Proceedings
of the
Steklov Institute of Mathematics

Number 134 (1975)

Theory of functions and its applications

DEDICATED
to
Academician Sergeĭ Mihaĭlovič Nikol'skiĭ

edited by

L. S. Pontrjagin

Providence, Rhode Island
AMERICAN MATHEMATICAL SOCIETY
1977

Академия Наук
Союза Советских Социалистических Республик

ТРУДЫ
ордена Ленина
МАТЕМАТИЧЕСКОГО ИНСТИТУТА имени В. А. СТЕКЛОВА
CXXXIV

ТЕОРИЯ ФУНКЦИЙ И ЕЕ ПРИЛОЖЕНИЯ

Редколлегия юбилейного сборника:
академик Л. С. ПОНТРЯГИН (главный редактор),
Д. Д. КУДРЯВЦЕВ (зам. главного редактора),
О. В. БЕСОВ и Е. А. ВОЛКОВ

Издательство "Наука"
Москва 1975

AMS (MOS) subject classifications (1970). Primary 00A70, 26A15, 26A42, 26A63, 26A82, 30A16, 30A82, 41A10, 41A25, 41A50, 41A55, 41A60, 42A04, 42A08, 42A20, 42A24, 42A32, 35B99, 49B30, 54C25, 54C35, 22D12, 35D99, 46E15, 46E30, 46E35, 54E99, 58E05, 93E20, 10G05, 46G05, 35H05, 47H15, 35J60, 65N15, 35Q05;
Secondary 26A72, 26A86, 30A06, 30A18, 41A25, 42A16, 42A56, 42A68, 42A88, 10B15, 10B45, 34B15, 35C15, 40C05, 22D15, 90D25, 35E99, 46E30, 46E35, 35G30, 10H15, 47H15, 35J25, 35J67, 35K35, 65N10.

ABSTRACT. The collection begins with a review of the scientific activity of Academician S. M. Nikol′skiĭ. This review reflects his fundamental results in the theory of approximation of functions, functional analysis, imbedding of function spaces and its applications, and Nikol′skiĭ's outstanding role in training scientific personnel is evidenced.

The collection contains papers by colleagues, students and disciples of Nikol′skiĭ, representing original scientific research in the theory of functions of one and several variables, as well as applications to differential equations.

Library of Congress Cataloging in Publication Data CIP

Main entry under title:

Theory of functions and its applications.

(Proceedings of the Steklov Institute of Mathematics; no. 134, (1975))

Translation of Teoriia͡ funkt͡siĭ i ee prilozheniia͡.

Bibliography: p.

1. Functions—Congresses. 2. Approximation theory—Congresses. 3. Functional analysis—Congresses. 4. Nikol′skiĭ Sergei Mikhaĭlovich. I. Nikol′skiĭ, Sergei Mikhaĭlovich. II. Pontri͡agin, Lev Semenovich, 1908– III. Series: Akademiia͡ nauk SSSR. Matematicheskiĭ institut. Proceedings; no. 134, (1975)

QA1.A413 no. 134 [QA331] 510′.8s [515] 77-10017

PROCEEDINGS OF THE STEKLOV INSTITUTE OF MATHEMATICS
IN THE ACADEMY OF SCIENCES OF THE USSR

(Труды математического института им. В. А. Стеклова, т. CXXXIV, 1975)

TABLE OF CONTENTS

Sergeĭ Mihaĭlovič
Nikol'skiĭ

Trudy Mat. Inst. Steklov.
134 (1975)

Proc. Steklov Inst. Math.
134 (1975)

SERGEĬ MIHAĬLOVIČ NIKOL'SKIĬ

This year marks the seventieth anniversary of the birth of one of the leading Soviet mathematicians, Academician Sergeĭ Mihaĭlovič Nikol'skiĭ.

Sergeĭ Mihaĭlovič Nikol'skiĭ was born on April 30 (O. S. 17), 1905, in the town of Talica in the Perm province (now in the Sverdlovsk region) in the family of an assistant chief forester. His childhood was spent in the west of what was then Russia, in the Suwalki province. During 1913–1914 he attended the gymnasium at Suwalki, and later a gymnasium at Chernigov. In 1918, in connection with the assignment of his father to new work in the Thorn Forest of the region of Voronezh, Nikol'skiĭ left school and continued his education under the tutelage of his father, who instructed him in mathematics, physics and the natural sciences. During this period Nikol'skiĭ was already working in experimental forestry and later at a state farm as an assistant gardener. In 1921 the Nikol'skiĭ family returned to Chernigov, where Sergeĭ Mihaĭlovič began work in provincial political education and attended at the same time a technical school.

In 1925, under the sponsorship of a trade union, Nikol'skiĭ entered the People's Educational Institute of Ekaterinoslav, where he majored in physics and mathematics. He graduated in 1929 and was retained at the institute (which shortly after became Dnepropetrovsk University) as a lecturer in mathematical analysis. In 1932 Sergeĭ Mihaĭlovič became chairman of the Department of Mathematics at the Transportation Institute,while at the same time continuing to hold a demanding position at Dnepropetrovsk University.

In 1933 Nikol'skiĭ entered graduate work and was sent to Moscow State University, where he studied under A. N. Kolmogorov. In 1935 he successfully defended his master's thesis in the field of functional analysis and, returning to Dnepropetrovsk University, headed the work there in the theory of functions. Under the influence of Kolmogorov, Nikol'skiĭ continued his studies in functional analysis and began to work in the field of approximation theory. His very first results won high marks from the specialists in this field. At Dnepropetrovsk, Sergeĭ Mihaĭlovič conducted a seminar in approximation theory, which contributed much to the creation at Dnepropetrovsk of a strong mathematical school in the field of approximation theory.

In 1940 Nikol'skiĭ entered the Steklov Institute of Mathematics as a doctoral candidate, and, after a successful defense in January of 1942 of his doctor's thesis on the theory of approximation of functions by polynomials, was retained at the institute as a senior scientist. He has remained at this institute ever since: from 1953 to 1962 he was its deputy director; since 1962, he has headed the theory of functions section.

AMS (MOS) subject classifications (1970). Primary 00A70.

In 1968 Nikol'skiĭ was elected a corresponding member, and in 1972 a member, of the Academy of Sciences of the USSR.

Nikol'skiĭ constantly has devoted and devotes much of his time to pedagogic work. From 1943 to 1947 he was chairman of the Department of Mathematics of the Moscow Highway Institute; since 1947 he has been a professor at the Moscow Institute of Physics and Technology (during 1950–1954 he was chairman of the Department of Higher Mathematics). His lectures are extremely popular and invariably successful with the students.

Nikol'skiĭ has written the course, *Mathematical Analysis*, in two volumes, which was published in 1973.

Nikol'skiĭ is one of the founders of the Soviet mathematical review journal *Referativniĭ Žurnal Matematika,* and was its first editor-in-chief. Sergeĭ Mihaĭlovič takes a very active part in the cultural life of the scientific community. As an example of his many-sided cultural activities, we note that he has been assistant executive editor for more than thirty years, and lately executive editor, of the journal *Trudy Matematičeskogo Instituta Imeni V. A. Steklova.* He has been for some time a member, and then chairman, of the mathematics committee of VAK*; for the past few years he has been a member of the plenum of VAK. Sergeĭ Mihaĭlovič was recently head of the organization of the Soviet-Hungarian mathematical journal *Analysis Mathematica.* He is the chief editor of the Soviet section of the editorial board of this journal.

Nikol'skiĭ devotes much drive, energy and enthusiasm to the education of young scientific cadres, sharing his knowledge and experience and infusing young people with a love for science and with high principles. He has taught many scientists now working in various fields of mathematics. He has trained at various periods of time 34 candidates of science. Among his former students are nine doctors of science. Three of them, T. I. Amanov, V. K. Dzjadyk and N. P. Korneĭčuk, are corresponding members of the academies of sciences of their respective Soviet republics.

More than 20 years ago, S. M. Nikol'skiĭ organized at the Steklov Institute of Mathematics of the Academy of Sciences of the USSR a seminar on the theory of differentiable functions of several real variables and its applications. This seminar has since been successfully and intensively conducted under his leadership (shared for several years with V. I. Kondrašev and L. D. Kudrjavcev, and more recently with L. D. Kudrjavcev), attracting many mathematicians to its sessions. A long list of participants have become candidates and doctors of science. Among these actively working in the seminar are O. V. Besov, Ja. S. Bugrov, P. I. Lizorkin, V. I. Burenkov, A. D. Džabrailov and L. N. Jakovlev.

Nikol'skiĭ has published more than 130 papers. His principal investigations belong to approximation theory, the theory of differentiable functions of several variables, the theory of direct methods of the calculus of variations and functional analysis.

**Editor's note.* In Cyrillic the complete acronym is BAK CCCP, literally, Higher Academic Courses in the USSR.

The first investigations of Nikol'skiĭ were devoted to the theory of linear operators in Banach spaces. The initial results in this direction, concerning compact operators, are due to F. Riesz and J. Schauder. Nikol'skiĭ has enriched this field with a new perspective by formulating necessary and sufficient conditions under which the Fredholm theory is valid for a linear equation. Within the limits of the so-called Fredholm domain, he has exhaustively studied the conditions under which the spectrum of a bounded linear operator is continuous or discrete. These results of Nikol'skiĭ received wide application in the theory of singular integral equations, while his investigations in connection with the new perspective mentioned above provided the foundation for the development of a wholly new direction in functional analysis in a series of papers by Soviet and foreign authors.

An important cycle of papers of Nikol'skiĭ pertains to the theory of approximation of functions. In the mid-1930s, following a paper of Kolmogorov, there arose a tendency to obtain sharp and asymptotically sharp estimates of approximations. Nikol'skiĭ was the first to undertake a comprehensive study of such problems and in the case of trigonometric approximations successively solved a number of protoproblems. Each new method of approximation and new class of functions considered by him revealed significant difficulties requiring in a number of cases the development of special methods. The follow-up problem of obtaining estimates of the same kind in the nonperiodic case required the overcoming of new obstacles. Nikol'skiĭ managed to obtain the first asymptotically sharp result in this area: an estimate of the best approximation by algebraic polynomials of functions with bounded first derivative. It should be noted that Nikol'skiĭ obtained asymptotically sharp estimates depending not only on the degree of the approximating polynomial but also on the location of a point in the interval on which the approximation is carried out. These subjects have received subsequent development both here and abroad. Of related interest are the more recent papers of Nikol'skiĭ on estimates of the approximation by polynomials of differentiable functions of several variables.

A second cycle of papers, written by Nikol'skiĭ during the same period, contains a number of subtle asymptotic estimates for approximations of individual functions having given singularities. In this series of papers Nikol'skiĭ obtained a number of fundamentally new results, among which we note the following:

a) He discovered that the functions satisfying a Lipschitz condition (with constant equal to unity) include some whose best approximations, at least for a subsequence of the degrees of the approximating polynomials, have an asymptotic behavior with constant greater than the corresponding constant for the function $|x|$, and determined the exact values of the greatest of the indicated constants.

b) He developed a method for obtaining asymptotic estimates of the best approximation of a function having a countable number of algebraic singularities.

c) He obtained the first asymptotic estimates in a number of cases of approximations in the metric of L_p.

We note in addition two articles of Nikol'skiĭ and an original book on the theory of quadrature formulas. In them a theory of best quadrature formulas for classes of functions is given (in the one-dimensional case), and best possible estimates are obrained for a number of concrete classes.

A third important cycle of papers of Nikol'skiĭ pertains to the theory of approximation of functions of several variables and to imbedding theorems. He obtained new sharp (in the sense of order) inequalities for trigonometric polynomials and entire functions of exponential type. These inequalities permit one to study with the use of the methods of approximation theory (i) the connections between the different differential properties of functions of several variables in different metrics and (ii) the connections between the differential properties of a function considered on spaces of different dimensions. On the basis of these methods, Nikol'skiĭ essentially developed an imbedding theory for classes of differentiable functions of several variables, the foundations of which were laid by S. L. Sobolev.

Nikol'skiĭ was the first to obtain sharp direct and converse imbedding theorems for classes of differentiable functions satisfying the Hölder-Zygmund integral condition. This showed that the indicated classes form a closed system relative to imbedding theorems. Imbedding theory was significantly extended in his papers to the case of functions having different properties in different variables. The basic principles determined by him in this direction and the corresponding imbedding-theorem formulas for Hölder classes were subsequently extended by other mathematicians to other function classes that are important in mathematical analysis and its applications. We note here a later paper of S. M. Nikol'skiĭ on the representation of functions with dominant mixed derivative in the form of a series in functions of exponential type.

A series of investigations of Nikol'skiĭ is devoted to the problem of extending a function outside its domain of definition (when the boundary is piecewise smooth) with preservation of its differential properties. He also obtained important results on extending functions given on smooth or piecewise-smooth manifolds and on domains of higher dimension. The cycle of papers of Nikol'skiĭ on approximation theory was awarded a State Prize in 1952. His investigations in the theory of approximation of functions of several variables and imbedding theorems were summarized by him in a monograph (*The approximation of functions of several variables and imbedding theorems*, "Nauka", Moscow, 1969; English transl., Springer-Verlag, Berlin and New York, 1973) that was highly acclaimed by specialists and immediately became a standard reference book for mathematicians working in this field. Nikol'skiĭ was awarded the Čebyšev Prize for this monograph in 1972.

Nikol'skiĭ also investigated the direct methods of the calculus of variations. With the use of his theorems on the extension of functions, he established conditions on a function given on a piecewise-smooth boundary of a domain that are sufficient for the solvability of the first boundary value problem for a polyharmonic equation. These conditions coincide to within an arbitrary $\epsilon > 0$ with corresponding necessary conditions obtained earlier by Sobolev. Thanks to this, the criteria for deciding whether or not to look for a solution of a boundary value problem by the variational method in the indicated case have reached a definitive form, inasmuch as it has consequently become possible in a sufficiently wide number of cases to judge the applicability of the variational method from the properties of the boundary function alone.

Subsequently, in a number of cases Nikol'skiĭ and his students have determined in the terminology of H-classes the exact correspondence (necessary and sufficient conditions) between the difference-differential properties of the solution of a boundary value problem and its boundary conditions. A series of papers of Nikol'skiĭ is devoted to proving that the variational method leads to a solution of the first boundary value problem for sufficiently general equations of elliptic type and even some that are not of elliptic or hypoelliptic type. Adjoining this series is a cycle of papers on the variational problem of Hilbert in n-dimensional space. An extensive literature, based on the use of the theory of the Cauchy integral (the school of N. I. Mushelišvili), which is inapplicable in the case of several variables, is devoted to this question in the two-dimensional case.

Recently, Nikol'skiĭ (initially in collaboration with P. I. Lizorkin) conducted a series of investigations of the differential properties right up to the boundary of the solutions of the Dirichlet problem for a linear equation of elliptic type of second (and higher) order with a strong degeneracy on the boundary. These investigations are based on his original method of "regular bridges".

His seventieth birthday finds S. M. Nikol'skiĭ in the full flowering of his creative powers, intensively and successfully conducting scientific investigations in various fields of mathematics, actively participating in the scientific-cultural life of the country, and tirelessly teaching and training new generations of students and young scientists.

Translated by S. SMITH

Trudy Mat. Inst. Steklov.
134 (1975)

Proc. Steklov Inst. Math.
134 (1975)

OPTIMAL IMPULSE AND CONTINUOUS CONTROL: METHOD OF NONLINEAR QUASI-VARIATIONAL INEQUALITIES *

UDC 519.3: 517.9

A. BENSOUSSAN AND J.-L. LIONS

Abstract. In previous papers [1]–[4] the authors have introduced the method of quasi-variational inequalities in order to formulate and to solve problems of optimal impulse control. In this type of problems one considers a dynamic stochastic system. From time to time, the state of the system is modified instantaneously (it receives an impulse). The controller chooses the impulse times and the intensity of these impulses in order to optimize a payoff. In this paper we consider the same problem, but we add now the possibility of a continuous control between successive impulse times. When there is no impulse control, the solution is well known and is obtained through solving the Hamilton-Jacobi-Bellman equation (see W. H. Fleming [5], W. H. Fleming–R. Rishel [6]), which is a nonlinear partial differential equation of parabolic type. For a purely impulse control problem we have introduced, in the papers mentioned, a "linear" quasi-variational inequality.(1) In this paper, we will use a "nonlinear" quasi-variational inequality, which reduces to the "linear" one when there is no continuous control.

Bibliography: 11 titles.

§ 1. Assumptions — notations and setting of the problem

We denote by $(\Omega, \mathfrak{A}, P)$ a probability space. Let $b(t)$ be a standart brownian motion in R^n. We denote by $\mathfrak{B}_t$ the σ-algebra generated by $b(s)$, $0 \leqslant s \leqslant t$. One considers a mapping $\sigma(x)$: $R^n \to R^n$ such that

$$|\sigma(x)| \leqslant C; \tag{1.1}$$

$$|\sigma_x| \leqslant C; \tag{1.2}$$

$$|\sigma^{-1}(x)| \leqslant C. \tag{1.3}$$

Let $\mathfrak{U}_{ad}$ be a convex compact subset of R^m $(0 \in \mathfrak{U}_{ad})$, and $g(x, t, d)$ a mapping from $R^n \times [0, T] \times \mathfrak{U}_{ad} \to R^n$ such that

$$g \text{ is continuous}; \tag{1.4}$$

$$|g| \leqslant C. \tag{1.5}$$

AMS (MOS) subject classifications (1970). Primary 93E20.

Editor's note. The body of this article is an unaltered reprinting of the original paper. Details of the bibliography have been checked and put in standard form, and MR citations have been added.

(1) As it will be seen, a quasi-invariational inequality is always a nonlinear problem, related to a linear or nonlinear partial differential operator.

Let now $L(x, t, d)$ be a functional: $R^n \times [0, T] \times \mathfrak{U}_{ad} \to R$ such that

$$L \text{ is continuous;} \tag{1.6}$$

$$0 \leqslant L \leqslant C; \tag{1.7}$$

$$L(x, t, 0) \in L^2(R^n \times]0, T[). \tag{1.8}$$

One defines for $p \in R^n$

$$H(x, t, p) = \min_{d \in \mathfrak{U}_{ad}} [L(x, t, d) + (p, g(x, t, d))]. \tag{1.9}$$

One easily checks the following properties

$$|H| \leqslant L(x, t, 0) + C|p|; \tag{1.10}$$

$$|H(x, t, p) - H(x, t, p')| \leqslant C|p - p'|. \tag{1.11}$$

1.1. Impulse controls

Let $\theta^1, \theta^2, \ldots$ be a chain of stopping times (with respect to the family $\mathfrak{B}_t$) such that

$$0 \leqslant \theta^1 \leqslant \theta^2 \leqslant \ldots \leqslant \theta^i \leqslant \theta^{i+1} \leqslant \ldots \leqslant T. \tag{1.12}$$

We denote by $\mathfrak{B}_{\theta^i}$ the σ-algebra of events previous to θ^i, i. e.

$$A \in \mathfrak{B}_{\theta^i} \Leftrightarrow A \cap \{\theta^i \leqslant t\} \subset \mathfrak{B}_t, \quad \forall t.$$

Furthermore, one defines a sequence

$$\xi^1, \ldots, \xi^i, \ldots$$

such that

$$\xi^i \in K = \text{compact subset of } R^n, \quad \xi^i \geqslant 0, \tag{1.13}$$

$$\xi^i \text{ is } \mathfrak{B}_{\theta^i} \text{ measurable.}$$

The set $\{\ldots \theta^i, \xi^i, \ldots\}$ is an *impulse control*, denoted by V_I. Let us consider an impulse control V_I for which

$$\theta^i \to T \text{ a. s.,} \quad i \to \infty.$$

One may then define a stochastic process $y(t) = y(t; \omega)$, $0 \leqslant t < T$, by the set of relations

$$\begin{aligned} dy &= \sigma(y)\, db(t), \quad \theta^i \leqslant t < \theta^{i+1}, \ \forall i, \\ y(\theta^i) &= y(\theta^{i-}) + \xi^i, \\ y(0) &= x_0. \end{aligned} \tag{1.14}$$

If $\theta^1 = 0$, we set $\theta^{1-} = 0$. If $\theta^{i+1} = \theta^i$, we set $\theta^{(i+1)-} = \theta^i$.

1.2. Continuous control

A stochastic process $d(t)$ with values in $\mathfrak{U}_{ad}$, which is adapted to $\mathfrak{B}_t$ is called a *continuous control.*

Knowing V_I (therefore also $y(.)$ and $d(.)$), we introduce a new process called $\tilde{b}(t)$ by setting

$$\tilde{b}(t) = b(t) - \int_0^t \sigma^{-1}(y(s))\, g(y(s), s, d(s))\, ds. \tag{1.15}$$

The process $\tilde{b}(t)$ is adapted to $\mathfrak{B}^t$.

We shall provide $(\Omega, \mathfrak{A})$ with a new probability measure $\tilde{P}$, which is absolutely continuous with respect to P, and which is given by the well known Girsanov's formula:

$$\frac{d\tilde{P}}{dP} = \exp\left\{\int_0^T (\sigma^{-1}g, \ db) - \frac{1}{2}\int_0^T ((\sigma\sigma^*)^{-1} g, \ g)\, dt\right\}. \tag{1.16}$$

Since $|\sigma^{-1}g| \leqslant C$, the process $\tilde{b}(t)$ becomes a standard brownian motion for the measure $\tilde{P}$ (see for instance, Gichman—Skorokhod [7]).

Let $k > 0$. We define a pay off

$$J(V_I, \ d(\cdot)) = \tilde{E}\left\{\int_0^T L(y(t), \ t, \ d(t))\, dt + k \sum_{i \geqslant 1} \mathrm{X}_{[0, \, T[}(\theta^i)\right\} \tag{1.17}$$

where $\mathrm{X}_{[a, \, b[}(s) = 1$ if $a \leqslant s < b$ and 0 otherwise.

In (1.17), the symbol $\tilde{E}$ means that we are considering the mathematical expectation with respect to $\tilde{P}$.

The right hand side of (1.17) is unambiguously defined if V_I is such that $\theta^i \to T$, $i \to \infty$. If V_I does not satisfy this property, we set $J(V_I, \ d(\cdot)) = +\infty$. Our problem consists in finding a pair $(\hat{V}_I, \ \hat{d}(\cdot))$ which minimizes J.

We shall finally introduce a subclass of controls, called the class of *Markov Controls*. This class is interesting by itself, and it turns out that there exists an optimal control which belongs to this class. This result will be proved in the next paragraph. A Markov control is defined by a set $\{C, \ \xi, \ d\}$ where: C is an open subset of $Q = R^n \times \,]0, \ T[$, $\xi(x, \ t)$ is a measurable mapping from $R^n \times \,]0, \ T[\, \to K$, $d(x, \ t)$ is a measurable mapping from $R^n \times \,]0, \ T[\, \to \mathfrak{U}_{ad}$. To the set $\{C, \ \xi, \ d\}$ we can associate an impulse control and a continuous control. Let S be the complement set of C in $R^n \times [0, \ T]$. Considering the Ito equation

$$\begin{aligned} dy &= \sigma(y)\, db, \\ y(0) &= x_0, \end{aligned} \tag{1.18}$$

we set

$$\theta^1 = \inf_{T \geqslant t > 0} t \qquad (\text{if } (y(t), \ t) \in S, \quad \forall t > 0, \quad \theta^1 = T) \tag{1.19}$$

$$(y(t), \ t) \in S,$$

$$\xi^1 = \xi(y(\theta^{1-}), \ \theta^1).$$

Knowing θ^i, ξ^i and $y(t)$ for $t < \theta^i$, one solves for $t \geqslant \theta^i$

$$dy = \sigma(y)\, db(t), \qquad t > \theta^i,$$

$$y(\theta^i) = y(\theta^{i-}) + \xi^i,$$

and one sets

$$\theta^{i+1} = \inf_{T \geqslant t \geqslant \theta^i} t \, (y(t), \ t) \in S,$$

$$\xi^{i+1} = \xi(y(\theta^{i+1-}), \ \theta^{i+1})$$

and so on. This procedure defines an impulse control and a process $y(t)$ satisfying (1.14). We then set

$$d(t) = d(y(t), \ t)$$

which defines a continuous control, at least if

$$\theta^i \to T \ \text{a. s.}$$

If $\lim \theta^i = \theta < T$, then $d(t)$ is well defined on $[0, \theta[$; we define it as an arbitrary deterministic constant on $[\theta, T[$.

§ 2. Characterization of an optimal solution

Let us introduce the differential operators

$$A\varphi(x) = -\sum_{i,j=1}^{n} \frac{1}{2} a_{ij}(x) \frac{\partial^2 \varphi}{\partial x_i \partial x_j}, \tag{2.1}$$

where

$$a_{ij}(x) = (\sigma\sigma^*)_{ij}(x)$$

and

$$B(t)\varphi(x) = H\left(x, t, \frac{\partial \varphi}{\partial x}\right). \tag{2.2}$$

We consider the following problem: Find a function $u(x, t)$ such that

$$u \text{ is continuous and bounded on } R^n \times [0, T]; \tag{2.3}$$

$$u \in L^2(0, T; H^1(R^n)), \quad \frac{\partial u}{\partial t} - Au \in L^2(0, T; L^2(R^n)); \tag{2.4}$$

$$-\frac{\partial u}{\partial u} + Au - B(t)u \leqslant 0 \text{ a. e.}; \tag{2.5}$$

$$u \leqslant Mu; \tag{2.6}$$

$$(u - Mu)\left(-\frac{\partial u}{\partial t} + Au - B(t)u\right) = 0 \text{ a. e.}; \tag{2.7}$$

$$u(x, T) = 0, \tag{2.8}$$

where we have set

$$Mu(x, t) = k + \min_{\substack{\xi \geqslant 0 \\ \xi \in K}} u(x + \xi, t). \tag{2.9}$$

If a function u satisfies (2.3), ..., (2.9), we set

$$C_u = \{x, t \in Q \mid u(x, t) < Mu(x, t)\}. \tag{2.10}$$

$\xi_u(x, t)$ selection which realizes the minimum in (2.9), such that $x, t \to \xi_u(x, t)$ is measurable. (2.11)

$d_u(x, t)$ selection which realizes the minimum of $L(x, t, d) + \left(\frac{\partial u}{\partial x}(x, t), g(x, t, d)\right)$ for $d \in \mathfrak{U}_{ad}$ such that $x, t \to d_u(x, t)$ is measurable. (2.12)

We notice that the function

$$x, t \to \min_{\substack{\xi \geqslant 0 \\ \xi \in K}} u(x + \xi, t)$$

is continuous; hence C_u is open. Furthermore, there exist selections ξ_u and d_u satisfying (2.11), (2.12). With respect to ξ_u, this assertion is clear since u is continuous and K compact. With respect to d_u, one first proves that for any compact subset of Q, say S, there exists $d_S : S \to u_{ad}$ measurable which realizes the minimum in (2.12) for $(x, t) \in S$. To prove this, one uses Lusin's theorem: there exists an increasing sequence of compacts $\Sigma_n \subset S$ such that

$$\operatorname{Meas}(S-\Sigma_n) < \frac{1}{n}$$

and $\partial u/\partial x$ is continuous on Σ_n. Hence, one can define $d_{\Sigma_n}(x, t): \Sigma_n \to \mathfrak{U}_{ad}$ as a measurable function, realizing the min in (2.12). Setting by induction

$$d'_{\Sigma_n}(x, t) = \begin{cases} d'_{\Sigma_{n-1}}(x, t) & \text{if } (x, t) \in \Sigma_{n-1}, \\ d_{\Sigma_n}(x, t) & \text{if } (x, t) \in \Sigma_n - \Sigma_{n-1}, \\ d_0 \in \mathfrak{U}_{ad} & \text{if } (x, t) \in S - \Sigma_n \end{cases}$$

the sequence $d'_{\Sigma_n}(x, t)$ converges for any (x, t) towards $d_S(x, t)$ satisfying the required properties. One then considers a sequence $S_n \to Q$ and one proceeds as above to define a selection $d_u(x, t)$.

We shall denote by V_I^u and $d^u(\cdot)$ the impulse control and continuous control associated with (C_u, ξ_u, d_u) as explained in the end of paragraph 1.

Theorem 2.1. *Under the assumptions and notations of § 1, and assuming that there exists a function u satisfying (2.3), ..., (2.8), the pair V_I^u, $d^u(\cdot)$ is optimal and furthermore*

$$u(x_0, 0) = J(V_I^u, d^u(\cdot)) = \inf_{V_I, d(\cdot)} J(V_I, d(\cdot)). \tag{2.13}$$

Proof:

We shall proceed eventually as in Bensoussan—Lions [4], with some new-technical lemmas.

Lemma 2.1. *Let $\Phi(x, t) \in L^2(Q)$, $\Phi \geqslant 0$. We set $\theta = \theta^i$, $\theta' = \theta^{i+1}$ (where θ^i, θ^{i+1} belong to some arbitrary impulse control) and for $h > 0$, we set $\theta'_h = \min(\theta', \theta + h)$; then*

$$\tilde{E} \int_{\theta'_h}^{\theta'} \Phi(y(s), s)\, ds \leqslant C(h) \|\Phi\|_{L^2(Q)}. \tag{2.14}$$

It is enough to consider the case when $\Phi \in \mathscr{D}(Q)$ (space of infinitely differentiable functions in Q with compact support). Defining $\Phi = 0$ for $t \geqslant T$, we obtain $\Phi \in \mathscr{D}(R^n \times R^+)$.

Setting $\chi_a(h) = 1$ if $h < a$ and 0 otherwise we have

$$E \int_{\theta'_h}^{\theta'} \Phi(y(s), s)\, ds = \int_h^T E \left\{ \chi_{\theta'-\theta}(h) \chi_{\theta'-\theta}(s) \Phi(y(\theta - s), \theta + s) \exp\left[\int_0^T (\sigma^{-1} g, db) \right.\right.$$
$$\left.\left. - \frac{1}{2} \int_0^T (g, (\sigma\sigma^*)^{-1} g)\, dt \right] \right\} ds \leqslant \left(E \exp\left[\int_0^T \gamma (\sigma^{-1} g, db) \right.\right.$$
$$\left.\left. - \frac{\gamma}{2} \int_0^T (g, (\sigma\sigma^*)^{-1} g)\, dt \right] \right)^{\frac{1}{\gamma}} \left[\int_h^T ds \left(E\left[\chi_{\theta'-\theta}(h) \chi_{\theta'-\theta}(s) \Phi^\beta(y(\theta + s), \theta + s) \right] \right)^{\frac{1}{\beta}} \right], \tag{2.15}$$

where $\beta, \gamma > 1$, $1/\beta + 1/\gamma = 1$.

But

$$E\exp\left[\int_0^T \gamma(\sigma^{-1}g,\ db)-\frac{\gamma}{2}\int_0^T (g,\ (\sigma\sigma^*)^{-1}g)\,dt\right]$$

$$=E\left\{\exp\left[\int_0^T \gamma(\sigma^{-1}g,\ db)-\frac{\gamma}{2}\int_0^T (g,\ (\sigma\sigma^*)^{-1}g)\,dt\right]\right.$$

$$\left.\times\exp\frac{\gamma(\gamma-1)}{2}\int_0^T (g,\ (\sigma\sigma^*)^{-1}g)\,dt\right\}$$

and

$$E\exp\left[\int_0^T \gamma(\sigma^{-1}g,\ db)-\frac{\gamma}{2}\int_0^T (g,\ (\sigma\sigma^*)^{-1}g)\,dt\right]=1; \tag{2.16}$$

$$\int_0^T (g,\ (\sigma\sigma^*)^{-1}g)\,dt\leqslant C. \tag{2.17}$$

We define for $s\geqslant 0$ the process $y_\theta(s)$ as the solution of

$$\begin{aligned} dy_\theta(s)&=\sigma(y_\theta(s))\,db_\theta(s),\\ y_\theta(0)&=\zeta=y(\theta),\end{aligned} \tag{2.18}$$

where

$$b_\theta(s)=b(s+\theta);$$

it is a brownian motion with respect to the family $\mathfrak{G}_s=\mathfrak{B}_{s+\theta}$, $s\geqslant 0$. Since for $0\leqslant s\leqslant\theta'-\theta$ $y_\theta(s)=y(\theta+s)$, it follows from (2.15)—(2.17) that

$$\tilde{E}\int_{\theta'_h}^{\theta'}\Phi(y(s),\ s)\,ds\leqslant\exp\frac{(\gamma-1)\,C}{2}\int_h^T ds\,(E\Phi^\beta(y_\theta(s),\ \theta+s))^{\frac{1}{\beta}}. \tag{2.19}$$

Let now $y_{\theta z}(s)$ be the solution of

$$\begin{aligned} dy_{\theta z}(s)&=\sigma(y_{\theta z}(s))\,db_\theta(s),\qquad s>0,\\ y_{\theta z}(0)&=z,\qquad z\in R^n.\end{aligned} \tag{2.20}$$

One has clearly

$$y_\theta(s)=y_{\theta\zeta}(s).$$

Now $y_{\theta z}(s)$ has a probability density which does not depend on θ; let us denote by $\pi(z,\ s,\ x)$ this probability density; then for fixed z, t

$$\Phi^\beta(y_{\theta z}(s),\ s+t)\quad\text{is independant of } \mathfrak{G}_0=\mathfrak{B}_\theta,$$

$$E\Phi^\beta(y_{\theta z}(s),\ s+t)=\int_{R^n}\Phi^\beta(x,\ s+t)\,\pi(z,\ s,\ x)\,dx.$$

Since ζ, θ are $\mathfrak{B}_\theta$ measurable, we can write

$$E\Phi^\beta(y_\theta(s),\ s+\theta)=\int_{R^n}E\Phi^\beta(x,\ s+\theta)\,\pi(\zeta,\ s,\ x)\,dx. \tag{2.21}$$

By virtue of (1.1)—(1.3) we have the well known estimate for Green functions of parabolic equations

$$0\leqslant\pi(z,\ s,\ x)\leqslant\frac{C}{s^{n/2}}\exp-\frac{C\,|x-z|^2}{s}. \tag{2.22}$$

Using (2.21) and (2.22) in (2.19) we get

$$\tilde{E}\int_{\theta'_h}^{\theta'}\Phi(y(s),\ s)\,ds$$

$$\leqslant \frac{C^{\frac{1}{\beta}}}{h^{\frac{n}{\beta}}}\exp\frac{(\gamma-1)C}{2}\int_h^T ds\left(\int_{R^n}E\Phi^\beta(x,\ s+\theta)\exp\left[-\frac{C}{T}|x-\zeta|^2\right]dx\right)^{\frac{1}{\beta}}. \tag{2.23}$$

Taking $1<\beta<2$ and using Hölder's inequality, one easily gets:

$$\tilde{E}\int_{\theta'_h}^{\theta'}\Phi(y(s),\ s)\,ds\leqslant C(h)\|\Phi\|_{L^2(Q)} \tag{2.24}$$

which completes the proof of the Lemma.

From lemma 2.1, we get an Ito's formula as follows: if $u\in L^2(0,\ T;\ H^1(R^n))$, $\frac{\partial u}{\partial t}-Au\in L^2(Q)$, u is continuous and bounded on Q, then

$$\tilde{E}u(y(\theta'-),\ \theta')=\tilde{E}u(y(\theta'_h),\ \theta'_h)+\tilde{E}\int_{\theta'_h}^{\theta'}\left(\frac{\partial u}{\partial t}-Au\right)(y(s),\ s)\,ds$$

$$+\tilde{E}\int_{\theta'_h}^{\theta'}\left(\frac{\partial u}{\partial x}(y(s),\ s),\ g(y(s),\ s,\ d(s))\right)ds. \tag{2.25}$$

The proof of (2.25) is similar to Bensoussan—Lions [4], it consists in approximating u by a sequence $u_n\in\mathscr{D}(\bar{Q})$ such that

$$u_n\to u\quad\text{in}\quad C(\bar{Q}),$$

$$\frac{\partial u_n}{\partial t}-Au_n\to\frac{\partial u}{\partial t}-Au\quad\text{in}\quad L^2(Q),$$

$$\frac{\partial u_n}{\partial x}\to\frac{\partial u}{\partial x}\quad\text{in}\quad L^2(Q)$$

and next applying Ito's formula to u_n and to the process y and then passing to the limit by using lemma 2.1.

We can now prove theorem 2.1. Let V_I and $d(\cdot)$ be any pair such that $\theta^i\to T$ as $i\to\infty$. Actually, if N denotes the maximum index n such that $\theta^n<T$, we can assume, without any loss of generality, that $N<+\infty$, a. s., because otherwise $J(V_I,\ d(.))=+\infty$.

We apply formula (2.25) for $\theta=\theta^i$ and $\theta'=\theta^{i+1}$, hence

$$\tilde{E}u(y(\theta^{i+1-}),\ \theta^{i+1})=\tilde{E}u(y(\theta_h^{i+1-}),\ \theta_h^{i+1})+\tilde{E}\int_{\theta_h^{i+1}}^{\theta^{i+1}}\left(\frac{\partial u}{\partial t}-Au\right)(y(s),\ s)\,ds$$

$$+\tilde{E}\int_{\theta_h^{i+1}}^{\theta^{i+1}}\left(\frac{\partial u}{\partial x}(y(s),\ s),\ g(y(s),\ s,\ d(s))\right)ds,\qquad i\geqslant 0, \tag{2.26}$$

where

$$\theta_h^{i+1}=\min(\theta^{i+1},\ \theta^i+h),\qquad\theta^0=0.$$

But

$$\frac{\partial u}{\partial t}-Au\geqslant -B(t)u=-H\left(x,\ t,\ \frac{\partial u}{\partial x}\right)\ \text{a. e.},$$

i. e.

$$\frac{\partial u}{\partial t} - Au + H\left(x,\ t,\ \frac{\partial u}{\partial x}\right) \geqslant 0 \quad \text{and} \in L^2(Q).$$

From lemma 2.1 it follows that

$$0 \leqslant \tilde{E} \int_{\theta_h^{i+1}}^{\theta^{i+1}} \left(\frac{\partial u}{\partial t} - Au + H\right)(y(s),\ s)\, ds < +\infty,$$

hence

$$\tilde{E} \int_{\theta_h^{i+1}}^{\theta^{i+1}} \left(\frac{\partial u}{\partial t} - Au\right)(y(s),\ s)\, ds \geqslant -\tilde{E} \int_{\theta_h^{i+1}}^{\theta^{i+1}} H(y(s),\ s)\, ds \tag{2.27}$$

since both integrals exist (we have set $H(x,\ t) = H\left(x,\ t,\ \frac{\partial u}{\partial x}(x,\ t)\right)$. On the other hand

$$-H(x,\ t,\ p) \geqslant -L(x,\ t,\ d) - (p,\ g(x,\ t,\ d)), \quad \forall x,\ t,\ p,\ d.$$

Let us consider a representative of the equivalence class $\partial u/\partial x$ still denoted by $\partial u/\partial x$. We have

$$-H\left(x,\ t,\ \frac{\partial u}{\partial x}\right) \geqslant -L(x,\ t,\ d) - \left(\frac{\partial u}{\partial x},\ g(x,\ t,\ d)\right), \quad \forall x,\ t,\ d.$$

We may take $x = y(t;\ \omega)$ and $d = d(t;\ \omega)$ and integrate in $(t,\ \omega)$. It follows that

$$-\tilde{E} \int_{\theta_h^{i+1}}^{\theta^{i+1}} H(y(t),\ t)\, dt \geqslant -\tilde{E} \int_{\theta_h^{i+1}}^{\theta^{i+1}} L(y(t),\ t,\ d(t))\, dt$$
$$- \tilde{E} \int_{\theta_h^{i+1}}^{\theta^{i+1}} \left(\frac{\partial u}{\partial x}(y(t),\ t),\ g(y(t),\ t,\ d(t))\right) dt. \tag{2.28}$$

Each of the above integrals exists and does not depend on the representative which has been chosen for $\partial u/\partial x$. From (2.27), (2.28) and (2.26) it follows that

$$\tilde{E}u(y(\theta^{i+1-}),\ \theta^{i+1}) \geqslant \tilde{E}u(y(\theta_h^{i+1-}),\ \theta_h^{i+1}) - \tilde{E} \int_{\theta_h^{i+1}}^{\theta^{i+1}} L(y(t),\ t,\ d(t))\, dt. \tag{2.29}$$

By adding these relations when i runs from 0 to $n-1$ we get

$$\sum_{i=1}^{n} [\tilde{E}u(y(\theta^{i-}),\ \theta^i) - \tilde{E}u(y(\theta_h^{i-}),\ \theta_h^i)] \geqslant -\tilde{E} \int_0^T L(y(t),\ t,\ d(t))\, dt. \tag{2.30}$$

But

$$\theta_h^i \to \theta^{i-1}, \qquad \text{when } h \to 0$$

and

$$y(\theta_h^{i-}) \to y(\theta^{i-1}) \quad \text{a. s. } \tilde{P}, \quad \text{when } h \to 0.$$

Since $u \in C(\bar{Q})$, it follows from (2.30), by letting $h \to 0$

$$u(x_0,\ 0) \leqslant \tilde{E} \int_0^T L(y(t),\ t,\ d(t))\, dt + \tilde{E}u(y(\theta^{n-}),\ \theta^n)$$
$$+ \sum_{i=1}^{n-1} [\tilde{E}u(y(\theta^{i-}),\ \theta^i) - \tilde{E}u(y(\theta^i),\ \theta^i)]. \tag{2.31}$$

But, as in Bensoussan—Lions [4], we have

$$u(y(\theta^{i-}), \theta^i) - u(y(\theta^i), \theta^i) \leqslant k\chi_{[0,T[}(\theta^i)$$

$$\tilde{E}u(y(\theta^{n-}), \theta^n) \to 0, \quad n \to +\infty \quad (\text{since } J(V_I, d(\cdot)) < +\infty)$$

hence

$$u(x_0, 0) \leqslant J(V_I, d(\cdot)) \qquad \forall V_I, d(\cdot). \tag{2.32}$$

We shall now prove that

$$u(x_0, 0) = J(V_I^u, d^u(\cdot)). \tag{2.33}$$

Let us set:

$$V_I^u = \{\ldots\theta^i, \xi^i\ldots\}.$$

As in (2.26) we have

$$\tilde{E}u(y(\theta^{i+1-}), \theta^{i+1}) = \tilde{E}u(y(\theta_h^{i+1-}), \theta_h^{i+1}) + \tilde{E}\int_{\theta_h^{i+1}}^{\theta^{i+1}} \left(\frac{\partial u}{\partial t} - Au\right)(y(s), s)\,ds$$

$$+ \tilde{E}\int_{\theta_h^{i+1}}^{\theta^{i+1}} \left(\frac{\partial u}{\partial t}(y(s), s), g(y(s), s, d_u(y(s), s))\right) ds. \tag{2.34}$$

Let us set:

$$\chi_{C_u}(x, t) = \begin{vmatrix} 1 & \text{if } x, t \in C_u, \\ 0 & \text{otherwise.} \end{vmatrix}$$

We have

$$\chi_{C_u}(x, t)\left(\frac{\partial u}{\partial t} - Au + H\right)(x, t) = 0 \quad \text{a. e.}$$

hence, from lemma 2.1

$$\tilde{E}\int_{\theta_h^{i+1}}^{\theta^{i+1}} \chi_{C_u}(y(s), s)\left(\frac{\partial u}{\partial t} - Au + H\right)(y(s), s)\,ds = 0. \tag{2.35}$$

But for $s \in]\theta_h^{i+1}, \theta^{i+1}[$, $(y(s), s) \in C_u$, hence $\chi_{C_u}(y(s), s) = 1$. Therefore

$$\tilde{E}\int_{\theta_h^{i+1}}^{\theta^{i+1}} \left(\frac{\partial u}{\partial t} - Au + H\right)(y(s), s)\,ds = 0. \tag{2.36}$$

But

$$H(x, t) = L(x, t, d_u(x, t)) + \left(\frac{\partial u}{\partial x}, g(x, t, d_u(x, t))\right)$$

by definition of d_u. Since $H(x, t)$ and

$$\left(\frac{\partial u}{\partial x}, g(x, t, d_u(x, t))\right) \in L^2(Q) \qquad \text{also } L(x, t, d_u(x, t)) \in L^2(Q).$$

From (2.35) it follows that

$$\tilde{E}\int_{\theta_h^{i+1}}^{\theta^{i+1}} \left(\frac{\partial u}{\partial t} - Au\right)(y(s), s)\,ds + \tilde{E}\int_{\theta_h^{i+1}}^{\theta^{i+1}} L(y(s), s, d_u(y(s), s))\,ds$$

$$+ \int_{\theta_h^{i+1}}^{\theta^{i+1}} \left(\frac{\partial u}{\partial x}(y(s), s), g(y(s), s, d_u(y(s), s))\right) ds = 0$$

which together with (2.34) implies

$$\tilde{E}u(y(\theta^{i+1-}), \theta^{i+1}) = \tilde{E}u(y(\theta_h^{i+1}), \theta_h^{i+1}) - \tilde{E}\int_{\theta_h^{i+1}}^{\theta^{i+1}} L(y(s), s, d_u(s))\, ds. \tag{2.37}$$

One then lets $h \to 0$. Proceeding as above, one gets

$$u(x_0, 0) = \tilde{E}\int_0^{\theta^n} f(y(s), s)\, ds + \tilde{E}u(y(\theta^{n-}), \theta^n) + \tilde{E}k\sum_{i=1}^{n-1} \chi_{[0, T[}(\theta^i). \tag{2.38}$$

Since u is bounded, it follows that $J(V_1^u, d^u(.)) < +\infty$.

Therefore one has also

$$\tilde{E}u(y(\theta^{n-}), \theta^n) \to 0, \qquad n \to \infty,$$

hence (2.33) follows. This completes the proof of theorem 2.1.

§ 3. Nonlinear quasi variational inequalities

Orientation: We shall now consider the problem of existence of a solution of (2.3), ..., (2.8). As in Bensoussan—Lions [4], we will slightly generalize the problem.

3.1. Setting of the problem

We consider the second order elliptic operator

$$A\varphi = -\sum_{i,j=1}^{n} \frac{1}{2}\frac{\partial}{\partial x_i}\left(a_{ij}\frac{\partial\varphi}{\partial x_j}\right) + \sum_{j=1}^{n} a_j(x)\frac{\partial\varphi}{\partial x_j} + a_0\varphi, \tag{3.1}$$

where

$$a_{ij},\ a_j,\ a_0 \in L^\infty(R^n), \qquad a_0 \geqslant 0, \qquad a_{ij} = a_{ji}; \tag{3.2}$$

$$\sum_{ij=1}^{n} a_{ij}(x)\xi_i\xi_j \geqslant \beta\sum_{i=1}^{n}\xi_i^2 \tag{3.3}$$

and the first order nonlinear operator

$$B(t)\varphi(x) = H\left(x, t, \frac{\partial\varphi}{\partial x}\right). \tag{3.4}$$

We will assume, besides (1.10), (1.11) that

$$H(x, t+h, p) \leqslant H(x, t, p), \qquad h \geqslant 0 \quad \forall x, t, p \tag{3.5}$$

which is obtained by taking

$$\begin{aligned} g(x, t, d) &= g(x, d), \\ L(x, t+h, d) &\leqslant L(x, t, d). \end{aligned} \tag{3.6}$$

We will not assume (1.8) any more, unless $\gamma = 0$ in the following. Setting for $\gamma \geqslant 0$

$$m_\gamma(x) = \exp(-\gamma|x|)$$

we introduce the spaces H_γ and V_γ as in Bensoussan—Lions [4]

$$H_\gamma = \{v \mid m_\gamma v \in L^2(R^n)\},$$

$$V_\gamma = \left\{v \mid m_\gamma v \in L^2(R^n),\ m_\gamma\frac{\partial v}{\partial x_i} \in L^2(R)^n,\ \forall i = 1 - n\right\}.\ (2)$$

* V_γ is a Sobolev space of orger 1 with «weight» m_γ cf. S. L. Sobolev [11].

We denote by

$$|v|=\left(\int_{R^n} m_\gamma^2 v^2 dx\right)^{1/2}$$

the norm in H_γ and by

$$\|v\|=\left(|v|^2+\sum_{i=1}^n\left|\frac{\partial v}{\partial x_i}\right|^2\right)^{1/2}$$

the norm in V_γ.

For $u, v \in V_\gamma$ we set

$$a(u, v)=\sum_{ij=1}^n \frac{1}{2}\int_{R^n} a_{ij}(x)\left(m_\gamma \frac{\partial u}{\partial x_j}\right)\left(m_\gamma \frac{\partial v}{\partial x_j}\right)dx$$

$$+\sum_{j=1}^n \int_{R^n}\left(a_j-\gamma\sum_{i=1}^n \frac{x_i}{|x|}a_{ij}\right)\left(m_\gamma \frac{\partial u}{\partial x_j}\right)(m_\gamma v)\,dx+\int_{R^n} a_0 m_\gamma u m_\gamma v dx; \tag{3.7}$$

$$a_0(u, v)=\sum_{ij=1}^n \frac{1}{2}\int_{R^n} a_{ij}\left(m_\gamma \frac{\partial u}{\partial x_j}\right)\left(m_\gamma \frac{\partial v}{\partial x_i}\right)dx. \tag{3.8}$$

One easily checks that there exists $\lambda \geqslant 0$ such that

$$a(u, u)+\lambda|u|^2 \geqslant \alpha\|u\|^2, \quad \alpha>0, \quad \forall u \in V_\gamma. \tag{3.9}$$

The problem is now stated as follows: to find u such that

$$u \in L^\infty(0, T; V_\gamma), \quad u \in L^\infty(Q), \quad \frac{\partial u}{\partial t} \in L^2(0, T; H_\gamma); \tag{3.10}$$

$$\frac{du}{dt} \leqslant 0 \text{ a. e.}, \ u \geqslant 0 \text{ a. e. } u \leqslant M(u) \text{ a. e.} \tag{3.11}$$

$$-\left(\frac{du}{dt}, v-u\right)+a(u, v-u)-(B(t)(u), v-u) \geqslant 0 \text{ a. e.}, \tag{3.12}$$

$$\forall v \leqslant M(u), \quad v \in V_\gamma;$$

$$u(T)=0. \tag{3.13}$$

The operator M is the same as in (2.9). As in Bensoussan—Lions [4], we can take a more general M, by axiomatizing the properties of M which will be used in the proof.

As in Bensoussan—Lions [4], we can justify that if we assume

$$\begin{aligned} &Au \in L^2(0, T, H_\gamma), \\ &v \in C(Q) \end{aligned} \tag{3.14}$$

then, for $\gamma=0$, u is solution of (2.3), ..., (2.8), where

$$a_{ij}=(\sigma\sigma^*)_{ij}, \quad a_j=\frac{1}{2}\sum_i \frac{\partial a_{ij}}{\partial x_j}, \quad a_0=0.$$

The regularity properties stated in (3.14) as well as uniqueness are an open problem. From the interpretation of the problem, we can however say that there exists at most one solution u of (3.10), ..., (3.13) satisfying the regularity conditions (3.14). We shall use the following property:

if for $u \in V_\gamma$ we define $\mathfrak{A}(t)u \in V'_\gamma$ by

$$(\mathfrak{A}(t)u, v)=a(u, v)+k(u, v)-e^{kt}(B(t)(e^{-kt}u), v), \quad \forall v \in V_\gamma, \tag{3.15}$$

then we can choose k such that

$$(\mathfrak{A}(t)u,\ u) \geqslant c\|u\|^2 - c|u|; \tag{3.16}$$

$$(\mathfrak{A}(t)u - \mathfrak{A}(t)v,\ u-v) \geqslant c\|u-v\|^2; \tag{3.17}$$

$$-(\mathfrak{A}(t)u,\ u^-) \geqslant c\|u^-\|^2; \tag{3.18}$$

$$-(\mathfrak{A}(t)u - \mathfrak{A}(t)v,\ (u-v)^-) \geqslant c\|(u-v)^-\|^2; \tag{3.19}$$

$$(\mathfrak{A}(t)u,\ v) \geqslant a(u,v) + k(u,v) - |v|\left(e^{kT}\sup L + \left|\frac{\partial u}{\partial x}\right| C\right). \tag{3.20}$$

Relation (3. 20) is true for any k and it follows from the properties of B. Aplying it with $v=u$, we get

$$(\mathfrak{A}(t)u,\ u) = a(u,\ u) + \lambda|u|^2 + (k-\lambda)|u|^2 - |u|e^{kT}\sup L - C|u|\left|\frac{\partial u}{\partial x}\right|$$

$$\geqslant \frac{\alpha}{2}\|u\|^2 + (k-\lambda)|u|^2 - |u|e^{kT}\sup L - \frac{C^2}{2\alpha}|u|^2$$

and $k \geqslant \lambda + C^2/2\alpha$ implies (3. 16).

Similary

$$(\mathfrak{A}(t)u - \mathfrak{A}(t)v,\ u-v) = a(u-v,\ u-v) + \lambda|u-v|^2$$
$$+ (k-\lambda)|u-v|^2 - e^{kt}(B(t)(e^{-kt}u) - B(t)(e^{-kt}v),\ u-v)$$
$$\geqslant \alpha\|u-v\|^2 + (k-\lambda)|u-v|^2 - C|u-v|\|u-v\|$$

hence (2. 17) follows. One has also:

$$-(\mathfrak{A}(t)u,\ u^-) = a(u^-,\ u^-) + (k-\lambda)|u^-|^2 + \lambda|u^-|^2 + e^{kt}(B(t)(e^{-kt}u),\ u^-).$$

But

$$e^{kt}(B(t)(e^{-kt}u),\ u^-) = \int_{R^n} m_\gamma^2 u^-\left[e^{kt}L(x,\ t,\ d_u) + \left(\frac{\partial u}{\partial x},\ g(x,\ d_u)\right)\right]dx \geqslant -C|u^-|\|u^-\| \tag{3.21}$$

hence

$$-(\mathfrak{A}(t)u,\ u^-) \geqslant \alpha\|u^-\|^2 + (k-\lambda)|u^-|^2 - C|u^-|\|u^-\|$$

i. e., (3. 18) for the above choice of k. Next

$$-(\mathfrak{A}(t)u - \mathfrak{A}(t)v,\ (u-v)^-) = a(u^- - v^-,\ u^- - v^-) + \lambda|u^- - v^-|^2$$
$$+ (k-\lambda)|u^- - v^-|^2 + e^{kt}(B(t)(ue^{-kt}) - B(t)(ve^{-kt}),\ (u-v)^-);$$

but since

$$e^{kt}(B(t)(e^{-kt}u) - B(t)(e^{-kt}v),\ (u-v)^-) \geqslant -C|(u-v)^-|\|(u-v)^-\| \tag{3.22}$$

hence (3. 19) follows. We now state the

Theorem 3. 1. *With assumptions* (1. 4), ..., (1. 7) *and* (3. 2), (3. 3), (3. 6) *there exists a solution u of* (3. 10), ..., (3. 13). *If* $\gamma=0$, *the same result holds if we add assumption* (1. 8).

3.2. Proof of theorem 3.1

We shall first consider an auxiliary problem as follows. Let us assume that:

$$\psi \in L^\infty(Q), \quad f \in L^\infty(Q), \quad \psi \geqslant 0, \quad f \geqslant 0,$$
$$\psi(t+h,\ x) \leqslant \psi(t,\ x) \text{ a. e. } h \geqslant 0, \tag{3.23}$$
$$f(t+h,\ x) \leqslant f(t,\ x) \text{ a. e. } h \geqslant 0.$$

We consider the problem

$$-\left(\frac{dw}{dt}, v\right)+a(w, v)+\lambda(w, v)-(B(t)w, v)+\mu((w-\psi)^+, v)=(f(t), v),\ w(T)=0; \tag{3.24}$$

$$w\in L^2(0, T; V_\gamma),\quad \frac{dw}{dt}\in L^2(0, T; H_\gamma),\quad \frac{dw}{dt}\leqslant 0,$$
$$0\leqslant w\leqslant C=\frac{\sup f+\sup L}{\lambda+\inf a_0}. \tag{3.25}$$

We shall show that there exists a unique solution w of (3.24), (3.25). Setting

$$w=e^{-kt}z$$

we see that z is solution of

$$-\frac{dz}{dt}+\mathfrak{A}(t)z+\lambda z+\mu e^{kt}(e^{-kt}z-\psi)^+=fe^{kt},$$
$$z(T)=0. \tag{3.26}$$

But the operator

$$\tilde{\mathfrak{A}}(t)z=\mathfrak{A}(t)z+\lambda z+\mu e^{kt}(e^{-kt}z-\psi)^+$$

is a monotonie bounded continuous coercive operator from $V_\gamma\to V'_\gamma$. Hence it follows that (see J. L. Lions [8]) there exists a unique $z\in L^2(0, T; V_\gamma)$ satisfying (3.26), and therefore there also exists a unique function w satisfying (3.24) and $w\in L^2(0, T; V_\gamma)$. We have to show that this solution satisfies

$$\frac{dw}{dt}\in L^2(0, T; H_\gamma); \tag{3.27}$$

$$0\leqslant w\leqslant c,\quad \frac{dw}{dt}\leqslant 0. \tag{3.28}$$

But w is solution of

$$-\left(\frac{dw}{dt}, v\right)+a(w, v)+\lambda(w, v)=(f(t), v)+(B(t)w, v)-\mu((w-\psi)^+, v),$$
$$w(T)=0$$

and

$$f(t)+B(t)w-\mu(w-\psi)^+\in L^2(0, T; H_\gamma).$$

By standard lineary theory (see Lions—Magenes [9]) (3.27) follows. Taking $v=w^-$ in (3.24), we get(3)

$$\frac{1}{2}\frac{d}{dt}|w^-|^2\quad a(w^-)-\lambda|w^-|^2-(B(t)w, w^-)=(f, w^-) \tag{3.29}$$

since $((w-\psi)^+, w)=0$. But

$$(B(t)w, w^-)\geqslant -C|w^-|\|w^-\|. \tag{3.30}$$

From (3.20), (3.30) it follows that $w^-=0$.

Let c be as in (3.25). Taking $v=(w-c)^+$ in (3.24) we get

$$-\frac{1}{2}\frac{d}{dt}|(w-c)^+|^2+a((w-c)^+)+\lambda|(w-c)^+|^2+c\inf a_0(1, (w-c)^+)+\lambda c(1, (w-c)^+)+$$
$$+\mu((w-\psi)^+, (w-c)^+)\leqslant(\sup f+\sup L)(1, (w-c)^+)+C|(w-c)^+|\|(w-c)^+\|$$

from which the first part of (3.28) follows, using Gronwall inequality. To prove the second part of (3.28) we discretize $(0, T)$ in

$$0, k, 2k, \ldots, Nk=T$$

(3) $a(z)=a(z, z)$.

and we set

$$f_n = \frac{1}{k}\int_{nk}^{(n+1)k} f(t)\,dt; \qquad \psi_n = \frac{1}{k}\int_{nk}^{(n+1)k} \psi(t)\,dt; \qquad B_n = \frac{1}{k}\int_{nk}^{(n+1)k} B(t)\,dt;$$

then we define

$$f_k(t) = f_n \quad \text{in }]nk, (n+1)k],$$
$$\psi_k(t) = \psi_n \quad \text{in }]nk, (n+1)k].$$

We notice that $\psi_{n+1} \leqslant \psi_n$, $f_{n+1} \leqslant f_n$ a. e. Let now w_n be a sequence defined by

$$-\left(\frac{w_{n+1}-w_n}{k}, v\right) + a(w_n, v) + \mu((w_n-\psi_n)^+, v) + \lambda(w_n, v) - (B_n w_n, v) = (f_n, v), \tag{3.31}$$
$$w_N = 0.$$

We shall show by induction that

$$0 \leqslant \ldots \leqslant w_{n+1} \leqslant w_n \leqslant \ldots \leqslant C \tag{3.32}$$

for k small enough.

We set

$$\tilde{a}(u, v) = a(u, v) + \lambda(u, v) + \frac{1}{k}(u, v).$$

Taking $v = (w_n - c)^+$ in $(3.31)_n$ and using the fact that $w_{n+1} \leqslant C$ we obtain

$$a((w_n-c)^+) + \lambda|(w_n-c)^+|^2 + \frac{1}{k}|(w_n-c)^+|^2 \leqslant C|(w_n-c)^+|\,\|(w_n-c)^+\|$$

which implies that for k small enough $(w_n - c)^+ = 0$. By difference between $(3.31)_n$ and $(3.31)_{n+1}$ and by taking $v = (w_n - w_{n+1})^-$, we obtain

$$\begin{aligned}
&\tilde{a}(w_n - w_{n+1}, (w_n-w_{n+1})^-) + \mu((w_n-\psi_n)^+ - (w_{n+1}-\psi_{n+1})^+, (w_n-w_{n+1})^-) \\
&\quad - (B_n w_n - B_n w_{n+1}, (w_n-w_{n+1})^-) + (B_{n+1}w_{n+1} - B_n w_{n+1}, (w_n-w_{n+1})^-) \\
&\quad = (f_n - f_{n+1}, (w_n-w_{n+1})^-) + \frac{1}{k}(w_{n+1}-w_{n+2}, (w_n-w_{n+1})^-).
\end{aligned} \tag{3.33}$$

But

$$B_{n+1}w_{n+1} - B_n w_{n+1} = \frac{1}{k}\int_{(n+1)k}^{(n+2)k} B(t)\,w_{n+1}dt - \frac{1}{k}\int_{nk}^{(n+1)k} B(t)\,w_{n+1}dt$$
$$= \frac{1}{k}\int_{(n+1)k}^{(n+2)k} H\left(x, t, \frac{\partial w_{n+1}}{\partial x}\right)dt - \frac{1}{k}\int_{nk}^{(n+1)k} H\left(x, t, \frac{\partial w_{n+1}}{\partial x}\right)dt \leqslant 0.$$

As in Bensoussan—Lions [4], we have

$$((w_n-\psi_n)^+ - (w_{n+1}-\psi_{n+1})^+, (w_n-w_{n+1})^-) \leqslant 0,$$
$$(f_n - f_{n+1}, (w_n-w_{n+1})^-) \geqslant 0,$$
$$(w_{n+1}-w_{n+2}, (w_n-w_{n+1})^-) \geqslant 0.$$

It follows from (3.33) that

$$\tilde{a}((w_n-w_{n+1})^-) + (B_n w_n - B_{n+1}w_{n+1}, (w_n-w_{n+1})^-) \leqslant 0; \tag{3.34}$$

but

$$(B_n w_n - B_{n+1}w_{n+1}, (w_n-w_{n+1})^-) \geqslant -C|(w_n-w_{n+1})^-|\,\|(w_n-w_{n+1})^-\|$$

hence it follows that

$$a((w_n-w_{n+1})^-) + \lambda|(w_n-w_{n+1})^-|^2 + \frac{1}{k}|(w_n-w_{n+1})^-|^2 \leqslant C|(w_n-w_{n+1})^-|\,\|(w_n-w_{n+1})^-\|$$

and for k small enough we get $(w_n - w_{n+1})^- = 0$. It remains to show that $w_{N-1} \geqslant 0$. But we have

$$\tilde{a}((w_{N-1})^-) + (B_{N-1}w_{N-1}, (w_{N-1})^-) \leqslant 0$$

and as before $(w_{N-1})^- = 0$. We now set

$$w_k(t) = w_n \quad \text{in }]nk, (n+1)k],$$

$$\tilde{w}_k(t) = \frac{(n+1)k - t}{k} w_n + \frac{t - nk}{k} w_{n+1}.$$

We shall show that

$$\begin{aligned} &\| w_k \|_{L^2(0, T; V_\gamma)} \leqslant C, \qquad \| w_k \|_{L^\infty(Q)} \leqslant C, \\ &\| \tilde{w}_k \|_{L^2(0, T; V_\gamma)} \leqslant C, \qquad \| \tilde{w}_k \|_{L^\infty(Q)} \leqslant C, \\ &\left\| \frac{d\tilde{w}_k}{dt} \right\|_{L^2(0, T; H_\gamma)} \leqslant C. \end{aligned} \tag{3.35}$$

Taking $v = w_n$ in $(3.31)_n$, we get

$$\frac{1}{2} | w_n |^2 - \frac{1}{2} | w_{n+1} |^2 + k\alpha \| w_n \|^2 \leqslant Ck | w_n | [1 + \| w_n \|] + k | f_n | | w_n |$$

from which the two first estimates (3.35) easily follow. On the other hand

$$\frac{d\tilde{w}_k}{dt} = -\frac{w_n - w_{n+1}}{k} \qquad \text{in }]nk, (n+1)k].$$

Taking

$$v = w_n - w_{n+1} \geqslant 0 \text{ in } (3.31)_n \text{ we get}$$

$$\frac{1}{k} | w_n - w_{n+1} |^2 + a(w_n, w_n - w_{n+1}) + \lambda (w_n, w_n - w_{n+1})$$
$$\leqslant (f_n, w_n - w_{n+1}) + (B_n w_n, w_n - w_{n+1}).$$

Using the decomposition $a = a_0 + a_1$

$$| (B_n w_n, w_n - w_{n+1}) | \leqslant C | w_n - w_{n+1} | [1 + \| w_n \|]$$

and reasoning as in Bensoussan—Lions [4], we get the last estimate in (3.35). Extracting a subsequence, we have

$$w_k \to w \text{ in } L^2(0, T; V_\gamma) \text{ weakly} \tag{3.36}$$

$\tilde{w}_k \to \tilde{w}$ in $L^2(0, T; V_\gamma)$ weakly and in $L^2(0, T; H_\gamma)$ strongly and a. e. x, t

$$\frac{d\tilde{w}_k}{dt} \to \frac{d\tilde{w}}{dt} \text{ in } L^2(0, T; H_\gamma) \text{ weakly.}$$

But $w = \tilde{w}$ (see Bensoussan—Lions [4]). Let $\varphi \in \mathscr{D}(0, T)$. We rewrite $(3.31)_n$ as follows:

$$-\int_0^T (\tilde{w}_k', v) \varphi(t)\, dt + \int_0^T a(w_k, v) \varphi dt + \lambda \int_0^T (w_k, v) \varphi dt + \mu \int_0^T ((w_k - \psi_k)^+, v) \varphi(t) dt$$
$$- \int_0^T (B_k(t)(w_k(t)), v) \varphi(t)\, dt = \int_0^T (f_k(t), v) \varphi(t)\, dt \tag{3.37}$$

where

$$B_k(t) = \frac{1}{k} \int_{nk}^{(n+1)k} B(t)\, dt \text{ for } t \in]nk, (n+1)k].$$

We may assume that

$$B_k(t)(w_k(t)) \to X(t) \text{ in } L^2(0, T; H_\gamma) \text{ weakly.} \tag{3.38}$$

Hence (3.37) implies that

$$-\left(\frac{dw}{dt}, v\right)+a(w, v)+\lambda(w, v)+\mu((w-\psi)^+, v)-(\chi(t), v)=(f(t), v). \qquad (3.39)$$

We also get from $(3.31)_n$

$$-\int_0^T(\tilde{w}_k', w_k)\,dt+\int_0^T a(w_k, w_k)\,dt+\lambda\int_0^T|w_k(t)|^2\,dt+\mu\int_0^T((w_k-\psi_k)^+, w_k)\,dt$$
$$-\int_0^T(B_k(t)(w_k(t)), w_k(t))\,dt=\int_0^T(f_k(t), w_k(t))\,dt.$$

Passing to the limit, using (3.36), (3.38) and the facts that $w_k(0)\to w(0)$ and $w_k(T)\to w(T)$ in H_γ (hence $w(T)=0$ follows), we get

$$\int_0^T a(w_k(t))\,dt\to-\frac{1}{2}|w(0)|^2-\lambda\int_0^T|w(t)|^2dt-\mu\int_0^T((w(t)-\psi(t))^+, w(t))\,dt$$
$$+\int_0^T(\chi(t), w(t))dt+\int_0^T(f(t), w(t))dt. \qquad (3.40)$$

But also from (3.39) it follows that

$$\frac{1}{2}|w(0)|^2+\int_0^T a(w(t))dt+\lambda\int_0^T|w(t)|^2\,dt+\mu\int_0^T((w(t)-\psi(t))^+, w(t))\,dt$$
$$-\int_0^T(\chi(t), w(t))\,dt=\int_0^T(f(t), w(t))\,dt.$$

Therefore

$$\int_0^T a(w_k(t))\,dt\to\int_0^T a(w(t))\,dt$$

hence $w_k\to w$ in $L^2(0, T; V_\gamma)$ strongly $B_k(t), (w_k(t))\to B(t)(w(t))$ in $L^2(0, T; H_\gamma)$ strongly i. e., $\chi(t)=B(t)(w(t))$; therefore $w(t)$ is the solution of (3.24). Since $d\tilde{w}_k/dt\leqslant 0$, we have $dw/dt\leqslant 0$. We can now proceed with the proof of theorem 3.1. We take an increasing sequense $\varphi(n)$, such that $\varphi(n)\to\infty$. We define u_0 as the solution of

$$-\left(\frac{du_0}{dt}, v\right)+a(u_0, v)-(B(t)u_0, v)=0, \qquad (3.41)$$
$$u_0(T)=0;$$

$$v_0\in L^2(0, T; V_\gamma), \quad \frac{du_0}{dt}\in L^2(0, T; H_\gamma), \quad u_0\geqslant 0, \quad u_0\in L^\infty(Q), \frac{du_0}{dt}\leqslant 0. \qquad (3.42)$$

The existence and uniqueness of u_0 solution of (3.41), (3.42) is not a direct consequence of (3.24), (3.25) since we cannot take $\lambda=0$. However, the same method can be applied to get (3.41), (3.42). However the value of c must be changed into

$$\frac{\sup Le^{kT}}{k+\inf a_0} \text{ for any } k>0,$$

i. e.

$$C = \inf_{k>0} \frac{\sup Le^{kT}}{k + \inf a_0}.$$

Knowing that $u_{n-1} \geqslant 0$, $u_{n-1} \in L^\infty(Q)$, $du_{n-1}/dt \leqslant 0$, we define u_n by solving

$$-\left(\frac{du_n}{dt}, v\right) + a(u_n, v) + \varphi(n)((u_n - M(u_{n-1}))^+, v) + \lambda(u_n, v) - (B(t)(u_n), v) = \lambda(u_{n-1}, v),$$

$$u_n(T) = 0, \tag{3.43}$$

$$u_n \in L^2(0, T; V_\gamma), \quad u_n \geqslant 0, \quad u_n \in L^\infty(Q),$$

$$\frac{du_n}{dt} \in L^2(0, T; H_\gamma); \quad \frac{du_n}{dt} \leqslant 0.$$

This is an application of (3. 24), (3. 25) with

$$\psi = M(u_{n-1}), \quad f = \lambda u_{n-1}, \quad \mu = \varphi(n).$$

Reasoning as in Bensoussan—Lions [4], we check that

$$0 \leqslant \ldots \leqslant u_n \leqslant u_{n-1} \leqslant u_{n-2} \leqslant \ldots \leqslant u_0. \tag{3.44}$$

As in Bensoussan—Lions [4], we can show that

$$\|u_n\|_{L^2(0, T; V_\gamma) \cap L^\infty(Q)} \leqslant C, \quad \left\|\frac{du_n}{dt}\right\|_{L^2(0, T; H_\gamma)} \leqslant C. \tag{3.45}$$

We can extract a subsequence such that

$$u_n \to u \text{ in } L^2(0, T; V_\gamma) \text{ weakly and } L^\infty(Q) \text{ weak star,}$$

$$u_n \to u \text{ in } L^2(0, T; H_\gamma) \text{ strongly and a. e.,} \tag{3.46}$$

$$\frac{du_n}{dt} \to \frac{du}{dt} \text{ in } L^2(0, T; H_\gamma) \text{ weakly.}$$

We get

$$u \geqslant 0, \quad u \in L^\infty(Q), \quad \frac{du}{dt} \leqslant 0, \quad u(T) = 0. \tag{3.47}$$

For $v \leqslant M(u_{n-1})$, it follows from (3. 43)

$$-\left(\frac{du_n}{dt}, v - u_n\right) + a(u_n, v - u_n) - (B(t)(u_n), v - u_n) \geqslant \lambda(u_{n-1} - u_n, v - u_n). \tag{3.48}$$

As in Bensoussan—Lions [4], we have

$$u \leqslant Mu. \tag{3.49}$$

Taking u in (3. 48) we get

$$-\int_0^T \left(\frac{du_n}{dt}, u - u_n\right) dt + \int_0^T a(u_n, u)\, dt - \int_0^T (B(t)(u_n), u - u_n)\, dt \geqslant \lambda \int_0^T (u_{n-1} - u_n, u - u_n)\, dt + \int_0^T a(u_n)\, dt,$$

hence

$$\limsup \int_0^T a(u_n)\, dt \leqslant \int_0^T a(u)\, dt,$$

i. e.,

$$\lim \int_0^T a(u_n)\,dt = \int_0^T a(u)\,dt$$

which means that

$$u_n \to u \qquad \text{in } L^2(0,\ T;\ V_\gamma) \text{ strongly,}$$
$$B(t)\,u_n \to B(t)\,u \qquad \text{in } L^2(0,\ T;\ H_\gamma) \text{ strongly.}$$

We can complete the proof as in Bensoussan—Lions [4].

Remark 3.1. Non linear quasi variational inequalities on an open subset $\mathfrak{O}$ of R^n can be studied similarly. One must add boundary conditions on $\partial\mathfrak{O}$ which may be

$$\left.\frac{\partial u}{\partial \nu_A}\right|_{\partial\mathfrak{O}} \leqslant 0,$$
$$u - Mu \leqslant 0,$$
$$(u - Mu)\left.\frac{\partial u}{\partial \nu_A}\right|_{\partial\mathfrak{O}} = 0$$

where $\partial/\partial\nu_A$ is the normal derivative with respect to the operator A.

Remark 3.2. We have only partial results in the study of Q. V. I. in L^p spaces, $p \neq 2$; no results are known in other spaces of the «interpolation type», such as those introduced in S. M. Nikolski, J. L. Lions, P. I. Lyzorkin [10].

BIBLIOGRAPHY

1. A. Bensoussan and J.-L. Lions a) *Nouvelle formulation de problèmes de contrôle impulsionnel et applications*, C. R. Acad. Sci. Paris Sér. A 276 (1973), A1189–A1192. MR 47 #5690.

b) *Contrôle impulsionnel et inéquations quasi-variationnelles d'evolution*, C. R. Acad. Sci. Paris Sér. A 276 (1973), A1333–A1338. MR 47 #5692.

2. ———, a) *Contrôle impulsionnel et contrôle continu. Méthode des inéquations quasi variationnelles non linéaires*, C. R. Acad. Sci. Paris Sér. A 278 (1974), 675–679. MR 49 #1273.

b) *Contrôle impulsionnel et systèmes d'inéquations quasi-variationnelles*, C. R. Acad. Sci. Paris Sér. A 278 (1974), 747–751. MR 49 #5996.

3. A. Bensoussan, M. Goursat and J.-L. Lions, *Contrôle impulsionnel et inéquations quasi-variationnelles stationnaires*, C. R. Acad. Sci. Paris Sér. A–B 276 (1973), A1279–A1284. MR 47 #5691.

4. A. Bensoussan and J.-L. Lions, *Nouvelles méthodes en contrôle impulsionnel*, Appl. Math. and Optimization 1 (1974/75), 289–312.

5. W. H. Fleming, *Optimal continuous-parameter stochastic control*, SIAM Rev. 11 (1969), 470–509. MR 41 #9633.

6. W. H. Fleming and R. W. Rishel, *Deterministic and stochastic optimal control*, Springer-Verlag, Berlin and New York, 1975.

7. I. I. Gihman and A. V. Skorohod, *Stochastic differential equations*, "Naukova Dumka", Kiev, 1968; English transl., Springer-Verlag, Berlin and New York, 1972. MR 41 #7777; 49 #11625.

8. J.-L. Lions, *Quelques méthodes de résolution des problèmes aux limites non linéaires*, Dunod; Gauthier-Villars, Paris, 1969. MR 41 #4326.

9. J.-L. Lions and E. Magenes, *Problèmes aux limites non homogènes et applications*. Vol. 1, Dunod, Paris, 1968; English transl., Springer-Verlag, Berlin and New York, 1972. MR 40 #512; 50 #2670.

10. S. M. Nikol′skiĭ, J.-L. Lions and P. I. Lizorkin, *Integral representation and isomorphism properties of some classes of functions*, Ann. Scuola Norm. Sup. Pisa (3) 19 (1965), 127–178. MR 32 #6203.

11. S. L. Sobolev, *Applications of functional analysis to mathematical physics*, Izdat. Leningrad. Gos. Univ., Leningrad, 1950; English transl., Amer. Math. Soc., Providence, R. I., 1963. MR 14, 565; 29 #2624.

Trudy Mat. Inst. Steklov.
134 (1975)

Proc. Steklov Inst. Math.
134 (1975)

DESCRIPTION OF MODULI OF CONTINUITY IN L_2

UDC 517.512

O. V. BESOV AND S. B. STEČKIN

Abstract. We give a description of the class of functions that are moduli of continuity in L_2. We show that it is smaller than the class of moduli of continuity in C.

Bibliography: 4 titles.

The function

$$\omega(\delta;\ f)_p = \sup_{|t|\leqslant\delta} \|\Delta(t) f\|_p,$$

where

$$\Delta(t) f(x) = f(x+t) - f(x),$$

is called the L_p-modulus of continuity $(1 \le p \le \infty)$ of the functions $f \in L_p = L_p(-\infty, \infty)$. The modulus of continuity in the space $C = C(-\infty, \infty)$ of uniformly continuous bounded functions is defined analogously.

It is well known that if $\omega(\delta)$ is the L_p-modulus of continuity for $1 \le p < \infty$ or the modulus of continuity in the space C, then the following properties hold:

$$\omega(\delta) \to 0 \quad \text{as} \quad \delta \to 0; \tag{1}$$

$$0 \leqslant \omega(\delta_2) - \omega(\delta_1) \leqslant \omega(\delta_2 - \delta_1) \leqslant M \quad \text{for} \quad 0 < \delta_1 < \delta_2. \tag{2}$$

Lebesgue [1] noted (in the periodic case) that every function $\omega(\delta)$ with the properties (1) and (2) is the modulus of continuity of some continuous periodic function. S. M. Nikol'skiĭ noted in [2] that (for $C(0, b)$) it is possible to take $\omega(x)$ as such a function. He also obtained a description of the set of moduli of continuity in C. It contains those and only those functions $\omega(\delta)$ for which the conditions (1) and (2) are satisfied. In this article we give the description of the set of L_2-moduli of continuity and we show that it is a proper subset of the set of moduli of continuity in C.

Let $L_2 = L_2(-\infty, \infty)$ denote the set of measurable real- or complex-valued functions for which the square of the modulus is integrable on $(-\infty, \infty)$. For $f \in L_2$ we introduce the function

---

AMS (MOS) subject classifications (1970). Primary 26A15; Secondary 42A16, 42A68, 42A88.

$$\Omega(h;\ f)=2\|f\|_2^2-\|\Delta(h)f\|_2^2,\quad -\infty<h<\infty. \tag{3}$$

The function $\Omega(h;\ f)$ is even, and vanishes in the limit as $h\to\infty$.

Denoting the Fourier transform of the function f by $\tilde{f}(\xi)=\mathfrak{F}[f](\xi)$, we have

$$\widetilde{\Delta(h)f}(\xi)=\tilde{f}(\xi)(e^{ih\xi}-1).$$

Hence, using Parseval's equality, we obtain

$$\|\Delta(h)f\|_2^2=2\|\tilde{f}\|_2^2-2\int_{-\infty}^{\infty}|\tilde{f}(\xi)|^2\cos h\xi d\xi$$

$$=2\|f\|_2^2-\int_{-\infty}^{\infty}(|\tilde{f}(\xi)|^2+|\tilde{f}(-\xi)|^2)\cos h\xi d\xi.$$

Thus,

$$\begin{gathered}\Omega(h;\ f)=\sqrt{2\pi}\mathscr{F}[\psi](h),\\ \psi(\xi)=|\tilde{f}(\xi)|^2+|\tilde{f}(-\xi)|^2\in L_1^+,\end{gathered} \tag{4}$$

where L_1^+ denotes the set of nonnegative even functions that are integrable on $(-\infty,\infty)$. The equality (4) states that

$$f\in L_2\Rightarrow\Omega(h;\ f)\in\mathscr{F}L_1^+. \tag{5}$$

We show that in fact

$$\{\Omega(h;\ f):f\in L_2\}=\mathscr{F}L_1^+. \tag{6}$$

Suppose that $\Omega(h)\in\mathfrak{F}L_1^+$. We consider the function

$$f(x)=\mathscr{F}\left[\sqrt{\frac{1}{\sqrt{2\pi}}\mathscr{F}[\Omega]}\right](x) \tag{7}$$

with the arithmetic value of the root. The function f is obviously even and real, and belongs to L_2. For it

$$|\tilde{f}(\xi)|^2+|\tilde{f}(-\xi)|^2=2|\tilde{f}(\xi)|^2=\frac{1}{\sqrt{2\pi}}\mathscr{F}[\Omega](\xi)=\psi(\xi),$$

so that, in view of (4),

$$\Omega(h;\ f)=\sqrt{2\pi}\mathscr{F}[\psi](h)=\Omega(h),$$

which implies (6).

We note that the set of functions $\Omega(h;\ f)$, $f\in L_2$, i.e. $\mathfrak{F}L_1^+$, makes up a subset of the set of positive definite functions. The latter are the Fourier-Stieltjes transforms of nondecreasing bounded functions, but they may be characterized by other properties, not connected with the Fourier transform (cf. [3], §8).

From (6) we also get in an obvious way the description of the class of L_2-moduli of continuity as functions of the form

$$\sup_{|h|\leq\delta}\sqrt{\varphi(0)-\varphi(h)},\qquad\varphi\in\mathscr{F}L_1^+. \tag{8}$$

A result analogous to (6) holds for the case of 2π-periodic functions. Let $f(x)$ be a measurable 2π-periodic function, and

$$\|f\|_{2^*}=\left\{\frac{1}{\pi}\int_{-\pi}^{\pi}|f(x)|^2\,dx\right\}^{\frac{1}{2}}.$$

We put

$$\omega(\delta,\ f)_{2^*}=\sup_{|h|\leqslant\delta}\|\Delta(h)f\|_{2^*}\quad(0\leqslant\delta\leqslant\pi),$$

so that $\omega(\delta, f)_{2^*}$ is the L_2^*-modulus of continuity of the function f.

A necessary and sufficient condition for $\omega(\delta)$ $(0\leq\delta\leq\pi)$ to be an L_2^*-modulus of continuity is that it have the representation

$$\omega(\delta)=\sup_{|h|\leqslant\delta}\sqrt{\varphi(0)-\varphi(h)},$$

where

$$\varphi(h)=\sum_{k=0}^{\infty}c_k\cos kh,\qquad c_k\geqslant 0,\qquad \sum_{k=0}^{\infty}c_k<\infty.$$

Indeed, if

$$f(x)\sim\frac{a_0}{2}+\sum_{k=1}^{\infty}(a_k\cos kx+b_k\sin kx),$$

then

$$\|\Delta(h)f\|_{2^*}^2=2\sum_{n=1}^{\infty}\rho_n^2-2\sum_{n=1}^{\infty}\rho_n^2\cos nh,$$

where $\rho_n^2=a_n^2+b_n^2\geq 0$, whence

$$\omega(\delta,\ f)_{2^*}=\sup_{|h|\leqslant\delta}\sqrt{\varphi(0)-\varphi(h)},$$

$$\varphi(h)=2\sum_{n=1}^{\infty}\rho_n^2\cos nh.$$

The converse statement is equally obvious.

The example given in [4], §12.4, shows that the function

$$(\sin\delta)^{\alpha}\qquad(0<\alpha\leqslant 1)$$

is an L_2^*-modulus of continuity.

We show that the set of L_2-moduli of continuity is a proper subset of the set of moduli of continuity in C. To see this we show that no L_2-modulus of continuity $\omega(\delta; f)_2$, $f\in L_2$, may coincide near zero with the linear function δ. Assuming the contrary and taking $\omega(\delta; f)$ to be an even function, we see that for some function $f\in L_2$, for all sufficiently small $t>0$,

$$\Delta^2(t)\,\Delta^2(-t)\,\omega^2(h;\ f)_2\big|_{h=0}=\Delta^2(t)\,\Delta^2(-t)\,h^2\big|_{h=0}=0. \tag{9}$$

On the other hand, by (4) and (3)

$$\Delta^2(t)\,\Delta^2(-t)\,\Omega(h;\ f) = -\Delta^2(t)\,\Delta^2(-t)\,\|\Delta(h)f\|_2^2 = \sqrt{2\pi}\mathscr{F}[\psi_t](h), \tag{10}$$

where

$$\psi_t(\xi) = |e^{-it\xi} - 1|^4\,\psi(\xi) \geqslant 0. \tag{11}$$

(9) and (10) imply that $\int_{-\infty}^{\infty}\psi_t(\xi)\,d\xi = 0$ for all sufficiently small $t > 0$, which, by virtue of (11), implies that $\psi(\xi) = 0$ almost everywhere. But then from (4) and (3) we have $f \equiv 0$, which contradicts the assumption and thus leads to the proof of our statement.

The corresponding results in the periodic case are established analogously.

BIBLIOGRAPHY

1. H. Lebesgue, *Sur le représentation trigonométrique approchée des fonctions satisfaisant à une condition de Lipschitz,* Bull. Soc. Math. France 38 (1910), 184–210.

2. S. M. Nikol'skiĭ, *La série de Fourier d'une fonction dont le module de continuité est donné,* C. R. (Dokl.) Acad. Sci. URSS 52 (1946), 191–194. MR 8, 149.

3. S. Bochner, *Monotone Funktionen, Stieltjessche Integrale und harmonische Analyse,* Math. Ann. 108 (1933), 378–410; English transl., *Lectures on Fourier integrals,* Princeton Univ. Press, Princeton, N. J., 1959, pp. 292–331. MR 21 #5851.

4. E. T. Whittaker and G. N. Watson, *A course of modern analysis,* 4th ed., Cambridge Univ. Press, New York, 1927.

Translated by JANE A. C. BURLAK

Trudy Mat. Inst. Steklov.
134 (1975)

Proc. Steklov Inst. Math.
134 (1975)

ON SOME CLASSES OF SOLUTIONS OF THE MAXWELL-EINSTEIN EQUATION

UDC 517.946.8

A. V. BICADZE AND V. I. PAŠKOVSKIĬ

Abstract. In the paper we construct three classes of solutions of the equation

$$\frac{\partial^2 w}{\partial z \partial \bar{z}} + \frac{1}{2(z+\bar{z})}\left(\frac{\partial w}{\partial z} + \frac{\partial w}{\partial \bar{z}}\right) - \frac{2}{w+\bar{w}}\frac{\partial w}{\partial z}\frac{\partial w}{\partial \bar{z}} = 0,$$

where

$$z = x + iy, \qquad \bar{z} = x - iy,$$
$$2\frac{\partial}{\partial z} = \frac{\partial}{\partial x} - i\frac{\partial}{\partial y}, \quad 2\frac{\partial}{\partial \bar{z}} = \frac{\partial}{\partial x} + i\frac{\partial}{\partial y}.$$

Solutions of the first class have the form $w = u + iu_1$, where

$$u_1 = x^2 \operatorname{Re} \int_0^1 f(y + ix - 2ixt)\sqrt{t(1-t)}\, dt,$$

$$u(x, y) = -\int_0^y \frac{\partial u_1(x, \tau)}{\partial x} d\tau + x \int_0^x \frac{\partial u(t, y)}{\partial y}\Big|_{y=0} \frac{dt}{t} + cx,$$

$f(\tau)$ is an arbitrary analytic function of the complex variable τ, while c is an arbitrary real constant.

The second and third classes comprise the functions

$$w = Ae^{\lambda v} + iB,$$
$$w = A(\operatorname{sech} \lambda v + i \tanh \lambda v),$$

where A, B and λ are arbitrary real constants, while $v(x, y)$ satisfies the equation

$$x\Delta v + v_x = 0.$$

As is well known, the solutions of this equation can be written in quadratures containing an arbitrary analytic function.

Bibliography: 3 titles.

The purpose of the present paper is to construct classes of solutions of the Maxwell-Einstein equation

$$\frac{\partial^2 w}{\partial z \partial \bar{z}} + \frac{1}{2(z+\bar{z})}\left(\frac{\partial w}{\partial z} + \frac{\partial w}{\partial \bar{z}}\right) - \frac{2}{w+\bar{w}}\frac{\partial w}{\partial z}\frac{\partial w}{\partial \bar{z}} = 0, \tag{1}$$

where

$$w(z) = u(x, y) + iu_1(x, y), \qquad z = x + iy, \qquad \bar{z} = x - iy,$$
$$2\frac{\partial}{\partial z} = \frac{\partial}{\partial x} - i\frac{\partial}{\partial y}, \qquad 2\frac{\partial}{\partial \bar{z}} = \frac{\partial}{\partial x} + i\frac{\partial}{\partial y}.$$

AMS (MOS) subject classifications (1970). Primary 35Q05; Secondary 35C15.

It is obvious that together with w the functions $\bar{w}$, $1/w$ and cw, where c is an arbitrary real constant, are solutions of (1).

We shall say that the function $w(z)$ belongs to class c^2 if $u(x, y) \in c^2$ and $u_1(x, y) \in c^2$. A function $w(z)$ of class c^2, satisfying equation (1), is called a regular solution of this equation.

Representing the left-hand side of (1) in the form

$$\frac{\partial^2 w}{\partial z \partial \bar{z}} - \frac{1}{2(z+\bar{z})}\left(\frac{\partial w}{\partial z} - \frac{\partial w}{\partial \bar{z}}\right) - \frac{2}{w+\bar{w}}\left(\frac{\partial w}{\partial \bar{z}} - \frac{w+\bar{w}}{2(z+\bar{z})}\right)\frac{\partial w}{\partial z}$$

and allowing for the fact that the equality

$$\frac{\partial^2 w}{\partial z \partial \bar{z}} = \frac{1}{2(z+\bar{z})}\left(\frac{\partial w}{\partial z} - \frac{\partial w}{\partial \bar{z}}\right)$$

stems from the equality

$$\frac{\partial w}{\partial \bar{z}} = \frac{w+\bar{w}}{2(z+z)}, \tag{2}$$

we arrive at the important conclusion that each solution $w(z)$ of class c^2 of the linear equation (2) is a solution of equation (1).

Writing (2) in the form of the system

$$x\left(\frac{\partial u}{\partial x} - \frac{\partial u_1}{\partial y}\right) - u = 0, \qquad \frac{\partial u}{\partial y} + \frac{\partial u_1}{\partial x} = 0, \tag{3}$$

we conclude that the class c^2 solutions $u(x, y)$and $u_1(x, y)$ of this system are the solutions, respectively, of the equations

$$\begin{aligned} &x^2 \Delta u - x\frac{\partial u}{\partial x} + u = 0, \\ &x \Delta u_1 - \frac{\partial u_1}{\partial x} = 0, \end{aligned} \tag{4}$$

where Δ is the Laplace operator.

Since for $x \neq 0$ the regular solutions of (4) are analytic functions of the variables x and y, in the notation

$$\xi = y + ix, \qquad \xi = y - ix,$$

$$u_1(x, y) = u_1\left(\frac{\xi - \xi}{2i}, \frac{\xi + \xi}{2}\right) = \omega(\xi, \xi)$$

this equation takes the form

$$\frac{\partial^2 \omega}{\partial \xi \partial \xi} + \frac{1}{2(\xi - \xi)}\frac{\partial \omega}{\partial \xi} - \frac{1}{2(\xi - \xi)}\frac{\partial \omega}{\partial \xi} = 0.$$

Using the well-known integral representation

$$\omega(x, y) = (y-x)^{1-\beta-\beta'} \int_0^1 f(x + (y-x)t)\, t^{-\beta}(1-t)^{-\beta'}\, dt$$

of the solutions of the Euler-Poisson-Darboux equation [1], [2]

$$\frac{\partial^2 \omega}{\partial x \partial y} - \frac{\beta'}{x-y}\frac{\partial \omega}{\partial x} + \frac{\beta}{x-y}\frac{\partial \omega}{\partial y} = 0, \tag{5}$$

we conclude that the formula

$$u_1(x, y) = x^2 \operatorname{Re} \int_0^1 f(y + ix - 2ixt) \sqrt{t(1-t)}\, dt, \tag{6}$$

where $f(\tau)$ is an arbitrary analytic function of the complex variable τ, yields the solutions of equation (4).

By a direct verification we easily see that the function $u_1(x, y)$ represented by formula (6) is a solution of (4).

On the basis of (6) we find from (3) that

$$u(x, y) = -\int_0^y \frac{\partial u_1(x, \tau)}{\partial x}\, d\tau + x \int_0^x \frac{\partial u_1(t, y)}{\partial y}\bigg|_{y=0} \frac{dt}{t} + cx, \tag{7}$$

where c is an arbitrary real constant.

Consequently the functions $w(z)$ whose real and imaginary parts are expressed by (7) and (6), respectively, constitute a class of regular solutions of (1).

In order to construct other classes of solutions of (1) we note that as a result of a change of the unknown solution $w = \phi(v)$, where $\phi(v)$ is an as yet unknown analytic function of the real variable v, this equation takes the form

$$\left[\frac{\partial^2 v}{\partial z \partial \bar{z}} + \frac{1}{2(z+\bar{z})}\left(\frac{\partial v}{\partial z} + \frac{\partial v}{\partial \bar{z}}\right)\right]\varphi' + \frac{\partial v}{\partial z}\frac{\partial v}{\partial \bar{z}}\left(\varphi'' - \frac{2}{\varphi + \bar{\varphi}}\varphi'^2\right) = 0.$$

Hence we see that if $\phi(v)$ and $v(z)$ are the solutions, respectively, of the equations

$$\varphi'' - \frac{2}{\varphi + \bar{\varphi}}\varphi'^2 = 0 \tag{8}$$

and

$$\frac{\partial^2 v}{\partial z \partial \bar{z}} + \frac{1}{2(z+\bar{z})}\left(\frac{\partial v}{\partial z} + \frac{\partial v}{\partial \bar{z}}\right) = 0, \tag{9}$$

then $w = \phi(v)$ is a solution of (1).

Equality (8) is a system of two ordinary second-order nonlinear differential equations in the real and the imaginary parts of ϕ.

A direct check easily convinces us that the functions

$$\varphi(v) = Ae^{\lambda v} + iB \tag{10}$$

and

$$\varphi(v) = A(\operatorname{sech} \lambda v + i \tanh \lambda v), \tag{11}$$

where A, λ and B are arbitrary real constants, are solutions of equation (8).

In the variables x and y we can write equation (9) as

$$x\Delta v + \frac{\partial v}{\partial x} = 0. \tag{12}$$

For solutions of the form $w = u(x, y) + iB$, equation (1) coincides with the equation

$$\Delta u + \frac{1}{x}\frac{\partial u}{\partial x} - \frac{1}{u}\left[\left(\frac{\partial u}{\partial x}\right)^2 + \left(\frac{\partial u}{\partial y}\right)^2\right] = 0,$$

which can be linearized by the substitution $u = e^{\lambda v}$, and by the same token we conclude that the formula $w = Ae^{\lambda v} + iB$, where $v(x, y)$ is a solution of (12), yields a second class of solutions of equation (1).

When expression (11) is chosen as $\phi(v)$, we have

$$w\bar{w} = u^2 + u_1^2 = A^2,$$

whence we find that

$$\frac{\partial^2 w}{\partial z \partial \bar{z}} + \frac{1}{2(z+\bar{z})}\left(\frac{\partial w}{\partial z} + \frac{\partial w}{\partial \bar{z}}\right) - \frac{2}{w+\bar{w}} \frac{\partial w}{\partial z} \frac{\partial w}{\partial \bar{z}}$$

$$= \left(1 - \frac{iu}{\sqrt{A^2 - u^2}}\right)\left\{\Delta u + \frac{1}{x}\frac{\partial u}{\partial x} - \frac{1}{u}\left(1 - \frac{u^2}{A^2 - u^2}\right)\left[\left(\frac{\partial u}{\partial x}\right)^2 + \left(\frac{\partial u}{\partial y}\right)^2\right]\right\}$$

and so, in the case being examined, equation (1) is equivalent to the equation

$$\Delta u + \frac{1}{x}\frac{\partial u}{\partial x} - \frac{1}{u}\left(1 - \frac{u^2}{A^2 - u^2}\right)\left[\left(\frac{\partial u}{\partial x}\right)^2 + \left(\frac{\partial u}{\partial y}\right)^2\right] = 0. \tag{13}$$

From what has been said above it is obvious that for sign-constant $u(x, y)$ the substitution $u = \operatorname{sech} \lambda v$ takes (13) into (12).

Consequently, functions of the form $w = A(\operatorname{sech} \lambda v + i \tanh \lambda v)$, where $v(x, y)$ is a solution of (12), constitute a third class of solutions of (1).

Thus, every time we have a solution $\phi(v)$ of equation (8), by substituting $w = \phi(v)$ equation (1) reduces to the linear equation (9), so that a new class of solutions of (1) is obtained as a result of the superposition $w = \phi[v(x, y)]$.

We now turn to equation (12). After the change of independent variables $\xi = y + ix$, $\bar{\xi} = y - ix$ in this equation, we obtain

$$\frac{\partial^2 \omega}{\partial \xi \partial \bar{\xi}} - \frac{1}{2(\xi - \bar{\xi})}\frac{\partial \omega}{\partial \xi} + \frac{1}{2(\xi - \bar{\xi})}\frac{\partial \omega}{\partial \bar{\xi}} = 0,$$

where

$$\omega(\xi, \bar{\xi}) = v\left(\frac{\xi - \bar{\xi}}{2i}, \frac{\xi + \bar{\xi}}{2}\right).$$

Since for $\beta + \beta' = 1$ the solutions of the Euler-Poisson-Darboux equation (5) is given by the formula [1]

$$\omega(x, y) = \int_0^1 f(x + (y - x)t)\, t^{-\beta}(1 - t)^{\beta - 1}\, dt$$

$$+ \int_0^1 f_1(x + (y - x)t)\, t^{-\beta}(1 - t)^{\beta - 1} \log[(y - x)t(1 - t)]\, dt,$$

it follows that the integral representation

$$v^{(1)}(x, y) = \operatorname{Re} \int_0^1 f(y + ix - 2ixt) \frac{dt}{\sqrt{t(1 - t)}}, \tag{14}$$

where $f(\tau)$ is an arbitrary analytic function of the complex variable τ, is valid for the solutions $v^{(1)}(x, y)$, analytic in x and y, of equation (12).

Analogously we conclude that the formula

$$v^{(2)}(x, y) = \operatorname{Re} \int_0^1 f(y + ix - 2ixt) \log[2ixt(1 - t)] \frac{dt}{\sqrt{t(1 - t)}} \tag{15}$$

yields a class of solutions $v^{(2)}(x, y)$, analytic in x and y in the halfplane $x > 0$, of equation (12) with a logarithmic singularity at $x = 0$.

Equating $f(\tau)$ to τ^n $(n = 0, 1, \dots)$, from (6), (14), and (15) we obtain a collection of solutions of equations (4) and (12):

$$u_1^{(n)}(x, y) = \sum_{k=0}^{n} \sum_{l=\left[\frac{k+1}{2}\right]}^{\left[\frac{n}{2}\right]} (-1)^{k+l} \binom{n}{k} \binom{n-k}{2l-k} 2^k \frac{\Gamma\left(k+\frac{3}{2}\right)\Gamma\left(\frac{3}{2}\right)}{(k+2)!} y^{n-2l} x^{2l},$$

$$v_n^{(1)}(x, y) = \sum_{k=0}^{n} \sum_{l=\left[\frac{k+1}{2}\right]}^{\left[\frac{n}{2}\right]} (-1)^{k+l} \binom{n}{k} \binom{n-k}{2l-k} 2^k \frac{\Gamma\left(k+\frac{1}{2}\right)\Gamma\left(\frac{1}{2}\right)}{k!} y^{n-2l} x^{2l},$$

$$v_n^{(2)}(x, y) = v_n^{(1)}(x, y) \log x + \sum_{k=0}^{n} \sum_{l=\left[\frac{k+1}{2}\right]}^{\left[\frac{n}{2}\right]} (-1)^{k+l} \binom{n}{k} \binom{n-k}{2l-k} \frac{2^k}{k!}$$
$$\times \Gamma\left(k+\frac{1}{2}\right)\Gamma\left(\frac{1}{2}\right)\left[\psi\left(k+\frac{1}{2}\right) - 2\psi(k+1) + \psi\left(\frac{1}{2}\right)\right] y^{n-2l} x^{2l}, \quad n = 0, 1, \dots$$

By allowing singularities in the analytic function $f(\tau)$ in (6), (14) and (15), we can construct solutions of (4) and (12) with specified singularities and, on the basis of (6), (7), (10) and (11), establish the nature of the corresponding solutions of equation (1).

The classes of solutions of (1) constructed above can prove useful both for the investigation of various boundary-value problems and for seeking the solutions of this equation with a specified asymptotic behavior as $z \to \infty$.

The authors announced the main conclusions of the present paper in the note [3].

BIBLIOGRAPHY

1. G. Darboux, *Leçons sur la théorie générale des surfaces et les applications géométriques du calcul infinitésimal.* Vol. II, 2nd ed., Gauthier-Villars, Paris, 1915, pp. 66–69.

2. I. N. Vekua, *New methods for solving elliptic equations*, OGIZ, Moscow, 1948, pp. 53–54. (Russian) MR 11, 598.

3. A. V. Bicadze and V. I. Paškovskiĭ, *On the theory of the Maxwell-Einstein equations*, Dokl. Akad. Nauk SSSR 216 (1974), 249–250 = Soviet Math. Dokl. 15 (1974), 762–764.

Translated by N. H. CHOKSY

Trudy Mat. Inst. Steklov.
134 (1975)

Proc. Steklov Inst. Math.
134 (1975)

CONDITIONS FOR THE UNIFORM BOUNDEDNESS OF TRIGONOMETRIC POLYNOMIALS IN THE L METRIC

UDC 517.52

Ja. S. BUGROV

Abstract. We obtain a new upper bound for a trigonometric cosine polynomial in the L metric, namely

$$\int_{-\pi}^{\pi}\left|\sum_{k=0}^{n}\lambda_k\cos kx\right|dx\leq c\sum_{k=0}^{N}|\Delta\lambda_{n_k}|+c\left\{\varkappa(n,\varphi)\sum_{k=0}^{N}|\Delta\lambda_{n_k}|^2\left[1+\sum_{\substack{l=0\\ l\neq k}}^{N}\varphi\left(\frac{c}{|n_k-n_l|}\right)\right]\right\}^{1/2},$$

where

$$\Delta\lambda_k=\lambda_k-\lambda_{k+1},\quad n_k=\psi(k)\geqslant k,\quad N=\psi^{-1}(n),\quad \varkappa(n,\varphi)=\int_{1/n}^{\pi}\frac{dx}{x^2\varphi'(x)}.$$

The function $\phi'(x)$ decreases to 0 as $x\to\infty$, and $\phi(0)=0$. It follows from this bound that all methods

$$\tau_{N\gamma}(f,x)=\frac{1}{N+1}\sum_{k=0}^{N}S_{k\gamma}(f,x),$$

where $S_k(f,x)$ are the Fourier partial sums of $f(x)$, are regular for all integers $\gamma\geq 1$.

Bibliography: 6 titles.

Let $f(x)$ be a continuous periodic function on $[-\pi,\pi]$, and

$$S(f)\equiv\frac{a_0}{2}+\sum_{k=1}^{\infty}A_k(x),\quad A_k(x)=a_k\cos kx+b_k\sin kx,$$

its Fourier series.

It is known that the Fourier series of some continuous functions do not converge pointwise. However, Fejér showed in 1904 that the Fourier arithmetic means

$$\sigma_n(f)=\frac{1}{n+1}\sum_{k=0}^{n}S_k(f),$$

where $S_k(f)$ are the partial sums of the Fourier series, converge uniformly to the continuous function $f(x)$.

If $\Lambda=\|\lambda_k^{(n)}\|$ is a lower triangular matrix, the functions

$$\tau_n(f)=\sum_{k=0}^{n}\lambda_k^{(n)}A_k(x)$$

AMS (MOS) subject classifications (1970). Primary 42A04, 42A24.

are called linear means. If $\tau_n(f)$ tends to a finite limit $s(x)$, we say that the Fourier series is summable to the limit $s(x)$ by the matrix Λ. A summation method is called regular if $\tau_n(f)$ converges uniformly to $f(x)$ for every $f(x) \in C(-\pi, \pi)$. If we insert the values of the Fourier coefficients we can write a linear mean in the form

$$\tau_n(f;\ x) = \frac{1}{\pi}\int_{-\pi}^{\pi} f(x+t) K_n(t)\, dt,$$

where

$$K_n(t) = \frac{\lambda_0^{(n)}}{2} + \sum_{k=1}^{n} \lambda_k^{(n)} \cos kt.$$

Thus each summation method generates a kernel $K_n(t)$, and our problem leads to the question of what properties of the kernel make the summation method regular.

It was shown by Lozinskiĭ and Nikol′skiĭ that necessary and sufficient conditions for a method to be regular are that

$$1)\ I_n = \frac{1}{\pi}\int_{-\pi}^{\pi} |K_n(t)|\, dt \leqslant H, \qquad 2)\ \lambda_k^{(n)} \to 1 \quad \text{as } n \to \infty,\ \forall k;\ \lambda_0^{(n)} = 1.$$

Condition 1) is not transparent, since it is not expressed in terms of the coefficients $\lambda_k^{(n)}$. It would be desirable to obtain a form of this condition expressed in terms of the $\lambda_k^{(n)}$. Hence we need to discover for which $\lambda_k^{(n)}$ the set of trigonometric cosine polynomials $K_n(t)$ have their norms bounded in the space L, uniformly in n.

Contributions to this problem have been made by many authors: Hille, Tamarkin, Sidon, Karamata, Nikol′skiĭ, Stečkin, Efimov, Teljakovskiĭ, Fomin, etc. These authors obtained upper and lower bounds for I_n in terms of the $\lambda_k^{(n)}$ and their differences

$$\Delta\lambda_k^{(n)} = \lambda_k^{(n)} - \lambda_{k+1}^{(n)}, \qquad \Delta^2\lambda_k^{(n)} = \Delta(\Delta\lambda_k^{(n)}).$$

We quote the following result:

$$I_n \leqslant c \sum_{k=0}^{n} |\Delta\lambda_k^{(n)}| + c\left(n^{p-1}\sum_{k=0}^{n} |\Delta\lambda_k^{(n)}|^p\right)^{\frac{1}{p}} \qquad (1 < p \leqslant 2). \tag{1}$$

For $p = 2$ this was proved by Fomin [1]. A proof of (1) for $1 < p < 2$ was presented by Stečkin in 1965 at a conference in Tartu and was mentioned later in Fomin's papers.

Our objective is to obtain an upper bound for I_n which is, in many cases, better than (1) in the sense that the right-hand side of (1) $\to \infty$ as $n \to \infty$, whereas the right-hand side of our inequality (see Theorem 1) is bounded in n. A short account of the present paper appeared in [3].

We shall write λ_k instead of $\lambda_k^{(n)}$, and we shall suppose that $\lambda_k = 0$ for $k \geq n+1$. We begin by proving two lemmas.

Lemma 1. *Let $\phi(x)$ increase on $[0, \infty)$ and have the properties that $\phi(0) = 0$ and $\phi'(x)$ decreases to zero as $x \to \infty$. Then*

$$\left|\int_0^T \cos ax\varphi'(x)\,dx\right| \leqslant \varphi\left(\frac{3\pi}{2a}\right)$$

for each $a > 0$ and all $T \geq \pi/2a$.

Proof. Since $T \geq \pi/2a$, there is an integer N such that

$$\frac{2N-1}{2a}\pi \leqslant T < \frac{2N+1}{2a}\pi.$$

Then

$$I = \int_0^T \cos ax\varphi'(x)\,dx = \int_0^{\frac{\pi}{2a}} \cos ax\varphi'(x)\,dx$$
$$+\sum_{k=1}^{N-1}\int_{\pi(2k-1)/2a}^{\pi(2k+1)/2a} \cos ax\varphi'(x)\,dx + \int_{\pi(2N-1)/2a}^{T} \cos ax\,\varphi'(x)\,dx.$$

The terms in the summation have decreasing moduli and alternate in sign with the first term negative. Hence for $T \geq 3\pi/2a$

$$|I| \leqslant \int_0^{\frac{\pi}{2a}} \cos ax\varphi'(x)\,dx + \left|\int_{\frac{\pi}{2a}}^{\frac{3\pi}{2a}} \cos ax\varphi'(x)\,dx\right| \leqslant \varphi\left(\frac{\pi}{2a}\right) + \varphi\left(\frac{3\pi}{2a}\right) - \varphi\left(\frac{\pi}{2a}\right) = \varphi\left(\frac{3\pi}{2a}\right).$$

On the other hand, if $T < 3\pi/2a$ $(N = 1)$ then

$$|I| \leqslant \int_0^{\frac{\pi}{2a}} \cos ax\varphi'(x)\,dx + \left|\int_{\frac{\pi}{2a}}^{T} \cos ax\varphi'(x)\,dx\right| \leqslant \varphi(T) \leqslant \varphi\left(\frac{3\pi}{2a}\right),$$

since $\phi(x)$ increases. This completes the proof of the lemma.

Remark. The functions $\phi(x) = x^{\alpha}$ $(0 < \alpha < 1)$ satisfy the hypotheses of the lemma.

Lemma 2. *Let $\phi(x)$ satisfy the hypotheses of Lemma 1; then*

$$M_n \equiv \int_0^{\pi}\left|\sum_{k=0}^{n} a_k \begin{Bmatrix}\cos(k+\alpha)x\\ \sin(k+\alpha)x\end{Bmatrix}\right|^2 d\varphi(x) \leqslant c\sum_{k=0}^{n} a_k^2\left[1 + \sum_{\substack{l=0\\ l\neq k}}^{n}\varphi\left(\frac{c}{|k-l|}\right)\right],$$

where a_k $(k = 0, 1, \ldots, n)$ are real; $\alpha \geq 0$.

Proof. If we square the sum in the integrand, we obtain

$$M_n=\int_0^\pi\left[\sum_{k=0}^n a_k^2\begin{Bmatrix}\cos(k+\alpha)x\\ \sin(k+\alpha)x\end{Bmatrix}^2\right.$$

$$\left.+2\sum_{k\neq l}\sum a_k a_l\begin{Bmatrix}\cos(k+\alpha)x\cos(l+\alpha)x\\ \sin(k+\alpha)x\sin(l+\alpha)x\end{Bmatrix}\right]d\varphi(x)$$

$$\leqslant c\sum_{k=0}^n a_k^2+c\sum_{k\neq l}\sum(a_k^2+a_l^2)\left|\int_0^\pi[\cos(k-l)x\pm\cos(k+l+2\alpha)x]\,d\varphi(x)\right|.$$

Since the expression in the double sum is symmetric in k and l, we have

$$M_n\leqslant c\sum_{k=0}^n a_k^2+c\sum_{k\neq l}\sum a_k^2\left(\left|\int_0^\pi\cos(k-l)x\,d\varphi(x)\right|+\left|\int_0^\pi\cos(k+l+2\alpha)x\,d\varphi(x)\right|\right).$$

Applying Lemma 1 and remembering that $\phi(x)$ increases, we have

$$M_n\leqslant c\sum_{k=0}^n a_k^2\left[1+\sum_{\substack{l=0\\ l\neq k}}^n\varphi\left(\frac{3\pi}{2|k-l|}\right)\right],$$

as was to be proved.

Remark. If the sequence $\{a_k\}$ is lacunary, i.e. its only nonzero elements are $a_{n_1},\dots,a_{n_N}$, then

$$M_n\leqslant c\sum_{k=0}^N a_{n_k}^2\left[1+\sum_{\substack{l=0\\ l\neq k}}^N\varphi\left(\frac{c}{|n_k-n_l|}\right)\right].$$

Theorem 1. *If the only nonzero differences* $\Delta\lambda_n$ *are* $\Delta\lambda_{n_k}$ $(k=0,1,\dots,N)$, *then the integral* $I_n=(1/n)\int_{-\pi}^{\pi}|K_n(t)|\,dt$ $(n\geq 1)$ *admits the estimate*

$$I_n\leqslant cM(\lambda,n)+c\left\{\varkappa(n,\varphi)\sum_{k=0}^N|\Delta\lambda_{n_k}|^2\left[1+\sum_{\substack{l=0\\ l\neq k}}^N\varphi\left(\frac{c}{|n_k-n_l|}\right)\right]\right\}^{1/2},$$

where $\phi(x)$ *is any function satisfying the hypotheses of Lemma* 1,

$$\varkappa(n,\varphi)=\int_{1/n}^{\pi}\frac{dt}{t^2\varphi'(t)},\qquad M(\lambda,n)=\min\left\{\sum_{k=0}^n|\Delta\lambda_k|,\ \sum_{k=0}^n\frac{|\lambda_k|}{n-k+1},\ \max_{0\leqslant k\leqslant n}|\lambda_k|\right\},$$

and c *is a constant.*

Proof. The kernel $K_n(t)$ $(n\geq 1)$ admits, for every t, the following estimates:

$$|K_n(t)|=\left|\frac{\lambda_0}{2}+\sum_{k=1}^n\lambda_k\cos kt\right|\leqslant cn\max_{0\leqslant k\leqslant n}|\lambda_k|;\tag{2}$$

$$|K_n(t)|=\left|\sum_{k=0}^n\frac{\lambda_k}{n-k+1}(n-k+1)\cos kt\right|\leqslant cn\sum_{k=0}^n\frac{|\lambda_k|}{n-k+1}.\tag{3}$$

If we apply Abel's transformation (cf. [2], vol. I, p. 3), we can write $K_n(t)$ in the form

$$K_n(t) = \sum_{k=0}^{n} \Delta\lambda_k D_k(t), \tag{4}$$

where

$$D_k(t) = \frac{1}{2} + \sum_{j=0}^{k} \cos jt = \frac{\sin [(2k+1)/2]\, t}{2 \sin (t/2)}$$

is the Dirichlet kernel. Since $|D_k(t)| \leq k+1$, we obtain from (4) that

$$|K_n(t)| \leqslant cn \sum_{k=0}^{n} |\Delta\lambda_k|. \tag{5}$$

Then, since $K_n(t)$ is even, we have

$$I_n \leqslant c \int_0^{\pi} |K_n(t)|\, dt \leqslant c \int_0^{\frac{1}{n}} |K_n(t)|\, dt + c \int_{\frac{1}{n}}^{\pi} |K_n(t)|\, dt.$$

We estimate the first integral by one of (2), (3) or (5), and the second integral by using the inequality $\sin(t/2) \geq t/\pi$ $(0 < t < \pi)$. Then

$$I_n \leqslant c \min \left\{ \max_{0 \leqslant k \leqslant n} |\lambda_k|,\ \sum_{k=0}^{n} \frac{|\lambda_k|}{n-k+1},\ \sum_{k=0}^{n} |\Delta\lambda_k| \right\}$$

$$+ c \int_{\frac{1}{n}}^{\pi} \left| \sum_{k=0}^{n} \Delta\lambda_k \sin\left(k + \frac{1}{2}\right) t \right| \frac{dt}{t}$$

$$\leqslant cM(\lambda, n) + c \int_{\frac{1}{n}}^{\pi} \left| \sum_{k=0}^{N} \Delta\lambda_{n_k} \sin\left(n_k + \frac{1}{2}\right) t \right| \frac{dt}{t}.$$

If we multiply and divide the integrand by $\sqrt{\phi'(t)}$, where $\phi(t)$ satisfies the hypotheses of Lemma 1, and apply Schwarz's inequality, we obtain

$$I_n \leqslant cM(\lambda, n) + c \left\{ \int_{\frac{1}{n}}^{\pi} \left| \sum_{k=0}^{N} \Delta\lambda_{n_k} \sin\left(n_k + \frac{1}{2}\right) \right|^2 d\varphi(t) \int_{\frac{1}{n}}^{\pi} \frac{dt}{t^2 \varphi'(t)} \right\}^{\frac{1}{2}}$$

$$\leqslant cM(\lambda, n) + c \left\{ \varkappa(n, \varphi) \int_0^{\pi} \left| \sum_{k=0}^{N} \Delta\lambda_{n_k} \sin\left(n_k + \frac{1}{2}\right) t \right|^2 d\varphi(t) \right\}^{\frac{1}{2}}.$$

Finally, applying Lemma 2, we obtain

$$I_n \leqslant cM(\lambda, n) + c \left\{ \varkappa(n, \varphi) \sum_{k=0}^{N} |\Delta\lambda_{n_k}|^2 \left[1 + \sum_{\substack{l=0 \\ l \neq k}}^{N} \varphi\left(\frac{c}{|n_k - n_l|} \right) \right] \right\}^{\frac{1}{2}},$$

and the theorem is established.

Corollary 1. *If* $\phi(t) = t^{\alpha}$ $(0 < \alpha < 1)$, *then*

$$I_n \leqslant cM(\lambda, n) + c\left\{n^{\alpha}\sum_{k=0}^{N}|\Delta\lambda_{n_k}|^2\left(1+\sum_{\substack{l=0\\ l\neq k}}^{N}\frac{1}{|n_k - n_l|^{\alpha}}\right)\right\}^{\frac{1}{2}}.$$

This follows from the fact that

$$\varkappa(n, \varphi) = c\int_{\frac{1}{n}}^{\pi}\frac{dt}{t^2 t^{\alpha-1}} = cn^{\alpha}.$$

Theorem 2. *If the only nonzero differences* $\Delta\lambda_n$ *are* $\Delta\lambda_{n_k}$ $(k = 0, 1, \ldots, N)$ *and*

$$\varkappa(n, \varphi)\sum_{k=0}^{N}|\Delta\lambda_{n_k}|^2\left[1+\sum_{\substack{l=0\\ l\neq k}}^{N}\varphi\left(\frac{c}{|n_k - n_l|}\right)\right] \leqslant c,$$

$M(\lambda, n) \leq c$, *then* $I_n \leq c$ *for all* n.

This follows from Theorem 1.

Corollary 2. *If* $M(\lambda, n) \leq c$ *and*

$$n^{\alpha}\sum_{k=0}^{N}|\Delta\lambda_{n_k}|^2\left(1+\sum_{\substack{l=0\\ l\neq k}}^{N}\frac{1}{|n_k - n_l|^{\alpha}}\right) \leqslant c,$$

then $I_n \leq c$ *for all* n.

Example. The methods

$$\tau_{N^{\gamma}}(f, x) = \frac{1}{N+1}\sum_{k=0}^{N} S_{k^{\gamma}}(f, x)$$

are regular for all $\gamma \geq 1$. (For $\gamma = 1$ we have Fejér's method. Consequently we shall suppose that $\gamma > 1$.) In fact,

$$\lambda_0 = 1, \quad \lambda_{k^{\gamma}+1} = \ldots = \lambda_{(k+1)^{\gamma}} = 1 - \frac{k+1}{N+1} \quad (k = 0, 1, \ldots, N-1);$$

$$\Delta\lambda_i = 0 \ \ (i = k^{\gamma}+1, \ldots, (k+1)^{\gamma} - 1); \quad \Delta\lambda_{k^{\gamma}} = \frac{1}{N+1}, \quad n = N^{\gamma}, \ N = n^{\frac{1}{\gamma}}.$$

In addition, $M(\lambda, n) \leq 1$; since we have $n_k = k^{\gamma}$, it follows that

$$n_k - n_l = k^{\gamma} - l^{\gamma} = (k-l)\cdot(k^{\gamma-1} + k^{\gamma-2}l + \ldots + l^{\gamma-1}).$$

Hence

$$\sum_{l=0}^{k-1}\frac{1}{|n_k - n_l|^{\alpha}} = \sum_{l=0}^{k-1}\frac{1}{|k-l|^{\alpha}(k^{\gamma-1} + \ldots + l^{\gamma-1})^{\alpha}}$$

$$\leqslant \frac{1}{k^{(\gamma-1)\alpha}} \sum_{l=0}^{k-1} \frac{1}{|k-l|^{\alpha}} = \frac{1}{k^{(\gamma-1)\alpha}} \sum_{l=1}^{k} \frac{1}{l^{\alpha}} \leqslant ck^{1-\alpha\gamma} \qquad (0<\alpha<1).$$

Similarly

$$\sum_{l=k+1}^{N} |n_k - n_l|^{-\alpha} \leqslant k^{(1-\gamma)\alpha} \sum_{l=k+1}^{N} |l-k|^{-\alpha} \leqslant k^{(1-\gamma)\alpha} \sum_{l=1}^{N-k} l^{-\alpha} \leqslant ck^{(1-\gamma)\alpha}(N-k)^{1-\alpha}.$$

Now

$$A \equiv n^{\alpha} \sum_{k=1}^{N} |\Delta\lambda_{k^{\gamma}}|^2 \left(1 + \sum_{l=0}^{k-1} |k^{\gamma} - l^{\gamma}|^{-\alpha} + \sum_{l=k+1}^{N} |l^{\gamma} - k^{\gamma}|^{-\alpha}\right)$$

$$\leqslant cn^{\alpha}N^{-2} \sum_{k=1}^{N} (1 + k^{1-\alpha\gamma} + k^{(1-\gamma)\alpha}(N-k)^{1-\alpha}).$$

Take $\alpha < 1/\gamma \leq 1$; then $n^{\alpha} = N^{\alpha\gamma}$ and

$$A \leqslant cN^{\alpha\gamma-2} \sum_{k=1}^{N} (k^{1-\alpha\gamma} + (N-k)^{1-\alpha} k^{(1-\gamma)\alpha})$$

$$\leqslant c + cN^{\alpha\gamma-2} \sum_{k=1}^{\left[\frac{N}{2}\right]} (N-k)^{1-\alpha} k^{(1-\gamma)\alpha} + cN^{\alpha\gamma-2} \sum_{k=\left[\frac{N}{2}\right]+1}^{N} (N-k)^{1-\alpha} k^{(1-\gamma)\alpha}$$

$$\leqslant c + cN^{\alpha\gamma-1-\alpha} \sum_{k=1}^{N} k^{(1-\gamma)\alpha} + cN^{\alpha-2} \sum_{k=1}^{N} (N-k)^{1-\alpha}$$

$$\leqslant c + cN^{\alpha-2} \sum_{k=1}^{N} k^{1-\alpha} \leqslant c.$$

Consequently we infer from Corollary 2 that $I_n \leq c$ uniformly in n, and consequently the methods $\tau_{N^{\gamma}}(f;\, x)$ are regular for $\gamma \geq 1$.

Remark. We note that when $\gamma > 1$ the Stečkin-Fomin theorem does not yield the regularity of $\tau_{N^{\gamma}}(f;\, x)$, since for $1 < p \leq 2$

$$n^{p-1} \sum_{k=1}^{N} |\Delta\lambda_{n_k}|^p = N^{\gamma(p-1)+1-p} = N^{(\gamma-1)(p-1)} \to \infty \quad \text{as} \quad N \to \infty.$$

This shows that when $\{\Delta\lambda_k\}$ has gaps our estimate in Theorem 1 is stronger than the estimate (1) of Stečkin and Fomin. It was shown in [4] ($\gamma = 2$) and [5] ($\gamma \geq 3$) that the linear means $\tau_{N^{\gamma}}(f;\, x)$ converge to $f(x)$ at all Lebesgue points. In particular, when $f(x)$ is continuous it can be shown that $\tau_{N^{\gamma}}(f;\, x)$ converges uniformly to $f(x)$, although this is not explicitly mentioned in [4] or [5].

The regularity of the method $\tau_{N^{\gamma}}(f;\, x)$ can also be established by using Trigub's [6] sufficient condition for the regularity of summation methods:

$$\sum_{k=0}^{n} |\Delta\lambda_k| \leqslant c, \quad \forall k_1,\, k_2, \quad |\lambda_{k_1} - \lambda_{k_2}| \leqslant c \left|\frac{k_1 - k_2}{n}\right|^{\alpha}, \quad \alpha > 0.$$

It is an open question whether there is a method whose regularity follows from our criterion but not from Trigub's. However, there exist methods whose regularity follows from Trigub's test but not from ours.

BIBLIOGRAPHY

1. G. A. Fomin, *On linear methods of summability of Fourier series*, Mat. Sb. 65 (**107**) (1964), 144–152. (Russian) MR **29** #6243.

2. A. Zygmund, *Trigonometric series*, Vols. 1, 2, 2nd rev. ed., Cambridge Univ. Press, New York, 1959. MR **21** #6498.

3. Ja. S. Bugrov, *On the regularity of linear summability methods for Fourier series*, Dokl. Akad. Nauk SSSR **217** (1974), 505–507 = Soviet Math. Dokl. **15** (1974), 1068–1071.

4. Z. Zalcwasser, *Sur la sommabilité des séries de Fourier*, Studia Math. 6 (1936), 82–88.

5. S. Izumi and T. Kawata, *Notes on Fourier series. X. Summability*, Tôhoku Math. J. 46 (1939), 154–158. MR **1**, 225.

6. R. M. Trigub, *Summability and absolute convergence of Fourier series in the large*, Metric Questions of the Theory of Functions and Mappings, no. 2, "Naukova Dumka", Kiev, 1971, pp. 173–266. (Russian) MR **45** #4057.

Translated by R. P. BOAS

Trudy Mat. Inst. Steklov.
134 (1975)

Proc. Steklov Inst. Math.
134 (1975)

ON INTEGRATION BY PARTS AND A PROBLEM ON COMPOSITION OF ABSOLUTELY CONTINUOUS FUNCTIONS WHICH ARISES IN THIS CONNECTION

UDC 517.51

V. I. BURENKOV

Abstract. The following (and also a somewhat more general) theorem is proved. If $f(x)$ has an absolutely continuous kth derivative $f^{(k)}(x)$ on $[a, b]$, and if $\Phi(x)$ is absolutely continuous on $[c, d]$, $c = \min_{[a,b]}f(x)$, $d = \max_{[a,b]}f(x)$, then the function $\Phi(f(x))f^{(k)}(x)$ is also absolutely continuous on $[a, b]$ (at the same time $\Phi(f(x))$ need not be absolutely continuous). This assertion enables us to integrate by parts in the integral $\int_a^b f^{(k+1)}(x)\Phi(f(x))dx$.

Bibliography: 8 titles.

§1. Introduction. On integration by parts

The following theorem is known from the theory of functions of a real variable (see [1], Russian p. 250).

Theorem 1. *If f and g are absolutely continuous functions on an interval $[a, b]$, then*

$$\int_a^b f'g\,dx = fg\,|_a^b - \int_a^b fg'\,dx. \tag{1}$$

The operation of integration by parts is used very often in the study of partial differential equations, and in many cases Theorem 1 turns out to be insufficient. We introduce an example in this direction. It is often necessary to integrate by parts the expression

$$\int_a^b u''v\,dx, \tag{2}$$

where $u = u(x)$ is a twice continuously differentiable function of one variable x, and the function $v = v(x)$ is chosen in a special way depending on u; for example $v = \Phi(u)$, where Φ is an absolutely continuous function (in particular, $\Phi(u) = (|u|^p)'$; here the case $1 < p < 2$ will be of interest to us).

We cannot apply Theorem 1 in the above situation since the composite $\Phi(u)$,

---

AMS (MOS) subject classifications (1970). Primary 26A42, 26A63; Secondary 26A72.

in general, need not be absolutely continuous. We note that this may happen even for a function $\Phi(u)$ of very simple form, for example for $\Phi(u) = |u|^\gamma$, $0 < \gamma < ½$. If for u we take $f_{\alpha,\beta}(x) = x^\alpha(\sin(1/x^\beta) + 2)$ for $0 < x \le 1$ and $f_{\alpha,\beta}(0) = 0$, where $\alpha > 0$ and $\beta > 0$, and we take $\beta > 2\gamma/(1 - 2\gamma)$ and $2 + 2\beta < \alpha \le \beta/\gamma$, then the function $f_{\alpha,\beta}(x)$ is twice continuously differentiable (since $\alpha > 2 + 2\beta$); at the same time the composite $\Phi(u) = x^{\alpha\gamma}(\sin(1/x^\beta) + 2)^\gamma$ is not absolutely continuous on $[0, 1]$ (since it follows from the condition $\alpha \le \beta/\gamma$ that the derivative of this function is not integrable on $[0, 1]$).

Nevertheless, it is possible to integrate by parts in (2), if the following considerations are used. We first relax the conditions of Theorem 1.

Theorem 2. *Suppose that the following conditions hold:*
1) *The derivatives f' and g' exist almost everywhere on $[a, b]$.*
2) *At least one of the products $f'g$ or fg' is integrable.*
3) *The product fg is absolutely continuous.*
Then (1) *holds.*

This theorem is not difficult to prove. In fact, we have $(fg)' = f'g + fg'$ almost everywhere on $[a, b]$. Since the product fg is absolutely continuous, the function $(fg)'$ is integrable, which means that by 2) the functions $f'g$ and fg' are also integrable. Integrating the equality $(fg)' = f'g + fg'$ and taking into account again that fg is absolutely continuous, we obtain the desired equality (1).

The following theorem, which was formulated by the author in [2], plays an essential role in the use of integration by parts in (2).

Theorem 3. *Suppose that a function $f(x)$ is differentiable everywhere on $[a, b]$ and that its derivative $f'(x)$ is absolutely continuous. In addition, suppose that the function $\Phi(x)$ is absolutely continuous on the interval $[c, d]$ ($c = \min_{[a,b]} f(x)$; $d = \max_{[a,b]} f(x)$). Then the function*

$$\Phi(f(x))\, f'(x) \tag{3}$$

is also absolutely continuous on $[a, b]$.

We note at once that under the conditions of Theorem 1 the factor $\Phi(f(x))$ is not necessarily absolutely continuous; thus, multiplication of $\Phi(f(x))$ by the absolutely continuous derivative $f'(x)$ improves the properties of the product in comparison with $\Phi(f(x))$ by making this product absolutely continuous.

In the example (2) considered above, $f = u'$ and $g = \Phi(u)$. By Theorem 3 the product $fg = \Phi(u)u'$ is absolutely continuous. In addition, the function $f'g = u''\Phi(u)$ is integrable, and $f' = u''$ exists everywhere. Thus,

$$\int_a^b u''\Phi(u)\,dx = \Phi(u)\,u'\Big|_a^b - \int_a^b (\Phi(u))'\,u'\,dx, \tag{4}$$

if the derivative $(\Phi(u))'$ exists almost everywhere.

Of the most interest from the point of view of applications is the case where $\Phi(0) = 0$ and $\Phi(u)$ fails to have a derivative only at $u = 0$.

If, in addition, it is known that $u(x) \neq 0$, $x \in (a, b)$, then it follows from (4) that

$$\int_a^b u''\Phi(u)\,dx = \Phi(u)\,u'\,|_a^b - \int_a^b \Phi'(u)\,(u')^2\,dx. \tag{4'}$$

If $u(x)$ may vanish on (a, b), we can argue in the following way. We denote by S the open set $S = \{x \in (a, b)\colon u(x) \neq 0\}$ and let $S = \bigcup_1^\infty (a_i, b_i)$, where the (a_i, b_i) are open intervals; moreover, if $u(a) \neq 0$ let $a_1 = a$, and if $u(b) \neq 0$ let $b_2 = b$. Taking into account that $u(b_1) = u(a_2) = 0$ and $u(a_i) = u(b_i) = 0$, $i = 3, 4, \dots$, we obtain

$$\int_a^b u''\Phi(u)\,dx = \int_a^{b_1} u''\Phi(u)\,dx + \int_{a_2}^b u''\Phi(u)\,dx + \sum_{i=3}^\infty \int_{a_i}^{b_i} u''\Phi(u)\,dx$$
$$= -\Phi(u(a))\,u'(a) + \Phi(u(b))\,u'(b) - \sum_{i=1}^\infty \int_{a_i}^{b_i} \Phi'(u)\,(u')^2\,dx.$$

Thus, in this case the formula for integration by parts reduces to the form

$$\int_a^b u''\Phi(u)\,dx = \Phi(u)\,u'\,|_a^b - \int_S \Phi'(u)\,(u')^2\,dx. \tag{5}$$

The proof of Theorem 3 is essentially based on the analogous theorem for functions of bounded variation.

Theorem 4. *Let $\Phi(x)$ be a function of bounded variation on $[c, d]$, and suppose that $f(x)$ is differentiable everywhere on $[a, b]$ and that its derivative $f'(x)$ is of bounded variation on $[a, b]$. Then the function $\Phi(f(x))f'(x)$ is also of bounded variation on $[a, b]$.*

Below, in §2 and 3, we prove Theorems 3 and 4, and generalize these theorems to the case where, for example, instead of (3) we consider the expression

$$\Phi(f(x))\,f^{(k)}(x). \tag{6}$$

§2. Theorems on composition of functions of bounded variation

The question of composition of absolutely continuous functions was studied in detail in papers by G. M. Fihtengol'c [3] and by N. K. Bari (see the detailed account in [4]); the latter investigated composition of functions of bounded variation [5].

It was established in [3], in particular, that for any absolutely continuous function $\Phi(x)$ the composite $\Phi(f(x))$ is absolutely continuous if and only if $f(x)$ is absolutely continuous and the solution of the equation $f(x) = y_0$ consists of isolated points and intervals, the total number of which does not exceed a number N that is independent of y_0. Hence it follows that, in general, we cannot make the composite $\Phi(f(x))$ absolutely continuous by improving the differential properties of $f(x)$. In particular, we can choose f and Φ so that $f(x)$ is infinitely differentiable, Φ absolutely continuous, but $\Phi(f(x))$ is not absolutely continuous. A similar example can be constructed also in the case where $\Phi(x)$ is a function of bounded variation.

On the other hand, it is known that the product $\Phi(f(x))f'(x)$ may have better properties than $\Phi(f(x))$. For example, if Φ is integrable, and $f(x)$ absolutely continuous and monotone, then $\Phi(f(x))f'(x)$ is integrable, even if $\Phi(f(x))$ is not integrable (see, for example, Natanson [1], Russian p. 245). We note further that if $\Phi(x)$ is of bounded variation and $f(x)$ is absolutely continuous, then $\Phi(f(x))f'(x)$ is integrable (de la Vallée-Poussin [6], Chapter VI, §2, paragraph 183).

We explain below the extent to which the properties of $\Phi(f(x))G(f^{(k)})$ are improved in comparison with $\Phi(f(x))$.

Suppose that $f(x)$ is defined on an interval $[a, b]$ and has a derivative of order k; we put

$$c = \inf_{[a,b]} f(x), \quad d = \sup_{[a,b]} f(x), \qquad \alpha = \inf_{[a,b]} f^{(k)}(x), \qquad \beta = \sup_{[a,b]} f^{(k)}(x). \tag{7}$$

We introduce the following norm:

$$\| f \|_{[a,b]} = \sup_{[a,b]} | f(x) | + \operatorname*{Var}_{[a,b]} f(x). \tag{8}$$

Theorem 5. *Suppose that the following conditions hold:*

1) *$f(x)$ has a derivative $f^{(k)}(x)$ of order k everywhere on $[a, b]$.*

2) *$\Phi(x)$ is of bounded variation on $[c, d]$.*

3) *$G(x)$ is defined on $[\alpha, \beta]$; moreover, $G(0) = 0$ if $0 \in [\alpha, \beta]$.*

4) *$G(f^{(k)}(x))$ is of bounded variation on $[a, b]$.*

Then the function

$$\Phi(f(x))\, G(f^{(k)}(x)) \tag{9}$$

is of bounded variation on $[a, b]$; moreover,

$$\| \Phi(f)\, G(f^{(k)}) \|_{[a,b]} \leqslant (k+1) \| \Phi \|_{[c,d]} \| G(f^{(k)}) \|_{[a,b]}. \tag{10}$$

We need the following lemma for a proof of this theorem.

Lemma 1. *Suppose that the functions $\phi(x)$ and $\psi(x)$ are given on $[a, b]$, and let*

$$S = \{x \in (a, b) : \psi(x) \neq 0\}.$$

Assume that $S \subset \bigcup_i (a_i, b_i) \subset [a, b]$. *Then*[1]

$$\operatorname*{Var}_{[a, b]} \varphi(x)\psi(x) \leqslant \sup_{[a, b]} |\varphi(x)| \operatorname*{Var}_{[a, b]} \psi(x) + \sum_i \operatorname*{Var}_{[a_i, b_i]} \varphi(x)\psi(x). \tag{11}$$

Proof. 1. Without loss of generality we may assume that the intervals (a_i, b_i) are pairwise disjoint and that in each of them there is at least one point of the set S. We put

$$M = [a, b] \setminus \bigcup_i (a_i, b_i)$$

($\psi(x) = 0$ on M after we exclude, possibly, the points a and b). We remove the interior points of M, which represent a finite or countably infinite union of pairwise disjoint intervals (a_i', b_i') such that $\psi(x) \equiv 0$ on $[a_i', b_i']$. We obtain the closed set

$$L = \Big([a, b] \setminus \bigcup_i (a_i, b_i)\Big) \setminus \bigcup_i (a_i', b_i')$$

which is nowhere dense in $[a, b]$.

We consider the following sets:

$$K = (a, b) \setminus L = \Big(\bigcup_i (a_i, b_i)\Big) \cup \Big(\bigcup_i (a_i', b_i')\Big) \equiv \bigcup_i (\alpha_i, \beta_i),$$

$$K_n = \bigcup_{\beta_i - \alpha_i \geqslant \frac{1}{n}} (\alpha_i, \beta_i) \equiv \bigcup_{i=1}^{i_n} (\alpha_i, \beta_i),$$

$$L_n = [a, b] \setminus K_n;$$

the set L_n consists of a finite number of intervals or isolated points:

$$L_n = \bigcup_{k=1}^{k_n} [\gamma_k, \delta_k], \qquad \gamma_k \leqslant \delta_k, \qquad k_n = i_n + 1.$$

2. Suppose we are given a partition of $[a, b]$: $a = x_1 < x_2 < \cdots < x_s = b$. The points of this partition, $x_2, \ldots, x_{s-1}$, belong either to L or to K. We choose n large enough so that 1) each $x_i \in K$ will belong to K_n, and 2) between two neighboring points $x_i, x_j \in L$ there will be at least one interval of the set K_n. This is possible since L is nowhere dense. In fact, in the interval (x_i, x_j) there exists an interval which does not belong to L, and hence there exists an interval of K and, consequently, an interval of K_n for sufficiently large n. Conditions

---

[1] If $\psi(x)$ is continuous, then for (a_i, b_i) it is natural to take the component intervals of S. This case is necessary for us in the proof of Theorem 5. If both $\phi(x)$ and $\psi(x)$ are continuous, then we have the following assertion, which is stronger than (11):

$$\operatorname*{Var}_{[a, b]} \varphi(x)\psi(x) = \sum_i \operatorname*{Var}_{[a_i, b_i]} \varphi(x)\psi(x). \tag{11'}$$

1) and 2) mean that in each of the intervals $[\gamma_k, \delta_k]$ $(\gamma_k \leq \delta_k)$ of the set L_n there is at most one point of the partition $x_1, \ldots, x_s$.

We now consider the partition $a = y_1 \leq y_2 \leq \cdots \leq y_t = b$ of $[a, b]$ by the points $x_1, \ldots, x_s$ and the endpoints of the intervals of K_n. Then, putting $\chi(x) = \phi(x)\psi(x)$, we obtain

$$\sum_{k=1}^{s-1} |\chi(x_{k+1}) - \chi(x_k)| \leqslant \sum_{k=1}^{t-1} |\chi(y_{k+1}) - \chi(y_k)|$$

$$= \sum_{i=1}^{i_n} \sum_{y_i, y_{i+1} \in [\alpha_i, \beta_i]} |\chi(y_{i+1}) - \chi(y_i)|$$

$$+ \sum_{k=1}^{k_n} \sum_{y_j, y_{j+1} \in [\gamma_k, \delta_k]} |\chi(y_{j+1}) - \chi(y_j)| \equiv \sum_i A_i + \sum_j B_j.$$

It is obvious that

$$\sum_i A_i \leqslant \sum_{i=1}^{i_n} \operatorname*{Var}_{[\alpha_i, \beta_i]} \chi(x) \leqslant \sum_i \operatorname*{Var}_{[\alpha_i, \beta_i]} \chi(x)$$

$$= \sum_i \operatorname*{Var}_{[a_i, b_i]} \chi(x) + \sum_i \operatorname*{Var}_{[a_i', b_i']} \chi(x) = \sum_i \operatorname*{Var}_{[a_i, b_i]} \chi(x).$$

Since $\chi(\gamma_k) = \chi(\delta_k) = 0$ and at most one of the points x_{ik} of the original partition lies in $[\gamma_k, \delta_k]$, we have either $B_j = 0$ or $(M = \sup_{[a, b]}|\phi(x)|)$

$$B_j = 2 |\varphi(x_{i_k}) \psi(x_{i_k})| \leqslant 2M |\psi(x_{i_k})| = M(|\psi(x_{i_k}) - \psi(\gamma_k)|$$

$$+ |\psi(\delta_k) - \psi(x_{i_k})|) \leqslant M \operatorname*{Var}_{[\gamma_k, \delta_k]} \psi(x).$$

Thus

$$\sum_j B_j \leqslant M \sum_{k=1}^{k_n} \operatorname*{Var}_{[\gamma_k, \delta_k]} \psi(x) \leqslant M \operatorname*{Var}_{[a, b]} \psi(x).$$

The desired inequality (11) follows from these bounds.

Proof of Theorem 5. Since $f^{(k)}(x)$ is of bounded variation, it is continuous (being a derivative, it has no discontinuity of the first kind; being of bounded variation it has no discontinuity of the second kind). We denote by (a_i, b_i) the component intervals of the open set $S = \{x \in (a, b): f^{(k)}(x) \neq 0\}$. By Lemma 1 it is sufficient to bound the expression

$$\sum_i \operatorname*{Var}_{[a_i, b_i]} \Phi(f(x)) G(f^{(k)}(x)).$$

Let $a_i = x_1 < x_2 < \cdots < x_s = b_i$ be any partition of $[a_i, b_i]$. Then

$$\sigma_i = \sum_{m=1}^{s-1} |\Phi(f(x_{m+1})) G(f^{(k)}(x_{m+1})) - \Phi(f(x_m)) G(f^{(k)}(x_m))|$$

$$\leqslant \sum_{m=1}^{s-1} |G(f^{(k)}(x_{m+1}))| |\Phi(f(x_{m+1})) - \Phi(f(x_m))|$$

$$+ \sum_{m=1}^{s-1} |\Phi(f(x_m))| |G(f^{(k)}(x_{m+1})) - G(f^{(k)}(x_m))| \leqslant$$

$$\leqslant \sup_m |G(f^{(k)}(x_{m+1}))| \sum_{m=1}^{s-1} |\Phi(f(x_{m+1})) - \Phi(f(x_m))|$$
$$+ M_\Phi \operatorname*{Var}_{[a_i, b_i]} G(f^{(k)}(x)),$$

where $M_\Phi = \sup_{[c,d]} |\Phi(y)|$.

Since $f^{(k)}(x) \neq 0$, the interval $[a_i, b_i]$ can be decomposed into r intervals, $r \leq k$, $[a_i^{(1)}, a_i^{(2)}], \ldots, [a_i^{(r)}, a_i^{(r+1)}]$, $a_i^{(1)} = a_i$, $a_i^{(r+1)} = b_i$, on which $f(x)$ is monotone. We now consider the partition $a = y_1 \leq y_2 \leq \cdots \leq y_t = b$ of $[a, b]$ by the points $x_1, \ldots, x_s$ and the endpoints of the intervals indicated above; then

$$\sum_{m=1}^{s-1} |\Phi(f(x_{m+1})) - \Phi(f(x_m))| \leqslant \sum_{l=1}^{r} \sum_{y_m, y_{m+1} \in [a_i^{(l)}, a_i^{(l+1)}]} |\Phi(f(y_{m+1})) - \Phi(f(y_m))|.$$

By monotonicity of f on $[a_i^{(l)}, a_i^{(l+1)}]$, the points $z_m = f(y_m)$ for $y_m \in [a_l^{(l)}, a_i^{(l+1)}]$ form a monotonically increasing sequence in $[c, d]$; therefore

$$\sum_{y_m, y_{m+1} \in [a_i^{(l)}, a_i^{(l+1)}]} |\Phi(f(y_{m+1})) - \Phi(f(y_m))| = \sum_m |\Phi(z_{m+1}) - \Phi(z_m)| \leqslant \operatorname*{Var}_{[c,d]} \Phi(z).$$

Thus,

$$\sigma_i \leqslant M_\Phi \operatorname*{Var}_{[a_i, b_i]} G(f^{(k)}(x)) + k \sup_m |G(f^{(k)}(x_{m+1}))| \operatorname*{Var}_{[c,d]} \Phi(x).$$

If $f^{(k)}(x) \neq 0$ on all of $[a, b]$, then $[a_i, b_i] = [a, b]$, and

$$\operatorname*{Var}_{[a,b]} \Phi(f) G(f^{(k)}) \leqslant M_\Phi \operatorname*{Var}_{[a,b]} G(f^{(k)}) + k M_{G(f^{(k)})} \operatorname*{Var}_{[c,d]} \Phi,$$

where $M_{G(f^{(k)})} = \sup_{[a,b]} G(f^{(k)})$, from which (10) follows.

If $f^{(k)}(x)$ vanishes anywhere on $[a, b]$, then either $f^{(k)}(a_i) = 0$ or $f^{(k)}(b_i) = 0$ (for example, $f^{(k)}(a_i) = 0$), and

$$\sup_m |G(f^{(k)}(x_{m+1}))| = \sup_m |G(f^{(k)}(x_{m+1})) - G(f^{(k)}(a_i))| \leqslant \operatorname*{Var}_{[a_i, b_i]} G(f^{(k)}(x)).$$

Hence

$$\operatorname*{Var}_{[a_i, b_i]} \Phi(f) G(f^{(k)}) - \sup \sigma_i \leqslant \operatorname*{Var}_{[a_i, b_i]} G(f^{(k)}) \left(M_\Phi + k \operatorname*{Var}_{[c,d]} \Phi\right).$$

Summing over i, we obtain

$$\operatorname*{Var}_{[a,b]} \Phi(f) G(f^{(k)}) \leqslant [M_\Phi + k \operatorname*{Var}_{[c,d]} \Phi] \operatorname*{Var}_{[a,b]} G(f^{(k)}),$$

from which the desired assertion follows. Theorem 5 is proved.

§3. A theorem on composition of absolutely continuous functions

Starting from Theorem 5, we now obtain a corresponding theorem for absolutely continuous functions which generalizes Theorem 3.

Theorem 6. *Suppose that the following conditions hold:*

1) $f(x)$ *has a derivative* $f^{(k)}$ *of order* k *everywhere on* $[a, b]$.

2) $\Phi(x)$ *is absolutely continuous on* $[c, d]$.

3) $G(x)$ *is defined on* $[\alpha, \beta]$; *moreover*, $G(0) = 0$ *if* $0 \in [\alpha, \beta]$.

4) $G(f^{(k)}(x))$ *is absolutely continuous on* $[a, b]$.

Then $\Phi(f(x))G(f^{(k)}(x))$ *is absolutely continuous on* $[a, b]$.

We say that a function $\phi(x)$ has *property* N $[a, b]$ if the image of every set of measure zero in $[a, b]$ has measure zero. We use a theorem of Banach and Zareckiĭ (see, for example, [1], Russian p. 232), in which it is asserted that a continuous function of bounded variation is absolutely continuous if and only if it has property N.

For a proof of Theorem 6 it is sufficient to take into account the Banach-Zareckiĭ theorem, Theorem 5, the obvious assertion that the composite of two functions with property N also has property N, and the following lemma.

Lemma 2. *If* $f(x)$ *and* $g(x)$ *are continuous functions which have property* N *on* $[a, b]$, *then the function* $f(x)g(x)$ *also has property* N *on* $[a, b]$.

Proof. Let $M \subset [a, b]$, $\operatorname{meas} M = 0$. Then $\operatorname{meas} f(M) = \operatorname{meas} g(M) = 0$. Let $\epsilon > 0$; then there exist open sets F and G such that $f(M) \subset F$, $g(M) \subset G$, $\operatorname{meas} F < \epsilon$ and $\operatorname{meas} G < \epsilon$. The preimages $f^{-1}(F)$ and $g^{-1}(G)$ are open and contain M. Let

$$L = f^{-1}(F) \cap g^{-1}(G) = \bigcup_i (\alpha_i, \beta_i),$$

where the (α_i, β_i) are the component intervals. We put

$$F_1 = f(L) = \bigcup_i f[(\alpha_i, \beta_i)] = \bigcup_i \{\alpha_i, \beta_i\},$$

where $\{a_i, b_i\}$ denotes any of the sets (a_i, b_i), $(a_i, b_i]$, $[a_i, b_i)$ or $[a_i, b_i]$ (in the last case it is possible that $a_i = b_i$);

$$G_1 = g(L) = \bigcup_i g[(\alpha_i, \beta_i)] = \bigcup_i \{c_i, d_i\}.$$

Since $F_1 \subset F$ and $G_1 \subset G$, we have $\operatorname{meas} F_1 < \epsilon$ and $\operatorname{meas} G_1 < \epsilon$.

Let $x \in M$; then $x \in L$, and hence $x \in (\alpha_{i_x}, \beta_{i_x})$. Thus $f(x) \in \{a_{i_x}, b_{i_x}\}$, $g(x) \in \{c_{i_x}, d_{i_x}\}$ and

$$f(x)\, g(x) \in \{a_{i_x}, b_{i_x}\} \times \{c_{i_x}, d_{i_x}\} \subset \bigcup_i \{a_i, b_i\} \times \{c_i, d_i\}.\text{(2)}$$

Thus

$$(fg)(M) \subset \bigcup_i \{a_i, b_i\} \times \{c_i, d_i\} = K.$$

We note that

$$\begin{aligned} \operatorname{mes} \{A_1, A_2\} \times \{B_1, B_2\} &= \max_{i, j, k, l = 1, 2} |A_i B_j - A_k B_l| \\ &= \max_{i, j, k, l = 1, 2} |(A_i - A_k) B_j + (B_j - B_l) A_k| \\ &\leqslant \max(|A_1|, |A_2|, |B_1|, |B_2|)[(A_2 - A_1) + (B_2 - B_1)], \end{aligned}$$

(2) Here $A \times B = \{x\colon x = yz,\ y \in A,\ z \in B\}$.

therefore

$$\text{meas}\, K \leqslant \max(M_f,\ M_g)\Big(\sum_i (b_i - a_i) + \sum_i (d_i - c_i)\Big) < 2\max(M_f,\ M_g)\,\varepsilon,$$

where $M_f = \max_{[a,b]} f(x)$ and $M_g = \max_{[a,b]} g(x)$. This means that $\text{meas}\,(fg)(M) = 0$, as required.

§4. An example

In §1 an example was given to show that under the assumptions of Theorem 3 the product $\Phi(f(x))f'(x)$ may be absolutely continuous while at the same time $\Phi(f(x))$ is not absolutely continuous.

We now introduce a less trivial example to show how bad the differential properties of the factor $\Phi(f(x))$ in Theorem 3 may be.

We note that by a theorem of Banach ([7], p. 169) the factor $\Phi(f(x))$ in Theorem 1 has a derivative on a set $E \subset [a, b]$ such that $\text{meas}\,(E \cap (\alpha, \beta)) > 0$ for every interval $(\alpha, \beta) \subset [a, b]$. On the other hand, for every perfect set $P \subset [a, b]$ that is nowhere dense in $[a, b]$ (the measure of P may be arbitrarily close to $b - a$) we can find a Φ and an f which satisfy the conditions of Theorem 3 and are such that the composite $\Phi(f(x))$ is not differentiable on P (in particular, if $\text{meas}\, P > 0$, then $\Phi(f(x))$ is a fortiori not absolutely continuous.

The functions Φ and f can be constructed in the following way. Let $\phi(x) \in C_0^\infty[0, 1]$, $0 \le \phi(x) \le 1$ for $x \in (0, 1)$, and $\phi(½) = 1$. Further, let $(a, b) - P = \Sigma_1^\infty (a_k, b_k)$ and $\delta_k = b_k - a_k$; suppose, moreover, that the intervals are numbered so that $\delta_1 \ge \delta_2 \ge \delta_3 \ge \cdots$. We denote by μ_n the length of the largest of the intervals of the set $[a, b] - \Sigma_1^n (a_k, b_k)$. Then $\mu_n \downarrow 0$.

We put $f(x) = 0$ on P and $f(x) = \psi_k(x) = e^{-1/\delta_k^n}\phi((x - a_k)/\delta_k)$ on (a_k, b_k). For $\Phi(x)$, $x \ge 0$, we take a function which is monotonically increasing, bounded and differentiable for all $x > 0$, and for which $\Phi(0) = 0$ and $\Phi(e^{-1/\delta_k}) = \mu_k + \delta_k$. The idea for the construction of this example was taken from a paper of N. K. Bari [4], p. 212; the difference is that it is necessary to choose f so that it is differentiable and its derivative is absolutely continuous.

The function $\Phi(x)$ is absolutely continuous on every interval $[0, A]$, since it is continuous on $[0, A]$ and has an integrable derivative there (this follows from monotonicity of $\Phi(x)$). The function $f(x)$ is infinitely differentiable on $[a, b]$, since $f(x) = \Sigma_1^\infty \psi_k(x)$ and the series $f^{(n)}(x) = \Sigma_1^\infty \psi_k^{(n)}(x)$ converges uniformly for every $n \ge 1$; in fact,

$$|\psi_k^{(n)}(x)| \leqslant \frac{1}{\delta_k^n}\, e^{-\frac{1}{\delta_k}} \leqslant c_n \delta_k \quad \left(\sum_{k=1}^{\infty} \delta_k \leqslant b - a\right).$$

On the other hand, $\Phi(f(x))$ is not differentiable at points of P. In fact, let $x_0 \in P$ and suppose that x_0 is not, for example, a left endpoint of any of the intervals (a_k, b_k). We denote by x_k the midpoint of (a_k, b_k). We construct a sequence x_{k_l} according to the following rule. We consider the particular one of the intervals (a_k, b_k) which lies between x_0 and b and has maximal length relative to this property (if there are several, choose the one that lies closest to x_0). Denote the length of this interval (a_{k_1}, b_{k_1}) by δ_{k_1}. Its midpoint will be x_{k_1}. Since there are no intervals of length greater than or equal to δ_{k_1} between x_0 and a_{k_1}, x_0 and a_{k_1} belong to exactly one interval of the set $[a, b] - \Sigma_1^{k_1}(a_k, b_k)$; consequently, $a_{k_1} - x_0 \le \mu_{k_1}$ and $x_{k_1} - x_0 \le \mu_{k_1} + \frac{1}{2}\delta_{k_1}$. In the same way we choose a point x_{k_2} between x_0 and x_{k_1}, etc. The sequence $x_{k_l} \to x_0$ as $l \to \infty$. In this connection,

$$\frac{\Phi(f(x_{k_l})) - \Phi(f(x_0))}{x_{k_l} - x_0} \geqslant \frac{\Phi\left(e^{-\frac{1}{\delta_{k_l}}}\right)}{\mu_{k_l} + \frac{1}{2}\delta_{k_l}} \geqslant 1,$$

at the same time as for the sequence $x_n \to x_0$, $x_n \in P$, this ratio equals zero, i.e., $(\Phi(f))'$ does not exist at x_0.

BIBLIOGRAPHY

1. I. P. Natanson, *Theory of functions of a real variable*, 3rd ed., "Nauka", Moscow, 1974; English transl. of 1st ed., Ungar, New York, 1955. MR 16, 804.

2. V. I. Burenkov, *On composition of absolutely continuous functions and composition of functions of bounded variation*, Dokl. Akad. Nauk SSSR 189 (1969), 234–236 = Soviet Math. Dokl. 10 (1969), 1379–1381. MR 41 #407.

3. G. M. Fihtengol'c, *On absolutely continuous functions*, Mat. Sb. 31 (1923), 285–295. (Russian)

4. N. K. Bari, *Mémoire sur la représentation finie des fonctions continues*. I, II, Math. Ann. 103 (1930), 185–248; 598–653.

5. ———, *Sur une classification des fonctions continues à partir de fonctions à variation bornée*, Mat. Sb. 40 (1933), 326–370.

6. Ch.-J. de la Vallée Poussin, *Cours d'analyse infinitésimale*, Vol. I, 7th ed., Librairie universitaire Louvain, 1930; reprint of the 8th ed., Dover, New York, 1946. MR 8, 15.

7. S. Banach, *Sur une classe de fonctions continues*, Fund. Math. 8 (1926), 166–172.

8. M. A. Zareckiĭ, *Sur un théorème des fonctions absolument continues*, Dokl. Ross. Akad. Nauk 1925, 88–90. (Russian)

Translated by C. WILDE

Trudy Mat. Inst. Steklov.
134 (1975)

Proc. Steklov Inst. Math.
134 (1975)

AN A POSTERIORI ERROR ESTIMATE FOR DIFFERENCE SOLUTIONS OF THE LAPLACE AND POISSON EQUATIONS

UDC 518: 517.944/947

E. A. VOLKOV

Abstract. A description is given of a method (algorithm) for calculating an a posteriori error estimate for difference solutions of the Dirichlet problem for the two-dimensional Laplace and Poisson equations under smoothness requirements on the boundary of the domain, the boundary values and the right-hand side of the equation, none of which can be substantially weakened. The convergence to zero of this estimate, which is of unimprovable second order in the grid step, is established.

Bibliography: 16 titles.

Introduction

The overwhelming majority of error estimates known in the literature for difference solutions of boundary problems are unsuitable for practical use, inasmuch as these estimates usually contain maxima of the derivatives at unknown points of the solution, unknown constants, etc. In connection with these determinations, a posteriori estimates of the error manifested in terms of an approximate solution acquire interest; although the algorithm for computing these estimates is not simple, it is completely determinate. The main idea on which a posteriori error estimates are based is that the difference solution can be extended in some way from the network to the entire domain, and then from the residual of the extended solution one finds for the original differential equation, for with the aid of the Green's function for example, the deviation of the approximate solution from the unknown exact solution on the whole domain, and in particular on the network. The first a posteriori error estimate for a difference solution of the Poisson equation was given by D. F. Davidenko [1]. In that paper, however, the constructions are actually carried out for a very limited number of network nodes (of the order unity), and questions concerning mesh refinement are not touched upon. In the present article we construct an a posteriori error estimate for difference solutions of the Dirichlet problem for the Laplace and Poisson equations by using a local piecewise-polynomial extension of the network function which has fixed degree independent of the step size. Under conditions of smoothness on the boundary of the domain, on the boundary values and on the right-hand side of the equation,

---

AMS (MOS) subject classifications (1970). Primary 65N15; Secondary 65N10.

which cannot be essentially weakened (see [2] and [3]), we establish convergence of this estimate with unimprovable second order in the step size.

A brief description of the method (algorithm) for computing the a posteriori error estimate is given in §2. The verification of the estimate, which requires considerably more space, occupies §§1 and 3.

§1. Definitions and auxiliary propositions

Consider the boundary problem

$$\Delta u = f \text{ on } \Omega, \qquad u = \varphi \text{ on } \gamma, \tag{1.1}$$

where $\Delta = \partial^2/\partial x^2 + \partial^2/\partial y^2$, γ is the boundary of a finite simply connected domain Ω, and f and ϕ are given continuous functions.

Definition 1.1. We say that $\gamma \in C_{2,\lambda}$, $0 < \lambda < 1$, if the boundary γ has curvature satisfying a Hölder condition with exponent λ and, moreover, any point of γ can be touched by the boundaries of two open disks of fixed radius, one of which does not have points in common with Ω and the other belongs entirely to Ω.

Definition 1.2. The function ϕ belongs to $C_{k,\lambda}(\gamma)$, $k \geq 0$, $0 < \lambda \leq 1$, if ϕ is k times differentiable along γ, where

$$\|\varphi\|_{k,\lambda} = \max_{\gamma} |\varphi| + \sup_{\substack{s,t\in\gamma \\ s\neq t}} \frac{|\varphi^{(k)}(s) - \varphi^{(k)}(t)|}{|s-t|^{\lambda}} < \infty. \tag{1.2}$$

We shall denote by D^q the differential operator of qth order with respect to x and y, and by $D^q f(P)$ the value of $D^q f$ at the point P. Let $\{D^q\}$ be the set of all differential operators of qth order with respect to x and y.

Definition 1.3. A function f belongs to the Hölder weight class $C^{m,\mu}_{k,\lambda}(\Omega)$ ($f \in C^{m,\mu}_{k,\lambda}(\Omega)$), where m and k are nonnegative integers, $0 < \mu \leq 1$, $0 < \lambda \leq 1$ and $k + \lambda \leq m + \mu$, if f is m times differentiable on Ω;

$$\max_{\{D^m\}} \sup_{\substack{P,Q\in\Omega \\ P\neq Q}} \rho_*^{\alpha} \frac{|D^m f(P) - D^m f(Q)|}{|P-Q|^{\mu}} = F^{m,\mu}_{k,\lambda} < \infty,$$

where $\alpha = m + \mu - k - \lambda$, $\rho_* = \min\{\rho_P, \rho_Q\}$, ρ_P is the distance from the point P to γ, and $|P - Q|$ is the greatest lower bound of the lengths of the curves joining the points P and Q and lying entirely in Ω.

Let us introduce a norm for $f \in C^{m,\mu}_{k,\lambda}(\Omega)$:

$$\|f\|^{m,\mu}_{k,\lambda} = F^{m,\mu}_{k,\lambda} + \sup_{P\in\Omega} |f(P)|.$$

In the particular case in which $m = k$ and $\mu = \lambda$, the weight classes under consideration are just the ordinary Hölder classes $C_{k,\lambda}(\Omega)$ consisting of the functions whose kth derivatives satisfy the usual Hölder condition on Ω with exponent λ. For the functions $f \in C_{k,\lambda}(G)$ we give a definition of the norm extended to the whole space:

$$\|f\|_{C_{k,\lambda}(G)} = \sup_{P\in G} |f(P)| + \max_{\{D^k\}} \sup_{\substack{P,Q\in G \\ P\neq Q}} \frac{|D^k f(P) - D^k f(Q)|}{|P-Q|^{\lambda}}.$$

. For the norm of the function ϕ on γ we keep the notation (1.2).

Let us denote for short

$$\|f, \varphi\|_{0,\lambda}^{2,\lambda} = \|f\|_{0,\lambda}^{2,\lambda} + \|\varphi\|_{2,\lambda}.$$

The following lemma is valid.

Lemma 1.1. *If $\gamma \in C_{2,\lambda}$, $\phi \in C_{2,\lambda}(\gamma)$ and $f \in C_{0,\lambda}^{2,\lambda}(\Omega)$, $0 < \lambda < 1$, then there exists a unique solution $u \in C_{2,\lambda}^{4,\lambda}(\Omega)$ of problem (1.1) such that*

$$\|u\|_{2,\lambda}^{4,\lambda} \leqslant c\|f, \varphi\|_{0,\lambda}^{2,\lambda},$$

where the constant c does not depend on f or ϕ.

From the imbedding Theorem 3.1 of [4] follows

Lemma 1.2. *If $\gamma \in C_{2,\lambda}$ and $f \in C_{k,\lambda}^{m,\lambda}(\Omega)$, $0 < \lambda < 1$, $m \geq k \geq 0$, then $f \in C_{k,\lambda}(\Omega)$, where*

$$\|f\|_{C_{k,\lambda}(\Omega)} \leqslant c\|f\|_{k,\lambda}^{m,\lambda},$$

and, moreover,

$$\max_{\{D^q\}} \sup_{P\in\Omega} \rho_P^{q-k-\lambda} |D^q f(P)| \leqslant c\|f\|_{k,\lambda}^{m,\lambda}, \tag{1.3}$$

in which $k < q \leq m$ and c is a constant not depending on f.

From Theorem 6 of [5] (see also [6], proof of Lemma 1.2) follows

Lemma 1.3. *Let $\gamma \in C_{2,\lambda}$ and $v \in C_{k,\lambda}(\Omega)$, $0 \leq k \leq 2$, $0 < \lambda < 1$. Then there exists a function $\tilde{v} \in C_{k,\lambda}(E^2)$, where E^2 is the xy-plane, such that $\tilde{v} = v$ on $\overline{\Omega}$ and*

$$\|\tilde{v}\|_{C_{k,\lambda}(E^2)} \leqslant c\|v\|_{C_{k,\lambda}(\Omega)},$$

in which the constant c does not depend on v.

As a consequence of the well-known Markov inequality (see [7], p. 245), we have

Lemma 1.4. *Let $\overline{R}_l = \{(x, y): |x| \leq l, |y| \leq l\}$, and let Q_n be an algebraic polynomial of degree n in y. Then*

$$\max_{\{D^q\}} \max_{\overline{R}_l} |D^q Q_n| \leqslant c_n l^{-q} \max_{\overline{R}_l} |Q_n|, \qquad 0 < q \leqslant n,$$

where the constant c_n depends only on n.

Let us consider the boundary problem

$$\Delta v = f \quad \text{on} \quad \Omega, \qquad v = 0 \quad \text{on} \quad \gamma. \tag{1.4}$$

Lemma 1.5. *If $\gamma \in C_{2,\lambda}$, $0 < \lambda < 1$, $f \in L_2(\Omega)$ and the inequality $f \leq 0$ holds* *almost everywhere on* Ω, *then there exists in the space* $W^2_{2,0}(\Omega)$(1) *a unique solution* v *of problem* (1.4) *continuous on* $\bar{\Omega}$ *and satisfying the inequality*

$$v \geqslant 0 \quad \text{on} \quad \bar{\Omega}. \tag{1.5}$$

Proof. The unique solvability of problem (1.4) in $W^2_{2,0}(\Omega)$ was established in [8]. From an a priori estimate of the solution v in $W^2_{2,0}(\Omega)$ ([8], Chapter III, formula (9.2)) and Sobolev's imbedding theorem [9] follows the continuity of v on $\bar{\Omega}$ as well as the inequality

$$\max_{\bar{\Omega}} |v| \leqslant c \|f\|_{L_2(\Omega)}, \tag{1.6}$$

where the constant c is independent of f. For any $\epsilon > 0$ it is known (see [9] or [10]) that the function $f \in L_2(\Omega)$ is representable on Ω in the form

$$f = f^* + f_\varepsilon,$$

where f^* is a function infinitely differentiable in the entire plane, obtained from f continued outside of Ω by zero and averaged with the aid of a nonnegative infinitely differentiable kernel with compact support, and $f_\epsilon \in L_2(\Omega)$ with $\|f_\epsilon\|_{L_2(\Omega)} < \epsilon$. It is obvious that under the conditions of the lemma

$$f^* \leqslant 0 \quad \text{on} \quad \bar{\Omega}. \tag{1.7}$$

Let us represent v in the form

$$v = v^*_\varepsilon + v_\varepsilon, \tag{1.8}$$

where v^*_ϵ (v_ϵ) is a solution of problem (1.4) when f is replaced by $f^* - \epsilon$ ($f^* + \epsilon$), $\epsilon > 0$. Since $v^*_\epsilon \in C_{2,\lambda}(\Omega)$ (see [11]) and, on the strength of (1.7), $\Delta v^*_\epsilon < 0$ on Ω, the function v^*_ϵ cannot achieve a minimum on Ω. Consequently,

$$v^*_\varepsilon \geqslant 0 \quad \text{on} \quad \bar{\Omega}. \tag{1.9}$$

On the basis of (1.6) we have

$$\max_{\bar{\Omega}} |v_\varepsilon| \leqslant c'\varepsilon,$$

where c' does not depend on ϵ. From this, and from (1.8) and (1.9), inequality (1.5) follows because ϵ is arbitrary. Lemma 1.5 is proved.

Let r_* be the radius of the largest open disk whose boundary can be tangent at any point of γ in such a way that this disk lies completely in Ω; $\Omega(\rho)$ is the subdomain consisting of all points of Ω at a distance more than ρ from γ; and γ_ρ is the boundary of $\Omega(\rho)$. If $\gamma \in C_{2,\lambda}$, $0 < \lambda < 1$, $0 < \rho < r_*$, then by Theorem 2.1 of [4] $\gamma_\rho \in C_{2,\lambda}$, and according to Lemma 2.2 of [4] there is a one-to-one correspondence between points of the disks γ and γ_ρ, in which the pairs of corresponding points at distance ρ lie on the common normals to γ and γ_ρ.

(1) $W^2_{2,0}(\Omega)$ is a subspace of $W^2_2(\Omega)$, a dense set which contains all twice continuously differentiable functions on $\bar{\Omega}$ vanishing on γ.

Let us agree to take the curvature of the boundary of the domain to be positive at points of convexity.

Lemma 1.6. *Let* $\gamma \in C_{2,\lambda}$, $0 < \lambda < 1$ *and* $0 < \rho < r_*$. *Then between the curvature* k *of the curve* γ *and the curvature* k_ρ *of the curve* γ_ρ *at corresponding points at distance* ρ, *we have the following relation*:

$$k_\rho = \frac{k}{1 - \rho k}. \tag{1.10}$$

Proof. For simplicity assume that the curve γ passes through the origin of a rectangular system of coordinates x, y, with the x-axis directed along the tangent to γ and the positive y-axis along the interior normal to γ. Then a piece of γ containing the origin is given by the equation

$$y = \sigma(x), \tag{1.11}$$

where on a certain interval containing the point $x = 0_\rho$ the function $\sigma(x)$ has a second derivative satisfying a Hölder condition with exponent λ, with $\sigma'(0) = 0$. The piece of the curve γ_ρ corresponding to this piece of the boundary γ can be represented parametrically:

$$x_\rho = x - \rho\sigma'(x)(1 + \sigma'^2(x))^{-1/2}; \tag{1.12}$$

$$y_\rho = \sigma(x) + \rho(1 + \sigma'^2(x))^{-1/2}. \tag{1.13}$$

Hence

$$\frac{dx_\rho}{dx} = 1 - \rho\sigma''(x)(1 + \sigma'^2(x))^{-3/2}; \tag{1.14}$$

$$\frac{dy_\rho}{dx} = \sigma'(x)(1 - \rho\sigma''(x)(1 + \sigma'^2(x))^{-3/2}), \tag{1.15}$$

and since $\sigma''(0) \leq r_*^{-1}$, $0 < \rho < r_*$, we have $dx_\rho/dx \neq 0$ in some neighborhood of $x = 0$. Consequently

$$\frac{dy_\rho}{dx_\rho} = \frac{dy_\rho/dx}{dx_\rho/dx} = \sigma'(x). \tag{1.16}$$

Using (1.14)–(1.16), we easily obtain $dy_\rho/dx_\rho|_{x_\rho = 0} = 0$ and

$$\left.\frac{d^2y_\rho}{dx_\rho^2}\right|_{x_\rho=0} = \left(\frac{dx_\rho}{dx}\right)^{-1} \left.\frac{d}{dx}\frac{dy_\rho}{dx_\rho}\right|_{x=0} = \sigma''(0)(1 - \rho\sigma''(0))^{-1},$$

i.e. (1.10) holds for the points $(0, 0) \in \gamma$ and $(0, \rho) \in \gamma_\rho$. (1.10) is obviously true also for any other pair of corresponding points of the curves γ and γ_ρ that lie on the common normals of these curves at a distance ρ from each other. Lemma 1.6 is proved.

Let Ω_* be the set of all points of the closed domain Ω which are at a distance less than r_* from γ, and let $\rho = \rho(x, y)$ be the distance from the variable point (x, y) to γ.

Lemma 1.7. *If* $\gamma \in C_{2,\lambda}$, $0 < \lambda < 1$, *then the function* $\rho(x, y)$ *is twice continuously differentiable on* Ω_*.

Proof. Let us assume that the curve γ is tangent to the x-axis at the origin and that the positive y-axis is in the direction of the interior normal to γ. We rewrite (1.12) and (1.13) in the following form:

$$F_1(x, y, x_\gamma, \rho) \equiv x - x_\gamma + \rho\sigma'(x_\gamma)(1+\sigma'^2(x_\gamma))^{-1/2} = 0;$$
$$F_2(x, y, x_\gamma, \rho) \equiv y - \sigma(x_\gamma) - \rho(1+\sigma'^2(x_\gamma))^{-1/2} = 0, \qquad (1.17)$$

where x, y are the coordinates of a typical point: $(x, y) \in \Omega_*$; x_γ is the abscissa of the point (whose distance is less than r_*) of γ from which issues the normal to γ passing through (x, y) (on the strength of Lemma 2.2 of [4] such a point of γ exists and (as a consequence of the proof of Lemma 2.2) is unique); ρ is the distance between the points (x, y) and $(x_\gamma, \sigma(x_\gamma))$, which, by virtue of Lemma 2.2 of [4], is the same as the distance from (x, y) to γ. Let $0 \le y_0 < r_*$. Then it is obvious that

$$F_1(0, y_0, 0, y_0) = F_2(0, y_0, 0, y_0) = 0,$$

where for $x = x_\gamma = 0$ and $y = \rho = y_0$ we have

$$\frac{\mathcal{D}(F_1, F_2)}{\mathcal{D}(x_\gamma, \rho)} = 1 - y_0\sigma''(0) \neq 0,$$

and the functions $F_i(x, y, x_\gamma, \rho)$, $i = 1, 2$, are continuously differentiable in a neighborhood of the point $(0, y_0, 0, y_0)$. Hence, according to a well-known theorem on implicit functions in a neighborhood of the arbitrary point $(0, y_0)$, $0 \le y_0 < r_*$, equations (1.17) implicitly yield the continuously differentiable functions $\rho = \rho(x, y)$(2) and $x_\gamma = x_\gamma(x, y)$, and, as simple computations show,

$$\frac{\partial\rho}{\partial x} = -\frac{\sigma'(x_\gamma)}{(1+\sigma'^2(x_\gamma))^{1/2}}, \quad \frac{\partial\rho}{\partial y} = \frac{1}{(1+\sigma'^2(x_\gamma))^{1/2}}, \qquad (1.18)$$

$$\frac{\partial x_\gamma}{\partial x} = \frac{1}{1+\sigma'^2(x_\gamma) - \rho\sigma''(x_\gamma)(1+\sigma'^2(x_\gamma))^{-1/2}}, \qquad (1.19)$$

$$\frac{\partial x_\gamma}{\partial y} = \frac{\sigma'(x_\gamma)}{1+\sigma'^2(x_\gamma) - \rho\sigma''(x_\gamma)(1+\sigma'^2(x_\gamma))^{-1/2}}. \qquad (1.20)$$

Since $\sigma''(x_\gamma)$ does not appear in expression (1.18) for the partial derivatives $\partial\rho/\partial x$ and $\partial\rho/\partial y$, there exist second partial derivatives $\partial^2\rho/\partial x^2$, $\partial^2\rho/\partial x\partial y$ and $\partial^2\rho/\partial y^2$, continuous in the same neighborhood of the point $(0, y_0)$, $0 \le y_0 < r_*$, as the one in which the partial derivatives (1.19), (1.20) and $\sigma''(x_\gamma)$ are continuous.

From Lemma 2.2 of [4] it follows that an arbitrary fixed point $(x_0, y_0) \in \Omega_*$ lies on a certain normal to γ issuing from the nearest point to it on γ at a distance less than r_*. Therefore, choosing a local system of rectangular coordinates, one of whose axes is directed along the indicated normal and the other along the

(2) Here the distance from γ of a point $(x, y) \notin \overline{\Omega}$ is taken to be negative.

corresponding tangent to γ, in complete analogy to what was done above, we establish continuity of the second derivatives of the distance to the boundary γ in a neighborhood of an arbitrary point $(x_0, y_0) \in \Omega_*$ in the appropriate local system of rectangular coordinates and consequently in any other rectangular coordinate system. Lemma 1.7 is proved.

Remark 1.1. If $\gamma \in C_{k,\lambda}$, $k \geq 2$, $0 < \lambda < 1$, i.e. if the curvature of the boundary has a $(k-2)$th derivative everywhere with respect to its arc that satisfies a Hölder condition with exponent λ, then, relying on formulas (1.18)–(1.20), we can easily establish that $\rho(x, y) \in C_{k,\lambda}(\Omega \setminus \bar{\Omega}(\rho_0))$, where $0 < \rho_0 < r_*$.

Lemma 1.8. *Let $y = y(x)$ and $\psi = \psi(x, y)$ be functions twice continuously differentiable in neighborhoods of the points $x = 0$ and $(0, 0)$, respectively, and let s be the arc measured from the point $x = 0$ along the curve $y = y(x)$. Then at $s = 0$*

$$\frac{d^2\psi}{ds^2} = \frac{\partial^2\psi}{\partial x^2} + k_0 \frac{\partial\psi}{\partial y}, \tag{1.21}$$

where k_0 is the curvature of the curve $y = y(x)$ at $x = 0$.

Proof. We have

$$\frac{d^2\psi}{ds^2} = \frac{\partial^2\psi}{\partial x^2}\left(\frac{dx}{ds}\right)^2 + \frac{\partial\psi}{\partial x}\frac{d^2x}{ds^2} + 2\frac{\partial^2\psi}{\partial x\partial y}\frac{dx}{ds}\frac{dy}{ds} + \frac{\partial^2\psi}{\partial y^2}\left(\frac{dy}{ds}\right)^2 + \frac{\partial\psi}{\partial y}\frac{d^2y}{ds^2},$$

where for $s = 0$ we have $dx/ds = 1$, $dy/ds = d^2x/ds^2 = 0$ and $d^2y/ds^2 = k_0$. Lemma 1.8 follows.

We shall now take up the Cauchy problem for the ordinary differential equation

$$z''_{tt} + a(t) z'_t = b(t), \qquad 0 \leqslant t \leqslant r_*, \tag{1.22}$$

$$z(0) = z'_t(0) = 0, \tag{1.23}$$

where $a(t)$ and $b(t)$ are given functions piecewise continuous on the interval $[0, r_*]$. As one can easily see, the function

$$z(t) = \int_0^t z'(\tau)\, d\tau, \tag{1.24}$$

where

$$z'(t) = e^{-\int_0^t a(\xi)\,d\xi} \int_0^t b(\eta)\, e^{\int_0^\eta a(\xi)\,d\xi}\, d\eta, \tag{1.25}$$

is the unique continuously differentiable solution of problem (1.22), (1.23).

We shall consider only the continuously differentiable solution of this problem below.

From (1.24) and (1.25) we immediately obtain

Lemma 1.9. *If* $b(t) \geq 0$ *on* $[0, r_*]$, *then the following inequalities hold for problem* (1.22), (1.23):

$$z(t) \geqslant 0, \quad z'(t) \geqslant 0, \quad t \in [0, r_*].$$

We have also

Lemma 1.10. *Let the functions* $a(t)$, $b(t)$, $\alpha(t)$ *and* $\beta(t)$ *be given, all piecewise continuous on the interval* $[0, r_*]$; *suppose a function* $z(t)$ *is the solution of problem* (1.22), (1.23) *and a function* $y(t)$ *is the solution of the Cauchy problem*

$$y''_{tt} + \alpha(t) y'_t = \beta(t), \quad 0 \leqslant t \leqslant r_*; \tag{1.26}$$

$$y(0) = y'_t(0) = 0. \tag{1.27}$$

If $\alpha(t) \leq a(t)$ *and* $\beta(t) \geq b(t) \geq 0$, $t \in [0, r_*]$, *then*

$$y(t) \geqslant z(t), \quad y'(t) \geqslant z'(t), \quad t \in [0, r_*]. \tag{1.28}$$

Proof. Let $v(t) = y(t) - z(t)$. By Lemma 1.9 we have

$$v''_{tt} + a(t) v'_t = \beta(t) - b(t) + (a(t) - \alpha(t)) y'_t(t) \geqslant 0, \quad 0 \leqslant t \leqslant r_*,$$
$$v(0) = y(0) - z(0) = 0, \quad v'_t(0) = y'_t(0) - z'_t(0) = 0.$$

Hence by this same lemma (1.28) is satisfied.

§2. A method of computing the a posteriori error estimate of the difference solution

Let us construct a square grid in the xy-plane by the lines $x = x_i = ih$ and $y = y_j = jh$, where $i, j = 0, \pm 1, \pm 2, \ldots$. Let Ω_h be the set of grid nodes on Ω such that all interior points of the segments joining them to the four neighboring nodes belong to Ω; let γ_h be the set of the remaining nodes on Ω, and $\tilde{\gamma}_h$ the set of points of intersection of the domain boundary γ with the lines that form the network.

We shall assume that a point (x_0^*, y_0^*) belongs to γ_h. Denote by $\mathbf{e}_k$ the vector with components $(\cos k(\pi/2), \sin k(\pi/2))$, and by ρ_k the distance from the point (x_0^*, y_0^*) to the nearest point (x_k^*, y_k^*) in the direction of $\mathbf{e}_k$, where $(x_k^*, y_k^*) \in \Omega_h \cup \tilde{\gamma}_h \cup \gamma_h \setminus (x_0^*, y_0^*)$. Also let $u_m = (x_m^*, y_m^*)$. Introduce operators l_h^3 and g_h^3 in the following way:

$$l_h^3 u_0 \equiv \sum_{i=1}^{4} \alpha_i u_i, \quad g_h^3 f_0 \equiv \frac{-\beta}{2h^2} f_0,$$

where

$$\alpha_i = \frac{\beta}{\rho_i(\rho_i + \rho_{i+2})}; \quad \beta = \frac{\rho_1\rho_2\rho_3\rho_4}{\rho_1\rho_3 + \rho_2\rho_4}.$$

Since $\alpha_1 + \alpha_2 + \alpha_3 + \alpha_4 = 1$, $\alpha_1\rho_1 = \alpha_3\rho_3$, $\alpha_2\rho_2 = \alpha_4\rho_4$ and $\alpha_1\rho_1^2 + \alpha_3\rho_3^2 = \alpha_2\rho_2^2 + \alpha_4\rho_4^2 = \beta$, it is an easy consequence of Taylor's formula that

$$\max_{\gamma_h} | u - l_h^3 u - h^2 g_h^3 \Delta u | \leqslant ch^3 \max_{\bar{\Omega}} \left(\left| \frac{\partial^3 u}{\partial x^3} \right| + \left| \frac{\partial^3 u}{\partial y^3} \right| \right),$$

where u is an arbitrary thrice continuously differentiable function on $\bar{\Omega}$, and c is a constant not depending on h or u. Moreover, it is clear that if u is linear in x and y, then $l_h^3 u \equiv u$ on γ_h. Finally, $0 < \min_{\gamma_h} \beta \le \max_{\gamma_h} \beta \le h^2/2$, and since the quantity $\rho_k(\rho_k + \rho_{k+2})$ has a minimum at the point $(x_{k_0}^*, y_{k_0}^*) \in \tilde{\gamma}_h \subset \gamma$, where $1 \le k_0 \le 4$ and $\alpha_i > 0$, $i = 1, 2, 3, 4$, we have

$$\max_{\gamma_h} \sum_{\substack{i=1 \\ i \neq k_0}}^{4} \alpha_i \leqslant \frac{3}{4}.$$

Thus the operators l_h^3 and g_h^3 satisfy all the requirements set forth in [6], §1, on the operators l_h^ν and g_h^ν for $\nu = 3$.

Consider the system of difference equations

$$u_h = Au_h - \frac{h^2 f}{4} \quad \text{on } \Omega_h; \tag{2.1}$$

$$u_h = l_h^3 u_h + h^2 g_h^3 f \quad \text{on } \gamma_h, \quad u_h = \varphi \text{ on } \tilde{\gamma}_h, \tag{2.2}$$

where $Au(x, y) \equiv (u(x + h, y) + u(x - h, y) + u(x, y + h) + u(x, y - h))/4$. This system, approximating problem (1, 1), has a unique solution (Lemma 1.3 of [6]).

Let us put $R_{ij} = \{(x, y)\colon x_i < x < x_{i+1}, y_j < y < y_{j+1}\}$ for a square; Ω_h^* for the set of vertices of all squares R_{ij} which have nonempty intersection with the domain Ω; Ω_h^0 for the set of nodes of Ω_h such that all interior points of the segments joining them to the four nodes at distance $\sqrt{2}h$ belong to Ω; r_* for the radius of the largest circle whose boundary can touch any point of γ and such that this circle lies completely in Ω.

Assume that $\Omega_h^0 \neq \varnothing$. To each point $(x_i, y_i) \in \Omega_h^0$ we make correspond a polynomial

$$\begin{aligned} P_{ij}(x, y) = {} & (1 + \bar{x}_i \delta_x (1 + h^2 \delta_y^2/6) + \bar{y}_j \delta_y (1 + h^2 \delta_x^2/6) \\ & + (\bar{x}_i^2 \delta_x^2 + \bar{y}_j^2 \delta_y^2)/2 + \bar{x}_i \bar{y}_j \delta_x \delta_y + ((3\bar{x}_i \bar{y}_j^2 - \bar{x}_i^3)\delta_x \delta_y^2 \\ & + (3\bar{x}_i^2 \bar{y}_j - \bar{y}_j^3)\delta_y \delta_x^2)/6) u_h(x_i, y_j) \\ & + (\bar{x}_i(\bar{x}_i^2 - h^2)\delta_x + \bar{y}_j(\bar{y}_j^2 - h^2)\delta_y) f(x_i, y_j)/6, \end{aligned} \tag{2.3}$$

where u_h is the solution of system (2.1), (2.2), $\bar{x}_i = x - x_i$, and

$$\bar{y}_j = y - y_j; \quad \delta_x f(x, y) = (f(x+h, y) - f(x-h, y))/2h;$$
$$\delta_x^2 f(x, y) \equiv (f(x-h, y) - 2f(x, y) + f(x+h, y))/h^2;$$

δ_y and δ_y^2 are defined in the same way. The polynomial P_{ij} that corresponds to the point $(x_i, y_j) \in \Omega_h^* \backslash \Omega_h^0$ we set equal to the arithmetic mean of the polynomials corresponding to all of the points of Ω_h^0 lying at the least distance from (x_i, y_j), and if $(x_i, y_j) \in \tilde{\gamma}_h \cup \gamma_h \cup \Omega_h$, we vary the free term to make $P_{ij}(x_i, y_j) = u_h(x_i, y_j)$.

Let us define a function $\tilde{u}_h$ on $\bar{\Omega}$ in the following way:

$$\tilde{u}_h(x,\ y)=\sum_{\mu=0}^{1}\sum_{\nu=0}^{1}P_{i+\mu,\ j+\nu}(x,\ y)\,\psi\left(\frac{\bar{x}_{i+\mu}}{h}\right)\psi\left(\frac{\bar{y}_{j+\nu}}{h}\right),\quad (x,\ y)\in\bar{\omega}_{ij},\tag{2.4}$$

where $\omega_{ij}=\Omega\cap R_{ij}$, $\psi(t)=1-2t^2$ for $|t|\le 1/2$, and $\psi(t)=2(1-|t|)^2$ for $t>1/2$.

Let $\gamma\in C_{2,\lambda}$, $0<\lambda<1$; let $2d$ be the width of the minimal strip containing Ω; $h_*=r_*/([r_*/h]+1)$; $N=r_*/h_*$; $\Omega(\rho)$ the set of points of Ω at a distance from γ greater than ρ; $\sigma_0=\Omega(r_*)$; $\sigma_n=\Omega(r_*-nh_*)\backslash\Omega(r_*-(n-1)h_*)$, $n=1,2,\dots,N$; $x_i^\alpha=x_i+\alpha h/2$; $y_j^\beta=y_j+\beta h/2$; $R_{ij}^{\alpha\beta}=\{(x,y)\colon x_i^\alpha<x<x_i^{\alpha+1},\ y_j^\beta<y<y_j^{\beta+1}\}$; $\omega_{ij}^{\alpha\beta}=\Omega\cap R_{ij}^{\alpha\beta}$, where α and β take the two values 0 and 1;

$$F_n=\max_{\substack{i,\ j,\ \alpha,\ \beta\\ \bar{\omega}_{ij}^{\alpha\beta}\cap\bar{\sigma}_n\neq\varnothing}}\ \max_{\bar{\omega}_{ij}^{\alpha\beta}}|\Delta\tilde{u}_h-f|;\tag{2.5}$$

$k_*=\inf_\gamma k(s)$, where $k(s)$ is the curvature of γ at the running point $s\in\gamma$ with the plus sign at points of convexity; $Z^*=z(r_*)$; $z(t)$ the solution of the Cauchy problem

$$z''_{tt}+a(t)\,z'_t=b(t),\qquad 0\leqslant t\leqslant r_*;\tag{2.6}$$

$$z(0)=z'_t(0)=0,\tag{2.7}$$

where $a(t)=1/(1/k_*-r_*+nh_*)$; $b(t)=F_n$, $(n-1)h_*\le t<nh_*$, $n=1,2,\dots,N$; and $a(r_*)=k_*$ and $b(r_*)=F_N$, computed with step h_* in elementary functions.

The following theorem is valid as an a posteriori result in [12]:

Theorem 2.1. *If* $\gamma\in C_{2,\lambda}$, $\phi\in C_{2,\lambda}(\gamma)$, $f\in C_{0,\lambda}^{2,\lambda}(\Omega)$,(3) $0<\lambda<1$, *and* $\Omega_h^0\neq\varnothing$, *then*

$$\max_{\Omega_h\cup\gamma_h}|u_h-u|\leqslant\max_{\bar{\Omega}}|\tilde{u}_h-u|\leqslant\max_{\gamma}|\tilde{u}_h-\varphi|+Z^*+F_0(d^2-r_*^2)/2,\tag{2.8}$$

with

$$\max_{\gamma}|\tilde{u}_h-\varphi|+Z^*+F_0(d^2-r_*^2)/2\leqslant ch^2\|f,\ \varphi\|_{0,\ \lambda}^{2,\ \lambda},\tag{2.9}$$

where u *is the solution of problem* (1.1), u_h *is the solution of system* (2.1), (2.2), *and* c *is a constant independent of* h, f *and* ϕ.

§3. Proof of Theorem 2.1.

We shall first prove the inequality

$$F_n\leqslant c_1h^\lambda(N-n+2)^{\lambda-2}\|f,\ \varphi\|_{0,\ \lambda}^{2,\ \lambda}\text{(4)}\tag{3.1}$$

where $n=0,1,\dots,N$, $N=[r_*/h]+1$.

On the basis of Lemmas 1.1–1.3, there exists a function $\tilde{u}(x,y)\in C_{2,\lambda}(E^2)$ such that $\tilde{u}=u$ on $\bar{\Omega}$, where u is the solution of problem (1.1) and

$$\|\tilde{u}\|_{C_{2,\ \lambda}(E^2)}\leqslant c_2\|f,\ \varphi\|_{0,\ \lambda}^{2,\ \lambda}.\tag{3.2}$$

(3) In order that $f\in C_{0,\lambda}^{2,\lambda}(\Omega)$ it is sufficient that $f\in C_{2,\lambda}(\Omega)$.

(4) Here and in what follows c_k, $k=1,2,\dots$, are constants independent of f, ϕ, h, n, i, j, p_{ij}, μ, ν, α, β.

By Lemma 1.2, $f \in C_{0,\lambda}(\Omega)$ and

$$\|f\|_{C_{0,\lambda}(\Omega)} \leqslant c_3 \|f, \varphi\|^{2,\lambda}_{0,\lambda}. \tag{3.3}$$

Let us denote by $Q^{(m)}_{ij}(x, y; u)$ the Taylor polynomial of degree m for the function $u(x, y)$ at the point (x_i, y_j). The expression

$$\frac{\partial^{p+q}}{\partial x^p \partial y^q} Q^{(m)}_{ij}(x, y; u), \qquad 0 \leqslant p+q \leqslant m,$$

is obviously a Taylor polynomial of degree $m - p - q$ for the derivative

$$\frac{\partial^{p+q} u(x, y)}{\partial x^p \partial y^q}$$

at (x_i, y_j). Consequently, with due regard for (3.2), for $0 \le p + q \le 2$ and $(x_i, y_j) \in \Omega^*_h$ we have

$$\max_{\substack{|x-x_i|\leqslant 7h \\ |y-y_j|\leqslant 7h}} \left| \frac{\partial^{p+q} Q^{(2)}_{ij}(x, y; \tilde{u})}{\partial x^p \partial y^q} - \frac{\partial^{p+q} \tilde{u}(x, y)}{\partial x^p \partial y^q} \right| \leqslant c_4 h^{2+\lambda-p-q} \|f, \varphi\|^{2,\lambda}_{0,\lambda}. \tag{3.4}$$

For definiteness we denote the polynomial (2.3) by $P_{ij}(x, y; u_h, f)$. From Taylor's formula together with (1.1), (3.2) and (2.3) it is easy to obtain the inequality

$$\max_{\substack{|x-x_i|\leqslant 7h \\ |y-y_j|\leqslant 7h}} |P_{ij}(x, y; u, f) - Q^{(2)}_{ij}(x, y; \tilde{u})| \leqslant c_5 h^{2+\lambda} \|f, \varphi\|^{2,\lambda}_{0,\lambda} \quad (x_i, y_j) \in \Omega^0_h. \tag{3.5}$$

Obviously any point of the set $\Omega^*_h \backslash \Omega^0_h$ is less than $\sqrt{2}h$ units from γ, and for $h \le r_*/3$ the distance from any point of the boundary γ to the set Ω^0_h is not greater than $4h$. For the proof of this last, less obvious, assertion it is sufficient to take an auxiliary point on the interior normal to γ at a distance $3h$ from γ; in a neighborhood of this point with radius h there is automatically at least one point belonging to Ω^0_h. Then, for $h \le r_*/3$ the distance from any point of the set $\Omega^*_h \backslash \Omega^0_h$ to Ω^0_h is less than $6h$. From this and from (3.4), (3.5), for $h \le r_*/3$, taking into account the given method of constructing the sets P_{ij} at points of $\Omega^*_0 \backslash \Omega^0_h$, follows the inequality

$$\max_{\bar{R}_{ij}} |P_{i+\mu, j+\nu}(x, y; u, f) - Q^{(2)}_{ij}(x, y; \tilde{u})| \leqslant c_6 h^{2+\lambda} \|f, \varphi\|^{2,\lambda}_{0,\lambda} \quad (x_i, y_j) \in \Omega^*_h, \tag{3.6}$$

where $\mu = 0, 1$ and $\nu = 0, 1$ when $\bar{\omega}_{ij} \neq \varnothing$. Because of (3.2), (3.3) and the boundedness of Ω, inequality (3.6) remains valid also for $h > r_*/3$ and $\Omega^0_h \neq \varnothing$, i.e. for any h under consideration, possibly with another constant c_6.

Since the difference $P_{i+\mu, j+\nu} - Q^{(2)}_{ij}$ is a polynomial in x and y of degree not greater than three, from (3.6) on the basis of Lemma 1.4 follows

$$\max_{\bar{R}_{ij}} \left| \frac{\partial^{p+q}}{\partial x^p \partial y^q} (P_{i+\mu, j+\nu}(x, y; u, f) - Q^{(2)}_{ij}(x, y, \tilde{u})) \right| \leqslant c_7 h^{2+\lambda-p-q} \|f, \varphi\|^{2,\lambda}_{0,\lambda}, \quad (x_i, y_j) \in \Omega^*_h, \tag{3.7}$$

where $0 \le p+q \le 2$, $\mu = 0, 1$, $\nu = 0, 1$ and $\bar{\omega}_{ij} \ne \varnothing$.

Evidently we also have

$$\max_{\bar{R}_{ij}^{\alpha\beta}} \left(\left| \frac{\partial^{p+q} \psi\left(\frac{\bar{x}_{i+\mu}}{h}\right)}{\partial x^p \partial y^q} \right| + \left| \frac{\partial^{p+q} \psi\left(\frac{\bar{y}_{j+\nu}}{h}\right)}{\partial x^p \partial y^q} \right| \right) \leqslant c_8 h^{-p-q}, \tag{3.8}$$

where $0 \le p+q \le 2$, $\alpha = 0, 1$, $\beta = 0, 1$, $\mu = 0, 1$ and $\nu = 0, 1$.

Denote by ρ_{ij} the distance from the point (x_i, y_j) to γ and by $\tilde{\Omega}_h$ the set of grid nodes that belong to $\bar{\Omega}$ and are not farther than $10h$ from γ. From Theorem 3.1 of [6] and Lemmas 2.3 and 2.7 of [6] we have

$$\max_{\tilde{\Omega}_h} |u_h - u| \leqslant c_9 h^{2+\lambda} \| f, \varphi \|_{0, \lambda}^{2, \lambda}. \tag{3.9}$$

From this it follows that

$$\max_{\bar{R}_{ij}} |P_{ij}(x, y; u_h - u, 0)| \leqslant c_{10} h^{2+\lambda} \| f, \varphi \|_{0, \lambda}^{2, \lambda}, \tag{3.10}$$

where $(x_i, y_j) \in \Omega_h^*$, $\rho_{ij} \le 8h$ and $\bar{\omega}_{ij} \ne \varnothing$. Also it is clear that

$$\max_{\bar{\omega}_{ij}^{\alpha\beta}} |f(x, y) - \Delta \tilde{u}_h(x, y)| \leqslant \max_{\bar{R}_{ij}^{\alpha\beta}} |\Delta \tilde{u}(x, y) - \Delta \tilde{u}_h(x, y)|. \tag{3.11}$$

Let us introduce the notation

$$S_{ij}(x, y; u, f) = \sum_{\mu=0}^{1} \sum_{\nu=0}^{1} P_{i+\mu, j+\nu}(x, y; u, f) \psi\left(\frac{\bar{x}_{i+\mu}}{h}\right) \psi\left(\frac{\bar{y}_{j+\nu}}{h}\right).$$

Since

$$\sum_{\mu=0}^{1} \sum_{\nu=0}^{1} \psi\left(\frac{\bar{x}_{i+\mu}}{h}\right) \psi\left(\frac{\bar{y}_{j+\nu}}{h}\right) \equiv 1, \quad (x, y) \in \bar{R}_{ij}, \tag{3.12}$$

we have

$$\begin{aligned} |\Delta \tilde{u}(x, y) - \Delta \tilde{u}_h(x, y)| &= |\Delta \tilde{u}(x, y) - \Delta S_{ij}(x, y; u_h, f)| \\ &\leqslant \left| \Delta \sum_{\mu=0}^{1} \sum_{\nu=0}^{1} (Q_{ij}^{(2)}(x, y; \tilde{u}) - P_{i+\mu, j+\nu}(x, y; u, f)) \psi\left(\frac{\bar{x}_{i+\mu}}{h}\right) \psi\left(\frac{\bar{y}_{j+\nu}}{h}\right) \right| \\ &\quad + |\Delta S_{ij}(x, y; u - u_h, 0)| + |\Delta(\tilde{u}(x, y) - Q_{ij}^{(2)}(x, y; \tilde{u}))|, \end{aligned}$$

where $(x, y) \in \bar{R}_{ij}^{\alpha\beta}$ and $\bar{\omega}_{ij}^{\alpha\beta} \ne \varnothing$. From this, as well as (3.4), (3.7)–(3.11) and Lemma 1.4, it follows that

$$\max_{\bar{\omega}_{ij}^{\alpha\beta}} |f(x, y) - \Delta \tilde{u}_h(x, y)| \leqslant c_{11} h^{\lambda} \| f, \varphi \|_{0, \lambda}^{2, \lambda}, \tag{3.13}$$

where $(x_i, y_j) \in \Omega_h^*$, $\rho_{ij} \le 6h$ and $\bar{\omega}_{ij}^{\alpha\beta} \ne \varnothing$.

Now let $(x_i, y_j) \in \Omega_h^*$ with $\rho_{ij} \ge 2h$. Then on the basis of Lemmas 1.1 and 1.2 we have

$$\max_{\substack{|x - x_i| \leqslant h \\ |y - y_j| \leqslant h}} \left| \frac{\partial^{p+q} Q_{ij}^{(3)}(x, y; u)}{\partial x^p \partial y^q} - \frac{\partial^{p+q} u(x, y)}{\partial x^p \partial y^q} \right| \leqslant c_{12} h^{4-p-q} \rho_{ij}^{\lambda-2} \| f, \varphi \|_{0, \lambda}^{2, \lambda}, \tag{3.14}$$

where $0 \le p+q \le 2$. With the help of Taylor's formula, relying on Lemmas 1.1

and 1.2 and taking into account that $f \in C^{2,\lambda}_{0,\lambda}(\Omega)$, we easily obtain the inequality

$$\max_{\substack{|x-x_i| \\ |y-y_j|}} |P_{ij}(x, y; u, f) - Q^{(3)}_{ij}(x, y; u)| \leqslant c_{13} h^4 \rho_{ij}^{\lambda-2} \|f, \varphi\|^{2,\lambda}_{0,\lambda}. \tag{3.15}$$

By virtue of (3.14) we have

$$\max_{\bar{R}_{ij}} |Q^{(3)}_{i+\mu,\, j+\nu}(x, y; u) - Q^{(3)}_{ij}(x, y; u)| \leqslant c_{14} h^4 \rho_{ij}^{\lambda-2} \|f, \varphi\|^{2,\lambda}_{0,\lambda},$$

where $\rho_{ij} \geq 4h$, $\mu = 0, 1$ and $\nu = 0, 1$. From this and (3.15) we find that

$$\max_{\bar{R}_{ij}} |P_{i+\mu,\, j+\nu}(x, y; u, f) - Q^{(3)}_{ij}(x, y; u)| \leqslant c_{15} h^4 \rho_{ij}^{\lambda-2} \|f, \varphi\|^{2,\lambda}_{0,\lambda}, \tag{3.16}$$

where $\rho_{ij} \geq 6h$, $\mu = 0, 1$ and $\nu = 0, 1$.

Let us denote by T^h_{ij} $(\bar{T}^h_{ij})$ the set of grid nodes belonging to the square

$$T_{ij} = \{(x, y) : |x - x_i| < 2h, \quad |y - y_j| < 2h\}$$

(the closure $\bar{T}_{ij}$ of this square). On the basis of Theorem 3.2 and Lemmas 2.1 and 2.2 of [6], when $\rho_{ij} \geq 6h$ we represent the solution of system (2.1), (2.2) on $\bar{T}^h_{ij}$ in the form

$$u_h = u + h^2 w + h^2 \rho^{\lambda}_{ij} \theta_h, \tag{3.17}$$

where u is the solution of problem (1.1), $w \in C_{2,\lambda}(T_{ij})$ is a completely determined function,

$$\max_{\bar{T}_{ij}} \left| \frac{\partial^{p+q} w}{\partial x^p \partial y^q} \right| \leqslant c_{16} \rho_{ij}^{\lambda-p-q} \|f, \varphi\|^{2,\lambda}_{0,\lambda}, \quad 0 \leqslant p+q \leqslant 2, \tag{3.18}$$

and θ_h is a function defined on $\bar{T}^h_{ij}$ and satisfying the condition

$$|\delta^2_x \theta_h| + |\delta^2_y \theta_h| + |\delta_y \delta_x \theta_h| \leqslant c_{17} \rho_{ij}^{-2} \|f, \varphi\|^{2,\lambda}_{0,\lambda} \text{ on } T^h_{ij}. \tag{3.19}$$

From (3.19) it follows that θ_h is representable on $\bar{T}^h_{ij}$ $(\rho_{ij} \geq 6h)$ in the form

$$\theta_h = l(x, y) + \theta^*_h, \tag{3.20}$$

where $l(x, y)$ is a function linear in x and y, and θ^*_h satisfies

$$\max_{T^h_{ij}} |\theta^*_h| \leqslant c_{18} h^2 \rho_{ij}^{-2} \|f, \varphi\|^{2,\lambda}_{0,\lambda}. \tag{3.21}$$

From (3.12) and (3.20) we have on $\bar{R}_{ij}$, when $\rho_{ij} \geq 6h$,

$$S_{ij}(x, y; \theta_h, 0) \equiv l(x, y) + S_{ij}(x, y; \theta^*_h, 0).$$

From this, using (3.8) and (3.21), we obtain

$$\max_{\bar{R}^{\alpha\beta}_{ij}} |\Delta S_{ij}(x, y; \theta_h, 0)| \leqslant c_{19} \rho_{ij}^{-2} \|f, \varphi\|^{2,\lambda}_{0,\lambda}, \tag{3.22}$$

where $\rho_{ij} \geq 6h$, $\alpha = 0, 1$ and $\beta = 0, 1$. Let us set

$$v(x, y) = w(x, y) - Q^{(1)}_{ij}(x, y; w). \tag{3.23}$$

By virtue of (3.18) we have on $\bar{T}_{ij}$

$$|v(x, y)| \leqslant c_{20} h^2 \rho_{ij}^{\lambda-2} \|f, \varphi\|_{0,\lambda}^{2,\lambda}, \qquad \rho_{ij} \geqslant 6h.$$

As in the preceding case, we find, on taking (3.23) into account, that

$$\max_{\bar{R}_{ij}^{\alpha\beta}} |\Delta S_{ij}(x, y; w, 0)| \leqslant c_{21} \rho_{ij}^{\lambda-2} \|f, \varphi\|_{0,\lambda}^{2,\lambda}, \tag{3.24}$$

where $\rho_{ij} \geq 6h$, $\alpha = 0, 1$ and $\beta = 0, 1$.

According to (1.1), (3.12) and (3.17) we have on $\overline{R}_{ij}^{\alpha\beta}$, when $\rho_{ij} \geq 6h$,

$$f(x, y) - \Delta\tilde{u}_h(x, y) = \Delta u(x, y) - \Delta Q_{ij}^{(3)}(x, y; u)$$
$$+ \Delta \sum_{\mu=0}^{1} \sum_{\nu=0}^{1} (Q_{ij}^{(3)}(x, y; u) - P_{i+\mu, j+\nu}(x, y; u, f)) \psi\left(\frac{\bar{x}_{i+\mu}}{h}\right) \psi\left(\frac{\bar{y}_{j+\nu}}{h}\right)$$
$$- h^2 \Delta S_{ij}(x, y; w, 0) - h^2 \rho_{ij}^{\lambda} \Delta S_{ij}(x, y; \theta_h, 0).$$

From this and from (3.8), (3.14), (3.16), (3.21), (3.24) and Lemma 1.4 follows the inequality

$$\max_{\bar{R}_{ij}^{\alpha\beta}} |f(x, y) - \Delta\tilde{u}_h(x, y)| \leqslant c_{22} h^2 \rho_{ij}^{\lambda-2} \|f, \varphi\|_{0,\lambda}^{2,\lambda}, \tag{3.25}$$

where $\rho_{ij} \geq 6h$, $\alpha = 0, 1$ and $\beta = 0, 1$.

By combining (3.13) and (3.25) we obtain the general inequality

$$\max_{\bar{\omega}_{ij}^{\alpha\beta}} |f(x, y) - \Delta\tilde{u}_h(x, y)| \leqslant c_{23} h^2 (\rho_{ij} + h)^{\lambda-2} \|f, \varphi\|_{0,\lambda}^{2,\lambda}, \tag{3.26}$$

where $(x_i, y_j) \in \Omega_h^*$, $\alpha = 0, 1$, $\beta = 0, 1$ and $\bar{\omega}_{ij}^{\alpha\beta} \neq \varnothing$. From (3.26) the inequality (3.1) being proved follows immediately.

Let π be the axis of the strip of minimal width $2d$ that contains the domain Ω, and, letting $\rho_\pi = \rho_\pi(x, y)$ be the distance from the variable point (x, y) to this axis, set

$$\Psi_1(x, y) = F_0 \begin{cases} \dfrac{d^2 - r_*^2 - \rho_\pi^2}{2}, & \rho_\pi \leqslant d - r_*, \\ (d - r_*)(d - \rho_\pi), & \rho_\pi > d - r_*. \end{cases}$$

It is obvious that $\Psi_1 \in W_2^2(\Omega) \cap C_{1,1}(\Omega)$,

$$\max_{\bar{\Omega}} \Psi_1 = F_0 \frac{d^2 - r_*^2}{2}, \qquad \min_{\bar{\Omega}} \Psi_1 = 0; \tag{3.27}$$

$$\Delta\Psi_1 = -F_0, \qquad (x, y) \in \sigma_0, \tag{3.28}$$

and the inequality

$$\Delta\Psi_1 \leqslant 0 \tag{3.29}$$

is satisfied almost everywhere on Ω.

Let us define on $\bar{\Omega}$ the function

$$\Psi_2(x, y) = \begin{cases} Z^* - z(r_* - \rho), & \rho \leqslant r_*, \\ Z^*, & \rho > r_*, \end{cases} \tag{3.30}$$

where $\rho = \rho(x, y)$ is the distance from the variable point $(x, y) \in \bar{\Omega}$ to γ, $z(t)$ is a solution of the Cauchy problem (2.6), (2.7), and $Z^* = z(r_*)$. By virtue of the fact that the function $z(t)$ is continuously differentiable on the interval $[0, r_*]$, wherein $z(0) = z'(0) = 0$, and from Lemma 1.7 and (1.18), where $\sigma'(0) = 0$, it follows that the first partial derivatives of $\rho(x, y)$ are continuous and bounded on the set

$$\Omega_* = \{(x, y): 0 \leqslant \rho(x, y) < r_*, \quad (x, y) \in \bar{\Omega}\},$$

$\rho(x, y)$ itself is continuous on $\bar{\Omega}$, and $\Psi_2(x, y)$ is continuously differentiable on $\bar{\Omega}$. Moreover, since the second derivative $z''(t)$ is piecewise continuous on $[0, r_*]$ ($z''(t)$ may have only discontinuities of the first kind at the points $t_m = mh_*$, $m = 1, 2, \ldots, N - 1$), and the function $\rho(x, y)$ is twice continuously differentiable on Ω_* by virtue of Lemma 1.7, it follows that the function $\Psi_2(x, y)$ is twice continuously differentiable everywhere on $\bar{\Omega}_*$, except possibly on a finite number of isolated smooth curves $\gamma_{\rho_m} = \{(x, y): \rho(x, y) = \rho_m, (x, y) \in \Omega\}$, $\rho_m = mh_*$, $m = 1, 2, \ldots, N - 1$ (according to Theorem 2.1 of [4], $\gamma_{\rho_m} \in C_{2,\lambda}$), and on the set $\gamma_* = \{(x, y): \rho(x, y) = r_*, (x, y) \in \Omega\}$ of two-dimensional Jordan measure zero.(5) From (1.18)–(1.20) and (3.30), with due regard for the fact that $\gamma \in C_{2,\lambda}$, $0 < \lambda < 1$, and that $z(t)$ is the solution of problem (2.6), (2.7), we easily establish the boundedness of the second partial derivatives of $\Psi_2(x, y)$ on the set

$$\Omega' = \bar{\Omega} \setminus \left(\bigcup_{m=1}^{N-1} \gamma_{\rho_m} \cup \gamma_* \right).$$

On the strength of what has been said concerning Theorem 2.1 of [4], Lemma 1.1 of [4] and the fact that $|z'(t)| \leq \tilde{c}t$, where $\tilde{c}$ is independent of t, it follows that $\Psi_2 \in W_2^2(\Omega)$.

Let us calculate the value of the Laplace operator of the function Ψ_2 at an arbitrary point $(x, y) \in \Omega$ at distance ρ from γ, $0 < \rho < r_*$, where ρ/h_* is not an integer. From Lemma 2.2 of [4] the quantity

$$\frac{\partial^2 \Psi_2}{\partial \rho^2} = -z''(r_* - \rho) \tag{3.31}$$

is the second derivative of Ψ_2 at the point (x, y) in the direction of the normal to the curve γ_ρ on which (x, y) lies. Let τ be the displacement along the tangent to γ_ρ at the point $(x, y) \in \gamma_\rho$. Since the function Ψ_2 has a constant value along γ_ρ, and by Theorem 2.1 of [4] $\gamma_\rho \in C_{2,\lambda}$, we have on the basis of Lemma 1.8 at the point (x, y) of tangency

$$\frac{\partial^2 \Psi_2}{\partial \tau^2} = -z'(r_* - \rho) k_\rho, \tag{3.32}$$

where k_ρ is the curvature of γ_ρ at this point. According to what was proved above, the function Ψ_2 has at the point (x, y) under consideration continuous

(5) The property that the Jordan measure of the set γ_* is zero readily follows from the fact that the curve $\gamma_{r_* - \epsilon}$ for any positive $\epsilon < r_*$ has finite length bounded by the length of γ, by virtue of Theorem 2.1 of [4] and Lemma 2.2 of [4].

second derivatives with respect to x and y; hence, according to (3.31) and (3.32),

$$\Delta\Psi_2(x, y) = -(z''(r_* - \rho) + z'(r_* - \rho)k_\rho), \tag{3.33}$$

where $(x, y) \in \Omega$, $0 < \rho = \rho(x, y) < r_*$, ρ/h_* is nonintegral, and k_ρ is the curvature of γ_ρ at the point (x, y).

Now let $(x, y) \in \Omega$, $r_* - nh_* < \rho(x, y) < r_* - (n-1)h_*$, n an integer, $1 \le n \le N$, k the curvature of the boundary γ at a point lying at a distance $\rho = \rho(x, y)$ from (x, y) and k_* the minimum of the curvature of γ. Since $\rho k < 1$, on the basis of Lemmas 1.8 and 1.9 we have

$$z'(r_* - \rho)k_\rho = z'(r_* - \rho)\frac{k}{1 - \rho k} \geqslant z'(r_* - \rho)\frac{k_*}{1 - \rho k_*} \geqslant \frac{z'(r_* - \rho)}{1/k_* - r_* + nh_*}.$$

From this, with regard for (2.6) and (3.33), we obtain

$$\Delta\Psi_2(x, y) \leqslant -F_n, \tag{3.34}$$

where $r_* - nh_* < \rho(x, y) < r_* - (n-1)h_*$, and F_n is given by (2.5), $n = 1, \dots, N$. It is also obvious that

$$\Delta\Psi_2(x, y) = 0, \quad \rho(x, y) > r_*. \tag{3.35}$$

Let

$$v_j(x, y) = \Psi_1(x, y) + \Psi_2(x, y) + c_0 + (-1)^j(\tilde{u}_h(x, y) - u(x, y)), \quad j = 0, 1, \tag{3.36}$$

where $\tilde{u}_h$ is the function (2.4), u is the solution of problem (1.1), and

$$c_0 = \max_\gamma |\tilde{u}_h - \varphi|.$$

In addition let $w_j(x, y)$ be a function harmonic on Ω, where

$$w_j|_\gamma = v_j|_\gamma \geqslant 0, \quad j = 0, 1. \tag{3.37}$$

Since $\Psi_i|_\gamma \in C_{1,1}(\gamma)$, $i = 1, 2$, $\tilde{u}_h|_\gamma \in C_{1,1}(\gamma)$ and $u|_\gamma = \phi \in C_{1,1}(\gamma)$, we have $w_j|_\gamma \in C_{1,1}(\gamma)$, and it follows from the estimate (2.8) of [13] that $w_j \in W_2^2(\Omega) \cap C_{1,2/3}(\Omega)$, $j = 0, 1$. Moreover, since $\tilde{u}_n \in W_2^2(\Omega) \cap C_{1,1}(\Omega)$ (by construction), $u \in C_{2,\lambda}(\Omega)$, and, as was established earlier, $\Psi_i \in W_2^2(\Omega) \cap C_{1,1}(\Omega)$, $i = 1, 2$, we have

$$g_j = (v_j - w_j) \in W_2^2(\Omega) \cap C_{1,2/3}(\Omega), \quad j = 0, 1.$$

Finally, according to (2.5), (3.28), (3.29) and (3.34)–(3.36), the following inequality is satisfied almost everywhere on Ω:

$$\Delta g_j \leqslant 0, \quad j = 0, 1. \tag{3.38}$$

We now prove that $g_j \in W_{2,0}^2(\Omega)$, $j = 0, 1$. Let us fix h for the time being. Continue the function g_j from the domain Ω across its boundary $\gamma \in C_{2,\lambda}$ to the entire plane E^2 so that $g_j \in W_2^2(E^2) \cap C_{1,2/3}(E^2)$ (cf. [14] and [15]), and the extended function g_j, for which we retain the previous notation, coincides with the original function on Ω and has compact support on E^2.

Let $\epsilon > 0$,

$$g_j^\varepsilon(x, y) = \frac{1}{\chi\varepsilon^2}\int_{-\infty}^{+\infty}\int_{-\infty}^{+\infty} g_j(\xi, \eta)\,\vartheta\left(\frac{(x-\xi)^2+(y-\eta)^2}{\varepsilon^2}\right)d\xi d\eta, \qquad (3.39)$$

where $j = 0, 1$;

$$\vartheta(t) = \begin{cases} e^{\frac{t}{t-1}}, & 0 \leqslant t < 1, \\ 0, & 1 \leqslant t; \end{cases}$$

and $\kappa = \pi\int_0^1 \vartheta(t)\,dt$. It is known (cf. [9] and [10]) that the function g_j^ϵ, $j = 0, 1$, is infinitely differentiable on the entire plane, with

$$\lim_{\varepsilon\to+0} \|g_j^\varepsilon - g_j\|_{W_2^2(\Omega)} = 0, \qquad j = 0, 1. \qquad (3.40)$$

Furthermore, since $\gamma \in C_{2,\lambda}$, $g_j \in C_{1,2/3}(\Omega)$ and $g_j|_\gamma \equiv 0$, elementary calculations carried out with the use of standard methods of differentiation and successive integration by parts of the integral (3.39) show that the function $\phi_{j\epsilon} = g_j^\epsilon|_\gamma$ belongs to the class $C_{2,\lambda}(\gamma)$, where

$$\max_{s\in\gamma} |\varphi_{j\varepsilon}^{(m)}(s)| < \tilde{c}_1 \varepsilon^{1+2/3-m}, \qquad j = 0, 1,^{(6)} \qquad (3.41)$$

s being the arc of γ, and $m = 0, 1, 2$. From (3.41) for $0 < \epsilon < 1$ we get the inequality (cf. [11], §33)

$$\|\varphi_{j\varepsilon}\|_{C_{1,5/9}(\gamma)} < \tilde{c}_2 \varepsilon^{1/9}, \qquad j = 0, 1. \qquad (3.42)$$

Furthermore, let w_j^ϵ be a solution of the Dirichlet problem

$$\Delta w_j^\varepsilon = 0 \text{ on } \Omega, \quad w_j^\varepsilon = \varphi_{j\varepsilon} \text{ on } \gamma.$$

Then (see [11]) $w_j^\epsilon \in C_{2,\lambda}(\Omega)$, and on the strength of (3.42) we have $\|w_j^\epsilon\|_{C_{1,5/9}(\Omega)} < \tilde{c}_3\epsilon$, where $0 < \epsilon < 1$; therefore, with regard for the estimates (29) and (30) of [16], for $0 < \epsilon < 1$ we have

$$\|w_j^\varepsilon\|_{W_2^2(\Omega)} < \tilde{c}_4 \varepsilon^{1/9}, \qquad j = 0, 1. \qquad (3.43)$$

Thus the function $\tilde{g}_j^\epsilon = g_j^\epsilon - w_j^\epsilon$ is twice continuously differentiable on $\bar{\Omega}$ and vanishes on γ, while by virtue of (3.40) and (3.43)

$$\lim_{\varepsilon\to+0} \|\tilde{g}_j^\varepsilon - g_j\|_{W_2^2(\Omega)} = 0, \qquad j = 0, 1.$$

Hence (see [8], Chapter I, §1) $v_j - w_j = g_j \in \overset{\circ}{W}{}_{2,0}^2(\Omega)$, $j = 0, 1$. The assertion that has been proved clearly is true for any h for which $\Omega_h^0 \neq \varnothing$. On the basis of (3.38) and Lemma 1.5, because of the nonnegativity of the harmonic function w_j (see (3.37)), we obtain from this

(6) Here and below $\tilde{c}_k$, $k = 1, 2, \ldots$, are constants independent of ϵ.

$$v_j(x, y) \geqslant v_j(x, y) - w_j(x, y) \geqslant 0, \qquad (x, y) \in \bar{\Omega}, \qquad j = 0, 1,$$

and the latter, when we take (3.27), (3.30) and (3.36) into account, implies that (2.8) is satisfied.

From (2.3), (2.4) and (3.9) follows the inequality

$$\max_{\gamma} |\tilde{u}_h - \varphi| \leqslant c_{24} h^{2+\lambda} \| f, \varphi \|_{0, \lambda}^{2, \lambda}, \tag{3.44}$$

and from (3.1) we immediately obtain

$$\frac{F_0 (d^2 - r_*^2)}{2} \leqslant c_{25} h^2 \| f, \varphi \|_{0, \lambda}^{2, \lambda}. \tag{3.45}$$

Let $\alpha(t) = -|k_*|$ and

$$\beta(t) = c_1 h^2 (r_* - t + h_*)^{\lambda - 2} \| f, \varphi \|_{0, \lambda}^{2, \lambda}, \qquad 0 \leqslant t \leqslant r_*.$$

Then according to (1.24) and (1.25) the solution of problem (1.26), (1.27) satisfies the conditions

$$0 \leqslant y'(t) \leqslant e^{|k_*| r_*} \int_0^t \beta(\eta)\, d\eta \leqslant c_1 e^{|k_*| r_*} \frac{h^2}{1-\lambda} (r_* - t + h_*)^{\lambda - 1} \| f, \varphi \|_{0, \lambda}^{2, \lambda},$$

$$0 \leqslant y(t) \leqslant c_1 e^{|k_*| r_*} \frac{h^2 (r_* + h_*)^{\lambda}}{\lambda (1 - \lambda)} \| f, \varphi \|_{0, \lambda}^{2, \lambda}, \tag{3.46}$$

where $0 \leq t \leq r_*$. On the basis of (3.1),

$$0 \leqslant b(t) \leqslant \beta(t), \qquad 0 \leqslant t \leqslant r_*, \tag{3.47}$$

where $b(t)$ is the right-hand side of equation (2.6). Furthermore, it is evident that $\alpha(t) \leq a(t)$, $0 \leq t \leq r_*$, where $a(t)$ is the coefficient of z' in (2.6). From this and from (3.46) and (3.47) according to Lemma 1.10 we have

$$Z^* \leqslant c_{26} h^2 \| f, \varphi \|_{0, \lambda}^{2, \lambda}. \tag{3.48}$$

From (3.44), (3.45) and (3.48) follows inequality (2.9)

Theorem 2.1 is completely proved.

BIBLIOGRAPHY

1. D. F. Davidenko, *An estimate of the error in the network solution of the Dirichlet problem for the Laplace equation*, Dokl. Akad. Nauk SSSR **138** (1961), 267–270 = Soviet Math. Dokl. 2 (1961), 574–577. MR 23 #B3157.

2. E. A. Volkov, *On conditions for convergence with speed h^2 of the net method for the Laplace equation*, Mat. Zametki 6 (1969), 669–679 = Math. Notes 6 (1969), 866–872. MR 41 #6418.

3. ———, *Weighted error estimates for the mesh method of solving the Laplace and Poisson equations*, Trudy Mat. Inst. Steklov. **117** (1972), 100–112 = Proc. Steklov Inst. Math. **117** (1972), 119–134.

4. ———, *Boundaries of subdomains, Hölder weight classes and solutions of the Poisson equation in these classes*, Trudy Mat. Inst. Steklov. **117** (1972), 75–99 = Proc. Steklov Inst. Math. **117** (1972), 89–118.

5. O. V. Besov and V. P. Il'in, *Natural extension of the class of regions in imbedding theorems*, Mat. Sb. **75** (117) (1968), 483–495 = Math. USSR Sb. 4 (1968), 445–456. MR **37** #1987.

6. E. A. Volkov, *Approximate solution of the Laplace and Poisson equations in weighted Hölder spaces*, Trudy Mat. Inst. Steklov **128** (1972), 76–112 = Proc. Steklov Inst. Math. **128** (1972), 85–129. MR **47** #9852.

7. V. L. Gončarov, *Theory of interpolation and the approximation of functions*, 2nd ed., GITTL, Moscow, 1954. (Russian) MR **16**, 803.

8. O. A. Ladyženskaja and N. N. Ural'ceva, *Linear and quasi-linear equations of elliptic type*, "Nauka", Moscow, 1964; English transl., Academic Press, New York, 1968. MR **35** #1955; **39** #5941.

9. S. L. Sobolev, *Applications of functional analysis in mathematical physics*, Izdat. Sibirsk. Otd. Akad. Nauk SSSR, Novosibirsk, 1962; English transl., Amer. Math. Soc., Providence, R. I., 1963. MR **29** #2624.

10. S. M. Nikol'skiĭ, *Approximation of functions of several variables and embedding theorems*, "Nauka", Moscow, 1969; English transl., Springer-Verlag, Berlin and New York, 1974. MR **46** #9714.

11. C. Miranda, *Equazioni alle derivate parziali di tipo ellittico*, Springer-Verlag, Berlin, 1955, English transl., 1970. MR **19**, 421.

12. E. A. Volkov, *A posteriori and weighted error estimates for solutions of Poisson's equation and their derivatives computed by the net method*, Dokl. Akad. Nauk SSSR **192** (1970), 717–720 = Soviet Math. Dokl. **11** (1970), 699–703; correction, ibid. **206** (1972), no. 4, VIII = ibid. **13** (1972), no. 5, iv. MR **42** #7085; **47** #2833.

13. ———, *The differential properties of the solutions of Laplace's equation, and the errors in the method of nets with boundary values in C_2 and $C_{1,1}$*, Ž. Vyčisl. Mat. i Mat. Fiz. **9** (1969), 573–584 = USSR Comput. Math. and Math. Phys. **9** (1969), no. 3, 97–112. MR **40** #8286.

14. S. M. Nikol'skiĭ, *On the solution of the polyharmonic equation by a variational method*, Dokl. Akad. Nauk SSSR **88** (1953), 409–411. (Russian) MR **15**, 425.

15. ———, *Extension of functions of several variables preserving differential properties*, Mat. Sb. **40** (82) (1956), 243–268: English transl., Amer. Math. Soc. Transl. (2) **83** (1969), 159–188. MR **19**, 398.

16. E. A. Volkov, *On the solution by the method of grids of the interior Dirichlet problem for the Laplace equation*, Vyčisl. Mat. **1** (1957), 34–61; English transl., Amer. Math. Soc. Transl. (2) 24 (1963), 279–307. MR **23** #B1665.

Translated by R. N. GOSS

Trudy Mat. Inst. Steklov.
134 (1975)

Proc. Steklov. Inst. Math.
134 (1975)

ON THE THEORY OF APPROXIMATION OF FUNCTIONS ON CLOSED SETS OF THE COMPLEX PLANE (WITH REFERENCE TO A PROBLEM OF S. M. NIKOL'SKIĬ)

UDC 517.51:517.53

V. K. DZJADYK

Abstract. A concise review of basic results on the subject for the past two decades is presented in §1 of this paper. For continuous functions $f(z)$ defined on arbitrary sets $\mathfrak{M}$ of the complex plane C the definition of the second modulus of continuity is introduced and its properties are investigated in detail in §2. Definition. Let $f\colon \mathfrak{M} \to C$, and let $\omega(t)$ be an arbitrary function of the type of second modulus of continuity (i.e. $\omega(0) = 0$, $\omega(t) \in C$, $t \geq 0$, $\omega(t)\uparrow$ and $\omega(\lambda t) \leq (\lambda + 1)^2\omega(t)$). Then we shall say that $f(z) \in H_2^{\omega(t)}$ if there exists a positive constant A such that for all $z_1, z_2 \in \mathfrak{M}$ and $z \in \mathfrak{M} \cap [U(z_1; h) \cup U(z_2; h)]$

$$|f(z) - P_1(z)| \leqslant At \int_t^{2h} \frac{\omega(s)}{s^2}\, ds,$$

where $h = |z_2 - z_1|$, $U(z_i; h) = \{z\colon |z - z_i| \leq h\}$, $i = 1, 2$, $t = \min\{|z - z_1|, |z - z_2|\}$ and $P_1(z) = P_1(z; z_1; z_2; f)$ is a polynomial of degree 1 which interpolates the function $f(z)$ at the points z_1 and z_2. It is established that for the case of convex sets $\mathfrak{M}$ this definition coincides with the usual definition of the classes $H_2^{\omega(t)}$ in terms of upper bounds of the second differences. In §3 a generalization of admissible sets of type B_k introduced earlier by the author is presented. In accordance with the results of V. I. Belyĭ and V. M. Mikljukov the structure of these sets can be revealed with the help of the theory of quasiconformal mappings. A very special case is furnished by arbitrary bounded sets with simply connected complement, whose boundary consists of finitely many sufficiently smooth arcs which meet to form nonzero exterior angles. A study of the kernels $K_{r,n,k,m}(z)$ introduced by the author as polynomials in z is carried out in §4; they approximate the Cauchy kernel $1/(\zeta - z)$ on sets of type B_k and are needed for subsequent purposes. These kernels play the same role in direct theorems as the Jackson kernels do in the periodic case. In §5 direct theorems on approximation of functions continuous on $\mathfrak{M}$, analytic in Int $\mathfrak{M}$ and defined on various admissible sets $\mathfrak{M}$ of type B_k are established in terms of second moduli of continuity. In particular, the problems of simultaneous approximation of a function and its derivatives, approximation of integrals of Cauchy type, approximation at interior points of a set, refinement of inequalities for the modulus of the derivative of an algebraic polynomial which approximates a given function "well", etc., are considered. In §6 converse theorems on arbitrary bounded continua with simply-connected complement are obtained in terms of second moduli of continuity; these theorems generalize earlier results of the author and subsequent results of N. A. Lebedev and P. M. Tamrazov. A consequence of the direct and converse theorems is a constructive

AMS (MOS) subject classifications (1970). Primary 30A82; Secondary 30A06.

characterization on various sets of type B_k, of the classes $H_2^{\omega(t)}$ and $W^r H_2^{\omega(t)}$ of functions, under the requirement that the function $\omega(t)$ satisfy the condition

a) $$\int_h^1 \frac{\omega(s)}{s^3}\,ds = O\left[\frac{1}{h}\int_h^1 \frac{\omega(s)}{s^2}\,ds\right],$$

for $r = 0$, while for $r \geq 1$ that it satisfy condition a) and the condition

b) $$\int_0^t \frac{\omega(s)}{s}\,ds = O\,[\omega(t)]$$

Bibliography: 41 titles.

§1. Introduction

1°. One of the most significant problems in approximation theory is to determine a relationship between smoothness of functions and the maximum rate of their approximation by polynomials when the degrees of these polynomials increase. In this connection theorems which establish estimates for the rate of convergence to zero of magnitudes of approximation of a particular function (or class of functions) in terms of its smoothness are called direct theorems, while theorems in which the degree of smoothness of a function is deduced from its approximations are called converse theorems. In the case when interlocking direct and converse theorems have been established on a class A of functions, we say that a constructive characterization has been obtained for A.

In 1911 D. Jackson obtained, in the most complete form, direct theorems on approximation of periodic continuous functions of the classes $C^{(r)}$, $r = 0, 1, 2, \ldots$ ($C^0 \overset{\text{def}}{=\!=} C$). According to this result, for any function $f(t) \in C^{(r)}$ there exists a trigonometric polynomial $T_n(t)$ of degree n for every $n = 1, 2, \ldots$, such that

$$|f(t) - T_n(t)| \leqslant A\,\frac{\omega(f^{(r)};\ 1/n)}{n^r}, \tag{1.1}$$

where $\omega(f^{(r)};\ t)$ is the modulus of continuity of the function $f^{(r)}(t)$; $A = A(r)$ is a constant.

In the sequel, to avoid repetition, the degree or order of a polynomial will be denoted by an index as subscript, while various constants will be denoted by the letters A and A_1.

A year later, S. N. Bernstein established inequalities for the modulus of the derivative of a trigonometric polynomial:

$$\text{if}\quad |T_n(t)| \leqslant A,\quad \text{then}\ |T_n'(t)| \leqslant An, \tag{1.2}$$

and with their help he obtained for the first time converse theorems for some important classes of functions. In 1919, improving somewhat upon Bernstein's method de la Vallée-Poussin established that if for some integer $r \geq 0$ and $0 < \alpha < 1$ there exist polynomials $T_n(t)$ for $n = 1, 2, \ldots$ for a function $f(t)$ such that

$$|f(t) - T_n(t)| \leqslant \frac{A}{n^{r+\alpha}},\quad \text{then}\ f(t) \in C^{(r)}H^\alpha, \tag{1.3}$$

i.e. $f(t) \in C^{(r)}$ and $f^{(r)}(t) \in H^\alpha$.

Since, in view of (1.3) and (1.1), direct and converse theorems interlock for the functions (of Hölder classes) considered by de la Vallée-Poussin, as a result a constructive characterization was obtained for all functions of the Hölder classes $C^{(r)}H^{\alpha}$, where r is a nonnegative integer and $0 < \alpha < 1$. The case $\alpha = 1$ did not fit into the original scheme.

In 1945 A. Zygmund showed that in this case it was feasible to introduce, instead of the Hölder classes $C^{(r)}H^1$, somewhat broader classes of functions $C^{(r)}Z$ by taking second differences into consideration, so that these functions satisfy the condition $|\Delta_h^2(f^{(r)};\, t)| \leq Ah$ for any $h > 0$. For these classes he also obtained a constructive characterization.

In the work of Jackson and Bernstein the direct and converse theorems are also extended to the case of approximation of nonperiodic functions defined on an interval $[a, b]$ by means of ordinary polynomials $P_n(x)$; however, the results obtained for this case were incomplete, since the sufficient conditions for a function f belong to the class H^{α} $(0 < \alpha < 1)$ did not coincide with the necessary conditions. This is explained, first of all, by the fact that, in the nonperiodic case, which is more complicated than the periodic one, it is, as a rule, not possible to characterize continuity properties of a function and its derivatives completely by means of a sequence of uniform approximants of this function. In particular, in this case there does not exist any decreasing function $\phi(n)$ of natural argument such that the condition

$$\|f - P_n\|_C \leqslant A\varphi(n), \qquad A = \text{const}, \qquad n = 1,\ 2,\ \ldots,$$

turns out to be (in the same way as in the periodic case) both necessary and sufficient for $f(x)$ to belong to H^{α} $(0 < \alpha < 1)$.

The first essential progress in this problem, after the above-mentioned work of Jackson and Bernstein, was obtained only in 1946 by S. M. Nikol'skiĭ, who showed [1] that for functions which satisfy the condition $\operatorname{Lip} 1 = H^1$ the Jackson theorem can be strengthened in the nonperiodic case.

In 1951 A. F. Timan [2], while generalizing and somewhat strengthening Nikol'skiĭ's result, established that if $f(x) \in C^{(r)}[-1, 1]$, then there exist polynomials $P_n(x)$ such that

$$|f(x) - P_n(x)| \leqslant A\left(\frac{\sqrt{1-x^2}}{n} + \frac{1}{n^2}\right)^r \omega\left(f^{(r)};\ \frac{\sqrt{1-x^2}}{n} + \frac{1}{n^2}\right),$$
$$x \in [-1,\ 1], \qquad n = 1,\ 2,\ \ldots. \tag{1.4}$$

In 1956 the author [3], using the identity of M. Riesz for the derivative of a trigonometric polynomial, established the following inequalities for the modulus of the derivative of a polynomial: if

$$|P_n(x)| \leqslant \left[\frac{\sqrt{1-x^2}}{n} + \frac{1}{n^2}\right]^s, \quad s > 0,$$

then

$$|P_n'(x)| \leqslant A\left[\frac{\sqrt{1-x^2}}{n}+\frac{1}{n^2}\right]^{s-1}, \tag{1.5}$$

and by using these inequalities he obtained converse theorems for a function $f(x) \in C^{(r)}H^{\alpha}$, $r = 0, 1, \ldots, 0 < \alpha < 1$.

As a result of these theorems and Nikol'skiĭ's direct theorems, Timan obtained a constructive characterization for nonperiodic functions of the Hölder classes $C^{(r)}H^{\alpha}$, $0 < \alpha < 1$.

In 1958 the author [4] obtained a constructive characterization for the classes $C^{(r)}Z\,[-1, 1]$, and the same year he [5], and a year later G. Freud [6], strengthened the direct theorems by replacing the first modulus of continuity in (1.4) by the second. We also note that G. K. Lebed' [7], and independently Ju. A. Brudnyĭ [8], by using instead of M. Riesz's identity other identities for the derivative of a trigonometric polynomial, generalized the result of the author [3] in the sense that if

$$|P_n(x)| \leqslant \left[\frac{\sqrt{1-x^2}}{n}+\frac{1}{n^2}\right]^{r} \omega\left(\frac{\sqrt{1-x^2}}{n}+\frac{1}{n^2}\right),$$

then

$$|P_n'(x)| \leqslant A\left[\frac{\sqrt{1-x^2}}{n}+\frac{1}{n^2}\right]^{r-1} \omega\left(\frac{\sqrt{1-x^2}}{n}+\frac{1}{n^2}\right), \tag{1.5'}$$

where $\omega(t)$ is an arbitrary modulus of continuity. Using this inequality, Timan [9] generalized the results obtained by the author in [3] to the classes $C^{(r)}H^{\omega}$.

From this discussion it is clear that in the approximation of nonperiodic functions the results (1.4), (1.5) and others differ from the corresponding results (1.1)–(1.3) for the case of periodic functions in that in the latter the quantity $1/n$ is replaced by $\sqrt{1-x^2}/n + 1/n^2$, which decreases to $1/n^2$ as $x \to \pm 1$.

In 1956, at the Third All-Union Mathematical Congress, S.M. Nikol'skiĭ conjectured that this special role of the points -1 and $+1$ should be explained by the fact that in the complex plane these points are angular points for the interval $[-1, 1]$, and he posed the problem of explaining this phenomenon [10]. Essentially the entire theory discussed below was obtained in the process of investigating this problem and the many new problems arising out of it.

Let us denote by $\Gamma_{1+1/n}$ the ellipse with foci at the points -1 and $+1$ and sum of semiaxes equal to $1 + 1/n$. As is well known, such an ellipse is the image of the circle $|w| = 1 + 1/n$ under conformal mapping by the Joukowski function $z = \frac{1}{2}(w + 1/w)$ of the exterior of the unit disc $|z| < 1$ onto the exterior of the interval $[-1, 1]$; that is, it is the level curve $\{z\colon |\Phi(z)| = 1 + 1/n\}$ for the interval $[-1, 1]$, where $\Phi(z)$ denotes the inverse of the Joukowski function:

$$\Phi(z) = z + \sqrt{z^2 - 1}.$$

In 1957 the author discovered that

$$\frac{\sqrt{1-x^2}}{n}+\frac{1}{n^2}\asymp\rho_{1+\frac{1}{n}}(x),\qquad x\in[-1,\ 1],\qquad n=1,\ 2,\ \ldots, \tag{1.6}$$

where $\rho_{1+1/n}(x)$ denotes the distance of a point $x\in[-1,1]$ from the level curve $\Gamma_{1+1/n}$ and the symbol $\asymp$ means that

$$A\rho_{1+\frac{1}{n}}(x)\leqslant\frac{\sqrt{1-x^2}}{n}+\frac{1}{n^2}\leqslant A'\rho_{1+\frac{1}{n}}(x),$$

where A and A' are positive constants.

Since the concept of level curves, and hence also the quantity $\rho_{1+1/n}(x)$, is also meaningful for sets of the complex plane, the basic outline of constructive function theory stated below on closed sets of the complex plane was already apparent in 1957. The foundation of this theory was laid in the works of the author [11]–[18] during 1959–1967. In the development of this theory after 1966 a significant contribution was made, aside from the author, by G. A. Alibekov, V. I. Belyĭ and V. M. Mikljukov, N. N. Vorob'ev, N. A. Lebedev, P. M. Tamrazov, I. A. Ševčuk and others.

2°. We shall introduce the required definitions and briefly state in chronological order the main results in the complex domain that have been obtained since 1959.

For a bounded continuum $\mathfrak{M}$ in the z-plane which contains more than one point and whose complement is a domain $G_\infty = C\mathfrak{M}$ containing the point at infinity, let $w=\Phi(z)$ denote a function which maps G_∞ conformally and univalently onto the domain $|w|>1$ and which is normalized by the condition

$$0<\lim_{z\to\infty}\frac{\Phi(z)}{z}<\infty. \tag{1.7}$$

Let $z=\Psi(w)$ denote the inverse of $\Phi(z)$. In the present paper we shall obtain converse theorems for arbitrary bounded continua, and direct theorems only for certain classes of such continua for which $\Psi(w)$ can be continuously extended to the unit circle $K=\{w\colon |w|=1\}$ and is absolutely continuous on K. Such continua we shall henceforth call *admissible.*

Definition 1.1. The image $\Psi(K)$ of the circle K under the mapping $z=\Psi(w)$ will be called the *boundary* $\Gamma=\partial\mathfrak{M}$ of the admissible continuum $\mathfrak{M}$.

It is known that, in order that the boundary Γ be rectifiable, it is necessary and sufficient that the function $\Psi(w)$ be absolutely continuous on K.

By A^r (r a nonnegative integer) we shall denote the class of functions $f(z)$, each of which is analytic in Int $\mathfrak{M}$ and has continuous derivatives up to order r inclusive on $\mathfrak{M}$, while $A^rH^{\omega(t)}$ will denote the subclass of those functions in A^r whose rth derivatives have modulus of continuity $\omega(f^{(r)};\ t)\leq \text{const}\ \omega(t)$.

A. We shall dwell on the most important results obtained during the period 1959 through 1965.

a) For every $R\geq 1$ we shall denote by Γ_R the level curve $\{z\colon |\Phi(z)|=R\}$, and for all $z\in\Gamma=\partial\mathfrak{M}$ we shall denote by $\rho_R(z)$ the distance from z to Γ_R.

In 1959 the author [11] carried out an investigation of level curves of the sets $\mathfrak{M}^V$ with piecewise smooth boundary for the case when the individual arcs of the boundary have continuous curvature and at the junction points z_j satisfy some additional conditions. The following theorem was established:

Theorem 1.1. *The distance* $\rho_R(z)$ *from a point* $z \in \partial\mathfrak{M}^V$ *to the level curve* Γ_R $(R > 1)$ *of this set can be expressed by the relation*

$$\rho_R(z) \asymp (R-1)[|z-z_j| + (R-1)^{\alpha_j}]^{1-\frac{1}{\alpha_j}} \tag{1.8}$$

or

$$\rho_R[\Psi(w)] \asymp (R-1)[|w-w_j| + R-1]^{\alpha_j-1}, \tag{1.8'}$$

where $w = \Phi(z)$, z_j *is the junction point nearest to* z, α_j *denotes the exterior angle with respect to* $\mathfrak{M}^V$ *at the point* z_j *and* $w_j = \Phi(z_j)$.

In addition, it was proved that for every $z \in \partial\mathfrak{M}^V$

$$\rho_{1+\frac{1}{n}}(z) \asymp \tilde{\rho}_{1+\frac{1}{n}}(\tilde{z}) \asymp |z - \tilde{z}|, \tag{1.9}$$

where

$$\tilde{z} \overset{\text{def}}{=} \Psi\left[\left(1+\frac{1}{n}\right)\Phi(z)\right]; \quad \tilde{\rho}_{1+\frac{1}{n}}(\tilde{z}) \overset{\text{def}}{=} \inf_{z \in \partial\mathfrak{M}^V} |\tilde{z} - z|. \tag{1.10}$$

An analogue of the inequalities (1.5) was subsequently obtained for the sets $\mathfrak{M}^V$: If

$$\forall \zeta \in \partial\mathfrak{M}^V : |P_n(\zeta)| \leqslant \rho^r_{1+\frac{1}{n}}(\zeta),$$

then

$$\forall \zeta \in \partial\mathfrak{M}^V : |P'_n(\zeta)| \leqslant A\rho^{r-1}_{1+\frac{1}{n}}(\zeta), \tag{1.5^V}$$

and by means of these inequalities converse theorems for functions of the classes $A^{(r)}H^\alpha$ were established, where r is a nonnegative integer and $0 < \alpha < 1$.

We note that in the case when $\mathfrak{M} = [-1, 1]$ the relation (1.8) yields (1.6), while in the case when $\mathfrak{M}$ is the disc $|z| \leq 1$ it yields the equality $\rho_{1+1/n}(z) = 1/n$, $|z| = 1$.

Owing to this fact, for the first time the connection between the constructive characterization of periodic functions and that of functions defined on the interval $[-1, 1]$ (or $[a, b]$) was cleared up.

b) In order to obtain direct theorems the author constructed polynomials of the form [see [12], [18]]:

$$P_n(z) = \frac{1}{4\pi^2 i}\int_{-\pi}^{\pi} J_{kn}(t)\,dt \int_{\partial\mathfrak{M}} \frac{f(\zeta_t)}{\zeta - z}\,d\zeta = \frac{1}{2\pi i}\int_{\partial\mathfrak{M}} f(\zeta)\,\pi_n(\zeta, z)\,d\zeta, \tag{1.11}$$

where

$$\pi_n(\zeta, z) \overset{\text{def}}{=} \frac{1}{2\pi}\int_{-\pi}^{\pi} \frac{J_{kn}(t)}{\Psi[\Phi(\zeta)e^{-it}] - z}\,\frac{\Psi'[\Phi(\zeta)e^{it}]}{\Psi'[\Phi(\zeta)]}\,e^{it}\,dt = \sum_{\nu=0}^{k(n-1)} \frac{j_\nu F_\nu(z)}{[\Phi(\zeta)]^\nu\,\Psi'[\Phi(\zeta)]}; \tag{1.12}$$

$$J_{kn}(t)=\gamma_{kn}\left(\frac{\sin(nt/2)}{\sin(t/2)}\right)^{2k}=\sum_{-k(n-1)}^{k(n-1)} j_\nu e^{i\nu t};\ \gamma_{kn}\overset{\text{def}}{=}\left(\frac{1}{2\pi}\int_0^{2\pi} J_{kn}(t)\,dt\right)^{-1} \tag{1.13}$$

are generalized Jackson polynomials; $F_\nu(z)$ are Faber polynomials (see, for example, [19], 1st ed., Chapter V, §4.4, 2nd ed., Chapter 5, §4.5) and

$$\zeta_t\overset{\text{def}}{=}\Psi[\Phi(\zeta)e^{-it}],\qquad t\in[-\pi,\pi], \tag{1.14}$$

is a generalized rotation. There was also obtained a more general result, according to which the following analogue of the well-known theorem on convolution of periodic summable functions holds.

Theorem 1.2 (on convolution; see [16], [18]). *Let $\mathfrak{M}$ be a bounded continuum with rectifiable boundary $\Gamma=\partial\mathfrak{M}$, and suppose two functions $f(\zeta)$ and $K(\zeta)$ on Γ are given such that the functions generated by them on the unit circle*

$$\hat f(t)\overset{\text{def}}{=}f[\Psi(e^{it})],\qquad \hat K(t)\overset{\text{def}}{=}K[\Psi(e^{it})]$$

are summable on $[0,2\pi]$. Moreover, suppose the following conditions hold:

1)
$$\hat f(t)\sim\sum_{-\infty}^{\infty} c_k e^{ikt},\quad \hat K(t)\sim\sum_{-\infty}^{\infty}\lambda_k e^{ikt} \tag{1.15}$$

(and hence $f(z)\sim\sum_0^\infty c_kF_k(z)$ and $K(z)\sim\sum_0^\infty\lambda_kF_k(z)$).

2) *At least one of the functions $\hat f(t)$, $\hat K(t)$ or $\Psi'(e^{it})$ has bounded variation on $[0,2\pi]$.*

Then for every $z\in\operatorname{Int}\mathfrak{M}$

$$(K*f)(z)\overset{\text{def}}{=}\frac{1}{4\pi^2 i}\int_{-\pi}^{\pi}\hat K(t)\,dt\int_\Gamma\frac{f\{\Psi[\Phi(\zeta)e^{-it}]\}}{\zeta-z}\,d\zeta=\sum_0^\infty\lambda_kc_kF_k(z). \tag{1.16}$$

The principal new points in these results are the following:

1) Introduction of the concept of generalized rotation by formula (1.14).

2) Establishment of a method which permits us to apply many important results from the theory of Fourier series to the summation of series in Faber polynomials and, in particular, to proceed from Fourier-Faber series to the construction of complete analogues of the Dirichlet, Fejér, Jackson and de la Vallée-Poussin polynomials, and the Poisson integral, etc., and their integral representations.

c) By means of the polynomials (1.11) a direct theorem [12], [14] was established in 1962–1963 for domains $\mathfrak{M}^V$ with exterior angles greater than π. This theorem states that if a function $f(z)$ belongs to A^rH^α, $0<\alpha\le 1$, on the domain $\mathfrak{M}^V$ with exterior angles greater than π, then for every $n=1,2,\dots$ there exists a polynomial $P_n(z)$ such that

$$|f(z)-P_n(z)|\leqslant A\rho^{r+\alpha}_{1+\frac{1}{n}}(z),\qquad \forall z\in\partial\mathfrak{M}^V, \tag{1.17}$$

and if the exterior angles are greater than π in $\mathfrak{M}^V$, then for $f(z)\in A$

$$|f(z)-P_n(z)|\leqslant A\omega\left(f;\ \rho_{1+\frac{1}{n}}(z)\right),\quad \forall z\in\partial\mathfrak{M}^V. \tag{1.17'}$$

Owing to this result and the above-mentioned converse theorem for functions of the Hölder classes on closed domains with corners, a constructive characterization was obtained for the first time.

B. After the results mentioned in A above were established, the principal refinements were directed toward their generalization and extension, and also toward the simplification of their proofs. In this direction, besides the author, an active participation was assumed by his students R. N. Koval'čuk, V. I. Belyĭ, N. N.Vorob'ev, R.V. Poljakov, I. A. Ševčuk and others, and also by N. A. Lebedev, P. M. Tamrazov, V. M. Mikljukov, N. A. Širokov and others. Especially here, in our opinion, the following four results must be noted.

1. For the purpose of generalizing (1.5') to sets $\mathfrak{M}^V$ with corners by solving the exterior Dirichlet problem, in 1963 the author constructed a function $G(z)$ harmonic in $C\mathfrak{M}^V$ according to the condition

$$G(z)|_{\partial\mathfrak{M}^V}=\ln\left\{\rho^s_{1+\frac{1}{n}}(z)\,\omega_k\left[\rho_{1+\frac{1}{n}}(z)\right]\right\}, \tag{1.18}$$

where $\omega_k(t)$ is a function of the type of kth modulus of continuity (i.e. $\omega_k(0)=0$, $\omega_k(t)\in C$, $\omega_k(t)\uparrow$ and $\omega_k(\lambda t)\leq(\lambda+1)^k\omega_k(t)$, $\lambda>0$), and then by investigating the Poisson integral from the image of $G(z)$ onto the unit disc he studied the problem of a good majorant for $G(z)$ in the neighborhood of the set $\mathfrak{M}$ bounded by the level curve $\Gamma_{1+1/n}$. In this manner he proved that $G(\tilde{z})|_{\Gamma_{1+1/n}}\leq AG(z)|_{\partial\mathfrak{M}^V}$, and using this inequality he established on $\partial\mathfrak{M}$ that if

$$|P_n(z)|\leqslant\rho^s_{1+\frac{1}{n}}(z)\,\omega_k\left[\rho_{1+\frac{1}{n}}(z)\right],$$

then

$$|P_n'(z)|\leqslant A\rho^{s-1}_{1+\frac{1}{n}}(z)\,\omega_k\left[\rho_{1+\frac{1}{n}}(z)\right],\quad s\geqslant 0. \tag{1.$\tilde{5}$'}$$

In 1968 N. A. Lebedev and P. M. Tamrazov, by considering the Green's function for a suitably selected domain, instead of the Poisson integral, estimated the function $G(z)$ in the neighborhood $U(z,\ \rho_{1+1/n}(z))$ of every point $z\in\partial\mathfrak{M}$ for the case of very general sets $\mathfrak{M}$ and thereby established (see [20] and [21]) the validity of (1.$\tilde{5}$') for arbitrary bounded continua $\mathfrak{M}$ (more precisely, on compacta which are regular in the Dirichlet problem). Using the validity of (1.$\tilde{5}$') for such sets, they generalized the above-mentioned converse theorem of the author to such sets.

2. In 1967 the author, for the purpose of obtaining good direct theorems, proposed polynomial kernels of the form [see [22], [23]]

$$K_{m,n}(\zeta,\ z)=\frac{1-[1-(\zeta-z)\,\pi_n(\zeta,\ z)]^m}{\zeta-z}, \tag{1.19}$$

in which $\pi_n(\zeta,\ z)$ denotes a function which for each fixed $\zeta\in\partial\mathfrak{M}$ is a polynomial in z of degree n approximating the Cauchy kernel $1/(\tilde{\zeta} - z)$, $\tilde{\zeta} = \Psi[(1 + 1/n)\Phi(\zeta)]$, on $\partial\mathfrak{M}$, and established the estimate

$$\left|\frac{1}{\zeta - z} - K_{m,n}(\zeta, z)\right| \preccurlyeq \frac{1}{|\zeta - z|}\left[\frac{\rho_{1+\frac{1}{n}}(z)}{|\tilde{\zeta} - z|}\right]^m \tag{1.20}$$

(the sign $\preceq$ means that $L \leq AM$, $A = \text{const} > 0$).

N. N. Vorob'ev noted (see [24]) that the proof of (1.20) given by the author can be greatly simplified if the polynomials $\pi_n(\zeta, z)$ used in (1.19) are replaced by polynomials of the form

$$\pi_n(\zeta, z) = \frac{1}{2\pi}\int_{-\pi}^{\pi}\frac{J_{kn}(t)\,dt}{\Psi[\Phi(\tilde{\zeta})e^{-it}] - z} = \sum_{0}^{k(n-1)}\frac{j_\nu}{[\Phi(\tilde{\zeta})]^\nu}\Pi_\nu(z), \tag{1.12$'$}$$

where the $\Pi_\nu(z)$ are generalized Faber polynomials (see, for example, [25]) which are defined by the expansion

$$\frac{1}{\zeta - z} = \sum_{1}^{\infty}\frac{\Pi_\nu(z)}{[\Phi(\zeta)]^\nu}, \qquad z \in \mathfrak{M}, \qquad \zeta \in C\mathfrak{M}. \tag{1.21}$$

We note that the polynomials (1.21) were obtained (for an even more general case) by A. V. Stovbun [26] by passing from the author's polynomials to generalized Faber polynomials.

The significance of the polynomial kernels (1.19)–(1.12$'$) in approximation theory on various admissible sets and their properties will be considered in detail in §4.

3. In 1968, in a joint paper of the author and G. A. Alibekov [27], for all $s > 0$, with the exception of a discrete sequence, a complete characterization on sets $\mathfrak{M}^V$ with corners was obtained for the classes B_s of functions f for which we can construct a sequence of polynomials $P_n(z)$ such that the inequality

$$|f(z) - P_n(z)| \leqslant \frac{1}{n^s}$$

is satisfied for each $z \in \mathfrak{M}^V$.

4. The most general properties of admissible closed domains $\mathfrak{M}$ for which the theory we have developed permits us to obtain estimates, definitive with respect to order, in the direct theorems, were studied by the author in [28]. The sets which possess the indicated properties will be called sets of type B. We note that in [28] all properties were formulated in terms of properties of the functions $\Phi(\zeta)$, $\Psi(\omega)$ and $\rho_{1+1/n}(z)$. Recently V. I. Belyĭ and V. M. Mikljukov, by using the refined apparatus of the theory of quasiconformal mapping, have succeeded in expressing the basic properties of sets of type B in rather simple geometrical terms [29].

C. Since so far the most definitive results have been obtained in terms of the first modulus of continuity, the present paper is basically devoted to the extension of all results obtained so far to the case of functions whose properties can be characterized by the second modulus of continuity. For this purpose a definition is introduced in §2 for functions defined on arbitrary sets of the complex plane, and the properties of the second modulus of continuity are investigated.

§3 is devoted to a generalization of the class of sets of type B on closed sets which are not closed domains, and to a brief description of the results of Belyĭ and Mikljukov.

A thorough investigation of properties of polynomial kernels is carried out in §4.

Direct theorems are established in terms of the second modulus of continuity in §5.

Converse theorems which include as corollaries on one hand, the results of the author established in [3] in terms of second modulus of continuity for the interval $[-1, 1]$, and on the other the results of Lebedev and Tamrazov [21], are established in §6.

§2. Moduli of continuity of second order for functions defined on arbitrary sets of the complex plane

Let a continuous function $f(z)$ be defined on some (for definiteness, bounded) set $\mathfrak{M}$ in the complex plane, and let a function $\omega(t)$ of the type of second modulus of continuity (i.e. $\omega(0) = 0$, $\omega(t) \in C$, $\omega(t)\uparrow$ and $\omega(\lambda t) \leq (\lambda + 1)^2\omega(t)$ $\forall \lambda > 0$) be given. The question arises as to how the class $H_2^{\omega(t)}$ can be defined on $\mathfrak{M}$.

It is clear that in the general case we cannot, generally speaking, construct second differences of the form $\Delta_h^2 f(z) = f(z) - 2f((z + z_1)/2) + f(z_1)$, where $h \overset{\text{def}}{=} (z_1 - z)/2$, for the condition $z, z_1 \in \mathfrak{M}$ need not imply that $(z + z_1)/2 \in \mathfrak{M}$.

In view of a well-known theorem of Whitney [30], which for the case $k = 2$ states that $\omega_2(f; (b - a)/2) \leq 4E_1(f)_{C[a,b]}$ on any interval $[a, b]$ for every function $f(x) \in C_{[a,b]}$, it can be shown that, in order that the function $f(z)$ belong to the class $H_2^{\omega(t)}$ on $\mathfrak{M}$, it is sufficient to demand that for an arbitrary $t > 0$ the function f can be approximated by a polynomial $P(z) = az + b$ with accuracy up to the order $\omega_2(t)$ on every set $\mathfrak{M} \cap U(\xi, t)$, where ξ is any point of $\mathbf{C}$ and

$$U(\xi, t) \overset{\text{def}}{=} \{z : |z - \xi| \leqslant t\}. \tag{2.1}$$

We shall show that we cannot get anything better by this method. In fact, let

$$\mathfrak{M} = \{0\} \cup \{x_n\}_{n=1}^{\infty} \cup \{\check{x}_n\}_{n=1}^{\infty},$$

where $x_n \overset{\text{def}}{=} 1/n$, $\check{x}_n \overset{\text{def}}{=} 1/n + 1/2^n$ and $f(z) = 0$ for $z = 0$ or $z = x_n$ and $f(z) = 2^{-n/2}$ for $z = \check{x}_n$, $n = 1, 2, \ldots$.

It is easy to see that $f(z) \in H^{1/2}$ and at the same time the function $f(z)$ can be approximated on any set $\mathfrak{M} \cap U(\xi; t)$ by a polynomial of first degree with an accuracy of $2^{-1/2\sqrt{t}} = 0(t^2)$ for all $t > 0$. In this connection we give a definition of the class $H_2^{\omega(t)}$ in which, along with the conditions of Whitney's theorem, it is required that Marchaud's inequality for the first modulus of continuity in terms of the second modulus be satisfied, i.e.

$$\omega(f;\ t) \leqslant At \int_t^1 \frac{\omega_2(f;\ u)}{u^2}\, du.$$

We shall denote such a class by $\breve{H}_2^{\omega(t)}$. More precisely, we have the following

Definition 2.1. Let $\mathfrak{M}$ be an arbitrary compact set in the complex plane $\mathbf{C}$, $f\colon \mathfrak{M} \to \mathbf{C}$ and $\omega(t)$ an arbitrary function of the type of second modulus of continuity, which we shall regard as defined anywhere it is needed. We say that $f(z) \in \breve{H}_2^{\omega(t)}$ if there exists a positive constant A such that for arbitrary $z_1, z_2 \in \mathfrak{M}$ the inequality

$$|f(z) - P_1(z)| \leqslant At \int_t^{2h} \frac{\omega(u)\, du}{u^2}, \tag{2.2}$$

holds at all points $z \in \mathfrak{M} \cap [U(z_1, h) \cup U(z_2, h)]$, where

$$h \overset{\text{def}}{=} |z_2 - z_1|, \quad U(z_i, h) = \{z : |z - z_i| \leqslant h\}, \quad i = 1,\ 2;$$

$$t = \min\{|z - z_1|;\ |z - z_2|\}$$

and

$$P_1(z) = P_1(f;\ z_1,\ z_2;\ z) \overset{\text{def}}{=} \frac{z_2 - z}{z_2 - z_1} f(z_1) + \frac{z - z_1}{z_2 - z_1} f(z_2). \tag{2.3}$$

Example. The function $\overset{*}{f}(z) = z^2$ belongs to the class $\breve{H}_2^{t^2}$ on an arbitrary compact set $\mathfrak{M} \subset \mathbf{C}$.

In fact, under the conditions and notation of Definition 2.1 we get

$$|\overset{*}{f}(z) - P_1(\overset{*}{f};\ z_1,\ z_2;\ z)| = |z^2 - P_1(z)|$$

$$= |(z - z_1)(z - z_2)| \leqslant 2th \leqslant 2t \int_t^{2h} \frac{u^2}{u^2}\, du.$$

Remark. For points $z \in \mathfrak{M}$ which do not satisfy the conditions of Definition 2.1, i.e. for $z \subset \mathfrak{M} \setminus [U(z_1, h) \cup U(z_2, h)]$, we have

$$|f(z) - P_1(f;\ z_1,\ z_2;\ z)| \leqslant At \int_h^{2t} \frac{\omega(u)}{u^2}\, du \leqslant 8A \frac{t^2}{h^2}\, \omega(h),$$

$$t = \min_{i=1,\,2} \{|z - z_i|\}. \tag{2.2'}$$

In fact, putting $f(z) - P_1(f;\ z_1,\ z_2;\ z) \overset{\text{def}}{=} \overset{\circ}{f}(z)$ and $t = |z - z_1|$, and noting that $\overset{\circ}{f}(z_1) = \overset{\circ}{f}(z_2) = 0$ according to (2.2), we get (on replacing therein z_2)

$$Ah\int_h^{2t}\frac{\omega(u)}{u^2}\,du \geqslant |\overset{0}{f}(z_2)-P_1(\overset{0}{f};\ z_1,\ z;\ z_2)|$$

$$=\left|\frac{z-z_2}{z-z_1}\overset{0}{f}(z_1)+\frac{z_2-z_1}{z-z_1}\overset{0}{f}(z)\right|=\frac{h}{|z-z_1|}|f(z)-P_1(f;\ z_1,\ z_2;\ z)|.$$

Here, since $|z-z_1|\asymp|z-z_2|\asymp t$, we obtain the first of the inequalities (2.2′). The second inequality follows from the first, since for $u>h$

$$\omega(u)\leqslant 4\frac{u^2}{h^2}\omega(h).$$

The major considerations in favor of Definition 2.1 consist in the validity of the following statements:

1) For any first modulus of continuity $\omega(t)$

$$H_1^{\omega(t)}\subset \breve{H}_2^{\omega(t)} \tag{2.4}$$

and simultaneously there exist moduli of continuity $\omega^*(t)$ of first order such that

$$\breve{H}_2^{\omega^*(t)}\not\subset H_1^{\omega^*(t)}. \tag{2.5}$$

2) There exists a method which permits us to start from a function $f(z)$ and find with accuracy up to a constant the minimal class $\breve{H}_2^{\omega(t)}$ to which this function belongs.

3) In the case when the set $\mathfrak{M}$ is an interval $[a,\ b]$, the classes $\breve{H}_2^{\omega(t)}$ coincide with the usual classes $H_2^{\omega(t)}$, i.e. the classes of those functions $f(z)$ for which the inequality

$$\left|f(z')-2f\left(\frac{z'+z''}{2}\right)+f(z'')\right|\leqslant A\omega(t)$$

is satisfied at any two points $z',\ z''\in[a,\ b]$ such that $|z'-z''|\leq t$.

4) In the case of sufficiently "good" sets with piecewise smooth boundary the classes $\breve{H}_2^{\omega(t)}$ coincide with the classes $\breve{H}_2^{\omega(t)}$ which were introduced in [31] and from which a significant strengthening of the Plemelj-Privalov theorem on limit values of an integral of Cauchy type was obtained in that paper, while a rather important addition to the author's direct theorems on approximation of analytic functions in regions with piecewise smooth boundary was obtained in [32].

We shall prove these statements.

1) a) First we prove (2.4). In fact, if $f(z)\in H_1^{\omega(t)}$ on some compact $\mathfrak{M}$, then for any three points $z_1,\ z_2,\ z\in\mathfrak{M}$, assuming the notation of Definition 2.1 and (for the sake of definiteness) that $|z-z_1|\leq|z-z_2|$, we get

$$|f(z)-P_1(z)|=\left|\frac{z_2-z}{z_2-z_1}[f(z)-f(z_1)]+\frac{z-z_1}{z_2-z_1}[f(z)-f(z_2)]\right|$$

$$\leqslant\left|\frac{z_2-z}{z_2-z_1}\right|\omega(t)+\frac{t}{h}\,\omega(2h)\leqslant 2\omega(t)+\frac{t}{h}\left(\frac{2h}{t}+1\right)\omega(t) \tag{2.6}$$

$$\leqslant 5\omega(t)\leqslant 10t\int_t^{2t}\frac{\omega(u)}{u^2}\,du\leqslant 10t\int_t^{2h}\frac{\omega(u)}{u^2}\,du,$$

which means that $f(z) \in \breve{H}_2^{\omega(t)}$, and hence, in view of the arbitrariness of the function $f(z) \in H_1^{\omega(t)}$, we have $H_1^{\omega(t)} \subset \breve{H}_2^{\omega(t)}$.

b) Now we give a nontrivial example showing that there exist first moduli of continuity $\omega(t)$ such that $\breve{H}_2^{\omega(t)} \not\subset H_1^{\omega(t)}$. For this purpose, on an arbitrary compact $\mathfrak{M}$ which does not contain points of any positive angle Ω included between the ray $\rho e^{-i\alpha}$, $0 < \rho < \infty$, $0 < \alpha < \pi$, $\alpha = \text{const}$, and the axis OX, and for which the point $z = 0$ is a limit point, we consider for definiteness that branch of the function

$$\overset{*}{f}(z) \overset{\text{def}}{=} z \ln z, \tag{2.7}$$

for which $\overset{*}{f}(-1) = -i\pi$.

In view of the condition imposed on the set $\mathfrak{M}$, it is quite clear that

$$\overline{\lim_{t\to 0}} \frac{\omega_1(f;\ t)}{t \ln \frac{1}{t}} = 1.$$

We shall show that $\overset{*}{f}(z) \in H_2^t$, i.e.

$$|\overset{*}{f}(z) - P_1(z)| \leqslant At \int_t^{2h} \frac{u}{u^2}\, du = At \ln \frac{2h}{t}, \qquad A = \text{const}. \tag{2.7'}$$

This will imply, e.g., that $\breve{H}_2^t \not\subset H_1^t$.

In fact, according to (2.3), for any three points z_1, z_2, $z \in \mathfrak{M} \setminus \{0\}$, in the notation of Definition 2.1, and assuming for definiteness that $|z - z_1| \leq |z - z_2|$, we get

$$\begin{aligned} |\overset{*}{f}(z) - P_1(z)| &= \left| \int_{z_1}^{z} [\overset{*}{f}{}'(\zeta) - P_1'(\zeta)]\, d\zeta \right| = \left| \int_{z_1}^{z} \left[\ln \zeta + 1 + \frac{z_1}{z_2 - z_1} \ln z_1 \right.\right. \\ &\left.\left. - \frac{z_2}{z_2 - z_1} \ln z_2 \right] d\zeta \right| = \left| \int_{z_1}^{z} \left[\ln \frac{\zeta}{z_1} + 1 - \frac{z_2}{z_2 - z_1} \ln \frac{z_2}{z_1} \right] d\zeta \right| \\ &= \left| \int_{z_1}^{z} \left[\ln \frac{\zeta}{z_2} + 1 - \frac{z_1}{z_1 - z_2} \ln \frac{z_1}{z_2} \right] d\zeta \right|. \end{aligned} \tag{2.8}$$

We shall verify that one of the following relations holds.

$$\frac{z_2}{z_2 - z_1} \ln \frac{z_2}{z_1} = O(1) \qquad \text{or} \qquad \frac{z_1}{z_1 - z_2} \ln \frac{z_1}{z_2} = O(1). \tag{2.9}$$

Since both of these relations are symmetric with respect to z_1 and z_2, without loss of generality we assume that, for example,

$$|z_1| \leqslant |z_2| \tag{2.10}$$

and consider the following three possible cases:

a) Let

$$|z_2| \geqslant 4|z_1|. \tag{2.11}$$

In this case we at once obtain

$$\left|\frac{z_1}{z_1-z_2}\ln\frac{z_2}{z_1}\right|\leqslant 2\left|\frac{z_1}{z_2}\right|\left|\ln\frac{z_1}{z_2}\right|\leqslant 2\left|\frac{z_1}{z_2}\right|\left(\ln\left|\frac{z_2}{z_1}\right|+2\pi\right)=O(1). \quad (2.9')$$

b) Let

$$|z_2|\geqslant 4|z_2-z_1|=4h, \quad (2.11')$$

so that $|z_1|\geq 3|z_2|/4\geq 3h$. In this case

$$\left|\frac{z_2}{z_2-z_1}\ln\frac{z_2}{z_1}\right|=\left|\frac{z_2}{z_2-z_1}\right|\left|\ln\left(1-\frac{z_2-z_1}{z_2}\right)\right|=O(1). \quad (2.9'')$$

c) Let

$$|z_2|<4|z_1| \quad (2.11'')$$

and simultaneously $|z_2|<4\,|z_2-z_1|=4h$. In this case

$$\left|\frac{z_2}{z_2-z_1}\ln\frac{z_2}{z_1}\right|<4(\ln 4+2\pi)=O(1). \quad (2.12)$$

This proves (2.9). Moreover, we obtain, in view of (2.8), (2.11′)–(2.11″) and (2.9)–(2.9″),

$$|\overset{*}{f}(z)-P_1(z)|\leqslant\left|\int_{z_1}^{z}\ln\frac{\zeta}{z_1}\,d\zeta\right|+\left|\int_{z_2}^{z}\ln\frac{\zeta}{z_2}\,d\zeta\right|+O(t). \quad (2.13)$$

From (2.13) we deduce the following facts:

i) If $h=|z_2-z_1|<|z_2|/4$, so that $3|z_2|/4<|z_1|\leq|z_2|$, and if the interval $[z,z_1]\overset{\text{def}}{=\!=}\{\zeta\colon \zeta=\alpha z+(1-\alpha)z_1,\ 0\leq\alpha\leq 1\}\subset\mathbf{C}\backslash\Omega$, then $|\zeta-z_1|\leq t\leq h$ for any $\zeta\in[z,z_1]$, so that $\zeta=z_1(1+\beta)$, where $|\beta|\leq|h/z_1|\leq 4|h/z_2|/3<1/3$, and thus, in view of (2.13),

$$|\overset{*}{f}(z)-P_1(z)|\leqslant A|z-z_1|\left(\ln\left|\frac{z_1}{z_1(1-1/3)}\right|\right)+O(t)\leqslant A_1t\ln\frac{2h}{t}. \quad (2.7')$$

But if $h=|z_2-z_1|<|z_2|/4$ and $[z,z_1]\not\subset C\backslash\Omega$, then obviously $|z_1|\asymp|z-z_1|\asymp t\asymp h\asymp|z_2|\asymp|z|$, and hence, again,

$$\begin{aligned}|\overset{*}{f}(z)-P_1(z)|&\leqslant A\left(\left|\int_0^z\ln\frac{\zeta}{z_i}\,d\zeta\right|+\left|\int_0^{z_1}\ln\frac{\zeta}{z_i}\,d\zeta\right|\right)+O(t)\\&\leqslant A\left\{|z|\ln\frac{2h}{|z|}+|z_1|\ln\frac{2h}{|z_1|}\right\}+O(t)\leqslant A_1t\ln\frac{2h}{t}.\end{aligned} \quad (2.7'')$$

ii) If $t=|z-z_1|<|z|/4$, $|z_2|\leq 4h$ and $[z,z_1]\subset C\backslash\Omega$, then $|\zeta-z_1|\leq t\leq h$ for any $\zeta\in[z,z_1]$, so that $\zeta=z_1(1+\beta)$, where $|\beta|<1/3$, and thus, whether $|z_2|<4|z_1|$ or $|z_2|\geq 4|z_1|$, we easily obtain

$$|\overset{*}{f}(z)-P_1(z)|\leqslant A|z-z_1|\ln\frac{2h}{t}+O(t)\leqslant A_1t\ln\frac{2h}{t}, \quad (2.7')$$

but if $t=|z-z_1|<|z|/4$ and $[z,z_1]\not\subset C\backslash\Omega$, then again $|z|\asymp t\asymp|z_1|=O(h)$, and hence, as in the second part of case i),

$$|\overset{*}{f}(z)-P_1(z)|\leqslant At\ln\frac{2h}{t}. \quad (2.7'')$$

iii) Now let $h=|z_2-z_1|\geq|z_2|/4$ and $t=|z-z_1|\geq|z|/4$. In this case we set $\min\{|z|, |z_1|\} = r$; for the sake of definiteness we assume that $|z| = r$, and denote by $\tilde{z}$ a point such that $|\tilde{z}| = r$ and $\arg \tilde{z} = \arg z_1$, and by γ the arc of the circle $|z| = r$ which joins the points z and $\tilde{z}$ but does not contain points of the angle Ω.

Then for $i = 1$ and $i = 2$, noting that $t \ln 1/t \uparrow$ for small $t > 0$, we obtain

$$\left|\int_z^{z_1} \ln\frac{\zeta}{z_i}\,d\zeta\right| \leqslant \int_\gamma \left(\ln\frac{|z_i|}{|z|} + 2\pi\right) ds + \left|\int_{|\tilde{z}|}^{|z_1|} \ln\frac{s}{|z_i|}\,ds\right| \leqslant 2\pi|z|\left(\ln\frac{|z_i|}{|z|} + 2\pi\right)$$

$$+|z_i|\left|\int_{\left|\frac{z}{z_i}\right|}^{\left|\frac{z_1}{z_i}\right|} \ln u\,du\right| \leqslant At\ln\frac{2h}{t} + z_i\omega\left(u\ln u - u;\ \frac{|z_1|-|z|}{|z_i|}\right) \qquad (2.7''')$$

$$\leqslant At\ln\frac{2h}{t} + (|z_1|-|z|)\ln\frac{|z_2|}{|z_1|-|z|} \leqslant A_1 t\ln\frac{2h}{t}.$$

On the basis of (2.7′)–(2.7‴) the validity of (2.7), and along with it the statement 1), is verified.

2) In order to establish the statement 2) we shall show that the following theorem holds.

Theorem 2.1. *For each function $\omega_1(t)$: $[0, a] \to R$ which is a first modulus of continuity, one can construct a function $\omega_2(t) = \omega_2(a; t)$ of the type of a second modulus of continuity which satisfies the inequalities*

$$\max\left\{\frac{t}{10}\int_t^a \frac{\omega_2(u)}{u^2}\,du;\ \frac{1}{8}\,\omega_2(t)\right\} \leqslant \omega_1(t) \leqslant t\int_t^a \frac{\omega_2(u)}{u^2}\,du \qquad (2.14)$$

for all $t \in [0, a/2]$.

Proof. First we construct the function

$$\omega_1^*(t) \overset{\text{def}}{=} \sup_{0\leqslant x\leqslant t\leqslant y\leqslant a} \frac{(t-x)\,\omega_1(y) + (y-t)\,\omega_1(x)}{y-x}.$$

It can easily be verified that this function is a convex-up modulus of continuity and satisfies the relations

$$\omega_1(t) \leqslant \omega_1^*(t) \leqslant 2\omega_1(t) \quad \text{and} \quad \omega_1^*(a) = \omega_1(a). \qquad (2.15)$$

Being monotone increasing and convex up, $\omega_1(t)$ is absolutely continuous and has a nonnegative monotone decreasing derivative $\omega_1^{*\prime}(t)$ almost everywhere. Starting from the values of $\omega_1^{*\prime}(x)$ on the set $X \subset [0, a]$ on which it is defined, we construct a new function

$$\lambda(t) \overset{\text{def}}{=} \inf_{0\leqslant x\leqslant t\leqslant y\leqslant a} \frac{(t-x)\,\omega_1^{*\prime}(y) + (y-t)\,\omega_1^{*\prime}(x)}{y-x}, \qquad x,\ y \in X. \qquad (2.16)$$

Obviously this function is the maximum convex-down minorant of the function $\omega_1^{*\prime}(x)$. We need the following of its properties, each of which holds at least on the set of those points $t \in [0, a]$ where the derivatives $\omega_1^{*\prime}(t)$ and $\lambda'(t)$ exist:

$$\text{a) } \lambda(t)\downarrow; \qquad \text{c) } -\lambda'(t)=|\lambda'(t)|\downarrow;$$

$$\text{b) } \lambda(t)\leqslant\omega_1^{*\prime}(t); \qquad \text{d) } \omega_1^*(t)\geqslant\int_0^t\lambda(u)\,du\geqslant\frac{\omega_1^*(t)}{2}. \tag{2.17}$$

Moreover, the function $\lambda(t)$ is obviously nonnegative and continuous.

All of these properties except the second inequality in d) are obvious. The inequality $\int_0^t\lambda(u)du\geq\frac{1}{2}\int_0^t\omega_1^{*\prime}(u)du$ follows, on the basis of elementary geometric considerations, from the fact that $\omega_1^{*\prime}(u)\downarrow$ and $\lambda(u)$ is the maximum convex-down minorant of the function $\omega_1^{*\prime}(u)$.

Now we shall put $\omega_2(0)=0$, and for $t>0$

$$\omega_2(t)=2\int_0^t[\lambda(\xi)-\lambda(t)]\,d\xi+Ct^2=2\left[\int_0^t\lambda(\xi)\,d\xi-t\lambda(t)\right]+Ct^2,\quad C\overset{\text{def}}{=}4\frac{\omega_1(a)}{a^2}, \tag{2.18}$$

and verify that this function satisfies (2.14) and is a function of the type of a second modulus of continuity.

In fact, on one hand, as a result of (2.18), (2.17), c) and (2.15), for all $t\in[0,a/2]$ we find that

$$\omega_2(t)\leqslant2\int_0^t\lambda(u)\,du+Ct^2\leqslant2\omega_1^*(t)+C\frac{a}{2}t\leqslant4\omega_1(t)+4\frac{\omega_1(a)}{a^2}\,\frac{a}{2}\,\frac{2a\omega_1(t)}{\omega_1(a)}=8\omega_1(t),$$

while on the other hand, taking account of (2.18) and putting $\int_0^u\lambda(\xi)d\xi=v(u)$, we get

$$t\int_0^a\frac{\omega_2(u)}{u^2}\,du=t\left[2\int_t^a\frac{v(u)}{u^2}\,du-2\int_t^a\frac{dv(u)}{u}\,du+C\int_t^a du\right]=2t\left[\frac{v(t)}{t}-\frac{v(a)}{a}\right]+C(a-t)t,$$

whence, in view of c) and d) and well-known properties of the first modulus of continuity, it follows that

$$t\int_t^a\frac{\omega_2(u)}{u^2}\,du\leqslant2\int_0^t\lambda(\xi)\,d\xi-\frac{2t}{a}\int_0^a\lambda(\xi)\,d\xi+Cat$$

$$\leqslant2\omega_1^*(t)+\left[\frac{\omega_1^*(a)}{a}+\frac{4\omega_1(a)}{a^2}\,a\right]\frac{2a}{\omega_1(a)}\,\omega_1(t)=10\omega_1(t),$$

and

$$t\int_t^a\frac{\omega_2(u)}{u^2}\,du\geqslant2\int_0^t\lambda(\xi)\,d\xi-2\frac{t}{a}\int_0^a\lambda(\xi)\,d\xi+\frac{Ca}{2}t$$

$$\geqslant\omega_1^*(t)-\frac{2t}{a}\,\omega_1^*(a)+\frac{4\omega_1(a)}{a_2}\,\frac{a}{2}t\geqslant\omega_1(t).$$

Thus the inequalities (2.14) are proved.

We shall establish that $\omega_2(t)$ is a function of the type of a second modulus of continuity:

1) $\omega_2(0)=0$ by definition;

2) $\omega_2(t)\uparrow$, since the function $\lambda(t)$ is nonnegative and monotone decreasing;

3) continuity of $\omega_2(t)$ for $t > 0$ is obvious, and for $t = 0$ it follows from the fact that $\lambda(t) \geq 0$, $\lambda(t)$ is integrable and $\lambda(t)\downarrow$, so that $0 \leq \omega_2(t) \leq 2\int_0^t \lambda(\xi)d\xi + Ct^2$;

4) we shall prove that

$$\omega_2(kt) = 2\int_0^{kt} [\lambda(\xi) - \lambda(kt)]\, d\xi + C(kt)^2 \leqslant k^2\omega_2(t) = k^2 2\int_0^t [\lambda(\xi) - \lambda(t)]\, d\xi + k^2ct^2$$

or equivalently

$$\int_0^{kt} [\lambda(\xi) - \lambda(kt)]\, d\xi \leqslant k^2 \int_0^t [\lambda(\xi) - \lambda(t)]\, d\xi.$$

In order to prove this inequality we must consider the following relations, which in view of the downward convexity of the curve $y = \lambda(t)$ are easily seen from the figure:

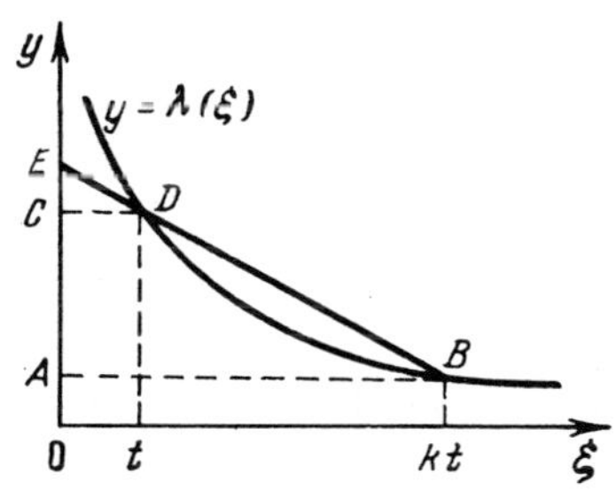

$$\int_0^{kt} [\lambda(\xi) - \lambda(kt)]\, d\xi \leqslant S_{ABE} + S = k^2 S_{CDE} + S \leqslant k^2(S_{CDE} + S)$$

and

$$\int_0^t [\lambda(\xi) - \lambda(t)]\, d\xi = S_{CDE} + S,$$

where S_{ABE} and S_{CDE} denote the areas of the triangles ABE and CDE, and S denotes the area bounded by the segment ED, the OY axis and the curve $y = \lambda(\xi)$.

Thus Theorem 2.1 is completely proved.

Now let $f(z)$ be a continuous function defined on an arbitrary compact set $\mathfrak{M} \subset C$, d the diameter of $\mathfrak{M}$ and $\omega(t)$ any first modulus of continuity for which $f(z) \in H_1^{\omega(t)}$. We denote by H the set of all $h \in (0, d]$, for each of which there exist at least two points $z_1, z_2 \in \mathfrak{M}$ such that $|z_2 - z_1| = h$, and by A_h the set of all possible pairs of points $z_1, z_2 \in \mathfrak{M}$ such that $|z_1 - z_2| = h$; then for each $t \in [0, h]$, $h \in H$, we put

$$\Delta(h;\, t) = \sup_{\substack{(z_1, z_2)\in A_h,\ h\in H \\ \min\limits_{i=1,2} \{|z - z_i|\} \leqslant t}} |f(z) - P_1(z)|, \tag{2.19}$$

where $P_1(z)$ is defined by (2.3). The function $\Delta(t) = \Delta(h;\ t)$, being monotone in t, is, generally speaking, neither continuous nor semiadditive. However, the function

$$\omega(h;\ t) \overset{\text{def}}{=} \sup_{0 \leqslant x \leqslant t \leqslant y \leqslant h} \frac{(t-x)\,\Delta(h;\ y) + (y-t)\,\Delta(h;\ x)}{y-x} \tag{2.20}$$

will obviously be the minimal convex-up function possessing all the properties of a first modulus of continuity and also satisfying the inequality

$$\omega(h;\ t) \leqslant A\omega(t), \tag{2.21}$$

for any $h \geq t$, $h \in H$, where the constant A depends neither on $h \in H$ nor on $t \leq h$.

Starting from $\omega(h;\ t)$, we construct, using Theorem 2.1, a function $\omega_2(h;\ t)$ of the type of a second modulus of continuity, such that the inequalities

$$\max\left\{\frac{t}{10}\int_t^{2h} \frac{\omega_2(h;\ u)}{u^2}\,du;\ \frac{1}{8}\,\omega_2(h;\ t)\right\} \leqslant \omega(h;\ t) \leqslant t\int_t^{2h} \frac{\omega_2(h;\ u)}{u^2}\,du \tag{2.14$'$}$$

are satisfied, and finally we put

$$\omega_2(t) = \sup_{h \geqslant t,\ h \in H} \omega_2(h;\ t). \tag{2.22}$$

As can easily be verified, this function $\omega_2(t)$ will be the required one; that is, up to a constant it will be the minimal function of the type of a second modulus of continuity determining the class $H_2^{\omega_2(t)}$ to which the function $f(z)$ belongs.

Definition. The function $\omega_2(t) = \omega_2(f;\ \mathfrak{M};\ t)$ constructed by (2.22) according to the above process will be called the *second modulus of continuity of the function $f(z)$ on the set $\mathfrak{M}$.*

Example. We take the function $\overset{*}{f}(z) = z \ln z$, defined by (2.7). Then under the same conditions on the set $\mathfrak{M}$ as above we get for any $h \in [0,\ d] \cap H$ and $t \leq h$, in view of (2.19) and (2.7),

$$\Delta(h;\ t) = \sup_{\substack{(z_1,\ z_2) \in A_h,\ h \in H \\ \min\limits_{i=1,\,2} \{|z - z_i|\} \leqslant t}} |\overset{*}{f}(z) - P_1(z)| \leqslant At \ln \frac{2h}{t} \overset{\text{def}}{=} \omega(h;\ t);\quad A = \text{const.}$$

Since in this case in view of (2.2$'$) and (2.16) we have

$$\omega_1^*(h;\ t) = \omega_1(h;\ t) = \omega(h;\ t) = At \ln \frac{2h}{t}$$

and

$$\frac{\partial}{\partial t}\,\omega_1^*(h;\ t) = A\left[\ln \frac{2h}{t} - 1\right] = \lambda(h;\ t),$$

we obtain, because of (2.18),

$$\begin{aligned}\omega_2(h;\ t) = 2\int_0^t [\lambda(h;\ \xi) - \lambda(h;\ t)]\,d\xi &= \omega^*(h;\ t) - t\lambda(h;\ t) \\ &= A\left[t \ln \frac{2h}{t} - t\left(\ln \frac{2h}{t} - 1\right)\right] = At,\end{aligned}$$

and hence, in view of (2.22),

$$\omega_2(t)=\sup_{h\geq t,\ h\in H}\omega_2(h;\ t)=At,$$

which means that up to a constant the class $H_2^{At}=H_2^t$ is the minimal class which contains the function $\overset{*}{f}(z)=z\ln z$. Thus the statement 2) is established.

3) Obviously it is sufficient to restrict attention to the case when $f(x)$ takes real values on $[a,\ b]$. It is required to prove that for any function $\omega(t)$ of the type of a second modulus of continuity there holds

$$H_2^{\omega(t)}=\breve{H}_2^{\omega(t)}. \tag{2.23}$$

i) First we establish that $\breve{H}_2^{\omega(t)}\subset H_2^{\omega(t)}$. In fact, if $f(x)\in\breve{H}_2^{\omega(t)}$, then for arbitrary fixed $x_1,\ x_2\in[a,\ b]$, $x_1<x_2$, and any $x\in[x_1,\ x_2]$ we have in view of (2.2)

$$|f(x)-P_1(x)|\leqslant t\int_t^{2h}\frac{\omega(u)}{u^2}\,du\leqslant\omega(2h)\leqslant 4\omega(h), \tag{2.24}$$

$$t=\min\{x-x_1,\ x_2-x\},\ h=x_2-x_1.$$

Hence, taking Whitney's theorem into account, we get

$$\Delta_h^2(f;\ x_1)\leqslant 4\omega_2(f;\ (x_2-x_1)/2)\leqslant 16E_1(f)_{C[x_1,\ x_2]}$$

$$\leqslant 16\|f(x)-P_1(x)\|_{C[x_1,\ x_2]}\leqslant 64\omega(h),$$

which means that $f(x)\in H_2^{64\omega(h)}=H_2^{\omega(h)}$.

Thus the inclusion $\breve{H}_2^{\omega(t)}\subset H_2^{\omega(t)}$ is proved.

ii) Now we shall prove that $H_2^{\omega(t)}\subset\breve{H}_2^{\omega(t)}$. In fact, if $f(x)\in H_2^{\omega(t)}$, then for any points x_1 and $x_2=x_1+h\in[a,\ b]$, taking into account the fact that $\omega_2(f;\ t)=\omega_2(f-P_1;\ t)$ and that, according to Marchaud's theorem,

$$\omega_1(f-P_1;\ t)\leqslant At\int_t^{2h}\frac{\omega_2(u)}{u^2}\,du,\qquad A=\mathrm{const},$$

for all $t\in[0,\ h]$, at any point $x\in[a,\ b]$ such that, e.g., $|x-x_1|=t\leq h$, we obtain

$$|f(x)-P_1(f;\ x_1,\ x_2;\ x)|=|[f(x)-P_1(x)]-[f(x_1)-P_1(x_1)]|$$

$$\leqslant\omega_1(f-P_1;\ |x-x_1|)\leqslant At\int_t^{2h}\frac{\omega(u)}{u^2}\,du.$$

This proves the inclusion $H_2^{\omega(t)}\subset\breve{H}_2^{\omega(t)}$, and together with it the equality (2.23), i.e. the statement 3) is proved.

Theorem 2.2. *Each class $\breve{H}_2^{\omega(t)}$ of functions is an algebra over the field of complex numbers such that if $f_1(z)$ and $f_2(z)\in\breve{H}_2^{\omega(t)}$, then*

(i) $\alpha_1f_1(z)+\alpha_2f_2(z)\in\breve{H}_2^{\omega(t)}$ *for arbitrary complex* α_1 *and* α_2, *and*

(ii)

$$f_1(z)\,f_2(z)\in\breve{H}_2^{\omega(t)}. \tag{2.25}$$

Proof. Since (i) is obvious, we shall prove only (ii). We denote by d the diameter of the set $\mathfrak{M}$ and on $[0, d]$ consider the function

$$\beta(\tau) = \tau \int_\tau^d \frac{\omega(u)}{u^2}\left[1 - \left(\frac{\tau}{u}\right)^2\right] du, \qquad \beta(0) \overset{\text{def}}{=} 0. \tag{2.26}$$

We show that for all $h \in [0, d/2]$

$$\omega_2(\beta;\ h) \leqslant A\omega(h). \tag{2.27}$$

For this purpose, at every point $\tau \in [0, d-2_t]$, where $t \leq h$, we consider the second difference

$$\Delta_h^2\beta(\tau) = \beta(\tau) - 2\beta(\tau+t) + \beta(\tau+2t).$$

First let $\tau \in [t, d-2t]$. Then, as is not difficult to see,

$$\Delta_t^2\beta(\tau) = \int_0^t\int_0^t \beta''(r+t_2+t_1)\,dt_1dt_2 \leqslant t^2 \max_{r\in[t,\ d-2t]} |\beta''(\tau)|$$

$$= t^2 \max_{\tau\in[t,\ d-2t]} \left|\int_\tau^d \frac{\omega(u)}{u^3}\left[20\left(\frac{\tau}{u}\right)^3 - 12\frac{\tau}{u}\right]du\right| \leqslant t^2 \max_{\tau\in[t,\ d-2t]} \left|\int_\tau^d \frac{\omega(u)}{u^3}\, 32\frac{\tau}{u}\,du\right|$$

$$\leqslant t^2 \max_{\tau\in[t,\ d-2t]} 32\cdot 4\cdot \frac{\omega(t)}{t^2}\tau \int_\tau^\infty \frac{du}{u^2} = 128\omega(t).$$

Let $\tau \in [0, t]$. We define the following functions:

$$\begin{aligned}
\beta_1(\tau) &\overset{\text{def}}{=} \beta(\tau) - \tau\int_\tau^d \frac{\omega(u)}{u^2}\,du,\\
\beta_2(\tau) &\overset{\text{def}}{=} \tau\int_\tau^{2t} \frac{\omega(u)}{u^2}\,du,\\
\beta_3(\tau) &\overset{\text{def}}{=} \tau\int_{2t}^d \frac{\omega(u)}{u^2}\,du,
\end{aligned} \tag{2.28}$$

so that $\beta(\tau) = \beta_1(\tau) + \beta_2(\tau) + \beta_3(\tau)$.

Now we shall estimate the second differences $\Delta_t^2\beta_i(\tau)$ of the functions $\beta_i(\tau)$ at the point $\tau \in [0, 2t]$, with increment t. We have

$$\Delta_t^2\beta_1(\tau) \leqslant 4\max_{\tau\in[0,\ 2t]} |\beta_1(\tau)| \leqslant 4\max_{\tau\in[0,\ 2t]} \left|\tau\int_\tau^d \frac{\omega(u)}{u^2}\left[\left(\frac{\tau}{u}\right)^4 - 2\left(\frac{\tau}{u}\right)^2\right]du\right|$$

$$\leqslant 4\max_{\tau\in[0,\ 2t]} \tau\cdot 4\frac{\omega(\tau)}{\tau^2}\int_\tau^d 3\left(\frac{\tau}{u}\right)^2 du \leqslant 48\max_{\tau\in[0,\ 2t]} \omega(\tau) \leqslant 48\omega(2t) \leqslant 192\omega(t);$$

$$\Delta_t^2\beta_2(\tau) \leqslant 4\max_{\tau\in[0,\ 2t]} \left|\tau\int_\tau^{2t} \frac{\omega(u)}{u^2}\,du\right| \leqslant 4\omega(2t)\,\tau\int_\tau^\infty \frac{du}{u^2} = 4\omega(2t) \leqslant 16\omega(t); \quad \Delta_t^2\beta_3(\tau) = 0.$$

Hence in view of (2.28) we find that for any $\tau \in [0, t]$

$$\Delta_t^2\beta(\tau) \leqslant \Delta_t^2\beta_1(\tau) + \Delta_t^2\beta_2(\tau) + \Delta_t^2\beta_3(\tau) \leqslant (192+16)\,\omega(t) = 208\omega(t).$$

The relation (2.27) is proved.

We shall also establish that for all $h \in [0, d/2]$

$$\left(h\int_h^d \frac{\omega(u)}{u^2}\,du\right)^2 = \left(h\int_{\sqrt{2}h}^d \frac{\omega(u)}{u^2}\,du + h\int_h^{\sqrt{2}h}\frac{\omega(u)}{u^2}\,du\right)^2$$

$$= \left(h\int_{\sqrt{2}h}^d \frac{\omega(u)}{u^2}\,du + \frac{h}{\sqrt{2}}\int_{\sqrt{2}h}^{2h}\frac{\omega(u/\sqrt{2})}{u^2/2}\,du\right)^2 \leqslant \left(h\int_{\sqrt{2}h}^d \frac{\omega(u)}{u^2}\,du + \sqrt{2}h\int_{\sqrt{2}h}^d \frac{\omega(u)}{u^2}\,du\right)^2$$

$$\leqslant (1+\sqrt{2})^2\, 16\left(h\int_{\sqrt{2}h}^d \frac{\omega(u)}{u^2}\left(1-\left(\frac{h}{u}\right)^2\right)^2 du\right)^2$$

$$\leqslant 96\left(h\int_h^d \frac{\omega(u)}{u^2}\left(1-\left(\frac{h}{u}\right)^2\right)^2 du\right)^2 = 96\beta^2(h). \tag{2.29}$$

On the basis of the bound $\omega_1^2(t) \le A\omega_2(t)$ (see [33]) the relations (2.27) and (2.29) imply that for all $h \in [0, d/2]$

$$\left(h\int_h^d \frac{\omega(u)}{u^2}\,du\right)^2 \leqslant 100\beta^2(h) \leqslant A\omega_1^2(\beta;\ h) \leqslant A_1\omega_2(\beta;\ h) \leqslant A_2\omega(h) \leqslant A_3 h\int_h^{2h}\frac{\omega(u)}{u^2}\,du. \tag{2.30}$$

Now let z_1 and z_2 be arbitrary points of $\mathfrak{M}$ for which $|z_1 - z_2| \overset{\text{def}}{=} h \le d/2$, and let the point $z^* \in \mathfrak{M}$ possess the property that $|z^* - z| \ge d/2$. Then, in view of (2.2),

$$|f_i(z_2) - P_1(f_i;\ z_1,\ z^*;\ z_2)|$$

$$= \left|\frac{z^*-z_2}{z^*-z_1}[f_i(z_2)-f_i(z_1)] + \frac{z_2-z_1}{z^*-z_1}[f_i(z_2)-f_i(z^*)]\right| \leqslant Ah\int_h^d \frac{\omega(u)}{u^2}\,du$$

and since

$$\left|\frac{z_2-z_1}{z^*-z_1}[f_i(z_2)-f_i(z^*)]\right| \leqslant \frac{2h}{d}\,2\|f\| \leqslant A_1 h,$$

it follows that

$$\left|\frac{z^*-z_2}{z^*-z_1}[f_i(z_2)-f_i(z_1)]\right| \leqslant Ah\int_h^d \frac{\omega(u)}{u^2}\,du + A_1 h \leqslant A_2 h\int_h^d \frac{\omega(u)}{u^2}\,du. \tag{2.31}$$

Thus, noting that by (2.3) each of the polynomials $P_1(f_i;\ z_1,\ z_2;\ z) \overset{\text{def}}{=} \overset{i}{P}_1(z)$ has the form

$$\overset{i}{P}_1(z) = \frac{f_i(z_2)-f_i(z_1)}{z_2-z_1}\,z + C_i, \qquad C_i = \text{const}, \qquad i = 1,\ 2,$$

putting $P_1(z) \overset{\text{def}}{=} P_1(f_1 f_2;\ z_1,\ z_2;\ z) - P_1(\overset{1}{P}_1\overset{2}{P}_1;\ z_1,\ z_2;\ z)$ and taking (2.31) and (2.30) into consideration, we get

$$|\overset{1}{P}_1(z)\overset{2}{P}_1(z)-P_1(z)|=\left|\frac{f_1(z_2)-f_1(z_1)}{z_2-z_1}\,\frac{f_2(z_2)-f_2(z_1)}{z_2-z_1}\right||z-z_1||z-z_2|$$

$$\leqslant 2A_1A_2\left(\int_h^d\frac{\omega(u)}{u^2}\,du\right)^2 th\leqslant A't\int_h^{2h}\frac{\omega(u)}{u^2}\,du\leqslant A't\int_t^{2h}\frac{\omega(u)}{u^2}\,du.$$

Using this inequality, we obtain

$$|f_1(z)f_2(z)-P_1(f_1f_2;\ z_1,\ z_2;\ z)|\leqslant|f_1(z)||f_2(z)$$
$$-\overset{2}{P}_1(z)|+|\overset{2}{P}_1(z)||f_1(z)-\overset{1}{P}_1(z)|+|\overset{1}{P}_1(z)\overset{2}{P}_1(z)-P_1(z)|\leqslant\bar{A}t\int_t^{2h}\frac{\omega(u)}{u^2}\,du.$$

for any $z_1,\ z_2\in\mathfrak{M}$ such that $|z_1-z_2|=h\leq d/2$ and $\min\{|z-z_1|,\ |z-z_2|\}=t$.

Thus inequality (2.25), and together with it Theorem 2.2 is proved.

§3. Admissible sets

In this section we shall define and study the most important classes of closed sets $\mathfrak{M}$ on which the results proved in §§4 and 5 on approximation of Cauchy kernels by polynomial kernels and the basic direct theorems on approximation of functions of the classes A^r, $r=0,\ 1,\ 2,\ \ldots$, will be valid.

Definition 3.1. We shall denote by $(\mathfrak{L})$ the class of admissible continua $\mathfrak{M}$ with the following properties:

1) The boundary Γ of each of the sets $\mathfrak{M}$ consists of finitely many smooth arcs Γ_j which form positive exterior angles $\alpha_j\pi$, $0<\alpha_j\leq 2$, at their junction points z_j, $j=1,\ \ldots,\ l$, $l=l(\mathfrak{M})$, and possess the property that for any pair of points $z',\ z''\in\Gamma_j$ the arc-length $s(z',\ z'')$ between these points satisfies the inequality

$$s(z',\ z'')\leqslant A|z'-z''|; \tag{3.1}$$

2) There exists a positive number $a=a(\mathfrak{M})$ such that in each of the closed regions $U_j\overset{\text{def}}{=}U(w_j,\ a)\cap\{w\colon |w|\geq 1\}$, where $w_j=\Phi(z_j)$ and $U(w_j,\ a)=\{w\colon |w-w_j|\leq a\}$, the function $\Psi(w)$ has the representation

$$\Psi(w)=\Psi(w_j)+A_j(w)(w-w_j)^{\alpha_j}, \tag{3.2}$$

where $A_j(w)$ is continuous on U_j and differentiable on $U_j\setminus\{w_j\}$, and $A_j(w_j)\neq 0$.

3) The modulus of the derivative of $\Psi(w)$ satisfies

$$C_1|w-w_j|^{\alpha_j-1}\leqslant|\Psi'(w)|\leqslant C_2|w-w_j|^{\alpha_j-1}, \tag{3.3}$$

for $w\in U_j\setminus\{w_j\}$, while for $w\in\{w\colon |w|>1\}\setminus(\bigcup_1^l U_j)$

$$C_1\leqslant|\Psi'(w)|\leqslant C_2, \tag{3.4}$$

where $C_i=C_i(\mathfrak{M})$, $i=1,\ 2$, are positive constants.

Definition 3.2. We shall say that an admissible continuum $\mathfrak{M}$ belongs to the class B_k, where k is a fixed natural number, if for any $n=1,\ 2,\ \ldots$ it satisfies the following conditions:

1) The boundary Γ, taking its folds into account, can be divided into finitely many arcs Γ_j, $j = 1, \ldots, \kappa$, on each of which certain relations hold: specifically, for arbitrary z and $\zeta \in \Gamma_j$ and $t \in [-\pi, \pi]$, having set

$$\tilde{\xi} = \tilde{\xi}\left(\frac{1}{n}\right) \stackrel{\text{def}}{=} \Psi\left[\left(1+\frac{1}{n}\right)\Phi(\xi)\right], \tag{3.5}$$

$$\tilde{\xi}_t = \tilde{\xi}_t\left(\frac{1}{n}\right) \stackrel{\text{def}}{=} \Psi\left[\left(1+\frac{1}{n}\right)\Phi(\xi)\,e^{-it}\right] \tag{3.6}$$

for $\xi \in \Gamma$, we have

$$\text{a) } |\tilde{z} - z| \asymp \rho^{(j)}_{1+\frac{1}{n}}(z), \qquad \forall z \in \Gamma_j, \tag{3.7}$$

where $\rho^{(j)}_{1+1/n}(z)$ denotes the distance from the point z to that arc $\Psi[(1 + 1/n)\Phi(\Gamma_j)]$ which contains the point $\tilde{z}$;

$$\text{b) } |\zeta - z| \leqslant A(1 + n|t|)^k |\tilde{\zeta}_t - z|, \qquad \forall \zeta,\ z \in \Gamma, \tag{3.8}$$

and thus also

$$\text{b}') \ |\tilde{\zeta}_t - z| \leqslant A(1 + n|t|)^k |\tilde{\zeta} - z|; \tag{3.9}$$

$$\text{c) } |\tilde{\zeta} - \zeta|^k \leqslant A|\tilde{\zeta} - z|^{k-1}|\tilde{z} - z|, \qquad \forall \zeta,\ z \in \Gamma_j. \tag{3.10}$$

2) For arbitrary ζ, $z \in \Gamma$ the arc-length $\check{s}(\zeta, z)$ of that part of the curve Γ which lies in the disk $U(z; |\zeta - z|)$ with center at z and radius $|\zeta - z|$ satisfies the inequality

$$\check{s}(\zeta, z) \leqslant A|\tilde{\zeta} - z|. \tag{3.11}$$

Remark. The above definition is somewhat more general than the one introduced by the author in [28].

Definition 3.3. We shall say that $\mathfrak{M}$ is a set of type (B) if $\mathfrak{M} \in B_k$ for some k.

We leave it to the reader to verify that a set of type $(\mathfrak{L})$ is also a set of type (B).

A set of type (B) is general. Theorem 3.1, in which sufficient conditions for regions to be of type (B) are presented in geometrical terms, is corroboration of this assertion. In order to formulate this theorem we introduce some concepts about quasiconformal mappings [34], [29].

A plane set which consists of not more than countably many continuous arcs (open or closed) and continuous closed curves will be called a *curve*. A function $\rho(z) \geq 0$ defined in the complex plane will be called *admissible* if it is square summable on the plane and

$$A(\rho) = \iint \rho^2(z)\,dx\,dy \neq 0.$$

Let Γ be a family of locally rectifiable curves in the plane, P the class of all admissible functions, and

$$L_\rho(\Gamma) = \inf_{\gamma \in \Gamma} \int_\gamma \rho\,|dz|.$$

Then the quantity

$$m(\Gamma)=\inf_{\rho\in P}\frac{A(\rho)}{[L_\rho(\Gamma)]^2}$$

is called the *modulus* of the family Γ, while $\lambda(\Gamma)=[m(\Gamma)]^{-1}$ is called the *extremal length* of Γ.

In the case when ρ is not measurable on γ, we shall, by definition, take $\int_\gamma \rho|dz|=\infty$.

For example, let Q be a sector of the circular annulus of angle $\theta\pi$, $0<\theta\leq 2$, and radii $0<r_1<r_2<\infty$; let Γ be the family of continuous locally rectifiable curves in Q which separate curvilinear arcs of the boundary. Then

$$m(\Gamma)=[\lambda(\Gamma)]^{-1}=\frac{1}{\theta\pi}\ln\frac{r_2}{r_1}.$$

An arbitrary Jordan region on whose boundary two closed nonintersecting Jordan arcs l_1 and l_2 are marked off is commonly called a *quadrilateral*. The arcs l_1 and l_2 are called distinguished sides of the quadrilateral. We shall denote quadrilaterals by the symbol $Q(l_1, l_2)$. By the modulus $m(Q)$ of a quadrilateral we shall understand the modulus of the family Γ of locally rectifiable continuous arcs in Q which separate its distinguished sides. In particular, we can speak of the modulus of a doubly connected region, regarding it as a quadrilateral whose distinguished sides are components of its boundary.

Let $\phi(z)$ be a topological map of a region Ω onto a region Ω' which preserves orientation, and let $\phi_z=\frac{1}{2}(\phi_x-i\phi_y)$ and $\phi_{\bar z}=\frac{1}{2}(\phi_x+i\phi_y)$ be generalized derivatives. We say that $\phi(z)$ is *K-quasiconformal* if ϕ_z and $\phi_{\bar z}$ are locally integrable and satisfy the inequality

$$|\varphi_{\bar z}|\leqslant\frac{K-1}{K+1}|\varphi_z|$$

for fixed $K\geq 1$. A mapping which is K-quasiconformal with some $K\geq 1$ is said to be *quasiconformal*.

Consider an arbitrary quasiconformal mapping of the extended plane onto itself. A plane curve that is the image of a circle (or straight line) under such a mapping is called a quasiconformal curve. A complete description of such curves is one of the noteworthy results in the theory of quasiconformal mappings.

Let L be a closed Jordan curve, $\infty\notin L$. We fix an arbitrary point $z^*\in L$. If z_1, z_2, z_3 are arbitrary successive points of the arc $L-z^*$, then the condition

$$|z_1,\ z^*,\ z_2,\ z_3|=\left|\frac{(z_1-z_2)(z^*-z_3)}{(z_1-z_3)(z^*-z_2)}\right|\leqslant C<\infty \qquad \text{(C)}$$

is necessary and sufficient for the curve L to be quasiconformal.

We consider an arbitrary closed set $\mathfrak{M}$ of the complex plane whose complement $C\mathfrak{M}$ relative to the extended plane is a simply connected domain $C\mathfrak{M}\ni\infty$, $\operatorname{diam}\mathfrak{M}>0$. Let $l\in\partial C\mathfrak{M}$ be an arbitrary prime end of $C\mathfrak{M}$, and z_l a principal point of l. We shall introduce the following notation: $s(x)$, $x\in(0, \frac{1}{2}\operatorname{diam}\mathfrak{M})$ is an arc of the circle $\{\zeta: |\zeta - \zeta_l| = x\}$ which belongs to $C\mathfrak{M}$ and separates l from ∞; $Q(z; \delta, \Delta) = Q[z_l; s(\delta), s(\Delta)]$ is the quadrilateral bounded by the arcs $s(\delta)$ and $s(\Delta)$ and the arcs of the boundary of $C\mathfrak{M}$ joining their endpoints ($Q(z_l, \delta, \Delta) \subset C\mathfrak{M}$; $0 < \delta < \Delta \leq \frac{1}{2} \operatorname{diam} \mathfrak{M}$); $\{\gamma\}$ is the family of all locally rectifiable curves in $Q(z_l, \delta, \Delta)$ which separate the sides $s(\delta)$ and $s(\Delta)$; $\lambda(\{\gamma\}) = \lambda(z_l, \delta, \Delta)$ is the extremal length of $\{\gamma\}$; and $m[Q(z_l, \delta, \Delta)]$ is the modulus of the quadrilateral $Q(z_l, \delta, \Delta)$. The function

$$A(z, \delta, \Delta) \overset{\text{def}}{=} \frac{1}{\pi} \lambda(z, \delta, \Delta) \ln \frac{\Delta}{\delta},$$

$z \in C\mathfrak{M}$, will be called the opening of the quadrilateral $Q(z, \delta, \Delta)$ [29]. It is easy to verify that if $Q(z, \delta, \Delta)$ is a rectilinear annular sector with angle $\theta\pi$, then $A(z, \delta, \Delta) = \theta$. Thus the function $A(z, \delta, \Delta)$ generalizes the concept of the opening of an annular sector.

Lemma 3.1 [29]. *Let $z \in \partial C\mathfrak{M}$, where $\partial C\mathfrak{M}$ is a bounded quasiconformal curve; let $w = \Phi(z)$ and $U_t = \{\tau: |\tau| \geq 1; |\tau - w| = t > 0\}$; let $\Psi(U_t)$ be the image of the arc U_t under the mapping $z = \Psi(w)$, and ζ an arbitrary point of the quadrilateral*

$$Q_t = Q[z; s(\min_{\tau \in U_t} |\Psi(\tau) - z|), s(\max_{\tau \in U_t} |\Psi(\tau) - z|)].$$

Then for all sufficiently small t

$$A(z, |z - \zeta|, \Delta) = \alpha(z, t, \Delta) + \frac{\varepsilon(z, \zeta, \Delta)}{\ln t}, \tag{3.12}$$

where $\alpha(z, t, \Delta)$ is a positive continuous function in t depending only on its arguments, and $\epsilon(z, \zeta, \Delta) \asymp O(1)$.

The function $\alpha(z, t, \Delta)$, $0 < t < \Delta$, will be called the *generalized (t, Δ) exterior angle* of the set $\mathfrak{M}$ at the point z.

We note that in [29] the relation

$$|z - \zeta| \asymp |\Phi(z) - \Phi(\zeta)|^{\alpha(z, |\Phi(z) - \Phi(\zeta)|, \Delta)}, \tag{3.13}$$

is established for rather general sets, and for $z = \Psi(w)$ and $\zeta = \tilde{z} = \Psi[(1 + 1/n)w]$ this yields

$$\rho_{1+\frac{1}{n}}(z) \asymp |z - \tilde{z}| \asymp \left(\frac{1}{n}\right)^{\alpha\left(z, \frac{1}{n}, \Delta\right)}. \tag{3.14}$$

This relation gives us the possibility of finding (up to a constant) the value of $\rho_{1+1/n}(z)$ for a rather large class of sets in terms of the generalized exterior angle. The results of [29] easily give

Theorem 3.1. *In order that a closed bounded region $\mathfrak{M}$ belong to the class (B) it is sufficient that the following conditions should be satisfied:*

1) The boundary Γ of $\mathfrak{M}$ is a rectifiable quasiconformal curve.

2) There exists a $\Delta \in (0, \frac{1}{2} \operatorname{diam} \mathfrak{M})$ such that for all $\delta \in (0, \Delta)$ and all $z \in \partial\mathfrak{M}$

$$A(z, \delta, \Delta) \geqslant c = \text{const} > 0; \tag{3.15}$$

3) *Condition* c) (*inequality* (3.10)) *and condition* 2) *of Definition* 3.2 *for sets of type* (B_k) *are satisfied for the region* $\mathfrak{M}$.

Remark 1. Condition 1) of this theorem is equivalent to the fulfilment of the condition (C).

Remark 2. Condition 2), roughly speaking, means that the region $\mathfrak{M}$ does not have zero exterior angles.

§4. Approximation of Cauchy kernels by polynomial kernels

An overwhelming majority of methods of approximating various functions by polynomials on admissible sets are based in the final analysis on good approximation by polynomials of the form $\Sigma_0^n a_k(\zeta)z^k$ of the Cauchy kernel $1/(\zeta - z)$. As we shall see in this section, such polynomials play the same role in the complex case as the polynomial kernels, Faber, Valleé-Poussin, Jackson kernels and others play in the real case.

We shall study the polynomial kernels mentioned in §1, which in the theory of Faber polynomials at present possess the strongest properties in comparison with other kernels.

Theorem 4.1 (on approximation of the Cauchy kernel). *Let* $\mathfrak{M}$ *be a set which belongs to the class* B_k *for some natural* k. *Starting from this set and an arbitrarily fixed point* $\zeta \in \partial\mathfrak{M}$ *and natural* r *and* m, *we construct for every natural* n *a polynomial kernel* $K_{r,m,k,n}(\zeta, z)$ *in* z *of degree equal to*

$$(l+1)(n-1)km - 1 \overset{\text{def}}{=} n' \asymp n, \quad l \overset{\text{def}}{=} [{}^1/_2 (r+2)k] + 2,$$

of the form

$$K_{r,m,k,n}(\zeta, z) = \frac{1 - [1 - (\zeta - z)\pi_{ln}(\zeta, z)]^{km}}{\zeta - z}, \tag{4.1}$$

in which $\pi_{ln}(\zeta, z)$ *is a polynomial in* z *of degree* $(l+1)(n-1) - 1$, *which is constructed from the formula*

$$\pi_{ln}(\zeta, z) \overset{\text{def}}{=} \sum_{\nu=1}^{(l+1)(n-1)} \frac{j_\nu \Pi_\nu(z)}{(1+1/n)^\nu \Phi^\nu(\zeta)}, \tag{4.2}$$

where $j_\nu = j_\nu(ln)$ *denote the* ν*th Fourier coefficients of the generalized Jackson kernel*

$$J_{ln}(t) = \frac{1}{\gamma_{ln}} \left(\frac{\sin \frac{nt}{2}}{\sin \frac{t}{2}} \right)^{2(l+1)} = \frac{1}{2} \sum_{\nu=-(l+1)(n-1)}^{(l+1)(n-1)} j_{|\nu|} e^{i\nu t}, \tag{4.3}$$

$\Phi(\zeta) = w$ *denotes a function which maps* $C\mathfrak{M}$ *conformally and univalently onto* $|w| > 1$ *so that at the same time* $\lim_{\zeta\to\infty} \zeta^{-1}\Phi(\zeta) > 0$, *and* $\Pi_\nu(z)$ *denote generalized Faber polynomials defined by the relation* (1.21).

Then, for all $\zeta \in \partial\mathfrak{M}$ *and* $z \in G_{n'}$, *where*

$$G_n \overset{\text{def}}{=} \mathfrak{M} \cup \Big(\bigcup_{\zeta \in \partial\mathfrak{M}} g(n, \zeta) \Big), \quad g(n, \zeta) \overset{\text{def}}{=} \Big\{ z : d(z, \zeta) \leqslant \frac{1}{2} \rho_{1+\frac{1}{n}}(\zeta) \Big\}$$

and $d(z, \zeta) \overset{\text{def}}{=} \inf_\gamma \int_\gamma |d\xi|$, *where* γ *are all possible arcs in* $C\overline{\mathfrak{M}}$ *with end points* z *and* ζ, *the following relations hold for the kernel* $K_{r,m,k,n}(\zeta, z)$:

1) *For all* $j = 0, 1, \ldots, r$

$$\left| \frac{\partial^j}{\partial z^j} \left[\frac{1}{\zeta - z} - K_{r,m,k,n}(\zeta, z) \right] \right| \leqslant \frac{A \rho^m_{1+\frac{1}{n}}(z^*)}{|\zeta - z|^{1+j} \left(|\zeta - z| + \rho_{1+\frac{1}{n}}(\zeta) \right)^m}, \tag{4.4}$$

where z^* *denotes the point on* Γ_j (*the* j*th piece of the boundary* $\partial\mathfrak{M}$) *nearest to* z.

2) $$|K_{r,m,k,n}(\zeta, z)| \leqslant \frac{A}{\rho_{1+\frac{1}{n}}(z^*) + |\zeta - z|}, \tag{4.5}$$

where z^* *denotes the same point as in* 1).

3) $$\frac{1}{2\pi i} \int_{\partial\mathfrak{M}} K_{r,m,k,n}(\zeta, z)\, d\zeta = 1 - \left(1 - \frac{j_1(ln)}{1 + \frac{1}{n}}\right)^{km} - 1 - \frac{\alpha_{lm}}{n^{km}} \overset{\text{def}}{=} \gamma'_n,$$
$$0 < \alpha_{lm} < A; \tag{4.6}$$

4) $$\frac{1}{2\pi i} \int_{\partial\mathfrak{M}} (\zeta - z) \frac{\partial^j}{\partial z^j} K_{r,m,k,n}(\zeta, z)\, d\zeta = \begin{cases} O(n^{-km+1}), & \textit{if} \quad j = 0, \\ 1 + O(n^{-km}), & \textit{if} \quad j = 1, \\ 0, & \textit{if} \quad j = 2, 3, \ldots, \end{cases} \tag{4.7}$$

$$\frac{1}{2\pi i} \int_{\partial\mathfrak{M}} \frac{\partial^j}{\partial z^j} K_{r,m,k,n}(\zeta, z)\, d\zeta = 0, \qquad j = 1, 2, \ldots. \tag{4.7'}$$

Here A *everywhere denotes constants which depend neither on* n *nor on* $z \in G_n$.

Proof. 1) Let ζ be an arbitrary point of $\partial\mathfrak{M}$. We put

$$w = \Phi(\zeta), \qquad \tilde{w} = \left(1 + \frac{1}{n}\right) w, \qquad \tilde{\zeta} = \Psi(\tilde{w}), \qquad \zeta_t = \Psi(\tilde{w}e^{-it})$$

and, starting from a kernel of Jackson type (4.3) and a representation (which follows from (1.21) of the function $[\Psi(\tilde{w}e^{-it}) - z]^{-1}$ in the form of the series (see, for example, [25])

$$\frac{1}{\Psi(\tilde{w}e^{-it}) - z} = \sum_{j=1}^{\infty} \frac{\Pi_j(z)}{\tilde{w}^j e^{-ijt}},$$

which converges uniformly in $z \in G_n$ and $t \in [0, 2\pi]$, where the $\Pi_j(z)$ denote the generalized Faber polynomials figuring in (1.21), we form the following polynomials of degree $(l + 1)(n - 1) - 1$:

$$\pi_{ln}(z) = \pi_{ln}(\zeta, z) = \frac{1}{\pi} \int_{-\pi}^{\pi} J_{ln}(t) \frac{dt}{\Psi(\tilde{w}e^{-it}) - z}$$
$$= \sum_{\nu=1}^{(l+1)(n-1)} \frac{j_\nu}{\tilde{w}^\nu} \Pi_\nu(z) = \sum_{\nu=1}^{(l+1)(n-1)} C_\nu \Pi_\nu(z), \tag{4.8}$$

where

$$C_\nu = C_\nu(\zeta) = \frac{j_\nu}{\left(1+\frac{1}{n}\right)^\nu [\Phi(\zeta)]^\nu}. \tag{4.9}$$

Hence, noting that, in view of (1.13), $(1/\pi)\int_{-\pi}^{\pi} J_{ln}(t)dt = 1$, we obtain

$$\left|\frac{1}{\tilde\zeta - z} - \pi_{ln}(z)\right| = \frac{1}{\pi}\left|\int_{-\pi}^{\pi} J_{ln}(t)\left[\frac{1}{\tilde\zeta - z} - \frac{1}{\tilde\zeta_t - z}\right]dt\right|$$

$$\leqslant \frac{1}{\pi}\frac{1}{|\tilde\zeta - z|^2}\int_{-\pi}^{\pi} J_{ln}(t)\left|\frac{\tilde\zeta - z}{\tilde\zeta_t - z}\right| |\tilde\zeta_t - \tilde\zeta|\, dt. \tag{4.10}$$

We estimate the third and second factors under the integral sign in the above inequality.

Taking (3.9) and (3.7) into account, we find that

$$|\tilde\zeta_t - \tilde\zeta| \leqslant |\tilde\zeta_t - \zeta| + |\zeta - \tilde\zeta| \preccurlyeq A(1+n|t|)^k |\tilde\zeta - \zeta|$$
$$\asymp (1+n|t|)^k \rho_{1+\frac{1}{n}}(\zeta). \tag{4.11}$$

Since $\tilde\zeta = (\tilde\zeta_t)_{-t}$, in view of (3.9) we have

$$\left|\frac{\tilde\zeta - z}{\tilde\zeta_t - z}\right| = \left|\frac{(\tilde\zeta_t)_{-t} - z}{\tilde\zeta_t - z}\right| \preccurlyeq A(1+n|t|)^k. \tag{4.12}$$

Substituting (4.11) and (4.12) into the right side of (4.10) and then noting that for a kernel of Jackson type (4.3)

$$\int_{-\pi}^{\pi} J_{ln}(t)|t|^i dt \asymp \frac{1}{n^i}, \qquad 0 \leqslant i \leqslant 2l-2, \tag{4.13}$$

and, in particular,

$$|1 - j_1| \asymp \frac{1}{n^2}, \tag{4.14}$$

which will be required in the sequel, we get

$$\left|\frac{1}{\tilde\zeta - z} - \pi_{ln}(\zeta, z)\right| \preccurlyeq \frac{\rho_{1+\frac{1}{n}}(\zeta)}{|\tilde\zeta - z|^2}\int_{-\pi}^{\pi} J_{ln}(t)(1+n|t|)^{2k} dt \preccurlyeq A\frac{\rho_{1+\frac{1}{n}}(\zeta)}{|\tilde\zeta - z|^2}. \tag{4.15}$$

We shall now establish that for all $z \in G_n$ the following relations hold for the denominator $|\tilde\zeta - z|$ on the right side of the above inequality:

$$|\tilde\zeta - z| \succcurlyeq |\zeta - z| + \rho_{1+\frac{1}{n}}(\zeta); \tag{4.16}$$

$$|\tilde\zeta - z| \succcurlyeq |\zeta - z^*|. \tag{4.16'}$$

In fact, if $|\zeta - z| \geq 2|\tilde\zeta - z|$, then

$$|\tilde\zeta - z| \geqslant |\zeta - z| - |\zeta - \tilde\zeta| \geqslant \frac{1}{2}|\zeta - z| \geqslant \frac{1}{4}(|\zeta - z| + |\zeta - \tilde\zeta|)$$
$$\succcurlyeq |\zeta - z| + \rho_{1+\frac{1}{n}}(\zeta)$$

and (4.16) is established.

But if

$$|\zeta - z| < 2|\tilde{\zeta} - \zeta|, \tag{4.17}$$

then the inequality

$$|\tilde{\zeta} - z| \geqslant \frac{1}{2}\tilde{\rho}_{1+\frac{1}{n}}(\tilde{\zeta}), \tag{4.18}$$

must hold for all $z \in G_n$, where $\tilde{\rho}_{1+1/n}(\tilde{\zeta})$ denotes the distance from the point $\tilde{\zeta} \in \Gamma_{1+1/n}$ to the set $\mathfrak{M}$. Since the validity of this inequality is obvious for $z \in \mathfrak{M}$, we shall consider the case when $z \in G_n \setminus \mathfrak{M}$. Denoting in this case by $\check{\xi}$ a point on ∂G_n for which the equality $|\tilde{\zeta} - \check{\xi}| = d(\tilde{\zeta}, G_n)$ is satisfied, and by ξ a point on $\partial\mathfrak{M}$ for which $|\xi - \check{\xi}| = \frac{1}{2}\rho_{1+1/n}(\xi)$, we get

$$\tilde{\zeta} - z| \geqslant |\tilde{\zeta} - \check{\xi}| \geqslant |\tilde{\zeta} - \xi| - |\xi - \check{\xi}| = |\xi - \tilde{\zeta}| - \frac{1}{2}\rho_{1+\frac{1}{n}}(\xi)$$

$$\geqslant \frac{1}{2}|\xi - \tilde{\zeta}| \geqslant \frac{1}{2}\tilde{\rho}_{1+\frac{1}{n}}(\tilde{\zeta}).$$

Thus (4.18) is proved for all $z \in G_n$.

Using this inequality and also (3.7), we find that

$$|\tilde{\zeta} - z| \geqslant \frac{1}{2}\tilde{\rho}_{1+\frac{1}{n}}(\tilde{\zeta}) \asymp |\tilde{\zeta} - \zeta| \asymp |\tilde{\zeta} - \zeta| + |\zeta - z| \asymp |\zeta - z| + \rho_{1+\frac{1}{n}}(\zeta).$$

Thus (4.16) is also proved for the second case.

In order to establish (4.16′) we consider two cases:

a) If $z \in \mathfrak{M}$, then, taking (1.16) into account, we at once obtain

$$|\tilde{\zeta} - z| \geqslant \frac{1}{2}(|\tilde{\zeta} - z| + |z - z^*|) \geqslant \frac{1}{2}|\tilde{\zeta} - z^*| \succcurlyeq |\zeta - z^*|;$$

b) If $z \in G_n \setminus \mathfrak{M}$, then, denoting by ξ_1 a point on $\partial\mathfrak{M}$ nearest to ξ_1 and taking (4.18) into account, we find

$$|\tilde{\zeta} - z| \geqslant \frac{1}{4}[|\tilde{\zeta} - z| + (|z - \tilde{\zeta}| + |\tilde{\zeta} - \xi_1|)] \geqslant \frac{1}{4}[|\tilde{\zeta} - z| + |z - \xi_1|]$$

$$\geqslant \frac{1}{4}(|\tilde{\zeta} - z| + |z - z^*|) \geqslant \frac{1}{4}|\tilde{\zeta} - z^*| \succcurlyeq \frac{1}{4}|\zeta - z^*|.$$

The relations (3.7) and (4.15) yield

$$\left|\frac{1}{\tilde{\zeta} - z} - \pi_{ln}(\zeta, z)\right| \preccurlyeq \frac{\rho_{1+\frac{1}{n}}(\zeta)}{\left[|\zeta - z| + \rho_{1+\frac{1}{n}}(\zeta)\right]^2}$$

$$\preccurlyeq \frac{|\tilde{\zeta} - \zeta|}{\left[|z - \zeta| + \rho_{1+\frac{1}{n}}(\zeta)\right](|\zeta - z| + |\tilde{\zeta} - \zeta|)} \tag{4.15″}$$

and hence

$$\left|\frac{1}{\zeta - z} - \pi_{ln}(\zeta, z)\right| \leqslant \frac{|\zeta - \tilde{\zeta}|}{|\zeta - z||\tilde{\zeta} - z|} + \left|\frac{1}{\tilde{\zeta} - z} - \pi_{ln}(\zeta, z)\right|$$

$$\preccurlyeq \frac{|\tilde{\zeta} - \zeta|}{|\zeta - z||\tilde{\zeta} - z|}. \tag{4.15′}$$

Multiplying both sides of this inequality by $|\zeta - z|$, then raising them to power

km and taking (3.10), (3.7), (4.16) and (4.16′) into account, we get

$$\begin{aligned}|1-(\zeta-z)\pi_{ln}(\zeta, z)|^{km} &\preccurlyeq \frac{|\tilde{\zeta}-\zeta|^{km}}{|\tilde{\zeta}-z|^{km}} \preccurlyeq \left(\frac{|\tilde{\zeta}-z^*|^{k-1}|\tilde{z}^*-z^*|}{|\tilde{\zeta}-z|^{k}}\right)^m \\ &\preccurlyeq \left(\frac{|\tilde{z}^*-z^*|}{|\tilde{\zeta}-z|}\right)^m \preccurlyeq \frac{\rho^m_{1+\frac{1}{n}}(z^*)}{|\tilde{\zeta}-z|^m} \preccurlyeq \frac{\rho^m_{1+\frac{1}{n}}(z^*)}{\left[|\zeta-z|+\rho_{1+\frac{1}{n}}(\zeta)\right]^m}.\end{aligned} \tag{4.19}$$

Dividing both sides by $|\zeta - z|$ and denoting by $K_{r,m,k,n}(\zeta, z)$ the polynomial (4.1), we obtain (4.4) for the case $j = 0$.

In order to prove (4.4) for $j \geq 1$, we establish first that

$$\left|\frac{\partial^\nu}{\partial z^\nu}\left[\frac{1}{\zeta-z}-\pi_{ln}(\zeta, z)\right]\right| \preccurlyeq \frac{|\tilde{\zeta}-\zeta|}{|\zeta-z|^{\nu+1}|\tilde{\zeta}-z|}, \qquad \nu=0, 1, \ldots, r. \tag{4.20}$$

Differentiating with respect to z the equality in (4.10) taken without absolute values, we obtain

$$\frac{\partial^\nu}{\partial z^\nu}\left[\frac{1}{\tilde{\zeta}-z}-\pi_{ln}(\zeta, z)\right]=\frac{\nu!}{\pi}\int_{-\pi}^{\pi} J_{ln}(t)\left[\frac{1}{(\tilde{\zeta}-z)^{\nu+1}}-\frac{1}{(\tilde{\zeta}_t-z)^{\nu+1}}\right]dt.$$

Hence, as a result of (4.16), (4.11) and (4.12), we get

$$\begin{aligned}\left|\frac{\partial^\nu}{\partial z^\nu}\left[\frac{1}{\zeta-z}-\pi_{ln}(\zeta, z)\right]\right| &= \nu!\left|\frac{1}{(\zeta-z)^{\nu+1}}-\frac{1}{(\tilde{\zeta}-z)^{\nu+1}}\right| \\ &+\frac{\nu!}{\pi}\left|\int_{-\pi}^{\pi} J_{ln}(t)\left[\frac{1}{(\tilde{\zeta}-z)^{\nu+1}}-\frac{1}{(\tilde{\zeta}_t-z)^{\nu+1}}\right]dt\right| \leqslant \frac{|\tilde{\zeta}-\zeta|\sum_{s=0}^{\nu}|\tilde{\zeta}-z|^{\nu-s}|\zeta-z|^s}{|\tilde{\zeta}-z|^{\nu+1}|\zeta-z|^{\nu+1}} \\ &+\frac{2}{\pi}\nu!\int_0^{\pi} J_{ln}(t)\frac{|\tilde{\zeta}_t-\zeta|\sum_{s=0}^{\nu}|\tilde{\zeta}_t-z|^{\nu-s}|\tilde{\zeta}-z|^{s+1}}{|\tilde{\zeta}-z|^{\nu+2}|\tilde{\zeta}_t-z|^{\nu+1}}\,dt \\ &\preccurlyeq \frac{|\tilde{\zeta}-\zeta|}{|\zeta-z|^{\nu+1}|\tilde{\zeta}-z|}+\frac{\rho_{1+\frac{1}{n}}(\zeta)}{|\tilde{\zeta}-z|^{\nu+2}} \preccurlyeq \frac{|\tilde{\zeta}-\zeta|}{|\zeta-z|^{\nu+1}|\tilde{\zeta}-z|}.\end{aligned}$$

Now assuming in an induction that (4.10) for some natural $j \geq 1$ is satisfied for all $\nu = 0, 1, \ldots, j-1$, and taking (4.1) into account, we obtain

$$\begin{aligned}\left|\frac{\partial^j}{\partial z^j}\left\{\frac{1}{(\zeta-z)^{km-1}}\left[\frac{1}{\zeta-z}-K_{r,m,k,n}(\zeta, z)\right]\right\}\right| &= \left|\frac{\partial^j}{\partial z^j}\left[\frac{1}{\zeta-z}-\pi_{ln}(\zeta, z)\right]^{km}\right| \\ &\preccurlyeq \sum_{s=1}^{j}\left|\frac{1}{\zeta-z}-\pi_{ln}(\zeta, z)\right|^{km-s}\left|\sum\nolimits_s\right|,\end{aligned} \tag{4.21}$$

where the sum Σ_s consists of finitely many terms $A_{\nu_1, \ldots, \nu_s}$ of the form

$$A_{\nu_1,\nu_2,\ldots,\nu_s} \overset{\text{def}}{=} \frac{\partial^{\nu_1}}{\partial z^{\nu_1}}\left(\frac{1}{\zeta-z}-\pi_{ln}(\zeta, z)\right)\frac{\partial^{\nu_2}}{\partial z^{\nu_2}}\left(\frac{1}{\zeta-z}-\pi_{ln}(\zeta, z)\right) \cdots \frac{\partial^{\nu_s}}{\partial z^{\nu_s}}\left(\frac{1}{\zeta-z}-\pi_{ln}(\zeta, z)\right),$$

where the ν_τ are, generally speaking, distinct natural numbers such that $\sum_1^s \nu_\tau = 1$. Since, as a result of (4.20),

$$|A_{\nu_1,\nu_2,\ldots,\nu_s}| \leqslant \prod_{\tau=1}^{s} \frac{|\tilde{\zeta}-\zeta|}{|\zeta-z|^{\nu_\tau+1}|\tilde{\zeta}-z|} = \left(\frac{|\tilde{\zeta}-\zeta|}{|\zeta-z||\tilde{\zeta}-z|}\right)^s \frac{1}{|\zeta-z|^j},$$

on the basis of (4.21), (4.20) and (3.10) we obtain

$$\left|\frac{\partial^j}{\partial z^j}\left\{\frac{1}{(\zeta-z)^{km-1}}\left[\frac{1}{\zeta-z}-K_{r,m,k,n}(\zeta,z)\right]\right\}\right| \preccurlyeq \sum_{s=1}^{j}\left(\frac{|\tilde{\zeta}-\zeta|}{|\zeta-z||\tilde{\zeta}-z|}\right)^{km-s+s}\frac{1}{|\zeta-z|^j}$$

$$\asymp \frac{j}{|\zeta-z|^j}\left(\frac{|\tilde{\zeta}-\zeta|}{|\zeta-z||\tilde{\zeta}-z|}\right)^{km} \preccurlyeq \frac{1}{|\zeta-z|^{km+j}}\frac{\rho^m_{1+\frac{1}{n}}(z^*)}{|\tilde{\zeta}-z|^m}$$

and, therefore, taking into account the assumption that (4.4) holds for $\nu = 0, 1, \ldots, j-1$, we also get the inequality

$$\left|\frac{1}{(\zeta-z)^{km-1}}\frac{\partial^j}{\partial z^j}\left[\frac{1}{\zeta-z}-K_{r,m,k,n}(\zeta,z)\right]\right| \preccurlyeq \frac{1}{|\zeta-z|^{km+j}}\frac{\rho^m_{1+\frac{1}{n}}(z^*)}{|\tilde{\zeta}-z|^m}$$

$$+\sum_{\nu=0}^{j-1}\frac{1}{|\zeta-z|^{km-1+(j-\nu)}}\left|\frac{\partial^\nu}{\partial z^\nu}\left[\frac{1}{\zeta-z}-K_{r,m,k,n}(\zeta,z)\right]\right| \preccurlyeq \frac{1}{|\zeta-z|^{km+j}}\frac{\rho^m_{1+\frac{1}{n}}(z^*)}{|\tilde{\zeta}-z|^m}.$$

Thus, in view of (4.16) the inequality (4.4) is proved for the case $j \geq 1$.

2) Since by virtue of (4.15), (4.15°) and (4.16)

$$|\pi_{ln}(\zeta,z)| \leqslant \frac{1}{|\tilde{\zeta}-z|}+\left|\frac{1}{\tilde{\zeta}-z}-\pi_{ln}(\zeta,z)\right|$$

$$\leqslant \frac{1}{|\zeta-z|+\rho_{1+\frac{1}{n}}(\zeta)+\rho_{1+\frac{1}{n}}(z^*)} \leqslant \frac{1}{|\zeta-z|+\rho_{1+\frac{1}{n}}(z^*)},$$

we get by using (4.1)

$$|K_{r,m\ k,n}(\zeta,z)| \leqslant \sum_{j=1}^{km}\binom{km}{j}\frac{|\zeta-z|^{j-1}}{\left[|\zeta-z|+\rho_{1+\frac{1}{n}}(z^*)\right]^j} \preccurlyeq \frac{1}{|\zeta-z|+\rho_{1+\frac{1}{n}}(z^*)}.$$

Thus (4.5) is proved.

3) Assuming that

$$\Psi(\tau) = c_1\tau + c_0 + \frac{c_{-1}}{\tau} + \ldots, \qquad \Psi'(\tau) = c_1 - \frac{c_{-1}}{\tau^2} - \ldots, \tag{4.22}$$

in the neighborhood of ∞, and noting that, firstly, $c_1 = \Psi'(\infty)$, and, secondly (see, for example, [28]), $\Pi_1(z) = 1/\Psi'(\infty) = 1/c_1$, we get on the basis of (4.1) and (4.8)

$$\gamma_n' = \frac{1}{2\pi i} \int\limits_{\partial\mathfrak{M}} K_{r,m,k,n}(\zeta, z)\, d\zeta = \sum_{s=1}^{km} (-1)^{s-1} \binom{km}{s} \frac{1}{2\pi i} \int\limits_{\partial\mathfrak{M}} (\zeta - z)^{s-1} [\pi_{ln}(\zeta, z)]^s\, d\zeta$$

$$= \sum_{s=1}^{km} (-1)^{s-1} \binom{km}{s} \frac{1}{2\pi i} \int\limits_{\partial\mathfrak{M}} (\zeta - z)^{s-1} \left[\sum_{\nu=1}^{(l+1)(n-1)} \frac{j_\nu}{\left(\left(1+\frac{1}{n}\right)\Phi(\zeta)\right)^\nu} \Pi_\nu(z) \right] d\zeta$$

$$= \sum_{s=1}^{km} (-1)^{s-1} \binom{km}{s} \frac{1}{2\pi i} \int\limits_{|\tau|=1} \left[c_1\tau + c_0 - z + \frac{c_{-1}}{\tau} + \dots \right]^{s-1} \left[\frac{j_1}{\left(1+\frac{1}{n}\right)\tau} \Pi_1(z) \right.$$

$$\left. + \frac{j_2}{\left(1+\frac{1}{n}\right)^2 \tau^2} \Pi_2(z) + \dots \right]^s \Psi'(\tau)\, d\tau = \sum_{s=1}^{km} (-1)^{s-1} \binom{km}{s} \frac{1}{2\pi i} \int\limits_{|\tau|=1} [c_1^{s-1}\tau^{s-1}$$

$$+ (s-1) c_1^{s-2} (c_0 - z) \tau^{s-2} + \dots] \left[\left(\frac{j_1}{\left(1+\frac{1}{n}\right) c_1} \right)^s \frac{1}{\tau^s} \right.$$

$$\left. + \frac{s j_1^{s-1} j_2}{\left(1+\frac{1}{n}\right)^{s+1} c_1^{s-1}} \Pi_2(z) \frac{1}{\tau^{s+1}} + \dots \right] c_1 d\tau$$

$$= \sum_{s=1}^{km} (-1)^{s-1} \binom{km}{s} \left(\frac{j_1}{1+\frac{1}{n}} \right)^s = 1 - \left(1 - \frac{j_1}{1+\frac{1}{n}} \right)^{km}.$$

Hence (4.6) follows by (4.14).

4) Reasoning in the same manner as above, we obtain [32]

$$\frac{1}{2\pi i} \int\limits_{\partial\mathfrak{M}} (\zeta - z) K_{r,m,k,n}(\zeta, z)\, d\zeta = \sum_{s=1}^{km} (-1)^{s-1} \binom{km}{s} \frac{1}{2\pi i} \int\limits_{|\tau|=1} \left(c_1\tau + c_0 - z \right.$$

$$\left. + \frac{c_{-1}}{\tau} + \dots \right)^s \left[\frac{j_1 \Pi_1(z)}{\left(1+\frac{1}{n}\right)\tau} + \frac{j_2 \Pi_2(z)}{\left(1+\frac{1}{n}\right)^2 \tau^2} + \dots \right]^s \Psi'(\tau)\, d\tau$$

$$= \sum_{s=1}^{km} (-1)^{s-1} \binom{km}{s} \frac{1}{2\pi i} \int\limits_{|\tau|=1} (c_1^s \tau^s + s c_1^s (c_0 - z) \tau^{s-1} + \dots) \left[\left(\frac{j_1}{\left(1+\frac{1}{n}\right) c_1} \right)^s \frac{1}{\tau^s} \right.$$

$$\left. + \frac{s j_1^{s-1} j_2 \Pi_2(z)}{\left(1+\frac{1}{n}\right)^{s+1} c_1^{s-1}} \frac{1}{\tau^{s+1}} + \dots \right] c_1 dt$$

$$= \sum_{s=1}^{km} (-1)^{s-1} \binom{km}{s} s \left(\frac{j_1}{1+\frac{1}{n}} \right)^{s-1} \left[\frac{c_1^2 j_2 \Pi_2(z)}{\left(1+\frac{1}{n}\right)^2} + \frac{c_1 (c_0 - z) j_1}{1+\frac{1}{n}} \right]$$

$$= \left[\frac{c_1^2 j_2 \Pi_2(z)}{\left(1+\frac{1}{n}\right)^2} + \frac{c_1 (c_0 - z) j_1}{\left(1+\frac{1}{n}\right)} \right] km \sum_{s=0}^{km-1} (-1)^s \binom{km-1}{s} \left(\frac{j_1}{1+\frac{1}{n}} \right)^s$$

$$= [\text{idem}]\, km \left(1 - \frac{j_1}{1+\frac{1}{n}} \right)^{km-1} \asymp \left(\frac{1}{n} \right)^{km-1}.$$

Thus (3.7) is proved for the case $j = 0$.

In the case $j = 1$, we get by using (4.1)

$$\frac{\partial}{\partial z} K_{r,m,k,n}(\zeta, z) = \frac{1}{(\zeta - z)^2} \left\{ -km(\zeta - z)[1 - (\zeta - z)\pi_{ln}(\zeta, z)]^{km-1} \right.$$
$$\left. \times \left[\pi_{ln}(\zeta, z) - (\zeta - z) \frac{\partial}{\partial z} \pi_{ln}(\zeta, z) \right] + 1 - [1 - (\zeta - z)\pi_{ln}(\zeta, z)]^{km} \right\} \tag{4.23}$$

and hence, reasoning in the same way as in 3) and taking (4.15) into account, we find

$$\frac{1}{2\pi i}\int_{\partial\mathfrak{M}}(\zeta-z)\frac{\partial}{\partial z}K_{r,m,k,n}(\zeta, z)\,d\zeta=\frac{-km}{2\pi i}\int_{\partial\mathfrak{M}}[1-(\zeta-z)\pi_{ln}(\zeta, z)]^{km-1}$$

$$\times\left[\pi_{ln}(\zeta, z)-(\zeta-z)\frac{\partial}{\partial z}\pi_{ln}(\zeta, z)\right]d\zeta+\frac{1}{2\pi i}\int_{\partial\mathfrak{M}}K_{r,m,k,n}(\zeta, z)\,d\zeta$$

$$=-\frac{km}{2\pi i}\int_{|\tau|=1}\left\{1-[\Psi(\tau)-z]\sum_{\nu=1}^{(l+1)(n-1)}\frac{j_\nu\Pi_\nu(z)}{\left(1+\frac{1}{n}\right)^\nu\tau^\nu}\right\}^{km-1}\left\{\sum_{\nu=1}^{(l+1)(n-1)}\frac{j_\nu\Pi_\nu(z)}{\left(1+\frac{1}{n}\right)^\nu\tau^\nu}\right.$$

$$\left.-[\Psi(\tau)-z]\sum_{\nu=1}^{(l+1)(n-1)}\frac{j_\nu\Pi'_\nu(z)}{\left(1+\frac{1}{\pi}\right)^\nu\tau^\nu}\right\}\Psi'(\tau)\,d\tau+1-\frac{a_{lm}}{n^{km}}$$

$$=-\frac{km}{2\pi i}\int_{|\tau|=1}\left\{1-c_1\tau\frac{j_1\Pi_1(z)}{\left(1+\frac{1}{n}\right)\tau}\right\}^{km-1}\left\{\frac{j_1\Pi_1(z)}{\left(1+\frac{1}{n}\right)\tau}-c_1\tau\frac{j_2\Pi'_2(z)}{\left(1+\frac{1}{n}\right)^2\tau^2}\right\}c_1d\tau$$

$$+1-\frac{a_{lm}}{n^{km}}=1-\frac{a_{lm}}{n^{km}}-km\left(1-\frac{j_1}{1+\frac{1}{n}}\right)^{km-1}\left\{\frac{j_1}{1+\frac{1}{n}}-c_1^2\Pi'_2(z)\frac{j_2}{\left(1+\frac{1}{n}\right)^2}\right\}.$$

Thus, since, as is easy to see, $c_1^2\Pi_2'(z)=1$, the second relation in (4.7) follows.

The case $j\geq 2$ is not required in the sequel. It is established by exactly the same reasoning using the relation

$$\frac{\partial^j}{\partial z^j}K_{r,m,k,n}(\zeta, z)=\sum_{s=0}^{j-1}\binom{j-1}{s}\frac{\partial^s}{\partial z^s}\left[\frac{1}{(\zeta-z)^2}\right]\frac{\partial^{j-1-s}}{\partial z^{j-1-s}}\left\{-km(\zeta-z)[1-(\zeta-z)\right.$$

$$\left.\times\pi_{ln}(\zeta, z)]^{km-1}\left[\pi_{ln}(\zeta, z)-(\zeta-z)\frac{\partial}{\partial z}\pi_{ln}(\zeta, z)\right]+1-[1-(\zeta-z)\pi_{ln}(\zeta, z)]^{km}\right\}.$$

which follows from (4.23). The relations (4.7$'$) are established analogously to the proof of (4.7). Thus Theorem 4.1 is completely proved.

We note that the relation (4.7) was obtained for $j=0$ by I. A. Ševčuk [32]. Shortly after the paper [35] by the author and A. I. Švaĭ was published, which contained the kernels (1.19)–(4.1) introduced by the author, and a simplification of the polynomials $\pi_n(\zeta, z)$ in this paper had appeared (see R. V. Poljakov [36]), N. A. Lebedev and N. A. Širokov [37], and at the same time the author [28], introduced into consideration, instead of the kernels

$$K_{r,m,k,n}(\zeta, z)=\frac{1-[1-(\zeta-z)\pi_{ln}(\zeta, z)]^{km}}{\zeta-z}\tag{4.1}$$

$$=\frac{1-\left[1-\frac{1}{\pi}\int_{-\pi}^{\pi}J_{ln}(t)\frac{\zeta-z}{\Psi[(1+n^{-1})\Phi(\zeta)e^{it}]-z}\,dt\right]^{km}}{\zeta-z}$$

the formally simpler kernels

$$\check{K}_{r,m,k,n}(\zeta, z) = \frac{1}{\pi}\int_{-\pi}^{\pi} J_{ln}(t)\frac{1-\left[1-\frac{\zeta - z}{\Psi[(1+n^{-1})\Phi(\zeta)e^{it}]-z}\right]^{km}}{\zeta - z}\,dt, \quad (4.1')$$

which, as is easy to see, are obtained from (4.1) by interchanging the operations of multiplication by the Jackson kernel $(1/\pi)J_{ln}(t)$ and integration, and which represent an immediate generalization of the kernel (4.2″) to the case $km > 1$. However, it has turned out that the kernels $\check{K}_{r,m,k,n}(\zeta, z)$ do not have as good properties as the kernels $K_{r,m,k,n}(\zeta, z)$, namely the relations (4.7), (4.8) for $j \geq 0$ and (4.4) for $j \geq 1$ do not hold for them. Thus we shall not discuss properties of these kernels here, and refer the readers to the papers of the author [28] and Lebedev and Širokov [37].

Finally, we note that for the approximation of functions defined on subsets of $\mathfrak{M}$ which are at a positive distance from the boundary, the kernels

$$K_{D_n}(\zeta, z) \overset{\text{def}}{=} \frac{1}{\pi}\int_{-\pi}^{\pi} D_n(t)\frac{dt}{\Psi[\Phi(\zeta)e^{-it}]-z},$$

$$K_{V_n}(\zeta, z) \overset{\text{def}}{=} \frac{1}{\pi}\int_{-\pi}^{\pi} V_n(t)\frac{dt}{\Psi[\Phi(\zeta)e^{-it}]-z},$$

are very useful, where $D_n(t)$ and $V_n(t)$ denote the Dirichlet and de la Vallée-Poussin kernels respectively.

§5. Direct theorems

In this section we shall establish basic theorems on approximation of functions on various classes of sets of type B_k.

Definition 5.1. A set $\mathfrak{M}$ of type B_k will be called a set of type $\overset{\circ}{B}{}_k^{(n)}$ for some n if, no matter what the point $z \in \partial\mathfrak{M}$ is, the relation

$$\rho_{1+\frac{1}{n}}(\zeta) \asymp \rho_{1+\frac{1}{n}}(z) \quad (5.1)$$

is satisfied at every point ζ of the set $\gamma_0 \overset{\text{def}}{=} \partial\mathfrak{M} \cap U(z, \rho_{1+1/n}(z))$.

A set $\mathfrak{M}$ of type $\overset{\circ}{B}{}_k^{(n)}$ is called a set of type $\overset{*}{B}{}_k^{(n)}$ if (3.10) is satisfied for it for arbitrary $\zeta, z \in \partial\mathfrak{M}$.

We shall say that $\mathfrak{M} \in \overset{\circ}{B}_k$ or $\mathfrak{M} \in \overset{*}{B}_k$ if $\mathfrak{M} \in \overset{\circ}{B}{}_k^{(n)}$, or $\mathfrak{M} \in \overset{*}{B}{}_k^{(n)}$ simultaneously for all $n = 1, 2, \ldots$.

Theorem 5.1 (on estimating the approximation of a function in terms of its second modulus of continuity on the boundary and within a set). *For some natural n let $f(z)$ be a function on a set $\mathfrak{M}$ of type $\overset{\circ}{B}{}_k^{(n)}$, with $f(z)$ analytic at all points $z \in \operatorname{Int}\mathfrak{M}$ and continuous on $\mathfrak{M}$. Then we can construct a polynomial $P_n(z) = P_n(f; \mathfrak{M}; z)$ of degree not greater than n such that at any point $z \in \mathfrak{M}$ the inequality*

$$|f(z) - P_n(z)| \leqslant A \min_{r>0} \Omega(r, z), \quad (5.2)$$

is satisfied, where

$$\Omega(r, z) \overset{\text{def}}{=} \max_{\zeta\in\Gamma}\left\{\omega_2(r)\left[\frac{\rho_{1+\frac{1}{n}}(\zeta)\left(|\zeta-z|+\rho_{1+\frac{1}{n}}(\zeta)+r\right)}{r\left(|\zeta-z|+\rho_{1+\frac{1}{n}}(\zeta)\right)}\right]^2\left[\frac{\rho_{1+\frac{1}{n}}(\zeta)}{|\zeta-z|+\rho_{1+\frac{1}{n}}(\zeta)}\right]^{m-2}\right\} \tag{5.2'}$$

and $\Gamma \overset{\text{def}}{=} \partial\mathfrak{M}$ *denotes the boundary of* $\mathfrak{M}$, $\omega_2(t) = \omega_2(f;\, t)$ *denotes the second modulus of continuity of* $f(z)$ *on* $\mathfrak{M}$, m *denotes an arbitrary fixed natural number* ≥ 4 *and* A *denotes a constant which depends on* $\mathfrak{M}$ *but not on* $z \in \mathfrak{M}$ *or* $f(z)$.

Remark 5.1. The complicated form of the function $\Omega(r, z)$ which appears on the right side of (5.2) is explained by the fact that if the boundary $\Gamma = \partial\mathfrak{M}$ of $\mathfrak{M}$ is sufficiently complicated, then the measure of approximation on the right side of (5.2) will depend in a rather complicated manner on the position of the point z in $\mathfrak{M}$.

Remark 5.2. The fact that the magnitude of the deviation decreases at points $z \in \partial\mathfrak{M}$ far from the boundary was first pointed out by N. A. Širokov [39].

Proof of Theorem 5.1. We denote by $P_n(z)$ a polynomial of degree n' of the form

$$P_n(z) = \frac{1}{2\pi i\gamma_n'}\int_\Gamma f(\zeta)\, K_{n'}(\zeta, z)\, d\zeta, \tag{5.3}$$

where $K_{n'}(\zeta, z) \overset{\text{def}}{=} K_{r,m,k,n}(\zeta, z)$ denotes a polynomial kernel in z of degree $n' \overset{\text{def}}{=} [\frac{1}{2}((r+2)k+4)km(n-1)] - 1$, defined by (4.1), r, k and m denote fixed natural numbers which are chosen so that, by Theorem 4.1, the properties of $K_{n'}(\zeta, z)$ necessary for the sequel are satisfied, and γ_n' denotes the number defined by (4.6). In order to investigate the magnitude of the deviation of this polynomial from $f(z)$, we shall denote, for arbitrary fixed $r \in (1/n^2, \frac{1}{2}\operatorname{diam}\mathfrak{M})$ (which figures in (5.2) and (5.2')), by z° a point of $\mathfrak{M}$ which is at a distance r from z ($z^\circ \in \mathfrak{M}$, $|z^\circ - z| = r$), and by z^* a fixed point which coincides with z if $z \in \operatorname{Int}\mathfrak{M}$, and belongs to $C\mathfrak{M}$ and is at a distance $\leq n^{-2km}$ from z if $z \in \partial\mathfrak{M} = \Gamma$; we put

$$p(\zeta) \overset{\text{def}}{=} q(\zeta - z), \qquad q = \frac{f(z^0) - f(z)}{z^0 - z}. \tag{5.4}$$

In such a case, taking into account Theorem 4.1 on the approximation of the Cauchy kernel (see (4.7), the case $j = 0$), we obtain

$$\begin{aligned} |f(z) - P_n(z)| &= \left|\frac{1}{2\pi i\gamma_n'}\int_\Gamma [f(z) - f(\zeta)]\, K_{n'}(\zeta, z)\, d\zeta\right| \\ &= \frac{1}{2\pi\gamma_n'}\left|\int_\Gamma [f(\zeta) - f(z)]\left[K_{n'}(\zeta, z) - \frac{1}{\zeta - z^*}\right]d\zeta\right| \\ &\leqslant \frac{1}{2\pi\gamma_n'}\left\{\left|\int_\Gamma [f(\zeta) - f(z) - p(\zeta)]\left[K_{n'}(\zeta, z) - \frac{1}{\zeta - z^*}\right]d\zeta\right|\right. \\ &\left.+ |q|\left|\int_\Gamma (\zeta - z)\, K_{n'}(\zeta, z)\, d\zeta\right|\right\} = \frac{1}{2\pi\gamma_n'}|I| + O(|q|\, n^{-km+1}); \end{aligned} \tag{5.5}$$

$$I = I(z) = \int_{\Gamma} \Delta(\zeta, z)\left[K_{n'}(\zeta, z) - \frac{1}{\zeta - z^*}\right] d\zeta; \tag{5.6}$$

$$\Delta(\zeta, z) \overset{\text{def}}{=} f(\zeta) - f(z) - p(\zeta) = f(z) - P_1(f;\ z,\ z_0;\ \zeta). \tag{5.7}$$

For the purpose of estimating the modulus $I = I(z)$ we proceed as follows:

a) If $z \in \bigcup_{\zeta \in \Gamma} U(\zeta, \rho_{1+1/n}(\zeta))$, $z \in \mathfrak{M}$, then, denoting by $\zeta_0 \in \Gamma$ the point (or one of the points) of Γ which is the center of the largest disk $U(\zeta_0, \rho_{1+1/n}(\zeta_0))$ whose closure contains z:

$$|z - \zeta_0| = \sup_{\zeta \in \Gamma,\ \overline{U\left(\zeta, \rho_{1+\frac{1}{n}}(\zeta)\right)} \ni z} |z - \zeta|,$$

we see that the disk $U_0 \overset{\text{def}}{=} \overline{U(\zeta_0, 2\rho_{1+1/n}(\zeta_0))}$ will contain all those points $\zeta \in \Gamma$ for which $\overline{U(\zeta, \rho_{1+1/n}(\zeta))} \ni z$.

In this case, on denoting by γ_0 and σ the arcs

$$\gamma_0 \overset{\text{def}}{=} (\Gamma \cap U_0) \cup (\mathfrak{M} \cap \partial U_0), \qquad \Gamma = \partial\mathfrak{M}, \qquad \sigma \overset{\text{def}}{=} \mathfrak{M} \cap \partial U_0$$

and using the fact that, according to Definition 3.2, $\Gamma = \bigcup_1^{\varkappa} \Gamma_j$, we estimate $|I|$, starting from the inequality

$$|I| \leqslant \left|\int_{\gamma_0} \frac{\Delta(\zeta, z)}{\zeta - z^*} d\zeta\right| + \int_{\gamma_0} |\Delta(\zeta, z)|\,|K_{n'}(\zeta, z)|\,|d\zeta|$$

$$+ \int_{\sigma} |\Delta(\zeta, z)| \left|K_{n'}(\zeta, z) - \frac{1}{\zeta - z^*}\right| |d\zeta| + \sum_{j=1}^{\varkappa} \int_{\Gamma_j \setminus \gamma_0} |\Delta(\zeta, z)| \left|\frac{1}{\zeta - z^*} \right. \tag{5.8}$$

$$\left. - K_{n'}(\zeta, z)\right| |d\zeta| = 0 + i_0 + i_0' + \sum_{j=1}^{\varkappa} I_j,$$

where

$$I_j \overset{\text{def}}{=} \int_{\Gamma_j \setminus \gamma_0} |\Delta(\zeta, z)| \left|\frac{1}{\zeta - z^*} - K_{n'}(\zeta, z)\right| |d\zeta|. \tag{5.8$'$}$$

b) If $z \in \mathfrak{M} \setminus \bigcup_{\zeta \in \Gamma} U(\zeta, \rho_{1+1/n}(\zeta))$, we shall use the simpler inequality

$$|I| \leqslant \sum_{j=1}^{\varkappa} \int_{\Gamma_j} |\Delta(\zeta, z)| \left|\frac{1}{\zeta - z^*} - K_{n'}(\zeta, z)\right| |d\zeta| = \sum_{j=1}^{\varkappa} I_j'. \tag{5.9}$$

Besides these inequalities, we take the following facts into consideration:

i) The inequalities (2.2) and (2.2$'$) imply the estimates

$$|\Delta(\zeta, z)| = |f(\zeta) - P_1(f;\ z,\ z_0;\ \zeta)| \leqslant \begin{cases} \omega_2(r), & \text{if } |\zeta - z| \leqslant r, \\ \dfrac{|\zeta - z|^2}{r^2}\,\omega_2(r), & \text{if } |\zeta - z| \geqslant r; \end{cases} \tag{5.10}$$

ii) Since by assumption $\mathfrak{M} \in \overset{\circ}{B}{}_k^{(n)}$, in view of Definition 3.3 we have

$$\check{s}(\zeta', \zeta'') \leqslant A|\zeta' - \zeta''| \tag{5.11}$$

for any $\zeta', \zeta'' \in \Gamma$.

On each of the arcs Γ_j we denote by ζ^j the point (or one of the points) closest to z, and consider the following four subcases of location of z on $\mathfrak{M}$ for fixed $r > 0$:

1) Let $|\zeta^j - z| < \rho_{1+1/n}(\zeta^j)$ and $r < \rho_{1+1/n}(\zeta^j)$. In this case, taking into account (5.6), (5.10) and (5.11) and properties (4.5) and (4.4) of the kernel $K_n{}'(\zeta, z)$, we find that

$$i_0,\ i_0' \leqslant A\frac{\omega_2(r)}{r^2}\rho^2_{1+\frac{1}{n}}(\zeta^j) \leqslant A\Omega(r, z);$$

$$I_j = \int\limits_{\Gamma_j\setminus\gamma_0} |\Delta(\zeta, z)|\left|\frac{1}{\zeta - z^*} - K_{n'}(\zeta, z)\right||d\zeta| \qquad (5.12^1)$$

$$\leqslant A\frac{\omega_2(r)}{r^2}\int\limits_{\rho_{1+\frac{1}{n}}(\zeta^j)}^{\infty}\frac{\rho^m_{1+\frac{1}{n}}(\zeta^j)}{\sigma^{m-1}}d\sigma \leqslant A\Omega(r, z).$$

2) Let $|z - \zeta^j| < \rho_{1+1/n}(\zeta^j)$ and $\rho_{1+1/n}(\zeta^j) \le r$. In this case, reasoning as in 1), we get

$$I_j = \int\limits_{\Gamma_j\setminus\gamma_0} |\Delta(\zeta, z)|\left|\frac{1}{\zeta - z^*} - K_{n'}(\zeta, z)\right||d\zeta| \qquad (5.12^2)$$

$$\leqslant A\int\limits_{\rho_{1+\frac{1}{n}}(\zeta^j)}^{r}\omega_2(r)\frac{\rho^m_{1+\frac{1}{n}}(\zeta^j)}{\sigma^{m+1}}d\sigma + A\frac{\omega_2(r)}{r^2}\int\limits_{r}^{\infty}\frac{\rho^m_{1+\frac{1}{n}}(\zeta^j)}{\sigma^{m-1}}d\sigma \leqslant A\Omega(r, z).$$

3) If $|z - \zeta^j| \ge \rho_{1+1/n}(\zeta^j)$ and $r \le |z - \zeta^j|$, then

$$I_j' = \int\limits_{\Gamma_j} |\Delta(\zeta, z)|\left|\frac{1}{\zeta - z^*} - K_{n'}(\zeta, z)\right||d\zeta| \leqslant A\frac{\omega_2(r)}{r^2}\int\limits_{|z-\zeta^j|}^{\infty}\frac{\rho^m_{1+\frac{1}{n}}(\zeta^j)}{\sigma^{m-1}}d\sigma \qquad (5.12^3)$$

$$\leqslant A\omega_2(r)\left[\frac{\rho_{1+\frac{1}{n}}(\zeta^j)}{r}\right]^2\left[\frac{\rho_{1+\frac{1}{n}}(\zeta^j)}{|z-\zeta^j|+r}\right]^{m-2} \leqslant A\Omega(r, z).$$

4) If $|z - \zeta^j| \ge \rho_{1+1/n}(\zeta^j)$ and $|z - \zeta^j| < r$, then

$$I_j' \leqslant A\omega_2(r)\int\limits_{|z-\zeta^j|}^{r}\frac{\rho^m_{1+\frac{1}{n}}(\zeta^j)}{\sigma^{m+1}}d\sigma + \frac{A\omega_2(r)}{r^2}\int\limits_{r}^{\infty}\frac{\rho^m_{1+\frac{1}{n}}(\zeta^j)}{\sigma^{m-1}}d\sigma \qquad (5.12^4)$$

$$\leqslant A\omega_2(r)\left\{\left[\frac{\rho_{1+\frac{1}{n}}(\zeta^j)}{|z-\zeta^j|}\right]^m + \left[\frac{\rho_{1+\frac{1}{n}}(\zeta^j)}{r}\right]^m\right\} \leqslant A\Omega(r, z).$$

Substituting the estimates (5.12) in (5.8) and (5.9) and taking into consideration (5.5) and the fact that $|q|n^{-km+1} \le A\omega_2(1/n^4) \le A\Omega(r, z)$ for $km \ge 4$, we obtain Theorem 5.1.

Theorem 5.1 yields

Corollary. *Under the conditions of Theorem* 5.1,

$$|f(z)-P_n(z)|\leqslant A\max_{\zeta\in\Gamma}\left\{\omega_2\left[\rho_{1+\frac{1}{n}}(\zeta)\right]\left[\frac{\rho_{1+\frac{1}{n}}(\zeta)}{|\zeta-z|+\rho_{1+\frac{1}{n}}(\zeta)}\right]^{m-2}\right\}. \quad (5.2'')$$

Theorem 5.2 (on approximation on the boundary of a set of type $\overset{*}{B}{}_k^{(n)}$). *If $\mathfrak{M}\in\overset{*}{B}{}_k^{(n)}$, then at all boundary points $z\in\partial\mathfrak{M}$*

$$|f(z)-P_n(z)|\leqslant A\omega_2\left[\rho_{1+\frac{1}{n}}(z)\right], \quad (5.13)$$

where the polynomial $P_n(z)$ is defined by (5.3), *$\omega_2(t)$ denotes the second modulus of continuity of the function f on $\mathfrak{M}$, and A denotes a constant which depends neither on f nor on $z\in\partial\mathfrak{M}$.*

Proof. In fact, assuming for definiteness that the right side of (5.2$'$) attains a maximum at the point $\zeta_0\in\partial\mathfrak{M}$ and that $m\geq 2k+2$, we obtain, on putting $r=\rho_{1+1/n}(\zeta_0)$ in (5.2) and taking (3.7) and (3.10) into account, that

$$|f(z)-P_n(z)|\leqslant A\omega_2\left[\rho_{1+\frac{1}{n}}(\zeta_0)\right]\cdot 2^2\frac{|\tilde{\zeta}_0-\zeta_0|^{2k}}{|\tilde{\zeta}_0-z|^{2k}}$$

$$\leqslant A\omega_2\left[\rho_{1+\frac{1}{n}}(z)\right]\left[\frac{\rho_{1+\frac{1}{n}}(\zeta_0)}{\rho_{1+\frac{1}{n}}(z)}+1\right]^2\frac{|\tilde{z}-z|^2}{\left[|\zeta_0-z|+\rho_{1+\frac{1}{n}}(\zeta_0)\right]^2}\leqslant A\omega_2\left[\rho_{1+\frac{1}{n}}(z)\right],$$

as was to be proved.

We note that a particular case of this theorem is the similar theorem in the real case (very complicated to prove) for the interval $[a, b]$ ($\in\overset{*}{B}_k$). We also note that for construction of the polynomial (5.3) one uses only those values of $f(z)$ which we approximate, and not those values which would be obtained from the original function by means of a (nonunique, in general) continuation.

Definition 5.2. A point $z\in\partial\mathfrak{M}$ is said to be of *multiplicity* s, $s\geq 1$, if there exist s and only s distinct points τ_j of the form $\tau_j=e^{i\alpha_j\pi}$, $0\leq\alpha_j<2$, which are preimages of z under the mapping $\eta=\Psi(\tau)$: $\Psi(\tau_j)=z$, $j=1,\ldots,s$. For $s=1$ the point z is said to be *simple*; for $s>1$, *multiple*.

In the case when a set $\mathfrak{M}$ of type B_k does not belong to the class $\overset{\circ}{B}{}_k^{(n)}$, the estimates carried out in Theorem 5.1 for the moduli of the integrals i_0, i_0' and I_j in (5.8) become less precise, because in this case for any choice of the arc γ_0 the estimate of at least one of the quantities i_0, i_0' or I_j is worse than is required by approximately $\ln n$. In this connection the following Theorems 5.3 and 5.3$'$, in which $\mathfrak{M}$, generally speaking, does not belong to the class $\overset{\circ}{B}{}_k^{(n)}$, are of interest.

Theorem 5.3 (on approximation on sets with piecewise smooth boundary). *Let the boundary of a bounded set $\mathfrak{M}$ with simply connected complement consist of finitely many sufficiently smooth arcs (for example, having bounded curvature), which do not form zero exterior angles. Then for any function $f(z)$ which is* *analytic in* Int $\mathfrak{M}$ *and continuous on* $\mathfrak{M}$, *for any natural* n *we can construct a polynomial* $P_n(z)$ *of degree not greater than* n *such that for all* $z \in \partial\mathfrak{M}$

$$|f(z) - P_n(z)| \leqslant A \min_{r>0} \Omega(r, z), \tag{5.14}$$

where the function $\Omega(r, z)$ *is defined by* (5.2$'$) *and* A *is a constant which depends on* $\mathfrak{M}$ *but not on* $f(z)$, $z \in \mathfrak{M}$ *or* n.

An important corollary of this theorem is

Theorem 5.3$'$ (on approximation on arcs). *Let the set* Γ *be an open piecewise smooth curve without cusps, which consists of finitely many sufficiently smooth arcs (for example, having bounded curvature). Then for any function* $f(z)$ *continuous on* Γ *and any natural* n *there exists a polynomial* $P_n(z)$ *of degree not greater than* n *such that for all* $z \in \Gamma$

$$|f(z) - P_n(z)| \leqslant A\omega_2\left[f;\ \overset{*}{\rho}_{1+\frac{1}{n}}(z)\right], \tag{5.14$'$}$$

where $\overset{*}{\rho}_{1+1/n}(z)$ *denotes at every double point* $z \in \Gamma$ *the larger of the distances to the two branches of the level curve of* $\Gamma_{1+1/n}$ *and* A *denotes a constant which does not depend on* z, f *or* n.

In order to establish the validity of Theorems 5.3 and 5.3$'$ we shall explain by examples only the principle of their proof, leaving the details to the reader.

Example 1. Let a continuous function $f(z)$ be given on a set $\mathfrak{M}$ which consists of three rectilinear segments joining the origin of coordinates with the points 1, $e^{i\pi/6}$ and $e^{i\pi} = -1$. In order to approximate the function $f(z)$ by a suitable method, we shall construct an auxiliary set $\mathfrak{M}^*$ obtained by adding to $\mathfrak{M}$ two closed regions bounded by arcs $\gamma_1 = \{z\colon z = e^{it},\ t \in [\pi/3, 5\pi/6]\}$ and $\gamma_2 = \{z\colon z = e^{it},\ t \in [7\pi/6, 11\pi/6]\}$ of the unit circle and by rays joining the origin with the points $e^{i\pi/3}$, $e^{i5\pi/6}$, $e^{i7\pi/6}$ and $e^{i11\pi/6}$, and, taking for convenience $f(0) = 0$, we define an auxiliary function $f^*(z)$ on $\mathfrak{M}^*$ by the formula

$$f^*(z) = \begin{cases} f(z), & z \in \mathfrak{M}, \\ p(z), & z \in \mathfrak{M}^* \setminus \mathfrak{M}, \end{cases}$$

where

$$p(z) \overset{\text{def}}{=} \frac{f\left[\overset{*}{\rho}_{1+\frac{1}{n}}(0)\right]}{\overset{*}{\rho}_{1+\frac{1}{n}}(0)}\, z$$

is a linear function which for a given n interpolates the function $f(z)$ at the points 0 and $\overset{*}{\rho}_{1+1/n}(0)$. We construct a polynomial $P_n^*(z) = P_n(\mathfrak{M}^*; f^*; z)$ such that the difference $f^*(z) - P_n^*(z)$ can be represented by (5.6). Then for z and $\zeta \in \partial\mathfrak{M}$ all estimates are carried out as in the proof of Theorem 5.1. But if, for example, the point $z \in [2\overset{*}{\rho}_{1+1/n}(0); \tfrac{1}{2}]$ and $\zeta = re^{11i\pi/6}$, $0 < r \leq 1$, then putting

$$\begin{aligned} p_1(\zeta) &= P_1(f;\ \ z;\ \ z + \overset{*}{\rho}_{1+1/n}(z);\ \ \zeta) \\ &= \frac{z + \overset{*}{\rho}_{1+1/n}(z) - \zeta}{\overset{*}{\rho}_{1+1/n}(z)} f(z) + \frac{\zeta - z}{\overset{*}{\rho}_{1+1/n}(z)} f[z + \overset{*}{\rho}_{1+1/n}(z)], \end{aligned}$$

we get

$$\Delta(\zeta, z)=\left\{f(0)-\left[\frac{z+\overset{*}{\rho}_{1+\frac{1}{n}}(z)}{\overset{*}{\rho}_{1+\frac{1}{n}}(z)}f(z)+\frac{-z}{\overset{*}{\rho}_{1+\frac{1}{n}}(z)}f\left[z+\overset{*}{\rho}_{1+\frac{1}{n}}(z)\right]\right]\right\}$$
$$+\zeta\left\{\frac{f\left[\overset{*}{\rho}_{1+\frac{1}{n}}(0)\right]}{\overset{*}{\rho}_{1+\frac{1}{n}}(0)}+\frac{f(z)}{\overset{*}{\rho}_{1+\frac{1}{n}}(z)}-\frac{f\left[z+\overset{*}{\rho}_{1+\frac{1}{n}}(z)\right]}{\overset{*}{\rho}_{1+\frac{1}{n}}(z)}\right\}.$$

Hence, taking (2.2), (2.3) and (2.2′) into account, we find

$$|\Delta(\zeta, z)|\leqslant\left|f(0)-P_1\left(f;\ z;\ z+\overset{*}{\rho}_{1+\frac{1}{n}}(z);\ 0\right)\right|+|\zeta|\left|\left[\frac{f\left(\overset{*}{\rho}_{1+\frac{1}{n}}(0)\right)}{\overset{*}{\rho}_{1+\frac{1}{n}}(0)}-\frac{f(z)}{z}\right]\right.$$
$$\left.-\frac{1}{\overset{*}{\rho}_{1+\frac{1}{n}}(z)}\left[f\left(z+\overset{*}{\rho}_{1+\frac{1}{n}}(z)\right)-\left(\frac{z+\overset{*}{\rho}_{1+\frac{1}{n}}(z)}{z}f(z)+\frac{\overset{*}{\rho}_{1+\frac{1}{n}}(z)}{z}f(0)\right)\right]\right|$$
$$\leqslant\left|f(0)-P_1\left(f;\ z;\ z+\overset{*}{\rho}_{1+\frac{1}{n}}(z);\ 0\right)\right|+\frac{|\zeta|}{\overset{*}{\rho}_{1+\frac{1}{n}}(0)}\left|f\left[\overset{*}{\rho}_{1+\frac{1}{n}}(0)\right]\right.$$
$$\left.-P_1\left(f;\ z;\ 0;\ \overset{*}{\rho}_{1+\frac{1}{n}}(0)\right)\right|+\frac{|\zeta|}{\overset{*}{\rho}_{1+\frac{1}{n}}(z)}\left|f\left[z+\overset{*}{\rho}_{1+\frac{1}{n}}(z)\right]\right.$$
$$\left.-P_1\left(f;\ z;\ 0;\ z+\overset{*}{\rho}_{1+\frac{1}{n}}(z)\right)\right|\leqslant A\left(\frac{|z|}{\overset{*}{\rho}_{1+\frac{1}{n}}(z)}\right)^2\omega_2\left[f;\ \overset{*}{\rho}_{1+\frac{1}{n}}(z)\right]$$
$$+A\frac{|\zeta|}{\overset{*}{\rho}_{1+\frac{1}{n}}(0)}\overset{*}{\rho}_{1+\frac{1}{n}}(0)\int\limits_{\overset{*}{\rho}_{1+\frac{1}{n}}(0)}^{2|z|}\frac{\omega_2(f;\ u)}{u^2}\,du$$
$$+A\frac{|\zeta|}{\overset{*}{\rho}_{1+\frac{1}{n}}(z)}\overset{*}{\rho}_{1+\frac{1}{n}}(z)\int\limits_{\overset{*}{\rho}_{1+\frac{1}{n}}(z)}^{2|z|}\frac{\omega_2(f,\ u)}{u^2}\,du\leqslant A\left(\frac{|\zeta|+|z|}{\overset{*}{\rho}_{1+\frac{1}{n}}(z)}\right)^2\omega_2\left[f;\ \overset{*}{\rho}_{1+\frac{1}{n}}(z)\right].$$

Using this inequality, we easily establish that for all $z\in\partial\mathfrak{M}$

$$|f(z)-P_n^*(z)|\leqslant A\min_{r>0}\Omega(r,\ z).$$

Example 2. Let a continuous function $f(z)$ be given on a set $\mathfrak{M}$ which consists of two rectilinear segments joining the origin of coordinates to the points $e^{i\pi/6}$ and $e^{i\pi}=-1$. In this case we construct three auxiliary sets $\mathfrak{M}_1^*$, $\mathfrak{M}_2^*$ and $\mathfrak{M}_3^*$, the first of which is obtained by adding to the interval $[-1, 0]$ the closed region bounded by line segments joining the origin with the points $e^{i\pi/6}$ and $e^{11i\pi/6}$ and the arc of the circle with center at the point $(\cos\pi/6)^{-1}=2/\sqrt{3}$ which, together with the indicated segments, forms a region with exterior angles at the points $e^{i\pi/6}$ and $e^{-i\pi/6}$ equal to 2π. The set $\mathfrak{M}_2^*$ is obtained in an analogous manner, starting from the segment $re^{i\pi/6}$, $0\leq r\leq 1$, and line segments which join the origin to the points -1 and $e^{4i\pi/3}$. The boundary of the set $\mathfrak{M}_3^*$ consists of the two above-mentioned arcs, the line segments joining the origin with the points $e^{i\pi/6}$, $e^{i\pi}$, $e^{i4\pi/3}$ and $e^{i11\pi/6}$, and the segment joining the points $e^{i4\pi/3}$ and $e^{i11\pi/6}$. The function $f^*(z)$ is constructed on the set $\mathfrak{M}_3^* \setminus \mathfrak{M}_1^* \cup \mathfrak{M}_2^*$ in the same way as in Example 1, and on the two arcs of the set $\partial\mathfrak{M}_3^*$ in such a way that in some fixed neighborhoods of the points -1 and $e^{i\pi/6}$ the condition $\omega_2(f^*; t) \le A\omega_2(f; t)$, $t > 0$, is satisfied, and then in such a way that on the remaining parts of the arcs the function $f^*(z)$ is linear, and on $\partial\mathfrak{M}_3^*$ it is continuous. After this we put

$$P_n^*(z) \overset{\text{def}}{=} P_n(\partial\mathfrak{M}_1^*;\ f^*;\ z) + P_n(\partial\mathfrak{M}_2^*;\ f^*;\ z) - P_n(\partial\mathfrak{M}_3^*;\ f^*;\ z),$$

where $P_n(\partial\mathfrak{M}_i^*; f^*; z)$ denotes polynomials which are constructed on corresponding curves by (5.3).

Remarks. a) As the sets $\mathfrak{M}_i^*$ we can also consider sets in which, instead of the arcs, we take rectilinear segments joining the corresponding points and then, starting from the polynomials $P_n^*(z)$ and $P_n(\partial\mathfrak{M}; f; z)$ so obtained, we find the required polynomial using the method of partition of unity into a sum of polynomials of fixed degree which possess the required roots at the vertices of the polygon $\partial\mathfrak{M}_3^*$, as in the author's paper [12].

b) The reasoning carried out in these examples bears a sufficiently general character and can be easily adapted for very many sets of type B_k.

Theorem 5.4 (on simultaneous approximation of a function and its derivatives). *If on a set* $\mathfrak{M}$ *of type* $\overset{*}{B}_k$ *a function which is continuous on* $\mathfrak{M}$ *together with all derivatives up to order* r *(*r *a positive integer) and analytic in* Int $\mathfrak{M}$ *is given, then for every* $n \ge r$ *we can construct a polynomial* $P_n(z) = P_n(f; \mathfrak{M}; r; z)$ *of degree not greater than* n *such that simultaneously for all* $j = 0, 1, \ldots, r$, *the inequality*

$$|f^{(j)}(z) - P_n^{(j)}(z)| \leqslant A_j\left[\rho_{1+\frac{1}{n}}(z)\right]^{r-j}\omega_2\left[f^{(r)};\ \rho_{1+\frac{1}{n}}(z)\right], \tag{5.15}$$

is satisfied at all points $z \in \partial\mathfrak{M}$, *where the* A_j *are constants which depend only on* $\mathfrak{M}$, r *and* j; $\rho_{1+1/n}(z)$ *denotes the distance from the point* $z \in \Gamma$ *to the level curve* $\Gamma_{1+1/n}$ *and* $\omega_2(f^{(r)}; t)$ *denotes the second modulus of continuity of* $f^{(r)}(z)$ *on* $\mathfrak{M}$.

Proof. For $r = 0$, Theorem 5.4 follows from Theorem 5.2. We assume in an induction that for some integer $r \ge 1$ the inequality (5.15) (or, equivalently, Theorem 5.4) is true for the case when $f(z) \in C^{(r-1)}$, i.e. when $f(z)$ is continuous on $\mathfrak{M}$ together with its derivatives up to order $r-1$, and we prove that in such a case it will also hold when $f(z) \in C^{(r)}$.

Since $f'(z) \in C^{(r-1)}$ and $(f')^{(r-1)} = f^{(r)}(z)$, according to the assumption on $f'(z)$ we can construct a polynomial $\hat{P}_n(z)$ of degree not greater than n such that for all $j = 0, 1, \ldots, r-1$

$$|f^{(j+1)}(z)-\hat{P}_n^{(j)}(z)|\leqslant A\left[\rho_{1+\frac{1}{n}}(z)\right]^{r-1-j}\omega_2\left[f^{(r)};\ \rho_{1+\frac{1}{n}}(z)\right],$$

$$j=0,\ 1,\ 2,\dots,r-1;\ r\geqslant 1. \tag{5.15'}$$

Now we carry out the proof of the theorem in two steps.

a) First we prove that there exists a polynomial $P_n(z)$ for which

$$|f(z)-P_n(z)|\leqslant A\rho^r_{1+\frac{1}{n}}(z)\,\omega_2\left(f^{(r)};\ \rho_{1+\frac{1}{n}}(z)\right), \tag{5.16}$$

i.e. we prove the theorem for the class $C^{(r)}$ in the particular case when $j=0$.

For this purpose we fix a point $z_0\in\Gamma$ and, putting

$$U(z)\overset{\text{def}}{=}\int_{z_0}^{z}[f'(\zeta)-\hat{P}_n(\zeta)]\,d\zeta=f(z)-\left[f(z_0)+\int_{z_0}^{z}\hat{P}_n(\zeta)\,d\zeta\right], \tag{5.17}$$

we approximate $f(z)$ by a polynomial $P_n(z)$ such that $f(z)-P_n(z)$ can be represented in the form

$$f(z)-P_n(z)=\frac{1}{2\pi i\gamma_n'}\int_\Gamma[U(\zeta)-U(z)]\left[\frac{1}{\zeta-z^*}-K_n(\zeta,\ z)\right]d\zeta, \tag{5.18}$$

where z^* is defined as in (5.5). Hence, noting that in view of (5.17) and (5.15′)

$$\begin{aligned}|U(\zeta)-U(z)|&=\left|\int_z^\zeta[f'(\zeta)-\hat{P}_n(\zeta)]\,d\zeta\right|\\&\leqslant A|\zeta-z|\max_{\xi\in\widetilde{z\zeta}}\left\{\rho^{r-1}_{1+\frac{1}{n}}(\xi)\,\omega_2\left[f^{(r)},\ \rho_{1+\frac{1}{n}}(\xi)\right]\right\}\\&\leqslant A|\zeta-z|\left[\rho_{1+\frac{1}{n}}(z)+|\zeta-z|\right]^{r-1}\omega_2\left[f^{(r)};\ \rho_{1+\frac{1}{n}}(z)+|\zeta-z|\right]\\&\leqslant A\left[|\zeta-z|+\rho_{1+\frac{1}{n}}(z)\right]^{r+1}|\zeta-z|\frac{\omega_2\left[f^{(r)};\ \rho_{1+\frac{1}{n}}(z)\right]}{\rho^2_{1+\frac{1}{n}}(z)},\end{aligned} \tag{5.19}$$

we obtain

$$\begin{aligned}&|f(z)-P_n(z)|\\&\leqslant A_1\frac{\omega_2\left[f^{(r)};\ \rho_{1+\frac{1}{n}}(z)\right]}{\rho^2_{1+\frac{1}{n}}(z)}\int_\Gamma\left[|\zeta-z|+\rho_{1+\frac{1}{n}}(z)\right]^{r+1}|\zeta-z|\frac{\rho^m_{1+\frac{1}{n}}(z)}{|\zeta-z|\left(|\zeta-z|+\rho_{1+\frac{1}{n}}(z)\right)^m}|d\zeta|\\&\leqslant A\rho^r_{1+\frac{1}{n}}(z)\,\omega_2\left[f^{(r)};\ \rho_{1+\frac{1}{n}}(z)\right].\end{aligned}$$

Thus, (5.16), and hence Theorem 5.4, is proved for the case $j=0$.

b) We shall prove that, for the polynomial $P_n(z)$ appearing in (5.18), inequality (5.15), and so Theorem 5.4, will also be valid for $j=1,\dots,r$.

In fact, differentiating (5.18), and taking into account the method by which the point z^* has been determined, we get

$$\begin{aligned}|f'(z)-P_n'(z)|&=-\frac{U'(z)}{2\pi i\gamma_{n'}}\int_\Gamma\left[\frac{1}{\zeta-z^*}-K_n(\zeta,\ z)\right]d\zeta\\&+\frac{1}{2\pi i\gamma_{n'}}\int_\Gamma[U(\zeta)-U(z)]\left[\frac{1}{(\zeta-z^*)^2}-\frac{\partial}{\partial z}K_n(\zeta,\ z)\right]d\zeta=I_1+I_2.\end{aligned} \tag{5.20}$$

Since on the basis of (5.17), (5.15 ') and (4.6)

$$|I_1| \preccurlyeq |f'(z) - \hat{P}_n(z)|(1-\alpha) \leqslant A \frac{\rho^{r-1}_{1+\frac{1}{n}}(z)\,\omega_2\left[f^{(r)};\ \rho_{1+\frac{1}{n}}(z)\right]}{n^{km}} \preccurlyeq \rho^{r-1}_{1+\frac{1}{n}}(z)\,\omega_2\left[f^{(r)};\ \rho_{1+\frac{1}{n}}(z)\right],$$

and on the basis of (5.19) and (4.4)

$$|I_2| \preccurlyeq A \int\limits_{\Gamma\setminus\gamma_0} \left[|\zeta - z| + \rho_{1+\frac{1}{n}}(z)\right]^{r+1} |\zeta - z| \frac{\omega_2\left[f^{(r)};\ \rho_{1+\frac{1}{n}}(z)\right]}{\rho^2_{1+\frac{1}{n}}(z)} \frac{\rho^m_{1+\frac{1}{n}}(z)}{\left(|\zeta - z| + \rho_{1+\frac{1}{n}}(z)\right)^{m+2}} |d\zeta|$$

$$+ A \left| \int\limits_{\gamma+\sigma} - \int\limits_{\sigma} [U(\zeta) - U(z)] \left[\frac{1}{(\zeta - z^*)^2} - \frac{\partial}{\partial z} K_n(\zeta, z)\right] d\zeta \right|$$

$$\preccurlyeq \omega_2\left[\rho_{1+\frac{1}{n}}(z)\right] \rho^{m-2}_{1+\frac{1}{n}}(z) \int\limits_{\rho_{1+\frac{1}{n}}(z)}^{\infty} \frac{ds}{s^{m-r}} + |U'(z)|$$

$$+ \int\limits_{\sigma} \rho^{r+2}_{1+\frac{1}{n}}(z) \frac{\omega_2\left[f^{(r)};\ \rho_{1+\frac{1}{n}}(z)\right]}{\rho^2_{1+\frac{1}{n}}(z)} |d\zeta| \preccurlyeq \omega_2\left[f^{(r)};\ \rho_{1+\frac{1}{n}}(z)\right] \rho^{r-1}_{1+\frac{1}{n}}(z),$$

where $\sigma \overset{\text{def}}{=} \mathfrak{M} \cap \partial U_0$ and $U_0 y_0$ are defined as in (5.8), the inequality (5.20) yields the estimate

$$|f'(z) - P'_n(z)| \leqslant A\rho^{r-1}_{1+\frac{1}{n}}(z)\,\omega_2\left[f^{(r)};\ \rho_{1+\frac{1}{n}}(z)\right]. \tag{5.21}$$

From this inequality and (5.15 ') for the case $j = 0$, putting

$$\pi_n(z) \overset{\text{def}}{=} \hat{P}_n(z) - P'_n(z) = [f'(z) - P'_n(z)] - [f(z) - \hat{P}_n(z)],$$

we obtain for the polynomial $\pi_n(z)$ the estimate

$$|\pi_n(z)| \leqslant A\rho^{r-1}_{1+\frac{1}{n}}(z)\,\omega_2\left[f^{(r)};\ \rho_{1+\frac{1}{n}}(z)\right] \tag{5.22}$$

for all $z \in \Gamma$. Hence, on the basis of (5.21), (5.15) and (1.5 ') it follows that for any $j - 1, 2, \ldots, r-1$

$$|f^{(j+1)}(z) - P_n^{(j+1)}(z)| \leqslant |f^{(j+1)}(z) - \hat{P}_n^{(j)}(z)| + |\hat{P}_n^{(j)}(z) - P_n^{(j+1)}(z)|$$

$$\leqslant A\rho^{r-j-1}_{1+\frac{1}{n}}(z)\,\omega_2\left[f^{(r)};\ \rho_{1+\frac{1}{n}}(z)\right] + |\pi_n^{(j)}(z)| \leqslant A_1\rho^{r-j-1}_{1+\frac{1}{n}}(z)\,\omega_2\left[f^{(r)};\ \rho_{1+\frac{1}{n}}(z)\right].$$

These inequalities coupled with (5.15 ') and (5.21) establish the validity of the theorem for the class $C^{(r)}$. Thus Theorem 5.4 is proved.

We note that particular cases of Theorem 5.4 when $\mathfrak{M}$ is a region and $j = 0$ were obtained earlier in [40] under the condition that $\partial\mathfrak{M} \in C^{(2)}$, and also in [32] (by using the kernel (4.1)) under the conditions that $\mathfrak{M} \in B_k$ and $t\int_t^1 s^{-2}\omega(s)ds \leq A\omega(t)$.

Theorem 5.5 (on approximation of integrals of Cauchy type). *If the region G belongs to the class $\overset{*}{B}_k$, then for an integral of Cauchy type determined by a function $f(\zeta)$ continuous on $\Gamma = \partial G$ there exists an algebraic polynomial $P_n(z)$ of degree not greater than n, for every natural n, such that for any point $z_0 \in \Gamma$ and all $z \in \gamma \overset{\text{def}}{=} G \cap U[z_0; \frac{1}{2}\rho_{1+1/n}(z_0)]$*

$$\left|\frac{1}{2\pi i}\int_\Gamma \frac{f(\zeta)}{\zeta - z}\,d\zeta - P_n(z)\right| \leqslant \left|\frac{1}{2\pi i}\int_{\partial\gamma} \frac{f(\zeta)-f(z_0)}{\zeta - z}\,d\zeta\right| + A\omega\left[f;\ \rho_{1+\frac{1}{n}}(z_0)\right], \quad (5.23)$$

$$\left|\frac{1}{2\pi i}\int_\Gamma \frac{f(\zeta)}{\zeta - z}\,d\zeta - P_n(z)\right| \leqslant \left|\frac{1}{2\pi i}\int_{\partial\gamma} \left[\frac{f(\zeta)-f(z)}{\zeta - z} - q\right]d\zeta\right| + A\omega_2\left[f;\ \rho_{1+\frac{1}{n}}(z_0)\right], \quad (5.23')$$

where q is defined by (5.4); $\omega(f;\,t)$ and $\omega_2(f;\,t)$ denote the first and second moduli of continuity of the function $f(\zeta)$ on Γ, $U[z_0; \frac{1}{2}\rho_{1+1/n}(z_0)]$ denotes the disk with center at z_0 and radius $\frac{1}{2}\rho_{1+1/n}(z_0)$, and A denotes constants which depend neither on z_0 and z, nor on n.

Proof. Defining the polynomial $P_n(z)$ by means of (5.3) and using (4.6) and (4.7), we obtain

$$\frac{1}{2\pi i}\int_\Gamma \frac{f(\zeta)}{\zeta - z}\,d\zeta - P_n(z) = \frac{1}{2\pi i}\int_\Gamma [f(\zeta) - f(z)]\left[\frac{1}{\zeta - z} - K_{n'}(\zeta,\ z)\right]d\zeta + O\left(\frac{1}{n^{km}}\right); \quad (5.24)$$

$$\frac{1}{2\pi i}\int_\Gamma \frac{f(\zeta)}{\zeta - z}\,dz - P_n(z) = \frac{1}{2\pi i}\int_\Gamma [f(\zeta) - f(z) - q(\zeta - z)]\left[\frac{1}{\zeta - z} - K_{n'}(\zeta,\ z)\right]d\zeta + O(|q|\,n^{-km+2}). \quad (5.24')$$

Starting, for example, from the second of the above relations, and taking into account the second of the inequalities (5.10) for $r = \rho_{1+1/n}(z_0)$ and the inequalities (4.4) and (4.5), we find that

$$\left|\frac{1}{2\pi i}\int_\Gamma \frac{f(\zeta)}{\zeta - z}\,d\zeta - P_n(z)\right| \leqslant \left|\frac{1}{2\pi i}\int_{\partial\gamma} \left[\frac{f(\zeta)-f(z)}{\zeta - z} - q\right]d\zeta\right|$$
$$+ A\omega_2\left[f;\ \rho_{1+\frac{1}{n}}(z_0)\right]\int_{\partial\gamma\setminus\Gamma}\left(|K_{n'}(\zeta,\ z)| + \frac{1}{|\zeta - z|}\right)|d\zeta|$$
$$+ A\,\frac{\omega_2\left[f;\ \rho_{1+\frac{1}{n}}(z_0)\right]}{\rho^2_{1+\frac{1}{n}}(z_0)}\int_{\Gamma\setminus\partial\gamma} |\zeta - z|^2\,\frac{\rho^m_{1+\frac{1}{n}}(z_0)}{|\zeta - z|^{m+1}}\,|d\zeta| + O\left(\frac{1}{n^{km-2}}\right)$$
$$\leqslant \left|\frac{1}{2\pi i}\int_{\partial\gamma}\left[\frac{f(\zeta)-f(z)}{\zeta - z} - q\right]d\zeta\right| + A\omega_2\left[f;\ \rho_{1+\frac{1}{n}}(z_0)\right].$$

Thus (5.23′) is proved; (5.23) is proved analogously though more easily.

The reasoning carried out above shows that for approximation of integrals of Cauchy type it is not difficult to obtain analogues of Theorems 5.1 and 5.4 also.

To conclude this section we make the following remarks.

1. The method by which our direct theorems have been established permit us in some cases to improve upon the estimates for the moduli of derivatives of poly-

nomials which we gave in the introduction. For example, we have

Theorem 5.6. *On a set $\mathfrak{M}$ of type $\overset{*}{B}_k$ let a function $f(z)$ which is continuous on $\mathfrak{M}$ and analytic at all interior points of $\mathfrak{M}$ be given. Then for the moduli of the derivatives of the polynomials $P_n(z)$ of the form*

$$P_n(z) = \frac{1}{2\pi i \gamma_{n'}} \int_\Gamma f(\zeta) K_{n'}(\zeta, z)\, d\zeta \tag{5.3}$$

(which, as we have seen, realizes a good approximation of the function $f(z)$ for all $z \in \Gamma$) we have

$$|P_n'(z_0)| \leqslant A \frac{\omega\left[f;\ \rho_{1+\frac{1}{n}}(z_0)\right]}{\rho_{1+\frac{1}{n}}(z_0)}, \qquad |P_n^{(j)}(z_0)| \leqslant A \frac{\omega_2\left[f;\ \rho_{1+\frac{1}{n}}(z_0)\right]}{\rho^j_{1+\frac{1}{n}}(z_0)}, \quad j \geqslant 2, \tag{5.25}$$

for all $z_0 \in \partial\mathfrak{M}$.

Proof. In fact, since by (4.7 ′) and (4.7)

$$P_n'(z) = \frac{1}{2\pi i \gamma_{n'}} \int_\Gamma f(\zeta) \frac{\partial}{\partial z}\left[K_{n'}(\zeta, z) - \frac{1}{\zeta - z}\right] d\zeta$$

$$= \frac{1}{2\pi i \gamma_{n'}} \int_\Gamma [f(\zeta) - f(z_0)] \frac{\partial}{\partial z}\left[K_{n'}(\zeta, z) - \frac{1}{1 - z}\right] d\zeta$$

and for $j \geq 2$

$$P_n^{(j)}(z) = \frac{1}{2\pi i \gamma_{n'}} \int_\Gamma f(\zeta) \frac{\partial^j}{\partial z^j}\left[K_{n'}(\zeta, z) - \frac{1}{\zeta - z}\right] d\zeta$$

$$= \frac{1}{2\pi i \gamma_{n'}} \int_\Gamma [f(\zeta) - f(z_0) - p(\zeta)] \frac{\partial^j}{\partial z^j}\left[K_{n'}(\zeta, z) - \frac{1}{\zeta - z}\right] d\zeta$$

at all points z which lie outside $\mathfrak{M}$ but sufficiently close to z_0, where $p(\zeta)$ is defined by (5.4), we obtain for $j \geq 2$, taking (5.5) into account and using the same notation as in the proof of (5.12),

$$|P_n^{(j)}(z)| \leqslant \frac{1}{2\pi \gamma_{n'}} \int_{\gamma_0} \omega_2\left[\rho_{1+\frac{1}{n}}(z_0)\right] \left|\frac{\partial^j K_{n'}(\zeta, z)}{\partial z^j}\right| |d\zeta|$$

$$+ \frac{j!}{2\pi\gamma_{n'}} \int_\sigma \omega_2\left[\rho_{1+\frac{1}{n}}(z_0)\right] \frac{d\zeta}{|\zeta - z|^{j+1}} + \int_{\Gamma \setminus \gamma_0} \frac{\omega_2\left[\rho_{1+\frac{1}{n}}(z_0)\right]}{\rho^2_{1+\frac{1}{n}}(z_0)} |\zeta - z|^2 \frac{\rho^m_{1+\frac{1}{n}}(z_0)}{|\zeta - z|^{m+1+j}} |d\zeta|$$

$$\leqslant A \frac{\omega_2\left[f;\ \rho_{1+\frac{1}{n}}(z_0)\right]}{\rho^j_{1+\frac{1}{n}}(z_0)} + A\omega_2\left[f;\ \rho_{1+\frac{1}{n}}(z_0)\right] \int_{\gamma_0} \left|\frac{\partial^j K_{n'}(\zeta, z)}{\partial z^j}\right| |d\zeta|. \tag{5.26}$$

Finally, since for each fixed $\zeta \in \gamma$

$$\left|\frac{\partial^j K_{n'}(\zeta, z)}{\partial z^j}\right| = \left|\frac{j!}{2\pi i} \int_{|\xi - z| = \frac{1}{4}\rho_{1+\frac{1}{n}}(z_0)} \frac{K_{n'}(\zeta, \xi)}{(\xi - z)^{j+1}} d\xi\right| \leqslant \frac{A}{\rho^j_{1+\frac{1}{n}}(z_0)},$$

by (4.5), we find, in view of (5.26), that

$$|P_n^{(j)}(z)| \leqslant A \frac{\omega_2\left[f;\ \rho_{1+\frac{1}{n}}(z_0)\right]}{\rho^j_{1+\frac{1}{n}}(z_0)}.$$

Thus the inequalities (5.25) are proved for $j \geq 2$. The first of these inequalities is proved analogously.

2. In a recent paper [29] V. I. Belyĭ and V. M. Mikljukov have considered problems of approximating functions which are defined on a set $\mathfrak{M}$ not in the class B_k but in a somewhat modified class $\widetilde{B}_k$, which differs from B_k in that the condition (3.10) is replaced by

$$\alpha(z, u) - \alpha(z, u^*) \leqslant \frac{A}{\ln \frac{1}{n}}, \qquad x \in \left[\frac{c}{3}, 1\right], \qquad u > 0,\ A = \text{const}, \tag{5.27}$$

where the function $\alpha(z, t) = \alpha(z, t, \Delta)$ is defined in the same way as in (3.21) and the constant c has the same sense as in the inequality (C) of §3. For regions of type $\widetilde{B}_k$ they obtained a direct theorem similar to Theorem 5.2, with the difference that on the right side of (5.13) the first modulus of continuity appears instead of the second.

If the methods applied in this section and the discussions of [29] are used, it is easy to obtain the estimates (5.13) also for the case of sets of type $\widetilde{B}_k$. However, the question as to what extent the condition (5.27) and condition c) of §3 are independent, and whether they are different in general, is still open. Belyĭ and Mikljukov have established in [29] that, for example, any convex region belongs to the class $\widetilde{B}_k$.

Finally, we note that, by using the polynomials $K_{D_n}(\zeta, z)$ introduced by the author at the end of §4, A. S. Prypik has recently obtained a very simple proof of Bernstein's theorem on a constructive characterization of functions which are analytic on bounded closed sets with simply-connected complements.

§6. Converse theorems and the constructive characterization of functions

Theorem 6.1. *If for some integer $r > 0$ there exists for a function $f(z)$ defined on the boundary Γ of an admissible set $\mathfrak{M}$ a sequence of algebraic polynomials $P_n(z)$ which approximate $f(z)$ on Γ for every natural n in such a way that the inequality*

$$|f(z) - P_n(z)| \leqslant \rho^r_{1+\frac{1}{n}}(z)\, \omega\left[\rho_{1+\frac{1}{n}}(z)\right], \qquad z \in \Gamma, \tag{6.1}$$

holds, where $\omega(t)$ is a function of the type of a second modulus of continuity, which in the case when $r \geq 1$ possesses the property that

$$\int_0^1 \frac{\omega(u)}{u}\, du < \infty, \tag{6.1$'$}$$

and $\rho_{1+1/n}(z)$ denotes the distance from the point $z \in \Gamma$ to the level curve $\Gamma_{1+1/n}$, then $f(z)$ has a continuous rth derivative on $\mathfrak{M}$ for which the inequality

$$|f^{(r)}(z)-p_1(f^{(r)};\ z)|\leqslant A\left\{th\int\limits_h^{\frac{1}{2}d}\frac{\omega(u)}{u^3}\,du+t\int\limits_t^{2h}\frac{\omega(u)}{u^2}\,du\right\}$$
$$+A\delta_r\left[\int\limits_0^t\frac{\omega(u)}{u}\,du+\frac{t}{h}\int\limits_0^h\frac{\omega(u)}{u}\,du\right], \tag{6.2}$$

holds for any $z_1,\ z_2\in\Gamma$ *and* $z\in\Gamma\cap\{U(z_1;\ |z_2-z_1|)\cup U(z_2;\ |z_2-z_1|)\}$, *where*

$$p_1(\varphi;\ z)=p_1(\varphi_1;\ z_1;\ z_2;\ z)\overset{\text{def}}{=}\frac{z-z_1}{z_2-z_1}\varphi(z_2)+\frac{z_2-z}{z_2-z_1}\varphi(z_1),$$
$$h\overset{\text{def}}{=}|z_2-z_1|,\qquad t\overset{\text{def}}{=}\min\{|z-z_1|,\ |z-z_2|\}; \tag{6.3}$$

A is an absolute constant and $\delta_r=1$ *for* $r\geq 1$; $\delta_0\overset{\text{def}}{=}0$; $d\overset{\text{def}}{=}\operatorname{diam}\mathfrak{M}$.

Remark. Taking Definition 2.1 into consideration, we see that (6.2) is equivalent to the fact that $f(z)\in W^rH_2^{\widetilde{\omega}(t)}$, where $\widetilde{\omega}_2(t)$ denotes any function of the type of a second modulus of continuity for which the inequality

$$\Omega(t)\leqslant At\int\limits_t^{2h}\frac{\widetilde{\omega}_2(u)}{u^2}\,du,\quad A=\text{const},$$

holds for any $h>0$ and $t\subset[0,\ h]$, where $\Omega(t)=\Omega(t;\ h;\ \widetilde{\omega};\ r)$ denotes the right side of (6.2).

Proof. I. First we consider the case $r=0$. Let z_1 and z_2 be arbitrary points of Γ and $z\in\Gamma\cap\{U(z_1;\ |z_2-z_1|)\cup U(z_2;\ |z_2-z_1|)\}$. For definiteness we shall take $z\in\Gamma\cap U(z_1;\ |z_2-z_1|)$.

No matter what the monotone increasing sequence of natural numbers $\{n_\nu\}_1^\infty$ is, the difference $f(z)-p_1(f;\ z)$ can be represented in the form

$$\begin{aligned}f(z)-p_1(f;\ z)&=P_{n_1}(z)-p_1(P_{n_1};\ z)+\sum_{\nu=1}^{N_1-1}[U_\nu(z)-p_1(U_\nu;\ z)]\\&+f(z)-P_{n_{N_1}}(z)-p_1(f-P_{n_{N_1}};\ z)=P_{n_1}(z)-p_1(P_{n_1};\ z)\\&+\sum_{\nu=1}^{N_1-1}[U_\nu(z)-p_1(U_\nu;\ z)]+f(z)-P_{n_{N_2}}(z)\\&+\sum_{\nu=N_1}^{N_2-1}[U_\nu(z)-U_\nu(z_1)]+P_{n_{N_2}}(z_1)-f(z_1)\\&+f(z_1)-P_{n_{N_1}}(z_1)-p_1(f-P_{n_{N_1}};\ z),\end{aligned} \tag{6.4}$$

where

$$U_\nu(z)\overset{\text{def}}{=}P_{n_{\nu+1}}(z)-P_{n_\nu}(z); \tag{6.5}$$

the numbers N_1 and N_2 are chosen in such a way that

$$\rho_{n_{N_1}}(z_1)\leqslant h<\rho_{n_{N_1-1}}(z_1),\qquad \rho_{n_{N_2}}(z_1)\leqslant t<\rho_{n_{N_2-1}}(z_1). \tag{6.6}$$

Since

$$\varphi(z)=\varphi(a)+\varphi'(a)(z-a)+\int\limits_a^z(z-\zeta)\varphi''(\zeta)\,d\zeta,$$

for any function $\phi(z)$ possessing a second derivative $\phi''(z)$ which is integrable on some interval $[a,\ z]$, we can represent each of the polynomials $U_\nu(z)-p_1(U_\nu;\ z_1;\ z_2;\ z)$ with zeros at the points z_1 and z_2 in the form

$$U_\nu(z)-p_1(U_\nu;\ z)=[U'_\nu(z_1)-p'_1(U_\nu;\ z_1)](z-z_1)+\int_{z_1}^{z}(z-\zeta)U''_\nu(\zeta)\,d\zeta$$

$$=-\frac{z-z_1}{z_2-z_1}\int_{z_1}^{z_2}(z_2-\zeta)U''_\nu(\zeta)\,d\zeta+\int_{z_1}^{z}(z-\zeta)U''_\nu(\zeta)\,d\zeta.$$

Substituting the expressions obtained for $U_\nu(z)-p_1(U_\nu;\ z)$ in the right side of (6.4), we get

$$f(z)-p_1(f;\ z)=-\frac{z-z_1}{z_2-z_1}\int_{z_1}^{z_2}(z_2-\zeta)\left\{P''_{n_1}(\zeta)+\sum_{\nu=1}^{N_1-1}U''_\nu(\zeta)\right\}d\zeta$$

$$+\int_{z_1}^{z}(z-\zeta)\left\{P''_{n_1}(\zeta)+\sum_{\nu=1}^{N_1-1}U''_\nu(\zeta)\right\}d\zeta+[f(z)-P_{N_2}(z)]$$

$$+\sum_{\nu=N_1}^{N_2-1}\int_{z_1}^{z}U'_\nu(\zeta)\,d\zeta-[f(z_1)-P_{N_2}(z_1)]-\frac{z-z_1}{z_2-z_1}\{[f(z_2)-P_{N_1}(z_2)]$$

$$-[f(z_1)-P_{N_1}(z_1)]\}.$$

Assuming that the integrals here can be taken along rectilinear segments joining the corresponding points, putting

$$|z_2-z_1|=h,\qquad |z-z_1|=t,\qquad \rho_{1+\frac{1}{n_\nu}}(z)\overset{\text{def}}{=}\rho_\nu(z)$$

and taking into account the fact that in view of (1.6)

$$\rho_\nu(z)\leqslant|z-z_1|+\rho_\nu(z_1)\leqslant 2\rho_\nu(z_1),\qquad \nu=1,\ 2,\ \ldots,\ N_2-1, \tag{6.7}$$

for any $z\in\Gamma\cap\overline{U(z_1,\ h)}$, we find that

$$|f(z)-p_1(f;\ z)|\leqslant 2t\int_{z_1}^{z_2}\left\{\|P''_{n_1}\|+\sum_{\nu=1}^{N_1-1}|U''_\nu(\zeta)|\right\}|d\zeta|$$
$$+4\omega[\rho_{N_2}(z_1)]+\sum_{\nu=N_1}^{N_2-1}\int_{z_1}^{z}|U'_\nu(\zeta)||d\zeta|+\omega[\rho_{N_2}(z_1)]+\frac{8t}{h}\omega[\rho_{N_1}(z_1)]. \tag{6.8}$$

Since for every natural $\nu\leq N_2-1$ and all $\zeta\in\Gamma$

$$|U_\nu(\zeta)|\leqslant|P_{n_{\nu+1}}(\zeta)-f(\zeta)|+|f(\zeta)-P_{n_\nu}(\zeta)|\leqslant 2\omega[\rho_\nu(\zeta)], \tag{6.9}$$

we see, in view of $(1.\tilde{5}')$ and (6.7), that

$$\zeta\in\bigcup_{z_1\prec\zeta\prec z,\ \zeta\in\Gamma}U(\zeta,\ \rho_\nu(\zeta))\ \text{for all}\ |U_\nu(\zeta)|\leqslant A\omega[\rho_\nu(z_1)]$$

and that therefore, in the case $\nu=1,\ 2,\ \ldots,\ N_1-1$ for all $\zeta\in[z_1,\ z_2]$ and in the case $\nu=N_1,\ N_1+1,\ \ldots,\ N_2-1$ for all $\zeta\in[z_1,\ z]$, we have

$$U_\nu(\zeta)\leqslant A\omega[\rho_\nu(z_1)]. \tag{6.10}$$

If it now turns out that the sequence $\{n_\nu\}$ can be chosen in such a way that

$$\rho_{\nu+1}(z_1)\in\left[\frac{1}{4}\rho_\nu(z_1);\ \frac{1}{2}\rho_\nu(z_1)\right], \tag{6.11}$$

for each $\nu=1,\ 2,\ \ldots,$ then in view of the properties of the second modulus of continuity $\omega(t)$ we get

$$| U_\nu^{(j)}(z) | \prec \frac{\omega[\rho_\nu(z_1)]}{\rho_{\nu+1}^{(j)}(z_1)} \leqslant 16 \frac{\omega[\rho_\nu(z_1)]}{\rho_\nu^{(j)}(z_1)}$$
$$\leqslant 16 \cdot 8 \int\limits_{\rho_\nu(z_1)}^{2\rho_\nu(z_1)} \frac{\omega(t)}{t^{1+j}} dt \leqslant A \int\limits_{\rho_{\nu+1}(z_1)}^{\rho_\nu(z_1)} \frac{\omega(t)}{t^{1+j}} dt, \ A = \text{const} \tag{6.12}$$

for each $\nu = 1, 2, \ldots$, and $j = 0, 1, 2, \ldots$. Taking these inequalities into account, we find as a result of (6.8) that

$$|f(z) - p_1(f; z)| \prec th + th \left[\sum_{\nu=1}^{N_1-2} \int\limits_{\rho_{\nu+1}(z_1)}^{\rho_\nu(z_1)} \frac{\omega(t)}{t^3} dt + \int\limits_{\rho_{N_1-1}(z_1)}^{2\rho_{N_1-1}(z_1)} \frac{\omega(t)}{t^3} dt \right]$$
$$+ \omega[\rho_{N_2}(z_1)] + t \left[\sum_{\nu=N_1}^{N_2-2} \int\limits_{\rho_{\nu+1}(z_1)}^{\rho_\nu(z_1)} \frac{\omega(t)}{t^2} dt + \int\limits_{\rho_{N_2-1}}^{2\rho_{N_2-1}} \frac{\omega(t)}{t^2} dt \right] + \frac{t}{h} \omega[\rho_{N_1}(z_1)]$$
$$\prec th + \omega(t) + \frac{t}{h} \omega(h) + 2th \int\limits_h^d \frac{\omega(u)}{u^3} du + 2t \int\limits_t^h \frac{\omega(u)}{u^2} du \tag{6.13}$$
$$\prec th \int\limits_h^d \frac{\omega(u)}{u^3} du + t \int\limits_t^h \frac{\omega(u)}{u^2} du,$$

where $d \overset{\text{def}}{=} \operatorname{diam} \mathfrak{M}$.

Thus, when (6.11) holds the theorem is proved for the case $r = 0$. It can be verified that (6.11) holds, for example, for admissible continua with piecewise smooth boundary of type ($\mathfrak{L}$). In the general case it is not known whether (6.11) holds or not. Thus in the general case in choosing the sequence $\{n_\nu\}$ we carry out some additional reasoning which is a modification of the reasoning employed by Lebedev in [21] for the first modulus of continuity.

Specifically, we choose the sequence $\{n_\nu\}$ inductively as follows. If $n_1, \ldots, n_\nu$ have already been chosen, we choose $n_{\nu+1}$ by the following rule:

1. If there exists at least one natural number n for which

$$\rho_{1+\frac{1}{n}}(z_1) \in \left[\frac{1}{4} \rho_\nu(z_1), \ \frac{1}{2} \rho_\nu(z_1) \right], \tag{6.11$'$}$$

then $n_{\nu+1}$ will denote the least such n.

2. If none of the numbers $\rho_{1+1/n}(z_1)$ is contained in the interval $[\frac14 \rho_\nu(z_1), \frac12 \rho_\nu(z_1)]$, then we consider the following two possibilities:

a) If

$$\rho_{1+\frac{1}{n_\nu+1}}(z) < \frac{1}{4} \rho_\nu(z_1), \tag{6.14}$$

then we put $n_{\nu+1} = n_\nu + 1$;

b) if

$$\rho_{1+\frac{1}{n_\nu+1}}(z_1) > \frac{1}{2} \rho_\nu(z_1), \tag{6.15}$$

then $n_{\nu+1}$ will denote the largest of the integers for which

$$\rho_{\nu+1}(z_1) = \rho_{1+\frac{1}{n_{\nu+1}}}(z_1) > \frac{1}{2}\rho_\nu(z_1), \tag{6.16}$$

and we shall call such a number $n_{\nu+1}$ a *singular number*. In this case, after the (singular) number $n_{\nu+1}$ we at once define the number $n_{\nu+2}$ by putting $n_{\nu+2} \stackrel{\text{def}}{=} n_{\nu+1} + 1$. In this connection the number $n_{\nu+2}$ is certainly not singular, because in view of (6.16) and the fact that by assumption $\rho_{1+1/n}(z_1) \notin [\frac{1}{4}\rho_\nu(z_1), \frac{1}{2}\rho_\nu(z_1)]$ for every natural n we have

$$\rho_{\nu+2}(z_1) < \frac{1}{4}\rho_\nu(z_1) < \frac{1}{2}\rho_{\nu+1}(z_1),$$

which means that two singular numbers cannot occur in succession in the above method of constructing the sequence $\{n_\nu\}$.

After having constructed the sequence with the properties described above, we consider the following possible cases:

1) If $n_{\nu+1}$ is such that (6.11) holds, then, as in the first case analyzed, (6.12) will be satisfied.

2) If $n_{\nu+2} = n_{\nu+1} + 1$, then, using (6.9) and $(1.\tilde{5}')$, we obtain instead of (6.12) the inequality

$$\begin{aligned} |U_{\nu+1}^{(j)}(z)| = |P_{n_{\nu+1}+1}^{(j)}(z) - P_{n_{\nu+1}}^{(j)}(z)| \leqslant A\frac{\omega[\rho_{\nu+1}(z_1)]}{\rho_{\nu+1}^j(z_1)} \\ \leqslant 8A\int_{\frac{1}{2}\rho_{\nu+1}(z_1)}^{\rho_{\nu+1}(z_1)} \frac{\omega(u)}{u^{j+1}}\,du \leqslant 8A\int_{\frac{1}{2}\rho_{\nu+2}(z_1)}^{\rho_{\nu+1}(z_1)} \frac{\omega(u)}{u^{j+1}}\,du. \end{aligned} \tag{6.17}$$

3) If the number $n_{\nu+1}$ is singular, then, taking (6.10) and (6.16) into account, we get

$$|U_\nu^{(j)}(\zeta)| \leqslant 4A\frac{\omega[\rho_\nu(z_1)]}{\rho_\nu^j(z_1)} \leqslant 32A\int_{\frac{1}{2}\rho_\nu(z_1)}^{\rho_\nu(z_1)} \frac{\omega(u)}{u^{1+j}}\,du \leqslant 32A\int_{\frac{1}{2}\rho_{\nu+1}(z_1)}^{\rho_\nu(z_1)} \frac{\omega(u)}{u^{1+j}}\,du \tag{6.18}$$

for all $\zeta \in [z_1, z_2]$ and $j = 1, 2$. Now using (6.12), (6.17) and (6.16) and noting that two singular numbers in the sequence $\{n_\nu\}$ cannot occur in succession, we verify (6.13) in the same way as when (6.10) holds.

Thus Theorem 6.1 is completely proved in the case $r = 0$.

II. We consider the case $r > 0$. First we establish that, if $(6.1')$ is satisfied, then the derivatives

$$f^{(\mu)}(z_0) \stackrel{\text{def}}{=} \lim_{z \to z_0,\ z_0 \in \Gamma} \frac{f^{(\mu-1)}(z) - f^{(\mu-1)}(z_0)}{z - z_0}, \quad \mu = 1, 2, \ldots, r, \tag{6.19}$$

exist "along" Γ at each point $z_0 \in \Gamma$, and that if $f(z)$ is represented in the form

$$f(z) = P_{n_1}(z) + \sum_{\nu=1}^{\infty} U_\nu(z), \tag{6.20}$$

where the sequence n_ν is chosen with respect to the point z_0 in the same way as it was chosen with respect to the point z_1 in case I, then

$$f^{(\mu)}(z_0) = P_{n_1}^{(\mu)}(z_0) + \sum_{\nu=1}^{\infty} U_\nu^{(\mu)}(z_0), \quad \forall z_0 \in \Gamma. \tag{6.21}$$

We carry out the proof only for the case $r = 1$, since it is similar in the remaining cases.

For arbitrary $z \neq z_0$, $z \in \Gamma$, and natural $N > n_1$ we represent the difference

$$\Delta = \frac{f(z) - f(z_0)}{z - z_0} - \left[P'_{n_1}(z_0) + \sum_{\nu=1}^{\infty} U'_\nu(z_0) \right]$$

in the form

$$\Delta = \frac{1}{z - z_0} \int_{z_0}^{z} [P'_{n_1}(\zeta) - P'_{n_1}(z_0)]\, d\zeta + \frac{1}{z - z_0} \sum_{\nu=1}^{N-1} \int_{z_0}^{z} [U'_\nu(\zeta) - U'_\nu(z_0)]\, d\zeta$$

$$+ \frac{1}{z - z_0} \sum_{\nu=N}^{\infty} [U_\nu(z) - U_\nu(z_0)] - \sum_{\nu=N}^{\infty} U'_\nu(z_0).$$

Assuming that $t \overset{\text{def}}{=} |z - z_0|$ is sufficiently small, the path from z_0 to z is a line segment, and choosing N so that $\rho_N(z_0) \leq t \leq \rho_{N-1}(z_0)$, we get on the basis of (6.1) and inequalities of the type (6.12), (6.17) and (6.18)

$$|\Delta| \leqslant t \| P''_{n_1} \| + At \sum_{\nu=1}^{N+1} | U''_\nu(z_0) | + \frac{A}{t} | f(z) - P_n(z) | + \frac{1}{t} | f(z_0) - U_\nu(z_0) |$$

$$+ \sum_{N}^{\infty} | U'_\nu(z_0) | \leqslant t \| P''_{n_1} \| + At \int_{t}^{d} \frac{\omega(s)}{s^2}\, ds + \omega[\rho_N(z_0)] + \frac{2\rho_N(z_0)}{t} \omega[\rho_N(z_1)]$$

$$+ \int_{0}^{t} \frac{\omega(s)}{s}\, ds \leqslant O(1) + At \int_{t}^{\sqrt{t}} \frac{\omega(s)}{s^2}\, ds + A_1 t \int_{\sqrt{t}}^{d} \frac{1}{s^2}\, ds \to 0$$

as $t \to 0$. Thus (6.21) is proved.

Having proved (6.21), we repeat the arguments which helped us establish (6.8), and obtain

$$| f^{(r)}(z) - p_1(f^{(r)}; z) | \leqslant 2t \int_{z_1}^{z_2} \left\{ \| P^{(r+2)}_{n_1} \| + \sum_{\nu=1}^{N_1 - 1} | U^{(r+2)}_\nu(\zeta) | \right\} | d\zeta |$$

$$+ \sum_{\nu=N_1}^{N_2-1} \int_{z_1}^{z} | U^{(r+1)}_\nu(\zeta) | | d\zeta | + 5 | f^{(r)}(z_1) - P^{(r)}_{N_2}(z_1) | + \frac{8t}{h} | f^{(r)}(z_1) - P^{(r)}_{N_1}(z_1) |. \qquad (6.8')$$

In view of (6.21), (6.5), (6.1) and (1.5'), arguing in the same way as in case I we find that

$$| f^{(r)}(z_1) - P^{(r)}_{N_i}(z_1) | \leqslant \sum_{\nu=N_i}^{\infty} | U^{(r)}_\nu | \leqslant A \int_{0}^{\rho_{N_i}(z_1)} \frac{\omega(s)}{s}\, ds,$$

Then, applying the same arguments to the first two terms in (6.8') as in case I, we obtain (6.2) for the case $r > 1$. Thus Theorem 6.1 is completely proved.

As a corollary of Theorem 6.1 we have the following theorem, mentioned in the Introduction.

Theorem 6.2 [21]. *If, under the conditions of Theorem 6.1, the function $\omega(t)$ in the inequality*

$$|f(z)-P_n(z)|\leqslant \rho^r_{1+\frac{1}{n}}\omega\left[\rho_{1+\frac{1}{n}}(z)\right],\quad z\in\Gamma, \tag{6.1}$$

is the first modulus of continuity, then $f(z)$ will have a continuous rth derivative $f^{(r)}(z)$ on Γ whose modulus of continuity $\omega(f^{(r)};\,t)$ will satisfy the inequality

$$\omega(f^{(r)};\,t)\leqslant A\left[t\int_t^d\frac{\omega(u)}{u^2}\,du+\delta_r\int_0^t\frac{\omega(u)}{u}\,du\right],\quad d\overset{\mathrm{def}}{=}\operatorname{diam}\mathfrak{M}, \tag{6.22}$$

where A is an absolute constant, and $\delta_r=1$ if $r\geq 1$ and $\delta_0=0$.

In fact, if z and z_1 are arbitrary points of $\mathfrak{M}$, the distance $|z-z_1|\overset{\mathrm{def}}{=}t$ between which is, for example, less than $0.1d$, then, taking any point z_2 in $\mathfrak{M}$ for which $|z_1-z_2|\geq d/2$, and putting $|z_1-z_2|=h$, we get, in view of the properties of the first modulus of continuity,

$$\begin{aligned}
|f^{(r)}(z)-p_1(f^{(r)};\,z)|&\leqslant A_1\left\{t\frac{\omega(t)}{t}\int_{\frac{d}{4}}^{\frac{d}{2}}\frac{du}{u^2}+t\int_t^{\frac{d}{2}}\frac{\omega(u)}{u^2}\,du\right\}\\
&+A\delta_r\left[\int_0^t\frac{\omega(u)}{u}\,du+\frac{4t}{d}\left(\int_0^t\frac{\omega(u)}{u}\,du+\frac{\omega(t)}{t}\int_t^{\frac{d}{4}}du\right)\right]\\
&\leqslant A_2\left\{\omega(t)+t\int_t^{\frac{d}{2}}\frac{\omega(u)}{u^2}\,du\right\}+A_2\delta_r\left[\int_0^t\frac{\omega(u)}{u}\,du+\omega(t)\right]\\
&\leqslant A_3t\int_t^{\frac{d}{2}}\frac{\omega(u)}{u}\,du+A_3\delta_3\int_0^t\frac{\omega(u)}{u}\,du.
\end{aligned}$$

From this, noting that $|p_1'(f^{(r)};\,z)|\leq 4\|f^{(r)}\|/d$, we obtain

$$\begin{aligned}
|f^{(r)}(z)-f^{(r)}(z_1)|&\leqslant|f^{(r)}(z)-p_1(f^{(r)};\,z)|+|p_1(f^{(r)};\,z)-p_1(f^{(r)};\,z_1)|\\
&\leqslant A_3t\int_t^{\frac{d}{2}}\frac{\omega(u)}{u}\,du+A_3\delta_3\int_0^t\frac{\omega(u)}{u}\,du+\frac{4\|f^{(r)}\|}{d}\,t.
\end{aligned}$$

Hence (6.22) follows in view of the arbitrariness of the points z and $z_1\in\mathfrak{M}$.

To conclude this section we formulate two theorems on a constructive characterization of functions in certain function classes $W^0H^{\omega(t)}$ and $W^rH_2^{\omega(t)}$, which contain the Hölder and Zygmund classes. Both of these theorems are simple corollaries of the direct Theorem 5.1 and Theorems 6.1 and 6.2.

Theorem 6.3. *Suppose the function $\omega(t)$ is a first modulus of continuity, and in the case of the function classes $W^0H^{\omega(t)}=H^{\omega(t)}$ suppose it satisfies the condition*

$$t\int_t^1 \frac{\omega(u)}{u^2}\,du = O[\omega(t)], \tag{6.23}$$

while in the case of the classes $W^rH^{\omega(t)}$, $r = 1, 2, \ldots$, *suppose it satisfies the additional condition*

$$\int_0^t \frac{\omega(s)}{s}\,ds = O[\omega(t)]. \tag{6.23$'$}$$

Then, in order that $f(z)$ *belong to* $W^rH^{\omega(t)}$ *on a set* $\mathfrak{M}$ *of type* $\overset{0}{B}{}_k^{(n)}$ *for some integer* $r \geq 0$, *it is necessary and sufficient that for each* $n = r, r+1, \ldots$ *there exist an algebraic polynomial* $P_n(z)$ *of degree not greater than* n *such that*

$$|f(z) - P_n(z)| \leqslant A\rho^r_{1+\frac{1}{n}}(z)\,\omega\left[\rho_{1+\frac{1}{n}}(z)\right], \tag{6.24}$$

for all $z \in \mathfrak{M}$, *where the constant* A *depends neither on* $z \in \partial\mathfrak{M}$ *nor on* n.

Theorem 6.4. *Suppose the function* $\omega(t)$ *is a second modulus of continuity, and suppose it satisfies the condition*

$$\int_h^{\frac{d}{2}} \frac{\omega(u)}{u^3}\,du = O\left[\frac{1}{h}\int_0^h \frac{\omega(u)}{u^2}\,du\right], \tag{6.25}$$

for any $h \in (0, d/4)$, *where* $d = \operatorname{diam} \mathfrak{M}$, *in the case of the functions classes* $W^0H_2^{\omega(t)} = H_2^{\omega(t)}$, *while the classes* $W^rH_2^{\omega(t)}$, $r = 1, 2, \ldots$, *suppose it satisfies the additional condition*

$$\int_0^t \frac{\omega(s)}{s}\,ds = O[\omega(t)]. \tag{6.23$'$}$$

Then, in order that $f(z)$ *belong to* $W^rH^{\omega(t)}$ *on a set* $\mathfrak{M}$ *of type* $\overset{0}{B}{}_k^{(n)}$ *for some integer* $r > 0$, *it is necessary and sufficient that for each* $n = r+1, r+2, \ldots$ *there exist an algebraic polynomial* $P_n(z)$ *of degree not greater than* n *such that*

$$|f(z) - P_n(z)| \leqslant A\rho^r_{1+\frac{1}{n}}(z)\,\omega\left[\rho_{1+\frac{1}{n}}(z)\right], \tag{6.26}$$

for all $z \in \partial\mathfrak{M}$, *where the constant* A *depends neither on* $z \in \partial\mathfrak{M}$ *nor on* n.

We note that Theorems 6.3 and 6.4 contain as corollaries the corresponding results mentioned in the Introduction for the interval $[a, b]$.

BIBLIOGRAPHY

1. S. M. Nikol'skiĭ, *On the best approximation of functions satisfying Lipschitz conditions by polynomials*, Izv. Akad. Nauk SSSR Ser. Mat. 10 (1946), 295–322. (Russian) MR 8, 153.

2. A. F. Timan, *A strengthening of Jackson's theorem on the best approximation of continuous functions by polynomials on a finite segment of the real axis*, Dokl. Akad. Nauk SSSR 78 (1951), 17–20. (Russian) MR 12, 823.

3. V. K. Dzjadyk, *Constructive characterization of functions satisfying the condition* Lip $\alpha(0 < \alpha < 1)$ *on a finite segment of the real axis*, Izv. Akad. Nauk SSSR Ser. Mat. 20 (1956), 623–642. (Russian) MR 18, 392.

4. ———, *Approximation of functions by ordinary polynomials on a finite interval of the real axis*, Izv. Akad. Nauk SSSR Ser. Mat. 22 (1958), 337–354. (Russian) MR 20 #3406.

5. V. K. Dzjadyk, *A further strengthening of Jackson's theorem on the approximation of continuous functions by ordinary polynomials*, Dokl. Akad. Nauk SSSR 121 (1958), 403–406. (Russian) MR 21 #249.

6. G. Freud, *Über die Approximation reeller stetigen Funktionen durch gewöhnliche Polynome*, Math. Ann. 137 (1959), 17–25. MR 21 #250.

7. G. K. Lebed′, *Inequalities for polynomials and their derivatives*, Dokl. Akad. Nauk SSSR 117 (1957), 570–572. (Russian) MR 20 #1165.

8. Ju. A. Brudnyĭ, *Approximation by entire functions on the exterior of a segment or on a semi-axis*, Dokl. Akad. Nauk SSSR 124 (1959), 739–742. (Russian) MR 21 #255.

9. A. F. Timan, *Converse theorems in the constructive theory of functions given on a finite segment of the real axis*, Dokl. Akad. Nauk SSSR 116 (1957), 762–765. (Russian) MR 19, 1175.

10. S. M. Nikol′skiĭ, *Approximation of functions of several variables by polynomials*, Proc. Third All-Union Math. Congr. (Moscow, 1956), vol. 3, Izdat. Akad. Nauk SSSR, Moscow, 1956, pp. 226–231. (Russian)

11. V. K. Dzjadyk, *On a problem of S. M. Nikol′skiĭ in a complex region*, Izv. Akad. Nauk SSSR Ser. Mat. 23 (1959), 697–736. (Russian) MR 22 #865.

12. ———, *On the approximation of continuous functions in closed domains with corners and on a problem of S. M. Nikol′skiĭ*. I, Izv. Akad. Nauk SSSR Ser. Mat. 26 (1962), 797–824; English transl., Amer. Math. Soc. Transl. (2) 53 (1966), 221–252. MR 27 #313.

13. ———, *Inverse theorems in the theory of approximation of functions in complex domains*, Ukrain. Mat. Ž. 15 (1963), 365–375. (Russian) MR 35 #4437.

14. ———, *On the theory of approximation of analytic functions continuous on closed domains and on a problem of S. M. Nikol′skiĭ*. II, Izv. Akad. Nauk SSSR Ser. Mat. 27 (1963), 1135–1164; English transl., Amer. Math. Soc. Transl. (2) 53 (1966), 253–284. MR 27 #4937b.

15. ———, *Theorems on the transformation and approximation of analytic functions*, Dokl. Akad. Nauk SSSR 151 (1963), 269–272 = Soviet Math. Dokl. 4 (1963), 983–986. MR 27 #4937a.

16. ———, *The approximation by polynomials of functions of the classes H^{ω} on closed regions*, Studies Contemporary Problems Constructive Theory of Functions (Proc. Second All-Union Conf., Baku, 1962), Izdat. Akad. Nauk Azerbaĭdžan. SSR, Baku, 1965, pp. 273–283. (Russian) MR 33 #5906.

17. ———, *On the approximation of analytic functions in regions with smooth and piecewise smooth boundary*, Third Math. Summer School (Kaciveli, 1965): Constructive Theory of Functions, "Naukova Dumka", Kiev, 1966, pp. 29–83. (Russian) MR 37 #1625.

18. ———, *Analytic and harmonic transformations and the approximation of harmonic functions*, Ukrain. Mat. Ž. 19 (1967), no. 5, 33–57 = Ukrainian Math. J. 19 (1967), 531–552 (1969). MR 36 #1684.

19. A. I. Markuševič, *Theory of analytic functions*. Vol. 2, "Nauka", Moscow, 1968; English transl. of 1st ed., Prentice-Hall, Englewood Cliffs, N. J., 1965. MR 31 #5965; 38 #304.

20. N. A. Lebedev and P. M. Tamrazov, *Inverse theorems of approximation on closed sets of the complex plane*, Dokl. Akad. Nauk SSSR 179 (1968), 1046–1049 = Soviet Math. Dokl. 9 (1968), 498–501. MR 37 #4271.

21. ———, *Inverse approximation theorems on regular compacta of the complex plane*, Izv. Akad. Nauk SSSR Ser. Mat. 34 (1970), 1340–1390 = Math. USSR Izv. 4 (1970), 1355–1405. MR 45 #2387.

22. A. I. Švaĭ, *On the approximation of functions of the classes W^rH^{ω} given on closed sets with arbitrary corners different from 0 and 2π*, Candidate's Dissertation, Kiev, 1967. (Russian)

23. V. K. Dzjadyk and O. I. Švaĭ, *A constructive characterization of functions of Hölder classes in regions with piecewise smooth boundary and arbitrary positive exterior angles*, Dopovidī Akad. Nauk Ukraïn. RSR Ser. A 1967, 1012–1025. (Ukrainian) MR 36 #5360.

24. R. V. Poljakov, *Approximation of continuous functions (that are analytic in the interior) in domains with a smooth boundary*, Ukrain. Mat. Ž. 24 (1972), 57–68 = Ukrainian Math. J. 24 (1972), 45–53. MR 45 #7077.

25. V. I. Smirnov and N. A. Lebedev, *The constructive theory of functions of a complex variable*, "Nauka", Moscow, 1964; English transl., *Functions of a complex variable: Constructive theory*, M. I. T. Press, Cambridge, Mass., 1968. MR 30 #2152; 37 #5369.

26. A. V. Stovbun, *Uniform approximation in a closed region by generalized Faber polynomials*, Ukrain. Mat. Ž. 21 (1969), 849–857 = Ukrainian Math. J. 21 (1969), 711–717. (1970). MR 40 #5876.

27. V. K. Dzjadyk and G. A. Alibekov, *Uniform approximation of functions of a complex variable on closed sets with corners*, Mat. Sb. 75 (117) (1968), 502–557 = Math. USSR Sb. 4 (1968), 463–518. MR 37 #428.

28. V. K. Dzjadyk, *The application of generalized Faber polynomials to the approximation of integrals of Cauchy type and functions of the classes A^r in domains with smooth and piecewise smooth boundary*, Ukrain. Mat. Ž. 24 (1972), 3–19 = Ukrainian Math. J. 24 (1972), 1–13. MR 47 #3686.

29. V. I. Belyi and V. M. Mikljukov, *Some properties of conformal and quasiconformal mappings and direct theorems of the constructive theory of functions*, Izv. Akad. Nauk SSSR Ser. Mat. 38 (1974), 1343–1361 = Math. USSR Izv. 8 (1974), 1323–1341. MR 52 #8443.

30. H. Whitney, *On functions with bounded n-th differences*, J. Math. Pures Appl. (9) 36 (1957), 67–95. MR 18, 889.

31. V. K. Dzjadyk and I. A. Ševčuk, *The limit values of a Cauchy type integral for functions of Zygmund classes*, Ukrain. Mat. Ž. 24 (1972), 601–617 = Ukrainian Math. J. 24 (1972), 483–496 (1973). MR 47 #2065.

32. I. A. Ševčuk, *Constructive characterization of functions of the classes $D^rH^{\omega_2(t)}$ on closed sets with a piecewise smooth boundary*, Ukrain. Mat. Ž. 25 (1973), 81–90 = Ukrainian Math. J. 25 (1973), 66–73. MR 47 #5491.

33. R. M. Trigub, *Approximation of functions with a given modulus of smoothness on the exterior of a segment and on a half-axis*, Dokl. Akad. Nauk SSSR 132 (1960), 303–306 = Soviet Math. Dokl. 1 (1960), 568–571. MR 23 #A460.

34. L. V. Ahlfors, *Lectures in quasiconformal mappings*, Van Nostrand, Princeton, N. J., 1966. MR 34 #336.

35. V. K. Dzjadyk and O. Ī. Švaĭ, *The approximation of functions of Hölder classes on closed sets with acute exterior angles*, Metric Questions of the Theory of Functions and Mappings, no. 2, "Naukova Dumka", Kiev, 1971, pp. 74–164. (Russian) MR 45 #5372.

36. R. V. Poljakov, *On the approximation of continuous functions given on bounded closed sets*, Candidate's Dissertation, Kiev, 1969. (Russian)

37. N. A. Lebedev and N. A. Širokov, *The uniform approximation of functions on closed sets that have a finite number of angular points with non-zero exterior angles*, Izv. Akad. Nauk Armjan. SSR Ser. Mat. 6 (1971), 311–341. (Russian) MR 45 #5375.

38. V. K. Dzjadyk, *Investigations in the theory of approximations of analytic functions conducted at the Institute of Mathematics of the Ukrainian Academy of Sciences*, Ukrain. Mat. Ž. 21 (1969), 173–192 = Ukrainian Math. J. 21 (1969), 143–159. MR 39 #2996.

39. N. A. Širokov, *On uniform approximation of functions on closed sets having a finite number of corner points with nonzero exterior angles*, Dokl. Akad. Nauk SSSR 205 (1972), 798–800 = Soviet Math. Dokl. 13 (1972), 1041–1044. MR 46 #9362.

40. V. K. Dzjadyk and D. M. Galan, *The approximation of analytic functions in domains with smooth boundary*, Ukrain. Mat. Ž. 17 (1965), no. 1, 26–38. (Russian) MR 33 #3007.

41. A. S. Prypik, *On the application of generalized Faber polynomials to the approximation of functions analytic on closed sets*, Theory of the Approximation of Functions and its Applications, Izdanie Inst. Mat. Akad. Nauk Ukrain. SSR, Kiev, 1974, pp. 114–118. (Russian)

Translated by P. K. KULSHRESTHA

Trudy Mat. Inst. Steklov.
134 (1975)

Proc. Steklov Inst. Math.
134 (1975)

ON SOME RELATIONS BETWEEN APPROXIMATION CHARACTERISTICS OF CLASSES OF PERIODIC FUNCTIONS

UDC 517.51

N. P. KORNEĬČUK

Abstract. Quantities characterizing the approximation characteristics of a class $\mathfrak{M}$ of periodic functions in the metric C are considered (best approximation by trigonometric polynomials T_n, best linear approximation by these polynomials, upper bound of a function in $\mathfrak{M}$ orthogonal to T_n). For the class W^rH^ω, given by a convex modulus of continuity $\omega(t)$, results connecting with the cases of coincidence or noncoincidence of these characteristics are found.

Bibliography: 14 titles.

1. In this paper X stands for one of the following classes of 2π-periodic functions: the space (C) of continuous functions, the space (M) of functions measurable and essentially bounded, or the space (L_p) of functions pth power summable over a period, with the corresponding norms

$$\|f\|_C = \max_t |f(t)|, \qquad \|f\|_M = \operatorname{ess\,sup}_t |f(t)|,$$

$$\|f\|_{L_p} = \left(\int_0^{2\pi} |f(t)|^p\,dt\right)^{\frac{1}{p}} \qquad (1 \leqslant p < \infty).$$

In problems of approximation to a class $\mathfrak{M} \subset X$ the following characteristics of the class are frequently related to one another:

$$E_n(\mathfrak{M})_X = \sup_{f\in\mathfrak{M}} \inf_{T_n} \|f - T_n\|_X \tag{1}$$

the best approximation of the class $\mathfrak{M}$ by trigonometric polynomials $T_n(t)$ of order $n-1$;

$$\gamma_n(\mathfrak{M})_X = \sup_{f\in\mathfrak{M}_n^\perp} \|f\|_X, \tag{2}$$

where $\mathfrak{M}_n^\perp$ is the set of functions $f \in \mathfrak{M}$ such that

$$\int_0^{2\pi} f(t) \begin{matrix}\sin kt\\ \cos kt\end{matrix}\, dt = 0 \qquad (k = 0, 1, \ldots, n-1);$$

$$\beta_n(\mathfrak{M})_X = \inf_{A\in\mathfrak{L}_n} \sup_{f\in\mathfrak{M}} \|f - Af\|_X, \tag{3}$$

where $\mathfrak{L}_n$ is the set of all linear operators mapping functions $f \in X$ to trigonometric polynomials of order $n-1$. The most detailed study has been carried out for operators A_n^λ given by a sequence $\{\lambda_1, \ldots, \lambda_{n-1}\}$ of coefficients and taking the function $f \in X$ into a polynomial

---

AMS (MOS) subject classifications (1970). Primary 42A08; Secondary 41A25.

$$A_n^\lambda(f, t) = \frac{a_0}{2} + \sum_{k=1}^{n-1} \lambda_k (a_k \cos kt + b_k \sin kt),$$

where a_k and b_k are the Fourier coefficients of f in the trigonometric system. Linear operators $A_n^{\mu,\nu}$ of the more general form

$$A_n^{\mu,\nu}(f, t) = \frac{a_0}{2} + \sum_{k=1}^{n-1} \{\mu_k (a_k \cos kt + b_k \sin kt) + \nu_k (a_k \sin kt - b_k \cos kt)\}$$

have also been considered. We shall therefore introduce the following quantities in addition to $\beta_n(\mathfrak{M})_X$:

$$\beta_n^\lambda(\mathfrak{M})_X = \inf_{A_n^\lambda} \sup_{f \in \mathfrak{M}} \|f - A_n^\lambda f\|_X, \tag{4}$$

$$\beta_n^{\mu,\nu}(\mathfrak{M})_X = \inf_{A_n^{\mu,\nu}} \sup_{f \in \mathfrak{M}} \|f - A_n^{\mu,\nu} f\|_X. \tag{5}$$

It follows at once from the definitions that

$$E_n(\mathfrak{M})_X \leqslant \beta_n(\mathfrak{M})_X \leqslant \beta_n^{\mu,\nu}(\mathfrak{M})_X \leqslant \beta_n^\lambda(\mathfrak{M})_X; \quad \gamma_n(\mathfrak{M})_X \leqslant \beta_n^{\mu,\nu}(\mathfrak{M})_X.$$

The last relation follows from the fact that if $f \in \mathfrak{M}_n^\perp$ then $A_n^{\mu,\nu}(f, t) \equiv 0$, and for any μ_k and ν_k

$$\sup_{f \in \mathfrak{M}_n^\perp} \|f\|_X \leqslant \sup_{f \in \mathfrak{M}} \|f - A_n^{\mu,\nu} f\|_X.$$

2. In a number of important cases all the characteristics introduced above coincide for concrete classes of functions. Let W_X^r be the class of 2π-periodic functions $f(t)$ for which $f^{(r-1)}(t)$ is absolutely continuous and $\|f^{(r)}\|_X \leq 1$.

In 1936 and 1937 Favard [1]–[3] showed that

$$E_n(W_M^r)_C = \gamma_n(W_M^r)_C = \beta_n^\lambda(W_M^r)_C = \frac{K_r}{n^r} \qquad (n, r = 1, 2, \ldots),$$

where

$$K_r = \frac{4}{\pi} \sum_{\nu=0}^{\infty} \frac{(-1)^{\nu(r+1)}}{(2\nu+1)^{r+1}}.$$

The values of $E_n(W_M^r)_C$ and $\beta_n^\lambda(W_M^r)_C$ were also calculated by Ahiezer and Kreĭn [4]. In 1946 Nikol'skiĭ calculated the quantities (1)–(5) and proved that they coincide in the metrics of C and $L = L_1$ for certain classes of functions representable in the form of a convolution [5]. In particular, Nikol'skiĭ showed that

$$E_n(W_L^r)_L = \gamma_n(W_L^r)_L = \beta_n^\lambda(W_L^r)_L = \frac{K_r}{n^r} \qquad (n, r = 1, 2, \ldots).$$

Later, in articles of Dzjadyk [6], [7] Stečkin [8] and Sun Yung-sheng [9] similar results were obtained in the spaces C and L for classes of functions $f(t)$ having bounded fractional derivatives (in the sense of Weyl) $f^{(r)}(t)$ $(r > 0)$, and also for some more general classes. We note also the equation $\gamma_n(W_M^r)_L = E_n(W_M^r)_L = 4K_{r+1}/n^r$ that follows from the results of Taĭkov [10].

In 1961 the first case in which the quantities (1) and (4) do not coincide was found. Let H^ω be the class of 2π-periodic functions $f(x)$ satisfying the condition

$$|f(x')-f(x'')|\leqslant\omega(|x'-x''|),$$

where $\omega(t)$ is a given modulus of continuity. It was proved in the author's article [11] that for $\omega(t)$ convex

$$E_n(H^\omega)_C=\frac{1}{2}\omega\left(\frac{\pi}{n}\right).$$

At the same time, if $g_n(x)$ is a function of period $2\pi/n$, even and equal to $\omega(x) - (n/\pi)\int_0^{\pi/n}\omega(t)dt$ on $[0, \pi/n]$, then $g_n \in H^\omega$, $g_n \perp T_n$, and so

$$\beta_n^\lambda(H^\omega)_C\geqslant\beta_n^{\mu,\nu}(H^\omega)_C\geqslant\gamma_n(H^\omega)_C\geqslant\|g_n\|_C=\frac{n}{\pi}\int_0^{\frac{\pi}{n}}\omega(t)\,dt>\frac{1}{2}\omega\left(\frac{\pi}{n}\right),$$

where $\omega(t)$ is nonlinear on $[0, \pi/n]$. The last condition, as was proved subsequently, also implies that $\sup_{f\in H^\omega}\|f - Af\|_C > E_n(H^\omega)_C$ for any $A \in \mathfrak{L}_n$.

Naturally the problem of calculating and comparing the characteristics (1)–(5) also arises for the classes W^rH^ω $(r = 1, 2, \ldots)$ of functions $f \in C$ for which $f^{(r)} \in H^\omega$. In [13] the following relation is obtained on the condition that $\omega(t)$ is convex:

$$E_n(W^rH^\omega)_X=\|f_{nr}\|_X\qquad(n,\ r=1,\ 2,\ \ldots),\tag{6}$$

where X is C or L and $f_{nr}(x) = f_{nr}(\omega, x)$ is a function of period $2\pi/n$ with mean value zero over a period for which $f_{nr}^{(r)}$ is odd and

$$f_{nr}^{(r)}(x)=f_{n0}(x)=\begin{cases}\frac{1}{2}\omega(2x) & \left(0\leqslant x\leqslant\frac{\pi}{2n}\right),\\ \frac{1}{2}\omega\left(\frac{2\pi}{n}-2x\right) & \left(\frac{\pi}{2n}\leqslant x\leqslant\frac{\pi}{n}\right).\end{cases}$$

We do not know the values of the quantities (2)–(5) for the classes W^rH^ω. Here we note only some facts about these characteristics.

3. **Theorem.** *If $\omega(t)$ is a modulus of continuity convex on $[0, \pi/n]$ and $\omega(t)$ satisfies the condition of strict convexity on some segment $[\alpha, \beta] \subset [0, \pi/n]$:*

$$\omega(t')+\omega(t'')<2\omega\left(\frac{t'+t''}{2}\right)\qquad(\alpha\leqslant t'<t''\leqslant\beta),$$

then for any $r = 1, 2, \ldots$ *and* $n = 2, 3, \ldots$

$$\beta_n^{\mu,\nu}(W^r H^\omega)_C > E_n(W^r H^\omega)_C. \tag{7}$$

Proof. There is a best method for the class $W^r H^\omega$ in the set of linear methods $A_n^{\mu,\nu}$ (for fixed n); that is, there is one that realizes the infimum in (5). Hence, taking (6) into account, it is enough for us to prove that for any operator $A_n^{\mu,\nu}$

$$\sup_{f\in W^r H^\omega} \| f - A_n^{\mu,\nu} f\|_C > \| f_{nr}(\omega)\|_C. \tag{8}$$

The polynomial $A_n^{\mu,\nu}(f, x)$ can be written in the form

$$A_n^{\mu,\nu}(f, x) = \frac{a_0}{2} + \frac{1}{\pi}\int_0^{2\pi} \tau_n(x-t) f^{(r)}(t)\,dt,$$

where $\tau_n(u)$ is a trigonometric polynomial of order $n-1$ with zero mean over a period, uniquely determined by the coefficients μ_k and ν_k. Since

$$f(x) = \frac{a_0}{2} + \frac{1}{\pi}\int_0^{2\pi} D_r(x-t) f^{(r)}(t)\,dt,$$

where

$$D_r(u) = \sum_{k=1}^{\infty} \frac{\cos(ku - \pi r/2)}{k^r},$$

we have

$$f(x) - A_n^{\mu,\nu}(f, x) = \frac{1}{\pi}\int_0^{2\pi} \Delta(x-t) f^{(r)}(t)\,dt,$$

where $\Delta(u) = D_r(u) - \tau_n(u)$. Noting that $\int_0^{2\pi}\Delta(t)dt = 0$ and that the class H^ω contains $\phi(x-t)$ for any x as well as $\phi(t)$, we have

$$\sup_{f\in W^r H^\omega} \| f - A_n^{\mu,\nu} f\|_C = \sup_{\varphi\in H^\omega} \frac{1}{\pi}\int_0^{2\pi} \Delta(t)\varphi(t)\,dt. \tag{9}$$

On the other hand, it follows from the definition of $f_{nr}(\omega, x)$ on taking into account its $(2\pi/n)$-periodicity that

$$\| f_{nr}\|_C = \frac{1}{\pi}\int_0^{2\pi} D_r(t)\varphi_n(t)\,dt = \frac{1}{\pi}\int_0^{2\pi} \Delta(t)\varphi_n(t)\,dt, \tag{10}$$

where

$$\varphi_n(t) = \begin{cases} (-1)^\nu f_{n0}(t) & (r = 2\nu+1,\ \nu = 0, 1, 2, \ldots), \\ (-1)^\nu f_{n0}\left(t + \frac{\pi}{2n}\right) & (r = 2\nu,\quad \nu = 1, 2, \ldots). \end{cases} \tag{11}$$

It is now clear from (9) and (10) that (8) will be proved if we construct a function $\phi_* \in H^\omega$ for which

$$\int_0^{2\pi} \Delta(t)[\varphi_*(t) - \varphi_n(t)]\,dt > 0. \tag{12}$$

Let σ_i $(i = 1, \ldots, 2n)$ denote the interval

$$\left(\frac{2i-1}{2n}\pi,\ \frac{2i+1}{2n}\pi\right)$$

for r odd and

$$\left(\frac{i-1}{n}\pi,\ \frac{i}{n}\pi\right)$$

for even r. The function $\phi_n(t)$ is monotonic on each of the intervals σ_i, its graph is symmetric about the point z_i (the mean of σ_i), and it is convex where $\phi_n(t) > 0$ and concave where $\phi_n(t) < 0$. Because of the conditions imposed on $\omega(t)$ and the hypotheses of the theorem, $\phi_n(t)$ is strictly concave or strictly convex on each of the intervals

$$\sigma_i' = \left(z_i - \frac{\beta}{2},\ z_i - \frac{\alpha}{2}\right) \quad \text{and} \quad \sigma_i'' = \left(z_i + \frac{\alpha}{2},\ z_i + \frac{\beta}{2}\right).$$

It is essential that for any point t' of the interval σ_i' there is a unique point $t'' \in \sigma_i''$, namely $t'' = 2z_i - t'$ (the reflection of t' in z_i), such that

$$|\varphi_n(t') - \varphi_n(t'')| = \omega(|t' - t''|).$$

If the functions $\Delta(t)$ and $\phi_n(t)$ have opposite signs on some interval (t_1, t_2) lying inside any one of the σ_i' or σ_i'', then we take for $\phi_*(t)$ the continuous function coinciding with $\phi_n(t)$ everywhere (for the period) outside (t_1, t_2) and linear on $[t_1, t_2]$. It is clear that $\phi_* \in H^\omega$ and for $t_1 < t < t_2$ the sign of the difference $\phi_*(t) - \phi_n(t)$ will coincide with that of $\Delta(t)$, so that (12) will hold.

We now assume that $\operatorname{sgn} \Delta(t) = \operatorname{sgn} \phi_n(t)$ on all the intervals σ_i' and σ_i'' $(i = 1, \ldots, 2n)$. Then $\Delta(t)$ changes sign exactly once on each segment $[z_i - \alpha/2, z_i + \alpha/2]$ $(i = 1, \ldots, 2n)$. We shall rely essentially on the fact that

$$\Delta(t) \not\equiv -\Delta(2z_1 - t) \qquad (t \in \sigma_1'). \tag{13}$$

For, supposing that $\Delta(t) \equiv -\Delta(2z_1 - t)$ on σ_1', we find that

$$\Delta^{(r-1)}(t) \equiv (-1)^r \Delta^{(r-1)}(2z_1 - t) \qquad (t \in \sigma_1');$$

that is, on σ_1'

$$\tau_n^{(r-1)}(t) + (-1)^{r-1}\tau_n^{(r-1)}(2z_1 - t) \equiv D_1(t) + (-1)^{r-1} D_1(2z_1 - t)$$
$$= \begin{cases} \pi - \dfrac{\pi}{n} & (r = 2\nu - 1), \\[2ex] \dfrac{\pi}{2n} - t & (r = 2\nu), \end{cases}$$

which is impossible for the trigonometric polynomial $\tau_n(t)$ of order $n-1\geq 1$.

It follows from (13) that for some interval (y_0, u_0) with center at z_1 and such that $\alpha/2 < z_1 - y_0 = u_0 - z_1 < \beta/2$, we shall have

$$\int_{y_0}^{u_0} \Delta(t)\,dt \neq 0.$$

For definiteness let the integral on the left be greater than zero and let $\Delta(t)$ change sign on (y_0, u_0) from negative to positive. Then there is a point u_1, $z_1 < u_1 < u_0$, such that

$$\int_{y_0}^{u_1} \Delta(t)\,dt = 0.$$

Let the function $\tilde{\phi}(t)$ be defined on $[y_0, u_1]$ by

$$\sup_{\varphi\in H^\omega} \int_{y_0}^{u_1} \Delta(t)\,\varphi(t)\,dt = \int_{y_0}^{u_1} \Delta(t)\,\tilde{\varphi}(t)\,dt,$$

where we can suppose that $\tilde{\phi}(y_0) = \phi_n(y_0)$. It follows from Lemma 5.1 of [13] that $\tilde{\phi}(u_1) - \tilde{\phi}(y_0) = \omega(u_1 - y_0)$; and since $\phi_n(u_1) - \phi_n(y_0) < \omega(u_1 - y_0)$, we have $\tilde{\phi}(u_1) > \phi_n(u_1)$. We continue $\tilde{\phi}(t)$ to the interval $(u_1, u_0]$, putting $\tilde{\phi}(t) = \phi_n(y_0) + \omega(t - y_0)$. Then for all $t \in (u_1, u_0)$ we have $\tilde{\phi}(t) > \phi_n(t)$ and $\Delta(t) > 0$, and so

$$\int_{y_0}^{u_0} \Delta(t)\,[\tilde{\varphi}(t) - \varphi_n(t)]\,dt > 0.$$

If now $\phi_*(t)$ and $\tilde{\phi}(t)$ coincide on $[y_0, u_0]$ and $\phi_*(t) = \phi_n(t)$ at the remaining points of the period, then clearly $\phi_* \in H^\omega$ and (12) holds. The theorem is proved.

In particular, for the classes W^rKH^α $(r = 1, 2, \ldots)$ of functions $f \in C$ for which $|f^{(r)}(x') - f^{(r)}(x'')| \leq K|x' - x''|^\alpha$, the following relations hold if $0 < \alpha < 1$:

$$\beta_n^{\mu,\nu}(W^rKH^\alpha)_C > E_n(W^rKH^\alpha)_C \qquad (r = 1, 2, \ldots;\ n = 2, 3, \ldots).$$

For the case $n = 1$ (that is, approximation by constants) we note that the results of [14] imply that

$$\beta_1^\lambda(W^{2\nu-1}H^\omega)_C = E_1(W^{2\nu-1}H^\omega)_C \qquad (\nu = 1, 2, \ldots)$$

holds for any convex modulus of continuity $\omega(t)$. For even r the assertion of the theorem holds for $n = 1$.

4. Although for $r = 0$ nonlinearity of $\omega(t)$ on $[0, \pi/n]$ implies that $\beta_n^{\mu,\nu}(H^\omega)_C > E_n(H^\omega)_C$, for odd $r = 2\nu - 1$ $(\nu = 1, 2, \ldots)$ the quantities $\beta_n^{\mu,\nu}(W^rH^\omega)_C$ and $E_n(W^rH^\omega)_C$ can coincide for $n = 2, 3, \ldots$ even for certain nonlinear moduli of continuity. Let $0 < \gamma < \pi/2n$,

$$\omega(t) = \begin{cases} Kt & (0 \leqslant t \leqslant 2\gamma), \\ 2K\gamma & (t \geqslant 2\gamma) \end{cases}$$

and let W^rH^ω $(r = 1, 2, \ldots)$ be the classes defined in terms of this modulus of continuity. If r is odd we can choose a polynomial of the form

$$\tau_n(t) = \sum_{k=1}^{n-1} \beta_k \sin kt \tag{14}$$

so that for all $i = 1, \ldots, 2n$

$$\int_{\delta_i} [D_r(t) - \tau_n(t)]\, dt = 0, \tag{15}$$

where $\delta_i = (i\pi/n - \gamma,\ i\pi/n + \gamma)$. Indeed, it is only necessary to verify that these relations hold for $i = 1, \ldots, n-1$, because $D_r(t)$ is an odd function. We obtain for β_k the system of equations

$$2\sum_{k=1}^{n-1} \frac{\beta_k}{k} \sin\frac{ki\pi}{n} \sin k\gamma = D_{r+1}\left(\frac{i\pi}{n} + \gamma\right) - D_{r+1}\left(\frac{i\pi}{n} - \gamma\right) \tag{16}$$

$$(i = 1, 2, \ldots, n-1),$$

whose determinant is nonzero. The polynomial (14), whose coefficients β_k are the solution of (16), defines a linear method A_n^λ such that for $f \in W^rH^\omega$

$$A_n^\lambda(f,\ x) = \frac{a_0}{2} + \frac{1}{\pi}\int_0^{2\pi} \tau_n(x-t) f^{(r)}(t)\, dt.$$

It follows from (15) that the difference $\Delta(t) = D_r(t) - \tau_n(t)$ changes sign on each interval $\delta_i (i = 1, \ldots, 2n)$, and this at one point only, while $\Delta(t)$ has no other zeros on a period. Hence $\Delta(t)$ is of constant sign on each of the complementary segments $\alpha_i = [i\pi/n + \gamma,\ (i+1)\pi/n - \gamma]$ $(i = 1, \ldots, 2n)$, and $\Delta(t)$ takes opposite signs on α_i and α_{i+1}. Let us write α_i^+ or α_i^- according to whether $\Delta(t) > 0$ or $\Delta(t) < 0$ on α_i. Any two segments α_i and α_j are separated by at least one interval δ_k of length 2γ. The set $A = \bigcup_1^{2n} \alpha_i$ can be partitioned into n disjoint subsets e_i $(i = 1, \ldots, n)$ in such a way that $\int_{e_i}\Delta(t)dt = 0$ and e_i is the union of a segment α_i^+ and a finite number of nonintersecting intervals from the set $\bigcup_1^n \alpha_i^-$. On taking the form of the modulus of continuity $\omega(t)$ into consideration and applying Lemma 5.1 from [13] to each of the δ_i and e_i we find that

$$\sup_{f\in W^rH^\omega} \|f - A_n^\lambda f\|_C = \sup_{\varphi\in H^\omega} \frac{1}{\pi}\int_0^{2\pi} \Delta(t)\varphi(t)\, dt$$

$$\leqslant \frac{1}{\pi}\left\{\sum_{i=1}^{2n} \sup_{\varphi\in H^\omega}\int_{\delta_i} \Delta\varphi dt + \sum_{i=1}^{n} \sup_{\varphi\in H^\omega}\int_{e_i}\Delta\varphi dt\right\} = \frac{1}{\pi}\int_0^{2\pi}\Delta(t)\varphi_n(t)\, dt,$$

where $\phi_n(t) = \pm K(t - i\pi/n)$ on δ_i, $\phi_n(t) = \pm K\gamma$ on e_i and the signs are so chosen that

$$\operatorname{sgn} \varphi_n(t) = \operatorname{sgn} \Delta(t) \qquad (t \in e_i),$$

$$\operatorname{sgn} \varphi_n'(t) = -\operatorname{sgn} \int_{\frac{i\pi}{n}-\gamma}^{t} \Delta(u)\, du \qquad (t \in \delta_i).$$

It is easy to see that ϕ_n belongs to H^ω, has period $2\pi/n$ and is expressed in terms of $f_{n0}(\omega, t)$ by (11). Hence

$$\frac{1}{\pi}\int_0^{2\pi} \Delta(t)\varphi_n(t)\, dt = \frac{1}{\pi}\int_0^{2\pi} D_r(t)\varphi_n(t)\, dt = \| f_{nr} \|_C,$$

and so for $r = 1, 3, 5, \ldots$

$$\sup_{f \in W^r H^\omega} \| f - A_n^\lambda f \|_C = E_n(W^r H^\omega)_C.$$

5. Finally we show that if $\omega(t)$ is a modulus of continuity convex and nonlinear on $[0, \pi/n]$ then

$$\gamma_n(W^{2\nu}H^\omega)_C > E_n(W^{2\nu}H^\omega)_C \qquad (\nu,\ n = 1, 2, \ldots). \tag{17}$$

Let $W_0^r H_n^\omega$ be the class of functions $f \in W^r H^\omega$ of period $2\pi/n$ with zero mean over a period. We show that under the assumptions made concerning $\omega(t)$

$$\sup_{f \in W_0^{2\nu} H_n^\omega} \| f \|_C > \| f_{nr} \|_C, \tag{18}$$

whence (17) will follow.

If $f \in W_0^r H_n^\omega$, then

$$f(x) = \frac{1}{\pi n^{r-1}} \int_0^{\frac{2\pi}{n}} D_{nr}(x - t) f^{(r)}(t)\, dt,$$

where $D_{nr}(u) = D_r(nu)$. For $r = 2\nu$ the function $D_{nr}(u)$ is even and Lemma 5.2 of [13] allows one to write

$$\sup_{f \in W_0^r H_n^\omega} \| f \|_C = \frac{2}{\pi n^{r-1}} \sup_{\varphi \in H^\omega} \left| \int_0^{\frac{\pi}{n}} D_{nr}(t)\varphi(t)\, dt \right| = \frac{2}{\pi n^r} \int_0^{\frac{\pi}{n}} \bar{D}_{nr+1}(t)\, \omega'(t)\, dt,$$

where $\bar{D}_{nr+1}(t)$ is the nonincreasing rearrangement of the function $|D_{nr+1}(t)|$ on the interval $[0, \pi/n]$.

On the other hand, since $f_{nr}(\omega, t) \in W_0^r H_n^\omega$ for $r = 2\nu$, we have

$$\|f_{nr}\|_C=\frac{(-1)^\nu}{2}\left[f_{nr}\left(\frac{\pi}{2n}\right)-f_{nr}\left(-\frac{\pi}{2n}\right)\right]$$

$$=\frac{2}{\pi n^{r-1}}\int_0^{\frac{\pi}{n}}\delta_{nr}(t)\,f_{n0}\left(t+\frac{\pi}{n}\right)dt=\frac{2}{\pi n^r}\int_0^{\frac{\pi}{n}}\bar{\delta}_{nr+1}(t)\,\omega'(t)\,dt,$$

where

$$\delta_{nr}(t)=\frac{(-1)^\nu}{2}\left[D_{nr}(t)-D_{nr}\left(t+\frac{\pi}{n}\right)\right]=\sum_{i=1}^{\infty}\frac{\cos(2i-1)\,nt}{(2i-1)^r},$$

and $\bar{\delta}_{nr+1}(t)$ is the nonincreasing rearrangement over $[0, \pi/n]$ of the function

$$\delta_{nr+1}(t)=\frac{(-1)^\nu}{2}\left[D_{nr+1}(t)-D_{nr+1}\left(t+\frac{\pi}{n}\right)\right]=\sum_{i=1}^{\infty}\frac{\sin(2i-1)\,nt}{(2i-1)^{r+1}}.$$

Thus

$$\sup_{f\in W_0^r H_n^\omega}\|f\|_C-\|f_{nr}\|_C=\frac{2}{\pi n^r}\int_0^{\frac{\pi}{n}}[\bar{D}_{nr+1}(t)-\bar{\delta}_{nr+1}(t)]\,\omega'(t)\,dt,\quad r=2\nu.$$

We put $A(t)=\bar{D}_{nr+1}(t)-\bar{\delta}_{nr+1}(t)$. The graph of $\delta_{nr+1}(t)$ is symmetric about the line $t=\pi/2n$, and

$$\max_t\delta_{nr+1}(t)=\delta_{nr+1}\left(\frac{\pi}{2n}\right)=\left|D_{nr+1}\left(\frac{\pi}{2n}\right)\right|<\max_t|D_{nr+1}(t)|=\bar{D}_{nr+1}(0),$$

and so $A(0)>0$. Moreover, $A(\pi/n)=0$ and

$$\int_0^{\frac{\pi}{n}}A(t)\,dt=\int_0^{\frac{\pi}{n}}|D_{nr+1}(t)|\,dt-\int_0^{\frac{\pi}{n}}\delta_{nr+1}(t)\,dt=0.$$

It is easy to verify, on studying the behavior of the functions $D_{nr+1}(t)$ and $\delta_{nr+1}(t)$, that $A(t)$ changes sign (from positive to negative) at a unique point t_0, $0<t_0<\pi/n$. But then, if the modulus of continuity $\omega(t)$ is nonlinear on $[0, \pi/n]$ and consequently $\omega'(t)\not\equiv\text{const}$, it follows that $\int_0^{\pi/n}A(t)\omega'(t)dt>0$, and relations (18) and with them (17) are proved.

The more exact relations

$$\gamma_n(W^{2\nu}KH^\alpha)_C-E_n(W^{2\nu}KH^\alpha)_C=\frac{2K\alpha}{\pi n^r}\int_0^{\frac{\pi}{n}}A(t)\,t^{\alpha-1}dt>\frac{c}{n^{2\nu+\alpha}},$$

where $c>0$ is a constant dependent only on ν and α, hold for the class $W^{2\nu}KH^\alpha$ for $0<\alpha<1$.

BIBLIOGRAPHY

1. J. Favard, *Application de la formule sommatoire d'Euler à la démonstration de quelques propriétés extrémales des intégrales des fonctions périodiques ou presque-périodiques*, Mat. Tidskr. (B) 1936, 81–94.

2. ———, *Sur l'approximation des fonctions périodiques par des polynomes trigonométriques*, C. R. Acad. Sci. Paris **203** (1936), 1122–1124.

3. ———, *Sur les meilleurs procédés d'approximation de certaines classes de fonctions par des polynomes trigonométriques*, Bull. Sci. Math. (2) 61 (1937), 209–224.

4. N. I. Ahiezer and M. G. Kreĭn, *Sur la meilleure approximation des fonctions périodiques dérivables au moyen des sommes trigonométriques*, C. R. (Dokl.) Acad. Sci. URSS **15** (1937), 107–111.

5. S. M. Nikol'skiĭ, *Approximation of functions in the mean by trigonometric polynomials*, Izv. Akad. Nauk SSSR Ser. Mat. **10** (1946), 207–256; English transl., Amer. Math. Soc. Transl. (2) **102** (1973), 53–94. MR **8**, 149.

6. V. K. Dzjadyk, *On best approximation in the class of periodic functions having a bounded s-th derivative* $(0 < s < 1)$, Izv. Akad. Nauk SSSR Ser. Mat. 17 (1953), 135–162. (Russian) MR **14**, 867.

7. ———, *Best approximation on classes of periodic functions defined by kernels which are integrals of absolutely monotone functions*, Izv. Akad. Nauk SSSR Ser. Mat. 23 (1959), 933–950. (Russian) MR 22 #2844.

8. S. B. Stečkin, *On best approximation of certain classes of periodic functions by trigonometric polynomials*, Izv. Akad. Nauk SSSR Ser. Mat. **20** (1956), 643–648. (Russian) MR **18**, 303.

9. Sun Yung-sheng, *Best approximation of periodic differentiable functions by trigonometric polynomials*, Izv. Akad. Nauk SSSR Ser. Mat. **23** (1959), 67–92. (Russian) MR 21 #2155.

10. L. V. Taĭkov, *On approximating some classes of periodic functions in the mean*, Trudy Mat. Inst. Steklov. 88 (1967), 61–70 = Proc. Steklov Inst. Math. 88 (1967), 65–74. MR 36 #5595.

11. N. P. Korneĭčuk, *The best uniform approximation on certain classes of continuous functions*, Dokl. Akad. Nauk SSSR 140 (1961), 748–751 = Soviet Math. Dokl. 2 (1961), 1254–1257. MR 24 #A2181.

12. ———, *On methods of investigating extremal problems in the theory of best approximation*, Uspehi Mat. Nauk **29** (1974), no. 3 (177), 9–42 = Russian Math. Surveys 29 (1974), no. 3, 7–44.

13. ———, *Extreme values of functionals and best approximation on classes of periodic functions*, Izv. Akad. Nauk SSSR Ser. Mat. 35 (1971), 93–124 = Math. USSR Izv. 5 (1971), 97–130. MR 43 #3721.

14. ———, *Extremal properties of periodic functions*, Dopovīdī Akad. Nauk Ukraïn. RSR 1962, 993–998. (Ukrainian) MR 27 #2774.

Translated by J. L. B. COOPER

Trudy Mat. Inst. Steklov.
134 (1975)

Proc. Steklov Inst. Math.
134 (1975)

ON THE STABILIZATION OF FUNCTIONS AT INFINITY AND ON A HYPERPLANE

UDC 517.51

L. D. KUDRJAVCEV

Abstract. Conditions are given for the stabilization of a function $f(x)$, $x = (x_1, \ldots, x_n)$, to a trigonometric polynomial or a harmonic function when $x_n \to a$ ($a = 0$ or $+\infty$). The conditions are formulated in terms of the summability with weight of certain derivatives. The only restrictions imposed on the weight are of the type of its summability or nonsummability to some power.

Bibliography: 16 titles.

In the present paper we give conditions under which a function of n variables ($n = 1, 2, \ldots$) stabilizes to either a trigonometric polynomial or a harmonic function. We consider the case of stabilization when one of the coordinates tends to zero or infinity. The conditions are formulated in terms of the p-power summability with weight of certain derivatives. No restrictions are imposed on the weight other than summability or nonsummability to a specific power. By derivatives are understood generalized derivatives (see the monograph of S. M. Nikol'skiĭ [1], §4.1).

S. V. Uspenskiĭ [2] considered the question of stabilization of a function at infinity to a constant, S. L. Sobolev [3] studied the question of stabilization of functions at infinity to a polynomial, while T. S. Pigolkina [4], V. R. Portnov [5], V. N. Sedov [6] and F. G. Bulaevskaja [16] studied the question of stabilization to algebraic polynomials in the weight case. The present paper is a continuation of the author's papers [7]–[13].

We will say that a function $f(x)$, $x = (x_1, \ldots, x_n)$, stabilizes to a function $\phi(x)$ for $x_n \to a$ if

$$\lim_{x_n\to a} [f(x) - \varphi(x)] = 0.$$

In this case we will write

$$\operatorname*{stab}_{x_n\to a} f(x) - \varphi(x)$$

and say that $\phi(x)$ stabilizes $f(x)$.

AMS (MOS) subject classifications (1970). Primary 41A60.

§1. Auxiliary lemmas

Let $\alpha = \alpha(x) \geq 0$ denote a function defined on the interval $(0, 1)$. The condition

$$\int_0^1 [\alpha(x)]^{\frac{p}{1-p}} dx < +\infty, \qquad p > 1, \tag{1}$$

is called the weak degeneracy condition for $\alpha(x)$ at zero.

Let $\|F\|_p$, $p \geq 1$, denote the norm of a function F in the space $L_p(0, 1)$.

Lemma 1. *If a function f is absolutely continuous on each interval $[\epsilon, 1]$, $\epsilon > 0$, and*

$$\left\| \alpha \frac{df}{dx} \right\|_p < +\infty, \tag{2}$$

where $\alpha(a)$ satisfies the weak degeneracy condition (1) at zero, then the limit

$$\lim_{x \to +0} f(x) = A \tag{3}$$

exists and is finite, and

$$\|f\|_p \leqslant \left\| \alpha \frac{df}{dx} \right\|_p \|\alpha^{-1}\|_q + A, \qquad q = \frac{p}{p-1}. \tag{4}$$

This fact is well known (see [14] or [9]).

The limit (3) is called the trace of $f(x)$ at $x = 0$. The trace of a function f at $x = 0$ will also be called the trace at $x = 0$ of any function f^* that is equivalent to f on $(0, 1)$. In what follows a function will sometimes be replaced by an equivalent function without explicit mention.

Lemma 2. *Suppose a function $f(x, y)$ is absolutely continuous in $x \in (a, b)$ for almost all $y \in (0, 1)$, and suppose the limit*

$$\lim_{y \to +0} f(x, y) \tag{5}$$

exists and is finite for almost all x. If

$$\int_a^b dx \int_1^{+\infty} \left| \beta(y) \frac{\partial f(x, y)}{\partial x} \right|^p dy < +\infty, \tag{6}$$

where $\beta(y) \geq 0$ satisfies the condition

$$\int_0^t [\beta(y)]^{\frac{p}{1-p}} dy = O\left(t^{\frac{p}{p-1}}\right) \quad \text{for } t \to +0, \tag{7}$$

then the limit (5) does not depend on x.

Proof. For almost all $y \in (0, 1)$, viz. for those y for which $f(x, y)$ is by assumption absolutely continuous in x,

$$f(x'', y)-f(x', y)=\int_{x'}^{x''}\frac{\partial f(t, y)}{\partial x}\,dt, \qquad a<x'<b, \ a<x''<b.$$

Taking the absolute values of both sides of this equality, replacing the interval of integration in the right side by the whole interval (a, b), and integrating the left side with respect to x' and x'' over (a, b), we get

$$\int_a^b dx'\int_a^b |f(x'', y)-f(x', y)|\,dx''\leqslant (b-a)^2\int_a^b\left|\frac{\partial f(x, y)}{\partial x}\right|dx. \tag{8}$$

We now fix a number $M\in(0, 1)$, take the infimum of the left side of (8) over the almost all $y\in[M/2, M]$ for which (8) holds, integrate the right side with respect to y from $M/2$ to M and apply Hölder's inequality:

$$\begin{aligned}\inf_{\frac{1}{2}M\leqslant y\leqslant M}&\int_a^b dx'\int_a^b|f(x'', y)-f(x', y)|\,dx''\leqslant\frac{2(b-a)^2}{M}\int_{\frac{1}{2}M}^{M}dy\int_a^b\left|\frac{\partial f(x, y)}{dx}\right|dx\\
&\leqslant\frac{2(b-a)^2}{M}\left[\int_{\frac{1}{2}M}^{M}dy\int_a^b\left|\beta(y)\frac{\partial f(x, y)}{\partial x}\right|^p dx\right]^{\frac{1}{p}}\left\{\int_{\frac{1}{2}M}^{M}[\beta(y)]^{-q}\,dy\int_a^b dx\right\}^{\frac{1}{q}}\\
&\leqslant 2(b-a)^{2+\frac{1}{q}}\left[\int_{\frac{1}{2}M}^{M}dy\int_a^b\left|\beta(y)\frac{\partial f(x, y)}{\partial x}\right|^p dx\right]^{\frac{1}{p}}\left\{M^{-q}\int_0^M[\beta(y)]^{-q}\,dy\right\}^{\frac{1}{q}},\end{aligned}\tag{9}$$

$$\frac{1}{p}+\frac{1}{q}=1.$$

As $M\to+0$, the right side of (9) tends to zero by virtue of conditions (6) and (7) of the lemma. We choose a sequence $M_k\to+0$, $k=1, 2, \ldots$, such that the right side of (9) is less than $1/k$ for $M=M_k$. Then

$$\inf_{\frac{1}{2}M_k<y<M_k}\int_a^b dx'\int_a^b|f(x'', y)-f(x', y)|\,dx''<\frac{1}{k}.$$

There therefore exists a sequence $y_k\in[\frac{1}{2}M_k, M_k]$, $k=1, 2, \ldots$, such that

$$\int_a^b dx'\int_a^b|f(x'', y_k)-f(x', y_k)|\,dx''<\frac{2}{k}$$

and, consequently,

$$\lim_{k\to\infty}y_k=0, \qquad \lim_{k\to\infty}\int_a^b dx'\int_a^b|f(x'', y_k)-f(x', y_k)|\,dx''=0. \tag{10}$$

We assume, further, that the points x' and x'' have been chosen according to condition (5) so that the limit

$$\lim_{k\to\infty} [f(x'', y_k) - f(x', y_k)]$$

exists. Then

$$\int_a^b dx' \int_a^b \lim_{k\to\infty} |f(x'', y_k) - f(x', y_k)|\, dx''$$
$$\leqslant \lim_{k\to\infty} \int_a^b dx' \int_a^b |f(x'', y_k) - f(x', y_k)|\, dx'' = 0$$

by virtue of Fatou's lemma and conditions (10). Therefore

$$\lim_{k\to\infty} [f(x'', y_k) - f(x', y_k)] = 0$$

for almost all points (x', x'') in the sense of the two-dimensional measure. It follows that this equality is also valid for almost all points x' and x'' from the interval (a, b), which implies the independence from x of the limit (5) for almost all $x \in (a, b)$. The lemma is proved.

If $\alpha(x) = x^\lambda$, then the weak degeneracy condition (1) means that $\lambda < 1 - 1/p$. If $\beta(y) = y^\mu$, then condition (7) has the form $\mu \leq -1/p$.

Assuming that $\alpha(x)$ satisfies condition (1), we put

$$\varphi(x) = [\alpha(x)]^{\frac{1}{1-p}} \left\{ \int_0^x [\alpha(t)]^{\frac{p}{1-p}}\, dt \right\}^{-1}. \tag{11}$$

Then the inequality

$$\left\{ \int_0^\eta \left[\varphi(x) \int_0^x |F(t)|\, dt \right]^p dx \right\}^{\frac{1}{p}} \leqslant c \left\{ \int_0^\eta |\alpha(x) F(x)|^p\, dx \right\}^{\frac{1}{p}}$$

(the constant c does not depend on the function F), which is usually called the generalized Hardy inequality (see [14] or [9]), is valid for any measurable function F on the interval $(0, \eta)$.

Lemma 3. *If $f(x)$ is an absolutely continuous function on the interval $[\epsilon, 1]$, the limit $\lim_{x\to+0} f(x) = A$ exists and*

$$\int_0^1 \left| \alpha(x) \frac{df(x)}{dx} \right|^p dx < +\infty,$$

where the weight function $\alpha(x)$ satisfies the weak degeneracy condition (1), *then the inequality*

$$\int_0^\eta |\varphi(x)[f(x) - A]|^p\, dx \leqslant c \int_0^\eta \left| \alpha(x) \frac{df(x)}{dx} \right|^d dx, \tag{12}$$

where the constant c does not depend on $f(x)$, is valid for any $\eta \in (0, 1)$.

Inequality (12) follows at once from the generalized Hardy inequality for $df(x)/dx$:

$$\int_0^{\eta} |\varphi(x)[f(x)-A]|^p\,dx \leqslant \int_0^{\eta}\left[\varphi(x)\int_0^{x}\left|\frac{df(t)}{dt}\right|dt\right]^p dx \leqslant c\int_0^{\eta}\left|\alpha(x)\frac{df(x)}{dx}\right|^p dx.$$

Lemma 4. *If $f(x)$ is an absolutely continuous function on the interval $[\epsilon, 1]$ for any $\epsilon > 0$ and there exists a function $\gamma(x) \geq 0$ such that*

$$\int_0^1 \left|[\gamma(x)]^{\frac{1}{1-p}} f(x)\right|^p dx < +\infty, \tag{13}$$

$$\int_0^1 \left|\gamma(x)\frac{df(x)}{dx}\right|^p dx < +\infty, \tag{14}$$

then $\lim_{x\to+0} f(x)$ exists and is finite.

Corollary. *If, in addition to the conditions of the lemma,*

$$\int_0^1 [\gamma(x)]^{\frac{p}{1-p}}\,dx = +\infty, \tag{15}$$

then

$$\lim_{x\to+0} f(x) = 0. \tag{16}$$

Proof. We note that

$$|f(x)|^p = |f(1)|^p - \int_x^1 \frac{d}{dt}|f(t)|^p\,dt. \tag{17}$$

Applying Hölder's inequality, we get by virtue of (13) and (14) that

$$\int_x^1 \left|\frac{d}{dt}|f(t)|^p\right| dt = p\int_x^1 |f(t)|^{p-1}\left|\frac{df(t)}{dt}\right| dt$$

$$\leqslant \left\{\int_0^1 \left|[\gamma(t)]^{\frac{1}{1-p}} f(t)\right|^p dt\right\}^{\frac{p-1}{p}} \left\{\int_0^1 \left|\gamma(t)\frac{df(t)}{dt}\right|^p dt\right\}^{\frac{1}{p}}; \qquad x\in(0,\ 1).$$

Thus, under the assumption of the lemma, $d|f(x)|^p/dx$ is summable on $[0, 1]$, and hence the right side of (17) tends to a finite limit as $x \to +0$, which implies the existence of $\lim_{x\to+0} f(x)$. The lemma is proved. If the limit is different from zero, then the integral (15) can diverge only if (13) diverges; therefore condition (15) implies (16).

If $\alpha(x) = x^{\lambda}$, condition (15) has the form $\lambda \geq 1 - 1/p$.

Lemmas 1–4 carry over to the case $x \to +\infty$. Their proofs are carried out analogously; we will therefore confine ourselves to their formulations.

The condition

$$\int_1^{+\infty} [\alpha(x)]^{\frac{p}{1-p}} dx < +\infty, \quad p > 1, \tag{18}$$

is called the weak degeneracy condition for $\alpha(x)$ at infinity. If $\alpha(x) = x^\lambda$, this condition means that $\lambda > 1 - 1/p$.

Lemma 1′. *If $f(x)$ is an absolutely continuous function on each interval $[1, \eta]$, $\eta > 1$, and*

$$\left\| \alpha \frac{df}{dx} \right\|_p < +\infty, \tag{19}$$

where $\alpha(x)$ satisfies the weak degeneracy condition (18) *at infinity, then the limit*

$$\lim_{x \to +\infty} f(x) = A$$

exists and is finite.

By the norm $\|F\|_p$ here is meant, of course, $\{\int_1^{+\infty} |F(x)|^p dx\}^{1/p}$.

Lemma 2′. *Suppose a function $f(x, y)$ is absolutely continuous in $x \in (a, b)$ for almost all $y > 1$ and suppose $\lim_{y \to +\infty} f(x, y)$ exists and is finite for almost all x. If*

$$\int_a^b dx \int_1^{+\infty} \left| \beta(y) \frac{\partial f(x, y)}{\partial x} \right|^p dy < +\infty, \tag{20}$$

where

$$\int_1^t [\beta(y)]^{\frac{p}{1-p}} dy = O\left(t^{\frac{p}{p-1}}\right) \quad \textit{for } t \to +\infty, \tag{21}$$

then $\lim_{y \to +\infty} f(x, y)$ does not depend on x.

If $\beta(y) = y^\mu$, condition (21) means $\mu \geq -1/p$ (see [13]).

Assuming that $\alpha(x)$ satisfies condition (18), we put

$$\psi(x) = [\alpha(x)]^{\frac{1}{1-p}} \left\{ \int_1^x [\alpha(t)]^{\frac{p}{1-p}} dt \right\}^{-1}. \tag{22}$$

Then the inequality

$$\left\{ \int_\eta^{+\infty} \left[\psi(x) \int_x^{+\infty} |F(t)| \, dt \right]^p dx \right\}^{\frac{1}{p}} \leq c \left[\int_\eta^{+\infty} |\alpha(x) f(x)|^p dx \right]^{\frac{1}{p}} \tag{23}$$

(here $\eta \geq 1$ and the constant c does not depend on F), which is also usually called the generalized Hardy inequality, is valid for any measurable function F on the halfline $x > 1$.

Lemma 3′. *If $f(x)$ is an absolutely continuous function on each interval*

$[1, \eta]$, $1 < \eta < +\infty$, if $\lim_{x\to+\infty} f(x) = A$ *exists, and if*

$$\int_1^{+\infty} \left|\alpha(x)\frac{df(x)}{dx}\right|^p dx < +\infty,$$

where the weight function $\alpha(x)$ *satisfies the weak degeneracy condition* (18) *at infinity, then*

$$\int_\eta^{+\infty} |\psi(x)[f(x) - A]|^p\, dx \leqslant c \int_\eta^{+\infty} |\alpha(x) f'(x)|^p\, dx \tag{24}$$

for any $\eta \geq 1$; *here* $\psi(x)$ *is defined by* (22) *and the constant* c *does not depend on* f.

Lemma 4′. *Suppose* $f(x)$ *is an absolutely continuous function on the interval* $[1, \eta]$ *for any* $\eta > 1$. *If there exists a function* $\gamma(x) \geq 0$ *such that*

$$\int_1^{+\infty} \left|[\gamma(x)]^{\frac{1}{1-p}} f(x)\right|^p dx < +\infty, \tag{25}$$

$$\int_1^{+\infty} \left|\gamma(x)\frac{df(x)}{dx}\right|^p dx < +\infty, \tag{26}$$

then the limit

$$\lim_{x\to+\infty} f(x) \tag{27}$$

exists and is finite.

Corollary. *If, in addition to the conditions of the lemma,*

$$\int_1^{+\infty} [\gamma(x)]^{\frac{p}{1-p}}\, dx = +\infty, \tag{28}$$

then $\lim_{x\to+\infty} f(x) = 0$.

It follows, in particular, when $\gamma(x) \equiv 1$ that if $f(x) \in W_p^1(1, +\infty)$, then

$$\lim_{x\to+\infty} f(x) = 0.$$

If $\gamma(x) = x^\lambda$, condition (28) has the form $\lambda \leq 1 - 1/p$.

An additional useful lemma can be proved for the case $x \to +\infty$.

Lemma 5. *Suppose* $f(x)$ *has an absolutely continuous derivative on the interval* $[1, \eta]$ *for any* $\eta > 1$, *and suppose* $\alpha(x)$ *and* $\psi(x)$ (*see* (22)) *satisfy the weak degeneracy condition at infinity. If*

$$\int_1^{+\infty} \left|\alpha(x)\frac{d^2f(x)}{dx^2}\right|^p dx < +\infty, \tag{29}$$

then the limits

$$\lim_{x\to+\infty}\frac{f(x)}{x}=\lim_{x\to+\infty}\frac{df(x)}{dx}=A \tag{30}$$

and

$$\lim_{x\to+\infty}[f(x)-Ax]=B \tag{31}$$

exist (i.e. $f(x)$ has an asymptote), and consequently $f(x)/x-A=O(1/x)$ for $x\to+\infty$.

In fact, since $\alpha(x)$ satisfies (18), it follows from (29) according to Lemma 1′ that the limit

$$\lim_{x\to+\infty}\frac{df(x)}{dx}=A$$

exists and is finite. By virtue of Lemma 3′ we have

$$\int_1^{+\infty}\left|\psi(x)\left[\frac{df(x)}{dx}-A\right]\right|^p dx\leqslant c\int_1^{+\infty}\left|\alpha(x)\frac{d^2f(x)}{dx^2}\right|^p dx,$$

or, equivalently,

$$\int_1^{+\infty}\left|\psi(x)\frac{d}{dx}[f(x)-Ax]\right|^p dx\leqslant c\int_1^{+\infty}\left|\alpha(x)\frac{d^2f(x)}{dx^2}\right|^p dx.$$

It follows according to Lemma 1′ again that the limit (31) exists. Finally,

$$\frac{f(x)}{x}-A=\frac{f(x)-Ax-B}{x}+\frac{B}{x}=o\left(\frac{1}{x}\right)+O\left(\frac{1}{x}\right)=O\left(\frac{1}{x}\right),\quad x\to+\infty$$

and therefore the limit

$$\lim_{x\to+\infty}\frac{f(x)}{x}=A$$

exists. Equality (30) and, consequently, Lemma 5 are proved.

When $\alpha(x)=x^\lambda$, $\lambda>1-1/p$, we have $\psi(x)=x^{\lambda-1}$, and the weak degeneracy condition for $\alpha(x)$ and $\psi(x)$ means $\lambda>2-1/p$.

§2. Stabilization to trigonometric polynomials

Consider a trigonometric polynomial $T=T(x)$ in n variables, $x=(x_1,\dots,x_n)$, of order m_k in the variable x_k, defined by generally complex frequencies ω_{j_k}, $j_k=1,\dots,m_k$, $k=1,\dots,n$, i.e.

$$T(x)=\sum_{j\in J}a_je^{i(\omega_j,x)}, \tag{32}$$

where

$$\begin{gathered}J=\{j=(j_1,j_2,\dots,j_n):j_k=0,\ \pm1,\ \pm2,\dots,\pm m_k;\quad k=1,2,\dots,n\},\\ \omega_{0k}=0,\quad \omega_{-j_k}=-\omega_{j_k},\quad k=1,2,\dots,n,\\ \omega_j=(\omega_{j_1},\omega_{j_2},\dots,\omega_{j_n}),\quad j=(j_1,j_2,\dots,j_n)\in J.\end{gathered} \tag{33}$$

Thus

$$T(x) = T(x_1, x_2, \ldots, x_n) = \sum_{j_1=-m_1}^{m_1} \sum_{j_2=-m_2}^{m_2} \cdots \sum_{j_n=-m_n}^{m_n} a_{j_1 j_2 \ldots j_n} e^{i(\omega_{j_1} x_1 + \omega_{j_2} x_2 + \ldots + \omega_{j_n} x_n)}. \tag{34}$$

In this section we will give conditions under which the stabilization

$$\underset{x_i \to a}{\operatorname{stab}} f(x) = T(x)$$

(the ith coordinate axis is fixed) will hold in the case of real frequencies ω_{j_k}.

Suppose given a (generally complex) number λ, and consider a trigonometric polynomial $T_\lambda(t)$ in one variable t of the form

$$T_\lambda(t) = a_0 + a_\lambda e^{i\lambda t} + a_{-\lambda} e^{-i\lambda t}.$$

We introduce for any complex μ the differential operators

$$\begin{aligned} \underset{t}{A_\mu^*} &= \frac{1}{2\mu^2} e^{-i\mu t} \left(\frac{d^2}{dt^2} + i\mu \frac{d}{dt} \right), \quad \mu \neq 0, \\ \underset{t}{L_\mu^*} &= \frac{d^2}{dt^2} + \mu^2. \end{aligned} \tag{35}$$

Then, for an arbitrary complex number μ,

$$\begin{aligned} \underset{t}{A_\mu^*} T_\lambda(t) &= a_\mu \quad \text{for} \quad \mu = \pm\lambda, \quad \underset{t}{A_\mu^*} a_0 = 0, \\ \underset{t}{L_\mu^*} T_\lambda(t) &= \mu^2 - \lambda^2, \quad \underset{t}{L_\mu^*} a_0 = \mu^2 a_0. \end{aligned} \tag{36}$$

Suppose given now m generally complex frequencies $\omega_k \neq 0$, $k = 1, \ldots, m$. Let $\omega_{-k} = -\omega_k$ and $\omega_0 = 0$. We introduce the differential operators

$$\begin{aligned} \underset{t}{A_k} &= \underset{t}{A_{\omega_k}^*}, \quad \underset{t}{L_k} = \underset{t}{L_{\omega_k}^*}, \\ \underset{t}{L^{(m)}} &= \prod_{k=1}^{m} \underset{t}{L_k}, \\ \underset{t}{L_k^{(m)}} &= \prod_{\substack{j=1 \\ j \neq k}}^{m} \underset{t}{L_j}, \quad k = 1, 2, \ldots, m. \end{aligned} \tag{37}$$

We also put

$$\varkappa = \prod_{j=1}^{m} \omega_j^2, \quad \varkappa_k = \prod_{\substack{j=1 \\ j \neq k}}^{m} (\omega_j^2 - \omega_k^2), \quad k = 1, 2, \ldots, m. \tag{38}$$

Now for the trigonometric polynomial

$$T(t) = \sum_{k=-m}^{m} a_k e^{i\omega_k t} \tag{39}$$

we will have

$$a_0 = \frac{1}{\varkappa} \underset{t}{L}^{(m)} T(t),$$
$$a_k = \frac{1}{\varkappa_{\hat{k}}} \underset{t}{A_k} \underset{t}{L}_{\hat{k}}^{(m)} T(t), \quad k = \pm 1, \pm 2, \ldots, \pm m \tag{40}$$

by virtue of (35) and (36).

We return again to the trigonometric polynomial (32). Let

$$\varkappa_\nu = \prod_{j=1}^{m_\nu} \omega_{j_\nu}^2, \quad \varkappa_{\hat{k}_\nu} = \prod_{\substack{j_\nu = 1 \\ j_\nu \neq k_\nu}}^{m_\nu} (\omega_{j_\nu}^2 - \omega_{k_\nu}^2),$$

$$D_{j_\nu} = \begin{cases} \dfrac{1}{\varkappa_\nu} \underset{x_\nu}{L}^{(m_\nu)} & \text{for } j_\nu = 0, \\ \dfrac{1}{\varkappa_{\hat{j}}} \underset{x_\nu}{A_{j_\nu}} \underset{x_\nu}{L}_{\hat{j}_\nu}^{(m_\nu)} & \text{for } j_\nu \neq 0, \end{cases} \tag{41}$$

and (see (33))

$$D_j = \prod_{\nu=1}^{n} D_{j_\nu}, \quad j = (j_1, \ldots, j_\nu, \ldots, j_n) \in J, \tag{42}$$

or, in more detail (see (42), (41), (38), (37) and (35)),

$$D_j = \prod_{j_\nu = 0} \frac{1}{\varkappa_\nu} \underset{x_\nu}{L}^{(m_\nu)} \prod_{j_\nu \neq 0} \frac{1}{\varkappa_{\hat{j}_\nu}} \underset{x_\nu}{A_{j_\nu}} \underset{x_\nu}{L}_{\hat{j}_\nu}^{(m_\nu)}, \quad j \in J. \tag{43}$$

Then the formulas

$$a_j = D_j T(x) \tag{44}$$

for the coefficients a_j of the trigonometric polynomial (32) will be valid by virtue of (40).

We will now prove a lemma, needed for the sequel, on determinants of a special form.

Consider the square matrix

$$\| \lambda_{j_1}^{k_1} \lambda_{j_2}^{k_2} \ldots \lambda_{j_\nu}^{k_\nu} \|, \tag{45}$$

where the j_s are indices and the k_s are powers, $j_s, k_s = 1, \ldots, \mu_s$; $s = 1, \ldots, \nu$. The multi-indices $j = (j_1, \ldots, j_\nu)$ and $k = (k_1, \ldots, k_n)$ will be regarded as the row and column indices respectively. We denote the determinant of the matrix (45) by

$$\Delta_{\mu_1\mu_2\ldots\mu_\nu} = \Delta(\lambda_{1_1}, \ldots, \lambda_{\mu_1}; \lambda_{1_2}, \ldots, \lambda_{\mu_2}; \ldots; \lambda_{1_\nu}, \ldots, \lambda_{\mu_\nu}).$$

Lemma 6. *The following recurrence formula is valid*:

$$\Delta_{\mu_1\mu_2\ldots\mu_\nu} = \lambda_{1_\nu}^{\mu_1\mu_2\ldots\mu_{\nu-1}} \prod_{s=2}^{\mu_\nu} (\lambda_{s_\nu} - \lambda_{1_\nu})^{\mu_1\mu_2\ldots\mu_{\nu-1}} \Delta_{\mu_1\mu_2\ldots\mu_{\nu-1}} \Delta_{\mu_1\mu_2\ldots\mu_{\nu-1}(\mu_\nu - 1)}, \tag{46}$$

where

$$\Delta_{\mu_1\mu_2\ldots\mu_{\nu-1}} = \Delta(\lambda_{1_1}, \ldots, \lambda_{\mu_1}; \ldots; \lambda_{1_{\nu-1}}, \ldots, \lambda_{\mu_{\nu-1}});$$
$$\Delta_{\mu_1\mu_2\ldots\mu_{\nu-1}(\mu_\nu-1)} = \Delta(\lambda_{1_1}, \ldots, \lambda_{\mu_1}; \ldots; \lambda_{\mu_{\nu-1}}; \lambda_{2_\nu}, \ldots, \lambda_{\mu_\nu}).$$

Corollary. *If* $\lambda_{j_s} \neq 0$ *and* $\lambda_{j_s} \neq \lambda_{k_s}$ *for* $j_s \neq k_s$, $j_s, k_s = 1, \ldots, \mu_s$, $s = 1, \ldots, \nu$, *then*

$$\Delta_{\mu_1\mu_2\ldots\mu_\nu} \neq 0. \tag{47}$$

Proof. We order the rows and columns of the matrix (45) lexicographically by writing the first row, for example, in the form

$$\lambda_{1_1}\lambda_{1_2}\ldots\lambda_{1_\nu},\ \lambda_{1_1}\lambda_{1_2}\ldots\lambda_{1_\nu}^2,\ \lambda_{1_1}\lambda_{1_2}\ldots\lambda_{1_\nu}^3,\ \ldots.$$

We subtract from all of the columns except those with indices

$$l = k\mu_\nu + 1, \qquad k = 0, 1, \ldots, \mu_1\mu_2\ldots\mu_{\nu-1} - 1, \tag{48}$$

the preceding column multiplied by λ_{1_ν}. As a result, in all of the columns with indices (48) all of the elements except those in the columns with indices (48) will be equal to zero. Applying the Laplace expansion theorem for determinants and extracting from the two determinants obtained in this way the factors $\lambda_{1_\nu}^{\mu_1\cdots\mu_{\nu-1}}$ and $\Pi_{k=2}^{\mu_\nu}(\lambda_{k_\nu} - \lambda_{1_\nu})^{\mu_1\cdots\mu_{\nu-1}}$, respectively, we obtain (46).

Let us prove the corollary. In the case $\nu = 1$ the determinant of the matrix (45) is a Vandermonde determinant and hence equal to the product of factors of the form $\lambda_{k_1} - \lambda_{j_1}$, $k > j$. Therefore, by virtue of the recurrence formula (46), the determinant $\Delta_{\mu_1\cdots\mu_\nu}$ can be represented as a product of powers of the λ_{j_s} and the differences $\lambda_{k_s} - \lambda_{j_s}$, $k_s > j_s$, $s = 1, \ldots, \nu$, and the corollary follows.

We will now prove a theorem on the "trigonometric polynomial representation" of functions of several variables. In this connection, we will use formula (44) for the coefficients of a trigonometric polynomial as a heuristic tool for obtaining the corresponding coefficients of an arbitrary function (see (49), below).

Theorem 1. *Suppose a function* $f(x) = f(x_1, \ldots, x_n)$ *has the partial derivatives*

$$D^l f(x) = \frac{\partial^{l_1+l_2+\ldots+l_n}}{\partial x_1^{l_1}\partial x_2^{l_2}\ldots\partial x_n^{l_n}} f(x), \quad l = (l_1, l_2, \ldots, l_n),\ l_\nu = 1, 2, \ldots, 2m_\nu,$$

suppose $\omega_{j_\nu} \neq 0$ *and* $\omega_{j_\nu} \neq \omega_{k_\nu}$ *for* $j_\nu \neq k_\nu$, $j_\nu, k_\nu = 1, \ldots, m_\nu$, $\nu = 1, \ldots, n$, *and suppose*

$$a_j(x) = D_j f(x) \tag{49}$$

for each $j = (j_1, \ldots, j_n) \in J$ *(see (43)). Then*

$$f(x) = \sum_{j\in J} a_j(x)\, e^{i(\omega_j, x)}. \tag{50}$$

Proof. We first note that by virtue of (41)–(43) we have

$$a_0(x) = D_0 f(x) = \prod_{\nu=1}^{n} \frac{1}{\varkappa_\nu} L_{x_\nu}^{(m_\nu)} f(x) = f(x) + \sum_{k \in K} b_k D^{2k} f(x),$$

where

$$D^{2k} = \frac{\partial^{2(k_1+k_2+\ldots+k_n)}}{\partial x_1^{k_1} \partial x_2^{k_2} \ldots \partial x_n^{k_n}},$$

$$k \in K = \left\{ k = (k_1, k_2, \ldots, k_n) : k_s = 0, 1, \ldots, m_s, \sum_{s=1}^{n} k_s^2 > 0 \right\}.$$

We next note that by virtue of (35), (37), (42) and (43) the terms with derivatives of odd order in the expressions for the coefficients $a_{j_1 \ldots j_n}$ and $a_{-j_1 \ldots j_n}$ will be equal in absolute value but opposite in sign. The same is true, of course, for any index j_k, $k = 1, \ldots, n$. Therefore

$$\sum_{j \in J} a_j(x) e^{i(\omega_j, x)} = f(x) + \sum_{k \in K} c_k D^{2k} f(x), \tag{51}$$

where the c_k are certain coefficients.

If any trigonometric polynomial of form (32) is taken as $f(x)$ in (51), the identity

$$\sum_{k \in K} c_k D^{2k} T(x) = 0 \tag{52}$$

will hold by virtue of (32) and (44).

We now fix a multi-index $j \in K$, and let

$$T_j(x) = e^{i(\omega_{j_1} x_1 + \omega_{j_2} x_2 + \ldots + \omega_{j_n} x_n)}, \quad j = (j_1, j_2, \ldots, j_n).$$

Substituting $T_j(x)$ in (52), we get

$$\sum_{k \in K} (-1)^{|k|} \omega_{j_1}^{2k_1} \omega_{j_2}^{2k_2} \ldots \omega_{j_n}^{2k_n} c_k = 0, \quad |k| = k_1 + k_2 + \ldots + k_n. \tag{53}$$

This is a linear equation for the coefficients c_k, the number of which is equal to the number of elements in the set K, viz. $(m_1 + 1)(m_2 + 1) \cdots (m_n + 1) - 1$. An equal number of equations of form (53) can be obtained by substituting all of the possible polynomials $T_j(x)$, $j \in K$, in (52).

We will show that the determinant of the system of equations (53) obtained in this way is not equal to zero. In the matrix

$$\left\| \omega_{j_1}^{2k_1} \omega_{j_2}^{2k_2} \ldots \omega_{j_n}^{2k_n} \right\|$$

of system (53) the multi-indices $j = (j_1, \ldots, j_n)$ and $k = (k_1, \ldots, k_n)$ will be regarded as the row and column indices respectively. For the sake of definiteness we order the rows and columns of the determinant of system (53) lexicographically, in the same way as was done in the proof of Lemma 6. As a result of the differentiation $D^{2k} T_j(x)$, it is obvious that nonzero coefficients are obtained for only those c_k, $k = (k_1, \ldots, k_n)$, for which $\omega_{j_\nu} \neq 0$ and $k_\nu \neq 0$, $\nu = 1, \ldots, n$. Suppose

$$T_j(x)=e^{i(\omega_{js_1}x_{s_1}+\omega_{js_2}x_{s_2}+\ldots+\omega_{js_\nu}x_{s_\nu})},$$
$$\omega_{js_\mu}\neq 0, \quad \mu=1,\ 2,\ldots,\nu. \tag{54}$$

Subtracting the appropriate preceding rows from the row of the determinant of system (53) obtained by substituting this polynomial in (52), we obtain a row whose only nonzero elements are those with column indices $k=(k_1,\ldots,k_n)$ such that $k_l=0$ if and only if $l\neq s_\mu$ for $\mu=1,\ldots,\nu$.

If we now take all of the possible $T_j(x)$ of form (54) for fixed indices $s_1, \ldots, s_\nu$, we will obtain $\Pi_{i=1}^{\nu} m_{s_i}$ rows, in which the only nonzero elements are on the indicated Πm_{s_i} columns, forming a matrix of the form

$$\left\|(-1)^{p_{s_1}+p_{s_2}+\cdots+p_{s_\nu}}\omega_{js_1}^{2p_{s_1}}\omega_{js_2}^{2p_{s_2}}\ldots\omega_{js_\nu}^{2p_{s_\nu}}\right\|, \tag{55}$$
$$p_{s_\mu}=1,\ 2,\ldots,m_{s_\mu}, \quad \mu=1,\ 2,\ldots,\nu,$$

i.e. a matrix of form (45) "to within sign." According to the corollary of Lemma 6, the determinant of this matrix is not zero by virtue of the assumptions made concerning the frequencies ω_{j_ν}. Inasmuch as after the indicated transformations the determinant of system (53) can be represented by means of the Laplace expansion theorem as a product of determinants of matrices of form (55), it is also not equal to zero.

Thus (53) is a system of linear equations having a nonvanishing determinant and hence only the trivial solution $c_k=0$, $k=(k_1,\ldots,k_n)\in K$. Substituting these values for the c_k in (51), we obtain (50). The theorem is proved.

Remark. Theorem 1 was proved for $n=1$ in [12]. In this case the solutions c_k of system (53) are found in explicit form. This system had been previously solved in explicit form for integral frequencies ω_j in [11]. In both cases the proof makes use of (52). V. I. Burenkov has proposed a direct method for finding the solutions of (53) when $n=1$, which is based on the formula

$$\sum_{\{\nu_1,\nu_2,\ldots,\nu_j\}_{\hat{k}}^{(m)}}\lambda_{\nu_1}\lambda_{\nu_2}\ldots\lambda_{\nu_j}=\sum_{\{\nu_1,\nu_2,\ldots,\nu_j\}^{(m)}}\lambda_{\nu_1}\lambda_{\nu_2}\ldots\lambda_{\nu_j}+\sum_{\mu=1}^{m-1}A_\mu\lambda_k^\mu; \tag{56}$$

here $\lambda_1,\ldots,\lambda_m$ are given complex numbers, the coefficients A_μ do not depend on k, the symbol $\Sigma_{\{\nu_1,\ldots,\nu_j\}^{(m)}}$ indicates a sum over all possible systems consisting of j elements chosen from the m natural numbers $1,\ldots,m$, and the symbol $\Sigma_{\{\nu_1,\ldots,\nu_j\}_{\hat{k}}^{(m)}}$ indicates a sum over all possible systems consisting of j elements chosen from the $m-1$ natural numbers $1,\ldots,\hat{k},\ldots,m$ (a "cap" means that the expression under it is missing). Formulas (56) permit us to reduce system (53) when $n=1$ to the system

$$\sum_{k=1}^{m}c_k=1, \quad \sum_{k=1}^{m}c_k\lambda_k^\mu=0, \quad \mu=1,\ 2,\ldots,m-1.$$

The solution of this system has the form (see [12])

$$c_k = \frac{\lambda_1}{\lambda_1 - \lambda_k} \cdots \widehat{\frac{\lambda_k}{\lambda_k - \lambda_k}} \cdots \frac{\lambda_m}{\lambda_m - \lambda_k}, \qquad k = 1, 2, \ldots, m.$$

In the case of integral frequencies, Burenkov has found a derivation of the formulas for the roots c_k of system (53) in the one-dimensional case on the basis of the formula

$$\sum_{\nu=0}^{m} (-1)^\nu C_m^\nu \nu^\mu = 0$$

for the binomial coefficients C_m^ν.

We introduce the additional notation

$$B_j^{(\mu)} = \prod_{\substack{j_\nu=0\\ \nu\neq\mu}} \frac{1}{\varkappa_\nu} \underset{x_\nu}{L^{(m_\nu)}} \prod_{\substack{j_\nu\neq0\\ \nu\neq\mu}} \frac{1}{\varkappa_{\hat{j}_\nu}} \underset{x_\nu}{A} j_\nu \underset{x_\nu \hat{j}_\nu}{L^{(m_\nu)}}, \tag{57}$$

$$\mu = 1, 2, \ldots, n, \qquad j = (j_1, j_2, \ldots, j_n) \in J.$$

Lemma 7. *Suppose a function $f(x)$ has the partial derivatives $D^l f(x)$, $l = (l_1, \ldots, l_n)$, $l_\nu = 1, \ldots, 2m_\nu$, $\nu = 1, \ldots, \hat{\mu}, \ldots, n$, $l_\mu = 2m_\mu + 1$, all of the frequencies ω_{j_k} are real and for definiteness*

$$0 = \omega_{0_k} < \omega_{1_k} < \cdots < \omega_{jk} < \cdots < \omega_{mk}, \qquad k = 1, 2, \ldots, \hat{n}. \tag{58}$$

Then for any $j = (j_1, \ldots, j_n) \in J$

$$\left|\frac{\partial a_j(x)}{\partial x_\mu}\right| \leqslant \frac{1}{\varkappa_\mu} \left| B_j^{(\mu)} \frac{\partial}{\partial x_\mu} \underset{x_\mu}{L^{(m_\mu)}} f(x) \right|, \qquad j_\mu = 0, \tag{59}$$

$$\left|\frac{\partial a_j(x)}{\partial x_\mu}\right| \leqslant \frac{1}{2\varkappa_{\hat{j}_\mu} \omega_{j_\mu}^2} \left| B_j^{(\mu)} \frac{\partial}{\partial x_\mu} \underset{x_\mu}{L^{(m_\mu)}} f(x) \right|, \qquad j_\mu \neq 0. \tag{60}$$

Proof. Inequality (59) follows at once from (42), (43) and (49). In order to prove (60), we note that (see (35))

$$\frac{\partial}{\partial x_\mu} \underset{x_\mu}{A_{j_\mu}} = \frac{1}{2\omega_{j_\mu}^2} \frac{\partial}{\partial x_\mu} e^{-i\omega_{j_\mu} x_\mu} \left(\frac{\partial^2}{\partial x_\mu^2} + i\omega_{j_\mu} \frac{\partial}{\partial x_\mu} \right) = \frac{1}{2\omega_{j_\mu}^2} e^{-i\omega_{j_\mu} x_\mu} \frac{\partial}{\partial x_\mu} \left(\frac{\partial^2}{\partial x_\mu^2} + \omega_{j_\mu}^2 \right). \tag{61}$$

It therefore follows from (49), (43) and (57) that

$$\begin{aligned}
\frac{\partial a_j(x)}{\partial x_\mu} &= B_j^{(\mu)} \frac{\partial}{\partial x_\mu} \frac{1}{\varkappa_{\hat{j}_\mu}} \underset{x_\mu}{A_{j_\mu}} \underset{x_\mu \hat{j}_\mu}{L^{(m_\mu)}} f(x) \\
&= B_j^{(\mu)} \frac{1}{2\varkappa_{\hat{j}_\mu} \omega_{j_\mu}^2} e^{-i\omega_{j_\mu} x_\mu} \frac{\partial}{\partial x_\mu} \left(\frac{\partial^2}{\partial x_\mu^2} + \omega_{j_\mu}^2 \right) \underset{x_\mu \hat{j}_\mu}{L^{(m_\mu)}} f(x) \\
&= \frac{1}{2\varkappa_{\hat{j}_\mu} \omega_{j_\mu}^2} e^{-i\omega_{j_\mu} x_\mu} \frac{\partial}{\partial x_\mu} \underset{x_\mu}{L^{(m_\mu)}} f(x)
\end{aligned}$$

when $j_\mu \neq 0$, which implies (60).

We now put

$$D_{j_{\hat{n}}} = \prod_{\mu=1}^{n-1} D_{j_\mu}; \tag{62}$$

$$a_{j_{\hat{n}}}(x) = D_{j_{\hat{n}}} f(x), \qquad j_{\hat{n}} \in J_{\hat{n}}; \tag{63}$$

$$J_{\hat{n}} = \{j_{\hat{n}} = (j_1, j_2, \ldots, j_{n-1}) : j_\mu = 0, \pm 1, \ldots, \pm m_\mu;\ \mu = 1, 2, \ldots, n-1\}. \tag{64}$$

By virtue of Theorem 1, any function $f(x)$ having the partial derivatives $D^{l_{\hat{n}}} f(x)$, where

$$l_{\hat{n}} \in L_{\hat{n}} = \{l_{\hat{n}} = (l_1, l_2, \ldots, l_{n-1}, 0) : l_\nu = 1, 2, \ldots, 2m_\nu;\ \nu = 1, 2, \ldots, n-1\}, \tag{65}$$

satisfies the identity

$$f(x) = \sum_{j_{\hat{n}} \in J_{\hat{n}}} a_{j_{\hat{n}}}(x) f(x) \tag{66}$$

(the variable x_n plays the role of a parameter here).

Lemma 8. *Suppose a function $f(x)$ has the partial derivatives $D^{l_{\hat{n}}} D_{x_n} f(x)$, $l_{\hat{n}} \in L_{\hat{n}}$, and condition* (58) *is satisfied. Then*

$$\left| \frac{\partial}{\partial x_n} a_{j_{\hat{n}}}(x) \right| \leqslant c \sum_{l_{\hat{n}} \in L_{\hat{n}}} |D^{l_{\hat{n}}} D_{x_n} f(x)|, \qquad j_{\hat{n}} \in J_{\hat{n}}, \tag{67}$$

where the constant c does not depend on f.

This follows from (63), (62) and (42).

We now proceed to the question of stabilization of functions. We will consider the cases of stabilization of a function for $x_n \to 0$ and for $x \to +\infty$ when the stabilizing function is a trigonometric polynomial depending either on all of the variables $x_1, \ldots, x_n$ or on just the variables $x_1, \ldots, x_{n-1}$. Let $x^{(n-1)}$ be a point of the $(n-1)$-dimensional space E^{n-1}: $x^{(n-1)} = (x_1, \ldots, x_{n-1})$, and let P^{n-1} be an $(n-1)$-dimensional parallelepiped (finite or infinite):

$$P^{n-1} = \{x^{(n-1)} : |x_k| < a_k \leqslant +\infty,\ k = 1, 2, \ldots, n-1\},$$

$$P^n = P^{n-1} \times (0, 1) \text{ and } D_{x_\nu} = \frac{\partial}{\partial x_\nu}, \qquad \nu = 1, 2 \ldots, n.$$

Theorem 2. *Suppose $f(x)$ has in P^n the locally summable generalized derivatives*

$$D^l D_{x_n} f(x), \qquad l \in L,$$

where

$$L = \{l = (l_1, l_2, \ldots, l_n) : l_\nu = 0, 1, \ldots, 2m_\nu;\ \nu = 1, 2, \ldots, n\}, \tag{68}$$

and suppose the frequencies ω_{j_ν}, $j_\nu = 1, \ldots, m_\nu$, $\nu = 1, \ldots, n$, are real. Suppose, further, $\alpha(x_n)$ satisfies the weak degeneracy condition (1) *at zero and $\beta(x_n)$ satisfies* (7). *If for almost all $x^{(n-1)} \in P^{n-1}$*

$$\int_0^1 \left| \alpha(x_n) B_j^{(n)} \frac{\partial}{\partial x_n} L_{x_n}^{(m_n)} f(x) \right|^p dx_n < +\infty, \quad j \in J, \tag{69}$$

and, in addition,

$$\int dE^{n-1} \int_0^1 \left| \beta(x_n) B_j^{(\mu)} \frac{\partial}{\partial x_\mu} L_{x_\mu}^{(m_\mu)} f(x) \right|^p dx_n < +\infty, \tag{70}$$
$$\mu = 1, 2, \ldots, n-1, \quad j \in J,$$

then there exists a trigonometric polynomial

$$T(x) = \sum_{j \in J} a_j e^{i(\omega_j, x)},$$

stabilizing $f(x)$ *for* $x_n \to +0$:

$$T(x) = \operatorname*{stab}_{x_n \to +0} f(x). \tag{71}$$

In this connection,

$$a_j = \lim_{x_n \to +0} a_j(x), \quad j \in J, \tag{72}$$

where the $a_j(x)$ *are defined by* (49).

Proof. It follows from (49) by virtue of our assumptions that the functions $a_j(x)$ have locally summable partial derivatives with respect to all of the variables x_ν, $\nu = 1, \ldots, n$. Therefore, according to a theorem of C. B. Morrey ([15], Theorem 3.1.8), each function $a_j(x)$ is equivalent to a function that is absolutely continuous in each variable x_ν for almost all of the values of the remaining variables (in the sense of $(n-1)$-dimensional measure) on any interval lying in P^n. Taking these representatives of the functions $a_j(x)$, we get according to Lemma 1 that the limits (72) exist by virtue of (59), (60) and (69). And by virtue of (70) and Lemma 2 these limits are constant, i.e. do not depend on $x_1, \ldots, x_{n-1}$.

We now take the identity (50) for $f(x)$ and pass to the limit in it for $x_n \to +0$. We then get (71) by virtue of (72). The theorem is proved.

Let $Q^n = P^{n-1} \times (1, +\infty)$.

Theorem 3. *Suppose* $f(x)$ *has in* Q^n *the locally summable generalized derivatives*

$$D^l D_{x_n} f(x), \quad l \in L$$

(see (68)), the frequencies ω_{j_μ}, $j_\mu = 1, \ldots, m_\mu$; $\mu = 1, \ldots, n$, *are real,* $\alpha(x_n)$ *satisfies the weak degeneracy condition at infinity (see (18)) and* $\beta(x_n)$ *satisfies (21). If for almost all* $x^{(n-1)} \in P^{n-1}$

$$\int_1^{+\infty} \left| \alpha(x_n) B_j^{(n)} \frac{\partial}{\partial x_n} L_{x_n}^{(m_n)} f(x) \right|^p dx_n < +\infty, \quad j \in J,$$

and, in addition,

$$\int dE^{n-1} \int_1^{+\infty} \left| \beta(x_n) B_j^{(\mu)} \frac{\partial}{\partial x_\mu} L_{x_\mu}^{(m_\mu)} f(x) \right|^p dx_n < +\infty,$$
$$\mu = 1, 2, \ldots, n-1, \quad j \in J,$$

then there exists a trigonometric polynomial (32) *stabilizing* $f(x)$ *for* $x_n \to +\infty$; *in this connection, formulas* (72) *hold.*

This theorem is proved in the same way as Theorem 2, except that Lemmas 1′ and 2′ are used in place of Lemmas 1 and 2 respectively.

We now take up the question of stabilization to trigonometric polynomials not depending on x_n. Let (see (64))

$$T(x^{(n-1)}) = \sum_{j_{\hat{n}} \in J_{\hat{n}}} a_{j_{\hat{n}}} e^{i(\omega_{j_{\hat{n}}}, x^{(n-1)})}, \tag{73}$$

where

$$\begin{gathered} \omega_{j_{\hat{n}}} = (\omega_{j_1}, \omega_{j_2}, \ldots, \omega_{j_{n-1}}), \\ \omega_{0_\mu} = 0 < \omega_{1_\mu} < \ldots < \omega_{m_\mu}, \\ \mu = 1, 2, \ldots, n-1. \end{gathered} \tag{74}$$

We introduce in addition the operator $B_{j_{\hat{n}}}^{(\mu)}$, which is defined by (57) with n replaced by $n-1$.

Theorem 4. *Suppose* $f(x)$ *has in* P^n *the locally summable generalized derivatives*

$$D^{l_{\hat{n}}} D_{x_\nu} f(x), \quad l_{\hat{n}} \in L_{\hat{n}}, \quad \nu = 1, 2, \ldots, n$$

(*see* (65)), *the frequencies* ω_{j_μ}, $j_\mu = 1, \ldots, m_\mu$; $\mu = 1, \ldots, n-1$, *are real,* $\alpha(x_n)$ *satisfies* (1) *and* $\beta(x_n)$ *satisfies* (7). *If for almost all* $x^{(n-1)} \in P^{n-1}$

$$\int_0^1 |\alpha(x_n) D^{l_{\hat{n}}} D_{x_n} f(x)|^p \, dx_n < +\infty, \quad l_{\hat{n}} \in L_{\hat{n}}, \tag{75}$$

and, in addition,

$$\int dE^{n-1} \int_0^1 \left| \beta(x_n) B_{j_{\hat{n}}}^{(\mu)} \frac{\partial}{\partial x_\mu} L_{x_\mu}^{(m_\mu)} f(x) \right|^p dx_n < +\infty, \tag{76}$$
$$\mu = 1, 2, \ldots, n-1, \quad j_{\hat{n}} \in J_{\hat{n}},$$

then there exists a trigonometric polynomial (73) *stabilizing* $f(x)$ *for* $x_n \to +0$; *in this connection,*

$$a_{j_{\hat{n}}} = \lim_{x_n \to +0} a_{j_{\hat{n}}}(x). \tag{77}$$

We note that under the conditions of Theorem 4 the polynomial (73) is the trace of $f(x)$ on the hyperplane $x_n = 0$ both in the sense of convergence almost everywhere and in the sense of convergence in the mean.

The proof of Theorem 4 is analogous to the proof of Theorem 2. In fact, the limits (77) exist according to Lemma 1 by virtue of (67) and (75), and, according to Lemma 2, these limits do not depend on $x^{(n-1)} \in P^{n-1}$ by virtue of inequalities (59) and (60) for $\mu = 1, \ldots, n-1$ and conditions (76). Passing to the limit for $x_n \to +0$ in (66), we therefore get

$$\lim_{x_n \to +0} f(x) = T(x^{(n-1)}).$$

The theorem is proved.

A theorem on the existence of a trigonometric-polynomial trace for $x_n \to +\infty$ of a function defined on Q^n can be formulated and proved analogously to Theorem 4. Conditions (1) and (7) on the functions $\alpha(x_n)$ and $\beta(x_n)$ should be replaced by (18) and (21) respectively, and Lemmas 1′ and 2′ should be used in the proof in place of Lemmas 1 and 2.

Lemmas 3 and 3′ permit one to estimate the integral rate of stabilization (see [12] for the case of functions of two variables and power weights). It can be shown that the conditions of Theorems 2–4 imply stabilization not only of the function itself but also of a series of differential expressions derived from this function. This was also shown in [12] for the case $n = 2$ when the stabilizing trigonometric polynomial depends on only one variable and the weight is a power.

§3. Stabilization to harmonic functions

In this section we consider another method of obtaining stabilization theorems, which is based on Lemmas 4 and 4′.

Consider a function $f(x, y)$ defined on the $(n+1)$-dimensional parallelepiped $P^{n+1} = P^n \times (0, 1)$, where

$$P^n = \{x : |x_j| < a_j \leqslant +\infty,\ j = 1, \ldots, n\},\quad x = (x_1, \ldots, x_n) \in P^n,\ y \in (0, 1).$$

Let $\Delta_x = \sum_{k=1}^n (\partial^2/\partial x_k^2)$ be the Laplace operator with respect to the variables $x_1, \ldots, x_n$.

Theorem 5. *Suppose $f(x, y)$ has the locally summable generalized derivatives $\partial^{k+1} f(x, y)/\partial x_\mu^k \partial y$, $k = 0, 1, 2$; $\mu = 1, \ldots, n$, on P^{n+1}, suppose $\alpha(y)$ satisfies the weak degeneracy condition* (1) *at zero and suppose $\gamma(y)$ satisfies* (15). *If*

$$\int_0^1 \left| \alpha(y) \frac{\partial^{\nu+1} f(x, y)}{\partial x_\mu^\nu \partial y} \right|^p dy < +\infty, \qquad \nu = 0, 1, 2, \tag{78}$$

and

$$\begin{gathered} \int_0^1 \left| [\gamma(y)]^{\frac{1}{1-p}} \Delta_x f(x, y) \right|^p dy < +\infty, \\ \int_0^1 \left| \gamma(y) \frac{\partial}{\partial y} \Delta_x f(x, y) \right|^p dy < +\infty, \end{gathered} \tag{79}$$

then $f(x, y)$ *has a trace on the hyperplane* $y = 0$, *and it is a harmonic function.*

Before proving the theorem, we note that condition (78) can be weakened by requiring that it hold only for $\nu = 0$ and 2, since it will then follow that the corresponding integrals for $\nu = 1$ are also finite.

Theorem 5 follows from Lemmas 1 and 4. In fact, the functions $f(x, y)$, $\partial f(x, y)/\partial x_\mu$ and $\partial^2 f(x, y)/\partial x_\mu^2$ have traces for $y = 0$ by virtue of condition (78) and Lemma 1; and if

$$\lim_{y\to+0} f(x, y) = f(x, 0), \quad \lim_{y\to+0} (\partial f(x, y)/\partial x_\mu) = g_\mu(x)$$

and

$$\lim_{y\to+0} \partial^2 f(x, y)/\partial x_\mu^2 = h_\mu(x),$$

then $\partial f(x, 0)/\partial x_\mu = g_\mu(x)$ and $\partial^2 f(x, 0)/\partial x_\mu^2 = h_\mu(x)$. Therefore

$$\lim_{y\to+0} \Delta_x f(x, y) = \Delta_x f(x, 0).$$

Conditions (79) and Lemma 4 imply

$$\lim_{y\to+0} \Delta_x f(x, y) = 0. \tag{80}$$

Thus $f(x, 0)$ is a function having locally summable second-order derivatives that satisfies the equation $\Delta_x f(x, 0) = 0$ by virtue of (80), and hence is a harmonic function.

Certain additional restrictions must be imposed on the weight function $\alpha(y)$ in order to guarantee that the traces of a function and its derivatives are locally summable (concerning this, see [9]).

Conditions similar to those of Theorem 5 can be obtained ensuring that the trace of a function for $y \to +0$ or for $y \to +\infty$ is a solution of some differential equation

$$F(x, u, Du, \ldots, D^m u) = 0, \tag{81}$$

where F is a continuous function of its arguments. Here (78) should be replaced by conditions that according to Lemmas 1 and 1$'$ guarantee the existence of the limits of all derivatives up to order m inclusively, while the Laplace operator in (79) should be replaced by the left side of (81).

BIBLIOGRAPHY

1. S. M. Nikol'skiĭ, *Approximation of functions of several variables and embedding theorems*, "Nauka", Moscow, 1969; English transl., Springer-Verlag, Berlin and New York, 1974. MR 46 #9714.

2. S. V. Uspenskiĭ, *Imbedding theorems for weighted classes*, Trudy Mat. Inst. Steklov. 60 (1961), 282–303; English transl., Amer. Math. Soc. Transl. (2) **87** (1970), 121–145. MR 25 #440; **41** #3202.

3. S. L. Sobolev, *The density of compactly supported functions in the space* $L_p^{(m)}(E_n)$, Sibirsk. Mat. Ž. **4** (1963), 673–682. (Russian) MR **30** #5156.

4. T. S. Pigolkina, *Density of compactly supported functions in weighted classes*, Mat. Zametki **2** (1967), 53–60 = Math. Notes **2** (1967), 518–522. MR **36** #3121.

5. V. R. Portnov, *A theorem on the density of compactly supported functions in weighted classes*, Dokl. Akad. Nauk. SSSR **160** (1965), 545–548 = Soviet Math. Dokl. **6** (1965), 149–152. MR **31** #597.

6. V. N. Sedov, *Functions that tend to a polynomial at infinity*, Imbedding Theorems and their Applications (Proc. Sympos., Baku, 1966), "Nauka", Moscow, 1970, pp. 204–212. (Russian) MR **47** #7416.

7. L. D. Kudrjavcev, *On the traces of functions of several variables and on direction fields regular at infinity for the Laplace operator*, Trudy Mat. Inst. Steklov. **112** (1971), 256–270 = Proc. Steklov Inst. Math. **112** (1971), 266–280. MR **48** #12035.

8. ———, *On parametric inequalities for functions of several variables*, Dokl. Akad. Nauk SSSR **202** (1972), 1261–1264 = Soviet Math. Dokl. **13** (1972), 276–280. MR **45** #9126.

9. ———, *On polynomial traces and moduli of smoothness of functions of several variables*, Trudy Mat. Inst. Steklov. **117** (1972), 180–211 = Proc. Steklov Inst. Math. **117** (1972), 215–250.

10. ———, *On a space of functions whose limiting values are trigonometric polynomials*, Dokl. Akad. Nauk SSSR **216** (1974), 491–494 = Soviet Math. Dokl. **15** (1974), 826–830.

11. ———, *Certain properties of squares of natural numbers*, Mat. Zametki **16** (1974), 33–39 = Math. Notes **16** (1974), 604–607 (1975). MR **50** #9822.

12. ———, *Functions whose limit values are trigonometric polynomials*, Differencial'nye Uravnenija **10** (1974), 2223–2241 = Differential Equations **10** (1974), 1720–1734.

13. ———, *On properties of traces of functions belonging to weight spaces*, Internat. Conf. (Madras, 1973), Functional Analysis and its Applications, Lecture Notes in Math., vol. 399, Springer-Verlag, Berlin and New York, 1974, pp. 292–312.

14. G. N. Jakovlev, *The traces of functions whose derivatives are summable with a certain weight*, Imbedding Theorems and their Applications (Proc. Sympos., Baku, 1966), "Nauka", Moscow, 1970, pp. 225–233. (Russian) MR **48** #4716.

15. C. B. Morrey, Jr., *Multiple integrals in the calculus of variations*, Springer-Verlag, Berlin and New York, 1966. MR **34** #2380.

16. F. G. Bulaevskaja, *Constant traces and the moduli of smoothness of the traces of functions of several variables on hyperplanes of different dimensions*, Differencial'nye Uravnenija **9** (1973), 487–493 = Differential Equations **9** (1973), 371–375. MR **47** #9265.

Translated by S. SMITH

Trudy Mat. Inst. Steklov.
134 (1975)

Proc. Steklov Inst. Math.
134 (1975)

SOME INEQUALITIES FOR TRIGONOMETRIC AND ALGEBRAIC POLYNOMIALS AND THEIR DERIVATIVES

G. K. LEBED′

Abstract. This work consists of three sections. In the first two of them, in the metric of $L_p(-1, 1)$, $1 \le p \le \infty$, a detailed proof is given of the inequalities for trigonometric and algebraic polynomials and their derivatives which were presented in the author's candidate's dissertation and earlier published without proof in Dokl. Akad. Nauk SSSR **117** (1957), 570–572 (MR **20** #1165).

In the third section, several new inequalities of a more general form are established.

Bibliography: 17 items.

The results of this work generalize the well-known, now becoming classical inequalities of A. A. Markov, S. N. Bernšteĭn, A. Zygmund and S. M. Nikol′skiĭ for trigonometric and algebraic polynomials and their derivatives, as well as certain important generalizations thereof obtained by N. K. Bari [1], [2], V. K. Dzjadyk [3] and M. K. Potapov [4].

The report is made up of three sections. In the first two of them, detailed proof is supplied of results which the author presented in his dissertation [5] and published without proof in [6]. In §3 some new theorems of a more general form are proved.

Matters akin to these in angular complex regions were considered later by Dzjadyk and his students [7], [8].

In those works, by means of other methods there were derived several inequalities including, in particular, the results of Theorems 3 and 6 and certain results of Theorem 8 of §2 of the present work. Subsequently, the methods and results of [7] were developed and generalized in works of N. A. Lebedev and P. M. Tamrazov [15]–[17]. Finally, note that Theorems 1 and 6 with proofs which in no essential way differ from the author's proof (see [5]) were also treated in [9], [10].

§1. Ancillary lemmas

In order to establish the main results of this paper we need the following lemmas.

AMS (MOS) subject classifications (1970). Primary 26A82, 42A04.

Lemma 1. *Let*

$$t_n = \frac{\sin (nx/2)}{\sin (x/2)}, \qquad \rho_n(x) = \sqrt{1-x^2} + \frac{1}{n},$$

where n *is an arbitrary natural number. Then for any real* x

$$|t_n(x)| \leqslant \frac{2}{\rho_n(\cos x/2)}, \qquad |t_n'(x)| \leqslant \frac{2n}{\rho_n(\cos x/2)}.$$

Proof. We have

$$|t_n(x)| \leqslant n, \qquad |t_n(x)| \leqslant \frac{1}{|\sin x/2|}, \qquad |t_n'(x)| \leqslant \frac{n}{|\sin x/2|},$$

$$|t_n'(x)| = \left|\left(\frac{e^{\frac{inx}{2}} - e^{-\frac{inx}{2}}}{e^{\frac{ix}{2}} - e^{-\frac{ix}{2}}}\right)'\right| = \left|\left(e^{-i\frac{n-1}{2}x} \sum_{k=0}^{n-1} e^{ikx}\right)'\right|$$

$$= \left| i\frac{n-1}{2} t_n(x) - ie^{-i\frac{n-1}{2}x} \sum_{k=1}^{n-1} ke^{ikx} \right| \leqslant n(n-1).$$

It follows that for any x

$$|t_n(x)| \leqslant \min\left\{n, \frac{1}{|\sin x/2|}\right\}, \qquad |t_n'(x)| \leqslant n \min\left\{n, \frac{1}{|\sin x/2|}\right\}. \tag{1}$$

If we now note that

$$\left\{n, \frac{1}{|\sin x/2|}\right\} \leqslant \frac{2}{\rho(\cos x/2)},$$

on the basis of (1) we obtain

$$|t_n(x)| \leqslant \frac{2}{\rho_n(\cos x/2)}, \qquad |t_n'(x)| \leqslant \frac{2n}{\rho_n(\cos x/2)},$$

and Lemma 1 is proved.

Lemma 2. *Let*

$$K_{n,r}(x) = \frac{1}{2}\left[\frac{1}{n} t_n(x)\right]^{2r} t_{2n(r+1)-2r+1}(x) \text{ and } \tilde{K}_{n,r}(x) = \frac{1}{2}\left[\frac{1}{n} t_n(x)\right]^{2r} t'_{2n(r+1)-2r+1}(x).$$

Then for any value of x

$$|K_{n,r}(x)| \leqslant \frac{c_1(r)\, n}{[n\rho_n(\cos x/2)]^{1+2r}}, \qquad c_1(r) = (1+r)\, 2^{1+2r},$$

$$|\tilde{K}_{n,r}(x)| \leqslant \frac{c_2(r)\, n^2}{[n\rho_n(\cos x/2)]^{1+2r}}, \qquad c_2(r) = 2(1+r)\, c_1(r).$$

This lemma follows immediately from Lemma 1 and the inequality

$$\frac{1}{|\sin x/2| + 1/n_1} \leqslant \frac{2(r+1)}{|\sin x/2| + 1/n}, \qquad n_1 = 2n(r+1) - 2r + 1.$$

Lemma 3. *Let* $s \geq 0$, $p \geq 1$ *and* $2r + 1/p' - s > 0$. *Then*

$$J_n^{(1)}(s)=\left[\int_{-\pi}^{\pi}\left(\left|\sin\frac{t}{2}\right|^s|K_{n,r}(t)|\right)^p dt\right]^{1/p}\leqslant C(s,\ r)\,n^{1/p'-s},$$

$$J_n^{(2)}(s)=\left[\int_{-\pi}^{\pi}\left(\left|\sin\frac{t}{2}\right|^s|\tilde{K}_{n,r}(t)|\right)^p dt\right]^{1/p}\leqslant C(s,\ r)\,n^{1+1/p'-s},$$

where $1/p+1/p'=1$ *and* $C(s,\ r)$ *is a constant depending only on* r *and* s.

Proof. By Lemma 2 we have

$$J_n^{(1)}(s)\leqslant c_1(r)\,nA_n,\qquad J_n^{(2)}(s)\leqslant c_2(r)\,n^2A_n,$$

where

$$A_n=\left(\int_{-\pi}^{\pi}\frac{|\sin(t/2)|^{ps}\,dt}{(n|\sin(t/2)|+1)^{p(1+2r)}}\right)^{1/p}.$$

Noting now that

$$A_n\leqslant 2^{1/p}n^{-s}\left(\int_0^{\pi}\frac{dt}{(nt/\pi+1)^{p(1+2r-s)}}\right)^{1/p}\leqslant\left(\frac{2\pi}{p(1+2r-s)-1}\right)^{1/p}n^{-s-1/p},$$

we obtain

$$J_n^{(1)}(s)\leqslant C(r,\ s)\,n^{1/p'-s},\qquad J_n^{(2)}(s)\leqslant C(r,\ s)\,n^{1+1/p'-s},$$

where

$$C(r,\ s)=\left(\frac{2\pi}{p(1+2r-s)-1}\right)^{1/p}c_2(r),$$

as desired.

Definition. We will say that $\omega(t)\in c_0A_\gamma[0,\ a]$ if $\omega(t)$ is continuous on $[0,\ a]$ and if for some $\gamma\geq 0$ and any t_1 and t_2 $(0<t_1<t_2\leq a)$ the conditions

$$0\leqslant\omega(t_1)\leqslant c_0\omega(t_2),\qquad \omega(t_2)/t_2^\gamma\leqslant c_0\omega(t_1)/t_1^\gamma$$

hold. Note that these conditions are satisfied, for example, by any function $\omega_k(t)$ that is a modulus of continuity of kth order $(k=1,\ 2,\ldots)$.

Lemma 4. *If* $\omega(t)\in c_0A_\gamma[0,\ 2]$, *then for any* $p\geq 1$, $s\geq 0$, $r>(s+\gamma-1/p')/2$ *and* $\alpha\geq 0$

$$I_{n,p}^{(1)}(s,\ \theta)=\left(\int_{-\pi}^{\pi}\left|\rho_n^s[\cos(\theta+t)]\,\omega\left(\frac{\rho_n[\cos(\theta+t)]}{n^\alpha}\right)K_{n,r}(t)\right|^p dt\right)^{1/p}$$
$$\leqslant Ac_0n^{1/p'}\rho_n^s(\cos\theta)\,\omega\left(\frac{\rho_n(\cos\theta)}{n^\alpha}\right),$$

where A *is a constant depending only on* s, γ *and* r.

Proof. Denote by E_θ the set of t for which for fixed θ we have $|\sin(\theta+t)|\leq|\sin\theta|$. Then by Lemma 3

$$I^{(1)}_{n,p}(s,\ \theta)=\Big(\int_{E_\theta}+\int_{[-\pi,\pi]-E_\theta}\Big)^{1/p}$$

$$\leqslant\Bigg(c_0^p\rho_n^{ps}(\cos\theta)\,\omega^p\Big[\frac{\rho_n(\cos\theta)}{n^\alpha}\Big]\int_{-\pi}^{\pi}|K_{n,r}(t)|^p\,dt$$

$$+c_0^p n^{-p\gamma}\frac{\omega^p\Big[\frac{\rho_n(\cos\theta)}{n^\alpha}\Big]}{\Big[\frac{\rho_n(\cos\theta)}{n}\Big]^{p\gamma}}\int_{-\pi}^{\pi}|\rho_n^{s+\gamma}[\cos(\theta+t)]\,K_{n,r}(t)|^p\,dt\Bigg)^{1/p}$$

$$\leqslant c_0\rho_n^s(\cos\theta)\,\omega\Big[\frac{\rho_n(\cos\theta)}{n^\alpha}\Big]\Bigg((1+2^{(s+\gamma)p})\int_{-\pi}^{\pi}|K_{n,r}(t)|^p\,dt$$

$$+4^{(s+\gamma)p}\int_{-\pi}^{\pi}\Big(\Big|n\sin\frac{t}{2}\Big|^{s+\gamma}|K_{n,r}(t)|\Big)^p\,dt\Bigg)^{1/p}$$

$$\leqslant Ac_0\rho_n^s(\cos\theta)\,\omega\Big[\frac{\rho_n(\cos\theta)}{n^\alpha}\Big]n^{1/p'},$$

and the lemma is proved.

One proves similarly

Lemma 5. *If the conditions of Lemma 4 hold, then*

$$I^{(2)}_{n,p}(s,\ \theta)=\Bigg(\int_{-\pi}^{\pi}\Big(\rho_n^s[\cos(\theta+t)]\,\omega\Big(\frac{\rho_n[\cos(\theta+t)]}{n^\alpha}\Big)\Big|\tilde K_{n,r}(t)|^p\,dt\Bigg)^{1/p}$$

$$\leqslant Ac_0 n^{1+1/p'}\rho_n^s(\cos\theta)\,\omega\Big[\frac{\rho_n(\cos\theta)}{n^\alpha}\Big].$$

Lemma 6. *Let $T_n(\theta)$ be an arbitrary trigonometric polynomial of order $\le n$. Then for all real θ*

$$T_n(\theta)=\frac{1}{\pi}\int_{-\pi}^{\pi}T_n(\theta+t)\,K_{n,r}(t)\,dt,$$

$$T'_n(\theta)=-\frac{1}{\pi}\int_{-\pi}^{\pi}T_n(\theta+t)\,\tilde K_{n,r}(t)\,dt,$$

where $K_{n,r}(t)$ and $\tilde K_{n,r}(t)$ are the kernels defined in Lemma 2.

Proof. It is known that for any trigonometric polynomial $P_m(u)$ of order m and its derivative $P'_m(u)$ the following identities hold:

$$P_m(u)=\frac{1}{2\pi}\int_{-\pi}^{\pi}P_m(u+t)\,t_{2m+1}(t)\,dt,$$

$$P'_m(u)=-\frac{1}{2\pi}\int_{-\pi}^{\pi}P_m(u+t)\,t'_{2m+n}(t)\,dt,$$

where

$$t_{2m+1}(t)=\frac{\sin\frac{2m+1}{2}t}{\sin\frac{t}{2}}.$$

If we set here $P_m(u)=T_n(\theta+u)t_n^{2r}(u)$ and note that $P_m(u)$ with respect to u is a trigonometric polynomial of order $m\le n(1+r)-r$, we will have

$$T_n(\theta+u)\,t_n^{2r}(u)=\frac{1}{2\pi}\int_{-\pi}^{\pi}T_n(\theta+u+t)\,t_n^{2r}(u+t)\,t_{2n(1+r)-2r+1}(t)\,dt,$$

$$T_n'(\theta+u)\,t_n^{2r}(u)+T_n(\theta+u)\,2rt_n^{2r-1}(u)\,t_n'(u)$$
$$=-\frac{1}{2\pi}\int_{-\pi}^{\pi}T_n(\theta+u+t)\,t_n^{2r}(u+t)\,t'_{2n(1+r)-2r+1}(t)\,dt.$$

Hence for $u=0$, by the equalities $t_n(0)=n$ and $t_n'(0)=0$, we obtain, respectively, the following identities:

$$T_n(\theta)=\frac{1}{2\pi}\int_{-\pi}^{\pi}T_n(\theta+t)\left[\frac{t_n(t)}{n}\right]^{2r}t_{2n(1+r)-2r+1}(t)\,dt,$$

$$T_n'(\theta)=-\frac{1}{2\pi}\int_{-\pi}^{\pi}T_n(\theta+t)\left[\frac{t_n(t)}{n}\right]^{2r}t'_{2n(1+r)-2r+1}(t)\,dt,$$

and Lemma 6 is proved.

§2. Inequalities for trigonometric and algebraic polynomials and their derivatives

Theorem 1. *Let the trigonometric polynomial $T_n(\theta)$ of order $\le n$, for all real values θ, satisfy the inequality $|T_n(\theta)|\le t(\theta)$, and for $t(\theta)$ let*

$$|t(\theta+h)+t(\theta-h)-2t(\theta)|\leqslant\omega(|h|),\qquad 0<|h|\leqslant\pi.$$

If $\omega(t)\in c_0A_\gamma[0,\pi]$, then for any $k=1,2,\dots$

$$|T_n^{(k)}(\theta)|\leqslant[An]^k\left[t(\theta)+kc_0\omega\left(\frac{1}{n}\right)\right],$$

where A is a constant depending neither on n nor on k and θ.

Proof. By Lemma 6 we have

$$T_n'(\theta)=-\frac{1}{\pi}\int_{-\pi}^{\pi}T_n(\theta+u)\,\tilde{K}_{n,r}(u)\,du.$$

Thus from the inequality $u/\pi\le|\sin u/2|$, $0\le u\le\pi$, we obtain

$$|T'_n(\theta)| \leqslant \frac{1}{\pi}\int_{-\pi}^{\pi} t(\theta+u)\,|\tilde{K}_{n,r}(u)|\,du$$

$$\leqslant \frac{t(\theta)}{\pi}\int_{-\pi}^{\pi} |\tilde{K}_{n,r}(u)|\,du + \frac{1}{\pi}\int_{0}^{\pi}\omega(u)\,|\tilde{K}_{n,r}(u)|\,du$$

$$\leqslant \frac{t(\theta)}{\pi}\int_{-\pi}^{\pi} |\tilde{K}_{n,r}(u)|\,du + \frac{c_0}{\pi}\,\omega\left(\frac{1}{n}\right)\int_{0}^{1/n} |\tilde{K}_{n,r}(u)|\,du$$

$$+\frac{c_0}{\pi}\,\frac{\omega(1/n)}{(1/n)^{\gamma}}\int_{\frac{1}{n}}^{\pi} u^{\gamma}\,|\tilde{K}_{n,r}(u)|\,du$$

$$\leqslant\left[\frac{t(\theta)}{\pi}+\frac{c_0}{\pi}\,\omega\left(\frac{1}{n}\right)\right]\int_{-\pi}^{\pi}|\tilde{K}_{n,r}(u)|\,du$$

$$+\frac{c_0}{\pi^{1-\gamma}}\,\omega\left(\frac{1}{n}\right)n^{\gamma}\int_{0}^{\pi}\left|\sin\frac{u}{2}\right|^{\gamma}|\tilde{K}_{n,r}(u)|\,du. \tag{2}$$

Now suppose r is chosen so that $2r-\gamma>0$. Then, by Lemma 3, for any s $(0\le s\le\gamma)$ we will have

$$\int_{-\pi}^{\pi}\left|\sin\frac{u}{2}\right|^{s}|\tilde{K}_{n,r}(u)|\,du \leqslant c_1(s,\ r)\,n^{1-s}.$$

On the basis of this inequality it follows from (2) that

$$|T'_n(\theta)| \leqslant An\left[t(\theta)+c_0\,\omega\left(\frac{1}{n}\right)\right].$$

Hence the theorem is proved for $k=1$.

Now set $t^{(1)}(\theta)=t(\theta)+c_0\omega(1/n)$. Then, since for any u and θ

$$|t^{(1)}(\theta+u)+t^{(1)}(\theta-u)-2t^{(1)}(\theta)| = |t(\theta+u)+t(\theta-u)-2t(\theta)|,$$

and since $|T'_n(\theta)/An|\le t^{(1)}(\theta)$, we obtain similarly that

$$\left|\frac{T''_n(\theta)}{An}\right| \leqslant An\left[t(\theta)+2c_0\omega\left(\frac{1}{n}\right)\right].$$

Repeating these calculations $k-2$ more times, we reach the desired inequality

$$|T_n^{(k)}(\theta)| \leqslant (An)^k\,[t(\theta)+k\,c_0\omega(1/n)],$$

and Theorem 1 is proved.

In particular, if $t(\theta)$ has bounded second derivative, then $\omega(u)=O(u^2)$. Hence by Theorem 1 we will have

$$|T'_n(\theta)| \leqslant An\left[t(\theta)+c_0\frac{1}{n^2}\right], \tag{3}$$

where A is a constant not depending on n.

Let us note here that if we improve the properties of $t(\theta)$, for example, by requiring that it have bounded third derivative, this will not in general bring about an increase in order of smoothness of the second term in the square brackets in (3). As before, it will have order of smoothness $1/n^2$. In order to see this we adduce an example.

It is known that, for any n and θ, $|\sin n\theta| \leq n|\sin\theta|$. Hence for the polynomials $T_n(\theta) = \sin^2 n\theta/n^2$, $n = 1, 2, \ldots$, the inequalities $T_n(\theta) \leq \sin^2\theta$ hold for any θ. In this case the function $t(\theta)$ is equal to $\sin^2\theta$ and has continuous derivatives of any order. In particular, it has a continuous derivative of second order. Therefore by (3) we have

$$|T'_n(\theta)| \leqslant An\left[\sin^2\theta + \frac{1}{n^2}\right], \qquad n = 1, 2, \ldots. \tag{4}$$

Now let us suppose that for the derivatives $T'_n(\theta)$, $n = 1, 2, \ldots$, a stronger inequality than (4) holds, for example,

$$|T'_n(\theta)| \leqslant An\left[\sin^2\theta + \frac{\alpha_n}{n^2}\right], \tag{5}$$

where $\alpha_n \to 0$ as $n \to \infty$. Then, if we set $\theta = \sqrt{\alpha_n}/2n$ in (5), we note that for any n we must have $1 \leq A\pi/\sqrt{\alpha_n}$, which is impossible since by hypothesis $\alpha_n \to 0$ as $n \to \infty$. This contradiction tells us that inequality (3), and thus also Theorem 1, for any k in the sense of order of growth of the right-hand side, cannot be improved.

Theorem 2. *For the trigonometric polynomial* $T_n(\theta)$ *let*

$$|T_n(\theta)| \leqslant \omega\left[\frac{\rho_n(\cos\theta)}{n^\alpha}\right], \qquad \alpha \geqslant 0$$

for all real θ. *If* $\omega(t) \in c_0 A_\gamma[0, 2]$, *then for any* $k = 1, 2, \ldots$

$$|T_n^{(k)}(\theta)| \leqslant (Ac_0 n)^k\, \omega\left[\frac{\rho_n(\cos\theta)}{n^\alpha}\right],$$

where A *is a constant depending neither on* $T_n(\theta)$ *nor on* k.

Proof. By Lemma 6 we have

$$T'_n(\theta) = -\frac{1}{\pi}\int_{-\pi}^{\pi} T_n(\theta + u)\,\tilde{K}_{n,r}(u)\,du,$$

whence from Lemma 5 we find that

$$|T'_n(\theta)| \leqslant \frac{1}{\pi}\int_{-\pi}^{\pi} \omega\left[\frac{\rho_n(\cos(u+\theta))}{n^\alpha}\right]|\tilde{K}_{n,r}(u)|\,du \leqslant Ac_0 n\omega\left[\frac{\rho_n(\cos\theta)}{n^\alpha}\right].$$

If we now apply this inequality successively to the trigonometric polynomials

$$T'_n(\theta)/Ac_0 n,\ T''_n(\theta)/(Ac_0 n)^2,\ \ldots,\ T_n^{(k-1)}(\theta)/(Ac_0 n)^{k-1},$$

we obtain Theorem 2.

Theorem 3. *For the algebraic polynomial $P_n(x)$ of degree $\le n$, suppose that on the interval $[-a, a]$, where $a=\sqrt{1-c^2/n^2}$ and c is some positive constant, there holds the inequality*

$$|P_n(x)| \leqslant \varphi_n(x),$$

where

$$\varphi_n(x) = \rho_n^s(x)\,\omega\left[\frac{\rho_n(x)}{n}\right].$$

If $\omega(t) \in c_0 A_\gamma[0, 2]$, then $|P_n(x)| \le \bar{A}\phi_n(x)$, where $\bar{A}$ is some constant ≥ 1 depending neither on n nor on x, will hold on all the interval $[-1, 1]$.

Proof. Since the theorem is true for $n \le 2(c+1)$, we will assume in proving it that $n > 2(c+1)$. We consider three cases.

Case 1. Let $s=0$ and $\omega(t) \equiv 1$. Then, setting $x = at$, for $P_n(at)$ we will have $|P_n(at)| \le M$, $|t| \le 1$. Hence by A. A. Markov's inequality (see [9]) for derivatives of algebraic polynomials we obtain

$$|P_n^{(k)}(at)\,a^k| \leqslant n^{2k}M, \qquad |t| \leqslant 1. \tag{6}$$

Now let us expand $P_n(x)$ by Taylor's formula in a neighborhood of $x = \pm a$. Then, by (6), for x satisfying $|x \pm a| \le 1-a$ we will have

$$|P_n(x)| = \left|\sum_{k=0}^{n} \frac{P_n^{(k)}(\pm a)}{k\,!}(x \pm a)^k\right| \leqslant M\sum_{k=0}^{n}\frac{(n^2/a)^k}{k\,!}\left(\frac{c^2/n^2}{1+a}\right)^k \tag{7}$$
$$\leqslant Me^{\frac{c^2}{a(1+a)}} \leqslant Me^{2c^2}, \qquad n \geqslant \frac{4}{\sqrt{3}}c.$$

Hence the theorem is true in this case.

Case 2. Let $s+\gamma \le 0$. Then by the inequality

$$\varphi_n(x) = Mn^{-\gamma}\rho_n^{s+\gamma}(x)\frac{\omega\left[\frac{\rho_n(x)}{n}\right]}{\left(\frac{\rho_n(x)}{n}\right)^{\gamma}} \leqslant c_0\varphi_n(a), \qquad |x| \leqslant a,$$

we have $P_n(x)| \le c_0\phi_n(a)$, $|x| \le a$, whence by the assertion proved above we conclude that

$$|P_n(x)| \leqslant A_0c_0\varphi_n(a),\ |x| \leqslant 1,$$

where $A_0 = e^{2c^2}$. Noting now that $\phi_n(a) \le c_0\phi_n(x)$ provided $a \le |x| \le 1$, and that $|P_n(x)| \le \phi_n(x)$ provided $|x| \le a$, we obtain

$$|P_n(x)| \leqslant (1 + A_0c_0)\,\varphi_n(x) = A_1\varphi_n(x), \qquad |x| \leqslant 1.$$

Hence the theorem is true in this case.

Case 3. Let $s+\gamma > 0$. Then by the inequality

$$M_n(x) = \frac{\omega\left[\frac{1}{n}\rho_n(x)\right]}{\left(\frac{\rho_n(x)}{n}\right)^{\gamma}}n^{-\gamma} \leqslant c_0M_n(a), \qquad |x| \leqslant a,$$

we obtain $|P_n(x)| \le c_0 M_n(a)\rho_u^{s+\gamma}(x)$, $|x| \le a$. If here we replace x by $a\cos\theta$, we obtain

$$|P_n(a\cos\theta)| \leqslant c_0 M_n(a)\left(\sqrt{1-a^2+a^2\sin^2\theta}+\frac{1}{n}\right)^{s+\gamma}$$
$$\leqslant c_0(1+c)^{s+\gamma} M_n(a)\,\rho_n^{s+\gamma}(\cos\theta) = r_n(a)\,\rho_n^{s+\gamma}(\cos\theta).$$

Thus by Theorem 2,

$$\begin{gathered} |P_n'(a\cos\theta)\,a\sin\theta| \leqslant Anr_n(a)\,\rho_n^{s+\gamma}(\cos\theta), \\ |P_n''(a\cos\theta)\,a^2\sin^2\theta - a\cos\theta P_n'(a\cos\theta)| \leqslant (An)^2 r_n(a)\,\rho_n^{s+\gamma}(\cos\theta). \end{gathered} \tag{8}$$

Further, from the second inequality with the aid of the first we have

$$\begin{aligned} |P_n''(a\cos\theta)\,a^2\sin^3\theta| &\leqslant |a\sin\theta P_n'(a\cos\theta)| + (An)^2 r_n(a)\,\rho_n^{s+\gamma}(\cos\theta)\,|\sin\theta| \\ &\leqslant (An)^2 r_n(a)\,\rho_n^{s+\gamma+1}(\cos\theta) \leqslant [A(n+1)]^2 r_n(a)\left(\frac{n+1}{n}\right)^{s+\gamma+1}\rho_{n+1}^{s+\gamma+1}(\cos\theta). \end{aligned} \tag{9}$$

Now, applying the same Theorem 2 to the polynomial $P_n''(a\cos\theta)a^2\sin^3\theta$, we obtain

$$\begin{aligned} |P_n'''(a\cos\theta)\,a^3\sin^5\theta| &\leqslant 3\,|P_n''(a\cos\theta)\,a^2\sin^3\theta| \\ &\quad + [A(n+1)]^3 r_n(a)\left(\frac{n+1}{n}\right)^{s+\gamma+1}\rho_{n+1}^{s+\gamma+2}(\cos\theta) \\ &\leqslant 1\cdot 3\,[A(n+2)]^3 r_n(a)\left(\frac{n+2}{n}\right)^{s+\gamma+1}\rho_{n+2}^{s+\gamma+2}(\cos\theta). \end{aligned} \tag{10}$$

From (10) we find similarly that

$$|P_n^{\mathrm{IV}}(a\cos\theta)\,a^4\sin^7\theta| \leqslant 1\cdot 3\cdot 5\,[A(n+3)]^4 r_n(a)\left(\frac{n+3}{n}\right)^{s+\gamma+1}\rho_{n+3}^{s+\gamma+3}(\cos\theta). \tag{11}$$

Upon analyzing inequalities (8)–(11) we find that they may be written in the form

$$\begin{gathered} |P_n^{(k)}(a\cos\theta)\,a^k\sin^{2k-1}\theta| \\ \leqslant (2k-1)!!\,[A(n+k-1)]^k r_n(a)\left(\frac{n+k-1}{n}\right)^{s+\gamma+1}\rho_{n+k-1}^{s+\gamma+k-1}(\cos\theta), \\ k=1,\ \ldots,\ 4. \end{gathered} \tag{12}$$

Now let us prove that (12) is valid for any k. Suppose that it is already established for all derivatives up to $(k-1)$th order inclusive $(k \ge 2)$. Then, applying Theorem 2 to the polynomial $P_n^{(k-1)}(a\cos\theta)a^{k-1}\sin^{2k-3}\theta$, we obtain

$$\begin{aligned} |P_n^{(k)}(a\cos\theta)\,a^k\sin^{2k-1}\theta| &\leqslant (2k-3)\,|P_n^{(k-1)}(a\cos\theta)\,a^{k-1}\sin^{2k-3}\theta| \\ &\quad + A(n+k-1)(2k-3)!!\,[A(n+k-2)]^{k-1} r_n(a) \\ &\quad \times\left(\frac{n+k-2}{n}\right)^{s+\gamma+1}\rho_{n+k-2}^{s+\gamma+k-2}(\cos\theta)\,|\sin\theta| \\ &\leqslant (2k-1)!!\,[A(n+k-1)]^k r_n(a)\left(\frac{n+k-1}{n}\right)^{s+\gamma+1}\rho_{n+k-1}^{s+\gamma+k-1}(\cos\theta). \end{aligned}$$

By induction we conclude that (12) holds for any $k = 1, \ldots, n$. In particular, for $\cos\theta = \pm\sqrt{1-1/n^2}$ by (12) we have

$$|P_n^{(k)}(\pm a_1)| \leqslant \left(\frac{4A}{a}\right)^k 4^{s+\gamma}(2k-1)!!\, r_n(a)\, n^{2k-s-\gamma}, \tag{13}$$

where $a_1 = a\sqrt{1 - 1/n^2}$.

Further, if we denote by $\bar{k}$ the smallest value of k satisfying $s + \gamma - \bar{k} \leq 0$ and note that for $|x| \leq a_1$

$$\sqrt{1-\left(\frac{x}{a}\right)^2}+\frac{1}{n} \leqslant 2\sqrt{1-\left(\frac{x}{a}\right)^2} \text{ и } \sqrt{1-\left(\frac{1+c}{n}\right)^2} \leqslant a_1,$$

we obtain the following inequality from (12):

$$|P_n^{(\bar{k})}(x)| \leqslant \left(\frac{4A}{a}\right)^{\bar{k}} 4^{s+\gamma}(2k-1)!!\, r_n(a)\, n^{2\bar{k}-s-\gamma} = \bar{M}_n(a), \quad |x| \leqslant \sqrt{1-\frac{(1+c)^2}{n^2}}.$$

Hence by (7) we have

$$|P_n^{(\bar{k})}(x)| \leqslant \bar{A}_0 \bar{M}_n(a), \qquad |x| \leqslant 1, \tag{14}$$

where $\bar{A}_0 = e^{2(c+1)^2}$.

Let us now turn to the estimation of $P_n(x)$.

Let $|x \mp a_1| \leq 1 - a_1$. Then from the identity

$$P_n(x) \equiv P_n(\pm a_1) + \sum_{\nu=1}^{\bar{k}-1} \frac{P_n^{(\nu)}(\pm a_1)}{\nu!}(x \mp a_1)^\nu + \int_{\pm a_1}^{x} \frac{(x-t)^{\bar{k}-1}}{(\bar{k}-1)!} P_n^{(\bar{k})}(t)\, dt,$$

using (13) and (14), we will have

$$\begin{aligned} |P_n(x)| &\leqslant \varphi_n(a_1) + \left(\frac{4}{n}\right)^{s+\gamma} A_0 r_n(a) \sum_{\nu=1}^{\bar{k}} \frac{(2\nu-1)!!}{\nu!}\left(\frac{4A}{a}\right)^\nu \\ &\leqslant A(s, \gamma)\,[\varphi_n(a_1) + n^{-(s+\gamma)} r_n(a)], \qquad |x \mp a_1| \leqslant 1 - a_1. \end{aligned}$$

Hence, in view of the inequalities

$$|P_n(x)| \leqslant \varphi_n(x), \qquad |x| \leqslant a;$$

$$\begin{aligned} \varphi_n(a_1) &= M\rho_n^{s+\gamma}(a_1)\, n^{-\gamma} \frac{\omega\left[\frac{\rho_n(a_1)}{n}\right]}{\left[\frac{\rho_n(a_1)}{n}\right]^\gamma} \leqslant M\left(1+\sqrt{1+c^2}\right)^{s+\gamma} \rho_n^{s+\gamma}(x) \\ &\times c_0 n^{-\gamma} \frac{\omega\left[\frac{\rho_n(x)}{n}\right]}{\left[\frac{\rho_n(x)}{n}\right]^\gamma} = c_0\left(1+\sqrt{1+c^2}\right)^{\gamma+s} \varphi_n(x), \qquad a_1 \leqslant x \leqslant 1; \end{aligned}$$

$$\frac{r_n(a)}{n^{s+\gamma}} = c_0(1+c)^{s+\gamma} M n^{-\gamma} \frac{\omega\left[\frac{\rho_n(a)}{n}\right]}{\left[\frac{\rho_n(a)}{n}\right]^\gamma} \leqslant c_0^2(1+c)^{s+\gamma} \varphi_n(x), \quad a \leqslant |x| \leqslant 1,$$

we obtain the estimate $|P_n(x)| \leq \bar{A}\phi_n(x)$, $|x| \leq 1$, where $\bar{A} = \bar{A}(s, \gamma, c_0)$, and Theorem 3 is proved.

Theorem 4. *Let* $P_n(x)$ *be an algebraic polynomial of degree* $\leq n$, *and let*

$$\sigma_{p,n} = \left(\int_{-1}^{1} \left| \frac{P_n(x)}{\varphi_n(x)} \right|^p dx \right)^{1/p}, \quad 1 \leqslant p \leqslant \infty,$$

where $\phi_n(x) = \rho_n^s(x)\omega(\rho_n(x)/n)$ *and* s *is an arbitrary number. If* $\omega(t) \in c_0A_\gamma[0, 2]$, *then*

$$\left| \frac{P_n(x)}{\varphi_n(x)} \right| \leqslant c(s) \left(\frac{n}{\rho_n(x)} \right)^{1/p} \sigma_{n,p}, \quad |x| \leqslant 1.$$

Proof. Set $x = \cos\theta$. Then, by Lemma 6, for any natural ν we will have

$$P_n(\cos\theta)\sin^\nu\theta = \frac{1}{\pi} \int_{-\pi}^{\pi} P_n(\cos u)\sin^\nu u K_{n+\nu,r}(u-\theta)\,du.$$

Hence by Hölder's inequality we find that

$$|P_n(\cos\theta)\sin^\nu\theta| \leqslant \frac{1}{\pi} \left(\int_{-\pi}^{\pi} \left| \frac{P_n(\cos u)}{\varphi_n(\cos u)} \right|^p \sin u\,du \right)^{1/p} \left(\int_{-\pi}^{\pi} |\rho_n^{\nu-1/p}(\cos u)\varphi_n(\cos u) K_{n+\nu,r}(u-\theta)|^{p'} du \right)^{1/p'}. \tag{15}$$

Now let us suppose that ν and r are chosen so that $\nu \geq 1$, $\nu + s - 1/p > 0$ and $2r - s - \gamma + 1/p' > 0$. Then from (15) and Lemma 4 we will have

$$|P_n(\cos\theta)\sin^\nu\theta| \leqslant c_1(s)\, n^{1/p} \rho_n^{\nu-1/p}(\cos\theta)\,\varphi_n(\cos\theta)\,\sigma_{p,n}. \tag{16}$$

If we set $\cos\theta = x$ in (16), after straightforward calculations we obtain for $P_n(x)$ the following estimate:

$$|P_n(x)| \leqslant 2^\nu c_1(s)\, n^{1/p} \rho_n^{-1/p}(x)\,\varphi_n(x)\,\sigma_{p,n}, \quad |x| \leqslant \sqrt{1 - \frac{1}{n^2}}.$$

Hence by Theorem 3 we have

$$\left| \frac{P_n(x)}{\varphi_n(x)} \right| \leqslant c(s) \left(\frac{n}{\rho_n(x)} \right)^{1/p} \sigma_{p,n}, \quad |x| \leqslant 1,$$

as desired.

As a corollary of Theorem 4 we obtain

Theorem 5. *Let* $P_n(x)$ *be an arbitrary algebraic polynomial of degree* $\leq n$. *If* $\omega(t) \in c_0A_\gamma[0, 2]$, *then for any* q $(1 \leq p < q \leq \infty)$

$$I_{q,n} \leqslant A(s)\, n^{1/p-1/q} I_{p,n}, \text{ where } I_{r,n} = \left(\int_{-1}^{1} \left| \frac{\rho_n^{-1/r}(x)\,P_n(x)}{\varphi_n(x)} \right|^r dx \right)^{1/r}, \quad 1 \leqslant r \leqslant \infty.$$

Proof. By Theorem 4 we have

$$I_{q,n}=\left(\int_{-1}^{1}\left|\frac{\rho_n^{-1/p}(x)P_n(x)}{\varphi_n(x)}\right|^p\left|\frac{P_n(x)}{\varphi_n(x)}\right|^{q-p}dx\right)^{1/q}$$

$$\leqslant I_{p,n}^{p/q}\left[c(s)n^{1/p}\left(\int_{-1}^{1}\left|\frac{P_n(x)}{\rho_n^{1/p}(x)\varphi_n(x)}\right|^p dx\right)^{1/p}\right]^{\frac{q-p}{q}}$$

$$\leqslant c^{\frac{q-p}{q}}(s)n^{1/p-1/q}I_{p,n},$$

and the theorem is proved.

Theorem 6. *Let $P_n(x)$ be an arbitrary algebraic polynomial of degree $\leq n$, and let $|P_n(x)|\leq\phi_n(x)$, $|x|\leq 1$. If $\omega(t)\in c_0A_\gamma[0,2]$, then*

$$|\rho_n^k(x)P_n^{(k)}(x)|\leqslant[A_k(s,c_0)n]^k\varphi_n(x),\qquad |x|\leqslant 1.\text{(}^1\text{)}$$

Proof. Set $x=\cos\theta$. Then for the trigonometric polynomial

$$P_n(\cos\theta)\sin^\nu\theta,\ \nu+s\geqslant 0,$$

we have the estimate

$$|P_n(\cos\theta)\sin^\nu\theta|\leqslant\rho_n^\nu(\cos\theta)\varphi_n(\cos\theta)\leqslant 2^{s+\nu+2\gamma}c_0\left(\frac{\rho_{n+\nu}(\cos\theta)}{n+\nu}\right)^{s+\nu}\omega\left(\frac{\rho_{n+\nu}(\cos\theta)}{n+\nu}\right).$$

If we apply Theorem 2 to this polynomial, we obtain

$$|P_n'(\cos\theta)\sin^{\nu+1}\theta-\nu\sin^{\nu-1}\theta\cos\theta P_n(\cos\theta)|$$

$$\leqslant Ac_0(n+\nu)2^{s+\nu+2\gamma}c_0\rho_{n+\nu}^{s+\nu}(\cos\theta)\omega\left(\frac{\rho_{n+\nu}(\cos\theta)}{n+\nu}\right)\leqslant\bar{A}c_0^3\varphi_n(\cos\theta).$$

After straightforward calculations we find that

$$|P_n'(x)|\leqslant Bc_0^3n\frac{\varphi_n(x)}{\rho_n(x)},\tag{17}$$

if $|x|\leq\sqrt{1-1/n^2}$.

From (17) it follows by Theorem 3 that

$$|\rho_n(x)P_n'(x)|\leqslant A_1(c_0)n\varphi_n(x),\qquad |x|\leqslant 1.$$

Hence for $k=1$ Theorem 6 is established. Applying it successively to the polynomials

$$P_n'(x)/A_1n,\quad P_n''(x)/[A_2n]^2,\ \ldots,\ P_n^{(k-1)}(x)/[A_{k-1}n]^{k-1},$$

we obtain the assertion for $k=2,\ldots,n$.

(1) In the special case $\omega(t)\equiv 1$ this inequality was obtained in a different fashion by V. K. Dzjadyk [3].

Theorem 7. *Let $P_n(x)$ be an arbitrary algebraic polynomial of degree $\leq n$, and let*

$$u \quad \sigma_{p,n}=\left(\int_{-1}^{1}\left|\frac{P_n(x)}{\varphi_n(x)}\right|^p dx\right)^{1/p}, \quad 1\leqslant p\leqslant\infty.$$

If $\omega(t)\in c_0A_\gamma[0, 2]$, then for any $k = 1, \ldots, n$

$$\left|\frac{\rho_n^{k+1/p}(x)\, P_n^{(k)}(x)}{\varphi_n(x)}\right|\leqslant[A_kc_0n]^{k+1/p}\sigma_{p,n}, \qquad |x|\leqslant 1.$$

Proof. By Theorem 4 we have

$$|P_n(x)|\leqslant c(s)\, n^{1/p}\,\frac{\varphi_n(x)}{\rho_n^{1/p}(x)}\,\sigma_{p,n}.$$

Applying Theorem 6, we obtain

$$|\rho_n^{k+1/p}(x)\, P_n^{(k)}(x)|\leqslant[A_k(c_0)\, n]^{k+1/p}\sigma_{p,n}, \qquad |x|\leqslant 1, \qquad k=1, \ldots, n,$$

and the theorem is proved.

Theorem 8. *Let the conditions of Theorem 7 hold. Then for any $k = 1, \ldots, n$ and $p \geq 1$*

$$\sigma_{p,n}^{(k)}=\left(\int_{-1}^{1}\left|\frac{\rho_n^k(x)\, P_n^{(k)}(x)}{\varphi_n(x)}\right|^p dx\right)^{1/p}\leqslant[A_k(s)\, n]^k\,\sigma_{p,n}.$$

Proof. Consider the trigonometric polynomial

$$P_n(\cos\theta)\sin\theta\,(\sin\theta+1/n)^\nu.$$

By Lemma 6 we have for its derivative

$$-P_n'(\cos\theta)\sin^2\theta\left(\sin\theta+\frac{1}{n}\right)^\nu+P_n(\cos\theta)\left(\sin\theta+\frac{1}{n}\right)^{\nu-1}\left[(\nu+1)\sin\theta+\frac{1}{n}\right]\cos\theta$$

$$=-\frac{1}{\pi}\int_{-\pi}^{\pi}P_n(\cos(u+\theta))\sin(u+\theta)\left(\sin(u+\theta)+\frac{1}{n}\right)^\nu\tilde{K}_{n_1,r}(u)\,du,$$

where $n_1 = n + \nu + 1$. By the Minkowski inequality we obtain

$$\left(\int_0^\pi\left|\frac{\sin^2\theta P_n'(\cos\theta)}{\rho_n(\cos\theta)\,\varphi_n(\cos\theta)}\right|^p\sin\theta d\theta\right)^{1/p}\leqslant(\nu+1)\left(\int_0^\pi\left|\frac{P_n(\cos\theta)}{\rho_n(\cos\theta)\,\varphi_n(\cos\theta)}\right|^p\sin\theta d\theta\right)^{1/p}$$

$$+\frac{1}{\pi}\int_{-\pi}^{\pi}|\tilde{K}_{n_1,r}(u)|\left(\int_{-\pi}^{\pi}\left|\frac{P_n(\cos(\theta+u))}{\varphi_n(\cos(\theta+u))}\right|^p|\sin(\theta+u)|\,F^p(u,\theta)\,d\theta\right)^{1/p}du,$$

where

$$F(u,\theta)=\frac{\rho_n^{\nu+1/p'}(\cos(\theta+u))\,\varphi_n(\cos(\theta+u))}{\rho_n^{\nu+1/p'}(\cos\theta)\,\varphi_n(\cos\theta)}. \tag{18}$$

Let us now take ν so that $s+1/p'+\nu\geq 0$, and for this ν estimate the expression

$$F(u,\theta)=\left[\frac{\rho_n(\cos(\theta+u))}{\rho_n(\cos\theta)}\right]^{s+\nu+1/p'}\left[\frac{\omega\left(\frac{\rho_n(\cos(\theta+u))}{n}\right)}{\omega\left(\frac{\rho_n(\cos\theta)}{n}\right)}\right].$$

Let $|\sin(\theta+u)|\leq|\sin\theta|$. Then $\rho_n(\cos(\theta+u))\leq\rho_n(\cos\theta)$, and hence $F(u,\theta)\leq c_0$. But if $|\sin(\theta+u)|>|\sin\theta|$, then

$$\rho_n(\cos(\theta+u))=|\sin(\theta+u)|+\frac{1}{n}\leqslant 2\left(|\sin\theta|+\frac{1}{n}+\left|\sin\frac{u}{2}\right|\right),$$

$$\frac{\omega\left(\frac{\rho_n(\cos(\theta+u))}{n}\right)}{\left[\frac{\rho_n(\cos(\theta+u))}{n}\right]^{\gamma}}\leqslant c_0\frac{\omega\left(\frac{\rho_n(\cos\theta)}{n}\right)}{\left(\frac{\rho_n(\cos\theta)}{n}\right)^{\gamma}}.$$

Hence

$$F(u,\theta)\leqslant 4^{s+\nu+\gamma+1/p'}c_0\left[1+\left|n\sin\frac{u}{2}\right|^{s+\nu+\gamma+1/p'}\right].$$

Hence for any u and θ we have

$$F(u,\theta)\leqslant 4^{s+\nu+\gamma+1/p'}c_0\left[1+\left|n\sin\frac{u}{2}\right|^{s+\nu+\gamma+1/p'}\right]. \tag{19}$$

Now substituting (19) into (18) and using Lemma 5 we obtain

$$\begin{aligned}&\left(\int_{-1}^{1}\left|\frac{(1-x^2)P_n'(x)}{\rho_n(x)\varphi_n(x)}\right|^p dx\right)^{1/p}\\ &\leqslant\left[\nu+1+\frac{1}{\pi}4^{s+\nu+\gamma+1/p'}c_0\int_{-\pi}^{\pi}|\tilde{K}_{n,r}(u)|\left(1+\left|n\sin\frac{u}{2}\right|^{s+\nu+\gamma+1/p'}\right)du\right]\sigma_{p,n}\\ &\leqslant A_1 n\sigma_{p,n}.\end{aligned} \tag{20}$$

Here it is assumed that $2r-s-\nu-\gamma-1/p'>0$. Now let us estimate the integral

$$\sigma_{p,n}^{(1)}=\left(\int_{-1}^{1}\left|\frac{\left(\sqrt{1-x^2}+\frac{1}{n}\right)^2P_n'(x)}{\rho_n(x)\varphi_n(x)}\right|^p dx\right)^{1/p}.$$

Since $(\sqrt{1-x^2}+1/n)^2\leq 2(1-x^2+1/n^2)$, we have

$$\sigma_{p,n}^{(1)}\leqslant\left(\int_{-1}^{1}\left|\frac{(1-x^2)P_n'(x)}{\rho_n(x)\varphi_n(x)}\right|^p dx\right)^{1/p}+\frac{2}{n^2}\sup_{[-1,1]}\rho_n^{1+1/p}(x)|P_n'(x)|\left(\int_{-1}^{1}\frac{dx}{\rho_n^{2p+1}(x)}\right)^{1/p}.$$

Hence by Lemma 7 and the inequalities (20) and

$$\left(\int_{-1}^{1}\frac{dx}{\rho_n^{2p+1}(x)}\right)^{1/p}\leqslant\left(\frac{4}{2p-1}\right)^{1/p}n^{2-1/p}$$

we obtain

$$\sigma_{p,n}^{(1)}\leqslant A_1(c_0)\,n\sigma_{p,n}.$$

Thus Theorem 8 is proved for $k=1$. If we apply it successively to the polynomials $P_n'(x),\dots,P_n^{(k-1)}(x)$, we arrive at the desired inequality $\sigma_{p,n}^{(k)}\le [A_k(c_0)n]^k\sigma_{p,n}$.

Theorem 9. *If the conditions of Theorem 7 hold, then for any p and q $(1\le p\le q\le\infty)$*

$$\left(\int_{-1}^{1}\left|\frac{\rho_n^{k-1/q}(x)\,P_n^{(k)}(x)}{\varphi_n(x)}\right|^q dx\right)^{1/q}$$

$$\leqslant[A_k(s,c_0)\,n]^{k+1/p-1/q}\left(\int_{-1}^{1}\left|\frac{\rho_n^{-1/p}(x)\,P_n(x)}{\varphi_n(x)}\right|^p dx\right)^{1/p},\qquad k=1,\dots,n.$$

This theorem follows directly from Theorems 5 and 8.

§3. Generalization of the results of §2 and [4]

Theorem 10. *Let $\phi_n(x)=\rho_n^s(x)\omega(\rho_n(x)/n)$. If $\omega(t)\in c_0A_\gamma[0,2]$, then for any $\alpha<2$ we have $\sigma_{p,n}\sim\sigma_{p,n}(\alpha)$, where*

$$\sigma_{p,n}(\alpha)=\left(\int_{-1}^{1}\left|\frac{P_n(x)}{\varphi_n(x)}\right|^p\psi_n^\alpha(x)\,dx\right)^{1/p},\qquad \psi_n(x)=\frac{\rho_n(x)}{\sqrt{1-x^2}},\qquad \sigma_{p,n}=\sigma_{p,n}(0).$$

Remark. We write $\alpha_n\sim\beta_n$ if there are two positive constants A and B, independent of n and such that $A\alpha_n\le\beta_n\le B\alpha_n$ for any n.

Proof. We first show that there exists a constant $B_1=B_1(\alpha,s)$ such that $\sigma_{p,n}(\alpha)\le B_1\sigma_{p,n}$. To this end let us write $\sigma_{p,n}^p(\alpha)$ as follows:

$$\sigma_{p,n}^p(\alpha)=\int_{-\sqrt{1-1/n^2}}^{\sqrt{1-1/n^2}}\psi_n^\alpha(x)\left|\frac{P_n(x)}{\varphi_n(x)}\right|^p dx+\int_{-1}^{-\sqrt{1-1/n^2}}\psi_n^\alpha(x)\left|\frac{P_n(x)}{\varphi_n(x)}\right|^p dx$$

$$+\int_{\sqrt{1-1/n^2}}^{1}\psi_n^\alpha(x)\left|\frac{P_n(x)}{\varphi_n(x)}\right|^p dx=S_1+S_2+S_3$$

and estimate first S_1, then S_2+S_3.

For S_1 by the inequality $(\sqrt{1-x^2}+1/n)^\alpha\le 2^{|\alpha|}(\sqrt{1-x^2})^\alpha$, valid for $|x|\le\sqrt{1-1/n^2}$, we have

$$S_1 \leqslant 2^{|\alpha|} \int_{-1}^{1} \left| \frac{P_n(x)}{\varphi_n(x)} \right|^p dx = 2^{|\alpha|} \sigma_{p,n}^p.$$

Further, since

$$\int_{\sqrt{1-1/n^2}}^{1} \psi_n^\alpha(x)\,dx \leqslant \int_{\sqrt{1-1/n^2}}^{1} \psi_n^{|\alpha|}(x)\,dx \leqslant \left(\frac{2}{n}\right)^{|\alpha|} \int_{\sqrt{1-1/n^2}}^{1} (1-x)^{-\frac{|\alpha|}{2}} dx \leqslant \frac{2^{1+|\alpha|}}{2-|\alpha|} \frac{1}{n^2} = \frac{c(\alpha)}{n^2},$$

by Lemma 4 we have

$$S_2 + S_3 \leqslant 2 \sup_{[-1,1]} \left| \frac{P_n(x)}{\varphi_n(x)} \right|^p \frac{c(\alpha)}{n^2} \leqslant c^p(\alpha, s)\, \sigma_{p,n}^p.$$

In view of the estimates for S_1 and $S_2 + S_3$ we observe that there exists a constant $B_1(\alpha, s)$ such that for any n

$$\sigma_{p,n}(\alpha) \leqslant B_1(\alpha, s)\, \sigma_{p,n}.$$

To obtain the reverse inequality, consider the polynomial $(1 - x^2)^\nu P_n(x)$, where ν is a positive integer such that $2p\nu + \alpha \geq 0$. Applying Theorem 4 to it, we will have

$$\left| \frac{P_n(x)}{\varphi_{n+\nu}(x)} \frac{(1-x^2)^\nu}{\rho_n^{2\nu}(x)} \right| \leqslant c(s, \nu)\, n^{2/p} \left(\int_{-1}^{1} \left| \frac{P_n(x)\,(1-x^2)^\nu}{\varphi_{n+\nu}(x)\,\rho_{n+\nu}^{2\nu}(x)} \right|^p dx \right)^{1/p}, \quad |x| \leqslant 1.$$

Hence by the inequalities

$$2^{-(|s+\gamma|+\gamma)} c_0 \varphi_n(x) \leqslant \varphi_{n+\nu}(x) \leqslant 2^{|s|} c_0 \varphi_n(x), \quad \nu \leqslant n,$$

and Theorem 3 we find that

$$\begin{aligned} \left| \frac{P_n(x)}{\varphi_n(x)} \right| &\leqslant c_1(s, \nu)\, n^{2/p} \left(\int_{-1}^{1} \left| \frac{P_n(x)}{\varphi_n(x)} \right|^p \left(\frac{\rho_n(x)}{\sqrt{1-x^2}} \right)^{-2\nu p} dx \right)^{1/p} \\ &\leqslant c_1(s, \nu)\, n^{2/p} \left(\int_{-1}^{1} \left| \frac{P_n(x)}{\varphi_n(x)} \right|^p \left(\frac{\rho_n(x)}{\sqrt{1-x^2}} \right)^{\alpha} dx \right)^{1/p} \\ &= c_1(s, \nu)\, n^{2/p} \sigma_{p,n}(\alpha), \quad \text{if} \quad \alpha \leqslant 0. \end{aligned}$$

Further, by Theorem 4 we have

$$\left| \frac{P_n(x)}{\varphi_n(x)} \right| \leqslant \frac{c(s)}{n^{2/p}} \sigma_{p,n} \leqslant \frac{c(s)}{n^{2/p}} \sigma_{p,n}(\alpha), \quad |x| \leqslant 1, \quad \text{if} \quad 0 < \alpha < 2.$$

Hence it is established that

$$\left| \frac{P_n(x)}{\varphi_n(x)} \right| \leqslant \frac{A(s, \alpha)}{n^{2/p}} \sigma_{p,n}(\alpha)$$

for any $\alpha < 2$. Now let us estimate $\sigma_{p,n}$. We have

$$\sigma_{p,n}=\left(\int\limits_{-\sqrt{1-1/n^2}}^{\sqrt{1-1/n^2}}\left|\frac{P_n(x)}{\varphi_n(x)}\right|^p dx+\int\limits_{-1}^{-\sqrt{1-1/n^2}}\left|\frac{P_n(x)}{\varphi_n(x)}\right|^p dx\right.$$
$$\left.+\int\limits_{-\sqrt{1-1/n^2}}^{1}\left|\frac{P_n(x)}{\varphi_n(x)}\right|^p dx\right)^{1/p}=(I_1+I_2+I_3)^{1/p}.$$

Hence by the inequalities

$$I_1\leqslant 2^{|\alpha|}\int\limits_{-\sqrt{1-1/n^2}}^{\sqrt{1-1/n^2}}\left|\frac{P_n(x)}{\varphi_n(x)}\right|^p \psi_n^\alpha(x)\,dx\leqslant 2^{|\alpha|}\sigma_{p,n}^p(\alpha),\quad \alpha<2,$$

$$I_2+I_3\leqslant \sup_{-[1,1]}\left|\frac{P_n(x)}{\varphi_n(x)}\right|^p 2\left(1-\sqrt{1-\frac{1}{n^2}}\right)\leqslant 2A^p(s,\alpha)\,\sigma_{p,n}^p(\alpha),\quad \alpha<2,$$

we obtain $\sigma_{p,n}\leq B_2\sigma_{p,n}(\alpha)$, $\alpha<2$, or $A\sigma_{p,n}\leq\sigma_{p,n}(\alpha)$, $\alpha\leq 2$, where $A>1/B_2>0$.

Thus Theorem 10 is completely proved.

Corollary. *For any* α *and* β $(\alpha, \beta<2)$ *and any algebraic polynomial* $P_n(x)$ *of degree* $\leq n$, *there holds the relation* $\sigma_{p,n}(\alpha)\sim\sigma_{p,n}(\beta)$.

Indeed, by Theorem 10

$$\sigma_{p,n}(\alpha)\sim\sigma_{p,n},\quad \sigma_{p,n}(\beta)\sim\sigma_{p,n},$$

and hence $\sigma_{p,n}(\alpha)\sim\sigma_{p,n}(\beta)$, as asserted.

Theorem 11. *If* $\omega(t)\in c_0A_\gamma[0,2]$, *then for any algebraic polynomial* $P_n(x)$ *of degree* $\leq n$ *and any* α *and* β $(\alpha,\beta<2)$:

$$\sigma_{q,n}^{(k)}(\alpha)\leqslant[A_k n]^{k+1/p-1/q}\sigma_{p,n}(\beta),\qquad 1\leqslant p\leqslant q\leqslant\infty,$$

where $A_k=A_k(\alpha,\beta,s,c_0)$;

$$\sigma_{q,n}^{(k)}(\alpha)=\left(\int\limits_{-1}^{1}\left|\rho_n^{k-1/q}(x)\frac{P_n^{(k)}(x)}{\varphi_n(x)}\right|^q\psi_n^\alpha(x)\,dx\right)^{1/q},\qquad k=0,1,\ldots,n,$$

and

$$\sigma_{p,n}(\beta)=\left(\int\limits_{-1}^{1}\left|\rho_n^{-1/p}(x)\frac{P_n(x)}{\varphi_n(x)}\right|^p\psi_n^\beta(x)\,dx\right)^{1/p}.$$

Proof. By Theorems 9 and 10 we have

$$\sigma_{q,n}^{(k)}(0)\leqslant[A_k(c_0)\,n]^{k+1/p-1/q}\sigma_{p,n}(0),$$
$$\sigma_{q,n}^{(k)}(0)\sim\sigma_{q,n}^{(k)}(\alpha),\qquad \alpha<2,$$
$$\sigma_{p,n}(0)\sim\sigma_{p,n}(\beta),\qquad \beta<2.$$

The theorem follows from this.

Now let us prove two theorems for trigonometric polynomials which are analogs of Theorems 10 and 11.

Theorem 12. *Let $T_n(\theta)$ be an arbitrary trigonometric polynomial of order $\leq n$, and let $\alpha < 1$ and $\beta < 1$. If $\omega(t) \in c_0 A_\gamma[0, 2]$, then*

$$I^*_{n,p}(\alpha) \sim I^*_{n,p}(\beta),$$

where

$$I^*_{n,p}(s) = \left(\int_{-\pi}^{\pi} \left| \frac{T_n(\theta)}{\varphi_n(\cos\theta)} \right|^p \left(\frac{\rho_n(\cos\theta)}{|\sin\theta|} \right)^s d\theta \right)^{1/p}.$$

Proof. We first establish this theorem for even and odd trigonometric polynomials. For example, let $\tau_n(\theta)$ be an odd trigonometric polynomial of order $\leq n$. Then, as is known, it can be written in the form $\tau_n(\theta) = \sin\theta P_{n-1}(\cos\theta)$, where $P_{n-1}(x)$ is an algebraic polynomial of degree $\leq n-1$. Hence

$$I^*_{n,p}(\alpha) = \left(\int_{-\pi}^{\pi} \left| \frac{\tau_n(\theta)}{\varphi_n(\cos\theta)} \right|^p \left(\frac{\rho_n(\cos\theta)}{|\sin\theta|} \right)^\alpha d\theta \right)^{1/p}$$
$$= 2^{1/p} \left(\int_{-1}^{1} \left| \rho_n^{\frac{p-1}{p}}(x) \frac{P_{n-1}(x)}{\varphi_n(x)} \right|^p \left(\frac{\rho_n(x)}{\sqrt{1-x^2}} \right)^{-p+\alpha+1} dx \right)^{1/p}, \quad \alpha < 1.$$

Thus from the corollary of Theorem 10 we obtain

$$I^*_{n,p}(\alpha) \sim \left(\int_{-1}^{1} \left| \rho_n^{\frac{p-1}{p}}(x) \frac{P_{n-1}(x)}{\varphi_n(x)} \right|^p \left(\frac{\rho_n(x)}{\sqrt{1-x^2}} \right)^{-p+\beta+1} dx \right)^{1/p}$$
$$= \left(\int_{-\pi}^{\pi} \left| \frac{\tau_n(\theta)}{\varphi_n(\cos\theta)} \right|^p \left(\frac{\rho_n(\cos\theta)}{|\sin\theta|} \right)^\beta d\theta \right)^{1/p} = I^*_{n,p}(\beta), \quad \beta < 1.$$

Hence for odd polynomials we have $I^*_{n,p}(\alpha) \sim I^*_{n,p}(\beta)$, $\alpha, \beta < 1$. It can be similarly shown that this relation holds for even polynomials.

Now suppose that $T_n(\theta)$ is an arbitrary trigonometric polynomial of order $\leq n$. Then, noting that the theorem is already established for even and odd polynomials, we obtain

$$I^*_{n,p}(\alpha) = \left(\int_{-\pi}^{\pi} \left| \frac{T_n(\theta)}{\varphi_n(\cos\theta)} \right|^p \left(\frac{\rho_n(\cos\theta)}{|\sin\theta|} \right)^\alpha d\theta \right)^{1/p}$$
$$\leqslant \frac{1}{2} \left(\int_{-\pi}^{\pi} \left| \frac{T_n(\theta) + T_n(-\theta)}{\varphi_n(\cos\theta)} \right|^p \left(\frac{\rho_n(\cos\theta)}{|\sin\theta|} \right)^\alpha d\theta \right)^{1/p}$$
$$+ \frac{1}{2} \left(\int_{-\pi}^{\pi} \left| \frac{T_n(\theta) - T_n(-\theta)}{\varphi_n(\cos\theta)} \right|^p \left(\frac{\rho_n(\cos\theta)}{|\sin\theta|} \right)^\alpha d\theta \right)^{1/p}$$
$$\sim \frac{1}{2} \left(\int_{-\pi}^{\pi} \left| \frac{T_n(\theta) + T_n(-\theta)}{\varphi_n(\cos\theta)} \right|^p \left(\frac{\rho_n(\cos\theta)}{|\sin\theta|} \right)^\beta d\theta \right)^{1/p}$$
$$+ \frac{1}{2} \left(\int_{-\pi}^{\pi} \left| \frac{T_n(\theta) - T_n(-\theta)}{\varphi_n(\cos\theta)} \right|^p \left(\frac{\rho_n(\cos\theta)}{|\sin\theta|} \right)^\beta d\theta \right)^{1/p}$$
$$\leqslant c_1(\alpha, \beta) I^*_{n,p}(\beta), \quad \alpha, \beta < 1.$$

Interchanging α and β here, we have

$$I^*_{n,p}(\beta) \leqslant c_2(\alpha, \beta) I^*_{n,p}(\alpha), \qquad \alpha, \beta < 1.$$

Hence $I^*_{n,p}(\alpha) \sim I^*_{n,p}(\beta)$, $\alpha < \beta < 1$, and the theorem is proved.

Theorem 13. *If the conditions of Theorem 12 hold, then for any p and q, $1 \le p < q \le \infty$, and any trigonometric polynomial $T_n(\theta)$ of order $\le n$*

$$I^{(k)}_{n,q}(\alpha) \leqslant [A(\alpha, \beta) n]^{k+1/p-1/q} I^*_{n,p}(\beta), \qquad k = 0, 1, \ldots,$$

where

$$I^{(k)}_{n,q}(\alpha) = \left(\int_{-\pi}^{\pi} \left| \frac{T^{(k)}_n(\theta)}{\varphi_n(\cos\theta)} \right|^q \left(\frac{|\sin\theta|}{\rho_n(\cos\theta)} \right)^\alpha d\theta \right)^{1/q},$$

$$I^*_{n,p}(\beta) = I^{(0)}_{n,p}(\beta).$$

Proof. We first consider the case of even and odd polynomials for $k = 1$. Let $\sigma_n(\theta)$ be even and $\tau_n(\theta)$ be odd. Then they can be written in the form

$$\sigma_n(\theta) = P_n(\cos\theta), \qquad \tau_n(\theta) = P_{n-1}(\cos\theta)\sin\theta,$$

where $P_n(x)$ and $P_{n-1}(x)$ are algebraic of degree $\le n$ and $\le n-1$, respectively.

By Theorem 11 we obtain for $\sigma_n(\theta)$ and $\tau_n(\theta)$ that

$$\left(\int_{-\pi}^{\pi} \left| \frac{\sigma'_n(\theta)}{\varphi_n(\cos\theta)} \right|^q d\theta \right)^{1/q} = 2^{1/q} \left(\int_0^{\pi} \left| \frac{P'_n(\cos\theta)\sin\theta}{\varphi_n(\cos\theta)} \right|^q d\theta \right)^{1/q}$$

$$= 2^{1/q} \left(\int_{-1}^{1} \left| \frac{P'_n(x)}{\varphi_n(x)} \rho_n^{1-1/q}(x) \right|^q \left(\frac{\sqrt{1-x^2}}{\rho_n(x)} \right)^{q-1} dx \right)^{1/q}$$

$$\leqslant 2^{1/p} (A_1 n)^{1+1/p-1/q} \left(\int_{-1}^{1} \left| \rho_n^{-1/p}(x) \frac{P_n(x)}{\varphi_n(x)} \right|^p \left(\frac{\rho_n(x)}{\sqrt{1-x^2}} \right)^{1+\beta} dx \right)^{1/p}$$

$$= (A_1 n)^{1+1/p-1/q} \left(\int_{-\pi}^{\pi} \left| \frac{P_n(\cos\theta)}{\varphi_n(\cos\theta)} \right|^p \left(\frac{\rho_n(\cos\theta)}{|\sin\theta|} \right)^\beta d\theta \right)^{1/p}, \qquad \beta < 1,$$

$$\left(\int_{-\pi}^{\pi} \left| \frac{\tau'_n(\theta)}{\varphi_n(\cos\theta)} \right|^q d\theta \right)^{1/q} = 2^{1/q} \left(\int_0^{\pi} \left| \frac{P'_{n-1}(\cos\theta)\sin^2\theta - \cos\theta P_{n-1}(\cos\theta)}{\varphi_n(\cos\theta)} \right|^q d\theta \right)^{1/q}$$

$$= 2^{\frac{1}{q}} \left(\int_{1}^{1} \left| \frac{\frac{1}{2} P'_{n-1}(x)(1-x^2) - xP_{n-1}(x) + \frac{1}{2} P'_{n-1}(x)(1-x^2)}{\varphi_n(x)} \right|^q \frac{dx}{\sqrt{1-x^2}} \right)^{1/q}$$

$$\leqslant 2^{\frac{1}{q}} \left(\int_{-1}^{1} \left| \frac{\rho_n^{1-1/q}(x) \left[\frac{1-x^2}{2} P_{n-1}(x) \right]'}{\rho_n(x)\varphi_n(x)} \right|^q \frac{\rho_n(x)}{\sqrt{1-x^2}} dx \right)^{1/q}$$

$$+ 2^{\frac{1}{q}} \left(\int_{-1}^{1} \left| \frac{\rho_n^{1-1/q}(x) P'_{n-1}(x)}{\rho_n^{-1}(x)\varphi_n(x)} \right|^q \left(\frac{\rho_n(x)}{\sqrt{1-x^2}} \right)^{1-2q} dx \right)^{1/q} \leqslant$$

$$\leqslant (\bar{A}_1 n)^{1+1/p-1/q} 2^{\frac{1}{q}} \left\{ \left(\int_{-1}^{1} \left| \frac{(1-x^2) P_{n-1}(x)}{2\rho_n^{(1+1/p)}(x)\,\varphi_n(x)} \right|^p \left(\frac{\rho_n(x)}{\sqrt{1-x^2}} \right)^{1+\beta} dx \right)^{1/p} \right.$$

$$\left. + \left(\int_{-1}^{1} \left| \frac{\rho_n^{1-1/p}(x)\, P_{n-1}(x)}{\varphi_n(x)} \right|^p \left(\frac{\rho_n(x)}{\sqrt{1-x^2}} \right)^{1-2q+\beta_1} dx \right)^{1/p} \right\}$$

$$= (\bar{A}_1 n)^{1+1/p-1/q} \left\{ \left(\int_{-1}^{\pi} \left| \frac{\sin\theta P_{n-1}(\cos\theta)}{\varphi_n(\cos\theta)} \right|^p \left(\frac{|\sin\theta|}{2\rho_n(\cos\theta)} \right)^p \left(\frac{\rho_n(\cos\theta)}{|\sin\theta|} \right)^\beta d\theta \right)^{1/p} \right.$$

$$\left. + \left(\int_{-\pi}^{\pi} \left| \frac{\sin\theta P_{n-1}(\cos\theta)}{\varphi_n(\cos\theta)} \right|^p \left(\frac{\rho_n(\cos\theta)}{|\sin\theta|} \right)^{p-2q+\beta_1} d\theta \right)^{1/p} \right\},$$

$$\beta < 1, \quad p - 2q + \beta_1 < 2.$$

Hence, setting $\beta_1 = 2q + \beta - p$, $\beta < 1$ and observing that $|\sin\theta|/\rho_n(\cos\theta) \le 1$, we find, respectively,

$$\left\| \frac{\sigma_n'(\theta)}{\varphi_n(\cos\theta)} \right\|_{L_q} \leqslant (\bar{A}_1 n)^{1+1/p-1/q} \left\| \frac{\sigma_n(\theta)}{\varphi_n(\cos\theta)} \left(\frac{\rho_n(\cos\theta)}{|\sin\theta|} \right)^{\beta/p} \right\|_{L_p},$$

$$\left\| \frac{\tau_n'(\theta)}{\varphi_n(\cos\theta)} \right\|_{L_q} \leqslant (\bar{A}_1 n)^{1+1/p-1/q} \left\| \frac{\tau_n(\theta)}{\varphi_n(\cos\theta)} \left(\frac{\rho_n(\cos\theta)}{|\sin\theta|} \right)^{\beta/p} \right\|_{L_p}.$$

If we write $T_n(\theta)$ in the form

$$T_n(\theta) = \sigma_n(\theta) + \tau_n(\theta),$$

where

$$\sigma_n(\theta) = [T_n(\theta) + T_n(-\theta)]/2, \quad \tau_n(\theta) = [T_n(\theta) - T_n(-\theta)]/2,$$

then after straightforward calculations akin to those in the proof of Theorem 12 we obtain

$$I_{n,q}^{(1)}(0) \leqslant (Bn)^{1+1/p-1/q} I_{n,p}^{*}(\beta).$$

But, by Theorem 12, $I_{n,q}^{(1)}(0) \sim I_{n,q}^{(1)}(\alpha)$, $\alpha < 2$, and hence

$$I_{n,q}^{(1)}(\alpha) \leqslant (A_1 n)^{1+1/p-1/q} I_{n,p}^{*}(\beta), \qquad \alpha,\ \beta < 2. \tag{21}$$

Hence Theorem 13 is proved for $k = 1$.

This theorem may be proved similarly for $k = 0$.

Let $k = 2, \ldots$. Then, applying (21) successively to

$$I_{n,q}^{(1)}(\alpha),\ I_{n,q}^{(2)}(\alpha),\ \ldots,\ I_{n,q}^{(k-1)}(\alpha),$$

we obtain

$$I_{n,q}^{(2)}(\alpha) \leqslant (A_1 n) I_{n,q}^{(1)}(\alpha),\quad I_{n,q}^{(3)}(\alpha) \leqslant A_1 n I_{n,q}^{(2)}(\alpha),\ \ldots,\ I_{n,q}^{(k)}(\alpha) \leqslant A_1 n I_{n,q}^{(k-1)}(\alpha).$$

Thus we find that

$$I_{n,q}^{(k)}(\alpha) \leqslant (A_1 n)^{k-1} I_{n,q}^{(1)}(\alpha) \leqslant (A_1 n)^{k+1/p-1/q} I_{n,p}^{*}(\beta).$$

Thus for any $k = 0, 1, \ldots$ we have

$$I_{n,q}^{(k)}(\alpha) \leqslant (A_1 n)^{k+1/p-1/q} I_{n,p}^{*}(\beta), \qquad \alpha, \beta < 2,$$

and the theorem is proved.

In concluding this section we note that Theorems 11 and 13 generalize the analogous results of Lemma 3, the main theorem and the results of subsections 1b and 2b of [4], and that the method of proof of these theorems is different from that of [4], where the proofs make use of ideas akin to those of N. K. Bari in the papers [2], [3], and [10]; see also [4].

A comment on this work. In connection with the fact that the inequalities proved in this work have an asymptotic character, note that in the formulations of the theorems the condition $\omega(t) \in c_0 A_\gamma[0, 2]$ may be replaced by the condition $\omega(t) \in c_0 A_\gamma[0, b]$, where b is a positive number ≤ 2.

BIBLIOGRAPHY

1. N. K. Bari, *Generalization of inequalities of S. N. Bernšteĭn and A. A. Markov*, Dokl. Akad. Nauk SSSR 90 (1953), 701–702. (Russian) MR 15, 215.

2. ———, *Generalization of inequalities of S. N. Bernšteĭn and A. A. Markov*, Izv. Akad. Nauk SSSR Ser. Mat. 18 (1954), 159–176. (Russian) MR 15, 788.

3. V. K. Dzjadyk, *Constructive characterization of functions satisfying the condition* Lip α $(0 < \alpha < 1)$ *on a finite segment of the real axis*. Izv. Akad. Nauk SSSR Ser. Mat. 20 (1956), 623–642. (Russian) MR 18, 392.

4. M. K. Potapov, *Some inequalities for polynomials and their derivatives*, Vestnik Moskov. Univ. Ser. I Mat. Meh. 1960, no. 2, 10–20. (Russian) MR 23 #A3412.

5. G. K. Lebed′, *Some problems in the approximation of functions by algebraic polynomials*, Candidate's Dissertation, Dnepropetrovsk, 1959. (Russian)

6. ———, *Inequalities for polynomials and their derivatives*, Dokl. Akad. Nauk SSSR 117 (1957), 570–572. (Russian) MR 20 #1165.

7. V. K. Dzjadyk, *Inverse theorems in the theory of approximation of functions in complex domains*, Ukrain. Mat. Ž. 15 (1963), 365–375. (Russian) MR 35 #4437.

8. M. I. Andraško, *Inequalities for the derivative of an algebraic polynomial in the metric* L_p $(p \geq 1)$ *in regions with corners*, Ukrain. Mat. Ž. 16 (1964), 431–444. (Russian) MR 30 #3198.

9. Ju. A. Brudnyĭ, *Approximation by integral functions on the exterior of a segment or on a semi-axis*, Dokl. Akad. Nauk SSSR 124 (1959), 739–742. (Russian) MR 21 #255.

10. A. F. Timan, *Theory of approximation of functions of a real variable*, Fizmatgiz, Moscow, 1960; English transl., Macmillan, New York, 1963. MR 22 #8257; 33 #465.

11. A. A. Markov, *On a problem of D. I. Mendeleev*, Zap. Imp. Akad. Nauk St. Petersburg 62 (1890), 1–24; reprinted in *Selected papers on continued fractions and the theory of functions deviating least from zero*, OGIZ, Moscow, 1948, pp. 51–75. (Russian) Fortschritte 22, 283; MR 11, 150.

12. S. N. Bernšteĭn, *Collected works*. Vol. 1: *The constructive theory of functions* [1905–1930], Izdat. Akad. Nauk SSSR, Moscow, 1952. (Russian) MR 14, 2.

13. S. M. Nikol′skiĭ, *Inequalities for entire functions of exponential type and their application to the theory of differentiable functions of several variables*, Trudy Mat. Inst. Steklov. 38 (1961), 244–278; English transl., Amer. Math. Soc. Transl. (2) 80 (1969), 1–38. MR 14, 32.

14. A. Zygmund, *A remark on conjugate series*, Proc. London Math. Soc. 34 (1932), 332–400.

15. N. A. Lebedev, *Inverse theorems of uniform approximation*, Dokl. Akad. Nauk SSSR 171 (1966), 788–790 = Soviet Math. Dokl. 7 (1966), 1545–1548. MR 35 #386.

16. N. A. Lebedev and P. M. Tamrazov, *Inverse theorems of approximation on closed sets of the complex plane*, Dokl. Akad. Nauk SSSR 179 (1968), 1046–1049 = Soviet Math. Dokl. 9 (1968), 498–501. MR 37 #4271.

17. ———, *Inverse approximation theorems on regular compacta of the complex plane*, Izv. Akad. Nauk SSSR Ser. Mat. 34 (1970), 1340–1390 = Math. USSR Izv. 4 (1970), 1355–1405. MR 45 #2387.

Translated by J. MALEK

Trudy Mat. Inst. Steklov.
134 (1975)

Proc. Steklov Inst. Math.
134 (1975)

ON THE UNIQUENESS OF ASSIGNMENT OF THE DIRICHLET SERIES TO A FUNCTION

UDC 517.52

A. F. LEONT'EV

Abstract. Let $L(\lambda)$ be an entire function of exponential type with indicator $h(\phi)$, and $\overline{D}$ the smallest convex closed set containing all the singularities of the function associated with $L(\lambda)$ in the Borel sense. To a function $f(z)$, analytic on $\overline{D}$, there corresponds a Dirichlet series according to the usual rule

$$f(z) \sim \Sigma P_k(z) e^{\lambda_k z}. \tag{1}$$

It has been shown (MR 34 #7780) that if $P_k(z) \equiv 0$ then $f(z) \equiv 0$. Under the additional condition

$$|L(re^{i\phi})| < A \frac{e^{h(\phi)r}}{r^{\mu}}, \quad \mu > 1,$$

the correspondence (1) can be extended to functions analytic in D and continuous in $\overline{D}$. It is shown that if all $P_k(z) \equiv 0$ for $f(z)$ in this wider class, then $f(z) \equiv 0$. It follows from this theorem that if the λ_k are simple zeros of $L(\lambda)$ and $\{\psi_k(t)\}$ is a system biorthogonal to the system $e^{\lambda_k t}$, then $\{\psi_k(t)\}$ is complete with respect to the metric C outside D in the space of functions analytic outside $\overline{D}$, continuous outside D and vanishing at infinity. The completeness of the system $\{\psi_k(t)\}$ can be used for reconstructing $f(z)$ if we know the coefficients of (1).

Bibliography: 8 titles.

Let $L(\lambda)$ be an entire function of exponential type with an infinite number of zeros, and $\gamma(t)$ the function associated with $L(\lambda)$ in the Borel sense. We denote by $\overline{D}$ the smallest closed convex set containing all the singularities of $\gamma(t)$. We take an arbitrary function $f(z)$, analytic in D, and set

$$\omega(\mu) = \frac{1}{2\pi i} \int_C \left[\int_0^t f(t-\eta) e^{\mu\eta} d\eta \right] \gamma(t)\, dt$$

(assuming that the origin lies in $\overline{D}$), where C is a closed contour containing $\overline{D}$ in its interior, such that $f(z)$ is analytic inside and on C. Let $\lambda_1, \lambda_2, \ldots$ be the zeros of $L(\lambda)$ and $m_1, m_2, \ldots$ their respective multiplicities. With the function $f(z)$ we associate the series

$$f(z) \sim \sum P_\nu(z) e^{\lambda_\nu z}, \tag{1}$$

AMS (MOS) subject classifications (1970). Primary 30A16; Secondary 30A18.

where

$$P_\nu(z)e^{\lambda_\nu z} = \frac{1}{2\pi i}\int_{C_\nu}\frac{\omega(\mu)e^{\mu z}d\mu}{L(\mu)} \qquad (\nu = 1, 2, \ldots). \tag{2}$$

Here C_ν is a closed contour with λ_ν, and no other zero of $L(\lambda)$, inside it.

In [1] it was shown that if all the $P_\nu(z) \equiv 0$ then $f(z) \equiv 0$. This uniqueness theorem implies that in principle we can define $f(z)$ from the coefficients in (1) without imposing any further restrictions on $L(\lambda)$.

We note an interesting fact which follows immediately from the uniqueness theorem. Suppose that all the zeros of $L(\lambda)$ are simple and different from zero. In this case $P_\nu(z) \equiv a_\nu$. The coefficients a_ν, as was shown in [2], can be calculated from the formula

$$a_\nu = \frac{1}{2\pi i}\int_C f(t)\psi_\nu(t)\,dt \qquad (\nu \geqslant 1),$$

where

$$\psi_\nu(t) = \frac{1}{L'(\lambda_\nu)}\int_0^\infty \frac{L(\lambda)}{\lambda - \lambda_\nu}e^{-\lambda t}d\lambda. \tag{3}$$

Here the $\psi_\nu(t)$ are the functions associated with

$$\frac{1}{L'(\lambda_\nu)}\frac{L(\lambda)}{\lambda-\lambda_\nu}$$

in the Borel sense; they are regular outside $\overline{D}$, vanish at infinity, and the system $\{\psi_\nu(z)\}$ is biorthogonal to $\{e^{\lambda_\nu z}\}$, i.e.

$$\frac{1}{2\pi i}\int_C e^{\lambda_k t}\psi_\nu(t)\,dt = \delta_{k\nu}.$$

The above-mentioned fact is the property (cf. [2]) that $\{\psi_\nu(t)\}$ is always complete in the space of functions $F(z)$ analytic outside D and vanishing at infinity:

As was observed by V. V. Napalkov, this circumstance can be used in the following way to construct $f(z)$ from the coefficients a_ν of its Dirichlet series. We have

$$\frac{1}{t-z} = \lim_{n\to\infty}\sum_{k=1}^{p_n} A_k^{(n)}(z)\psi_k(t), \qquad z \in D, \qquad t \in c. \tag{4}$$

Hence

$$f(z) = \frac{1}{2\pi i}\int_C \frac{f(t)\,dt}{t-z} = \lim_{n\to\infty}\sum_{k=1}^{p_n} A_k^{(n)}(z)\frac{1}{2\pi i}\int_C f(z)\psi_k(t)\,dt$$
$$= \lim_{n\to\infty}\sum_{k=1}^{p_n} A_k^{(n)}(z)\,a_k. \tag{5}$$

We suppose now that $L(\lambda)$ satisfies the condition ($h(\phi)$ is the indicator of growth)

$$|L(re^{i\varphi})| < A\frac{e^{h(\varphi)r}}{r^\mu} \qquad (\mu > 1). \tag{6}$$

Under this condition the function $\gamma(t)$ is regular outside $\overline{D}$ and continuous outside D (D is the interior of $\overline{D}$), and consequently we can define the function $\omega(\mu)$ by the formula

$$\omega(\mu) = \frac{1}{2\pi i} \int_{\partial\overline{D}} \left[\int_0^t f(t-\eta) e^{\mu\eta} d\eta \right] \gamma(t)\, dt. \tag{7}$$

Furthermore, this holds not only for functions $f(z)$ analytic in $\overline{D}$, but also for functions which are analytic in D and continuous in $\overline{D}$. If $f(z)$ belongs to this wider class, we associate with it the series (1) with terms (2).

Theorem 1. *If all $P_\nu(z)$ in* (1) *are zero, then $f(z) \equiv 0$.*

If (6) is satisfied, then the functions $\psi_\nu(t)$ defined by (3) are analytic outside $\overline{D}$, continuous outside D and vanish at infinity.

Theorem 2. *Suppose that $\overline{D}$ is not a segment. If* (6) *holds, then the system $\{\psi_\nu(t)\}$ in the closed domain $\overline{E}$ (the exterior of D) is complete with respect to the metric of C in the space of functions which are analytic in E (the exterior of $\overline{D}$), continuous in $\overline{E}$ and vanish at ∞.*

In view of Theorem 2 we can use (5) to construct the function $f(z)$ at points $z \in D$, if we know the coefficients a_ν.

We can then establish a uniqueness theorem for the more general case when

$$L(\lambda) = \int_{\partial\overline{D}} e^{\lambda t} d\sigma(t), \tag{8}$$

where $\sigma(t)$ is a function of bounded variation on $\partial\overline{D}$. In this case we have

$$\omega(\mu) = \int_{\partial\overline{D}} \left[\int_0^t f(t-\eta) e^{\mu\eta} d\eta \right] d\sigma(t), \tag{9}$$

and $f(z)$ is analytic in D and continuous in $\overline{D}$. We associate $f(z)$ with the series (1) with terms (2).

Theorem 3. *Suppose that* (8) *holds. If all $P_\nu(z) \equiv 0$ in* (1) *then $f(z) \equiv 0$.*

The theorem is proved by reducing case (8) to case (6).

In case (8) we can form the formulas (2), where

$$\psi_\nu(t) = -\frac{e^{-\lambda_\nu t}}{L'(\lambda_\nu)} \int_{\alpha_\nu}^t e^{\lambda_\nu \eta} d\sigma(\eta), \qquad \alpha_\nu \in \partial\overline{D}, \qquad t \in \partial\overline{D}.$$

The functions $\psi_\nu(t)$ are not continuous on $\partial\overline{D}$, and so we may not pose the question of the completeness of $\{\psi_\nu(t)\}$ with respect to the C metric on $\partial\overline{D}$. However, in this case too we can solve the problem of constructing $f(z)$ from the coefficients a_ν of the Dirichlet series by reducing it to case (6).

§1. Proof of Theorem 1

For the proof we first give several lemmas, some of which concern the properties of the function (7). It is convenient to denote (7) in the following way:

$$\omega_L(\mu, f) = \frac{1}{2\pi i}\int_{\partial\bar{D}}\left[\int_0^t f(t-\eta)e^{\mu\eta}d\eta\right]\gamma(t)\,dt.$$

Lemma 1. *Let β be a zero of $L(\lambda)$, and if $L_1(\lambda) = L(\lambda)/(\lambda-\beta)$ let $\gamma_1(t)$ be the function associated with $L_1(\lambda)$ in the Borel sense. If $\psi(t)$ is regular in D and continuous in $\bar{D}$, then*

$$\omega_L(\mu, \psi) = (\mu-\beta)\,\omega_{L_1}(\mu, \psi) + \frac{1}{2\pi i}\int_{\partial\bar{D}}\psi(t)\gamma_1(t)\,dt. \tag{10}$$

This lemma is proved in the same way as Lemma 2 in [1].

Lemma 2. *Let*

$$\frac{\omega_L(\mu, f)}{L(\mu)} \tag{11}$$

be an entire function, and suppose that $L(\beta) = 0$. Then

$$\frac{\omega_L(\mu, f)}{L(\mu)} = \frac{\omega_{L_1}(\mu, f)}{L_1(\mu)}, \qquad L_1(\mu) = \frac{L(\mu)}{\mu-\beta}.$$

It follows from the condition that $\omega_L(\beta, f) = 0$. Setting $\mu = \beta$ and $\psi = f$ in (10), we obtain

$$\frac{1}{2\pi i}\int_{\partial\bar{D}} f(t)\gamma_1(t)\,dt = 0,$$

and the result follows from this and (10).

Lemma 3. *Let*

$$L(\lambda) = \lambda^p e^{\alpha\lambda+\beta}\prod_{k=1}^{\infty}\left(1-\frac{\lambda}{\lambda_k}\right)e^{\frac{\lambda}{\lambda_k}}, \qquad p \geqslant 0,$$

and

$$L_n(\lambda) = L(\lambda)\left\{\frac{p}{\lambda}+\alpha+\sum_{k=1}^{n}\left(\frac{1}{\lambda-\lambda_k}+\frac{1}{\lambda_k}\right)\right\} \qquad (n = 1, 2, \ldots).$$

Denote by D_h the exterior of the discs $|\lambda-\lambda_k| \le |\lambda_k|^{-h}$ $(k = 1, 2, \ldots)$ and of $|\lambda| \le 1$, where $\sum_{k=1}^{\infty}|\lambda_k|^{-h} < \infty$. Then

$$|L_n(\lambda)| < A|\lambda|^{h+2}|L(\lambda)| \qquad (\lambda \in D_h),$$

where A does not depend on n.

This is Lemma 5 from [1].

Lemma 4. *Let $\gamma_n(t)$ be the functions associated in the Borel sense with the functions $L_n(\lambda)$ (from the previous lemma), and $g(t)$ the function associated with the derivative $L'(\lambda)$ in the Borel sense. If $L(\lambda)$ satisfies*

$$|L(re^{i\varphi})| < A\frac{e^{h(\varphi)r}}{r^{\mu}}, \quad \mu > h+3, \quad \sum_{k=1}^{\infty} |\lambda_k|^{-h} < \infty, \tag{12}$$

then

$$\lim_{n\to\infty} \gamma_n(t) = g(t)$$

uniformly on the boundary.

We first note that if $L(\lambda)$ satisfies (6), then its derivative satisfies the analogous inequality

$$|L'(re^{i\varphi})| < B\frac{e^{h(\varphi)r}}{r^{\mu}}$$

(this was proved in [4]). It follows that $g(t)$ is continuous outside D.

We also need the inequality

$$|h(\varphi_1)r_1 - h(\varphi_2)r_2| \leqslant |z_2 - z_1| \max_{z\in \bar D}|z|, \qquad z_j = r_j e^{i\varphi_j} \quad (j=1,\ 2) \tag{13}$$

which was also established in [4].

Let

$$H = 2\left(\sum_{k=1}^{\infty} |\lambda_k|^{-h} + 1\right).$$

We take an arbitrary point $\tilde\lambda$, $|\tilde\lambda| > 4H$. Among all circles $|\lambda - \tilde\lambda| = r$, $0 < r < 2H$, there is at least one (which we denote by $|\lambda - \tilde\lambda| = \tau$) which lies entirely in D_h. By Lemma 3 we have

$$L_n(\tilde\lambda)| \leqslant \max_{|\lambda-\tilde\lambda|=\tau} |L_n(\lambda)| \leqslant A(|\tilde\lambda| + 2H)^{h+2} \max_{|\lambda-\tilde\lambda|=2H} |L(\lambda)|$$

$$= A(|\tilde\lambda| + 2H)^{h+2}|L(\lambda_1)|, \qquad |\lambda_1 - \tilde\lambda| = 2H.$$

Next, from (12) and (13) we obtain ($\phi_1 = \arg\lambda_1$ and $\tilde\phi = \arg\tilde\lambda$)

$$|L(\lambda_1)| < A\frac{e^{h(\varphi_1)|\lambda_1|}}{|\lambda_1|^{\mu}} \leqslant A\left|\frac{\tilde\lambda}{\lambda_1}\right|^{\mu} \exp\left(2H\max_{z\in\bar D}|z|\right)\frac{e^{h(\tilde\varphi)|\tilde\lambda|}}{|\tilde\lambda|^{\mu}}.$$

Since $|\tilde\lambda| > 4H$ we have $|\lambda_1| \geq |\tilde\lambda| - 2H > \frac{1}{2}|\tilde\lambda|$, and so

$$\left|\frac{\tilde\lambda}{\lambda_1}\right| < 2, \quad |\tilde\lambda| + 2H < 2|\tilde\lambda|.$$

Hence

$$|L_n(\tilde\lambda)| < B\frac{e^{h(\tilde\varphi)|\tilde\lambda|}}{|\tilde\lambda|^{\mu-h-2}}, \qquad |\tilde\lambda| \geqslant 4H, \tag{14}$$

where

$$B = A2^{\mu+h+2}\exp\left(2H\max_{z\in\bar D}|z|\right).$$

We observe that $\mu - h - 2 > 1$, since by hypothesis $\mu > h+3$.

We now take into consideration that, in any bounded domain,

$$\lim_{n\to\infty} L_n(\lambda) = L'(\lambda)$$

uniformly. In view of this and (14), and using

$$\gamma_n(t) - g(t) = \int_0^{\infty e^{i\varphi}} [L_n(\lambda) - L'(\lambda)] e^{-\lambda t}\, d\lambda,$$

we deduce that

$$\lim_{n\to\infty} \gamma_n(t) = g(t)$$

uniformly on the boundary $\partial\bar{D}$.

Lemma 5. *Suppose that* (12) *holds and that* (11) *is an entire function. Then*

$$\frac{\omega_{L'}(\mu, f)}{L'(\mu)}$$

is also an entire function.

The lemma is proved using Lemmas 2 and 4 in the same way as Lemma 7 in [1].

Lemma 6. *Suppose that* (12) *holds and* (11) *is an entire function. Then*

$$\frac{1}{2\pi i}\int_{\partial\bar{D}} t^k \gamma(t) f(t)\, dt = 0 \qquad (k = 0, 1, 2, \ldots).$$

It is proved in [2] that the coefficients of the polynomial

$$P_\nu(z) = \sum_{k=0}^{m_\nu - 1} \alpha_{\nu k} z^k$$

in (1) can be computed from the formulas

$$\alpha_{\nu k} = \frac{1}{2\pi i}\int_{\partial\bar{D}} f(t)\, \psi_{\nu k}(t)\, dt \qquad (k = 0, 1, \ldots, m_\nu - 1;\ \nu = 1, 2, \ldots),$$

where

$$\psi_{\nu k}(t) = \frac{1}{k!\,(m_\nu - k - 1)!} \sum_{q=0}^{m_\nu - k - 1} q!\, C^q_{m_\nu - k - 1} [M_\nu(\mu)]^{(m_\nu - k - q - 1)}_{\mu = \lambda_\nu} A_{\nu, q}(t),$$

$$M_\nu(\mu) = \frac{(\mu - \lambda_\nu)^{m_\nu}}{L(\mu)},$$

and the $A_{\nu,q}(t)$ are associated in the Borel sense with the functions

$$\frac{L(\lambda)}{(\lambda - \lambda_\nu)^{q+1}}.$$

By (12) the functions $\psi_{\nu k}(t)$ are regular outside $\bar{D}$ and continuous outside D. We observe that

$$\psi_{\nu, m_\nu - 1}(t) = \frac{1}{(m_\nu - 1)!} M_\nu(\lambda_\nu) A_{\nu, 0}(t).$$

Here

$$A_{\nu, 0}(t) = \int_0^{\infty} \frac{L(\lambda)}{\lambda - \lambda_\nu} e^{-\lambda t}\, d\lambda.$$

We obtain an asymptotic relation for $A_{\nu,0}(t)$ for large ν. We have

$$A_{\nu,0}(t)=-\frac{1}{\lambda_\nu}\int_0^\infty L(\lambda)e^{-\lambda t}d\lambda-\frac{1}{\lambda_\nu^2}\int_0^\infty \lambda L(\lambda)e^{-\lambda t}d\lambda+\frac{1}{\lambda^2}\int_0^\infty \frac{\lambda^2}{\lambda-\lambda_\nu}L(\lambda)e^{-\lambda t}d\lambda.$$

It was established in [4] that if (6) holds, then

$$\left|\frac{L(\lambda)}{\lambda-\lambda_\nu}\right|<C\frac{e^{h(\varphi)r}}{r^\mu},\qquad \lambda=re^{i\varphi},$$

where C does not depend on ν. By (12), $\mu>3$. Hence

$$\left|\int_0^\infty \lambda L(\lambda)e^{-\lambda t}d\lambda\right|<B,\qquad \left|\int_0^\infty \frac{\lambda^2}{\lambda-\lambda_\nu}L(\lambda)e^{-\lambda t}d\lambda\right|<B,\qquad t\in\partial\bar D,$$

where B does not depend on ν. Consequently

$$A_{\nu,0}(t)=-\frac{\gamma(t)}{\lambda_\nu^2}+\frac{\gamma_\nu(t)}{\lambda_\nu^2},\qquad |\gamma_\nu(t)|<2B,\qquad t\in\partial\bar D.$$

Hence

$$\alpha_{\nu,m_\nu-1}=\frac{1}{2\pi i}\int_{\partial\bar D} f(t)\psi_{\nu,m_\nu-1}(t)\,dt=\frac{1}{(m_\nu-1)!}M_\nu(\lambda_\nu)\times\left\{-\frac{1}{\lambda_\nu}\frac{1}{2\pi i}\int_{\partial\bar D} f(t)\gamma(t)\,dt+O\left(\frac{1}{\lambda_\nu^2}\right)\right\}. \tag{15}$$

Since (11) is an entire function, all the $\alpha_{\nu,k}$ vanish. From the asymptotic relation (15) it then follows that

$$\frac{1}{2\pi i}\int_{\partial\bar D} f(t)\gamma(t)\,dt=0. \tag{16}$$

Hence (16) holds for the entire functions (11).

We consider the derivative $L^{(n)}(\lambda)$. It satisfies a condition of type (12). By Lemma 5,

$$\frac{\omega_{L^{(k)}}(\mu,f)}{L^{(k)}(\mu)}\qquad (k=1,2,\ldots)$$

is an entire function. But then, as proved above, we have

$$\frac{1}{2\pi i}\int_{\partial\bar D} g_k(t)f(t)\,dt=0\qquad (k=1,2,\ldots),$$

where $g_k(t)$ are the functions associated with $L^{(k)}(\lambda)$. It remains only to note that $g_k(t)$ differs from $(-1)^k t^k\gamma(t)$ by an entire function. The lemma is now proved.

We now commence the proof of the theorem. We have to show that if (11) is an entire function, then $f(t)\equiv 0$.

We first observe, that if (6) holds, then $\bar D$ cannot reduce to a single point. Indeed, if this were false then $\gamma(t)$, being bounded in a neighborhood of $t_0=\bar D$, would be analytic at t_0. Then, however, $\gamma(t)$ is analytic in the whole plane, which means that $L(\lambda)\equiv 0$, and this is impossible. It follows that $L(\lambda)$ has infinitely many zeros. For otherwise $L(\lambda)$ has the form

$$L(\lambda)=P(\lambda)e^{a\lambda},$$

where $P(\lambda)$ is a polynomial, and $\overline{D}$ reduces to the single point a.

We note further that during the proof we may assume the number μ in (6) to be as large as we please, without loss of generality. Indeed, if we take the function

$$M(\lambda)=\frac{L(\lambda)}{(\lambda-\lambda_1)\dots(\lambda-\lambda_k)},$$

then it satisfies

$$|M(re^{i\varphi})|<K\frac{e^{h(\varphi)r}}{r^{\mu+k}}.$$

Furthermore, since (11) is an entire function, we deduce from successive applications of Lemma 2 that

$$\frac{\omega_M(\mu,t)}{M(\mu)}$$

is an entire function. Thus instead of $L(\lambda)$ we may take a function $M(\lambda)$ which satisfies (6) with $\mu+k$ replacing μ as the exponent in the denominator. This can be made as large as desired by suitable choice of k. It is enough for our purposes that it exceed $h+4$, where h is as in Lemma 3.

We consider the case when $\overline{D}$ is a segment (and D is the empty set). It is convenient to assume, without loss of generality, that $\overline{D}$ is an interval $[a, b]$ ($\operatorname{Im} a < \operatorname{Im} b$) of the imaginary axis. The points a and b are singular points of $\gamma(t)$.

By Lemma 6 we have

$$\frac{1}{2\pi i}\int_{\partial\overline{D}} t^k\gamma(t)f(t)\,dt=0 \qquad (k=0,1,\dots).$$

Let $\overset{+}{\gamma}(t)$ be the value of $\gamma(t)$ at one end of the segment $[a, b]$, and $\overline{\gamma}(t)$ its value at the other. We obtain

$$\int_a^b t^k[\overset{+}{\gamma}(t)-\overline{\gamma}(t)]f(t)\,dt=0 \qquad (k=0,1,\dots)$$

and hence

$$[\overset{+}{\gamma}(t)-\overline{\gamma}(t)]f(t)=0, \qquad t\in[a,b]. \tag{17}$$

Let $\gamma_{\nu,k}(t)$ be the function associated with

$$\frac{L(\lambda)}{(\lambda-\lambda_\nu)^k} \qquad (k=1,2,\dots,m_\nu)$$

in the Borel sense. Since we may take this function instead of $L(\lambda)$, we have, as above

$$[\overset{+}{\gamma}_{\nu,k}(t)-\overline{\gamma}_{\nu,k}(t)]f(t)=0, \quad t\in[a,b] \quad (k=1,2,\dots,m_\nu;\quad \nu=1,2,\dots). \tag{18}$$

We now assume that our result is false, i.e. that $f(t)\not\equiv 0$ on $[a, b]$. Let (a_1, b_1), $\operatorname{Im} a < \operatorname{Im} a_1 < \operatorname{Im} b_1 < \operatorname{Im} b$, be any interval in which $f(t)\neq 0$. From (17) and (18) we find that

$$\overset{+}{\gamma}(t)=\bar{\gamma}(t),\quad \overset{+}{\gamma}_{\nu,k}(t)=\bar{\gamma}_{\nu,k}(t),\quad t\in(a_1,b_1)\quad (k=1,2,\ldots,m_\nu;\ \nu=1,2\ldots). \tag{19}$$

We have to show that the relations (19) are in fact impossible.

We make use of the following theorems ([5], Chapter V, §4, Theorem II): If an entire function $f(\lambda)$ of exponential type is bounded on the real axis (in our case the set $\overline{D}$ is an interval $[a, b]$ of the imaginary axis), then the upper density Δ_f of the sequence of zeros of this function satisfies

$$\Delta_f=\overline{\lim_{r\to\infty}}\frac{n(r)}{r}=\frac{|b-a|}{\pi},$$

where $n(r)$ is the number of zeros of $f(\lambda)$ in $|\lambda|\leq r$.

Our function $L(\lambda)$ has the form

$$L(\lambda)=\frac{1}{2\pi i}\int_a^b[\overset{+}{\gamma}(t)-\bar{\gamma}(t)]e^{\lambda t}d\lambda.$$

We see from this that it is bounded on the real axis. Hence the upper density Δ_L of its sequence of zeros is

$$\Delta_L=\frac{|b-a|}{\pi}. \tag{20}$$

We now return to the equations (19). It follows from (19) that $\gamma(t)$ and $\gamma_{\nu,k}(t)$ can be analytically continued from the left half-plane over the segment (a_1, b_1) into the right half-plane, and further that on every contour Γ such that $[a, a_1]$ lies wholly inside Γ and $[b_1, b]$ outside Γ, these functions are single-valued. Let Γ be such a contour. We set

$$M(\lambda)=\frac{1}{2\pi i}\int_\Gamma\gamma(t)e^{\lambda t}dt=\frac{1}{2\pi i}\int_a^{a_1}[\overset{+}{\gamma}(t)-\bar{\gamma}(t)]e^{\lambda t}dt; \tag{21}$$

then $M(\lambda)\not\equiv 0$ since a is a singular point of $\gamma(t)$, We shall show that the points λ_ν are zeros of $M(\lambda)$ of multiplicity $\geq m_\nu$.

We have

$$\gamma_{\nu,k}(t)=\int_0^\infty\frac{L(\lambda)e^{-\lambda t}d\lambda}{(\lambda-\lambda_\nu)^k}.$$

Hence

$$\gamma_{\nu,k}'(t)+\lambda_\nu\gamma_{\nu,k}(t)=-\gamma_{\nu,k-1}(t)$$

(for $k=1$ we have $\gamma_{\nu,k-1}(t)$). Solving this equation by the method of varying an arbitrary constant* we obtain

$$\gamma_{\nu,k}(t)=-e^{-\lambda_\nu t}\int_{t_0}^t\gamma_{\nu,k-1}(\eta)e^{\lambda_\nu\eta}d\eta+Ce^{-\lambda_\nu t},\qquad t_0\in\Gamma,$$

where C is a constant (which depends on ν, k and t_0). Since the $\gamma_{\nu,k}(t)$ are single-valued on Γ, it is evident from this formula that the functions

$$v_k(t)=\int_{t_0}^t\gamma_{\nu,k-1}(\eta)e^{\lambda_\nu\eta}d\eta\qquad(k=1,\ 2,\ \ldots,\ m_\nu)$$

are also single-valued on Γ.

---

* *Translator's note.* This is a literal translation; it is the integrating factor method.

Bearing this in mind and using integration by parts, we obtain from (21)

$$M(\lambda)=\frac{1}{2\pi i}\int_\Gamma \gamma(t)\,e^{\lambda_\nu t}e^{(\lambda-\lambda_\nu)t}\,dt = v_1(t)\,e^{(\lambda-\lambda_\nu)t}\Big|_\Gamma - \frac{(\lambda-\lambda_\nu)}{2\pi i}\int_\Gamma v_1(t)\,e^{(\lambda-\lambda_\nu)t}\,dt.$$

The first term on the right vanishes since $v_k(t)$ is single-valued on Γ. Thus

$$M(\lambda)=(\lambda-\lambda_\nu)\frac{1}{2\pi i}\int_\Gamma \gamma_{\nu,1}(t)\,e^{\lambda t}\,dt.$$

In an analogous way, using the single-valuedness of $v_k(t)$ on Γ, we successively find that

$$M(\lambda)=(\lambda-\lambda_\nu)^k\frac{1}{2\pi i}\int_\Gamma \gamma_{\nu,k}(t)\,e^{\lambda t}\,dt \qquad (k=2,\ \ldots,\ m_\nu).$$

Thus λ_ν is a zero of $M(\lambda)$ of order $\geq m_\nu$ ($\nu = 1, 2, \ldots$).

The function $M(\lambda)$ is bounded on the real axis, as is evident from (21). The singular points of the function associated with $M(\lambda)$ are situated on $[a, a_1]$. By the density theorem stated above we conclude that the upper density of the sequence $\{\lambda_\nu\}$ (earlier denoted by Δ_L) satisfies

$$\Delta_L \leqslant \frac{|a_1-a|}{\pi}.$$

However, this contradicts (20). Consequently $f(t) = 0$ on $[a, b]$. Thus the uniqueness theorem is proved in the case when $\bar{D}$ is a segment. We note that this result for a segment is not new, being found in [6]. Here we have merely given a new, simpler proof.

Lemma 7. *Let* (11) *be an entire function. Then for some* H *the function*

$$\frac{\omega_L(\mu,\ F)}{L(\mu)},$$

where $F(z) = \int_0^z f(t)dt - H$, *is also entire.*

We set

$$\Phi(z)=\int_0^z f(t)\,dt, \qquad M(\lambda)=\lambda L(\lambda).$$

The function associated with $M(\lambda)$ in the Borel sense equals $-\gamma'(t)$. We have

$$\omega_M(\mu,\ \Phi)=-\frac{1}{2\pi i}\int_{\partial\bar{D}}\gamma'(t)\left[\int_0^t \Phi(t-\eta)\,e^{\mu\eta}\,d\eta\right]dt.$$

Integrating by parts, we obtain

$$\omega_M(\mu,\ \Phi)=-\gamma(t)\left(\int_0^t \Phi(t-\eta)\,e^{\mu\eta}\,d\eta\right)\Bigg|_{\partial\bar{D}}$$

$$+\frac{1}{2\pi i}\int_{\partial\bar{D}}\gamma(t)\left[\Phi(0)\,e^{\mu t}+\int_0^t f(t-\eta)\,e^{\mu\eta}\,d\eta\right]dt=\omega_L(\mu,\ f),$$

whence

$$\frac{\omega_M(\mu, \Phi)}{M(\mu)} = \frac{\omega_L(\mu, f)}{\mu L(\mu)}. \tag{22}$$

Now (11) is an entire function. Hence

$$\frac{1}{2\pi i}\int_{C_0} \frac{\omega_L(\mu, f)}{\mu L(\mu)} e^{\mu z} d\mu = H. \tag{23}$$

Here C_0 is a circle of sufficiently small radius with center the origin, and H is a constant. We compute $\omega_M(\mu, \Phi_1)$, where $\Phi_1 \equiv H$. We have

$$\omega_M(\mu, \Phi_1) = -\frac{H}{2\pi i}\int_{\partial \bar{D}} \gamma'(t)\left(\int_0^t e^{\mu\eta} d\eta\right) dt = \frac{H}{\mu}(M(\mu) - M(0)) = H\frac{M(\mu)}{\mu}.$$

Set

$$F(z) = \Phi(z) - H = \int_0^z f(t)\,dt - H.$$

We obtain

$$\omega_M(\mu, F) = \omega_M(\mu, \Phi) - H\frac{M(\mu)}{\mu}.$$

Using this and (22), we deduce that

$$\frac{\omega_M(\mu, F)}{M(\mu)} = \frac{\omega_L(\mu, f)}{\mu L(\mu)} - \frac{H}{\mu}. \tag{24}$$

It is evident from this equation that the left-hand side is regular everywhere except possibly at the origin. By (23) we obtain

$$\frac{1}{2\pi i}\int_{C_0} \frac{\omega_M(\mu, F)}{M(\mu)} e^{\mu z} d\mu = 0.$$

Consequently the origin is also not a singular point of the left-hand side. Hence the function (24) is entire. It now only remains to use Lemma 2, setting $\beta = 0$.

Lemma 8. *Let* (11) *be an entire function. If a and b are arbitrary vertices of the boundary $\partial\bar{D}$ (i.e. they are not interior points of linear segments contained in $\partial\bar{D}$), then*

$$\int_0^a f(t)\,dt = \int_0^b f(t)\,dt = H,$$

where H is the number of Lemma 7.

By Lemma 7 the function

$$\frac{\omega_L(\mu, F)}{L(\mu)} \quad \left(F(z) = \int_0^z f(t)\,dt - H\right)$$

is entire. From Lemma 6 we then have

$$\frac{1}{2\pi i}\int_{\partial\bar{D}} t^k \Phi(t)\,dt = 0, \qquad \Phi(t) = \gamma(t) F(t) \qquad (\kappa = 0, 1, 2, \ldots). \tag{25}$$

The function $\Phi(t)$ defined on $\partial\bar{D}$ is continuous and has a continuous derivative $\Phi'(t)$ ($F(t)$ has a continuous derivative by construction; $\gamma(t)$ has a continuous derivative by (6) if we suppose that $\mu > 2$). Hence

$$|\Phi(t_2)-\Phi(t_1)|=\left|\int_{t_1}^{t_2}\Phi'(t)\,dt\right|\leqslant K_1 s(t_1,\ t_2),\qquad K_1=\max_{t\in\partial\bar{D}}|\Phi'(t)|,$$

where $s(t_1, t_2)$ is the length of the arc of the boundary between t_1 and t_2. In the case of a convex domain there exists a constant K_2 such that

$$s(t_1,\ t_2)\leqslant K_2|t_2-t_1|.$$

Consequently

$$|\Phi(t_2)-\Phi(t_1)|<K|t_2-t_1|,\qquad t_1\in\partial\bar{D},\qquad t_2\in\partial\bar{D}. \tag{26}$$

We consider the Cauchy-type integral

$$A(z)=\frac{1}{2\pi i}\int_{\partial\bar{D}}\frac{\Phi(t)\,dt}{t-z},\qquad z\in D.$$

It is known ([7], Chapter III, §2.3b) that the difference

$$\frac{1}{2\pi i}\int_{\partial\bar{D}}\frac{\Phi(t)\,dt}{t-z_1}-\frac{1}{2\pi i}\int_{\partial\bar{D}}\frac{\Phi(t)\,dt}{t-z_2},\qquad z_1\in D,\qquad z_2\notin\bar{D},$$

tends to $\Phi(t)$ for almost all $t\in\partial\bar{D}$, when z_1 and z_2 tend to t along nontangential paths. By (25) the value of the integral containing $z_2\notin\bar{D}$ is zero. Thus for almost all $t\in\partial\bar{D}$ we have

$$\lim_{z\to t}A(z)=\Phi(t).$$

By (26) the integral $A(z)$ is a continuous function in $\bar{D}$ ([7], Chapter III, §4.4 2). Consequently $A(t)=\Phi(t)$ everywhere on $\partial\bar{D}$. Thus there is a function $A(z)$, analytic in D, continuous in $\bar{D}$, and coinciding with $\Phi(t)$ on $\partial\bar{D}$.

Let a be a vertex of $\bar{D}$. Then it is immediately seen to be a singular point of $\gamma(t)$ (and also singular for the functions associated with functions obtained by multiplication or division of $L(\lambda)$ by a polynomial). We show that $F(a)=0$.

Suppose, on the contrary, that $F(a)\neq 0$. Then there is a neighborhood K of a such that $F(t)\neq 0$ for $t\in K\cap\bar{D}$.

Consider the function

$$\frac{A(t)}{F(t)}.$$

It is regular in $K\cap D$ (the part of D adjacent to a) and continuous onto the arc $\Gamma=K\cap\partial\bar{D}$. The function $\gamma(t)$ is regular outside $\bar{D}$ and continuous onto the boundary. On Γ these functions coincide, since $A(t)=\gamma(t)F(t)$ everywhere on $\partial\bar{D}$. Hence the former function is an analytic continuation of $\gamma(t)$ over Γ into D. Consequently the point a is not a singular point of $\gamma(t)$, and this is impossible. Therefore $F(a)=0$.

If b is another vertex, then $F(b)=0$ too. We have

$$F(a)=\int_0^a f(t)\,dt-H=0, \qquad F(b)=\int_0^b f(t)\,dt-H=0,$$

whence

$$\int_0^a f(t)\,dt=\int_0^b f(t)\,dt\equiv H,$$

which is what we wished to prove.

We proceed to complete the proof of the theorem.

Let a and b be any two vertices of $\overline{D}$. According to Lemma 8 we have

$$\int_0^a f(t)\,dt=\int_0^b f(t)\,dt=H,$$

whence

$$\int_a^b f(t)\,dt=0.$$

We set

$$f_1(t)=\int_0^t f(t)\,dt-H=\int_0^t f(t)\,dt-\int_0^a f(t)\,dt=\int_a^t f(t)\,dt.$$

By Lemma 7

$$\frac{\omega_L(\mu, f_1)}{L(\mu)}$$

is an entire function. Thus we have shown that if

$$\frac{\omega_L(\mu, f)}{L(\mu)}$$

is an entire function, then

$$\frac{\omega_L(\mu, f_1)}{L(\mu)} \qquad \left(f_1=\int_a^t f(t)\,dt\right)$$

is also an entire function, and

$$\int_a^b f(t)\,dt=0.$$

Repeating the argument, we deduce that

$$\frac{\omega_L(\mu, f_2)}{L(\mu)} \qquad \left(f_2=\int_a^t f_1(t)\,dt\right)$$

is an entire function and

$$\int_a^b f_1(t)\,dt=0,$$

etc. Thus we have

$$\int_a^b f_k(t)\,dt = 0 \qquad (k = 0,\ 1,\ 2,\ \ldots),$$

where

$$f_0(t) = f(t), \qquad f_k(t) = \int_a^t f_{k-1}(t)\,dt \qquad (k \geqslant 1).$$

We write this equation in the form

$$\frac{1}{k!}\int_a^b (b-t)^k f(t)\,dt = 0 \qquad (k = 0,\ 1,\ \ldots).$$

Here the integration takes place along a segment $[a, b]$. Hence $f(t) = 0$ on $[a, b]$. Consequently $f(t) \equiv 0$. The theorem is proved.

§2. Proof of Theorem 2

We assume that all the zeros λ_k are simple and different from zero.

We denote by $A(\overline{E})$ the space of functions analytic in E, continuous in $\overline{E}$ and vanishing at infinity. For the proof of the completeness of the system $\{\psi_k(t)\}$ in the space $A(\overline{E})$ with the norm

$$\|f\| = \max_{t \in \overline{E}} |f(t)|$$

we have to show that if $l(f)$ is a linear functional in $A(E)$ satisfying the conditions

$$l(\psi_k) = 0 \qquad (k = 1,\ 2,\ \ldots),$$

then $l(f) = 0$ for every function $f \in A(\overline{E})$.

We extend the functional $l(f)$ to the space of functions $\psi(t)$ continuous on $\partial\overline{D}$, where

$$\|\psi\| = \max_{t \in \partial\overline{D}} |\psi(t)|.$$

A linear functional in this space has the form

$$l(\psi) = \int_{\partial\overline{D}} \psi(t)\,d\sigma(t),$$

where $\sigma(t)$ is a function of bounded variation on $\partial\overline{D}$. Thus

$$l(f) = \int_{\partial\overline{D}} f(t)\,d\sigma(t), \qquad f(t) \in A(\overline{E}). \tag{27}$$

Let

$$l(\psi_k) = \int_{\partial\overline{D}} \psi_k(t)\,d\sigma(t) = 0 \qquad (k = 1,\ 2,\ \ldots). \tag{28}$$

We have to show that for every $f(t)$ in $A(\overline{E})$

$$l(f) = \int_{\partial\overline{D}} f(t)\,d\sigma(t) = 0. \tag{29}$$

Integration by parts gives

$$l(\psi_k) = \sigma(t)\,\psi_k(t)\,|_{\partial\bar D} - \int_{\partial\bar D} \sigma(t)\,\psi_k'(t)\,dt.$$

The first term on the right vanishes. Hence we have

$$l(\psi_k) = -\int_{\partial\bar D} \sigma(t)\,\psi_k'(t)\,dt = 0.$$

We observe that it follows from

$$\psi_k(t) = \frac{1}{L'(\lambda_k)} \int_0^\infty \frac{L(\lambda)\,e^{-\lambda t}d\lambda}{\lambda - \lambda_k},$$

in accordance with condition (6), that the function $\psi'_k(t)$ is continuous in $\bar E$. From this it follows that

$$\psi_k'(t) = -\frac{\gamma(t)}{L'(\lambda_k)} - \lambda_k\psi_k(t), \tag{30}$$

and from this

$$\frac{1}{L'(\lambda_k)} \int_{\partial\bar D} \sigma(t)\,\gamma(t)\,dt + \lambda_k \int_{\partial\bar D} \sigma(t)\,\psi_k(t)\,dt = 0.$$

Without loss of generality we may assume that the first term on the left vanishes. We prove this. We have

$$L(\lambda) = \sum_0^\infty \frac{c_k}{k!}\lambda^k, \qquad \gamma(t) = \sum_0^\infty \frac{c_k}{t^{k+1}}.$$

Since $L(0) \neq 0$, it follows that $c_0 \neq 0$, and so

$$\int_{\partial\bar D} \gamma(t)\,dt \neq 0.$$

Consequently we can choose c so that

$$\int_{\partial\bar D} [\sigma(t) + c]\,\gamma(t)\,dt = \int_{\partial\bar D} \sigma(t)\,\gamma(t) + c\int_{\partial\bar D} \gamma(t)\,dt = 0,$$

and then we may take $\sigma(t) + c$ instead of $\sigma(t)$ in (27).

For such a choice of $\sigma(t)$ we obtain

$$\int_{\partial\bar D} \sigma(t)\,\psi_k(t)\,dt = 0 \qquad (k = 1,\ 2,\ \ldots). \tag{31}$$

We observe that if $L(0) = 0$, then it would follow from

$$\frac{L(\lambda)}{\lambda - \lambda_k} = \sum_{m=0}^\infty \frac{c_m^{(k)}}{m!}\lambda^m, \qquad c_0^{(k)} = 0, \qquad \psi_k(t) = \sum_{m=0}^\infty \frac{c_m^{(k)}}{t^{m+1}}$$

that the point at infinity is a zero of $\psi_k(t)$ of the second order. We could then discuss the completeness of the system $\{\psi_k(t)\}$ in the space of functions analytic in E, continuous in $\bar E$ with a zero of at most the second order at infinity.

We write (31) in the form

$$\int_{\partial\bar D} \sigma_1(t)\,\psi_k(t)\,dt = 0 \qquad (k = 1,\ 2,\ \ldots), \qquad \sigma_1(t) = \sigma(t) + \frac{a}{t},$$

where a is chosen so that the function

$$\varphi(t)=\int_{\alpha}^{t}\sigma_1(t)\,dt=\int_{\alpha}^{t}\sigma(t)\,dt+a\int_{\alpha}^{t}\frac{dt}{t},\qquad \alpha\in\partial\bar{D},$$

is single-valued on $\partial\bar{D}$. Integrating by parts we obtain

$$\varphi(t)\psi_k(t)\big|_{\partial\bar{D}}-\int_{\partial\bar{D}}\varphi(t)\psi_k'(t)\,dt=0.$$

The first term vanishes. We take note of (30) and see that

$$\frac{1}{L'(\lambda_k)}\int_{\partial\bar{D}}\varphi(t)\gamma(t)dt+\lambda_k\int_{\partial\bar{D}}\varphi(t)\psi_k(t)\,dt=0,$$

whence

$$\int_{\partial\bar{D}}\varphi(t)\psi_k(t)\,dt=-\frac{\mu}{\lambda_k L'(\lambda_k)},\qquad \mu=\int_{\partial\bar{D}}\varphi(t)\gamma(t)\,dt.$$

Note that if p is a constant, then

$$\int_{\partial\bar{D}}p\psi_k(t)\,dt=-\frac{2\pi i p L(0)}{\lambda_k L'(\lambda_k)}.$$

Setting $p=-\mu/2\pi i L(0)$, we obtain

$$\int_{\partial\bar{D}}F(t)\psi_k(t)\,dt=0\qquad(k=1,\,2,\,\ldots),\qquad F(t)=\varphi(t)+p. \tag{32}$$

Next, we have

$$F(t_2)-F(t_1)=\varphi(t_2)-\varphi(t_1)=\int_{t_1}^{t_2}\sigma_1(t)\,dt,$$

whence

$$|F(t_2)-F(t_1)|\leqslant As(t_1,\,t_2)\leqslant K|t_2-t_1|, \tag{33}$$

where $s(t_1,\,t_2)$ is the length of the arc of $\partial\bar{D}$ between t_1 and t_2.

In view of (33) we can represent $F(t)$ in the form ([7], Chapter III, §4.4)

$$F(t)=F_1(t)+F_2(t),$$

where $F_1(t)$ is regular in D and continuous in $\bar{D}$ and $F_2(t)$ is regular in E, continuous in $\bar{E}$ and $F_2(\infty)=0$.

From (32) we obtain

$$\int_{\partial\bar{D}}F_1(t)\psi_k(t)\,dt=0\qquad(k=1,\,2,\,\ldots).$$

Hence in accordance with the already-proved uniqueness theorem we deduce that $F_1(t)=0$. Thus $F(t)=\phi(t)+p$ is regular in E, continuous in $\bar{E}$ and $F(\infty)=0$. As a consequence of this we now show that (29) is satisfied.

Since $\phi(t)=F(t)-p$, we have

$$\int_{\partial\bar{D}} \frac{\varphi(t)}{t^m}\,dt = 0 \qquad (m \geqslant 2). \tag{34}$$

We recall that

$$\varphi(t) = \int_{\alpha}^{t} \sigma_1(t)\,dt, \qquad \alpha \in \partial\bar{D}.$$

Hence

$$\int_{\partial\bar{D}} \frac{\sigma_1(t)\,dt}{t^{m-1}} = \frac{\varphi(t)}{t^{m-1}}\Bigg| + (m-1)\int_{\partial\bar{D}} \frac{\varphi(t)\,dt}{t^m}.$$

Since the first term on the right vanishes, and (34) holds, we have

$$\int_{\partial\bar{D}} \frac{\sigma_1(t)\,dt}{t^{m-1}}\,dt = 0 \qquad (m \geqslant 2),$$

whence, reflecting that $\sigma_1(t) = \sigma(t) + a/t$, we obtain

$$\int_{\partial\bar{D}} \frac{\sigma(t)\,dt}{t^{m-1}} = 0 \qquad (m \geqslant 2). \tag{35}$$

Finally, we investigate the relation

$$\int_{\partial\bar{D}} \frac{d\sigma(t)}{t^k} = \frac{\sigma(t)}{t^k}\Bigg|_{\partial\bar{D}} + k\int_{\partial\bar{D}} \frac{\sigma(t)\,dt}{t^{k+1}} \qquad (k \geqslant 1).$$

By this and (35) we find that

$$\int_{\partial\bar{D}} \frac{d\sigma(t)}{t^k} = 0 \qquad (k \geqslant 1). \tag{36}$$

We can approximate a function $f(t) \in A(\bar{E})$ arbitrarily closely in $\bar{E}$, in particular on $\partial\bar{D}$, by a polynomial (without constant term) in $1/t$. Hence from (36) we obtain

$$\int_{\partial\bar{D}} f(t)\,d\sigma(t) = 0.$$

The theorem is proved.

§3. Proof of Theorem 3

We examine the functions (8) with zeros λ_k $(k \geq 1)$. Denote the function (9) by

$$\omega_L(\mu, f) = \int_{\partial\bar{D}} \left[\int_0^t f(t-\eta)\,e^{\mu\eta}\,d\eta\right] d\sigma(t).$$

With $f(z)$, analytic in D and continuous in $\bar{D}$, we associate the series (1) with terms calculated from (2). In order to emphasize that the series is formed using $L(\lambda)$, it is convenient to write (1) as

$$f(z) \sim (L)\sum_{v=1}^{\infty} P_v(z)\,e^{\lambda_v z}. \tag{37}$$

Consider the function

$$L_1(\lambda) = \frac{L(\lambda)}{\lambda - \lambda_1} = \int_{\partial\bar{D}} \psi(t)\,e^{\lambda t}dt, \qquad \text{where} \qquad \psi(t) = -e^{-\lambda_1 t}\int_{\alpha}^{t} e^{\lambda_1\eta}d\sigma(\eta), \qquad t \in \partial\bar{D}.$$

To this corresponds

$$\omega_{L_1}(\mu, f) = \int_{\partial \bar{D}} \left[\int_0^t f(t-\eta) e^{\mu\eta} d\eta \right] \psi(t)$$

and using this we construct a series of form (37) but defined using $L_1(\lambda)$, not $L(\lambda)$.

Lemma 9. *Let the series* (37) *correspond to* $f(z)$. *Then to the function*

$$\varphi(z) = f(z) - P_1(z) e^{\lambda_1 z}$$

corresponds the series

$$\varphi(z) \sim (L_1) \sum_{\nu=2}^{\infty} P_\nu(z) e^{\lambda_\nu z}.$$

The proof of the lemma is based on the following two lemmas.

Lemma 10. *Let* $\Phi(z)$ *be regular in* D *and continuous in* $\bar{D}$. *Then*

$$\omega_L(\mu, \Phi) = (\mu - \lambda_1) \omega_{L_1}(\mu, \Phi) + \int_{\partial \bar{D}} \Phi(t) \psi(t) dt. \tag{38}$$

Lemma 10 is analogous to Lemma 1, and is proved in the same way.

Lemma 11. *Let* $F(z) = z^p e^{\lambda_1 z}$, $0 \leq p < m_1$. *Then the residue of the function*

$$\frac{\omega_L(\mu, F)}{L(\mu)} e^{\mu z}$$

is $F(z)$ *at* $\mu = \lambda_1$ *and is zero at all other points.*

This lemma is analogous to Lemma 7 in [8], and is proved in the same way.

We now prove Lemma 9. Set $P_1(z) e^{\lambda_1 z} = F(z)$. By Lemma 11 we have

$$\frac{1}{2\pi i} \int_{C_1} \frac{\omega_L(\mu, F) e^{\mu z} d\mu}{L(\mu)} = F(z)$$

(where C_1 is a circle with center λ_1 and sufficiently small radius).

By definition

$$\frac{1}{2\pi i} \int_{C_1} \frac{\omega_L(\mu, f) e^{\mu z} d\mu}{L(\mu)} = P_1(z) e^{\lambda_1 z} = F(z).$$

Hence, since $\phi(z) = f(z) - F(z)$, we have

$$\frac{1}{2\pi i} \int_{C_1} \frac{\omega_L(\mu, \varphi) e^{\mu z} d\mu}{L(\mu)} = 0.$$

Consequently $\mu = \lambda_1$ is a zero of $\omega_L(\mu, \phi)$ of order $\geq m_1$. We now apply Lemma 10. Setting $\Phi(z) = \phi(z)$ and $\mu = \lambda_1$ in (38), we deduce that

$$\int_{\partial \bar{D}} \varphi(t) \psi(t) dt = 0,$$

in view of which

$$\frac{\omega_L(\mu, \varphi)}{L(\mu)} = \frac{\omega_{L_1}(\mu, \varphi)}{L_1(\mu)}. \tag{39}$$

We observe that the function (39) is regular at $\mu = \lambda_1$.

Further, we have from (39)

$$\frac{\omega_L(\mu, f) - \omega_L(\mu, F)}{L(\mu)} = \frac{\omega_L(\mu, \varphi)}{L_1(\mu)},$$

whence

$$\frac{1}{2\pi i}\int_{C_k} \frac{\omega_L(\mu, f)\, e^{\mu z} d\mu}{L(\mu)} - \frac{1}{2\pi i}\int_{C_k} \frac{\omega_L(\mu, F)\, e^{\mu z} d\mu}{L(\mu)} = \frac{1}{2\pi i}\int_{C_k} \frac{\omega_{L_1}(\mu, \varphi)\, e^{\mu z}}{L_1(\mu)}\, d\mu.$$

For $k \neq 1$ the second term on the left vanishes. Hence

$$\frac{1}{2\pi i}\int_{C_k} \frac{\omega_{L_1}(\mu, \varphi)\, d\mu}{L_1(\mu)} = P_k(z)\, e^{\lambda_k z} \qquad (k \geqslant 1),$$

where for $k = 1$ the corresponding integral is zero. Therefore

$$\varphi(z) \sim (L_1) \sum_{k=2}^{\infty} P_k(z)\, e^{\lambda_k z}.$$

The lemma is proved.

We now take the function

$$L_2(\lambda) = \frac{L_1(\lambda)}{\lambda - \lambda_2} = \int_{\partial \bar{D}} \tilde{\psi}(t)\, e^{\lambda t} dt, \qquad \tilde{\psi}(t) = -e^{-\lambda_2 t} \int_{\alpha}^{t} e^{\lambda_2 \eta} \psi(\eta)\, d\eta,$$

to which corresponds

$$\omega_{L_2}(\mu, \Phi) = \int_{\partial \bar{D}} \left[\int_0^t \Phi(t - \eta)\, e^{\mu \eta} d\eta \right] \tilde{\psi}(t)\, dt. \tag{40}$$

By a lemma analogous to Lemma 9 we have

$$\varphi(z) - P_2(z)\, e^{\lambda_2 z} \sim (L_2) \sum_{\nu=3}^{\infty} P_\nu(z)\, e^{\lambda_\nu z}.$$

Thus

$$f(z) - P_1(z)\, e^{\lambda_1 z} - P_2(z)\, e^{\lambda_2 z} \sim (L_2) \sum_{\nu=3}^{\infty} P_\nu(z)\, e^{\lambda_\nu z}. \tag{41}$$

The function $L(\lambda)$ satisfies

$$|L(re^{i\varphi})| < Ae^{h(\varphi) r},$$

and so $L_2(\lambda)$ satisfies the following condition:

$$|L_2(re^{i\varphi})| < B\frac{e^{h(\varphi) r}}{r^2}.$$

Let $\gamma_2(t)$ be the function associated with $L_2(\lambda)$ in the Borel sense. It is regular outside $\bar{D}$ and continuous outside D. It is easily seen that the function (40) can be written in the form

$$\omega_{L_2}(\mu, \Phi)=\frac{1}{2\pi i}\int_{\partial\bar{D}}\left[\int_0^t \Phi(t-\eta)e^{\mu\eta}d\eta\right]\gamma_2(t)\,dt$$

(this equation is proved first for $\Phi(z) = e^{\lambda z}$ and then in the general case).

Thus (41) is the series investigated in the previous sections. It permits results on case (8) to be reduced to case (6).

We prove the uniqueness Theorem 3. Let all $P_\nu(z) \equiv 0$. Then by Theorem 1 we have

$$f(z)-P_1(z)e^{\lambda_1 z}-P_2(z)e^{\lambda_2 t}=f(z)\equiv 0.$$

It was noted in the Introduction that we can construct the function using Theorem 2 if we know the coefficients of its Dirichlet series (in the case of simple roots). By this method we can consequently construct the function

$$F(z)=f(z)-a_1e^{\lambda_1 z}-a_2e^{\lambda_2 z},$$

but then

$$f(z)=F(z)+a_1e^{\lambda_1 z}+a_2e^{\lambda_2 z}.$$

BIBLIOGRAPHY

1. A. F. Leont′ev, *A uniqueness property*, Mat. Sb. **72** (**114**) (1967), 237–249 = Math. USSR Sb. **1** (1967), 209–220. MR **34** #7780.

2. ———, *The representation of functions by generalized Dirichlet series*, Uspehi Mat. Nauk 24 (1969), no. 2 (146), 97–164 = Russian Math. Surveys 24 (1969), no. 2, 101–178. MR **41** #471.

3. V. V. Napalkov, *On the recovery of a function from the coefficients of its Dirichlet series*, Mat. Zametki **17** (1975), 545–553 = Math. Notes **17** (1975), 321–326.

4. A. F. Leont′ev, *On conditions of expandibility of analytic functions in Dirichlet series*, Izv. Akad. Nauk SSSR Ser. Mat. 36 (1972), 1282–1295 = Math. USSR Izv. 6 (1972), 1265–1277. MR **49** #543.

5. B. Ja. Levin, *Distribution of zeros of entire functions*, GITTL, Moscow, 1956; English transl., Amer. Math. Soc., Providence, R. I., 1964. MR **19**, 402; **28** #217.

6. A. F. Leont′ev, *Properties of sequences of Dirichlet polynomials which are convergent on an interval of the imaginary axis*, Izv. Akad. Nauk SSSR Ser. Mat. **29** (1965), 269–328. (Russian) MR **31** #5039.

7. I. I. Privalov, *Boundary properties of analytic functions*, 2nd ed., GITTL, Moscow, 1950; German transl., VEB Deutscher Verlag, Berlin, 1956. MR **13**, 926; **18**, 727.

8. A. F. Leont′ev, *On the transformation of a functional equation to simpler form*, Mat. Sb. 67 (**109**) (1965), 541–560. (Russian) MR 32 #7986.

Translated by K. W. LUCAS

Trudy Mat. Inst. Steklov.
134 (1975)

Proc. Steklov Inst. Math.
134 (1975)

ON A FUNCTIONAL CHARACTERIZATION OF THE INTERPOLATION SPACES $(L_p(\Omega), W_p^1(\Omega))_{\theta,p}$

UDC 517.51:513.88

P. I. LIZORKIN

Abstract. We give a new functional characterization of the space of function arising from interpolation by Lions' method of traces between the space $L_p(\Omega)$ of functions $f(x)$ whose pth power is integrable over the region Ω and the Sobolev space $W_p^1(\Omega)$ of functions $f(x) \in L_p(\Omega)$ whose generalized derivatives df/dx_i, $i = 1, \dots, n$, belong to $L_p(\Omega)$.

Bibliography: 10 titles.

Introduction

We consider the problem of describing in functional terms the interpolation spaces between the space $L_p(\Omega)$ of functions $f(x)$ whose pth power is integrable over the region $\Omega \subset R^n$ and the Sobolev space $W_p^1(\Omega)$ of functions $f(x) \in L_p(\Omega)$ whose generalized derivatives $\partial f/\partial x_i$, $i = 1, \dots, n$, belong to $L_p(\Omega)$.

The interpolation spaces considered are constructed by the "method of traces" of Lions, whose scheme, in the case of interest to us, is as follows. Let X_0 and X_1 be Banach spaces such that there is a continuous embedding $X_0 \subset X_1$. We denote by $L_q^*(X)$ the space of functions $v(t)$, $0 < t < \infty$, which take values in X and are such that

$$\|v\|_{L_q^*(x)} = \left\{\int_0^\infty |v|_X^q \frac{dt}{t}\right\}^{\frac{1}{q}} < \infty.$$

Furthermore, let V be the set of functions $u(t)$ for which

$$t^\theta u(t) \in L_q^*(X_0), \qquad t^\theta \frac{du}{dt} \in L_q^*(X_1),$$
$$0 < \theta < 1, \qquad 1 \leqslant q \leqslant \infty$$

($u(t)$ takes values in X_0, and hence in X_1; and du/dt is the generalized derivative of an X_1-valued function).

We define

$$\|u\|_V = \left(\|t^\theta u(t)\|^q_{L_q^*(X_0)} + \left\|t^\theta \frac{du}{dt}\right\|^q_{L_q^*(X_1)}\right)^{\frac{1}{q}}.$$

AMS (MOS) subject classifications (1970). Primary 46E35; Secondary 26A86.

It is easy to see that for given values of the parameters we have the existence of the following limit in X_1:

$$\lim_{t\to+0} u(t) = u(0).$$

The set of all elements f which are representable in the form $f = u(0)$, $u \in V$, is called the "space of traces" $(X_0, X_1)_{\theta,q}$. Under the norm defined by

$$\|f\|_{(X_0, X_1)_{\theta,q}} = \inf_{u(0)=f} \|u\|_V$$

it becomes a Banach space. This space is intermediate between X_0 and X_1, i.e.

$$X_0 \subset (X_0, X_1)_{\theta,q} \subset X_1,$$

and has the interpolation property: a linear operator T which is bounded in X_0 and X_1 is bounded also in $(X_0, X_1)_{\theta,q}$.

If we set $X_0 = W_p^1(R^n)$ and $X_1 = L_p(R^n)$, with $q = p$, then it follows from the definitions that the space $(W_p^1(R^n), L_p(R^n))_{\theta,p}$ consists of those and only those functions $f(x)$, $x \in R^n$, which are representable in the form

$$f(x) = \lim_{t\to 0} u(x, t),$$

where the limit is understood in the sense of convergence in the norm of L_p, while the function $u(x, t)$ has generalized derivatives $\partial u/\partial x_i$, $i = 1, \ldots, n$, and $\partial u/\partial t$, and has a finite integral

$$\int_{R_+^{n+1}} t^{\theta p} \left(\sum_{i=1}^{n} \left|\frac{\partial u}{\partial x_i}\right|^p + \left|\frac{\partial u}{\partial t}\right|^p + |u|^p \right) dx \frac{\partial t}{t}.$$

The set of functions $u(x, t)$ with the indicated properties is called the weight space $W_{p,\alpha}^1(R_+^{n+1})$, $\alpha = p\theta - 1$.

An intrinsic characterization of the traces of functions in $W_{p,\alpha}^1(R_+^{n+1})$ was investigated by Aronszajn [1] ($p = 2$, $\alpha = 0$), Gagliardo [2] ($\alpha = 0$, $1 \le p < \infty$), Vašarin [3] ($p = 2$, $0 < \alpha < 1$), and, in the general case, in the work of Lions ([4], p. 401), where an interpolational interpretation of the problem was first given. This characterization has the form (1) (for $\Omega = R^n$ and $r = 1 - \theta$). Somewhat later the properties of traces of functions from more general weight classes were studied in [5].

In an analogous way one may consider the case $X_0 = W_p^1(\Omega)$, $X_1 = L_p(\Omega)$, where Ω is a region in R^n. In this case the interpolation treatment leads to a consideration of the weight classes $W_{p,\alpha}^1(\Pi)$ of functions $u(x, t)$ defined in the cylinder $\Pi = \Omega \times (0, \infty)$ which have generalized derivatives $\partial u/\partial x_i$ and $\partial u/\partial t$ and a finite integral

$$\|u\|_{W_{p,\alpha}^1(\Pi)} = \int_{\Pi} t^{\alpha} \left(\sum_1^n \left|\frac{\partial u}{\partial x_i}\right|^p + \left|\frac{\partial u}{\partial t}\right|^p + |u|^p \right) dx dt.$$

Lions and Magenes ([6], p. 53) observed that for a region Ω with smooth boundary the set of traces $u(x, 0) = f(x)$ of functions in $W_{p,\alpha}^1(\Omega)$ coincides with the familiar Sobolev-Slobodeckiĭ-Besov space $B_p^r(\Omega)$, $r = 1 + \theta$, with norm

$$\|f\|_{B_p^r(\Omega)} \overset{\text{def}}{=} \left(\|f\|_{L_p(\Omega)}^p + \int_\Omega\int_\Omega \frac{|f(x)-f(y)|^p}{(x-y)^{n+pr}}\,dxdy\right)^{\frac{1}{p}} < \infty. \tag{1}$$

Moreover, they suggested the possibility of extending f from Ω onto the whole space R^n with preservation of class (see [7]).

In this paper we obtain a new intrinsic characterization of a function f in the space $(W_p^1(\Omega), L_p(\Omega))_{\theta,p}$. It is expressed by the requirement of finiteness for the norm

$$\|f\|_{\breve{B}_p^r(\Omega)} = \left\{\|f\|_{L_p(\Omega)}^p + \int_\Omega dx \int_{|x-y|<d_x} \frac{|f(x)-f(y)|^p}{|x-y|^{n+pr}}\,dy\right\}^{\frac{1}{p}}, \tag{2}$$

where $d_x = \operatorname{dist}(x, \partial\Omega)$ is the distance from the point x to the boundary of Ω. A priori, this norm way serve for an adequate description of the given interpolation class in the case of more general regions Ω. Some of our estimates are obtained for arbitrary regions. Furthermore, we use a straightforward method: we prove directly an embedding theorem $W_{p,\alpha}^1(\Phi) \to \breve{B}_p^r(\Omega)$ and an extension theorem $\breve{B}_p^r(\Omega) \to W_{p,\alpha}^1(\Phi)$. We mention that we were able to get the extension theorem only for a region Ω with smooth boundary. It follows from above that for "nice" regions the expressions (1) and (2) are equivalent, which is not without interest in itself. In the course of the paper we also prove some weight estimates.

§1. One-dimensional case (guiding concepts). Fundamental definitions

Let R_+^1 denote the halfline $0 < x < \infty$, and let $0 < r < 1$. The collection of functions $f(x) \in L_p(R_+^1)$ which are p-integrable, $1 \le p < \infty$, on R_+^1 and for which the integral

$$\left\{\int_0^\infty\int_0^\infty \frac{|f(x)-f(y)|^p}{|x-y|^{1+pr}}\,dxdy\right\}^{\frac{1}{p}} \equiv b_p^r(f;\ R_+^1) \equiv b_p^r(f),(1)$$

is finite is called the space $B_p^r(R_+^1)$. This is a Banach space with the norm

$$\|f\|_{B_p^r(R_+^1)} = \|f\|_{L_p(R_+^1)} + b_p^r(f,\ R_+^1).$$

Let d_x denote the distance from the point $x \in R_+^1$ to the boundary of the region $\Omega = R_+^1$:

$$d_x = \operatorname{dist}(x,\ R_+^1) = x.$$

Then it is easy to obtain the estimate

$$\frac{1}{2}[b_p^r(f)]^p \le \int_0^\infty dx \int_{|x-y|<d_x} \frac{|f(x)-f(y)|^p}{|x-y|^{1+pr}}\,dy \equiv [\breve{b}_p^r(f,\ R_+^1)]^p. \tag{1}$$

Indeed, we have

(1) When the region Ω for which some seminorm (or norm) is given is clear from the context, we will not always indicate it in the notation.

$$[b_p^{(r)}(f)]^p=\int_0^\infty dx\int_0^x \frac{|f(x)-f(y)|^p}{|x-y|^{1+pr}}\,dy+\int_0^\infty dx\int_x^\infty \frac{|f(x)-f(y)|^p}{|x-y|^{1+pr}}\,dy=I_1+I_2.$$

We change the order of integration in I_2 and then rename the variables. We get

$$I_2=\int_0^\infty dy\int_0^y \frac{|f(x)-f(y)|^p}{|x-y|^{1+pr}}\,dx=\int_0^\infty d\eta\int_0^\eta \frac{|f(\xi)-f(\eta)|^p}{|\xi-\eta|^{1+pr}}\,d\xi.$$

Therefore

$$[b_p^{(r)}(f)]^p=2\int_0^\infty dx\int_0^x \frac{|f(x)-f(y)|^p}{|x-y|^{1+pr}}\,dy\leqslant 2\int_0^\infty dx\int_0^{2x} \frac{|f(x)-f(y)|}{|x-y|^{1+pr}}\,dy. \qquad (2)$$

Because $d_x = x$ in the given case, the inequality $|x-y| < d_x$ means that $0 < y < 2x$, and therefore (2) coincides with (1). Consequently, we can write

$$\left\{\int_0^\infty dx\int_{|x-y|<d_x} \frac{|f(x)-f(y)|^p}{|x-y|^{1+pr}}\,dy\right\}^{\frac{1}{p}}\leqslant b_p^{(r)}(f,\ R_+^1)$$

$$\leqslant 2^{\frac{1}{p}}\left\{\int_0^\infty dx\int_{|x-y|<d_x} \frac{|f(x)-f(y)|^p}{|x-y|^{1+pr}}\,dy\right\}^{\frac{1}{p}}.$$

Thus the quantities $b_p^{(r)}(f,\, R_+^1)$ and $\breve{b}_p^{(r)}(f,\, R_+^1)$ (see the notation of (1)) are equivalent:

$$b_p^{(r)}(R_+^1)\sim \breve{b}_p^{(r)}(R_+^1).$$

If we consider the space $\breve{B}_p^r(R_+^1)$ of functions $f(x)$ with the norm

$$\|f\|_{\breve{B}_p^r(R_+^1)}=\|f\|_{L_p(R_+^1)}+\breve{b}_p^{(r)}(f,\ R_+^1),$$

then the spaces $B_p^r(R_+^1)$ and $\breve{B}_p^r(R_+^1)$, as we showed, *coincide to within equivalence of norms* (and even of seminorms). It turns out that an analogous fact holds also in a more general situation. Namely, we prove the coincidence (to within equivalence of norms) of the spaces $B_p^r(\Omega)$ and $\breve{B}_p^r(\Omega)$, where Ω is a region in n-dimensional euclidean space R^n with smooth boundary $\partial\Omega$, and the spaces mentioned are defined in the following way.

Definition 1. A function $f(x)$ which is defined on the region Ω belongs to the space $B_p^r(\Omega)$ if it is p-integrable, $1 \le p < \infty$, on Ω and

$$\left\{\int_\Omega \frac{|f(x)-f(y)|^p}{|x-y|^{n+pr}}\,dy\right\}^{\frac{1}{p}}\in L_p(\Omega).$$

It is well known (see, for example, [8]) that $B_p^r(\Omega)$ is a Banach space with the norm

$$\|f\|_{B_p^r(\Omega)}=\|f\|_{L_p(\Omega)}+b_p^{(r)}(f,\ \Omega),$$

where

$$b_p^{(r)}(f,\ \Omega)=\left\{\int\limits_{\Omega} dx \int\limits_{\Omega} \frac{|f(x)-f(y)|^p}{|x-y|^{n+pr}}\, dy\right\}^{\frac{1}{p}}.$$

Definition 2. A function $f(x)$ defined in the region Ω belongs to the space $\breve{B}_p^r(\Omega)$ if it is p-integrable, $1 \leq p < \infty$, on Ω and

$$\Pi_p^r f(x) \equiv \left\{\int\limits_{|x-y|<d_x} \frac{|f(x)-f(y)|^p}{|x-y|^{n+pr}}\, dy\right\}^{\frac{1}{p}} \in L_p(\Omega),$$

where $d_x = \operatorname{dist}(x, \partial\Omega)$.

We define a norm in $B_p^r(\Omega)$ by the relation

$$\|f\|_{\breve{B}_p^r(\Omega)} = \|f\|_{L_p(\Omega)} + \|\Pi_p^r f\|_{L_p(\Omega)}.$$

In §2 we will prove a theorem on the boundary values of a function in the weight class $W_{p,\alpha}^1(\Pi(\Omega))$ which is important in the framework of our exposition. The proof is carried out for the case of an arbitrary region Ω. Various types of auxiliary estimates and propositions are proved in §3. We mention, in particular, an extension lemma which is an essential step for the interpolation treatment of the fractional classes $\breve{B}_p^r(\Omega)$ considered. A generalization of an inequality of Hardy to the n-dimensional case is of some interest.

We prove in §4 the coincidence of the spaces $B_p^r(R_+^n)$ and $\breve{B}_p^r(R_+^n)$. Essentially from this it follows directly that $B_p^r(\Omega)$ and $\breve{B}_p^r(\Omega)$ coincide (to within equivalence of norms) for a region Ω with simple structure and with smooth boundary. Namely, the use of a partition of unity with a subsequent local transformation of coordinates reduces the problem to the case of R_+^n. We do not give the details of this procedure.

In §5 we consider various weight estimates for functions from the spaces $\breve{B}_p^r(R_+^n)$ and $B_p^r(R^n)$. These estimates have the character of boundary embedding theorems with respect to the hyperplane R^m in the case when the existence of a trace on R^m is not guaranteed.

§2. Boundary values of functions from the weight class $W_{p,\alpha}^1$

Let Ω be an arbitrary region in R^n, and let $\Pi(\Omega)$ = the Π-cylinder "on Ω", i.e.

$$\Pi(\Omega)=\{(x,\ t);\ x\in\Omega,\ 0<t<\infty\}.$$

We denote by $W_{p,\alpha}^1(\Pi)$, $1 < p < \infty$, $-1 < \alpha < p-1$, the class of functions $u(x, t)$ defined and locally integrable in Π which have generalized (Sobolev) derivatives of first order and a finite integral

$$\|u\|_{W_{p,\alpha}^1(\Pi)}^p = \int\limits_{\Pi} t^\alpha \left(\sum_{i=1}^{n}\left|\frac{\partial u}{\partial x_i}\right|^p + \left|\frac{\partial u}{\partial t}\right|^p + |u|^p\right) dx dt. \tag{1}$$

$W_{p,\alpha}^1(\Pi)$ is a Banach space with the norm $\|\cdot\|_{W_{p,\alpha}^1(\Pi)}^1$.

We will prove that a function in $W_{p,\alpha}^1(\Pi)$ has a trace ($\equiv$ a boundary value) for $t = 0$, i.e.

$$\lim_{t\to 0} u(x, t) = f(x), \tag{2}$$

where the limit is understood in the sense of convergence for almost every $x \in \Omega$. Indeed, suppose the integral (1) exists. Then for almost every $x \in \Omega$ we have the existence of the integral

$$\int_0^\infty t^\alpha \left|\frac{\partial u}{\partial t}\right|^p dt.$$

Consequently, for any ϵ and T, $0 < \epsilon < T$, the integral

$$\int_\varepsilon^T \frac{\partial u}{\partial t} dt$$

exists, and therefore

$$u(x, T) - u(x, \varepsilon) = \int_\varepsilon^T \frac{\partial u}{\partial t} dt. \tag{3}$$

We calculate the estimate

$$\left|\int_\varepsilon^T \frac{\partial u}{\partial t} dt\right| \leqslant \left(\int_\varepsilon^T \left|\frac{\partial u}{\partial t}\right|^p t^\alpha dt\right)^{\frac{1}{p}} \left(\int_\varepsilon^T t^{-\alpha\frac{p'}{p}} dt\right)^{\frac{1}{p'}} \leqslant \left(\int_0^\infty \left|\frac{\partial u}{\partial t}\right|^p t^\alpha dt\right)^{\frac{1}{p}} \left(\int_\varepsilon^T t^{-\frac{\alpha}{p}p'} dt\right)^{\frac{1}{p'}}.$$

We have used Hölder's inequality, $1/p + 1/p' = 1$. Since $\alpha < p - 1$, it follows that $\alpha p'/p < 1$, and for $\epsilon \to 0$ the quantity on the right remains bounded. This means that $\partial u/\partial t$ is integrable on $(0, T)$, and in (3) we may pass to the limit for $\epsilon \to 0$. Thus $\lim_{\epsilon\to 0} u(x, \epsilon)$ exists for almost every $x \in \Omega$; denoting it by $f(x)$, we get (2).

It is easy to see that $f(x)$ is p-integrable over Ω. We have

$$f(x) = u(x, T) - \int_0^T \frac{\partial u}{\partial t} dt.$$

We raise both sides of this equality to the power p, apply Hölder's inequality for sums to the right side, and multiply everything by T^α and integrate over $\Omega \times (0, 1)$. For $k = \int_0^1 T^\alpha dT$ we get

$$k\int_\Omega |f(x)|^p dx \leqslant 2^{p-1}\left\{\int_0^1 T^\alpha dT \int_\Omega |u(x, T)|^p dx + \int_{\Omega\times(0,1)} T^\alpha \left|\int_0^T \frac{\partial u}{\partial t} dt\right|^p dx dT\right\}. \tag{4}$$

We denote the second term in the curly brackets on the right by I, and we apply to it Hölder's inequality; we get

$$I \leqslant \int_{\Omega\times(0,1)} T^\alpha \left(\int_0^T \left|\frac{\partial u}{\partial t}\right|^p t^\alpha dt\right) \left(\int_0^T t^{-\frac{\alpha}{p}p'} dt\right)^{\frac{p}{p'}} dx dT$$

$$\leqslant \int_0^1 T^\alpha \left(\int_0^T t^{-\frac{\alpha p'}{p}} dt\right)^{p-1} dt \int_\Omega dx \int_0^\infty \left|\frac{\partial u}{\partial t}\right|^p t^\alpha dt = C\int_Ц t^\alpha \left|\frac{\partial u}{\partial t}\right|^\alpha dt. \tag{5}$$

Putting together (4) and (5), we arrive at the estimate

$$\int_{\Omega} |f(x)|^p \leqslant \text{const} \int_{\Pi} t^{\alpha}\left(|u(x, t)|^p + \left|\frac{\partial u}{\partial t}\right|^p\right) dx dt.$$

Thus $f \in L_p(\Omega)$.

Now it is easy to see that $f(x)$ is the trace of $u(x, t)$ for $t \to 0$ also in the sense of convergence with respect to the norm of L_p. It turns out that f not only belongs to $L_p(\Omega)$, but even has in addition a certain smoothness property, which is expressed by the existence of an integral of a certain type of difference quotient.

Let the point x be in Ω and let $d_x = \text{dist}(x, \partial\Omega)$ denote the distance from x to the boundary $\partial\Omega$. We consider the integral

$$A(x) = \int_{|x-y|<d_x} \frac{|f(x) - f(y)|^p}{|x-y|^{n+pr}} dy, \qquad r = 1 - \frac{1+\alpha}{p}, \tag{6}$$

and some inequalities connected with it (having in view an estimate of $\int_{\Omega} A(x)dx$ as the final result). The difference $f(x) - f(y)$ can be described by the derivatives of u in the following way (we "rise" from the region Ω in the cylinder Π on Ω vertically from the points x and y to the height $t = |x - y|$, and we connect the resulting points with a segment L which is parallel to R^n):

$$f(x) - f(y) = -\int_0^t \frac{\partial u(x, \tau)}{\partial \tau} d\tau - \int_0^t \frac{\partial u}{\partial \omega}(\xi, t)\, ds + \int_0^t \frac{\partial u(y, \tau)}{\partial \tau} d\tau. \tag{7}$$

In the middle integral the derivative is taken with respect to the direction ω of the segment L ($\omega = (x - y)/|x - y|$, $\xi = x + s\omega$ and $\partial u/\partial\omega = \text{grad}_x u(\xi, t) \cdot \omega$). Thus for an estimate of the integral (6) we estimate the three integrals A_1, A_2, A_3 which correspond to the three terms in (7).(2)

We transform to spherical coordinates with center at the point

$$x - y = t\omega,$$

where ω is a unit vector which runs through the unit sphere $S = S_n$ in R^n (whose area element will be denoted by $d\omega$). Then we have

$$\begin{aligned} A_1(x) &= \int_{|x-y|<d_x} \frac{1}{|x-y|^{n+p\tau}} \left| \int_0^{|x-y|} \frac{\partial u(x, \tau)}{\partial \tau} d\tau \right|^p dy \\ &= \int_S d\omega \int_0^{d_x} \frac{1}{t^{n+pr}} \left| \int_0^t \frac{\partial u(x, \tau)}{\partial \tau} d\tau \right|^p t^{n-1} dt \\ &\leqslant \varkappa \int_S d\omega \int_0^{d_x} t^{p(1-r)-1} \left| \frac{\partial u}{\partial t} \right|^p dt, \quad p(1-r) - 1 = \alpha, \quad \varkappa - r^{-p}. \end{aligned} \tag{8}$$

For the last inequality we have used an inequality of Hardy (see §3, where this inequality is deduced and given its generalization to the n-dimensional case, which is used in subsequent calculations). Extending the integration in the inside integral in (8) from 0 to ∞ and considering that the integration with respect to ω

(2) As is clear from the subsequent consideration, $A \leq 3^{p-1}(A_1 + A_2 + A_3)$.

gives the area $|S|$ of the sphere S, we can write the inequality

$$\int_{\Omega} A_1(x)\,dx \leqslant \varkappa\,|S|\int_{\Pi} t^{\alpha}\left|\frac{\partial u}{\partial t}\right| dx dt. \tag{9}$$

We have a corresponding estimate for A_2:

$$A_2(x) = \int_{|x-y|<d_x} \frac{1}{|x-y|^{n+pr}}\left|\int_0^{|x-y|} \operatorname{grad}_x u\left(x+s\frac{x-y}{|x-y|},\ |x-y|\right)\frac{(x-y)}{|x-y|}\,ds\right|^p dy.$$

Performing the change of variables $x - y = z$ and using Hölder's inequality in the inside integral, we write

$$A_2(x) \leqslant \int_{|z|<d_x} \frac{1}{|z|^{n+pr}} \int_0^{|z|} \left|\operatorname{grad}_x u\left(x+s\frac{z}{|z|},\ |z|\right)\right|^p ds \left(\int_0^{|z|} ds\right)^{p-1} dz$$

$$= \int_{|z|<d_x} \frac{dz}{|z|^{n+1+p(r-1)}} \int_0^{|z|} \left|\operatorname{grad}_x u\left(x+s\frac{z}{|z|},\ |z|\right)\right|^p ds$$

$$= \int_0^{d_x} \frac{dt}{t^{2+p(r-1)}} \int_S d\omega \int_0^t |\operatorname{grad}_x u(x+s\omega,\ t)|^p\,ds = \int_0^{d_x} F(x,\ t)\,dt.$$

For the computation of the integral

$$\int_{\Omega} A_2(x)\,dx = \int_{\Omega} dx \int_0^{d_x} F(x,\ t)\,dt \tag{10}$$

we change the order of integration. The integral on the right can be considered as an integral over the "cone-like body" $\Lambda = \Lambda(\Omega)$:

$$\Lambda = \{(x,\ t);\quad x \in \Omega,\quad 0 < t < \operatorname{dist}(x,\ \partial\Omega)\}.$$

If we take the intersection of this body with the hyperplane $t = \text{const}$, then the projection of this cross-section on R^n gives the set of points x for which $\operatorname{dist}(x, \partial\Omega) > t$. This is in general some open set $\Lambda_t \subset R^n$:

$$\Lambda_t = \{x;\ x \in R^n,\ \operatorname{dist}(x,\ \partial\Omega) > t\}.$$

This set will be nonempty as long as

$$t < r_{\Omega} = \sup_{x\in\Omega} \operatorname{dist}(x,\ \partial\Omega).\text{(3)}$$

Thus,

$$\Lambda = \{(x,\ t);\ 0 < t < r_{\Omega},\ x \in \Lambda_t\}.$$

From these arguments, we can write

$$\int_{\Omega} dx \int_0^{d_x} F(x,\ t)\,dt = \int_0^{r_{\Omega}} dt \int_{\Lambda_t} F(x,\ t)\,dx =$$

(3) r_{Ω} is the radius of the largest ball entirely contained in Ω.

$$=\int_0^{r_\Omega} \frac{dt}{t^{2+p(r-1)}} \int_{\Lambda_t} dx \int_S d\omega \int_0^t |\operatorname{grad} u(x+s\omega,\ t)|^p\, ds.$$

Changing again the order of integration and passing from spherical coordinates to cartesian coordinates ($y = s\omega$), we have from (10)

$$\int_\Omega A_2(x)\,dx = \int_0^{r_\Omega} \frac{dt}{t^{2+p(r-1)}} \int_S d\omega \int_0^t ds \int_{\Lambda_t} |\operatorname{grad} u(x+s\omega,\ t)|^p\, dx$$

$$= \int_0^{r_\Omega} \frac{dt}{t^{2+p(r-1)}} \int_{|y|<t} \frac{dy}{|y|^{n-1}} \int_{\Lambda_t} |\operatorname{grad} u(x+y,\ t)|^p\, dx.$$

In the inside integral we make the substitution $x + y = v$. Taking into account that Λ_t consists of those points x for which $\operatorname{dist}(x, \partial\Omega) > t$, while $|y| < t$, we see that for $x \in \Lambda_t$ the point $x + y$ is not outside Ω. Therefore we only increase the expression under consideration if the integration with respect to v is carried out over Ω. Since in addition

$$\int_{|y|<t} \frac{dy}{|y|^{n-1}} - t \int_{|z|<1} \frac{dz}{|z|^{n-1}} - |s|\, t,$$

we obtain finally

$$\int_\Omega A_2(x)\,dx \leqslant |s| \int_0^{r_\Omega} t^{p(1-r)-1} dt \int_\Omega |\operatorname{grad} u(v,\ t)|^p\, dv$$
$$\leqslant |s| \int_\Pi t^\alpha |\operatorname{grad}_x u(x,\ t)|^p\, dx dt. \tag{11}$$

In conclusion we estimate the term with $A_3(x)$. We write it in the form

$$\int_\Omega A_3(x)\,dx = \int_\Omega dx \int_{|x-y|<d_x} \left| \int_0^{|x-y|} \frac{\partial u(y,\ \tau)}{\partial \tau} d\tau \right|^p \frac{dy}{|x-y|^{n+pr}}$$
$$= \int_\Omega dx \int_{|h|<d_x} \frac{dh}{|h|^{n+pr}} \left| \int_0^{|h|} \frac{\partial u(x+h,\ \tau)}{\partial \tau} d\tau \right|^p. \tag{12}$$

We make the change of variables $x + h = \alpha$, $h = \beta$. Then the volume element can be expressed in terms of the Jacobian:

$$dxdh = \left| \frac{D(x,\ h)}{D(\alpha,\ \beta)} \right| d\alpha d\beta = d\alpha d\beta.$$

It is clear that under the indicated substitution the variable $\alpha = x + h$ will run through the region Ω (since h runs through the ball of radius $\operatorname{dist}(x, \partial\Omega)$), and the domain of the variable β (for fixed α) is of course determined by the inequality

$$|h| < \operatorname{dist}(x,\ \partial\Omega) \Leftrightarrow |\beta| < \operatorname{dist}(\alpha - \beta,\ \partial\Omega)$$

Consequently the integral (12) can be written in the form

$$\int_\Omega A_3(x)\,dx = \int_\Omega d\alpha \int_{|\beta|<\operatorname{dist}(\alpha-\beta,\,\partial\Omega)} \frac{d\beta}{|\beta|^{n+pr}} \left| \int_0^{|\beta|} \frac{\partial u(\alpha,\tau)}{\partial\tau}\,d\tau \right|^p .$$

From this we clearly have the estimate

$$\int_\Omega A_3(x)\,dx \leqslant \int_\Omega d\alpha \int_{R^n} \frac{d\beta}{|\beta|^{n+pr}} \left| \int_0^{|\beta|} \frac{\partial u(\alpha,\tau)}{\partial\tau}\,d\tau \right|^p .$$

Passage to spherical coordinates $\beta = t\omega$ and subsequent use of Hardy's inequality gives

$$\int_\Omega A_3(x)\,dx \leqslant |s| \int_\Omega d\alpha \int_0^\infty \frac{dt}{t^{1+pr}} \left| \int_0^t \frac{\partial u(\alpha,\tau)}{\partial\tau}\,d\tau \right|^p$$

$$\leqslant \varkappa\,|s| \int_\Omega d\alpha \int_0^\infty t^{p(1-r)-1} \left| \frac{\partial u(\alpha,t)}{\partial t} \right|^p dt = \varkappa\,|s| \int_{Ц} t^\alpha \left| \frac{\partial u}{\partial t} \right|^p dx\,dt. \tag{13}$$

Putting together the estimates (8), (11) and (13), and considering (6) and (7), we arrive at the following estimate:

$$\int_\Omega A(x)\,dx = \int_\Omega \frac{|f(x)-f(y)|^p}{|x-y|^{n+pr}}\,dx \leqslant 3^{p-1}\left(\int_\Omega A_1(x)\,dx + \int_\Omega A_2(x)\,dx \right.$$

$$\left. + \int_\Omega A_3(x)\,dx \right) \leqslant 3^{p-1}\left(2\varkappa\,|s| \int_{Ц} t^\alpha \left| \frac{\partial u}{\partial t} \right|^p dx\,dt + |s| \int_{Ц} t^\alpha\, |\operatorname{grad}_x u|^p\,dx\,dt \right)$$

$$\leqslant C \int_{Ц} t^\alpha \left(\sum \left| \frac{\partial u}{\partial x_i} \right|^p + \left| \frac{\partial u}{\partial t} \right|^p \right) dx\,dt.$$

The above calculations can be summed up in the form of Theorem 2.1.

Theorem 2.1. *Suppose the function $u(x, t)$ lies in the weight spaces $W^1_{p,\alpha}(Ц)$, $1 < p < \infty$, $-1 < \alpha < p-1$, $Ц = Ц(\Omega) = \{(x, t);\ x \in \Omega \subset R^n,\ 0 < t < \infty\}$. Then in the sense of convergence for almost every $x \in \Omega$ the limit*

$$\lim_{t\to 0} u(x, t) = f(x)$$

exists, and the function $f(x)$ belongs to the space $\breve{B}^r_p(\Omega)$, i.e. it has a finite norm

$$\|f\|_{\breve{B}^r_p(\Omega)} = \|f\|_{L_p(\Omega)} + \left\{ \int_\Omega dx \int_{|x-y|<d_x} \frac{|f(x)-f(y)|^p}{|x-y|^{n+pr}}\,dy \right\}^{\frac{1}{p}}, \qquad r = 1 - \frac{1+\alpha}{p}.$$

Moreover,

$$\|f\|_{\breve{B}^r_p(\Omega)} \leqslant C\,\|u\|_{W^1_{p,\alpha}(\dot{Ц})}, \tag{14}$$

where the constant C does not depend on u or Ω, but only on p, α and n.

Remark 1. Theorem 2.1 means that there is a continuous embedding

$$(W^1_p(\Omega),\ L_p(\Omega))_{\theta,p} \to \breve{B}^r_p(\Omega).$$

Indeed, the constant C in (14) does not depend on u. By definition,

$$\|f\|_{\theta,p} \equiv |f;\ (W_p^1(\Omega),\ L_p(\Omega))_{\theta,p}| = \inf_{u(x,0)=f} \|u\|_{W^1_{p,\alpha}(\Pi)}.$$

Consequently, for any $\epsilon > 0$ there is a function $u_\epsilon \in W^1_{p,\alpha}(\Pi)$ such that

$$\|u_\varepsilon\|_{W^1_{p,\alpha}(\Pi)} \leqslant \|f\|_{\theta,p} + \varepsilon. \tag{15}$$

By (14) and (15),

$$\|f\|_{\check{B}^r_p(\Omega)} \leqslant C(\|f\|_{\theta,p} + \varepsilon).$$

Since ϵ is arbitrarily small, we get from this

$$\|f\|_{\check{B}^r_p(\Omega)} \leqslant C\|f\|_{\theta,p}.$$

§3. Some auxiliary estimates and an extension lemma

In the following we will need a generalization of a well-known inequality of Hardy to the n-dimensional case. Let $L_{p,\alpha}(0,\infty)$ denote the space of functions $f(t)$ which are p-integrable with weight t^α on the halfline $(0,\infty)$, with the norm

$$\|f\|_{L_{p,\alpha}(0,\infty)} = \left\{\int_0^\infty t^\alpha |f(t)|^p\,dt\right\}^{\frac{1}{p}}.$$

Theorem of Hardy (see [9], inequality (9.9.8)). *The averaging operator*

$$F(t) = \frac{1}{t}\int_0^t f(\tau)\,d\tau$$

is bounded in the space $L_{p,\alpha}(0,\infty)$, $1 \le p < \infty$, $\alpha < p-1$, *and its norm is equal to* $p/(p-1-\alpha)$. *In other words, the following inequality holds (with best possible constant):*

$$\int_0^\infty t^\alpha |F(t)|^p\,dt \leqslant \left(\frac{p}{p-1-\alpha}\right)^p \int_0^\infty r^\alpha |f(\tau)|^p\,d\tau. \tag{1}$$

The estimate (1) is called Hardy's inequality. The inequality clearly remains true also in the finite limit

$$\int_0^a t^\alpha |F(t)|^p\,dt \leqslant \left(\frac{p}{p-1-\alpha}\right)^p \int_0^a r^\alpha |f(\tau)|^p\,d\tau. \text{(4)}$$

The generalizations we mentioned above follow immediately from Hardy's inequality, and we formulate them in the form of lemmas. Here by the space $L_{p,\alpha}(R^n\backslash 0) = L_{p,\alpha}(R^n)$ we mean the space of functions $f(x)$, $x \subset R^n\backslash 0$, with norm

$$\|f(x)\|_{L_{p,\alpha}(R^n)} = \left\{\int_{R^n} |x|^\alpha |f(x)|^p\,dx\right\}^{\frac{1}{p}}.$$

We let $|x| = r$, and we again let $S = S_n$ be the unit sphere in R^n and $|S|$ the area of S. Then the expression $|S|r^n/n$ gives the volume of a ball of radius r in R^n.

Lemma 1. *The operator of "radial averaging in* R^n*"*

---

(4) Just set $f(t) = 0$ for $t > a$.

$$F(x)=F(r\omega)=\frac{1}{r^n}\int_0^r f(\rho\omega)\,\rho^{n-1}d\rho, \qquad \omega\in S, \tag{2}$$

acts boundedly from $L_{p,\beta}(R^n)$ *into* $L_{p,\beta}(R^n)$ *for* $1\le p<\infty$ *and* $\beta<n(p-1)$. *Moreover, the following inequality holds (with best possible constant):*

$$\int_{R^n}|x|^\beta|F(x)|^p\,dx\leqslant\left(\frac{p}{n(p-1)-\beta}\right)^p\int_{R^n}|x|^\beta|f(x)|^p\,dx. \tag{3}$$

Proof. We have

$$\|F(x)\|^p_{L_{p,\beta}(R^n)}=\int_{R^n}|x|^\beta|F(x)|^p\,dx=\int_S d\omega\int_0^\infty r^{n-1+\beta}|F(r\omega)|^p\,dr$$

$$=\int_S d\omega\int_0^\infty r^{(n-1)(1-p)+\beta}\left|\frac{1}{r}\int_0^r f(\rho\omega)\,\rho^{n-1}d\rho\right|^p dr\leqslant$$

(we use Hardy's inequality for $\alpha=(n-1)(1-p)+\beta<p-1$)

$$\leqslant\left(\frac{p}{(p-1)n-\beta}\right)^p\int_S d\omega\int_0^\infty|f(r\omega)r^{n-1}|^p\,r^{(n-1)(1-p)+\beta}dr$$

$$=\left[\frac{p}{n(p-1)-\beta}\right]^p\int_{R^n}|f(x)|^p|x|^\beta\,dx.$$

The sharpness of the constant follows from Hardy's theorem (one should treat the spherically symmetric case $f(\rho\omega)=f(\rho)$).

Lemma 2. *For* $\gamma<np-1$

$$\int_0^\infty h^\gamma\left|\frac{1}{h^n}\int_{|y|<h}f(y)\,d(y)\right|^p dh\leqslant C\int_{R^n}|y|^\delta|f(y)|^p\,dy, \tag{4}$$

where $\delta=-n+1+\gamma$, *and* C *does not depend on* f.

The proof follows from the following chain of relations:

$$\int_0^\infty h^\gamma\left|\frac{1}{h^n}\int_{|y|\leqslant h}f(y)\,dy\right|^p dh=\int_0^\infty h^{-(n-1)p+\gamma}\left|\frac{1}{h}\int_0^h r^{n-1}\left[\int_S f(r\omega)\,d\omega\right]dr\right|^p dh\leqslant$$

(by Hardy's inequality for $\alpha=-(n-1)p+\gamma<p-1$ $(\Rightarrow\gamma<np-1)$)

$$\leqslant\left(\frac{p}{np-1-\gamma}\right)^p\int_0^\infty r^{-(n-1)p+\gamma}\left|r^{n-1}\int_S f(r\omega)\,d\omega\right|^p dr$$

$$\leqslant\left(\frac{p}{np-1-\gamma}\right)^p\int_0^\infty r^\gamma\int_S|f(r\omega)|^p\,d\omega\left(\int_S d\omega\right)^{\frac{p}{p'}}$$

$$=C\int_0^\infty r^{\gamma-(n-1)}\left(\int_S|f(r\omega)|^p\,d\omega\right)r^{n-1}dr$$

$$=C\int_{R^n}|y|^\delta|f(y)|^p\,dy, \qquad C=|S|^{p-1}\left(\frac{p}{np-1-\gamma}\right)^p.$$

Remark to Lemma 2. It is clear that (4) holds also in the case when $f(x)$ vanishes at zero for $|y| > a$. Then, decreasing the left side, we can write

$$\int_0^a h^\gamma \left| \frac{1}{h^n} \int_{|y|<h} f(y)\,dy \right|^p dh \leqslant C \int_{|y|<a} |y|^\delta |f(y)|^p\,dy,$$

$$\gamma < np-1, \qquad \delta = -n+1+\gamma, \qquad C = |S|^{p-1}\left(\frac{p}{np-1-\gamma}\right)^p. \tag{5}$$

We proceed to the question of extension. We suppose that the function $f(x)$ is defined in Ω and belongs to $\breve{B}_p^r(\Omega)$. Our goal is to construct a function $u(x, t)$ having specific properties and defined in some region $\Lambda \subset R^{n+1}$ "adjacent" to Ω. We take as the region $\Lambda = \Lambda(\Omega)$ the cone-like body

$$\Lambda = \{(x, t);\ x \in \Omega,\ 0 < t < \operatorname{dist}(x, \partial\Omega)\}.$$

We give a precise formulation of the result.

Lemma on Extension. *Let the function $f(x) \in \breve{B}_p^r(\Omega)$, i.e.*

$$\|f\|_{\breve{B}_p^r(\Omega)} = \|f\|_{L_p(\Omega)} + \left\{\int_\Omega dx \int_{|x-y|<d_x} \frac{|f(x)-f(y)|^p}{|x-y|^{n+pr}}\,dy\right\}^{\frac{1}{p}} < \infty.$$

Then there exists a function $\overset{*}{u}(x, t) \in W^1_{p,\alpha}(\Lambda)$, $\alpha = p(1-r)-1$, such that

$$\lim_{t\to 0} \overset{*}{u}(x, t) = f(x) \quad (\textit{almost everywhere in } \Omega). \tag{6}$$

Moreover,

$$\|\overset{*}{u}\|_{W^1_{p,\alpha}(\Lambda)} \overset{\text{def}}{=} \int_\Lambda t^\alpha \left\{\sum_{i=1}^n \left|\frac{\partial \overset{*}{u}}{\partial x_i}\right|^p + \left|\frac{\partial \overset{*}{u}}{\partial t}\right|^p + |\overset{*}{u}|^p\right\} dx\,dt \leqslant C\|f\|_{\breve{B}_p^r(\Omega)}, \tag{7}$$

where the constant C does not depend on u or Ω.

Proof. We introduce the following notation: $\operatorname{dist}(x, \partial\Omega) = d$, $S = S_n$ is the unit sphere in R^n, and $Ш^t_x(R^n)$ $[Ш^t(R^n)]$ is the ball of radius t with center at the point x [at the origin]:

$$Ш^t_x(R^n) = \{y;\ y \in R^n,\ |x-\xi| < t\} \equiv Ш^t_x,$$

$|Ш^t_x| = |Ш^t|$ is the volume of the ball $Ш^t$, and $|Ш^t| = |S_n| t^n/n = a^{-1}t^n$.

We define the function $u(x, t)$ by the equation

$$u(x, t) = \frac{1}{|Ш^t|} \int_{Ш^t_x} f(y)\,dy = \frac{a}{t^n} \int_{|x-y|<t} f(y)\,dy, \qquad 0 < t < \operatorname{dist}(x, \partial\Omega).$$

It is clear that $u(x, t)$ is defined in Λ and

$$\lim_{t\to 0} u(x, t) = f(x) \quad (\text{almost everywhere in } \Omega).$$

There are estimates for the derivatives of $u(x, t)$. We begin with $\partial u/\partial t$. Passing to spherical coordinates with center at the point x: $y - x = \rho\omega$, $\omega \in S$, we have

$$u(x, t) = \frac{a}{t^n} \int_{\rho<t} f(x+\rho\omega)\rho^{n-1}d\rho d\omega = \frac{a}{t^n}\int_S d\omega \int_0^t f(x+\rho\omega)\rho^{n-1}d\rho,$$

$$\frac{\partial u(x, t)}{\partial t} = -\frac{an}{t^{n+1}}\int_S d\omega \int_0^t [f(x+\rho\omega)\rho^{n-1}d\rho + \frac{a}{t^n}\int_S f(x+t\omega)t^{n-1}d\omega$$

$$= \frac{an}{t^{n+1}}\int_S d\omega \int_0^t [f(x+t\omega) - f(x+\rho\omega)]\rho^{n-1}d\rho.$$

Consequently

$$\int_\Lambda t^\alpha \left|\frac{\partial u}{\partial t}\right|^p dxdt = \int_\Omega dx \int_0^d t^\alpha \left|\frac{\partial u}{\partial t}\right|^p dt$$

$$= (na)^p \int_\Omega dx \int_0^d t^\alpha \left|\frac{1}{t^{n+1}}\int_S d\omega \int_0^t [f(x+t\omega) - f(x) + f(x) - f(x+\rho\omega)]\, d\omega\right|^p dt$$

$$\leqslant 2^{p-1}(na)^p \int_\Omega dx \int_0^\alpha \frac{dt}{t^{p(n+1)-\alpha}} \left|\int_0^t \rho^{n-1}d\rho \int_S [f(x+\rho\omega) - f(x)]\, d\omega\right| \tag{8}$$

$$+ 2^{p-1}(na)^p \int_\Omega dx \int_0^\alpha \frac{dt}{t^{p(n+1)-\alpha}} \left|\int_0^t \rho^{n-1}d\rho \int_S [f(x+t\omega) - f(x)]\, d\omega\right|^p$$

$$= 2^{p-1}(na)^p (I_1 + I_2).$$

The integral I_1 is estimated by (5) (for $\gamma = \alpha - p$ and $\delta = -n + 1 + \alpha - p = -n - pr$):

$$I_1 = \int_\Omega dx \int_0^d \frac{dt}{t^{p-\alpha}} \left|\frac{1}{t^n}\int_{|y|\leqslant t} [f(x+y) - f(x)]\, dy\right|^p$$
$$\leqslant C \int_\Omega dx \int_{|y|<d} \frac{|f(x+y) - f(x)|^p}{|y|^{n+pr}}\, dy. \tag{9}$$

In I_2 we perform the inside integration with respect to ρ:

$$I_2 = \int_\Omega dx \int_0^d \frac{dt}{t^{p(n+1)-\alpha}} \left|\frac{t^n}{n}\int_S [f(x+t\omega) - f(x)]\, d\omega\right|^p$$
$$\leqslant \frac{1}{n^p}\int_\Omega dx \int_{|y|<d} \frac{|f(x+y) - f(x)|^p}{|y|^{n+pr}}\, dy. \tag{10}$$

Putting together the estimates (8)–(10), we obtain

$$\int_\Lambda t^\alpha \left|\frac{\partial u}{\partial t}\right|^p dxdt \leqslant C \int_\Omega dx \int_{|y|<d} \frac{|f(x+y) - f(x)|^p}{|y|^{n+pr}}\, dy. \tag{11}$$

We consider the corresponding estimates for the derivatives $\partial u/\partial x_i$. For definiteness let $i = n$. We write u in the form

$$u = \frac{a}{t^n}\int_{Ш_x^t(R^n)} f(y)\, dy = \frac{a}{t^n}\int_{Ш_{x'}^t(R^{n-1})} dy' \int_{x_n-\sqrt{}}^{x_n+\sqrt{}} f(y', y_n)\, dy_n,$$

where

$$y' = (y_1, \ldots, y_{n-1}), \qquad y = (y', y_n), \qquad dy' = dy_1 dy_2 \ldots dy_{n-1}.$$

$$\sqrt{} = \sqrt{t^2 - |x' - y'|^2} = \sqrt{t^2 - (x_1 - y_1)^2 - \ldots - (x_{n-1} - y_{n-1})^2}.$$

Then

$$\frac{\partial u}{\partial x_n} = \frac{a}{t^n} \int\limits_{|x'-y'|<t} [f(y', x_n + \sqrt{}) - f(y', x_n - \sqrt{})]\, dy'$$

$$= \frac{a}{t^n} \int\limits_{|x'-y'|<t} [f(y', x_n + \sqrt{}) - f(x)]\, dy'$$

$$+ \frac{a}{t^n} \int\limits_{|x'-y'|<t} [f(x) - f(y', x_n + \sqrt{})]\, dy' = \frac{a}{t^n}(A_1 + A_2).$$

If we introduce spherical coordinates ω, ρ $(y - x = \omega\rho)$ with center at x and polar axis with direction along the unit vector e_n (the axis x_n), then the first of the above integrals can be written as an integral over the upper hemisphere S^+:

$$A_1 = \int\limits_{S+} [f(x + \omega t) - f(x)]\, t^{n-1} \cos \varphi d\omega, \text{(5)}$$

where $\cos \phi = (e_n, (y - x)/|y - x|)$, and $d\omega$ is the surface element of the sphere $S = S^+ \cup S^-$ and

$$A_2 = \int\limits_{S-} [f(x) - f(x + \omega\rho)]\, \rho^{-n-1} (-\cos \varphi)\, d\omega.$$

Thus

$$\frac{\partial u}{\partial x_n} = \frac{a}{t^n} \int\limits_S [f(x + \omega t) - f(x)]\, t^{n-1} \cos \varphi d\omega.$$

From this we derive

$$\int\limits_\Lambda t^\alpha \left| \frac{\partial u}{\partial x_n} \right|^p dx dt = \int\limits_\Omega dx \int\limits_0^d t^\alpha \left| \frac{\partial u}{\partial x_n} \right|^p dt$$

$$= a^p \int\limits_\Omega dx \int\limits_0^d t^\alpha \left| \frac{1}{t} \int\limits_S [f(x + \omega t) - f(x)] \cos \varphi d\omega \right|^p dt$$

(we use Hölder's inequality)

$$\leqslant a^p \int\limits_\Omega dx \int\limits_0^d t^{\alpha-p} \int\limits_S |f(x + \omega t) - f(x)|^p d\omega \left(\int\limits_S |\cos \varphi|^{p'} d\omega \right)^{\frac{p}{p'}} dt$$

$$\leqslant a^n |S|^{p-1} \int\limits_\Omega dx \int\limits_{|u|<d} \frac{|f(x+u) - f(x)|^p}{|u|^{n-1+p-\alpha}} du, \qquad n - 1 + p - \alpha = n + pr.$$

The remaining derivatives are estimated in an analogous way:

$$\int\limits_\Lambda t^\alpha \left| \frac{\partial u}{\partial x_i} \right|^p dx dt \leqslant \frac{n^p}{|S|} \int\limits_\Omega dx \int\limits_{|y|<d} \frac{|f(x+y) - f(x)|^p}{|y|^{n+pr}} dy, \qquad i = 1, \ldots, n. \quad (12)$$

Finally, we consider an estimate for the function u itself:

(5) The substitution $y' - x' = z'$ gives $f(y', x_n + \sqrt{}) = f(x' + z', x_n + \sqrt{t^2 - |z'|^2})$; ωt is calculated from $y' - x' = z'$ and $t \cos \phi e_n = \sqrt{t^2 - |z'|^2} e_n$. Thus, $f(x' + z', x_n + \sqrt{t^2 - |z'|^2}) = f(x + \omega t)$.

$$\int_\Lambda t^\alpha |u|^p\,dxdt = \int_\Omega dx \int_0^d t^\alpha |u|^p\,dt = \int_\Omega dx \int_0^d t^\alpha \left| \frac{1}{t^n} \int_{|y|<t} f(x+y)\,dy \right|^p dt.$$

In the inside integral we use (5):

$$\int_0^d t^\alpha \left| \frac{1}{t^n} \int_{|y|<t} f(x+y)\,dy \right|^p dt \leqslant C \int_{|y|<d} |y|^{-n+1+\alpha} |f(x+y)|^p\,dy.$$

Then, after the substitution $x+y=\xi$,

$$\int_\Lambda t^\alpha |u|^p\,dxdt \leqslant C \int_\Omega dx \int_{|x-\xi|<\operatorname{dist}(x,\partial\Omega)} |f(\xi)|^p |x-\xi|^{-n+1+\alpha}\,d\xi = \qquad (*)$$

(we change the order of integration on the right; it is clear that for this the domain of ξ will be Ω, and $|x-\xi|$ does not exceed the diameter of Ω)

$$= C \int_\Omega |f(\xi)|^p\,d\xi \int_{|x-\xi|<\operatorname{dist}(x,\partial\Omega)} |x-\xi|^{-n+1+\alpha}\,dx \leqslant C \int_\Omega |f(\xi)|^p\,ds\,|S| \int_0^{\operatorname{Diam}\Omega} \rho^\alpha\,d\rho$$

$$= \frac{C|S|}{1+\alpha} (\operatorname{Diam}\Omega)^{1+\alpha} \int_\Omega |f(\xi)|^p\,d\xi.$$

Thus, in conclusion we get

$$\int_\Lambda t^\alpha |u|^p\,dxdt \leqslant \frac{C|S|}{1+\alpha} (\operatorname{Diam}\Omega)^{1+\alpha} \int_\Omega |f(\xi)|^p\,d\xi. \qquad (13)$$

Consequently (13) is proved, provided only that Ω is bounded. Simple examples show that in the general case (for unbounded Ω) $\int_\Lambda t^\alpha |u|^p\,dxdt = +\infty$.

The estimate (7) follows from (11)–(13):

$$\|u\|_{W^1_{p,\alpha}(\Lambda)} \leqslant C \|f\|_{B^r_p(\Omega)},$$

where the constant C does not depend on u. The dependence of C on the parameters p, α, n and Diam Ω can be explicitly shown (Ω a bounded region).

In the case of an unbounded region Ω we "adjust" the function $u(x,t)$ by multiplying by a function $\xi(t)$ such that

1) $\xi(t) \in C^\infty([0,\infty))$,
2) $0 \leqslant \xi(t) \leqslant 1$,
3) $\xi(t) = \begin{cases} 1, & 0 \leqslant t \leqslant \frac{1}{2}, \\ 0, & t \geqslant 1. \end{cases}$

Then it is easy to see that the function

$$\overset{*}{u}(x,t) = u(x,t)\,\xi(t)$$

will satisfy all the requirements of the lemma. For example, the last calculations leading to (13) are changed in the following way: $\min\{d,1\}$ appears instead of d; one should in general write $\leq$ in the step $(*)$; and finally, in place of Diam Ω should be $\min\{1, \operatorname{Diam}\Omega\}$. As a result we obtain

$$\int_\Lambda t^\alpha |\overset{*}{u}|^p\,dxdt \leqslant \frac{C|S|}{1+\alpha} \int_\Omega |f(\xi)|^p\,d\xi.$$

§4. Coincidence of the spaces $B_p^r(R_+^n)$ and $\breve{B}_p^r(R_+^n)$

Suppose the function $f(x)$ is defined in the halfspace $R_+^n = \{x;\ x = (x_1, \dots, x_n) \in R^n,\ x_n > 0\}$ and belongs to the class $\breve{B}_p^r(R_+^n)$, i.e.

$$\|f\|_{L_p(R_+^n)} + \left\{\int\limits_{R_+^n} dx \int\limits_{|x-y|<\operatorname{dist}(x,\partial R_+^n)} \frac{|f(x)-f(y)|^p}{|x-y|^{n+pr}}\, dy\right\}^{\frac{1}{p}} = \|f\|_{\breve{B}_p^r(R_+^n)} < \infty.$$

We remark that $\operatorname{dist}(x, \partial R_+^n) = x_n$. Let $Ш = Ц(R_+^n)$ be the "cylinder" on R_+^n:

$$Ц = \{(x, t),\quad x \in R_+^n,\quad 0 < t < \infty\}.$$

We construct the function $u(x, t)$ in the following way. As in the previous section, let $Ш_x^t$ be the open ball with radius t and center at x:

$$Ш_x^t = \{y;\ |x-y| < t\}.$$

For $x \in R_+^n$ and $0 < t < x_n$, we set

$$u(x, t) = \frac{a}{t^n} \int\limits_{Ш_x^t} f(y)\, dy = \frac{a}{t^n} \int\limits_{|x-y|<t} f(y)\, dy, \tag{1}$$

where $a = n|S_n|^{-1}$, $|S_n|$ being the area of the unit sphere in R^n. Thus $u(x, t)$ is equal to the average value of the function $f(y)$ on the ball $|x-y| < t$. Therefore $\lim_{t\to 0} u(x, t) = f(x)$ for almost every $x \in R_+^n$.

The function $u(x, t)$ is not defined in the whole cylinder $Ц$ but only in the cone-like body $\Lambda = \Lambda(R_+^n)$:

$$\Lambda = \{(x, t);\quad x \in R_+^n,\quad 0 < t < x_n\}.$$

We now set (denoting by e^n the unit vector of the nth axis in R^n):

$$\tilde{u}(x, t) = u(x + te^n, t) = u(x_1, \dots x_{n-1}, x_n + t, t) = u(x', x_n + t, t). \tag{2}$$

The function $\tilde{u}(x, t)$ is defined on all of $Ц$ and clearly has the property

$$\lim_{t\to 0} \tilde{u}(x, t) = f(x).$$

We estimate the derivatives of $\tilde{u}$. Since

$$\frac{\partial \tilde{u}}{\partial x_1} = \frac{\partial u}{\partial x_1}(x', x_n + t, t),$$

we have

$$\int\limits_{Ц} t^\alpha \left|\frac{\partial \tilde{u}}{\partial x_1}\right|^p dx dt = \int\limits_{R_+^n} dx \int\limits_0^\infty \left|\frac{\partial u}{\partial x}\right|(x', x_n + t, t)\Big|^p t^\alpha dt$$

$$= \int\limits_{R^{n-1}} dx' \int\limits_0^\infty t^\alpha dt \int\limits_0^\infty \left|\frac{\partial u}{\partial x_1}(x', x_n + t, t)\right|^p dx_n \qquad (x_n + t = y_n)$$

$$= \int\limits_{R^{n-1}} dx' \int\limits_0^\infty t^\alpha dt \int\limits_t^\infty \left|\frac{\partial u}{\partial x_1}(x', y_n, t)\right|^p dy_n =$$

$$= \int_{R^{n-1}} dx' \int_0^\infty dy_n \int_0^{y_n} t^\alpha \left| \frac{\partial u}{\partial x_1} (x', y_n, t) \right|^p dt \qquad (y_n \to x_n)$$

$$= \int_{R^{n-1}} dx' \int_0^\infty dx_n \int_0^{x_n} t^\alpha \left| \frac{\partial u}{\partial x_1} (x', x_n, t) \right|^p dt = \int_{R^n_+} dx \int_0^{x_n} t^\alpha \left| \frac{\partial u}{\partial x_1} (x, t) \right|^p dt.$$

Thus we have obtained an integral over the indicated cone-like body $\Lambda = \Lambda(R^n_+)$:

$$\int_{\Pi} t^\alpha \left| \frac{\partial \tilde{u}}{\partial x_1} \right|^p dx dt = \int_\Lambda t^\alpha \left| \frac{\partial u}{\partial x} \right|^p dx dt.$$

By the extension lemma from the preceding section, the last integral is bounded (see §3, (11)) by

$$\int_{R^n_+} dx \int_{|y| < x_n} \frac{|f(x+y) - f(x)|^p}{|y|^{n+pr}} dy.$$

An analogous calculation is valid for the derivatives $\partial \tilde{u}/\partial x_i$, $i = 2, \ldots, n-1$, and also for $(\partial \tilde{u}/\partial x_n)(x, t) = (D_n u)(x', x_n + t, t)$. It holds also for $\partial \tilde{u}/\partial t$, since

$$\frac{\partial \tilde{u}}{\partial t}(x, t) = (D_n u)(x', x_n + t, t) + (D_{n+1} u)(x', x_n + t, t).$$

Therefore we can write

$$\int_{\Pi(R^n_+)} t^\alpha \left\{ \sum_{i=1}^n \left| \frac{\partial \tilde{u}}{\partial x_i} \right|^p + \left| \frac{\partial \tilde{u}}{\partial t} \right|^p \right\} dx dt \leqslant C \int_{R^n_+} dx \int_{|y| < x_n} \frac{|f(x+y) - f(x)|^p}{|y|^{n+pr}} dy. \quad (3)$$

We now perform a "truncation" of $\tilde{u}(x, t)$. Let, as in §3,

$$\xi(t) \in C^\infty([0, \infty)), \qquad \xi(t) = \begin{cases} 1, & 0 \leqslant t \leqslant \frac{1}{2}, \\ 0, & t > 1. \end{cases}$$

$$0 \leqslant \xi(t) \leqslant 1.$$

We set $u^*(x, t) = \tilde{u}(x, t)\xi(t)$, and consider the integral

$$I = \int_\Pi t^\alpha |u^*(x, t)|^p dx dt = \int_{R^{n-1}_+} dx' \int_0^\infty dx_n \int_0^1 |\tilde{u}(x, t)\xi(t)|^p t^\alpha dt. \quad (4)$$

Carrying out some trivial transformations, we get

$$I = \int_{R^{n-1}_+} dx' \int_0^1 t^\alpha [\xi(t)]^p dt \int_0^\infty |u(x', x_n + t, t)|^p dx_n$$

$$= \int_{R^{n-1}_+} dx' \int_0^1 t^\alpha [\xi(t)]^p dt \int_t^\infty |u(x', y_n, t)|^p dy_n \quad (5)$$

$$= \int_{R^{n-1}_+} dx' \int_0^1 dy_n \int_0^{y_n} t^\alpha [\xi(t)]^p |u(x', y_n, t)|^p dt$$

$$+ \int_{R^{n-1}_+} dx' \int_1^\infty dy_n \int_0^1 t^\alpha [\xi(t)]^p |u(x', y_n, t)|^p dt = I_1 + I_2.$$

We rewrite I_1, replacing y_n by x_n, using the fact that $0 \leq \xi(t) \leq 1$, and substituting in place of $u(x, t)$ its expression (1):

$$I_1 \leqslant a^p \int\limits_{R_+^{n-1}} dx' \int\limits_0^1 dx_n \int\limits_0^{x_n} t^\alpha \left| \frac{1}{t^n} \int\limits_{|\xi|<t} f(x+\xi)\, d\xi \right|^p dt.$$

By the inequality (5) from §3, after simple transformations we get

$$I_1 \leqslant Ca^p \int\limits_{R_+^{n-1}} dx' \int\limits_0^1 dx_n \int\limits_{|y|<x_n} |y|^{-n+1+\alpha} |f(x+y)|^p dy$$

$$= Ca^p \int\limits_{R_+^{n-1}} dx' \int\limits_0^1 dx_n \int\limits_{|x-u|<x_n = \min\{x_n, 1\}} |f(u)|^p |x-u|^{-n+1-\alpha} du$$

$$\leqslant Ca^p \int\limits_{R_+^n} dx \int\limits_{|x-u|<\min\{x_n, 1\}} |f(u)|^p |x-u|^{-n+1-\alpha} du.$$

It is clear that for the last step we have increased the expression under consideration. We now change the order of integration. Since the variable point u runs through R_+^n, we have

$$I_1 \leqslant Ca^p \int\limits_{R_+^n} |f(u)|^p du \int\limits_{|x-u|<\min\{x_n, 1\}} |x-u|^{-n+1-\alpha} dx$$

$$\leqslant Ca^p \int\limits_{R_+^n} |f(u)|^p du \int\limits_{|x-u|<1} |x-u|^{-n+1-\alpha} dx$$

As a final result we arrive at the estimate

$$I_1 \leqslant Ca^p \frac{|S|}{1+\alpha} \int\limits_{R_+^n} |f(u)|^p du. \tag{6}$$

In estimating I_2 we use Hölder's inequality:

$$I_2 = \int\limits_{R_+^{n-1}} dx' \int\limits_1^\infty dx_n \int\limits_0^1 t^\alpha [\xi(t)]^p \left| \frac{1}{t^n} \int\limits_{|\xi|<t} f(x+\xi)\, d\xi \right|^p dt$$

$$\leqslant \int\limits_{x_n>1} dx \int\limits_0^1 \frac{dt}{t^{np-\alpha}} \int\limits_{|\xi|<t} |f(x+\xi)|^p d\xi \left(\int\limits_{|\xi|<t} d\xi \right)^{\frac{p}{p'}}$$

$$= \left(\frac{|S|}{n}\right)^{p-1} \int\limits_{x_n>1} dx \int\limits_0^1 \frac{dt}{t^{n-\alpha}} \int\limits_{|\xi|<t} |f(x+\xi)|^p d\xi$$

$$= \left(\frac{|S|}{n}\right)^{p-1} \int\limits_0^1 \frac{dt}{t^{n-\alpha}} \int\limits_{x_n>1} dx \int\limits_{|x-y|<t} |f(y)|^p dy.$$

We change the order of integration with respect to x and y. For fixed x the variable y is varied in the ball $|x-y| < t$. Since $0 < t < 1$ and x is of distance greater than one ($x_n > 1$) from ∂R_+^n, it follows that $y \in R_+^n$. Therefore

$$I_2 \leqslant \left(\frac{|S|}{n}\right)^{p-1} \int_0^1 \frac{dt}{t^{n-\alpha}} \int_{R_+^n} |f(y)|^p \, dy \int_{|x-y|<t} dx = \left(\frac{|S|}{n}\right)^p \frac{1}{1+\alpha} \|f\|^p_{L_p(R_+^n)}. \tag{7}$$

Putting (4)–(7) together, we get

$$\int_{R_+^n} t^\alpha |u^*(x, t)|^p \, dxdt \leqslant C \int_{R_+^n} |f(x)|^p \, dx. \tag{8}$$

It is easy to see also that for u^* we have an estimate of the form

$$\int_{\Pi(R_+^n)} t^\alpha \left\{ \sum_{i=1}^n \left| \frac{\partial u^*}{\partial x_i} \right|^p + \left| \frac{\partial u^*}{\partial t} \right|^p \right\} dxdt$$
$$\leqslant C_1 \left\{ \int_{R_+^n} dx \int_{|y|<x_n} \frac{|f(x+y) - f(x)|^p}{|y|^{n+pr}} dy + \int_{R_+^n} |f(x)|^p \, dx \right\}. \tag{9}$$

Indeed,

$$\frac{\partial u^*}{\partial x_i} = \frac{\partial \tilde{u}}{\partial x_i} \xi(t), \quad \frac{\partial u^*}{\partial t} = \frac{\partial \tilde{u}}{\partial t} \xi(t) + \tilde{u}\xi'$$

and substitution of these expressions in the right side reduces the matter to estimates already obtained (namely (3) and (8)).

Summing up, from (8) and (9) we get

$$\|u^*\|_{W^1_{p,\alpha}(\Pi)} \leqslant C^* \|f\|_{\breve{B}^r_p(R_+^n)}, \qquad r = 1 - \frac{1+\alpha}{p}. \tag{10}$$

Thus we have proved

Theorem 4.1. *Suppose the function $f(x)$ belongs to the space $\breve{B}^r_p(R_+^n)$, $1 \leq p < \infty$, $0 < r < \infty$. Then there exists a function $u^*(x, t)$ which is defined in $\Pi(R_+^n)$ and is such that*

$$u^* \in W^1_{p,\alpha}(\Pi), \qquad \alpha = p(1-r) - 1, \qquad \lim_{t \to 0} u^*(x, t) = f(x) \ (\text{a.e. on } \Omega).$$

Moreover, the inequality (10) *holds, where the constant C^* does not depend on f.*

Remark 1. Theorem 4.1 means that there is a continuous embedding

$$\breve{B}^r_p(R_+^n) \to (W^1_p(R_+^n), L_p(R_+^n))_{\theta, p},$$

where $r = 1 - \theta$. It is necessary to assume that the norm in $(W^1_p, L_p)_{\theta,p}$ is defined as

$$\inf \|u\|_{W^1_{p,\alpha}(R_+^n)},$$

where the inf runs over all $u \in W^1_{p,\alpha}(R_+^n)$ with $u(x, 0) = f$.

Remark 2. Theorem 4.1 and the preceding remark are easily generalized to the case of a region Ω with smooth boundary (of class C^1). For this, one can use a partition of unity to reduce the problem to consideration of a function which is different from zero only in a neighborhood of a boundary point. A substitution of variables then permits one to reduce the matter to the case of R_+^n. Finally, one should use the invariance of the class $\breve{B}^r_p(\Omega)$ with respect to smooth changes of variables.

§5. Some weight estimates

We first consider the case of a halfspace.

Theorem 5.1. *Suppose the function $f(x)$ belongs to the space $\breve{B}_p^r(R_+^n)$, $1 \le p < \infty$, $0 < r < 1/p$. Then*

$$\int\limits_{R_+^n} \frac{|f(x)|^p}{x_n^{pr}}\,dx \leqslant C \int\limits_{R_+^n} dx \int\limits_{|x-y|<d_x} \frac{|f(x)-f(y)|^p}{|x-y|^{n+pr}}\,dy \equiv C\,[\breve{b}_p^r(f,\ R_+^n)]^p, \tag{1}$$

where the constant C does not depend on f.

Remark. By the results of §4, Theorem 5.1 is true also for the space $B_p^r(R_+^n)$.

Proof. For simplicity of notation and clearness of the calculations we carry out the proof for $n = 2$. Thus $x = (x_1, x_2)$ and $y = (y_1, y_2)$. Since the smooth functions with bounded support are dense in $B_p^r(R_+^n)$,(6) we will take f to be such a function. Then we can write (using the triangle inequality for the norm)

$$\left\{\int\limits_{R_+^2} dx \int\limits_{\frac{x_2}{2}<|x-y|<x_2} \frac{|f(x)-f(y)|^p}{|x-y|^{2+pr}}\,dy\right\}^{\frac{1}{p}} \geqslant \left|\left\{\int\limits_{R_+^2} dx \int\limits_{\frac{x_2}{2}<|x-y|<x_2} \frac{|f(x)|^p}{|x-y|^{2+pr}}\,dy\right\}^{\frac{1}{p}} - \left\{\int\limits_{R_+^2} dx \int\limits_{\frac{x_2}{2}<|x-y|<x_2} \frac{|f(y)|^p}{|x-y|^{2+pr}}\,dy\right\}^{\frac{1}{p}}\right|. \tag{2}$$

In the first integral on the right $|f(x)|^p$ is taken out from under the inside integral sign, after which the integral is easily computed (in polar coordinates ρ, ϕ):

$$\int\limits_{\frac{x_2}{2}<|x-y|<x_2} \frac{dy}{|x-y|^{2+pr}} = 2\pi \int\limits_{\frac{x_2}{2}}^{x_2} \rho^{-pr}\,d\rho = \frac{C_1}{x_2^{pr}}, \qquad C_1 = 2\pi \int\limits_{\frac{1}{2}}^{1} \frac{dt}{t^{1+pr}}. \tag{3}$$

In the second integral we change the order of integration:

$$\int\limits_{R_+^2} dx \int\limits_{\frac{x_2}{2}<|x-y|<x_2} \frac{|f(y)|^p\,dy}{|x-y|^{2+pr}} = \int\limits_{R_+^2} |f(y)|^p\,dy \int\limits_{\frac{x_2}{2}<|x-y|<x_2} \frac{dx}{|x-y|^{2+pr}} = \int\limits_{R_+^2} |f(x)|^p\,dx \int\limits_{\frac{y_2}{2}<|y-x|<y_2} \frac{dy}{|y-x|^{2+pr}}. \tag{4}$$

Thus the problem is reduced to the calculation of the integral

$$I(x) = \int\limits_{\frac{y_2}{2}<|y-x|<y_2} \frac{dy}{|y-x|^{2+pr}}.$$

An easy computation shows that the curve $|y-x| = y_2$ is a parabola

$$y_1^2 = 2x_2\left(y_2 - \frac{x_2}{2}\right) \tag{5}$$

in the plane (y_1, y_2) (for fixed x_2), while the curve $|y-x| = y_2/2$ is the ellipse

---

(6) This fact follows, for example, from the results of §4: for regions with smooth boundaries the spaces $B_p^r(\Omega)$ and $\breve{B}_p^r(\Omega)$ coincide (C_0^∞ is dense in $B_p^r(\Omega)$).

$$y_1^2+\frac{3}{4}(y_2-x_2)^2=\frac{1}{3}x_2^2. \tag{6}$$

With respect to polar coordinates, (5) is written in the form

$$\rho=\frac{x_2}{1-\sin\varphi}, \tag{5'}$$

and (6) takes the form

$$\rho=\frac{x_2}{2-\sin\varphi}. \tag{6'}$$

Therefore

$$I(x)=2\int_{-\frac{\pi}{2}}^{\frac{\pi}{2}} d\varphi \int_{\frac{x_2}{2-\sin\varphi}}^{\frac{x_2}{1-\sin\varphi}} \frac{\rho\, d\rho}{\rho^{2+pr}}=\frac{C_2}{x_r^{pr}},\quad C_2=2\int_{-\frac{\pi}{2}}^{\frac{\pi}{2}} d\varphi \int_{\frac{1}{2-\sin\varphi}}^{\frac{1}{1-\sin\varphi}} \frac{dt}{t^{1+pr}}.$$

Consequently (2) is converted into the inequality

$$\breve{b}_p^{(r)}(f,\ R_+^2)\geqslant C^{\frac{1}{p}}\left\{\int_{R_+^2}\frac{|f(x)|^p\,dx}{x_2^{pr}}\right\}^{\frac{1}{p}},\qquad C^{\frac{1}{p}}=\left|C_1^{\frac{1}{p}}-C_2^{\frac{1}{p}}\right|. \tag{2'}$$

It can be verified that $C_1\neq C_2$ for $0<r<1/p$. Moreover, under this condition both terms on the right in (2) are finite, and hence all of the calculations carried out are valid. From the aforementioned density of the smooth functions of bounded support in $B_p^r(R_+^2)$ it follows that (2′) holds also in the general case. We remark that for arbitrary n the calculations are essentially unchanged (since there is spherical symmetry with respect to the axis x_n). The theorem is proved.

Theorem 5.1 is in some measure an analogue of a boundary embedding theorem, since, more precisely, it gives a representation of the behavior of the possible order of growth of the function on approach to the boundary. This analogy is justified by the fact that for $r<1/p$, when the embedding theorems guarantee the existence of a boundary value, it is possible to show that in place of (1) we have the estimate $(1/p<r<1+1/p)$

$$\left\{\int_{R_+^n}\frac{|f(x',\ x^n)-f(x',\ 0)|^p}{x_n^{pr}}\,dx\right\}^{\frac{1}{p}}\leqslant C\breve{b}_p^r(f,\ R_+^n).$$

We consider extension of Theorem 5.1 in another direction.

It is known [10] that the space $B_p^r(R^n)$, $1\leq p<\infty$, can be characterized as the collection of functions in $L_p(R^n)$ with finite norm

$$\|f\|^*_{B_p^r(R^n)}=\|f\|_p+\left\{\int_{R^n}dx\int_{R^m}\frac{|f(x',\ x'')-f(y',\ y'')|^p}{|x'-y'|^{m+pr}}\,dy'\right\}^{\frac{1}{p}}$$
$$+\left\{\int_{R^n}dx\int_{R^{n-m}}\frac{|f(x',\ x'')-f(y',\ y'')|^p}{|x''-y''|^{n-m+pr}}\,dy''\right\}^{\frac{1}{p}},$$

where

$$R^n = R^m \times R^{n-m},\ x = (x_1, \ldots, x_m, x_{m+1}, \ldots, x_n) = (x', x''),\ x' = (x_1, \ldots, x_m) \in R_m,$$

$$x'' = (x_{m+1}, \ldots, x_n) \in R^{n-m},\ x' = \sqrt{\sum_{i=1}^{m} x_i^2}\ ,\ \text{etc.}$$

We use the equality ($a > 1$; $a|x''|^{n-m}/k$ is the volume of the ball of radius $a|x''|$ in R^{n-m}):

$$f(x) = \frac{k}{(a\,|x''|)^{n-m}} \int\limits_{|y''|<a|x''|} f(x', y'')\,dy'' + \frac{k}{(a\,|x''|)^{n-m}} \int\limits_{|y''|<a|x''|} [f(x) - f(x'', y'')]\,dy''.$$

Applying Hölder's inequality, we get

$$\begin{aligned} |f(x)| \leqslant{} & \frac{k}{(a\,|x''|)^{n-m}} \Bigg(\int\limits_{|y''|<a|x''|} |f(x, y'')|^p\,dy'' \Bigg)^{\frac{1}{p}} \Bigg(\int\limits_{|y''|<a|x''|} dy'' \Bigg)^{\frac{1}{p'}} \\ & + \frac{1}{(a\,|x''|)^{n-m}} \Bigg(\int\limits_{|y''|<a|x''|} \frac{|f(x) - f(x', y'')|^p}{|x'' - y'')|^{n-m+pr}}\,dy'' \Bigg) \\ & \times \Bigg(\int\limits_{|y''|<a|x''} |x'' - y''|^{\left(r+\frac{n-m}{p}\right)p'}\,dy'' \Bigg)^{\frac{1}{p'}}. \end{aligned} \tag{7}$$

We compute that

1) $$\Bigg(\int\limits_{|y''|<a|x''|} dy'' \Bigg)^{\frac{1}{p'}} = \left[\frac{(a\,|x''|)^{n-m}}{k} \right]^{\frac{1}{p'}},$$

2) $$\begin{aligned} & \Bigg\{ \int\limits_{|y''|<a|x''|} |x'' - y''|^{\left(r+\frac{n-m}{p}\right)p'}\,dy'' \Bigg\}^{\frac{1}{p'}} \\ & = \Bigg\{ \int\limits_{|z|<1} |x'' - a\,|x''|\,z|^{\left(r+\frac{n-m}{p}\right)p'} (a\,|x''|)^{n-m}\,dz \Bigg\}^{\frac{1}{p'}} \\ & \leqslant (a|x''|)^{\frac{n-m}{p'}} \Bigg\{ \int\limits_{|z|<1} (|x''| + a\,|x''|\,|z|)^{\left(r+\frac{n-m}{p}\right)p'}\,dz \Bigg\}^{\frac{1}{p'}} \\ & \leqslant (a\,|x''|)^{r+n-m} \Bigg(\int\limits_{|z|<1} (1 + |z|)^{\left(r+\frac{n-m}{p}\right)p'}\,dz \Bigg)^{\frac{1}{p'}} = C\,(a\,|x''|)^{r+n-m}. \end{aligned} \tag{8}$$

Then (7) is rewritten in the strengthened form

$$\begin{aligned} |f(x)| \leqslant{} & \Bigg[\frac{k}{(a\,|x''|)^{n-m}} \int\limits_{|y''|<a|x''|} |f(x', y'')|^p\,dy'' \Bigg]^{\frac{1}{p}} \\ & + C\,(a\,|x''|)^r \Bigg(\int\limits_{|y''|<a|x''|} \frac{|f(x) - f(x', y'')|^p}{|x'' - y''|^{n-m+pr}}\,dy'' \Bigg)^{\frac{1}{p}}. \end{aligned}$$

We deivide both sides by $|x''|^r$ and take the $L_p(R^n)$-norm. We obtain

$$\left(\int_{R^n} \frac{|f(x)|^p\,dx}{|x''|^{pr}}\right)^{\frac{1}{p}} \leqslant \frac{k^{\frac{1}{p}}}{a^{\frac{n-m}{p}}}\left\{\int_{R^n} \frac{1}{|x''|^{pr}}\left[\frac{1}{|x''|^{n-m}} \int_{|y''|<a|x''|} |f(x', y'')|^p\,dy''\right]dx\right\}^{\frac{1}{p}}$$

$$+ Ca^r\left\{\int_{R^n} dx \int_{|y''|\leqslant a|x''|} \frac{|f(x)-f(x', y'')|^p}{|x''-y''|^{n-m+pr}}\,dy''\right\}^{\frac{1}{p}} = \frac{k^{\frac{1}{p}}}{a^{\frac{n-m}{p}}} I_1^{\frac{1}{p}} + Ca^r I_2^{\frac{1}{p}}. \tag{9}$$

We transform I_2, changing the order of integration:

$$I_2 = \int_{R^m} dx' \int_{R^{n-m}} \frac{dx''}{|x''|^{n-m+pr}} \int_{|y''|<a|x''|} |f(x', y'')|^p\,dy''$$

$$= \int_{R^m} dx' \int_{R^{n-m}} |f(x', y'')|^p\,dy'' \int_{|x''|>\frac{|y''|}{a}} |x''|^{-(n-m+pr)}\,dx''$$

$$= |S''| \int_{R^m} dx' \int_{R^{n-m}} |f(x', y'')|^p\,dy'' \int_{\frac{|y''|}{a}}^{\infty} \frac{d\rho}{\rho^{1+pr}} = \frac{|S''|}{pr} a^{pr} \int_{R^n} \frac{|f(x)|^p}{|x''|^{pr}}\,dx.$$

Here we denote the area of the unit sphere in R^{n-m} by S''. Using these calculations and the fact that $k = (n-m)/|S''|$, we rewrite (9) in the form

$$\left(\int_{R^n} \frac{|f(x)|^p\,dx}{|x''|^{pr}}\right)^{\frac{1}{p}} \leqslant \left(\frac{n-m}{pr}\right)^{\frac{1}{p}} \frac{1}{a^{\frac{n-m}{p}-r}} \left(\int_{R^n} \frac{|f(x)|^p}{|x''|^{pr}}\,dx\right)^{\frac{1}{p}} + Ca^r I_1^{\frac{1}{p}}. \tag{10}$$

If $(n-m)/p - r > 0$, then, choosing a, for example, from the condition

$$\left(\frac{n-m}{pr}\right)^{\frac{1}{p}} \frac{1}{a^{\frac{n-m}{p}-r}} = \frac{1}{2}, \tag{11}$$

we get from (10) the estimate

$$\left(\int_{R^n} \frac{|f(x)|^p dx}{|x''|^{pr}}\right)^{\frac{1}{p}} \leqslant 2Ca^r\left\{\int_{R^n} dx \int_{R^{n-m}} \frac{|f(x)-f(x', y'')|^p}{|x''-y''|^{n-m+pr}}\,dy''\right\}^{\frac{1}{p}}. \tag{12}$$

Speaking somewhat roughly, it is possible to formulate this result in the form of a theorem.

Theorem 5.2. *Suppose the function* $f(x)$ *belongs to the space* $B_p^r(R^n)$, $1 \le p < \infty$, $0 < r < 1$, $pr < n - m$, $0 \le m \le n-1$. *Then*

$$\left(\int_{R^n} \frac{|f(x)|^p}{|x''|^{pr}}\,dx\right)^{\frac{1}{p}} \leqslant C\|f\|_{B_p^r(R^n)}, \tag{13}$$

where $|x''| = \sqrt{x_{m+1}^2 + \cdots + x_n^2}$ *and the constant* C *does not depend on* f.

Remark. The estimate (13) (respectively, (12)) can be considered to be a certain analogue of a boundary embedding theorem for the space $B_p^r(R^n)$ with respect to the space R^m,(7) since it in some measure characterizes the growth of the function f on approach to R^m. From the fact that the space $B_p^r(R^n)$ is isotropic it follows that a similar estimate holds with respect to any hyperplane of

(7) We point out that Theorem 5.2 for $m = n = 1$ turns into Theorem 5.1 (for the space $B_p^r(R^n)$), which was proved by another method.

dimension m. All these circumstances are particularly clear in the case $m = 0$, which we isolate in the form of a separate assertion.

Theorem 5.2′. *Suppose* $f(x) \in B_p^r(R^n)$, $1 \leq p < \infty$, $0 < r < 1$, $pr < n$. *Then*

$$\sup_{\xi \in R^n} \left\{ \int_{R^n} \frac{|f(x)|^p}{|x-\xi|^{pr}} dx \right\}^{\frac{1}{p}} \leqslant 2Ca^r \left\{ \int_{R^n} dx \int_{R^n} \frac{|f(x)-f(y)|^p}{|x-y|^{n+pr}} dy \right\}^{\frac{1}{p}}, \tag{14}$$

where C *is determined by the final equality in* (8), *and* a *is determined by the equality* (11) *for* $m = 0$.

We mention that the region of variation of the parameter r in Theorems 5.1, 5.2 and 5.2′ can be extended, whereupon there appear weight integrals of differences on the left sides of the corresponding inequalities, while the right sides contain appropriate expressions for the norm (or seminorm) of the space $B_p^r(R^n)$ considered.

BIBLIOGRAPHY

1. N. Aronszajn, *Boundary value of functions with finite Dirichlet integral*, Technical Report no. 14, Univ. of Kansas, 1955, pp. 77–94.

2. E. Gagliardo, *Caratterizzazioni delle tracce sulla frontiera relative ad alcune classi di funzioni in n variabili*, Rend. Sem. Mat. Univ. Padova 27 (1957), 284–305. MR 21 #1525.

3. A. A. Vašarin, *The boundary properties of functions having a finite Dirichlet integral with a weight*, Dokl. Akad. Nauk SSSR **117** (1957), 742–744. (Russian) MR 20 #1113.

4. J.-L. Lions, *Théorèmes de trace et d'interpolation*. I, Ann. Scuola Norm. Sup. Pisa (3) 13 (1959), 389–403. MR 22 #9858.

5. P. I. Lizorkin, *Boundary properties of functions from "weight" classes*, Dokl. Akad. Nauk SSSR 132 (1960), 514–517 = Soviet Math. Dokl. 1 (1960), 589–592. MR 23 #A434.

6. J.-L. Lions and E. Magenes, *Problemi ai limiti non omogenei*. III, Ann. Scuola Norm. Sup. Pisa (3) 15 (1961), 41–103. MR 26 #4048.

7. S. V. Uspenskiĭ, *Properties of* W_p^r *classes with a fractional derivative on differentiable manifolds*, Dokl. Akad. Nauk SSSR 132 (1960), 60–62 = Soviet Math. Dokl. 1 (1960), 495–497. MR 23 #A2042b.

8. O. V. Besov et al., *Imbedding theory for classes of differentiable functions of several variables*, Partial Differential Equations (Proc. Sympos. S. L. Sobolev on the Occasion of His Sixtieth Birthday), "Nauka", Moscow, 1970, pp. 38–63; English transl., Amer. Math. Soc. transl. (2) 105 (1976), 57–94.

9. G. H. Hardy, J. E. Littlewood and G. Pólya, *Inequalities*, Cambridge Univ. Press, New York, 1934.

10. L. N. Slobodeckiĭ, *Generalized Sobolev spaces and their applications to boundary problems for partial differential equations*, Leningrad. Gos. Ped. Inst. Učen. Zap. **197** (1958), 54–112; English transl., Amer. Math. Soc. Transl. (2) 57 (1966), 207–276. MR 34 #3075.

Translated by H. McFADEN

Trudy Mat. Inst. Steklov.
134 (1975)

Proc. Steklov Inst. Math.
134 (1975)

PROPERTIES OF CESÀRO MEANS OF NEGATIVE ORDER FOR FOURIER SERIES OF SUMMABLE FUNCTIONS

D. E. MEN'ŠOV

Abstract. The following theorem is proved. *Suppose* $\alpha < 0$ *and* $\alpha \neq -1, -2, \ldots$. *Then any summable function can be altered outside any perfect nondense set in such a way that the Fourier series of the new function is summable* (C, α) *almost everywhere with respect to some sequence of its indices.*

Bibliography: 6 titles.

§1. Introduction

The following theorem is valid.

If $f(x)$ *is an arbitrary function that is summable on* $[-\pi, \pi]$ *and* P *is an arbitrary perfect nondense subset of* $[-\pi, \pi]$, *then* $f(x)$ *can be altered outside* P *in such a way that a summable function is obtained whose Fourier series converges almost everywhere.*(1)

The question arises as to whether this theorem is correct if, in place of ordinary convergence, one considers the Cesàro summation methods of negative order. The answer to this question is not known, but it is possible to prove a theorem that provides a partial answer.

Let us first recall the definition of the Cesàro summation method of arbitrary order α. Take a numerical series

$$\sum_{n=0}^{\infty} u_n. \tag{1.1}$$

One says that this series is summable by the Cesàro method of order α or, briefly, summable (C, α) to the limit S if

$$\lim_{m\to\infty} C_m^{(\alpha)} = S, \tag{1.2}$$

where

$$C_m^{(\alpha)} = \frac{1}{\binom{m+\alpha}{\alpha}} \sum_{k=0}^{m} \binom{m-k+\alpha-1}{\alpha-1} S_k, \tag{1.3}$$

$$S_k = \sum_{j=0}^{k} u_j, \tag{1.4}$$

$$\binom{\alpha}{\alpha} = 1, \quad \binom{m+\alpha}{\alpha} = \frac{(\alpha+1)(\alpha+2)\ldots(\alpha+m)}{m!} = \binom{m+\alpha}{m} \quad (m = 1, 2, \ldots).^{(2)} \tag{1.5}$$

AMS (MOS) subject classifications (1970). Primary 40G05, 42A24.

(1) [1]; see also [2], Chapter VI, §7.

(2) See, for example, [3], Chapter III, formula (1.20).

It is immediately seen that

$$\binom{m+\alpha}{\alpha} \neq 0 \qquad (m=0,\ 1,\ 2,\ \ldots), \tag{1.6}$$

if α is not equal to a negative integer. The Cesàro method is therefore defined for any $\alpha \neq -1, -2, \ldots$. The quantities (1.3) will be called the Cesàro means of order α of the series (1.1).

Let us now take the Fourier series

$$\frac{a_0}{2} + \sum_{n=1}^{\infty} (a_n \cos nx + b_n \sin nx) \tag{1.7}$$

of some function $G(x) \in L[-\pi, \pi]$ and denote its Cesàro means of order α by

$$C_n^{(\alpha)}(x, G) \qquad (n=0,\ 1,\ 2,\ \ldots). \tag{1.8}$$

We have

Theorem 1. *Suppose $\alpha < 0$ and $\alpha \neq -1, -2, \ldots$. Then for any summable function $f(x)$ on $[-\pi, \pi]$ and any perfect nondense set P satisfying the condition*

$$P \subset [-\pi, \pi], \tag{1.9}$$

it is possible to determine a summable function $g(x)$ on $[-\pi, \pi]$ and an increasing sequence of natural numbers

$$m_\nu \quad (\nu=0,\ 1,\ 2,\ \ldots), \tag{1.10}$$

that satisfy the following conditions:

α^0) $g(x) = f(x)$ $(x \in P)$; (1.11)

β^0) *the Cesàro means*

$$C_{m_\nu}^{(\alpha)}(x, g) \qquad (\nu=0,\ 1,\ 2,\ \ldots) \tag{1.12}$$

converge to $g(x)$ almost everywhere on $[-\pi, \pi]$.(3)

We will obtain Theorem 1 as a corollary of a more general theorem in which the methods (C, α) for

$$\alpha < 0 \qquad (\alpha \neq -1, -2, \ldots) \tag{1.13}$$

are replaced by the class of summation methods of type $R^{(\beta)}$, $\beta > 0$, defined in [5]. Let us recall the definition of these methods.

Definition *A*.(4) The method T determined by a matrix $\|C_{mk}\|$ $(m, k = 0, 1, 2, \ldots)$ is called a *method of type* $R^{(\beta)}$ if the following conditions are satisfied:

a^0) $C_{mk} = 0 \qquad (k > m,\ m=0,\ 1,\ 2,\ \ldots)$; (1.14)

b^0) $\lim\limits_{m\to\infty} C_{mk} = 0 \qquad (k=0,\ 1,\ 2,\ \ldots)$; (1.15)

c^0) $\lim\limits_{m\to\infty} \sum\limits_{k=0}^{\infty} C_{mk} = 1$; (1.16)

---

(3) Theorem 1 was formulated without proof in [4].

(4) [5], Definition *A*.

d⁰) $|C_{mm}| \leqslant Mm^{\beta}, \quad |C_{mk}| \leqslant M \frac{m^{\beta}}{(m-k)^{\beta+1}} \quad (m=1, 2, \ldots; 0 \leqslant k < m).$ (1.17)

If (1.7) is the Fourier series of a summable function $G(x)$ on $[-\pi, \pi]$ and if T is a summation method of type $R^{(\beta)}$ determined by a matrix $\|C_{mk}\|$, we will call the functions

$$T_m^{(\beta)}(x, G) = \sum_{k=0}^{m} C_{mk} S_k(x, G) \qquad (m = 0, 1, 2, \ldots), \tag{1.18}$$

where the $S_k(x, G)$ are the partial sums of (1.7), the *T-means* of (1.7). We have

Theorem 2. *Suppose $f(x)$ and P satisfy the conditions of Theorem 1 and suppose the method T determined by a matrix $\|C_{mk}\|$ is an arbitrary method of type $R^{(\beta)}$, where $\beta > 0$.*

Then there exist a summable function $g(x)$ on $[-\pi, \pi]$ and an increasing sequence of natural numbers (1.10) for which condition α^0) of Theorem 1 is satisfied and

$$\lim_{\nu \to \infty} T_{m_\nu}^{(\beta)}(x, g) = g(x) \tag{1.19}$$

almost everywhere on $[-\pi, \pi]$.(5)

The proof of Theorem 1 is obtained from Theorem 2 and the following lemma.

Lemma A. *If α satisfies conditions (1.13), then (C, α) is a method of type $R^{(|\alpha|)}$.*(6)

§2.

Here we recall a lemma from [5] and prove an additional lemma.

Lemma B. *Suppose given for all $x \in (-\infty, +\infty)$ a periodic function $F(x)$ of period 2π having continuous derivatives up to order*

$$p \geqslant 4 \tag{2.1}$$

inclusively. If

$$0 < \beta \leqslant p - 3, \tag{2.2}$$

then the Fourier series (1.7) of $F(x)$ is uniformly summable to this function on $(-\infty, +\infty)$ by any method of type $R^{(\beta)}$, i.e. the sequence

$$T_m^{(\beta)}(x, F) \qquad (m = 0, 1, 2, \ldots) \tag{2.3}$$

uniformly converges to $F(x)$ on $(-\infty, +\infty)$.(7)

Lemma C. *Suppose given a summable function $F(x)$ on $[-\pi, \pi]$. Then for any $\delta > 0$ it is possible to define a perfect set Q and a trigonometric polynomial*

(5) The method of proof of Theorem 2 in the present paper coincides to a considerable extent with the method of proof of Theorem 2 in [5].

(6) [5], §2, Lemma 1.

(7) [5], §2, Lemma 2.

$\Pi(x)$ *that satisfy the conditions*

$$1^\circ.\quad \operatorname{meas}\Omega > 2\pi - \delta \qquad (\Omega \subset [-\pi, \pi]); \tag{2.4}$$

$$2^\circ.\quad |F(x) - \Pi(x)| < \delta \qquad (x \in \Omega); \tag{2.5}$$

$$3^\circ.\quad \int_{-\pi}^{\pi} |F(x) - \Pi(x)|\, dx < \delta. \tag{2.6}$$

Proof. As is well known, the set of trigonometric polynomials is everywhere dense in the space $L(-\pi, \pi]$ (see, for example, [2], Chapter 1, §28). Therefore, since, by assumption, $F(x) \in L[-\pi, \pi]$, there exists a sequence of trigonometric polynomials

$$\Pi_n(x) \qquad (n = 1, 2, \ldots), \tag{2.7}$$

such that

$$\lim_{n\to\infty} \int_{-\pi}^{\pi} |F(x) - \Pi_n(x)|\, dx = 0, \tag{2.8}$$

i.e. the sequence of trigonometric polynomials (2.7) converges to $F(x)$ in the norm of $L[-\pi, \pi]$.

It follows that the sequence (2.7) converges in measure to $F(x)$ on $[-\pi, \pi]$ (see, for example, [6], Chapter VII, §6). There therefore exists an increasing sequence of natural numbers n_j, $j = 0, 1, 2, \ldots$, such that

$$\lim_{j\to\infty} \Pi_{n_j}(x) = F(x)$$

almost everywhere on $[-\pi, \pi]$, and hence by virtue of Egorov's theorem it is possible to define a perfect set Ω satisfying condition 1°, on which the sequence $\Pi_{n_j}(x)$, $j = 1, 2, \ldots$, converges uniformly to $F(x)$. In this connection, on the basis of (2.8),

$$\lim_{j\to\infty} \int_{-\pi}^{\pi} |F(x) - \Pi_{n_j}(x)|\, dx = 0.$$

It follows that there exists a natural number j_0 satisfying the conditions

$$|F(x) - \Pi_{n_0}(x)| < \delta \qquad (x \in \Omega), \qquad \int_{-\pi}^{\pi} |F(x) - \Pi_{n_0}(x)|\, dx < \delta,$$

where $n_0 = n_{j_0}$. Setting $\Pi_{n_0}(x) = \Pi(x)$ in these equalities, we obtain conditions 2° and 3° of Lemma C.

Inasmuch as the perfect set Ω satisfies condition 1°, Lemma C is proved.

§3. Proof of Theorem 2

Take the Fourier series (1.7) of an arbitrary function $f(x) \in L[-\pi, \pi]$ and an arbitrary summation method T of type $R^{(\beta)}$ determined by a matrix $\|C_{mk}\|$, where $\beta > 0$. Suppose p is a natural number satisfying the inequality

$$p > 3 + \beta, \tag{3.1}$$

which implies

$$p > 3. \tag{3.2}$$

Consider the T-mean $T_m^{(\beta)}(x, f)$ of (1.7). We have

$$T_m^{(\beta)}(x, f) = \sum_{k=0}^{m} C_{mk} S_k(x, f) \quad (m = 0, 1, 2, \ldots), \tag{3.3}$$

where the $S_k(x, f)$ are the partial sums of (1.7).

We will now recursively define an increasing sequence of natural numbers (1.10), perfect sets

$$\Omega_\nu \quad (\nu = 0, 1, 2, \ldots), \tag{3.4}$$

trigonometric polynomials

$$\Pi_\nu(x) \quad (\nu = 0, 1, 2, \ldots) \tag{3.5}$$

and periodic functions

$$\varphi_\nu(x) \quad (\nu = 0, 1, 2, \ldots), \tag{3.6}$$

of period 2π having continuous derivatives up to order p inclusively for all $x \in (-\infty, +\infty)$.

We put

$$m_0 = 1, \quad \Omega_0 = [-\pi, \pi], \tag{3.7}$$

$$\Pi_0(x) = 0, \quad \varphi_0(x) = 0 \tag{3.8}$$

for all $x \in (-\infty, +\infty)$ and assume that natural numbers

$$m_j \quad (0 \leqslant j \leqslant \nu - 1), \tag{3.9}$$

perfect sets

$$\Omega_j \subset [-\pi, \pi] \quad (0 \leqslant j \leqslant \nu - 1), \tag{3.10}$$

trigonometric polynomials

$$\Pi_j(x) \quad (0 \leqslant j \leqslant \nu - 1) \tag{3.11}$$

and functions

$$\varphi_j(x) \quad (0 \leqslant j \leqslant \nu - 1), \tag{3.12}$$

having the abovementioned properties have been defined for some natural number ν.

Let

$$F_\nu(x) = \sum_{j=0}^{\nu-1} \varphi_j(x) \tag{3.13}$$

and consider the T-means

$$T_m^{(\beta)}(x, F_\nu) = \sum_{k=0}^{m} C_{mk} S_k(x, F_\nu) \quad (\nu = 0, 1, 2, \ldots), \tag{3.14}$$

of the Fourier series of $F_\nu(x)$. Clearly, $F_\nu(x)$ has the properties assumed above of the functions (3.12), i.e. it is periodic of period 2π and has continuous derivatives up to order p inclusively for all $x \in (-\infty, +\infty)$.

Consequently, since the summation method T considered by us is of type $R^{(\beta)}$, where $\beta > 0$, it follows on the basis of Lemma B that the sequence of T-means (3.14) converges uniformly on $(-\infty, +\infty)$ to $F_\nu(x)$. Hence there exists a natural number m_ν satisfying the conditions

$$m_\nu > m_{\nu-1}, \quad |T_{m_\nu}^{(\beta)}(x, F_\nu) - F_\nu(x)| < \frac{1}{4^\nu} \tag{3.15}$$

for all $x \in (-\infty, +\infty)$.

Let

$$K_m^{(\beta)}(x) = \sum_{k=0}^{m} C_{mk} D_k(x) \qquad (m = 0, 1, 2, \ldots), \tag{3.16}$$

where the $D_k(x)$ are the Dirichlet kernels. Since the $D_k(x)$ are periodic functions of period 2π, so are the $K_m^{(\beta)}(x)$.

Further, let

$$L_\nu = \max_{0 \leqslant j \leqslant \nu} \max_{x \in (-\infty, +\infty)} |K_{m_j}^{(\beta)}(x)|, \tag{3.17}$$

which implies

$$|K_{m_\nu}^{(\beta)}(x)| \leqslant L_\nu \; (x \in (-\infty, +\infty)). \tag{3.18}$$

Since $f(x) \in L[-\pi, \pi]$, it is possible on the basis of Lemma C to define a perfect set Ω_ν and a trigonometric polynomial $\Pi_\nu(x)$ that satisfy the conditions

$$\operatorname{meas} \Omega_\nu > 2\pi - \delta_\nu, \qquad \Omega_\nu \subset [-\pi, \pi], \tag{3.19}$$

$$|f(x) - \Pi_\nu(x)| < \delta_\nu \qquad (x \in \Omega_\nu), \tag{3.20}$$

$$\int_{-\pi}^{\pi} |f(x) - \Pi_\nu(x)| \, dx < \delta_\nu, \tag{3.21}$$

where

$$\delta_\nu = \frac{1}{(L_\nu + 1) \cdot 4^\nu}, \tag{3.22}$$

which implies, since $L_\nu \geq 0$ (see (3.17)),

$$\sum_{\nu=1}^{\infty} \delta_\nu < +\infty. \tag{3.23}$$

We have already seen that the functions (3.16) are trigonometric polynomials for $m = 0, 1, 2, \ldots$. It follows that for any $m = 1, 2, \ldots$ there exist points d_{ms} $(0 \leq s \leq j_m)$ satisfying the condition

$$d_{m,0} = -\pi < d_{m,1} < \ldots < d_{m,s} < \ldots < d_{m,j_m} = \pi \tag{3.24}$$

and such that $K_m^{(\beta)}(x)$ is monotonic on each of the segments $[d_{m,s-1}, d_{m,s}]$ $(1 \le s \le j_m)$.

We now introduce the notation

$$\Phi_\nu(x) = \Pi_\nu(x) - \Pi_{\nu-1}(x). \tag{3.25}$$

Since $\Pi_\nu(x)$ and $\Pi_{\nu-1}(x)$ are trigonometric polynomials, so is $\Phi_\nu(x)$, which implies the existence of a finite number

$$\mu_\nu = \max_{x\in(-\infty,+\infty)} |\Phi_\nu(x)|. \tag{3.26}$$

Now let

$$i_\nu = \max_{1\leqslant m\leqslant m_\nu} j_m, \tag{3.27}$$

$$x_v^{(\nu)} = -\pi + \frac{2\pi v}{v_\nu} \quad (0 \leqslant v \leqslant v_\nu), \tag{3.28}$$

where v_ν is a natural number satisfying the inequality

$$v_\nu > 2\pi i_\nu (L_\nu + 1)(\mu_\nu + 1) 4^\nu. \tag{3.29}$$

From (3.28) and (3.29) we get

$$x_0^{(\nu)} = -\pi < x_1^{(\nu)} < \ldots < x_v^{(\nu)} < \ldots < x_{v_\nu}^{(\nu)} = \pi, \tag{3.30}$$

where $v = 2, 3, \ldots, v_\nu - 1$, and

$$x_v^{(\nu)} - x_{v-1}^{(\nu)} = \frac{2\pi}{v_\nu} < \frac{1}{i_\nu (L_\nu + 1)(\mu_\nu + 1)\cdot 4^\nu} \quad (1 \leqslant v \leqslant v_\nu). \tag{3.31}$$

In addition, on the basis of (3.26) and (3.31) we have

$$\int_{x_{v-1}^{(\nu)}}^{x_v^{(\nu)}} |\Phi_\nu(x)|\, dx \leqslant \mu_\nu \cdot (x_v^{(\nu)} - x_{v-1}^{(\nu)}) < \frac{1}{i_\nu (L_\nu + 1)\cdot 4^\nu} \quad (1 \leqslant v \leqslant v_\nu). \tag{3.32}$$

Since the set P is nondense on $[-\pi, \pi]$, by taking (3.31) into consideration we can define a number $\theta^{(\nu)}$ satisfying the conditions

$$y_v^{(\nu)} \notin P, \quad 0 < \theta^{(\nu)} < \frac{2\pi}{v_\nu} \quad (1 \leqslant v \leqslant v_\nu), \tag{3.33}$$

where

$$y_v^{(\nu)} = x_v^{(\nu)} - \theta^{(\nu)} \quad (1 \leqslant v \leqslant v_\nu). \tag{3.34}$$

In this connection, by virtue of the second of conditions (3.33),

$$x_{v-1}^{(\nu)} < y_v^{(\nu)} \quad (1 \leqslant v \leqslant v_\nu). \tag{3.35}$$

Since P is a perfect set, it follows from the first of conditions (3.33) that all of the points $y_v^{(\nu)}$ $(1 \le v \le v_\nu)$ lie at a positive distance from this set, and hence, by virtue of (3.35), we can define a number $\theta_0^{(\nu)}$ satisfying the conditions

$$\theta_0^{(\nu)} > 0, \quad x_{v-1}^{(\nu)} < z_v^{(\nu)}, \quad [z_v^{(\nu)}, y_v^{(\nu)}] \cap P = \varnothing \quad (1 \leqslant v \leqslant v_\nu), \tag{3.36}$$

where

$$z_v^{(\nu)} = y_v^{(\nu)} - \theta_0^{(\nu)} \quad (1 \leqslant v \leqslant v_\nu). \tag{3.37}$$

Taking (3.33), (3.34), (3.36) and (3.37) into consideration, we obtain

$$x_{v-1}^{(\nu)} < z_v^{(\nu)} < y_v^{(\nu)} < x_v^{(\nu)} \quad (1 \leqslant v \leqslant v_\nu). \tag{3.38}$$

We introduce the notation

$$J_{\nu v} = \int_{x_{v-1}^{(\nu)}}^{x_v^{(\nu)}} \Phi_\nu(x)\,dx \quad (1 \leqslant v \leqslant v_\nu), \tag{3.39}$$

where $\Phi_\nu(x)$ is given by (3.25). Comparing (3.39) with (3.32), we get

$$|J_{\nu v}| \leqslant \mu_\nu (x_v^{(\nu)} - x_{v-1}^{(\nu)}) < \frac{1}{i_\nu (L_\nu + 1)\cdot 4^\nu} \quad (1 \leqslant v \leqslant v_\nu). \tag{3.40}$$

Further, let

$$\rho_{\nu v}(x) = (x - z_v^{(\nu)})^{p+1} (y_v^{(\nu)} - x)^{p+1} \quad (1 \leqslant v \leqslant v_\nu), \tag{3.41}$$

$$\gamma_{\nu, v} = \int_{z_v^{(\nu)}}^{y_v^{(\nu)}} \rho_{\nu v}(x)\,dx \quad (1 \leqslant v \leqslant v_\nu), \tag{3.42}$$

where p is a natural number satisfying (3.1). It follows by virtue of (3.38) that

$$\rho_{\nu v}(x) > 0 \quad (z_v^{(\nu)} < x < y_v^{(\nu)},\ 1 \leqslant v \leqslant v_\nu), \tag{3.43}$$

$$\gamma_{\nu v} > 0 \quad (1 \leqslant v \leqslant v_\nu). \tag{3.44}$$

Taking into consideration the latter inequality, we can also introduce the notation

$$H_{\nu v} = \frac{1}{\gamma_{\nu v}} J_{\nu v} \quad (1 \leqslant v \leqslant v_\nu). \tag{3.45}$$

We define a function $\psi_\nu(x)$ from the conditions

$$\psi_\nu(x) = H_{\nu v}\rho_{\nu v}(x) \quad (z_v^{(\nu)} \leqslant x \leqslant y_v^{(\nu)},\quad 1 \leqslant v \leqslant v_\nu), \tag{3.46}$$

$$\psi_\nu(x) = 0 \quad \left(x \in [-\pi, \pi] - \sum_{v=1}^{v_\nu} [z_v^{(\nu)}, y_v^{(\nu)}]\right). \tag{3.47}$$

It follows on the basis of (3.30) and (3.38) that $\psi_\nu(x)$ is uniquely defined for all $x \in [-\pi, \pi]$, while the definition of the functions $\rho_{\nu v}(x)$ (see (3.41)) and (3.30), (3.38), (3.47) imply

$$\psi_\nu(y_v^{(\nu)}) = 0, \quad \psi_\nu(z_v^{(\nu)}) = 0 \quad (1 \leqslant v \leqslant v_\nu), \tag{3.48}$$

$$\psi_\nu(-\pi) = \psi_\nu(\pi) = 0. \tag{3.49}$$

Using the latter equality, we can define $\psi_\nu(x)$ for all $x \in (-\infty, +\infty)$ by extending it onto all of the x axis as a periodic function of period 2π.

We will prove $\psi_\nu(x)$ is continuous and has continuous derivatives up to order p inclusively for all x, where p is a natural number satisfying (3.1).

Take any integer r satisfying the inequalities

$$1 \leqslant r \leqslant p. \tag{3.50}$$

Using Leibniz' theorem, we get from (3.41) that

$$\rho_{\nu v}^{(r)}(x) = \sum_{s=0}^{r} \alpha_{rs} (x - z_v^{(\nu)})^{p+1-s} (y_v^{(\nu)} - x)^{p+1-r+s} \quad (1 \leqslant v \leqslant v_\nu), \tag{3.51}$$

where the α_{rs} are numerical constants, and since (3.50) implies

$$p+1-s \geqslant p+1-r \geqslant 1, \quad p+1-r+s \geqslant 1, \tag{3.52}$$

if $0 \le s \le r$, we find by virtue of (3.51) that

$$\rho_{\nu v}^{(r)}(y_\nu^{(v)}) = 0, \quad \rho_{\nu v}^{(r)}(z_\nu^{(v)}) = 0 \quad (1 \leqslant v \leqslant v_\nu, \quad 1 \leqslant r \leqslant p). \tag{3.53}$$

From (3.30) and (3.38) it follows that the segments

$$[z_v^{(\nu)}, y_v^{(\nu)}] \quad (1 \leqslant v \leqslant v_\nu) \tag{3.54}$$

are mutually disjoint and lie on the interval $(-\pi, \pi)$. Consequently, comparing (3.46), (3.47) with (3.48), (3.53) and taking into consideration the fact that $\psi_\nu(x)$ is a periodic function of period 2π, we see that $\psi_\nu(x)$ is continuous and has continuous derivatives up to order p inclusively for all real x. On the basis of (3.39), (3.42) and (3.45) we get

$$\int_{x_{v-1}^{(\nu)}}^{x_v^{(\nu)}} \Phi_\nu(x)\, dx = H_{\nu v} \gamma_{\nu v} = \int_{z_v^{(\nu)}}^{y_v^{(\nu)}} H_{\nu v} \rho_{\nu v}(x)\, dx \quad (1 \leqslant v \leqslant v_\nu), \tag{3.55}$$

which implies by virtue of (3.30), (3.38), (3.46) and (3.47) that

$$\int_{x_{v-1}^{(\nu)}}^{x_v^{(\nu)}} \Phi_\nu(x)\, dx = \int_{z_v^{(\nu)}}^{y_v^{(\nu)}} \psi_\nu(x)\, dx = \int_{x_{v-1}^{(\nu)}}^{x_v^{(\nu)}} \psi_\nu(x)\, dx \quad (1 \leqslant v \leqslant v_\nu) \tag{3.56}$$

and hence that

$$\int_{x_{v-1}^{(\nu)}}^{x_v^{(\nu)}} \{\Phi_\nu(x) - \psi_\nu(x)\}\, dx = 0 \quad (1 \leqslant v \leqslant v_\nu). \tag{3.57}$$

In addition, from (3.30), (3.38), (3.43), (3.46), (3.47) and (3.56) we see that

$$\left| \int_{x_{v-1}^{(\nu)}}^{x_v^{(\nu)}} \Phi_\nu(x)\, dx \right| = \int_{x_{v-1}^{(\nu)}}^{x_v^{(\nu)}} |\psi_\nu(x)|\, dx \quad (1 \leqslant v \leqslant v_\nu). \tag{3.58}$$

We put

$$\varphi_\nu(x) = \Phi_\nu(x) - \psi_\nu(x), \tag{3.59}$$

which implies on the basis of (3.57) that

$$\int_{x_{v-1}^{(\nu)}}^{x_v^{(\nu)}} \varphi_\nu(x)\, dx = 0 \quad (1 \leqslant v \leqslant v_\nu) \tag{3.60}$$

and, also, by virtue of (3.32), (3.58) and (3.59), that

$$\int_{x_{v-1}^{(\nu)}}^{x_v^{(\nu)}} |\varphi_\nu(x)|\, dx \leqslant \frac{2}{i_\nu (L_\nu + 1)\cdot 4^\nu} \qquad (1 \leqslant v \leqslant v_\nu). \tag{3.61}$$

Since $\Phi_\nu(x)$ is a trigonometric polynomial (see the remark following (3.25)) and since $\psi_\nu(x)$ is a periodic function of period 2π having continuous derivatives up to order p inclusively for all $x \in (-\infty, +\infty)$ (see the remark following (3.54)), it follows by virtue of (3.59) that $\phi_\nu(x)$ has the same properties as $\psi_\nu(x)$.

Thus we first defined m_0, Ω_0, $\Pi_0(x)$ and $\phi_0(x)$ from (3.7) and (3.8). We then assumed that we had already defined natural numbers (3.9), perfect sets Ω_j $(1 \le j \le \nu - 1)$, trigonometric polynomials (3.11) and p-times continuously differentiable periodic functions (3.12) of period 2π for all $x \in (-\infty, +\infty)$ (see the remark following (3.12)) for some natural number ν. After this, we defined a natural number m_ν satisfying the first inequality of (3.15), a perfect set Ω_ν, a trigonometric polynomial $\Pi_\nu(x)$ and a function $\phi_\nu(x)$ having the properties formulated after (3.61).

We can therefore recursively define an increasing sequence of natural numbers (1.10), a sequence of sets (3.4), a sequence of trigonometric polynomials (3.5) and a sequence of functions (3.6) satisfying the conditions formulated above (see the conditions following (3.6)). In the process, we define for all $\nu = 1, 2, \ldots$ and $m = 0, 1, 2, \ldots$ functions $F_\nu(x)$, $K_m^{(\beta)}(x)$, $\Phi_\nu(x)$, $\psi_\nu(x)$, natural numbers j_m, i_ν, v_ν and real numbers L_ν, δ_ν, $\theta^{(\nu)}$, $\theta_0^{(\nu)}$, $d_{m,s} (0 \le s \le j_m)$, as well as real numbers $x_v^{(\nu)}$, $y_v^{(\nu)}$, $z_v^{(\nu)}$, $J_{\nu v}$, $\gamma_{\nu v}$, $H_{\nu v}$ and functions $\rho_{\nu v}(x)$ $(1 \le v \le v_\nu)$, which together with m_ν, Ω_ν, $\Pi_\nu(x)$ and $\phi_\nu(x)$ satisfy the conditions from (3.13) to (3.61) inclusively.

In this connection, on the basis of (1.9), (3.36) and (3.47) we have

$$\psi_\nu(x) = 0 \qquad (x \in P,\ \nu = 1, 2, \ldots) \tag{3.62}$$

and hence, by virtue of (3.59),

$$\varphi_\nu(x) = \Phi_\nu(x) \qquad (x \in P,\ \nu = 1, 2, \ldots). \tag{3.63}$$

From (3.21), which is valid for any $\nu = 1, 2, \ldots$, we get

$$\int_{-\pi}^{\pi} |f(x) - \Pi_{\nu-1}(x)|\, dx < \delta_{\nu-1} \qquad (\nu = 2, 3, \ldots), \tag{3.64}$$

which implies by virtue of (3.25) that

$$\int_{-\pi}^{\pi} |\Phi_\nu(x)|\, dx < \delta_{\nu-1} + \delta_\nu \quad (\nu = 2, 3, \ldots). \tag{3.65}$$

Further, on the basis of (3.30) and (3.58) we have

$$\int_{-\pi}^{\pi} |\psi_\nu(x)|\,dx = \sum_{v=1}^{v_\nu} \int_{x_{v-1}^{(\nu)}}^{x_v^{(\nu)}} |\psi_\nu(x)|\,dx$$

$$= \sum_{v=1}^{v_\nu} \left| \int_{x_{v-1}^{(\nu)}}^{x_v^{(\nu)}} \Phi_\nu(x)\,dx \right| \leqslant \int_{-\pi}^{\pi} |\Phi_\nu(x)|\,dx \qquad (\nu = 1, 2, \ldots).$$

Consequently, from (3.65) we get

$$\int_{-\pi}^{\pi} |\psi_\nu(x)|\,dx < \delta_{\nu-1} + \delta_\nu \qquad (\nu = 2, 3, \ldots)$$

and hence, by virtue of (3.59) and (3.65),

$$\int_{-\pi}^{\pi} |\varphi_\nu(x)|\,dx < 2(\delta_{\nu-1} + \delta_\nu) \qquad (\nu = 2, 3, \ldots). \tag{3.66}$$

Comparing the latter inequality with (3.23), we see that

$$\sum_{\nu=1}^{\infty} |\varphi_\nu(x)| = g_0(x) < +\infty \tag{3.67}$$

almost everywhere on $[-\pi, \pi]$, where $g_0(x) \in L[-\pi, \pi]$.(8) Hence

$$\sum_{\nu=1}^{\infty} \varphi_\nu(x) = g(x) \tag{3.68}$$

almost everywhere on $[-\pi, \pi]$, with $g(x) \in L[-\pi, \pi]$. Since the set P satisfies (1.9) (see Theorems 1 and 2), it follows on the basis of (3.63) that

$$\sum_{\nu=1}^{\infty} \Phi_\nu(x) = g(x) \tag{3.69}$$

almost everywhere on P.

Further, by virtue of (3.8) and (3.25), which is valid for any $\nu = 1, 2, \ldots$, we get

$$\Pi_\nu(x) = \Pi_0(x) + \sum_{j=1}^{\nu} [\Pi_j(x) - \Pi_{j-1}(x)] = \sum_{j=1}^{\nu} \Phi_j(x) \qquad (\nu = 1, 2, \ldots), \tag{3.70}$$

and, consequently, we see from (3.69) that

$$\lim_{\nu\to\infty} \Pi_\nu(x) = g(x) \tag{3.71}$$

almost everywhere on P.

Let

$$\Omega = \varliminf_{\nu\to\infty} \Omega_\nu, \tag{3.72}$$

On the basis of (3.19) and (3.23) we have

(8) See, for example, [2], Introduction, §14, Theorem 5.

$$\operatorname{meas}\Omega = 2\pi, \qquad \Omega \subset [-\pi, \pi]. \tag{3.73}$$

Further, for an arbitrary point

$$x \in \Omega \tag{3.74}$$

there exists a natural number ν_x such that

$$x \in \Omega_\nu \qquad (\nu > \nu_x). \tag{3.75}$$

It consequently follows from (3.20) that

$$|f(x) - \Pi_\nu(x)| < \delta_\nu \qquad (x \in \Omega,\ \nu > \nu_x), \tag{3.76}$$

and hence, on the basis of (3.23) and (3.73), that

$$\lim_{\nu \to \infty} \Pi_\nu(x) = f(x) \tag{3.77}$$

almost everywhere on $[-\pi, \pi]$.

Since P is a subset of $[-\pi, \pi]$, (3.77) holds almost everywhere on P, which implies by virtue of (3.71) that

$$g(x) = f(x) \tag{3.78}$$

almost everywhere on P. Altering $g(x)$ on a set of measure zero, if necessary, we can assume that (3.78) is satisfied everywhere on P, i.e. condition α^0) of Theorem 1 is satisfied, with $g(x) \in L[-\pi, \pi]$ (see the remark following (3.68)). Therefore the proof of Theorem 2 will be complete if we prove that $g(x)$ and the increasing sequence of natural numbers (1.10) defined by us satisfy condition (1.19).

Since the partial sums of the Fourier series of a function $G(x) \in L[-\pi, \pi]$ are denoted by $S_k(x, G)$ $(k = 0, 1, 2, \ldots)$, we have

$$S_k(x, G) = \frac{1}{\pi} \int_{-\pi}^{\pi} G(t) D_k(t - x)\, dt \qquad (k = 0, 1, 2, \ldots), \tag{3.79}$$

where $D_k(x)$ is Dirichlet's kernel; and since the summation method T of type $R^{(\beta)}$ considered by us is determined by the matrix $\|C_{mk}\|$, we have from (1.18) and (3.16) that

$$T_m^{(\beta)}(x, G) = \frac{1}{\pi} \int_{-\pi}^{\pi} G(t) K_m^{(\beta)}(t - x)\, dt \qquad (m = 0, 1, 2, \ldots) \tag{3.80}$$

for any function $G(x) \in L[-\pi, \pi]$.

We put

$$B_\nu(x) = \sum_{u=\nu}^{\infty} \varphi_u(x) \qquad (\nu = 1, 2. \ldots), \tag{3.81}$$

noting that the series in the right side of this equality converges almost everywhere on $[-\pi, \pi]$. From (3.68) and (3.13) it follows that

$$g(x) = F_\nu(x) + B_\nu(x) \qquad (\nu = 1, 2, \ldots) \tag{3.82}$$

almost everywhere on $[-\pi, \pi]$, and hence, since (1.18) is valid for any function $G(x) \in L[-\pi, \pi]$, that

$$T_m^{(\beta)}(x, g) = T_m^{(\beta)}(x, F_\nu) + T_m^{(\beta)}(x, B_\nu) \quad (\nu = 1, 2, \ldots; \ m = 0, 1, 2, \ldots). \tag{3.83}$$

On the basis of (3.13) and (3.68) we have

$$\lim_{\nu\to\infty} F_\nu(x) = g(x)$$

almost everywhere on $[-\pi, \pi]$, which implies by virtue of the second inequality of (3.15) that

$$\lim_{\nu\to\infty} T_{m_\nu}^{(\beta)}(x, F_\nu) = g(x) \tag{3.84}$$

almost everywhere on $[-\pi, \pi]$. In addition, it follows from (3.80) that

$$T_m^{(\beta)}(x, \varphi_u) = \frac{1}{\pi} \int_{-\pi}^{\pi} \varphi_u(t) K_m^{(\beta)}(t - x)\, dt \quad (u = 1, 2, \ldots, m = 0, 1, 2, \ldots) \tag{3.85}$$

Let

$$b_{w,s}^{(\nu)} = d_{m_\nu, s} + 2\pi w, \quad j'_\nu = i_{m_\nu} \quad (0 \leqslant s \leqslant j'_\nu; \ \nu = 1, 2, \ldots; \ w = 0, 1, 2, \ldots, -1, -2, \ldots), \tag{3.86}$$

where the $d_{m,s} (0 \le s \le j_m)$ satisfy inequalities (3.24) and the condition following these inequalities. Then for any $\nu = 1, 2, \ldots$ the function $K_{m_\nu}^{(\beta)}(x)$ is monotonic on each of the segments

$$[d_{m_\nu, s-1}, d_{m_\nu, s}] \quad (1 \leqslant s \leqslant j'_\nu),$$

and, since it is periodic of period 2π, on each of the segments

$$[b_{w,s-1}^{(\nu)}, b_{w,s}^{(\nu)}] \quad (1 \leqslant s \leqslant j'_\nu, \ w = 0, 1, 2, \ldots, -1, -2, \ldots). \tag{3.87}$$

It follows for the same ν and for any x that $K_{m_\nu}^{(\beta)}(t - x)$ is monotonic as a function of t on each of the segments

$$[b_{w,s-1}^{(\nu)} + x, b_{w,s}^{(\nu)} + x] \quad (1 \leqslant s \leqslant j'_\nu; \ w = 0, 1, 2, \ldots, -1, -2, \ldots). \tag{3.88}$$

On the basis of (3.24) and (3.86) we have

$$b_{w,0}^{(\nu)} = \pi(2w - 1) < b_{w,1}^{(\nu)} < \ldots < b_{w,s}^{(\nu)} < \ldots < b_{w,j'_\nu}^{(\nu)} = \pi(2w + 1) \quad (1 < s \leqslant j'_\nu; \ \nu = 1, 2, \ldots; \ w = 0, 1, 2, \ldots, -1, -2, \ldots), \tag{3.89}$$

$$b_{w+1,s}^{(\nu)} - b_{w,s}^{(\nu)} = 2\pi, \quad b_{w,j'_\nu}^{(\nu)} = b_{w+1,0}^{(\nu)} = \pi(2w + 1) \quad (0 \leqslant s \leqslant j'_\nu; \ \nu = 1, 2, \ldots; \ w = 0, 1, 2, \ldots, -1, -2, \ldots), \tag{3.90}$$

which implies that for each $\nu = 1, 2, \ldots$ and for any $x \in (-\infty, +\infty)$ there exists an integer $w_\nu = w_\nu(x)$ satisfying

$$b_{w_\nu,0}^{(\nu)} + x \leqslant -\pi < b_{w_\nu, j'_\nu}^{(\nu)} + x = b_{w_\nu+1,0}^{(\nu)} + x \leqslant \pi < b_{w_\nu+1, j'_\nu}^{(\nu)} + x. \tag{3.91}$$

Taking (3.89)–(3.91) into consideration, we see that for each $\nu = 1, 2, \ldots$ and for any $x \in (-\infty, +\infty)$ the segments

$$[b^{(\nu)}_{w_\nu, s-1} + x,\ b^{(\nu)}_{w_\nu, s} + x], \quad [b^{(\nu)}_{w_\nu+1, s-1} + x,\ b^{(\nu)}_{w_\nu+1, s} + x] \quad (1 \leqslant s \leqslant j'_\nu) \tag{3.92}$$

have mutually disjoint interiors and cover the segment $[-\pi, \pi]$, with

$$\begin{gathered}[-\pi, \pi] \subset [b^{(\nu)}_{w_\nu, 0} + x,\ b^{(\nu)}_{w_\nu+1, j'_\nu} + x] = \bigcup_{s=1}^{j'_\nu} [b^{(\nu)}_{w_\nu, s-1} + x,\ b^{(\nu)}_{w_\nu, s}] \\ + \bigcup_{s=1}^{j'_\nu} [b^{(\nu)}_{w_\nu+1, s-1} + x,\ b^{(\nu)}_{w_\nu+1, s} + x] \quad (\nu = 1, 2, \ldots; -\infty < x < +\infty).\end{gathered} \tag{3.93}$$

Let

$$\Delta^{(\nu)}_j \quad (1 \leqslant j \leqslant j''_\nu) \tag{3.94}$$

denote the intersections of $[-\pi, \pi]$ and each segment (3.92) whose interior intersects $[-\pi, \pi]$. Then

$$[-\pi, \pi] = \bigcup_{j=1}^{j''_\nu} \Delta^{(\nu)}_j \quad (\nu = 1, 2, \ldots; -\infty < x < +\infty), \tag{3.95}$$

with the segments (3.94) having mutually disjoint interiors since the segments (3.92) have this property. Also, the number of segments (3.94) does not exceed the number of segments (3.92), which implies that

$$j''_\nu \leqslant 2j'_\nu \quad (\nu = 1, 2, \ldots; -\infty < x < +\infty). \tag{3.96}$$

We assume that the segments (3.94) have been renumbered from left to right. Then for each $\nu = 1, 2, \ldots$ we can determine points $\tau^{(\nu)}_j$, $0 \le j \le j''_\nu$, such that

$$\Delta^{(\nu)}_j = [\tau^{(\nu)}_{j-1}, \tau^{(\nu)}_j] \quad (1 \leqslant j \leqslant j''_\nu; -\infty < x < +\infty), \tag{3.97}$$

$$-\pi = \tau^{(\nu)}_0 < \ldots < \tau^{(\nu)}_j < \ldots < \tau^{(\nu)}_{j''_\nu} = \pi. \tag{3.98}$$

Comparing (3.85) with (3.98), we have

$$\begin{gathered}T^{(\beta)}_{m_\nu}(x, \varphi_u) = \frac{1}{\pi} \sum_{j=1}^{j''_\nu} \int_{\tau^{(\nu)}_{j-1}}^{\tau^{(\nu)}_j} \varphi_u(t) K^{(\beta)}_{m_\nu}(t - x)\, dt \\ (\nu = 1, 2, \ldots; u = 1, 2, \ldots; x \in (-\infty, +\infty)).\end{gathered} \tag{3.99}$$

Further, by virtue of their definition, each of the segments (3.94) is contained in one of the segments (3.92), and since $K^{(\beta)}_{m_\nu}(t - x)$ is monotonic as a function of t on each of the segments (3.88) for any $\nu = 1, 2, \ldots$, this function has the same property on each of the segments (3.94) for any fixed $x \in (-\infty, +\infty)$.

Consequently, on the basis of (3.97) and the second mean value theorem for integrals we have

$$\int_{\tau_{j-1}^{(\nu)}}^{\tau_j^{(\nu)}} \varphi_u(t)\,K_{m_\nu}^{(\beta)}(t-x)\,dt = K_{m_\nu}^{(\beta)}(\tau_{j-1}^{(\nu)}-x)\int_{\tau_{j-1}^{(\nu)}}^{\xi_j} \varphi_u(t)\,dt + K_{m_\nu}^{(\beta)}(\tau_j^{(\nu)}-x)\int_{\xi_j^{(\nu)}}^{\tau_j^{(\nu)}} \varphi_u(t)\,dt \tag{3.100}$$

$$(1\leqslant j\leqslant j_\nu''; \ \nu=1,\,2,\,\ldots;\ u=1,\,2,\,\ldots;\ x\in(-\infty,\,+\infty)),$$

where

$$\tau_{j-1}^{(\nu)} < \xi_j^{(\nu)} < \tau_j^{(\nu)} \tag{3.101}$$

for the same j, ν and x.

Let us estimate the integrals in the right side of (3.100). Take an arbitrary segment

$$[c,\ d]\subset[-\pi,\ \pi] \tag{3.102}$$

and an arbitrary number $\nu = 1, 2, \ldots$. From (3.30), which is valid for any $\nu = 1, 2, \ldots$, and (3.102), we can determine for any $u = 1, 2, \ldots$ two integers v_u' and v_u'' such that

$$0\leqslant v_u' < v_u'' \leqslant v_u, \quad x_{v_u'}^{(u)}\leqslant c < x_{v_u'+1}^{(u)}, \quad x_{v_u''-1}^{(u)} < d \leqslant x_{v_u''}^{(u)} \quad (u=1,\,2,\,\ldots). \tag{3.103}$$

These inequalities imply that for any $u = 1, 2, \ldots$ either $v_u' + 1 < v_u'' - 1$, or $v_u' + 1 = v_u'' - 1$ or $v_u' + 1 = v_u''$. Therefore, comparing (3.30) and (3.102) with (3.60), we see that

$$\left|\int_c^d \varphi_u(x)\,dx\right| \leqslant \int_{x_{v_u'}}^{x_{v_u'+1}} |\varphi_u(x)|\,dx + \int_{x_{v_u''-1}}^{x_{v_u''}} |\varphi_u(x)|\,du \quad (u=1,\,2,\,\ldots) \tag{3.104}$$

for any segment $[c, d]$ satisfying (3.102), which implies by virtue of (3.61) and (3.103) that

$$\left|\int_c^d \varphi_u(x)\,dx\right| \leqslant \frac{4}{i_u(L_u+1)\cdot 4^u} \quad (u=1,\,2,\,\ldots), \tag{3.105}$$

if (3.102) is satisfied.

Taking (3.98) and (3.101) into consideration, we get

$$[\tau_{j-1}^{(\nu)},\ \xi_j^{(\nu)}]\subset[-\pi,\ \pi],\ [\xi_j^{(\nu)},\ \tau_j^{(\nu)}]\subset[-\pi,\ \pi] \tag{3.106}$$

$$(1\leqslant j\leqslant j_\nu'';\ \nu=1,\,2,\,\ldots)$$

for any $x \in (-\infty, +\infty)$, from which we determine the points $\tau_j^{(\nu)}$ and $\xi_j^{(\nu)}$ and which implies on the basis of (3.105) that

$$\left|\int_{\tau_{j-1}^{(\nu)}}^{\xi_j^{(\nu)}} \varphi_u(x)\,dx\right| \leqslant \frac{4}{i_u(L_u+1)\cdot 4^u}, \quad \left|\int_{\xi_j^{(\nu)}}^{\tau_j^{(\nu)}} \varphi(u)\,du\right| \leqslant \frac{4}{i_u(L_u+1)\cdot 4^u} \tag{3.107}$$

$$(1\leqslant j\leqslant j_\nu'';\ \nu=1,\,2,\,\ldots;\ u=1,\,2,\,\ldots).$$

From (3.13), (3.99), (3.100) and (3.107) we have

$$|T^{(\beta)}_{m_\nu}(x,\ \varphi_u)| \leqslant \frac{4j''_\nu L_\nu}{i_u(L_u+1)\cdot 4^u} \quad (\nu=1,\ 2,\ \ldots;\ u=1,\ 2,\ \ldots;\ x\in(-\infty,\ +\infty)). \tag{3.108}$$

Further, taking into consideration the first inequality of (3.15), we get

$$m_\nu \leqslant m_u \quad (u \geqslant \nu;\ \nu=1,\ 2,\ \ldots), \tag{3.109}$$

which implies by virtue of (3.27) and the second equality of (3.86) that

$$i_u = \max_{1\leqslant m\leqslant m_u} j_m \geqslant j'_\nu \quad (u\geqslant \nu;\ \nu=1,\ 2,\ \ldots) \tag{3.110}$$

and hence, on the basis of (3.96), that

$$2i_u \geqslant j''_\nu \quad (u\geqslant\nu,\ \nu=1,\ 2,\ \ldots): \tag{3.111}$$

In addition, from (3.17) we get

$$0\leqslant L_\nu \leqslant L_u \quad (u\geqslant\nu;\ \nu=1,\ 2,\ \ldots). \tag{3.112}$$

Comparing (3.108) with (3.111) and (3.112), we have

$$|T^{(\beta)}_{m_\nu}(x,\ \varphi_u)| \leqslant \frac{8}{4^u} \quad (\nu=1,\ 2,\ \ldots;\ u\geqslant\nu;\ x\in(-\infty,\ +\infty)), \tag{3.113}$$

while from (3.17) we have

$$|\varphi_u(t)\, K^{(\beta)}_{m_\nu}(t-x)| \leqslant |\varphi_u(t)|\, L_\nu \quad (\nu=1,\ 2,\ \ldots,\ u\geqslant\nu)$$

for any real t and x. It follows from (3.67) that

$$\left|\sum_{u=\nu}^{N} \varphi_u(t)\, K^{(\beta)}_{m_\nu}(t-x)\right| \leqslant L_\nu g_0(x) \quad (\nu=1,\ 2,\ \ldots,\ N>\nu) \tag{3.114}$$

for any real t and x.

Since $g_0(x)\geq 0$ almost everywhere on $[-\pi,\ \pi]$ (see (3.67)) and $g_0(x)\in L[-\pi,\ \pi]$, it follows from (3.81) and (3.114) on the basis of a well-known theorem(9) that

$$\lim_{N\to\infty}\sum_{u=\nu}^{N}\frac{1}{\pi}\int_{-\pi}^{\pi}\varphi_u(t)\, K_{m_\nu}(t-x)\, dt = \frac{1}{\pi}\int_{-\pi}^{\pi} B_\nu(t)\, K_{m_\nu}(t-x)\, dt \tag{3.115}$$

$$(\nu=1,\ 2,\ \ldots,\ x\in(-\infty,\ +\infty)),$$

where $B_\nu(x)\in L[-\pi,\ \pi]$ for $\nu=1,\ 2,\ \ldots$, which implies on the basis of (3.80) that

$$T^{(\beta)}_{m_\nu}(x,\ B_\nu) = \frac{1}{\pi}\int_{-\pi}^{\pi} B_\nu(t)\, K_{m_\nu}(t-x)\, dt \quad (\nu=1,\ 2,\ \ldots;\ x\in(-\infty,\ +\infty)). \tag{3.116}$$

In addition, since the functions $\phi_u(x)$ ($u=0,\ 1,\ 2,\ \ldots$) are continuous on $[-\pi,\ \pi]$ (see (3.8) and the remark following (3.61)), we have

(9) See, for example, [2], Introduction, §14, Theorem 2 or [6], Chapter VI, §3, Theorem 1.

$$T_{m_\nu}^{(\beta)}(x, \varphi_u) = \frac{1}{\pi} \int_{-\pi}^{\pi} \varphi_u(t) K_{m_\nu}^{(\beta)}(t-x)\, dt \tag{3.117}$$

$$(\nu = 1, 2, \ldots; \ u = 0, 1, 2, \ldots; \ x \in (-\infty, +\infty)).$$

Comparing (3.115) with (3.116) and (3.117), we get

$$\lim_{N\to\infty} \sum_{u=\nu}^{N} T_{m_\nu}^{(\beta)}(x, \varphi_u) = T_{m_\nu}^{(\beta)}(x, B_\nu) \qquad (\nu = 1, 2, \ldots, x \in (-\infty, +\infty)), \tag{3.118}$$

which implies by virtue of (3.113) that

$$|T_{m_\nu}^{(\beta)}(x, B_\nu)| \leqslant 8 \cdot \sum_{u=\nu}^{\infty} \frac{1}{4^u} = \frac{32}{3} \cdot \frac{1}{4^\nu} \qquad (\nu = 1, 2, \ldots; \ x \in (-\infty, +\infty)) \tag{3.119}$$

and hence that

$$\lim_{\nu\to\infty} T_{m_\nu}^{(\beta)}(x, B_\nu) = 0 \tag{3.120}$$

for all real x.

From (3.83), (3.84) and (3.120) we see that (1.19) is satisfied almost everywhere on $[-\pi, \pi]$. By the same token the proof of Theorem 2 is complete (see the remark following (3.78)).

BIBLIOGRAPHY

1. D. Men′šov, *On Fourier series of summable functions*, Trudy Moskov. Mat. Obšč. **1** (1952), 5–38. (Russian) MR **14**, 161.

2. N. K. Bari, *Trigonometric series*, Fizmatgiz, Moscow, 1961; English transl., Macmillan, New York; Pergamon Press, Oxford, 1964. MR **23** #A3411; **30** #1347.

3. A. Zygmund, *Trigonometric series*. Vol. 1, 2nd rev. ed., Cambridge Univ. Press, New York, 1959. MR **21** #6498.

4. D. Men′šov, *The application of Cesàro summability methods of negative order to Fourier-Lebesgue series*, Dokl. Akad. Nauk SSSR **210** (1973), 271–273 = Soviet Math. Dokl. **14** (1973), 726–729. MR **48** #788.

5. ———, *Properties of Cesàro means of negative order and of certain other T-means for Fourier series of continuous functions*, Mat. Sb. **86** (**128**) (1971), 419–445 = Math. USSR Sb. **15** (1971), 415–442. MR **46** #5916.

6. I. P. Natanson, *Theory of functions of a real variable*, 2nd rev. ed., GITTL, Moscow, 1957; English transl. of the 1st ed., Ungar., New York, 1955. MR **16**, 804.

Translated by S. SMITH

Trudy Mat. Inst. Steklov.
134 (1975)

Proc. Steklov Inst. Math.
134 (1975)

HOMOGENEOUS REPRESENTATIONS OF A LOCALLY COMPACT GROUP ACTING ON A PAIR OF DUAL SPACES(1)

UDC 513.88

M. A. NAĬMARK

Abstract. By a representation T, $\hat{T}$ of a locally compact group G acting in a pair E, F of spaces dually related to each other by a bilinear form (ξ, η) we mean a pair of maps $g \to T(g)$, $g \to \hat{T}(g)$ into the groups of continuous linear operators in E and F satisfying the following conditions:

1) $T(e) = 1_E$, $T(g_1g_2) = T(g_1)T(g_2)$, $\hat{T}(e) = 1_F$, $\hat{T}(g_1g_2) = \hat{T}(g_2)\hat{T}(g_1)$;

2) $(T(g)\xi, \eta) = (\xi, \hat{T}(g)\eta)$;

3) $\{g, \xi\} \to T(g)\xi$, $\{g, \eta\} \to \hat{T}(g)\eta$ are maps of $G \times E$ and $G \times F$ into E and F respectively (E and F are assumed to have the $\sigma(E, F)$ and $\sigma(F, E)$ topologies).

It is shown that if $x(g)$ is a continuous function with compact support, then $\int x(g)T(g)dg$ and $\int x(g)\hat{T}(g)dg$ exist in the weak sense without any assumptions concerning the completeness of E and F in any sense whatsoever. Using this result, the auther carries over to representations acting on a pair the basic results (with appropriate changes) of the theory of homogeneous representations which he has developed previously (see MR 47 #396).

Bibliography: 8 titles.

§1. Introduction

In the papers [2]–[4] the present author has introduced the concept and investigated the properties of a homogeneous representation of a locally compact group G acting in a complete locally convex space E. The completeness of E was employed mainly for the purpose of ensuring the existence in E of integrals of E-valued functions used for the investigation of homogeneous representations. Later it became apparent that in a series of applications of the theory of homogeneous representations the requirement of the completeness of E is too restrictive and inconvenient. For instance, such a situation arises when, instead of a representation in E one considers a representation of the group G on a pair(2) of spaces E, F which are dual to each other with respect to a nondegenerate bilinear form (ξ, η) (see [5], Chapter 4). (Here and below we shall adhere to the terminology of [5].) The difficulty lies in the fact that in general the extension

AMS (MOS) subject classifications (1970). Primary 22D12; Secondary 22D15.

(1) A short exposition of the basic results of this paper was published by the author in [1].

(2) For the definition of a representation on a pair E, F, see §2.

of the form (ξ, η) to the completions of E and F may cease to be nondegenerate. And even if one assumes to begin with that the spaces E and F are complete in their σ-topologies, still a quotient space, for instance E/M, where M is closed subspace of E may be incomplete (see for instance [5], Fascicule de résultats, §3.5).

The concept of a representation on a pair of spaces is not new. It appears, for instance, in [6], but it is assumed there, to begin with, that the integrals of the vector-valued functions under consideration (right weighted means, in general) exist. In the present paper we prove the existence of the integrals of the vector-valued functions which are needed in the theory of homogeneous representations without making any assumptions whatsoever about the completeness of the spaces E and F. A quite general character is thus given to the theory of homogeneous representations, with suitable alterations, and this character appears to us to be final. This paper presents the theory of homogeneous representations with those changes.

§2. Representations in a pair of dual spaces and the associated integrals

Let E, F be a pair of vector spaces dual to each other with respect to the nondegenerate bilinear form (ξ, η) (for short, the pair $P = \{E, F, (\xi, \eta)\}$). We shall regard E and F as being equipped with the topologies $\sigma(E, F)$ and $\sigma(F, E)$ respectively, and all topological concepts in E and F (for instance closedness, the continuity of a mapping into E or F, etc.) will be understood, unless specified otherwise, in the sense of the σ-topology in E and F.(3)

In this paper G will always denote a locally compact group, U will be a compact subgroup of G, $c\colon u \to c(u)$ will be an irreducible unitary representation of the group U and $c_{jl}(u)$ will be the matrix elements of $c(u)$ in some fixed orthonormal basis.

By a representation of the group G in the pair $P = \{E, F, (\xi, \eta)\}$ we shall mean a pair of maps $T\colon g \to T(g)$, $\hat{T}\colon g \to \hat{T}(g)$ of the group G into the groups of linear operators $T(g)$ in E and $\hat{T}(g)$ in F which satisfy, in addition to the usual conditions

1) $T(e) = 1_E$, $T(g_1g_2) = T(g_1)T(g_2)$; $\hat{T}(e) = 1_F$, $\hat{T}(g_1g_2) = \hat{T}(g_2)\hat{T}(g_1)$

(1_E and 1_F are the identity operators in E and F respectively), also

2) $(T(g)\xi, \eta) = (\xi, \hat{T}(g)\eta)$,

3) $(g, \xi) \to T(g)\xi$ and $\{g, \eta\} \to \hat{T}(g)\eta$, are continuous maps of the topological products $G \times E$ and $G \times F$ into E and F respectively.

Condition 2) means that(4)

---

(3) It is evident that these σ-topologies may be replaced by any other topologies conforming with the duality between E and F.

(4) In this paper A' will always denote the operator adjoint to A, defined by $(A\xi, \eta) = (\xi, A'\eta)$.

$$\hat{T}(g) = (T(g))'. \tag{2.1}$$

The mapping $g \to \hat{T}(g)$ is usually called an *antirepresentation* of the group G.

Two representations $T_1, \hat{T}_1$ and $T_2, \hat{T}_2$ in the pairs

$$P_1 = \{E_1, F_1, (\xi, \eta)_1\}, \quad P_2 = \{E_2, F_2, (\xi, \eta)_2\}$$

respectively are called *equivalent* if there exists a pair of linear maps $S_1: E_1 \to E_2$, $S_2: F_1 \to F_2$ of E_1 onto E_2 and F_1 onto F_2 respectively which satisfies the condition

$$(S_1\xi_1, S_2\eta_1)_2 = (\xi_1, \eta_1)_1$$

and which carries $T_1, \hat{T}_1$ over into $T_2, \hat{T}_2$. A pair of maps S_1, S_2 of this kind is called an isomorphism of the pairs P_1, P_2 and an equivalence map of $T_1, \hat{T}_1$ on $T_2, \hat{T}_2$. Evidently, an isomorphism S_1, S_2 is also a homeomorphism of E_1 onto E_2 and F_1 onto F_2 in the corresponding σ-topologies.

A representation T in the space E is called irreducible if there does not exist in E a closed (with respect to the $\sigma(E, F)$-topology) subspace which is invariant under T and distinct from (0) and from E. The irreducibility of an antirepresentation $\hat{T}$ in F is defined analogously. The irreducibility of T in E is equivalent to the irreducibility of $\hat{T}$ in F. Indeed, let T be irreducible, but assume that $\hat{T}$ is reducible. Let M be a subspace of F which is closed in the $\sigma(F, E)$-topology, invariant with respect to $\hat{T}$ and distinct from (0) and F. We put

$$M^{\perp} = \{\xi;\ \xi \in E,\ (\xi, \eta) = 0 \quad \text{for all} \quad \eta \in M\}.$$

Then $M^{\perp}$ is invariant with respect to T, closed in the $\sigma(E, F)$ topology and different from (0) and E, which contradicts the irreducibility of T. The converse assertion is proved similarily.

The representation $T, \hat{T}$ in the pair $P = \{E, F, (\xi, \eta)\}$ is called irreducible if T is irreducible in E, and thus $\hat{T}$ is irreducible in F.

Let X denote the algebra of all continuous functions $x = x(g)$ on G with compact supports, pointwise addition and multiplication by numbers, and with the convolution product as multiplication.

I. *If $x \in X$, then for every $\xi \in E$ there exists a vector $\xi_1 \in E$ such that*

$$\int x(g)(T(g)\xi, \eta)\, dg = (\xi_1, \eta) \quad \textit{for all} \quad \eta \in F. \tag{2.2}$$

Proof. Let Q be the, by definition compact, support of the function $x(g)$. We put(5) (see (2.1))

$$l(\eta) = \int x(g)(T(g)\xi, \eta)\, dg = \int x(g)(\xi, \hat{T}(g)\eta)\, dg \tag{2.3}$$

---

(5) Here and below, dg denotes the differential of left-invariant Haar measure on G.

and for a fixed ξ

$$\varphi(g, \eta) = (\xi, \hat{T}(g)\eta). \tag{2.4}$$

The continuity of the form (ξ, η) with respect to η in the $\sigma(F, E)$-topology and condition 3) imply that *the correspondence* $\{g, \eta\} \to \phi(g, \eta)$ *is a continuous mapping of the topological product* $G \times F$ *into* C^1. It is evident that $l(\eta)$ is a linear functional on F. Let us prove that $l(\eta)$ is continuous. Due to the linearity of $l(\eta)$, it suffices to prove its continuity at the point $\eta = 0$. We put

$$V(\varepsilon) = \{\lambda, \lambda \in C^1, |\lambda| < \varepsilon\}; \tag{2.5}$$

$V(\epsilon)$ is an open subset of C^1. Then the set

$$W = \{\eta, \eta \in F, \varphi(g, \eta) \in V(\varepsilon) \quad \text{for every} \quad g \in Q\} \tag{2.6}$$

is open (see for example [7], Chapter 1, §2, V), and it contains the point 0. Hence W is a neighborhood of 0 in F, and by the definitions (2.5) and (2.6) of the set W we have $|\phi(g, \eta)| < \epsilon$ for $\eta \in W$ and all $g \in Q$. It follows now from (2.3) and (2.4) that

$$|l(\eta)| \leqslant \int_Q |x(g)| |\varphi(g, \eta)| \, dg \leqslant \varepsilon \int_Q |x(g)| \, dg,$$

for $\eta \in W$, and consequently $l(\eta)$ is continuous in the point $\eta = 0$. But then (see [5], Chapter IV, §1, Proposition 1) $l(\eta)$ can be uniquely represented in the form

$$l(\eta) = (\xi_1, \eta), \tag{2.7}$$

which concludes (see (2.3)) the proof of Proposition I.

We shall denote by $\int x(g)T(g)\xi dg$ the vector ξ_1 defined in this fashion, so that by definition

$$\int x(g)(T(g)\xi, \eta) \, dg = \left(\int x(g) T(g) \xi dg, \eta\right). \tag{2.8}$$

Exchanging in the preceding proof the roles of E and F, we conclude that for each $\eta \in F$ there exists an $\eta_1 \in F$ such that

$$\int x(g)(\xi, \hat{T}(g)\eta) \, dg = (\xi, \eta_1). \tag{2.9}$$

We shall denote this vector η_1 by $\int x(g)\hat{T}(g)\eta dg$, so that

$$\left(\xi, \int x(g) \hat{T}(g) \eta dg\right) = \int x(g)(\xi, \hat{T}(g)\eta) \, dg. \tag{2.10}$$

Here we have, by (2.3),

$$\left(\int x(g) T(g) \xi dg, \eta\right) = \left(\xi, \int x(g) \hat{T}(g) \eta dg\right). \tag{2.11}$$

It is evident that the map $\xi \to \xi_1$ is linear and thus defines a linear operator in E; let us denote this operator by $\int x(g)T(g)dg$, or for short by $T(x)$. Then, by definition,

$$T(x)\xi=\left(\int x(g)T(g)\,dg\right)\xi=\int x(g)T(g)\xi dg. \tag{2.12}$$

In an analogous way we define the operator $\hat{T}(x)=\int x(g)\hat{T}(g)dg$ in F by

$$\hat{T}(x)\eta=\left(\int x(g)\hat{T}(g)\,dg\right)\eta=\int x(g)\hat{T}(g)\eta dg, \tag{2.13}$$

so that (2.11) becomes

$$(T(x)\xi,\ \eta)=(\xi,\ \hat{T}(x)\eta). \tag{2.14}$$

It follows from (2.14) and [5], Chapter IV, §4.1, Proposition 1, that

II. *For every function $x\in X$, the operators $T(x)$ and $\hat{T}(x)$ are continuous in the $\sigma(E, F)$- and $\sigma(F, E)$-topologies respectively and*

$$\hat{T}(x)=(T(x))'. \tag{2.15}$$

Remark 1. The definition of the integrals $\int x(g)T(g)\xi dg$ and $\int x(g)T(g)dg$ may be easily generalized to a definition of $\int x(g)AT(g)\xi dg$ and $\int x(g)AT(g)dg$, where A is a continuous linear operator in E. Indeed, the continuity of the operator A, and thus of A', implies that the correspondence $\{g, \eta\}\to(AT(g)\xi, \eta)=(\xi, \hat{T}(g)A'\eta)$ is a continuous mapping of the space $G\times F$ into C^1. Hence all preceeding considerations will apply also in the present case. Furthermore,

$$\int x(g)AT(g)\,dg=A\int x(g)T(g)\,dg. \tag{2.16}$$

Indeed, for $\xi\in E$ and $\eta\in F$ we have

$$\begin{aligned}\left(\left(\int x(g)AT(g)\,dg\right)\xi,\ \eta\right)&=\left(\int x(g)AT(g)\xi dg,\ \eta\right)=\int x(g)(AT(g)\xi,\ \eta)\,dg\\&=\int x(g)(T(g)\xi,\ A'\eta)\,dg=\left(\left(\int x(g)T(g)\,dg\right)\xi,\ A'\eta\right)\\&=\left(A\left(\int x(g)T(g)\,dg\right)\xi,\ \eta\right),\end{aligned}$$

whence (2.16).

III. *Put*

$$x_{g_0}(g)=x(g_0^{-1}g),\qquad x^{g_0}(g)=x(gg_0); \tag{2.17}$$

then

$$T(x_{g_0})=T(g_0)T(x),\qquad T(x^{g_0})=\lambda(g_0^{-1})T(x)T(g_0^{-1}), \tag{2.18}$$

where

$$\lambda(g)=\frac{dg}{d_rg}. \tag{2.19}$$

Proof. (2.12) and (2.16) imply that for $\xi\in E$ and $\eta\in F$

$$(T(x_{g_0})\xi, \eta) = \int x_{g_0}(g)(T(g)\xi, \eta)\, dg = \int x(g_0^{-1}g)(T(g)\xi, \eta)\, dg$$
$$= \int x(g)(T(g_0 g)\xi, \eta)\, dg = \int x(g)(T(g_0)T(g)\xi, \eta)\, dg$$
$$= \Big(T(g_0)\Big(\int x(g)\, T(g)\, dg\Big)\xi, \eta\Big),$$

and thus $T(x_{g_0}) = T(g_0)T(x)$. Furthermore,

$$(T(x^{g_0})\xi, \eta) = \int x^{g_0}(g)(T(g)\xi, \eta)\, dg = \int x(gg_0)(T(gg_0)\xi, \eta)\, dg$$
$$= \int x(gg_0)(T(g)\xi, \eta)\lambda(g)\, d_r(g) = \int x(g)(T(gg_0^{-1})\xi, \eta)\lambda(gg_0^{-1})\, d_r g$$
$$= \int x(g)(T(g)\,T(g_0^{-1})\xi, \eta)\lambda(g_0^{-1})\, dg = \lambda(g_0^{-1})\, T(x)\, T(g_0^{-1}).$$

It is evident that the correspondence $x \to T(x)$ is linear. Moreover,

$$T(x_1 x_2) = T(x_1)\, T(x_2). \tag{2.20}$$

Indeed, $(x_1 x_2)(g) = \int x_1(g_1) x_2(g_1^{-1} g)\, dg_1$ holds by the definition of the convolution product, whence, due to (2.18), we have for $\xi \in E$ and $\eta \in F$ that

$$T(x_1 x_2) = \int \Big[\int x_1(g_1)\, x_2(g_1^{-1}g)\, dg_1\Big](T(g)\xi, \eta)\, dg$$
$$= \int x_1(g_1)\Big[\int x_2(g_1^{-1}g)(T(g)\xi, \eta)\, dg\Big] dg_1$$
$$= \int x_1(g_1)(T(g_1)\, T(x_2)\xi, \eta)\, dg_1$$
$$= \Big(\Big(\int x_1(g_1)\, T(g_1)\, dg_1\Big) T(x_2)\xi, \eta\Big) = (T(x_1)\, T(x_2)\xi, \eta).$$

This implies (2.20). Hence

IV. *The map $x \to T(x)$ is a homomorphism of the algebra X into the algebra of all continuous operators in E.*

By exactly analogous considerations we obtain

V. *The map $x \to \hat{T}(x)$ is an antihomomorphism of the algebra X into the algebra of all continuous operators on F; in particular,*

$$\hat{T}(x_1 x_2) = \hat{T}(x_2)\, \hat{T}(x_1).$$

This assertion follows directly from IV and (2.15).

§3. Certain auxiliary operators

Given T and c, we put (see, for instance, [4])

$$T^c_{jl} = r_c \int \overline{c_{jl}(u)}\, T(u)\, du, \tag{3.1}$$
$$T^c_j = T^c_{jj} = r_c \int \overline{c_{jj}(u)}\, T(u)\, du, \tag{3.2}$$

where $r_c = \dim c$ and du is normalized so that $\int du = 1$. The existence of the integrals (3.1) and (3.2) follows from I of §2, applied to the restriction $u \to T(u)$ of the representation T to the group U and to the continuous function $\overline{c_{jl}(u)}$ (in place of $x(g)$); moreover, II in §2 implies that T^c_{jl} and T^c_j are continuous operators on E.

I. *The following relationships hold for the operators* T^c_{jl} *and* T^c_j:

$$T^c_j T^{c}_{j'} = 0, \tag{3.3}$$

if c' *and* c *are nonequivalent or if* $j' \neq j$;

$$T^c_{jl} T^c_{ls} = T^c_{js}; \tag{3.4}$$

in particular

$$T^c_j T^c_{jl} = T^c_{jl} T^c_l = T^c_{jl} \tag{3.5}$$

and

$$T(u)\, T^c_{jl} = \sum_{s=1}^{r_c} c_{sj}(u)\, T^c_{sl}. \tag{3.6}$$

Proof. (2.16) implies

$$\begin{aligned} T(u)\,T^c_{jl} &= T(u)\, r_c \int \overline{c_{jl}(u_1)}\, T(u_1)\, du_1 = r_c \int \overline{c_{jl}(u_1)}\, T(u)\, T(u_1)\, du_1 \\ &= r_c \int \overline{c_{jl}(u_1)}\, T(uu_1)\, du_1 = r_c \int \overline{c_{jl}(u^{-1}u_1)}\, T(u_1) du_1 \\ &= r_c \int \sum_{s=1}^{r_c} \overline{c_{js}(u^{-1})}\, \overline{c_{sl}(u_1)}\, T(u_1)\, du_1 \\ &= \sum_{s=1}^{r_c} c_{sj}(u)\, r_c \int \overline{c_{sl}(u_1)}\, T(u_1)\, du_1 = \sum_{s=1}^{r_c} c_{sj}(u_1)\, T^c_{sl}; \end{aligned}$$

that is, (3.6) holds. Further, applying (3.6) and the orthogonality relations for the irreducible representations of the group U, we obtain

$$\begin{aligned} T^c_{jl} T^{c'}_{j'l'} &= r_c \int \overline{c_{jl}(u)}\, T(u)\, T^{c'}_{j'l'} du = r_c \int \overline{c_{jl}(u)} \left(\sum_{s=1}^{r_{c'}} c'_{sj'}(u)\, T^{c'}_{sl'} \right) du \\ &= \left(r_c \sum_{s=1}^{r_{c'}} \int \overline{c_{jl}(u)}\, c'_{sj'}(u)\, du \right) T^{c'}_{sl'} \\ &= \begin{cases} 0, & \text{when } c' \text{ is not equivalent to } c, \text{ or } l \neq j'; \\ \left(r_c \int \overline{c_{jl}(u)}\, c_{jl}(u)\, du \right) T^c_{jl'} = T^c_{jl'} & \text{for } c' = c \text{ or } l = j'. \end{cases} \end{aligned}$$

This implies (3.3)–(3.5).

II. T^c_{jl} *and* T^c_{lj}, *when regarded as operating only on* $T^c_l E$ *and* $T^c_j E$ *respectively, are mutually inverse maps of* $T^c_l E$ *onto* $T^c_j E$ *and* $T^c_j E$ *onto* $T^c_l E$, *and*

$$T^c_j E = T^c_{jl} E. \tag{3.7}$$

Proposition I shows that the proof is identical with the proof of Proposition II in §1 of [4].

Let us define in X the left and the right operators e_j^c by the formulas (see, for instance, [8])

$$(e_j^c x)(g) = r_c \int \overline{c_{jj}(u)}\, x(u^{-1}g)\, du, \qquad (xe_j^c)(g) = r_c \int c_{jj}(u)\, x(gu)\, du. \quad (3.8)$$

It is easy to show that (see, for instance, [8])

$$e_j^c e_{j'}^{c'} = 0, \quad (3.9)$$

if c and c' are nonequivalent, or $j \neq j'$;

$$e_j^c e_j^c = e_j^c; \quad (3.10)$$

$$T(e_j^c x) = T_j^c T(x), \qquad T(xe_j^c) = T(x)\, T_j^c. \quad (3.11)$$

Let us next define the operators $\hat{T}_j^c$ by

$$\hat{T}_j^c = r_c \int \overline{c_{jj}(u)}\, \hat{T}(u)\, du. \quad (3.12)$$

Then

$$\hat{T}_j^c = (T_j^c)'. \quad (3.13)$$

The properties of the operators T_j^c (see (3.3), (3.11) and (2.15)) imply that $\hat{T}_j^c$ is a continuous projection in F, and that

$$\hat{T}_j^c \hat{T}_{j'}^{c'} = 0, \quad (3.14)$$

if c and c' are nonequivalent, or if $j \neq j'$, and

$$\hat{T}(e_j^c x) = \hat{T}(x)\, \hat{T}_j^c, \qquad \hat{T}(xe_j^c) = \hat{T}_j^c \hat{T}(x). \quad (3.15)$$

§4. Representations in a subspace and in a quotient space

Let $C(E)$ and $C(F)$ denote the algebras of all continuous operators in E and F respectively, and let us give to $C(E)$ and $C(F)$ the locally convex topologies defined by the systems of seminorms

$$p_{\xi,\eta}(A) = |(A\xi, \eta)|, \qquad q_{\xi,\eta}(B) = |(\xi, B\eta)|, \; A \in C(E),\; B \in C(F),\; \xi \in E,\; \eta \in F.$$

Let $A \in C(E)$, and let M be a closed subspace of E which is invariant with respect to A; let us denote by A^M the restriction of the operator A to M and by A_M the operator on the quotient space E/M generated by A. Let us put further

$$N = M^\perp = \{\eta : \eta \in F,\ (\xi, \eta) = 0 \quad \text{for all} \quad \xi \in M\}; \quad (4.1)$$

then (see [5], §1.4, Proposition 4)

$$M = N^\perp = \{\xi,\ \xi \in E,\ (\xi, \eta) = 0 \quad \text{for all} \quad \eta \in N\}. \quad (4.2)$$

Further, let us put

$$\dot{F} = F/N, \quad (4.3)$$

and let us define a bilinear form $(\xi, \dot{\eta})$ on the pair M, $\dot{F}$ by putting

$$(\xi, \dot{\eta}) = (\xi, \eta) \qquad \text{for } \eta \in \dot{\eta}, \ \xi \in M. \tag{4.4}$$

It is easy to see that this definition does not depend on the choice of the representation $\eta \in \dot{\eta}$, and that $(\xi, \dot{\eta})$ is nondegenerate on M, $\dot{F}$, thus creating a duality relationship between M and $\dot{F}$.

According to Propositions 6 and 7 in [5], Chapter IV, §1.5, the $\sigma(M, \dot{F})$-topology in M coincides with the topology induced by the $\sigma(E, F)$-topology, and the $\sigma(\dot{F}, M)$-topology coincides with that of the quotient space $\dot{F} = F/M$ (the assertion concerning M is valid without the assumption of closedness). Here

$$(A^M\xi, \dot{\eta}) = (\xi, (A')_N\dot{\eta}) \qquad \text{for } \xi \in M, \ \dot{\eta} \in \dot{F} = F/N, \tag{4.5}$$

that is,

$$(A^M)' = (A')_N. \tag{4.6}$$

Indeed, if $\xi \in M$, then also $A^M\xi = A\xi \in M$, and thus, by (4.4), for $\eta \in \dot{\eta}$ we have

$$(A^M\xi, \dot{\eta}) = (A\xi, \eta) = (\xi, A'\eta) = (\xi, (A'\eta)^{\cdot}). \tag{4.7}$$

But $(A'\eta)^{\cdot} = (A')_N\dot{\eta}$, and thus (4.7) implies (4.5) and (4.6). Interchanging the roles of E and F, we conclude that also

$$(A_M)' = (A^N)'. \tag{4.8}$$

From this we deduce (see [5], Chapter IV, §4.1, Proposition 1) that if $A \in C(E)$ and M is invariant with respect to A, then

$$A^M \in C(M) \text{ and } (A')_M \in C(\dot{F}). \tag{4.9}$$

Let $C^M(E)$ denote the collection of all those $A \in C(E)$ for which M is an invariant subspace, and let us equip $C^M(E)$ with the subspace topology induced by the space $C(E)$. Then, as can be easily seen,

I. *$C^M(E)$ is a closed subalgebra of the algebra $C(E)$, and the maps $A \to A^M$ and $A \to A_M$ are homomorphisms*(6) *of the algebra $C^M(E)$ into $C(M)$ and $C(E/M)$ respectively.*

Suppose now that T, $\hat{T}$ is a representation of the group G in the pair $P = \{E, F, (\xi, \eta)\}$ and M is a closed subspace of E invariant under T, so that N is a closed subspace of F invariant under $\hat{T}$. Let T^M denote the restriction of T to M and let $\hat{T}_N$ denote the antirepresentation which is generated in the quotient space $\dot{F} = F/N$ by the antirepresentation $\hat{T}$. Then T^M, $\hat{T}_N$ is a representation of the group G in the pair $\dot{P} = \{M, \dot{F}, (\xi, \dot{\eta})\}$ in the sense of the definition of §2 (so that all conditions 1)–3) of §2 are fulfilled). Evidently conditions 1) and 2) hold,

(6) It is easy to see that these homomorphisms are continuous, but this fact will not be needed here.

so only condition 3) has to be verified. The condition concerning the map $\{g, \xi\} \to T(g)\xi$ is evidently preserved under taking a restriction to M; the condition regarding $\{g, \dot\eta\} \to \hat T_N(g)\eta$ is obtained as follows:

Let ϕ denote the canonical mapping of F onto $\dot F$, let $\dot U = \phi(U)$ be an open set in $\dot F$, where U is open in F, and let f denote the map $\{g, \eta\} \to \hat T(g)\eta$. Then the counterimage of the set $\dot U$ with respect to the map $\{g, \dot\eta\} \to \hat T_N(g)\eta$ is $\phi f^{-1}(U)$, and since f is by assumption continuous, this counterimage is open. Let us further remark that

$$(T^M)^c_j = (T^c_j)^M \text{ and } (T_M)^c_j = (T^c_j)_M. \tag{4.10}$$

Indeed, for $\xi \in M$, $\dot\eta \in \dot F$ and $\eta \in \dot\eta$ we have

$$\begin{aligned}((T^M)^c_j\xi, \dot\eta) = ((T^M)^c_j \xi, \eta) = \Big(\int \overline{c_{jj}(u)}\,(T^M(u)\xi, \eta)\,du \\ = \int \overline{c_{jj}(u)}\,(T(u)\xi, \eta)\,du = (T^c_j\xi, \eta).\end{aligned} \tag{4.11}$$

If $\eta \in N$, then $\int \overline{c_{jj}(u)}(T(u)\xi, \eta)du = 0$, whence $T^c_j\xi \in M$, due to (4.2). Thus M is invariant under T^c_j, and (4.11) implies that for $\xi \in M$, $\eta \in \dot F$ and $\eta \in \dot\eta$

$$((T^M)^c_j \xi, \dot\eta) = ((T^c_j)^M \xi, \eta) = ((T^c_j)^M \xi, \dot\eta),$$

which means that $(T^M)^c_j = (T^c_j)^M$. Further, for $\xi \in \dot\xi$, $\dot\xi \in E/M$ and $\eta \in N$

$$((T^c_j\xi)^\cdot, \eta) = (T^c_j\xi, \eta). \tag{4.12}$$

Since M is invariant under T^c_j and $\eta \in N = M^\perp$, we conclude that $\phi(T^c_j\xi) = (T^c_j\xi)^\cdot = (T^c_j)_M\dot\xi$, and we rewrite (4.12) in the form

$$((T^c_j)_M \dot\xi, \eta) = (T^c_j\xi, \eta). \tag{4.13}$$

But we also have

$$((T_M)^c_j \dot\xi, \eta) = r_c \int \overline{c_{jj}(u)}\,(T_M(u)\dot\xi, \eta)\,du = r_c \int \overline{c_{jj}(u)}\,(T(u)\xi, \eta)\,du = (T^c_j\xi, \eta) \tag{4.14}$$

for $\xi \in \dot\xi$ and $\eta \in F$. The comparison of (4.13) and (4.14) yields the second equality (4.10). In an analogous fashion one proves

$$(\hat T^N)^c_j = (\hat T^c_j)^N \text{ and } (\hat T_N)^c_j = (\hat T^c_j)_N. \tag{4.15}$$

§5. The definition and simple properties of a homogeneous representation

Given any sets $M \subset E$ and $N \subset F$, we shall denote by $[M]$ and $[N]$ the closed linear subspaces spanned by M and N with respect to the topologies $\sigma(E, F)$ and $\sigma(F, E)$ correspondingly.

A representation T of the group G in E is called quasi-homogeneous with respect to U and c if[7]

$$[T(x) T_j^c E, \ x \in X] = E.$$

By analogy, an antirepresentation $\hat{T}$ of the group G in F will be called quasi-homogeneous with respect to U and c if $[\hat{T}(x)\hat{T}_j^c F, \ x \in X] = F$.

A representation $T, \hat{T}$ in the pair $P = \{E, F, (\xi, \eta)\}$ is called homogeneous with respect to U and c if T is quasi-homogeneous in E with respect to U and c and $\hat{T}$ is quasi-homogeneous in F with respect to U and c.

Remark 1. *The following equalities hold*:

$$[T(x) T_j^c E, \ x \in X] = [T(g) T_j^c E, \ g \in G], \tag{5.1}$$

$$[\hat{T}(x) \hat{T}_j^c F, \ x \in X] = [\hat{T}(g) \hat{T}_j^c F, \ g \in G]. \tag{5.2}$$

Proof. Let M_1 and M_2 denote the left and the right sides of (5.1), and put $N_1 = M_1^\perp$ and $N_2 = M_2^\perp$. Taking $\xi \in E$ and $\eta \in N_1$, we then have

$$0 = (T(x) T_j^c \xi, \ \eta) = \int x(g) (T(g) T_j^c \xi, \ \eta) \, dg$$

for any $x(g) \in X$. Thus $(T(g)T_j^c \xi, \eta) = 0$ for all $g \in G$, whence $\eta \in N_2$, so that

$$N_1 \subset N_2. \tag{5.3}$$

We have, further, that for $\xi \in E$ and $\eta \in N_2$

$$0 = (T(g) T_j^c \xi, \ \eta)$$

for all $g \in G$, and thus

$$0 = \int x(g) (T(g) T_j^c \xi, \ \eta) \, dg = (T(x) T_j^c \xi, \ \eta),$$

so that $\eta \in N_1$ and

$$N_2 \subset N_1. \tag{5.4}$$

We conclude from (5.3) and (5.4) that $N_1 = N_2$, and thus $M_1 = N_1^\perp = N_2^\perp = M_2$ and (5.1) is proved. The proof of (5.2) is entirely similar.

I. *The representation T (antirepresentation $\hat{T}$) is quasi-homogeneous with respect to U and c if and only if $[T(g)T_j^c E; \ g \in G] = E$ (respectively $[\hat{T}(g)\hat{T}_j^c F, \ g \in G] = F$). In paritcular, a representation $T, \hat{T}$ in a pair $P = \{E, F, (\xi, \eta)\}$ is homogeneous if and only if $[T(g)T_j^c E, \ g \in G] = E$ and $[\hat{T}(g)\hat{T}_j^c F, \ g \in G] = F$.*

The assertion follows directly from Remark 1.

Remark 2. *Proposition I implies that if $\hat{T}$ is quasi-homogeneous with respect to U and c, then $\hat{T}_j^c \neq 0$, whence also $T_j^c = (\hat{T}_j^c)' \neq 0$.*

---

[7] We write $[\ ]$ in place of $[\{\ \}]$; in particular, $[T(x)T_j^c E, \ x \in X]$ instead of $[\{T(x)T_j^c E, \ x \in X\}]$.

This means that c belongs to the restriction $u \to T(u)$ of the representation T to the group U and $\dim T_j^c$ is the multiplicity $k(T, U, c)$ of c in the representation $u \to T(u)$.

II. *Suppose that* $\hat{T}$ *is a quasi-homogeneous representation in* F *with respect to* U *and* c, *and* M *is a closed subspace of* E, *invariant with respect to* T *and such that* $M \neq (0)$ *and* $M \neq E$. *Then the following assertions are true*:

1) $(T^M)'$ *is quasi-homogeneous in* $\dot{F} = F/M^\perp$ *with respect to* U *and* c.

2) $(T^M)_j^c \neq 0$; *that is, the restriction* $u \to T^M(u)$ *of* T^M *to* M *contains* c.

Proof. Let ϕ denote the canonical mapping of the space F onto $\dot{F} = F/N$, where $N = M^\perp$. Then, for every $A \in C_N(F)$,

$$A\varphi(\eta) = \varphi(A\eta). \tag{5.5}$$

Moreover, due to (4.15) we have

$$(\hat{T}_N)_j^c \varphi(\eta) = (\hat{T}_j^c)_N \varphi(\eta). \tag{5.6}$$

Since the map $A \to A_N$ is a homomorphism (see §4, Proposition I), we have, by (5.6) and (5.5),

$$\begin{aligned} \hat{T}_N(x)(\hat{T}_N)_j^c \varphi(\eta) &= \hat{T}_N(x)(\hat{T}_j^c)_N \varphi(\eta) = (\hat{T}(x)\hat{T}_j^c)_N \varphi(\eta) \\ &= (\hat{T}(x)\hat{T}_j^c)_N \varphi(\eta) = \varphi(\hat{T}(x)\hat{T}_j^c \eta). \end{aligned} \tag{5.7}$$

But then the continuity of ϕ and (5.7) imply that

$$\begin{aligned} [\hat{T}_N(x)(\hat{T}_N)_j^c(F/N),\ x \in X] &= [\hat{T}_N(x)(\hat{T}_j^c)_N \varphi(F),\ x \in X] \\ &= \varphi[\hat{T}(x)\hat{T}_j^c F,\ x \in X] = \varphi(F) = F/N. \end{aligned}$$

This proves assertion 1). Assertion 2) follows from Remark 2 and (4.5).

§6. Criteria for irreducibility and total irreducibility

Theorem 1. *Let* T, $\hat{T}$ *be a representation of the group* G *in the pair* $P = \{E, F, (\xi, \eta)\}$ *which is homogeneous with respect to* U *and* c. *Then each of the conditions*

1) $\dim T_j^c E = k(T, U, c) = 1$; (6.1)

2) *for every nonzero vector* $\xi \in T_j^c E$

$$[T_j^c T(x)\xi,\ x \in X] = T_j^c E, \tag{6.2}$$

implies the irreducibility of T, $\hat{T}$. *Condition* 2) *is also necessary for the irreducibility of* T, $\hat{T}$.

Proof. Suppose condition 1) is satisfied, and assume that there exists a closed subspace $M \subsetneqq E$, $M \neq (0)$, invariant under all $T(g)$, $g \in G$. According to II in §4, $(T^M)_j^c \neq 0$; and, moreover, M is invariant under T_j^c (see §4). Thus (4.10) implies

$$T^c_j M = (T^c_j)^M M = (T^M)^c_j M \neq (0) \quad \text{and} \quad T^c_j M \subset T^c_j E. \tag{6.3}$$

Thus (see (6.1))

$$0 < \dim T^c_j M \leqslant \dim T^c_j E = 1,$$

which implies

$$T^c_j M = T^c_j E. \tag{6.4}$$

But by assumption T is quasi-homogeneous with respect to U and c, and thus

$$M \supset [T(g)\, T^c_j M,\ g \in G] = [T(g)\, T^c_j E,\ \ g \in G] = E,$$

which means that $M = E$, contrary to what we have assumed. We conclude that T, and therefore also $\hat{T}$, are irreducible.

Suppose not that condition 2) is satisfied. We assume again that there exists a closed subspace $M \subsetneqq E$, $M \neq (0)$, which is invariant under T. Then (6.3) holds, whence there exists a nonzero vector $\xi \in T^c_j M \subset M$. Since M is invariant under $T(g)$, we also have $T^c_j T(g)\xi \in M$. Thus, by (6.2),

$$T^c_j E = [T^c_j T(g)\, \xi,\ \ g \in G] \subset M. \tag{6.5}$$

(6.5) and the quasi-homogeneity of T with respect to U and c imply that

$$E = [T(g)\, T^c_j E,\ \ g \in G] \subset [T(g)\, M,\ \ g \in G] \subset M,$$

due to the invariance of M under $T(g)$. Thus $M = E$, contrary to our assumption, and therefore T and hence also $\hat{T}$ are irreducible.

Let us prove the last assertion of the theorem. Assume that T is irreducible. Then

$$[T(g)\, \xi,\ \ g \in G] = E,$$

for every vector $\xi \neq 0$ in E, in particular, in $T^c_j E$, and therefore

$$[T^c_j T(g)\, \xi,\ \ g \in G] = T^c_j [T(g)\, \xi,\ \ g \in G] = T^c_j E,$$

which means that condition 2) is satisfied.

In the following we shall use the following lemmas:

Lemma 1. *Suppose that G is a locally compact group, U a compact subgroup of G and C a complete system of pairwise nonequivalent continuous representations of U. Then, for every $\xi_0 \in E$,*

$$\xi_0 \in \left[\sum_{j=1}^{r_c} T^c_j \xi_0,\ \ c \in C \right]. \tag{6.6}$$

Proof. Let

$$M = \left[\sum_{j=1}^{r_c} T^c_j \xi_0,\ \ c \in C \right], \qquad N = M^{\perp}, \tag{6.7}$$

and suppose that $\xi_0 \notin M$. Let, furthermore,

$$M_1 = [\xi_0, M] \tag{6.8}$$

and

$$N_1 = M_1^{\perp}. \tag{6.9}$$

Then $M_1 \supsetneqq M$, whence $N_1 \subsetneqq N$. Thus there will be a vector $\eta_0 \in N$ such that $\eta_0 \notin N_1$. Evidently

$$(\xi_0, \eta_0) \neq 0 \tag{6.10}$$

and

$$(\xi, \eta_0) = 0 \quad \text{for all} \quad \xi \in M. \tag{6.11}$$

We conclude from (6.11) that

$$0 = \sum_{j=1}^{r_c} (T_j^c \xi_0, \eta_0) = r_c \sum_{j=1}^{r_c} \int \overline{c_{jj}(u)}\, (T(u)\xi_0, \eta_0)\, du \\ = r_c \int \overline{\chi_c(u)}\, (T(u)\xi_0, \eta_0)\, du. \tag{6.12}$$

According to the lemma in [4], §1, (6.12) implies that

$$0 = (T(e)\xi_0, \eta_0) = (\xi_0, \eta_0),$$

and this contradicts (6.10).

Remark 1. By analogous considerations we obtain

$$\eta_0 \in \left[\sum_{j=1}^{r_c} \hat{T}_j^c \eta_0,\ c \in C \right]$$

for every $\eta_0 \in F$.

Lemma 2. *Suppose G is a locally compact group, U a compact subgroup of G with countable basis, $c^1, c^2, \ldots$ a complete system of pairwise nonequivalent irreducible representations of U, and T, $\hat{T}$ a representation of G in the pair $P = \{E, F, (\xi, \eta)\}$. Let $\mathfrak{G}$ be the linear span of all the operators*

$$P_n = \sum_{L=1}^{n} \sum_{j=1}^{r_{cL}} T_j^{cL}, \tag{6.13}$$

and $\overline{\mathfrak{G}}$ its closure in the space $C(E)$. Then $1 \in \overline{\mathfrak{G}}$.

Proof. Suppose

$$V(\xi_1^0, \ldots, \xi_n^0, \eta_1^0, \ldots, \eta_n^0, \varepsilon) = \{A,\ A \in C(E),\ |(A\xi_k^0 - \xi_k^0, \eta_k^0)| < \varepsilon, \\ k = 1, \ldots, n\},\ \xi_j^0 \in E,\ \eta_j^0 \in F,\ j = 1, \ldots, n,$$

is a neighborhood of the operator 1 in $C(E)$. It has to be shown that this neighborhood contains an element of $\mathfrak{G}$.

Let E^n and F^n denote the Cartesian products of n copies of the spaces E

and F respectively, and let us define on E^n and F^n the bilinear form $(\xi, \eta)^n$ by the formula

$$(\xi, \eta)^n = (\xi_1, \eta_1) + \ldots + (\xi_n, \eta_n) \qquad \text{for} \quad \xi = \{\xi_1 \ldots, \xi_n\}, \qquad \eta = \{\eta_1, \ldots, \eta_n\}.$$

It is easy to see that $(\xi, \eta)^n$ is nondegenerate on E^n and F^n, thus establishing a duality between E^n and F^n. Let us define a representation T^n, $\hat{T}^n$ in the pair $P^n = \{E^n, F^n, (\xi, \eta)^n\}$, putting $T^n(g)\{\xi_1, \ldots, \xi_n\} = \{T(g)\xi_1, \ldots, T(g)\xi_n\}$ and $\hat{T}^n(g)\{\eta_1, \ldots, \eta_n\} = \{\hat{T}(g)\eta_1, \ldots, \hat{T}(g)_n\}$. It is easy to verify that all axioms of a representation hold in this case. Let us apply Lemma 1 to the representation T^n and to the vector $\xi^0 = \{\xi_1^0, \ldots, \xi_n^0\}$. We obtain that

$$\{\xi_1^0, \ldots, \xi_n^0\} \in \left[\left\{\sum_{L=1}^{r_e L} T_j^{c_L}\xi_1^0, \ldots, \sum_{L=1}^{r_c L} T_j^{c_L}\xi_n^0\right\}, L = 1, 2, \ldots\right]. \tag{6.14}$$

Let

$$Q_1 = P_1, \quad Q_L = \sum_{j=1}^{r_c L} T_j^{c_L}; \tag{6.15}$$

then $Q_1 \in \mathfrak{G}$ and, by (6.13), $Q_L = P_L - P_{L-1}$ for $L > 1$, and thus $Q_L \in \mathfrak{G}$ and (6.14) can be rewritten in the form

$$\{\xi_1^0, \ldots, \xi_n^0\} \in [\{Q_L\xi_1^0, \ldots, Q_L\xi_n^0\}, L = 1, 2, \ldots]. \tag{6.16}$$

Now we observe that the set

$W = W(\xi_1^0, \ldots, \xi_n^0, \eta_1^0, \ldots, \eta_n^0; \varepsilon)$,

$$\{\{\xi_1, \ldots, \xi_n\} \in E_n, \ |(\xi_j - \xi_j^0, \eta_j^0)| < \varepsilon, \ j = 1, \ldots, n\}.$$

is a neighborhood of the point $\xi^0 = \{\xi_1^0, \ldots, \xi_n^0\}$ in the $\sigma(E^n, F^n)$ topology in E^n. By (6.16), there exists an operator $Q \in \mathfrak{G}$ such that W contains the point $\{Q\xi_1^0, \ldots, Q\xi_n^0\}$; that is,

$$|(\xi_j^0 - Q\xi_j^0, \eta_j^0)| < \varepsilon, \qquad j = 1, \ldots, n. \tag{6.17}$$

By the definition of the topology in $C(E)$, this means that $1 \in \overline{\mathfrak{G}}$, and Lemma 2 is proved.

Remark 2. One proves in an analogous way that

$$1 \in \overline{\hat{\mathfrak{G}}}, \tag{6.18}$$

where $\hat{\mathfrak{G}}$ is the set $\mathfrak{G}$ for $\hat{T}$ instead of T and $\overline{\mathfrak{G}}$ is the closure of $\mathfrak{G}$ in $C(F)$.

Lemma 3. *Let G, U and C be as in Lemma* 1. *Then, for every $j \leq r_c$, $c \in C$, $x \in X$ and $\xi_0 \in E$,*

$$T(x)\xi_0 \in [T(g)\xi_0, g \in G] \tag{6.19}$$

and

$$T_j^c \xi_0 \in [T(g)\xi_0, \ g \in G]. \tag{6.20}$$

Proof. We set

$$M = [T(g)\xi_0, \ g \in G], \quad N = M^{\perp}. \tag{6.21}$$

Then, for $\eta \in N$,

$$0 = \int x(g)(T(g)\xi_0, \ \eta)\, dg = (T(x)\xi_0, \ \eta); \tag{6.22}$$

whence $T(x)\xi_0 \in N^{\perp} = M$. By analogy one shows that $T_j^c \xi_0 \in M$.

Remark 3. Applying identical considerations to $\hat{T}$ instead of T, we obtain for every $j \leq r_c$, $c \in C$, $x \in X$ and $\eta_0 \in F$ that

$$\hat{T}(x)\eta_0 \in [\hat{T}(g)\eta_0, \ g \in G], \tag{6.23}$$

$$\hat{T}_j^c \eta_0 \in [\hat{T}(g)\eta_0, \ g \in G]. \tag{6.24}$$

Lemma 4. *Suppose that G, U and C are as in Lemma* 1; *let $\mathfrak{R}$ be the linear span of all the $T(g)$, $g \in G$, and $\overline{\mathfrak{R}}$ the closure of $\mathfrak{R}$ in $C(E)$. Then for every $j \leq r_c$, $c \in C$ and $x \in X$*

$$T_j^c \in \overline{\mathfrak{R}} \quad \textit{and} \quad T(x) \in \overline{\mathfrak{R}}. \tag{6.25}$$

The proof is similar to that of Lemma 2; it is based on the application of (6.19) and (6.20) (instead of (6.6) in Lemma 2) to the representation T^n, $\hat{T}^n$ in the pair $P^n = \{E^n, F^n, (\xi, \eta)^n\}$.

Remark 4. In a similar way one proves that

$$\hat{T}_j^c \in \overline{\hat{\mathfrak{R}}} \quad \text{and} \quad \hat{T}(x) \in \overline{\hat{\mathfrak{R}}}, \tag{6.26}$$

where $\hat{\mathfrak{R}}$ is the linear span of all $\hat{T}(g)$, $g \in G$, and $\overline{\hat{\mathfrak{R}}}$ is the closure of $\hat{\mathfrak{R}}$ in $C(F)$.

A representation T, $\hat{T}$ in the pair $P = \{E, F, (\xi, \eta)\}$ is called *totally irreducible* if $\overline{\mathfrak{R}} = C(E)$ holds for this representation, and thus $\overline{\hat{\mathfrak{R}}} = C(F)$. It is clear that total irreducibility implies irreducibility; the converse is in general false.

Theorem 2. *Let G be a locally compact group, U a compact subgroup of G with a countable basis and T, $\hat{T}$ an irreducible representation of G in the pair $P = \{E, F, (\xi, \eta)\}$. Then the condition*

a) $\dim T_j^c E = k(T, U, c) < \infty$ *for every irreducible representation c of U*

implies that T, $\hat{T}$ is totally irreducible.

Proof. Let $\mathfrak{R}$ be again the linear span of all the $T(g)$, $g \in G$, and $\overline{\mathfrak{R}}$ the closure of $\mathfrak{R}$ in $C(E)$. Evidently $\mathfrak{R}$, and hence $\overline{\mathfrak{R}}$, is a subalgebra of $C(E)$. Let c^k, P_n, $\mathfrak{G}$ and $\overline{\mathfrak{G}}$ have the same meaning as in the proof of Lemma 2. By Lemma 4,

$$P_n \in \overline{\mathfrak{R}}, \tag{6.27}$$

and thus

$$\overline{\mathfrak{G}} \subset \overline{\mathfrak{R}}. \tag{6.28}$$

We put

$$c_n = \sum_{k=0}^{n} \sum_{j=1}^{r_{c^k}} e_j^{c^k}. \tag{6.29}$$

From (3.3)–(3.5), (3.7) and (3.11) it follows that

$$T(e_n x) = P_n T(x), \quad T(x e_n) = T(x) P_n, \tag{6.30}$$

$$P_n P_m = P_n \qquad \text{for} \quad m \geqslant n. \tag{6.31}$$

Moreover, condition a) implies that

$$P_n E \text{ is of finite dimension for every } n = 1, 2, 3, \ldots . \tag{6.32}$$

We set

$$X_n = \{x, \; x \in X, \; e_n x = x e_n = x\}; \tag{6.33}$$

Then X_n is a subalgebra of X, and it follows from (6.30) and (6.33) that

$$T(x) = P_n T(x) = T(x) P_n \qquad \text{for all} \quad x \in X_n; \tag{6.34}$$

whence $P_n E$ is a subspace of E which is invariant under all $T(x)$, $x \in X_n$, and

$$T'(x) = 0 \text{ on } (1 - P_n) E \quad \text{for } x \in X_n. \tag{6.35}$$

Let $a(x)$ be the restriction of $T(x)$ to $P_n E$. Then $x \to a(x)$ is a representation of X_n in $P_n E$. Let us prove that this representation is irreducible. Assuming the contrary, there is a subspace M_n of $P_n E$ which is invariant under all $a(x)$, $x \in X_n$, and such that $M_n \neq (0)$ and $M_n \neq P_n E$. From $M_n \neq P_n E$ it follows that there exists a functional $\psi_0 \in (P_n E)'$ such that $\psi_0 \neq 0$ and

$$\psi_0(\xi) = 0 \qquad \text{for all} \quad \xi \in M_n. \tag{6.36}$$

But for $\xi \in M_n$ we have also $a(x)\xi \in M_n$, since M_n is invariant under $x \to a(x)$. Thus (6.36) implies that also

$$\psi_0(a(x)\xi) = 0 \qquad \text{for all} \quad \xi \in M_n \text{ and } x \in X_n. \tag{6.37}$$

We set

$$f_0(\xi) = \psi_0(P_n \xi) \qquad \text{for} \quad \xi \in E. \tag{6.37'}$$

The continuity of P_n implies the continuity on E of the linear functional f_0. Thus $f_0(\xi)$ is of the form

$$f_0(\xi) = (\xi, \eta_0), \qquad \eta_0 \in F, \; \xi \in E$$

([5], Chapter IV, §1, Proposition 1), and $\eta_0 \neq 0$ since $\psi_0 \neq 0$. Moreover, by (6.37′),

$$(\xi, P'_n\eta_0) = (P_n\xi, \eta_0) = \psi_0(P_n^2\xi) = \psi_0(P_n\xi) = (\xi, \eta_0)$$

for all $\xi \in E$, and thus

$$P'_n\eta_0 = \eta_0, \tag{6.38}$$

implying that (6.37) can be written as

$$0 = \psi_0(a(x)\xi) = \psi_0(P_n a(x)\xi) = f_0(a(x)\xi) = (a(x)\xi, \eta_0) \tag{6.39}$$

for all $x \in X_n$ and $\xi \in M_n$. By assumption, $M_n \neq (0)$, so that there is in M_n a vector $\xi_0 \neq 0$ and (6.39) holds for $\xi = \xi_0$; that is,

$$0 = (a(x)\xi_0, \eta_0) \quad \text{for all} \quad x \in X_n. \tag{6.40}$$

Suppose now that $x \in X$. Then $e_n x e_n \in X_n$, so that, by (6.40),

$$(T(e_n x e_n)\xi_0, \eta_0) = 0 \quad \text{for all} \quad x \in X. \tag{6.41}$$

But $P_n\xi_0 = \xi_0$ since $\xi_0 \in M_n \subset P_nE$, and $P'_n\eta_0 = \eta_0$ in view of (6.38). Thus (6.41) implies

$$0 = (P_n T(x) P_n\xi_0, \eta_0) = (T(x)\xi_0, P'_n\eta_0) = (T(x)\xi_0, \eta_0) \tag{6.42}$$

for all $x \in X$. By assumption, T is irreducible. Thus (see (5.1)) $[T(x)\xi_0, x \in X] = [T(g)\xi_0, g \in X] = E$, contradicting condition (6.42) in which $\eta_0 \neq 0$. Thus $x \to a(x)$ is an irreducible representation of the algebra X_n in the finite-dimensional space P_nE, and by Burnside's theorem(8)

$$\{a(x), x \in X_n\} = C(P_nE). \tag{6.43}$$

Suppose now that A is an arbitrary operator belonging to $C(E)$. We shall prove that $A \in \overline{\mathfrak{R}}$. This will imply that $\overline{\mathfrak{R}} = C(E)$, and the theorem will be proved.

We set

$$A_{nm} = P_n A P_m \tag{6.44}$$

and

$$A_n = A_{nn} = P_n A P_n \tag{6.45}$$

Evidently P_nE is invariant under A_n, and

$$A_n = 0 \text{ on } (1 - P_n)E. \tag{6.46}$$

---

(8)(6.43) follows also from the results of Fell ([6], p. 282, Lemma 10; see [4], Theorem 2, for another proof).

Let a_n denote the restriction of A_n to P_nE. (6.43) implies the existence of a function $x_n \in X_n$ such that $a(x_n) = a_n$; that is,

$$T(x_n) = A_n \text{ on } P_nE. \tag{6.47}$$

But (6.35) and (6.46) imply that both sides of (6.47) are null on $(1 - P_n)E$, and therefore

$$T(x_n) = A_n \text{ on the whole of } E, \tag{6.48}$$

whence

$$A_n \in \overline{\mathfrak{R}}, \quad n = 1, 2, 3, \ldots, \tag{6.49}$$

by Lemma 4. Moreover, (6.13) and Lemma 4 imply that $P_n \in \overline{\mathfrak{R}}$. Thus also

$$A_{nm} = P_nAP_m = P_nAP_nP_m = A_nP_m \in \overline{\mathfrak{R}} \quad \text{for } m \leqslant n$$

and

$$A_{nm} = P_nAP_m = P_nP_mAP_m = P_nA_m \in \overline{\mathfrak{R}} \quad \text{for } m \geqslant n,$$

because $\overline{\mathfrak{R}}$ is an algebra. But then, using the notation of Lemma 2, we have $Q_1AQ_2 \in \overline{\mathfrak{R}}$ for arbitrary $Q_1, Q_2 \in \mathfrak{G}$. Hence Lemma 2 and the continuity of the map $B \to QAB$ for $B \in C(E)$ and fixed Q and A imply that

$$QA = QA \cdot 1 \in \overline{\mathfrak{R}} \quad \text{for every} \quad Q \in \mathfrak{G}. \tag{6.50}$$

To conclude the proof of the theorem, it now suffices to use again Lemma 2, the continuity of the map $B \to BA$ for fixed A, and lastly (6.50).

Combining Theorems 1 and 2, we arrive at the following result:

Theorem 3. *Let G be a locally compact group, U a compact subgroup of G with a countable basis, c^0 a fixed irreducible representation of U and T, $\hat{T}$ a representation of G in the pair $P = \{E, F, (\xi, \eta)\}$ which is homogeneous with respect to U and c^0 and satisfies the condition*

1) $\dim(T_j^c E) = k(T, U, c) < \infty$ *for every irreducible representation of U*

and at least one of the conditions,

2) $\dim(T_{ji}^{c^0} E) = k(T, U, c^0) = 1$,

2') $[T_j^c T(g)\xi, g \in G] = T_j^{c^0}E$ *for every nonzero element* $\xi \in T_j^{c^0}E$.

Then T, $\dot{T}$ is completely irreducible in P.

BIBLIOGRAPHY

1. M. A. Naimark, *On homogeneous representations of a locally compact group acting in a pair of dual spaces*, Dokl. Akad. Nauk SSSR **220** (1975), 38–40 = Soviet Math. Dokl. **16** (1975), 35–38.

2. ———, *On homogeneous and quasi-homogeneous representations of a locally compact group*, Dokl. Akad. Nauk SSSR **205** (1972), 783–786 = Soviet Math. Dokl. **13** (1972), 1022–1025. MR **47** #396.

3. M. A. Naĭmark, *Criteria of irreducibility and discrete decomposability of a homogeneous representation of a locally compact group*, Dokl. Akad. Nauk SSSR **205** (1972), 1040–1042 = Soviet Math. Dokl. **13** (1972), 1077–1080. MR **47** #397.

4. ———, *Homogeneous and quasi-homogeneous representations of a locally compact group*, Izv. Akad. Nauk SSSR Ser. Mat. **36** (1972), 1020–1048 = Math. USSR Izv. **6** (1972), 1009–1036. MR **47** #396.

5. N. Bourbaki, *Espaces vectoriels topologiques*. Chaps. 1–5 and fascicule de résultats, Actualités Sci. Indust., nos. 1189, 1229, 1230, Hermann, Paris, 1953, 1955. MR **14**, 880; **17**, 1109.

6. J. M. G. Fell, *Non-unitary dual spaces of groups*, Acta Math. **114** (1965), 267–310. MR **32** #4210.

7. M. A. Naĭmark, *Normed rings*, 2nd rev. ed., "Nauka", Moscow, 1968; English transl., *Normed algebras*, Wolters-Noordhoff, Groningen, 1972. MR **50** #8075; #8076.

8. ———, *The decomposition of a unitary representation of a complex semisimple Lie group into its irreducible representations*, Studia Math. **31** (1968), 383–398. (Russian) MR **38** #4618.

Translated by S. SWIERCZKOWSKI

Trudy Mat. Inst. Steklov.
134 (1975)

Proc. Steklov Inst. Math.
134 (1975)

AN APPROXIMATE METHOD OF FINDING CRITICAL POINTS OF EVEN FUNCTIONALS

UDC 513.88

J. NEČAS

Abstract. It is shown that, in a separable Hilbert space, for even functionals $f(x)$ and $g(x)$ which are approximately quadratic the conditionally critical values of $g(x)$ with respect to the manifold $M_r(f) = \{x|f(x) = r\}$ obtained by the Ljusternik-Šnirel'man theorem can be obtained as limits of values $g(x(t))$ as $t \to \infty$, where $x(t)$ is the well-defined solution to the equation

$$\frac{dx}{dt} = g'(x) - \frac{(f'(x), g'(x))}{(f'(x), f'(x))} f'(x).$$

A method is given for the approximate location of the corresponding critical points.

Bibliography: 7 titles.

§1. Introduction

To study the problem of the critical points of a weakly continuous functional $g(x)$ with respect to the manifold $M_r(f) \overset{\text{def}}{=} \{x|\ f(x) = r\}$, the Courant extremum principle was generalized by Ljusternik and Šnirel'man [1] to general functionals. These authors introduced the concept of the category of a set and proved that the minimax principle can be used to establish the existence of a countable number of critical values $\gamma_1 \geq \gamma_2 \geq \cdots$ of the functional $g(x)$ with respect to the manifold $M_r(f)$ such that $\gamma_n \to 0$, i.e. the existence of $x_k \in M_r(f)$, $k = 1, 2, \ldots$, such that $\lambda_k f'(x_k) - g'(x_k) = 0$ and $g(x_k) = \gamma_k$. In a series of papers by the present author and his collaborators [2]–[4], the concept of the category of a set was replaced by the concept of the order of a set, and results of the same type were obtained. In addition, for the case when the functionals f and g are real and analytic and with certain additional conditions, it was shown that the set of critical points is no larger than countably infinite. This result is a corollary of the analogous theorem of Morse and Sard for analytic functions of several complex variables proved in [5], which asserts that locally the set of critical values of such a function is finite.

In this paper, we limit our consideration to a separable Hilbert space H. The critical points are obtained as limits of values as $t \to \infty$ of the solution of the ordinary differential equation

AMS (MOS) subject classifications (1970). Primary 58E05, 47H15; Secondary 34B15, 35G30.

$$\frac{dx}{dt}=g'(x)-f'(x)\frac{(f'(x),\ g'(x))}{(f'(x),\ f'(x))},\qquad x(0)=x_0,$$

and it is shown that under certain conditions one can select the initial (noncritical) point in such a way as to obtain the desired Ljusternik-Šnirel′man critical value.

§2. The Ljusternik-Šnirel′man theory and trajectories on the manifold $M_r(f)$

For the sake of simplicity in the argument, we shall not attempt to determine the weakest assumptions for the functionals f and g. We shall assume that f and g are twice Gâteaux differentiable. Set $f'(x, h) \equiv (f'(x), h)$ and $f''(x, h, k) \equiv (F(x)k, h)$, and assume that

$$f(\theta)=0,\qquad f'(\theta)=\theta, \tag{2.1}$$

$$\|F(x)\|\leqslant M, \tag{2.2}$$

$$(F(x)h,\ h)\geqslant\alpha\|h\|^2,\qquad \alpha>0, \tag{2.3}$$

$$f(-x)=f(x). \tag{2.4}$$

Set $g'(x, h) \equiv (g'(x), h)$ and $g''(x, h, k) \equiv (G(x)k, h)$, and assume that

$$g(\theta)=0,\qquad g'(\theta)=\theta, \tag{2.5}$$

$$\|G(x)\|\leqslant N, \tag{2.6}$$

$$(G(x)h,\ h)>0\quad \text{for}\quad h\neq\theta, \tag{2.7}$$

$$g(-x)=g(x), \tag{2.8}$$

$$x_n\rightharpoonup x\ \text{(weak convergence) implies}\quad g'(x_n)\to g'(x). \tag{2.9}$$

Here both $(F(x)k, h)$ and $(G(x)k, h)$ are assumed to be continuous for fixed k and h.

Now let K be a symmetric closed set in H. We will say that ord $K = 0$ if K is empty; that ord $K = 1$ if $K = K_1 \cup K_2$, where the K_i are closed subsets of K and neither K_1 nor K_2 contains antipodal points; that ord $K = n$ if $K = \bigcup_{i=1}^{n+1} K_i$, where the K_i are closed subsets of K not containing antipodal points and n is the least possible number; and that ord $K = \infty$ if no such n exists.

Let us introduce the autonomous differential equation

$$\dot{x}=g'(x)-f'(x)\frac{(f'(x),\ g'(x))}{(f'(x),\ f'(x))},\qquad x(0)=x_0\in M_r(f), \tag{2.10}$$

which we shall use both in the theoretical arguments and in the analysis of the approximate method. Since from (2.1)–(2.3) it follows that $\frac{1}{2}\alpha\|x\|^2 \le f(x) \le \frac{1}{2}M\|x\|^2$ and $\|f'(x)\|^2 \ge \alpha^2\|x\|^2$, we obtain, in view of (2.2) and (2.6), that (2.10) has a solution on the interval $[0, \infty)$, and that $x(t) \in M_r(f)$ since $(f'(x), \dot{x}) = 0$. It is also clear that x is a critical point of g with respect to the manifold $M_r(f)$ if and only if $T(x) = \theta$, where

$$T(x)=g'(x)-f'(x)\frac{(f'(x),\ g'(x))}{(f'(x),\ f'(x))}.$$

It is clear that if $T(x_0)=\theta$, then $x(t)\equiv x_0$. In what follows, we shall assume, with regard to the solution of (2.10), that x_0 is a noncritical point.

Theorem 2.1. *Suppose that conditions (2.1)–(2.9) hold. Then $T(x(t))\neq\theta$, $g(x(t))$ is a nondecreasing function for $t\in[0,\infty)$,*

$$\int_0^\infty \|\dot{x}\|^2\,dt=\lim_{t\to\infty} g(x(t))-g(x_0)<\infty, \tag{2.11}$$

$$\lim_{t\to\infty} T(x(t))=\theta \tag{2.12}$$

and from each sequence $t_n\to\infty$ one can select a subsequence t_{n_k} such that $x(t_{n_k})\to\bar{x}$, where $\bar{x}$ is a critical point. If N is the set of all such points $\bar{x}$, then N is a compact connected set.

Proof. If $T(x(t_0))=\theta$, then from the uniqueness of the solution to the Cauchy problem it follows that $T(x_0)=\theta$, which contradicts the assumption.

We have $dg(x(t))/dt=(g'(x(t)),\dot{x}(t))\geq 0$ and $\|\dot{x}(t)\|^2=(g'(x(t)),\dot{x}(t))$, from which (2.11) follows. Since $\|T(x(t_1))-T(x(t_2))\|\leq l|t_1-t_2|$ and $\int_0^\infty\|T(x(t))\|^2dt<\infty$, we obtain (2.12). From $x(t_n)$ it is possible to select a subsequence $x(t_{n_k})$ such that $x(t_{n_k})\to\bar{x}$. Set $x(t_{n_k})=x_k$. We can assume that

$$\frac{(f'(x_k),\ g'(x_k))}{(f'(x_k),\ f'(x_k))}\to c.$$

If $c=0$, then $g'(\bar{x})=\theta\Rightarrow\bar{x}=0\Rightarrow g(\bar{x})=0$, which is not possible since $g(x_0)>0$ and $g(x(t))$ is nondecreasing. But from (2.3) it follows that

$$(f'(x_n)-f'(\bar{x}),\ x_n-\bar{x})\geqslant\alpha\|x_n-x\|^2, \tag{2.13}$$

and in view of (2.9) and (2.12) we obtain that $(f'(x_n)-f'(\bar{x}),x_n-\bar{x})\to 0$, so that, by (2.13), $x_n\to\bar{x}$. Consequently $\bar{x}\in M_r(f)$, and from (2.12) it follows that $\bar{x}$ is a critical point. One can show that N is compact in the same way.

If N is not connected, then $N=N_1\cup N_2$ where $N_1\cap N_2=\emptyset$ and the N_i are compact sets, which is impossible in view of the connectedness of the trajectory. This completes the proof.

Corollary 2.1. *If the set of critical points corresponding to a critical value $\mu>0$ is isolated and $\lim_{t\to\infty}g(x(t))=\mu$, then $\lim_{t\to\infty}x(t)=\bar{x}$.*

Let $V_n=\{K|\ K\subset M_r(f)$ is a symmetric compact set and $\operatorname{ord}K\geq n\}$. Set $\gamma_k=\sup_{K\in V_k}\min_{x\in K}g(x)$.

Let us note in passing that the following fundamental theorem of Ljusternik and Šnirel'man is an immediate corollary of Theorem 2.1 (see the analogous proof in [6]):

Theorem 2.2. *Suppose that conditions* (2.1)–(2.9) *hold. Then there exist* $x_k \in M_r(f)$, $k = 1, 2, \ldots$, *such that*

$$g'(x_k) - f'(x_k)\frac{(f'(x_k), g'(x_k))}{(f'(x_k), f'(x_k))} = \theta,$$

$$g(x_k) = \gamma_k, \quad \gamma_k \searrow 0, \quad x_k \rightharpoonup \theta.$$

Proof. Since ord $M_n = n$, where M_n is the intersection of the manifold $M_r(f)$ with an $(n+1)$-dimensional subspace of the space H (cf. [7]), it follows that the set V_k is nonempty. Now suppose that $\delta > 0$ and consider a sequence $K_n \in V_k$ such that $\min_{x \in K_n} g(x) \to \gamma_k$. Let

$$Q_n = \{x \in K_n \mid \min_{x \in K_n} g(x) \leq g(x) \leq \min_{x \in K_n} g(x) + \delta\}.$$

We have

$$\lim_{n\to\infty} \min_{x \in Q_n} \| T(x) \| = 0. \tag{2.14}$$

Suppose on the contrary that $\min_{x \in Q_{n_m}} \|T(x)\|^2 \geq \epsilon > 0$. Consider all trajectories $x(t, x_0)$ for $x_0 \in Q_{n_m}$. For sufficiently small μ we will have, then, that

$$\| T(x(t, x_0)) \|^2 \geq \varepsilon/2$$

for $0 \leq t \leq \mu$. Since $dg(x(t))/dt = \|T(x(t))\|^2$, we have that

$$g(x(\mu, x_0)) \geq \min_{x \in K_n} g(x) + \frac{\varepsilon}{2}\mu$$

and consequently

$$\min_{x_0 \in K} g(x(\mu, x_0)) \geq \min\left(\min_{x \in K_n} g(x) + \frac{\varepsilon}{2}\mu, \ \min_{x \in K_n} g(x) + \delta\right).$$

But $x(\mu, x_0)$ is an odd mapping of the set K into the manifold $M_r(f)$. Since ord $x(\mu, K) \geq$ ord K (cf. [7]), a contradiction is obtained.

Now choose $\delta_n \to 0$. Select $K_n \in V_k$ such that $\min_{x \in K_n} g(x) \to \gamma_k$ and $\min_{x \in Q_n} \|T(x)\| \to 0$, where

$$Q_n = \{x \in K_n \mid \min_{x \in K_n} g(x) \leq g(x) \leq \min_{x \in K_n} g(x) + \delta_n\}.$$

Suppose $x_n \in Q_n$. If $x_{n_k} \rightharpoonup x_k$, then, arguing as in the proof of Theorem 2.1, we obtain (after selecting a subsequence) that $x_{n_{k_l}} \to x_k$, and consequently x_k is the desired point.

Standard methods can be used to show that $\gamma_k \searrow 0$ and $x_k \rightharpoonup \theta$ (cf. [6], [7]).

§3. Approximate method for finding critical values

Theorem 3.1. *Suppose that conditions* (2.1)–(2.9) *hold and that* γ_1 *is an isolated critical value of the functional* $g(x)$ *with respect to the manifold* $M_r(f)$. *Let* $x_0 \in M_r(f)$ *be such that* $g(x_0) > \gamma_1 - \delta$, *and let* $x(t)$ *be the trajectory from Theorem* 2.1. *Then*

$$\lim_{t\to\infty} g(x(t)) = \gamma_1.$$

The theorem is a direct corollary of Theorem 2.1.

Theorem 3.2. *Suppose that $\gamma_1 \geq \gamma_2 \geq \cdots \geq \gamma_k > \gamma_{k+1}$, that there are no critical points in the interval $(\gamma_k - \delta, \gamma_k)$, and that conditions (2.1)–(2.9) hold. Let $K \in V_k$ be such that $\gamma_k - \delta < \min_{x \in K} g(x) < \gamma_k$, and suppose that there are no critical points in K corresponding to the critical value γ_k. Then*

$$\lim_{t\to\infty} \min_{x_0 \in K} g(x(t, x_0)) = \gamma_k$$

and there exists an $x_0 \in K$ such that $\lim_{t\to\infty} g(x(t, x_0)) = \gamma_k$, where $x(t, x_0)$ is the trajectory of equation (2.10) leaving the point x_0.

Proof. Let $K_t = \{x \mid x = x(t, x_0), x_0 \in K\}$. Since the mapping $x_0 \to x(t, x_0)$ is odd and continuous, it follows that $\operatorname{ord} K_t \geq \operatorname{ord} K$ (cf. [7]). From this we obtain that

$$\lim_{t\to\infty} \min_{x_0 \in K} g(x(t, x_0)) \leqslant \gamma_k.$$

Let $\lim_{t\to\infty} \min_{x_0 \in K} g(x(t, x_0)) = \kappa$. There exist $x_{0n} \in K$ such that $g(x(t_n, x_{0n})) \to \kappa$ and $x_{0n} \to x_0$ for $t_n \to \infty$. But $\lim_{t\to\infty} g(x(t, x_0)) \geq \gamma_k$, so that for each $\epsilon > 0$ there exist n_0 and t_0 such that $g(x(t_0, x_{0n})) \geq \gamma_k - \epsilon$ for $n \geq n_0$. From this it follows that for some $n_1 \geq n_0$ and all $n > n_1$

$$g(x(t_n, x_{0n})) \geqslant g(x(t_0, x_{0n})) \geqslant \gamma_k - \varepsilon \Rightarrow \varkappa \geqslant \gamma_k,$$

and it is also clear that $\lim_{t\to\infty} g(x(t, x_0)) = \gamma_k$.

Remark 3.1. From the simple example of a trajectory in R_n: $\dot{x}(t) = T(x)$, where $T(x) = z'(x)$ and $z(x) = x_1^2 + x_2^2 + \cdots + x_i^2 - x_{i+1}^2 - \cdots - x_n^2$, it is clear that, in general, the location of x_0 from the previous theorem is an unreasonable problem.

Remark 3.2. As noted in the Introduction, the origin is the unique accumulation point of the set of critical values if f and g are real analytic functionals satisfying, in addition to (2.1)–(2.9), certain other conditions (cf. [7]).

BIBLIOGRAPHY

1. L. A. Ljusternik and L. G. Šnirel'man, *Topological methods in variational problems*, Gosudarstvennoe Izdat., Moscow, 1930; rev. French transl., Actualités Sci. Indust., no. 188, Hermann, Paris, 1934.

2. S. Fučík and J. Nečas, *Ljusternik-Schnirelmann theorem and nonlinear eigenvalue problems*, Math. Nachr. 53 (1972), 277–289. MR 48 #12185.

3. S. Fučík et al., *Upper bound for the number of critical levels for nonlinear operators in Banach spaces of the type of second order nonlinear differential operators*, J. Functional Analysis 11 (1972), 314–333. MR 49 #5974.

4. J. Nečas, *On the discreteness of the spectrum of a nonlinear Sturm-Liouville equation*, Dokl. Akad. Nauk SSSR **201** (1971), 1045–1048 = Soviet Math. Dokl. 12 (1971), 1779–1783. MR **45** #638.

5. J. Souček and V. Souček, *Morse-Sard theorem for real-analytic functions*, Comment. Math. Univ. Carolinae **13** (1972), 45–51. MR **46** #7459.

6. M. M. Vaĭnberg, *Variational methods for the study of non-linear operators*, GITTL, Moscow, 1956; English transl., Holden-Day, San Francisco, Calif., 1964. MR **19**, 567; **31** #638.

7. S. Fučík et al., *Spectral analysis of nonlinear operators*, Lecture Notes in Math., vol. 346, Springer-Verlag, Berlin and New York, 1973.

Translated by W. L. BYNUM

Trudy Mat. Inst. Steklov.
134 (1975)

Proc. Steklov. Inst. Math.
134 (1975)

AN ESTIMATE IN THE APPROXIMATION OF CONTINUOUS FUNCTIONS BY SUBSEQUENCES OF FOURIER SUMS

K. I. OSKOLKOV

Abstract. Let $\omega(\delta)$ be a modulus of continuity, H_ω the class of 2π-periodic continuous functions $f(x)$ whose moduli of continuity $\omega(f, \delta)$ satisfy the condition $\omega(f, \delta) \le \omega(\delta)$ for $\delta \in [0, \pi]$, and let $s_n(f, x) = s_n(f)$ be the Fourier sum of order n of f. Set

$$S_n(\omega) = \sup_{f \in H_\omega} \|f - s_n(f)\| \quad (n = 0, 1, \dots).$$

It is proved that

$$\max_{f \in H_\omega} \overline{\lim_{n\to\infty}} \frac{\|f - s_n(f)\|}{S_n(\omega)} = \frac{1}{2}$$

for an arbitrary class H_ω. If L_n are Lebesgue constants $(n = 0, 1, \dots)$, then

$$\max_{f \in L^\infty,\ \|f\|=1} \overline{\lim_{n\to\infty}} \frac{\|S_n(f)\|}{L_n} = \frac{1}{2}.$$

Bibliography: 24 titles.

Let C be the space of continuous 2π-periodic functions $f(x)$ with norm $\|f\| = \max_x |f(x)|$. We denote the modulus of continuity of a function $f \in C$ by $\omega(f, \delta)$ and the partial sum of degree n $(n = 0, 1, \dots)$ of its Fourier series by $s_n(f, x)$. Let $\omega(\delta) \not\equiv 0$ be a modulus of continuity and H_ω the class of all functions f from C whose moduli of continuity satisfy $\omega(f, \delta) \le \omega(\delta)$ whenever $\delta \in [0, \pi]$ (see [1]). We consider the upper bound of the norms of the deviations of the Fourier sums from the functions over the class H_ω; that is,

$$S_n(\omega) = \sup_{f \in H_\omega} \|f - s_n(f)\| \quad (n = 0, 1, \dots).$$

It is clear that if $f \in H_\omega$, then for all n

$$\|f - s_n(f)\| \leqslant S_n(\omega). \tag{1}$$

The question of the accuracy of the estimate (1) for an individual function f from the class H_ω as $n \to \infty$ reduces to the study of the sequence

$$\left\{\frac{\|f - s_n(f)\|}{S_n(\omega)}\right\}_{n=0}^{\infty},$$

AMS (MOS) subject classifications (1970). Primary 42A08.

and hence there arises the problem of estimating the quantities

$$\bar{S}(\omega)=\sup_{f\in H_\omega}\varlimsup_{n\to\infty}\frac{\|f-s_n(f)\|}{S_n(\omega)},$$

$$\underline{S}(\omega)=\sup_{f\in H_\omega}\varliminf_{n\to\infty}\frac{\|f-s_n(f)\|}{S_n(\omega)}.$$

In view of (1) neither of these quantities exceeds unity. The first of them, $\bar{S}(\omega)$, has already been studied (see below). It does not depend on the class H_ω and equals unity.

In the present article the size of $\underline{S}(\omega)$ will be computed. It will be proved that its value also does not depend on the class H_ω and always equals ½.

We mention some results which will be needed later concerning the estimation of the upper bounds of deviations of Fourier sums and the question of the accuracy of the estimate (1).

Lebesgue [2] proved that([1])

$$\|f-s_n(f)\|\leqslant c\omega\left(f,\frac{1}{n}\right)\log n \qquad (n=2,3,\ldots), \tag{2}$$

and consequently, for $n=2,3,\ldots,$

$$S_n(\omega)\leqslant c\omega\left(\frac{1}{n}\right)\log n. \tag{3}$$

The first result in the direction of better accuracy in the investigation of upper bounds on the deviations of Fourier sums from continuous functions was obtained by Fejér [3]. In [3] it is established that for the Lebesgue constant

$$L_n=\sup_{f,\,\|f\|\leqslant 1}\|s_n(f)\| \qquad (n=0,1,\ldots)$$

the estimate

$$\left|L_n-\frac{4}{\pi^2}\log n\right|\leqslant c \qquad (n=1,2,\ldots) \tag{4}$$

holds. The corresponding estimate for the upper bound of the partial sum of a Fourier series of power type was found by Landau [4].

Kolmogorov [5] obtained an asymptotic estimate of the upper bound of the deviations of Fourier sums for the class of differentiable functions (see also [6]).

S. M. Nikol'skiĭ [7], [1] investigated the asymptotics of the quantity $S_n(\omega)$ as $n\to\infty$ for arbitrary classes H_ω and proved that if $\omega(\delta)$ is a convex upwards function then for $n=1,2,\ldots$

([1]) Here and below the symbol C will stand for absolute constants which, in general, are distinct.

$$\left| S_n(\omega) - \frac{2\log n}{\pi^2} \int_0^{\frac{\pi}{2}} \omega\left(\frac{2t}{n}\right) \sin t dt \right| \leqslant c\omega\left(\frac{1}{n}\right). \tag{5}$$

If the modulus of continuity $\omega(\delta)$ is not convex, then, as Nikol'skiĭ established,

$$\left| S_n(\omega) - \frac{2\log n}{\pi^2} \theta_n \int_0^{\frac{\pi}{2}} \omega\left(\frac{2t}{n}\right) \sin t dt \right| \leqslant c\omega\left(\frac{1}{n}\right), \tag{6}$$

where θ_n is a number which satisfies the condition $\frac{1}{2} \leq \theta_n \leq 1$ for all n.

Efimov [8] obtained an estimate of the size of $S_n(\omega)$ in terms of the upper bound of the Fourier coefficients of the class H_ω:

$$\left| S_n(\omega) - \frac{C_n(\omega)}{\pi} \log n \right| \leqslant c\omega\left(\frac{1}{n}\right), \tag{7}$$

where

$$C_n(\omega) = \sup_{f \in H_\omega} \frac{1}{\pi} \int_{-\pi}^{\pi} f(t) \sin ntdt \qquad (n = 1, 2, \ldots).$$

The estimate (7) holds for an arbitrary class H_ω without assuming convexity for $\omega(\delta)$. If we suppose that $\omega(\delta)$ is convex upwards, then (see [2])

$$C_n(\omega) = \frac{2}{\pi} \int_0^{\frac{\pi}{2}} \omega\left(\frac{2t}{n}\right) \sin t dt;$$

consequently (5) follows from (7) in this case.(2)

The question of the accuracy of the estimate (2) for individual functions as $n \to \infty$ was considered by Lebesgue [2]. From his results it follows that in every class H_ω there is a function f such that

$$\overline{\lim_{n \to \infty}} \frac{\| f - s_n(f) \|}{\omega(1/n) \log n} \geqslant A > 0,$$

where A is an absolute constant. Hence from (3) we have

$$\bar{S}(\omega) \geqslant A' > 0. \tag{8}$$

Bohr [10] and Neder [11] solved the problem of the accuracy of Landau's estimate [4] for individual functions as $n \to \infty$. In [11] the following result for the Lebesgue constants L_n is mentioned. There exists a bounded 2π-periodic function $f(x)$ with norm $\|f\| = \operatorname{ess\,sup}_x |f(x)|$ for which

$$\overline{\lim_{n \to \infty}} \frac{\| s_n(f) \|}{L_n} = \overline{\lim_{n \to \infty}} \frac{s_n(f, 0)}{L_n} = 1, \tag{9}$$

(2) Efimov [9] refined the estimate for the numbers θ_n in (6) and proved that $2/3 \leq \theta_n \leq 1$ for all n.

Doronin [12] utilized the estimate (5) and proved (for the class Lipα) that

$$\overline{S}(\omega) = 1. \tag{10}$$

We note that the relations (8)–(10) give the lower bound for $\overline{S}(\omega)$; that is, there is a function $f \in H_\omega$ and a subsequence of Fourier sums $s_{n_i}(f)$ for which the norms of the deviations $\|f = s_{n_i}(f)\|$ are large. A problem of another kind, amounting to the estimation of $\underline{S}(\omega)$ (that is, studying all subsequences of the Fourier sums of a given function at once) was considered by Men'šov [13]. He proved there exists a continuous 2π-periodic function f such that every subsequence of its Fourier sums diverges at some point. In particular, it follows from this theorem that there exists a continuous function such that no subsequence of its Fourier sums is uniformly convergent.

Busko [14] strengthened this corollary of Men'šov's theorem and proved that for every sequence of positive numbers $\rho(n)$ tending to ∞ as $n \to \infty$ and satisfying $\rho(n) = o(\log n)$ $(n \to \infty)$ there exists a function $f \in C$ such that

$$\lim_{n\to\infty} \frac{\|s_n(f)\|}{\rho(n)} = \infty.$$

In addition, Busko proved that there exists a bounded function f for which

$$\|s_n(f)\| \geqslant \log n$$

for all $n = 1, 2, \ldots$.(3)

Korevaar (see [15], p. 185) posed the question of whether the quantity $\underline{S}(\omega)$ is positive or not (for the classes Lipα). The author [16] proved that this question has a positive answer, and, what is more (see [17]), that in every class H_ω there exists a function f such that for an arbitrary increasing sequence of natural numbers $\{n_i\}_1^\infty$ there exists a point x at which

$$\overline{\lim_{i\to\infty}} \frac{s_{n_i}(f, x) - f(x)}{\omega(1/n_i)\log n_i} \geqslant A > 0$$

and

$$\underline{\lim_{i\to\infty}} \frac{s_{n_i}(f, x) - f(x)}{\omega(1/n_i)\log n_i} \leqslant -A < 0,$$

where A is an absolute constant.

The results we quoted of Men'šov and Busko are corollaries of this theorem. The following result will be proved here.

Theorem 1. *For an arbitrary modulus of continuity* $\omega(\delta) \not\equiv 0$,

$$\underline{S}(\omega) = \max_{f\in H_\omega} \underline{\lim_{n\to\infty}} \frac{\|f - s_n(f)\|}{S_n(\omega)} = \frac{1}{2}. \tag{11}$$

---

(3) Unfortunately, Busko's results became known to the author only after the publication of [16] and [17].

The condition $\lim_{\delta \to 0} \omega(\delta) = 0$ which enters into the definition of the modulus of continuity will not be used in the proof of this theorem. Therefore the following result holds.

Theorem 2. *Let L^∞ be the space of 2π-periodic measurable functions which are bounded almost everywhere with norm $\|f\| = \operatorname{ess\,sup}_x |f(x)|$, and let L_n be the Lebesgue constants. Then*

$$\max_{f \in L^\infty, \|f\| \leqslant 1} \varlimsup_{n \to \infty} \frac{\|s_n(f)\|}{L_n} = \frac{1}{2}.$$

Before proving Theorem 1 we will prove that with its help it is possible to obtain a lower bound for the sequence of best approximations $E_n(f)$ of an individual function lying, for example, in the class Lip 1; that is, in the class H_ω, where $\omega(\delta) = \delta$ for $0 \leq \delta \leq \pi$.

Theorem 3. *There exists a function $f \in \text{Lip}\,1$ such that*

$$\varlimsup_{n \to \infty} nE_n(f) \geqslant \frac{1}{2}. \tag{12}$$

In order to deduce Theorem 3 from Theorem 1 we will use a well-known method; namely, we estimate from below the best approximation with the help of Lebesgue's inequality (see [18], Chapter III, formula (13.25)):

$$\|f - s_n(f)\| \leqslant (L_n + 1) E_n(f) \qquad (n = 0, 1, \ldots). \tag{13}$$

By Theorem 1 there exists a function $f \in \text{Lip}\,1$ such that

$$\varlimsup_{n \to \infty} \frac{\|f - s_n(f)\|}{S_n(\text{Lip}\,1)} = \frac{1}{2}, \tag{14}$$

and in view of Kolomogorov's estimate [5] (see also (5), where $\omega(\delta) = \delta$)

$$\left| S_n(\text{Lip}\,1) - \frac{4}{\pi^2} \frac{\log n}{n} \right| \leqslant \frac{c}{n} \qquad (n = 1, 2, \ldots).$$

Therefore, with the help of (13), (14) and (4) we find that for the function f

$$\varlimsup_{n \to \infty} nE_n(f) \geqslant \varlimsup_{n \to \infty} \frac{\|f - s_n(f)\|}{S_n(\text{Lip}\,1)} \frac{nS_n(\text{Lip}\,1)}{L_n + 1} = \frac{1}{2};$$

that is, (12) is satisfied.

We set

$$E_m = E_m(\text{Lip}\,1) = \sup_{f \in \text{Lip}\,1} E_m(f) \qquad (m = 0, 1, \ldots). \tag{15}$$

As Favard [19] and Ahiezer and Kreĭn [20] proved,

$$E_m = \frac{\pi}{2(m+1)},$$

and, in particular, for any fixed function $f \in \text{Lip}\,1$

$$\overline{\lim_{n\to\infty}}\, nE_n(f) \leqslant \frac{\pi}{2}. \tag{16}$$

In addition it is established that for every $m = 0, 1, \ldots$ the upper bound in the right-hand side of (15) is attained by the function

$$f_m = f_m(x) = \frac{4}{\pi} \sum_{\nu=0}^{\infty} \frac{\cos(2\nu+1)(m+1)x}{(2\nu+1)^2},$$

which is unique to within a shift in argument and the addition of an arbitrary constant.

We note (see [21], p. 394) that if we consider the sequence $\{nE_n(f_m)\}$ for a fixed m as $n \to \infty$, then it follows from results of Bernstein [22] that for every $m = 0, 1, \ldots$

$$\lim_{n\to\infty} nE_n(f_m) = 0.282 \tag{17}$$

to within an accuracy of 0.004; that is, the magnitude of the limit of the sequence $\{nE_n(f_m)\}$ as $n \to \infty$ is considerably smaller than the $\pi/2$ in (16). In connection with this there was posed the problem of obtaining for fixed functions f of the class Lip 1 both upper and lower estimates for the limits of the sequences $\{nE_n(f)\}$ as $n \to \infty$ (see [23], p. 297, and [21], p. 394).

S. M. Nikol'skiĭ [21] solved the problem of the upper limit and proved there exists a function $f \in \mathrm{Lip}\,1$ such that

$$\overline{\lim_{n\to\infty}}\, nE_n(f) = \frac{\pi}{2};$$

that is, the constant $\pi/2$ in (16) is exact. But from Theorem 3 it follows that in Lip 1 there exists a function f such that the lower bound of the sequence $\{nE_n(f)\}$ is no less than ½; that is, it is greater than the value (17) of the limit of the sequence $\{nE_n(f_m)\}$.

The question of the exact value of the quantity

$$\sup_{f\in \mathrm{Lip}\,1} \underline{\lim_{n\to\infty}}\, nE_n(f)$$

remains open.

For the proof of Theorem 1 we will use Efimov's estimate (7). We also require the following results.

Lemma 1. *Let* $\omega(\delta)$ *be an arbitrary modulus of continuity and set, for* $n = 1, 2, \ldots,$

$$C_n(\omega) = \sup_{f\in H_\omega} \frac{1}{\pi} \int_{-\pi}^{\pi} f(t) \sin nt\,dt, \quad D_n(\omega) = \frac{2}{\pi} \sup_{f\in H_\omega} \int_0^{\frac{\pi}{2}} \left[f\left(\frac{t}{n}\right) - f\left(-\frac{t}{n}\right)\right] \sin t\,dt.$$

Then

$$C_n(\omega) = D_n(\omega) \tag{18}$$

and in H_ω there exists a function $\phi_n(t)$ with a Fourier series of the form

$$\varphi_n(t) \sim \sum_{\nu=0}^{\infty} \hat{\varphi}_{\nu,n} \sin(2\nu+1)nt, \tag{19}$$

where for the coefficients $\hat{\phi}_{\nu,n}$ we have

$$\hat{\varphi}_{0,n} = C_n(\omega), \tag{20}$$

$$|\hat{\varphi}_{\nu,n}| \leqslant c\frac{\omega(1/n)}{\nu} \qquad \text{for } \nu \geqslant 1. \tag{21}$$

Lemma 2. *Suppose f is an element of H_ω, $x_0 \in [-\pi, \pi]$, I is an interval of length $|I| \leq 2\pi$ and center x_0, n is a natural number ≥ 3 and ξ is an arbitrary number satisfying*

$$|1 - n\xi| \leqslant \frac{c}{\log n}. \tag{22}$$

Let I' stand for the interval which is concentric with I and of length $|I'| = 2|I|$. Then for $x \in I$

$$\begin{aligned} |s_n(f, x) - f(x)| &\leqslant \frac{\log(1+n|I|)}{\log n} S_n(\omega) \\ &+ \frac{1}{2\pi}\left| \int\limits_{[-2\pi, 2\pi]\setminus I'} \frac{\sin n(t-x)}{t-x_0}[f(t) - f(t+\pi\xi)]\,dt \right| + c\omega\left(\frac{1}{n}\right). \end{aligned} \tag{23}$$

Proof of Lemma 1. We first prove that

$$C_n(\omega) \leqslant D_n(\omega). \tag{24}$$

Indeed, let $f(x)$ be that function from H_ω for which the upper bound in the definition of $C_n(\omega)$ is attained; that is,

$$\frac{1}{\pi}\int_{-\pi}^{\pi} f(t)\sin nt\,dt = C_n(\omega).$$

We set (see also [8], proof of Lemma 1)

$$f_1(t) = \frac{1}{4n}\sum_{k=0}^{2n-1}(-1)^k\left[f\left(\frac{k\pi}{n}+t\right) - f\left(\frac{k\pi}{n}-t\right)\right]. \tag{25}$$

The function $f_1(t)$ is odd, belongs to H_ω and satisfies

$$f_1\left(t+\frac{\pi}{n}\right) = -f_1(t), \tag{26}$$

and

$$\begin{aligned} \frac{1}{\pi}\int_{-\pi}^{\pi} f_1(t)\sin nt\,dt &= \frac{1}{4n\pi}\sum_{k=0}^{2n-1}\int_{-\pi}^{\pi}(-1)^k\left[f\left(\frac{k\pi}{n}+t\right) - f\left(\frac{k\pi}{n}-t\right)\right]\sin nt\,dt \\ &= \frac{1}{\pi}\int_{-\pi}^{\pi} f(t)\sin nt\,dt = C_n(\omega). \end{aligned} \tag{27}$$

From (26) it follows that

$$\frac{1}{\pi}\int_{-\pi}^{\pi} f_1(t)\sin nt\,dt=\frac{1}{\pi}\int_{-\pi}^{\pi} f_1\left(\frac{t}{n}\right)\sin t\,dt$$
$$=\frac{2}{\pi}\int_{0}^{\pi} f_1\left(\frac{t}{n}\right)\sin t\,dt=\frac{2}{\pi}\int_{0}^{\frac{\pi}{2}}\left[f_1\left(\frac{t}{n}\right)-f_1\left(-\frac{t}{n}\right)\right]\sin t\,dt. \tag{28}$$

Inequality (24) follows from (27) and (28).

Now let $f_2(t)$ be an element of H_ω for which the upper bound in the definition of $D_n(\omega)$ is attained. We set

$$f_3(t)=\frac{1}{2}[f_2(t)-f_2(-t)].$$

The function $f_3(t)$ also belongs to H_ω, is odd and satisfies

$$\frac{2}{\pi}\int_0^{\frac{\pi}{2}}\left[f_3\left(\frac{t}{n}\right)-f_3\left(-\frac{t}{n}\right)\right]\sin t\,dt=\frac{2}{\pi}\int_0^{\frac{\pi}{2}}\left[f_2\left(\frac{t}{n}\right)-f_2\left(-\frac{t}{n}\right)\right]\sin t\,dt=D_n(\omega). \tag{29}$$

Let $\phi(t)=\phi_n(t)$ be the odd function with period $2\pi/n$ and defined on $[0, \pi/n]$ by the equalities

$$\varphi(t)=\begin{cases}\max\limits_{0\leqslant\tau\leqslant t} f_3(\tau), & \text{if}\quad 0\leqslant t\leqslant\frac{\pi}{2n};\\ \varphi\left(\frac{\pi}{n}-t\right), & \text{if}\quad \frac{\pi}{2n}<t\leqslant\frac{\pi}{n}.\end{cases} \tag{30}$$

Then $\phi(t)$ is nondecreasing on $[-\pi/2n, \pi/2n]$, $\|\phi\|\leq \frac{1}{2}\omega(\pi/n)$ and $\phi(t+\pi/n)=-\phi(t)$. We will show that $\phi\subset H_\omega$. For this we fix x and y, $x<y$, and show that

$$|\varphi(y)-\varphi(x)|\leqslant\omega(y-x). \tag{31}$$

Since $\phi(t)$ is odd, has period $2\pi/n$ and satisfies the equation $\phi(t+\pi/n)=-\phi(t/n)$, it suffices to establish (31) for x and y satisfying the condition $-\pi/2n\leq x<y\leq\pi/2n$. We set

$$t_x=\max\left\{t\colon t\in\left[-\frac{\pi}{2n},\frac{\pi}{2n}\right],\ f_3(t)=\varphi(x)\right\},$$
$$t_y=\min\left\{t\colon t\in\left[-\frac{\pi}{2n},\frac{\pi}{2n}\right],\ f_3(t)=\varphi(y)\right\}.$$

Then $0\leq t_y-t_x\leq y-x$, and therefore

$$|\varphi(y)-\varphi(x)|=|f_3(t_y)-f_3(t_x)|\leqslant\omega(t_y-t_x)\leqslant\omega(y-x),$$

whence (31) follows.

From the estimate

$$\frac{2}{\pi}\int_0^{\frac{\pi}{2}}\left[\varphi\left(\frac{t}{n}\right)-\varphi\left(-\frac{t}{n}\right)\right]\sin t\,dt=\frac{4}{\pi}\int_0^{\frac{\pi}{2}}\varphi\left(\frac{t}{n}\right)\sin t\,dt=$$

$$=\frac{4}{\pi}\int_0^{\frac{\pi}{2}}\max_{0\leqslant\tau\leqslant t} f_3\left(\frac{\tau}{n}\right)\sin t\,dt\geqslant\frac{4}{\pi}\int_0^{\frac{\pi}{2}} f_3\left(\frac{t}{n}\right)\sin t\,dt=\frac{2}{\pi}\int_0^{\frac{\pi}{2}}\left[f_3\left(\frac{t}{n}\right)-f_3\left(-\frac{t}{n}\right)\right]\sin t\,dt$$

and from (29) it follows that

$$\frac{2}{\pi}\int_0^{\frac{\pi}{2}}\left[\varphi\left(\frac{t}{n}\right)-\varphi\left(-\frac{t}{n}\right)\right]\sin t\,dt=D_n(\omega).$$

From this equality, taking into account the last of the properties in (30), we obtain

$$D_n(\omega)=\frac{1}{\pi}\int_{-\pi}^{\pi}\varphi(t)\sin nt\,dt$$

and, in particular, $D_n(\omega)\leq C_n(\omega)$. Thus (18) follows from (24). But the relations (19)–(21) are consequences of (18) and (30). The lemma is proved.

Proof of Lemma 2. For the proof of the lemma we use the following estimate:

$$|s_n(f,\ x)-f(x)|\leqslant\frac{1}{2\pi}\left|\int_{-2\pi}^{2\pi}\frac{\sin n(t-x)}{t-x}[f(t)-f(t+\pi\xi)]\,dt\right|+c\omega\left(\frac{1}{n}\right),\tag{32}$$

whenever $f\in H_\omega$, $x\in[-\pi,\pi]$ and the number ξ satisfies condition (22). This estimate follows from results of Rogosinski and Bernstein (see [18], Chapter III, §12, and also [24]), and we will not cite its proof.

Let

$$J_1=\frac{1}{2\pi}\left|\int_{I'}\frac{\sin n(t-x)}{t-x}[f(t)-f(t+\pi\xi)]\,dt\right|,$$

$$J_2=\frac{1}{2\pi}\left|\int_{[-2\pi,\ 2\pi]\setminus I'}\frac{\sin n(t-x)}{t-x}[f(t)-f(t+\pi\xi)]\,dt\right|.$$

We first estimate J_1. Let s be an integer (positive or negative) with the properties

$$s\neq 0,\quad x+\frac{\pi s}{n}\in I',\tag{33}$$

and let $F_s(t)=F_s(x,t)=f(t+\pi s/n+x)$. Then, taking into account that $0\leq\xi\leq c/n$, we obtain

$$\left|\int_{I'}\frac{\sin n(t-x)}{t-x}[f(t)-f(t+\pi\xi)]\,dt-\frac{1}{\pi}\sum_s\frac{(-1)^s}{s}\int_0^{\frac{\pi}{2}}\sin t\left\{\left[F_s\left(\frac{t}{n}\right)-F_s\left(-\frac{t}{n}\right)\right]\right.\right.$$
$$\left.\left.-\left[F_s\left(\pi\xi+\frac{t}{n}\right)-F_s\left(\pi\xi-\frac{t}{n}\right)\right]\right\}dt\right|\leqslant c\omega\left(\frac{1}{n}\right),\tag{34}$$

where the sum extends over the integers s satisfying (33). Indeed,

$$(-1)^s\int_0^{\frac{\pi}{2}}\sin t\left\{\left[F_s\left(\frac{t}{n}\right)-F_s\left(-\frac{t}{n}\right)\right]-\left[F_s\left(\pi\xi+\frac{t}{n}\right)-F_s\left(\pi\xi-\frac{t}{n}\right)\right]\right\}dt$$

$$=n\int_{\frac{\pi}{n}\left(s-\frac{1}{2}\right)}^{\frac{\pi}{n}\left(s+\frac{1}{2}\right)}[f(t+x)-f(t+x+\pi\xi]\sin nt\,dt,$$

and

$$\left|\frac{n}{\pi}\sum_s\frac{1}{s}\int_{\frac{\pi}{n}\left(s-\frac{1}{2}\right)}^{\frac{\pi}{n}\left(s+\frac{1}{2}\right)}\sin nt\,[f(t+x)-f(t+x+\pi\xi)]\,dt-\int_{I'}\frac{\sin n(t-x)}{t-x}[f(t+x)-f(t+x+\pi\xi)]\,dt\right|\leqslant c\omega(\pi\xi),$$

whence (34) follows. Since $F_s(t)\in H_\omega$, from (34), (33) and (18) it follows that

$$J_1\leqslant\frac{C_n(\omega)}{\pi}\log(1+n|I|)+c\omega\left(\frac{1}{n}\right)$$

and consequently, in view of (7),

$$J_1\leqslant\frac{\log(1+n|I|)}{\log n}S_n(\omega)+c\omega\left(\frac{1}{n}\right). \tag{35}$$

We now estimate J_2. Let

$$J_3=\left|\int_{[-2\pi,2\pi]\setminus I'}\left(\frac{1}{t-x}-\frac{1}{t-x_0}\right)\sin n(t-x)\,[f(t)-f(t+\pi\xi)]\,dt\right|.$$

Since $|x-x_0|\leq\frac{1}{2}|I|$ and

$$\int_{[-2\pi,2\pi]\setminus I'}\left|\frac{1}{t-x}-\frac{1}{t-x_0}\right|dt\leqslant\frac{|I|}{2}\frac{c}{|I|}\leqslant c,$$

on taking into account the fact that $|\xi|\leq c/n$ we obtain

$$J_3\leqslant c\omega\left(\frac{1}{n}\right).$$

Hence

$$J_2\leqslant\frac{1}{2\pi}\left|\int_{[-2\pi,2\pi]\setminus I'}\frac{\sin n(t-x)}{t-x_0}[f(t)-f(t+\pi\xi)]\,dt\right|+c\omega\left(\frac{1}{n}\right)$$

and consequently (see (35) and (32)) the estimate (23) holds. Lemma 2 is proved.

Proof of Theorem 1. We first prove the estimate from above; that is, we establish that for every element f of H_ω

$$\varlimsup_{n\to\infty}\frac{\|f-s_n(f)\|}{S_n(\omega)}\leqslant\frac{1}{2}. \tag{36}$$

Let m be a sufficiently large natural number. We consider the quantities $\|f - s_n(f)\|$ for n satisfying the condition

$$m \leqslant n \leqslant m\left(1 + \frac{1}{\log m}\right). \tag{37}$$

In order to establish (36) it is sufficient to prove that for every sufficiently large m there exists a natural number n satisfying (37) and such that

$$\|f - s_n(f)\| \leqslant \frac{1}{2} S_n(\omega) + c\omega\left(\frac{1}{n}\right) \log\log n. \tag{38}$$

For every natural number n we select a single point y_n belonging to $[-\pi, \pi]$ at which the value $\|f - s_n(f)\|$ is assumed; that is,

$$\|f - s_n(f)\| = |f(y_n) - s_n(f, y_n)|. \tag{39}$$

We consider all possible segments I lying in $[-\pi, \pi]$ and having the same length;

$$|I| = 2\pi \sqrt{\frac{\log m}{m}},$$

and denote by $N(I)$ the number of all points y_n with indices n satisfying (37) which lie in I. From (37) it follows that there exists at least one segment I such that

$$N = N(I) \geqslant \sqrt{\frac{m}{\log m}} - 1. \tag{40}$$

Fixing I with property (40), we denote the center of this segment by x_0 and the points y_n lying in I by $x_1, \ldots, x_N$. We denote the numbers n corresponding to the points $x_i = y_n$ $(i = 1, \ldots, N)$ by n_i. The natural numbers n_i are pairwise distinct, and from (37) it follows that

$$m \leqslant n_i \leqslant m\left(1 + \frac{1}{\log m}\right). \tag{41}$$

Furthermore, let I' be the segment with center x_0 and length

$$|I'| = 2|I| = 4\pi \sqrt{\frac{\log m}{m}}, \tag{42}$$

and let $\xi = 1/m$ and

$$J_i = \frac{1}{2\pi} \int_{[-2\pi, 2\pi] \setminus I'} \frac{\sin n_i (t - x_i)}{t - x_0} [f(t) - f(t + \pi\xi)]\, dt,$$

$$\varepsilon_i = \operatorname{sgn} J_i \qquad \text{for} \quad i = 1, \ldots, N.$$

Since, in view of (41),

$$0 \leqslant n_i \xi - 1 \leqslant \frac{c}{\log n_i}$$

and (see (41) and (42))

$$\frac{\log(1 + n_i |I|)}{\log n_i} \leqslant \frac{1}{2} + c\frac{\log\log n_i}{\log n_i},$$

in view of Lemma 2 and (39), for the proof of (38) it remains to check that there exists an i, $1 \le i \le N$, such that

$$|J_i| \leqslant c\omega\left(\frac{1}{n}\right). \tag{43}$$

But

$$\frac{1}{N}\sum_{i=1}^{N}|J_i| = \frac{1}{2\pi N}\int\limits_{[-2\pi,\,2\pi]\setminus I'} \frac{f(t)-f(t+\pi\xi)}{t-x_0}\left(\sum_{i=1}^{N}\varepsilon_i \sin n_i(t-x_i)\right)dt,$$

and therefore, applying Cauchy's inequality, we obtain

$$\begin{aligned}\frac{1}{N}\sum_{i=1}^{N}|J_i| \\ \leqslant \frac{1}{2\pi N}\left\{\int\limits_{[-2\pi,\,2\pi]\setminus I'}\left[\frac{f(t)-f(t+\pi\xi)}{t-x_0}\right]^2 dt\right\}^{\frac{1}{2}}\left\{\int\limits_{-2\pi}^{2\pi}\left[\sum_{i=1}^{N}\varepsilon_i \sin n_i(\tau-x_i)\right]^2 d\tau\right\}^{\frac{1}{2}}.\end{aligned} \tag{44}$$

Furthermore,

$$\int\limits_{[-2\pi,\,2\pi]\setminus I'}\left[\frac{f(t)-f(t+\pi\xi)}{t-x_0}\right]^2 dt \leqslant c\,\frac{\omega^2(\pi\xi)}{|I|} \leqslant c\,\frac{\omega^2(1/m)}{|I|},$$

$$\int\limits_{-2\pi}^{2\pi}\left[\sum_{i=1}^{N}\varepsilon_i \sin n_i(\tau-x_i)\right]^2 d\tau = 2\pi\sum_{i=1}^{N}\varepsilon_i^2 \leqslant cN,$$

and therefore it follows from (44) that

$$\frac{1}{N}\sum_{i=1}^{N}|J_i| \leqslant c\,\frac{\omega(1/m)}{(N|I|)^{\frac{1}{2}}}.$$

Since, in view of (40) and (42), $(N|I|)^{-1/2} \le c$, from the last estimate we obtain that for some i, $1 \le i \le N$, (43) is fulfilled; that is, the estimate (36) is established.

We now prove the second part of the conclusion of the theorem; that is, we construct a function f in H_ω such that

$$\lim_{n\to\infty}\frac{\|f-s_n(f)\|}{S_n(\omega)} = \frac{1}{2}. \tag{45}$$

Let p and m be natural numbers, $1 \le p \le m$, and suppose the function $\phi_m(t)$ satisfies the conditions (19)–(21) of Lemma 1 for $n = m$. We define a 2π-periodic function $\phi_{p,m}(t)$ on $[-\pi, \pi)$ by

$$\varphi_{p,m}(t) = \begin{cases}\varphi_m(|t|), & \text{if } \ t \in \left(-\frac{\pi p}{m}, \frac{\pi p}{m}\right), \\ 0, & \text{if } \ t \in [-\pi, \pi)\setminus\left(-\frac{\pi p}{m}, \frac{\pi p}{m}\right).\end{cases} \tag{46}$$

It is not difficult to see that $\phi_{p,m}(t)$ belongs to the class H_ω together with $\phi_m(t)$. We estimate $s_n(\phi_{p,m}, 0)$ when $m \le n \le 2m$. In view of (46), (19) and (20) we have

$$\int_{-\pi}^{\pi} \frac{\sin nt}{t} \varphi_{p,m}(t)\,dt = 2\int_0^{\frac{\pi p}{m}} \frac{\sin nt}{t}\varphi_m(t)\,dt = 2\sum_{\nu=0}^{\infty}\hat{\varphi}_{\nu,m}\int_0^{\frac{\pi p}{m}} \frac{\sin nt \sin(2\nu+1)mt}{t}\,dt$$

$$= C_m(\omega)\int_{\pi p\left(\frac{n}{m}-1\right)}^{\pi p\left(\frac{n}{m}+1\right)} \frac{1-\cos t}{t}\,dt + \sum_{\nu=1}^{\infty}\hat{\varphi}_{\nu,m}\int_{\pi p\left(2\nu+1-\frac{n}{m}\right)}^{\pi p\left(2\nu+1+\frac{n}{m}\right)} \frac{1-\cos t}{t}\,dt.$$

From (21) and the condition $m \le n \le 2m$ obtain

$$\left|\int_{-\pi}^{\pi} \frac{\sin nt}{t}\varphi_{p,m}(t)\,dt - C_m(\omega)\log\min\left(p, \frac{m}{n-m}\right)\right| \leqslant c\omega\left(\frac{1}{m}\right),$$

and consequently, for $m \le n \le 2m$,

$$\left|s_n(\varphi_{p,m}, 0) - \varphi_{p,m}(0) - \frac{C_m(\omega)}{\pi}\log\min\left(p, \frac{m}{n-m}\right)\right| \leqslant c\omega\left(\frac{1}{m}\right). \tag{47}$$

Now let k be a natural number, $k \ge k_0$, where k_0 is a sufficiently large number, $m_k = k^2$, $p_k = [k/\log^3 k]$, $x_k = 1/\log k$ and $f_k(t) = \phi_{p_k,m_k}(t - x_k)$. The number k_0 is chosen so large that for $k \ge k_0$ the intervals

$$\Delta_k = \left(x_k - \frac{\pi p_k}{m_k},\ x_k + \frac{\pi p_k}{m_k}\right),$$

which are the carriers of the functions f_k (see (46)), lie in $[0, \pi]$ and are pairwise disjoint and, what is more, the conditions

$$\inf\{|x-y| : x \in \Delta_i,\ y \in \Delta_j\} \geqslant \max(|\Delta_i|, |\Delta_j|) \tag{48}$$

are satisfied for $i \ne j$, $i, j \ge k_0$.

Since for $m_k \le n \le m_{k+1}$

$$\left|\log\min\left(p_k, \frac{m_k}{n-m_k}\right) - \frac{1}{2}\log n\right| \leqslant c\log\log n$$

and

$$|C_n(\omega) - C_{m_k}(\omega)| \leqslant c\frac{\omega(1/n)}{\sqrt{n}},$$

from (47) and (7) it follows that

$$|S_n(f_k, x_k) - f_k(x_k) - \frac{1}{2}S_n(\omega)| \leqslant c\omega\left(\frac{1}{n}\right)\log\log n \tag{49}$$

for $m_k \le n \le m_{k+1}$. We set

$$f(t) = \sum_{k=k_0}^{\infty} f_k(t)$$

and prove that f is the function we seek. Since each of the functions f_k belongs to H_ω, with

$$f_k(t)=0, \quad \text{if} \quad t\in[-\pi,\pi)\setminus\Delta_k; \quad \max_t f_k(t) = -\min_t f_k(t), \tag{50}$$

and since the intervals Δ_k belong to $[0,\pi]$, are pairwise disjoint and satisfy (48), we have $f\in H_\omega$. Indeed, let $|x-y|\le\pi$. We estimate the difference $f(x)-f(y)$. It is easy to see that it suffices to consider the case where the points x and y belong to distinct intervals Δ_k; that is, $x\in\Delta_i$, $y\in\Delta_j$ and $i\neq j$. Then in view of (50) there is a point x' in Δ_j such that $f_j(x')=f_i(x)$, and therefore

$$|f(x)-f(y)|=|f_i(x)-f_j(y)|=|f_j(x')-f_j(x)|\leqslant\omega(|x'-x|).$$

But, in view of (48), $|x'-x|\le|y-x|$, whence it follows that $|f(y)-f(x)|\le\omega(|y-x|)$.

Fix a natural number $k\ge k_0$ and set

$$\Phi(t)=\Phi_k(t)=f(t)-f_k(t).$$

From (49) it follows that (45) will be proved if we establish that for all n

$$|s_n(\Phi, x_k)-\Phi(x_k)|\leqslant c\omega\left(\frac{1}{n}\right). \tag{51}$$

For the proof of this estimate we use the following result (see [17]):

Lemma 3. *Let $g\in H_\omega$ and suppose that $g(t)=0$ for $t\in[-\pi,\pi]\setminus G$, where G is a 2π-periodic measurable set. Denote the characteristic function of the set G by $\chi(G,t)$. Then for all x and $n>1$*

$$|s_n(g,x)-g(x)|\leqslant c\omega\left(\frac{1}{n}\right)\left[\int_{\frac{1}{n}}^{\pi}\frac{\chi(G,x+z)+\chi(G,x-z)}{z}\,dz+1\right].^{(4)} \tag{52}$$

We apply Lemma 3 to the function $g(t)=\Phi(t)$ at the point $x=x_k$. The set G corresponding to this function is the set

$$G=\bigcup_{i=k_0}^{\infty}\Delta_i\setminus\Delta_k,$$

extended with period 2π.

The intervals Δ_i belong to $[0,\pi]$, and, as is easily seen,

$$|x_k-x_i|\leqslant c|z-x_k|, \quad \text{if} \quad z\in\Delta_i,\ i\geqslant k_0,\ i\neq k.$$

Therefore, taking into account that the length of the interval Δ_i does not exceed $2\pi/i\log^3 i$, we obtain

---

(4) We note that (52) is a simple consequence of (32), where one must set $\xi=1/n$.

$$\int_0^{\pi} \frac{\chi(G, x_k + z) + \chi(G, x_k - z)}{z} dz \leqslant c \left(\sum_{i=k_0}^{k-1} \frac{|\Delta_i|}{x_i - x_k} + \sum_{i=k+1}^{\infty} \frac{|\Delta_i|}{x_k - x_i} \right)$$
$$\leqslant c \left(\sum_{i=k_0}^{k-1} \frac{\log k}{i \log^2 i (\log k - \log i)} + \sum_{i=k+1}^{\infty} \frac{\log k}{i \log^2 i (\log i - \log k)} \right). \tag{53}$$

To estimate the sums in the right-hand side of (53) we use the inequalities

$$\frac{1}{\log k - \log i} \leqslant \frac{2}{\log k}, \quad \text{if} \quad k_0 \leqslant i \leqslant \sqrt{k};$$
$$\frac{1}{\log k - \log i} \leqslant \frac{k}{k - i}, \quad \text{if} \quad \sqrt{k} < i \leqslant k - 1;$$
$$\frac{1}{\log i - \log k} \leqslant \frac{i}{i - k}, \quad \text{if} \quad k + 1 \leqslant i \leqslant k^2;$$
$$\frac{1}{\log i - \log k} \leqslant \frac{1}{\log k}, \quad \text{if} \quad i \geqslant k^2.$$

It follows from these inequalities that

$$\int_0^{\pi} \frac{\chi(G_k, x_k + z) + \chi(G, x_k - z)}{z} dz \leqslant c,$$

and therefore, using (52), we obtain (51). The theorem is proved.

I express my thanks to Academician S. M. Nikol′skiĭ for some very useful discussions of the results in this article.

BIBLIOGRAPHY

1. S. M. Nikol′skiĭ, *La série de Fourier d'une fonction dont le module de continuité est donné*, C. R. (Dokl.) Acad. Sci. URSS 52 (1946), 191–194. MR 8, 149.

2. H. Lebesgue, *Sur la représentation trigonométrique approchée des fonctions satisfaisant à une condition de Lipschitz*, Bull. Soc. Math. France 38 (1910), 184–210.

3. L. Fejér, *Lebesguesche Konstanten und divergente Fourierreihen*, J. Reine Angew. Math. 139 (1910), 22–53.

4. E. Landau, *Darstellung und Begründung einiger neuerer Ergebnisse der Funktionentheorie*, Springer, Berlin, 1916.

5. A. N. Kolmogoroff, *Zur Grössenordnung des Restgliedes Fourierscher Reihen differenzierbarer Funktionen*, Ann. of Math. (2) 36 (1935), 521–526.

6. S. B. Stečkin, *Estimate of the remainder of a Taylor series for certain classes of analytic functions*, Izv. Akad. Nauk SSSR Ser. Mat. 17 (1953), 461–472. (Russian) MR 15, 206.

7. S. M. Nikol′skiĭ, *Sur l'évaluation asymptotique du reste dans l'approximation au moyen des sommes de Fourier*, C. R. (Dokl.) Akad. Sci. URSS 32 (1941), 386–389. MR 5, 118.

8. A. V. Efimov, *Approximation of functions with given modulus of continuity by Fourier sums*, Izv. Akad. Nauk SSSR Ser. Mat. 23 (1959), 115–134; English transl., Amer. Math. Soc. Transl. (2) 22 (1962), 139–162. MR 22 #1787.

9. ———, *Approximation of continuous periodic functions by Fourier sums*, Izv. Akad. Nauk SSSR Ser. Mat. 24 (1960), 243–296. (Russian) MR 22 #9792b.

10. H. Bohr, *Über Koeffizientensumme einer beschränkten Potenzreihe*, Nachr. Konigl. Ges. Wiss. Göttingen **1916**, 276–291.

11. L. Neder, *Über Koeffizientensumme einer beschränkten Potenzreihe*, Math. Z. 11 (1921), 115–123.

12. G. Ja. Doronin, *Some inequalities for approximation by trigonometric polynomials*, Dokl. Akad. Nauk SSSR 69 (1949), 487–490. (Russian) MR **11**, 348.

13. D. E. Men′šov, *Sur les sommes partielles des séries de Fourier des fonctions continues*, Mat. Sb. **15** (57) (1944), 385–432. MR 6, 264.

14. E. Busko, *Fonctions continues et fonctions bornées non adhérentes dans* $L^{\infty}(T)$ *à la suite de leurs sommes partielles de Fourier*, Studia Math. 34 (1970), 319–337. MR 41 #8906.

15. *On approximation theory* (Proc. Conf. Oberwolfach, 1963), Internat. Ser. Numer. Math., vol. 5, Birkhäuser, Basel, 1964. MR 31 #2537.

16. K. I. Oskolkov, *Sharpness of the Lebesgue estimate for approximation of functions with given modulus of continuity by means of Fourier sums*, Trudy Mat. Inst. Steklov. 112 (1971), 337–345 = Proc. Steklov Inst. Math. 112 (1971), 348–357. MR 49 #970.

17. ———, *Subsequences of the Fourier sums of functions with a given modulus of continuity*, Mat. Sb. 88 (130) (1972), 447–469 = Math. USSR Sb. 17 (1972), 441–466. MR 48 #11874.

18. A. Zygmund, *Trigonometric series*. Vol. 1, 2nd rev. ed., Cambridge Univ. Press, New York, 1959. MR 21 #6498.

19. J. Favard, *Sur les meilleurs procédés d'approximation de certaines classes de fonctions par les polynomes trigonométriques*, Bull. Sci. Math. (2) 61 (1937), 209–224.

20. N. I. Ahiezer and M. G. Kreĭn, *Sur la meilleure approximation des fonctions périodiques dérivables au moyen de sommes trigonométriques*, C. R. (Dokl.) Acad. Sci. URSS 15 (1937), 107–111.

21. S. M. Nikol′skiĭ, *On interpolation and best approximation of differentiable periodic functions by trigonometric polynomials*, Izv. Akad. Nauk SSSR Ser. Mat. 10 (1946), 393–410. (Russian) MR 8, 262.

22. Serge Bernstein, *Sur la meilleure approximation de* $|x|$ *par des polynomes des degrés donnés*, Acta Math. 37 (1914), 1–57.

23. S. M. Nikol′skiĭ, *On the best approximation of functions satisfying Lipschitz conditions by polynomials*, Izv. Akad. Nauk SSSR Ser. Mat. 10 (1946), 295–322. (Russian) MR 8, 153.

24. S. N. Bernstein, *Sur un procédé de sommation des séries trigonométriques*, C. R. Acad. Sci. Paris 191 (1930), 976–979.

Translated by LE BARON O. FERGUSON

Trudy Mat. Inst. Steklov.
134 (1975)

Proc. Steklov Inst. Math.
134 (1975)

SUR CERTAINES FORMULES DE QUADRATURE DE DEGRE D'EXACTITUDE MAXIMUM*

T. POPOVICIU

Abstract. The problem of finding the coefficients and nodes of a quadrature formula of the form

$$\int_{-1}^{1} f(x)\,dx \approx \sum_{i=1}^{p} \sum_{j=0}^{k_i-1} c_{ij} f^{(j)}(z_i)$$

is considered, where the formula is exact for polynomials of maximum degree. The cases $p = 2$, $k_1 = k$, $k_2 = 1$ and $p = 3$, $k_1 = k$, $k_2 = k_3 = 1$ (k is an arbitrary odd integer) are investigated in detail. In the second case effective formulas are found for the c_{ij} and z_i as functions of k.

Bibliography: 6 titles.

1. Considérons la formule de quadrature

$$\int_{-1}^{1} f(x)\,dx = \sum_{i=1}^{p} \sum_{j=0}^{k_i-1} c_{i,j} f^{(j)}(z_i) + R(f) \tag{1}$$

où $z_1, z_2, \ldots, z_p$ sont p (> 0) points distincts de l'axe réel. Ces sont les noeuds de la formule et $k_1, k_2, \ldots, k_p$ sont p nombres naturels, les ordres de multiplicité des noeuds $z_1, z_2, \ldots, z_p$ respectifs. Nous pouvons supposer que k_i noeuds sont confondus dans le point z_i, donc que z_i est un noeud d'ordre k_i de multiplicité. Le nombre total des noeuds distincts ou non est égal à $k_1+k_2+\ldots+k_p=m$, où m est donc un nombre naturel tel que $m \geqslant p \geqslant 1$.

En ce qui concerne la fonction f nous pouvons supposer qu'elle est définie et continue sur un intervalle contenant les points -1, 1 et les noeuds z_i, $i=1, 2, \ldots, p$. De plus, f admet un certain nombre de dérivées, pour que le second membre de la formule (1), où les accents signifient des dérivations successives, ait un sens. Les $c_{i,j}$ sont des coefficients indépendants de la fonction f, les coefficients de la formule (1) et $R(f)$ est le reste de la formule de quadrature considérée. De nombreux auteurs dont nous-même avons étudié [3, 4] la structure de ce reste.

2. La formule (1), ou bien le reste $R(f)$ de cette formule, ont un degré d'exactitude. C'est le nombre entier $n \geqslant -1$ complètement déterminé par la condition que $R(P)$ soit nul sur tout polynome P de degré n(1) et que $R(x^{n+1}) \neq 0$. On dit aussi que la formule (1), ou le reste $R(f)$ de cette formule, est de

AMS (MOS) subject classifications (1970). Primary 41A55.

*Editor's note. The body of this article is an unaltered reprinting of the original paper. Details of the bibliography have been checked and put in standard form, and MR citations have been added.

(1) Un polynome de degré g est un polynome de degré effectif $\leq g$. Le polynome unique de degré -1 est le polynome nul.

degré d'exactitude n. $R(f)$ est, d'ailleurs, comme les deux autres termes de la formule (1), une fonctionnelle linéaire (additive et homogène) définie sur un certain ensemble linéaire des fonctions f. Dans la suite on peut toujours supposer $n \geqslant 0$, le cas $n = -1$ n'intervenant pas. La condition $n \geqslant 0$ est, d'ailleurs, équivalente à l'égalité $2 = \sum_{i=0}^{p} c_{i,0}$.

3. Les noeuds z_i et leurs ordres de multiplicité k_i, $i=1, 2, \ldots, p$, étant donnés, mais quelconques, la condition $n \geqslant m-1$ détermine complètement la formule de quadrature (1), donc les coefficients $c_{i,j}$, $j=0, 1, \ldots, k_i-1$, $i=1, 2, \ldots, p$, de cette formule. On peut alors obtenir facilement ces coefficients, en appliquant la formule (1) au polynome d'interpolation de Lagrange—Hermite de degré $m-1$ de la fonction f correspondant aux noeuds z_i, comptés avec leurs ordres de multiplicité respectifs. Il est inutil d'écrire ici ce polynome et les valeurs des coefficients $c_{i,j}$ qu'on en déduit. Dans quelques cas particuliers considérés plus loin nous donnerons leurs valeurs explicites.

Dans la suite nous supposerons toujours qu'il s'agit d'une telle formule de quadrature (1).

4. Le degré d'exactitude de la formule est alors égal à $n=m+q-1$, où l'entier non-négatif q est caractérisé par la propriété que le polynome

$$l(x) = \prod_{i=1}^{p} (x - z_i)^{k_i} \tag{2}$$

est orthogonal à tout polynome de degré $q-1$, donc que

$$\int_{-1}^{1} Q(x)\, l(x)\, dx = 0 \tag{3}$$

pour tout polynome Q de degré $q-1$ et que

$$\int_{-1}^{1} x^q l(x)\, dx \neq 0.$$

Pour la démonstration voir, par exemple, l'un des nos travaux cités [4].

Retenons la propriété exprimée par le

L e m m e 1. *Si la formule de quadrature* (1) *est de degré d'exactitude* $n = m+q-1$, *où* $q \geqslant 0$, *le polynome* (2) *change de signe, à l'intèrieur de l'intervalle* $[-1, 1]$, *au moins* q *fois.*

C'est une conséquence immédiate de l'orthogonalité.

5. Nous pouvons chercher à imposer aux noeuds z_i des conditions supplémentaires pour obtenir un ordre d'exactitude le plus grand possible. L'entier non-négatif q est au plus égal à p [4] et le maximun p de q peut être atteint. Nous dirons alors (donc si $q=p$) que la formule de quadrature (1) est de degré d'exactitude maximum ou encore qu'elle est du type Gauss. De telles formules ont été étudiées par plusieurs auteurs parmi lesquels nous signalons L. Tchakaloff [5], P. Turán [6], A. Ossicini [1, 2], nous les avons considérées dans nos travaux cités [3, 4]. Nous remarquons, en passant, qu'on a aussi considéré des problèmes analogues pour des fonctionnelles linéaires plus gé-

nérales que le premier membre de la formule (1).

Retenons, en particulier, la propriété exprimée par le

Théorème 1. *Pour qu'il existe au moins une formule du type Gauss, de la forme indiquée plus haut, il faut et il suffit que les ordres de multiplicité* $k_1, k_2, \ldots, k_p$, *supposés donnés, des noeuds* $z_1, z_2, \ldots, z_p$ *soient tous impairs.*

Nous pouvons résumer la démonstration de la façon suivante:

La condition est nécessaire d'après le lemme 1.

La condition est aussi suffisante. En effet, si les nombres k_i sont tous impairs, le polynome de p variables $z_1, z_2, \ldots, z_p$,

$$\Phi(z_1, z_2, \ldots, z_p) = \int_{-1}^{1} l(x) \prod_{i=1}^{p} (x - z_i)\, dx$$

est positif sur tout l'espace à p dimensions. Il a donc un minimum positif qui est atteint sur un point de minimum relatif. On peut voir facilement que pour un point de minimum (absolu) $(z_1, z_2, \ldots, z_p)$ les coordonnées $z_1, z_2, \ldots, z_p$ sont distinctes.

Sur un tel point on a

$$\frac{\partial \Phi}{\partial z_i} = 0, \quad i = 1, 2, \ldots, p. \tag{4}$$

Il suffit alors de remarquer que le système (4) est équivalent à l'orthogonalité du polynome (2) avec tout polynome de degré $p-1$.

6. Il résulte de ce qui précède, que les nombres k_i, $i=1, 2, \ldots, p$ étant supposés toujours impairs, dans toute formule (1) du type Gauss, les noeuds $z_1, z_2, \ldots, z_p$ sont distincts et à l'intérieur de l'intervalle $[-1, 1]$.

Mais, à ma connaissance, on ne sait pas encore quel est le nombre exact de formules distinctes du type Gauss, pour un système donné d'ordres de multiplicité (impairs). Il semble que ce nombre est égal au nombre des permutations distinctes des nombres $k_1, k_2, \ldots, k_p$, à toute permutation correspondant une formule du type Gauss bien déterminée. Ceci est vrai si $k_1 = k_2 = \ldots = k_p$. Il y a alors une seule formule du type Gauss.

Deux formules sont, d'ailleurs, distinctes si et seulement si les polynomes (2) correspondants sont différents.

Le but de ce petit travail est de résoudre complètement ce problème dans quelques cas particuliers, différents de celui où les ordres de multiplicité sont tous égaux. Nous complétons ainsi certains de nos résultats antérieurs [3].

7. Considérons le cas $p=2$, $k_1=k$, $k_2=1$. Le cas $k=3$ a été considéré antérieurement [3]. Supposons maintenant que k soit impair quelconque. Les noeuds z_1, z_2 sont alors donnés par le système

$$\int_{-1}^{1} (x - z_1)^{k+i} (x - z_2)\, dx = 0, \qquad i = 0, 1.$$

Si nous posons

$$\frac{1 - z_1}{1 + z_1} = y \tag{5}$$

le système peut s'écrire

$$\begin{aligned} &\frac{y^{k+2}+1}{k+2}+\frac{z_1-z_2}{1+z_1}\cdot\frac{y^{k+1}-1}{k+1}=0, \\ &\frac{y^{k+3}-1}{k+3}+\frac{z_1-z_2}{1+z_1}\cdot\frac{y^{k+2}+1}{k+2}=0. \end{aligned} \tag{6}$$

En éliminant z_2, le noeud z_1 est donné par (5) et par l'équation réciproque en y

$$(k+2)^2(y^{k+1}-1)(y^{k+3}-1)-(k+1)(k+3)(y^{k+2}+1)^2=$$
$$=y^{2k+4}-(k+2)^2y^{k+3}-2(k+1)(k+3)y^{k+2}-(k+2)^2y^{k+1}+1=0, \tag{7}$$

qui a exactement 2 racines réelles positives et distinctes (séparées par 1).

Il en résulte que dans le cas étudié nous avons exactement deux formules du type Gauss qui sont confondues pour $k=1$ dans la formule classique de Gauss, mais sont bien distincts pour $k>1$. En effet, le noeud z_2 s'obtient de l'une des équations (6) et la théorie précédente nous montre qu'il sera complétement déterminé, différent de z_1 et compris strictement entre -1 et 1.

Pour $k=3$ nous avons déjà donné la valeur explicite de z_1 (donc aussi de z_2) et de la formule (1) correspondante. z_1 et z_2 sont alors des irrationalités quadratiques. Pour $k>3$ la nature arithmétique de z_1 et z_2 semble être plus compliquée.

8. Pour calculer z_1 il faut résoudre d'abord l'équation réciproque (7). Pour cela nous posons

$$y+\frac{1}{y}=t. \tag{8}$$

Alors y^s+1/y^s est égal à un polynome $P_s(t)$ en t, de degré s, donné par la relation de récurrence

$$P_{s+1}=tP_s-P_{s-1},\quad s=1,2,\dots\qquad (P_0=2,\ P_1=t).$$

L'équation en t est donc

$$P_{k+2}(t)-(k+2)^2t-2(k+1)(k+3)=0. \tag{9}$$

De ce qui précède il résulte que cette équation en t a une racine réelle >2. On vérifie facilement qu'elle a aussi la racine double -2.

Pour $k=1$ et $k=3$ le premier membre de (9) est

$$t^3-12t-16=(t+2)^2(t-4),$$
$$t^5-5t^3+5t^2-25t-48=(t+2)^2(t-3)(t^2-t+4)$$

respectivement, ce qui donne les résultats signalés bien connus.

Pour $k=5$ l'équation (9) devient

$$t^7-7t^5+14t^3-56t-96=(t+2)^2(t^5-4t^4+5t^3-4t^2+10t-24)=0$$

et on vérifie que cette équation n'a plus de racines positives rationnelles.

9. Considérons encore le cas $p=3$, $k_1=k$, $k_2=k_3=1$. Le cas $k=3$ a déjà été considéré antérieurement [3]. Supposons que k soit impair quelconque. Les noeuds z_1, z_2, z_3 sont donnés par le système

$$\int_{-1}^{1}(x-z_1)^{k+i}(x-z_2)(x-z_3)\,dx=0,\qquad i=0,1,2. \tag{10}$$

En écrivant (5) et en éliminant z_2 et z_3, on trouve que y est donné par l'équation

$$\begin{vmatrix} \frac{y^{k+3}-1}{k+3} & \frac{y^{k+2}+1}{k+2} & \frac{y^{k+1}-1}{k+1} \\ \frac{y^{k+4}+1}{k+4} & \frac{y^{k+3}-1}{k+3} & \frac{y^{k+2}+1}{k+2} \\ \frac{y^{k+5}-1}{k+5} & \frac{y^{k+4}+1}{k+4} & \frac{y^{k+3}-1}{k+3} \end{vmatrix} = 0$$

ou, en faisant les calculs,

$$4y^{3k+9} - Ay^{2k+8} - By^{2k+7} - Cy^{2k+6} - Dy^{2k+5} - Ey^{2k+4} + Ey^{k+5} + Dy^{k+4} + Cy^{k+3} + By^{k+2} + Ay^{k+1} - 4 = 0 \qquad (11)$$

avec

$$A = (k+2)^2 (k+3)^2, \qquad B = 4(k+1)(k+3)^2(k+4),$$
$$C = 6(k+2)(k+4)(k^2+6k+7), \quad D = 4(k+2)(k+3)^2(k+5),$$
$$E = (k+3)^2(k+4)^2.$$

Remarquons que A, B, C, D, E sont positifs.

Si nous désignons par $\varphi(y)$ le premier membre de l'équation, nous avons $\varphi(0) < 0$ et $\varphi'(1) = 12(k+3) - (k+7)A - (k+5)B - (k+3)C - (k+1)D - -(k-1)E < 0$. La règle des signes de Descartes nous montre alors que l'équation réciproque (11) a exactement 3 racines réelles et positives qui sont distinctes, dont l'une est égale à 1.

De l'orthogonalité et du lemme 1 il résulte que pour un z_1 déterminé le polynome $(x-z_2)(x-z_3)$ est déterminé complètement, z_2, z_3 étant distincts, compris strictement entre -1 et 1 et étant différents de z_1.

Dans ce cas nous avons donc exactement 3 formules du type Gauss. Si $k=1$ ces trois formules sont confondues avec la formule classique de Gauss avec 3 noeuds. Mais si $k > 1$ les trois formules sont distinctes. L'une correspond à la valeur 0 de z_1 et les deux autres à deux valeurs de z_1 symétriques par rapport à l'origine.

Remarquons que, dans ce cas, le nombre des permutations distinctes des nombres k_1, k_2, k_3 est bien égale à 3.

10. Toujours dans le cas particulier considéré, la formule du type Gauss correspondant au noeud $z_1=0$ peut être obtenue explicitement sous une forme assez simple.

En posant $z_1 = 0$ dans le système (10) on obtient $(x-z_2)(x-z_3) = x^2 - \lambda^2$, où $\lambda = \sqrt{(k+2)/(k+4)}$. Nous avons donc $z_2 = \sqrt{(k+2)/(k+4)}$, $z_3 = -\sqrt{(k+2)/(k+4)}$.

Écrivons la formule de quadrature sous la forme

$$\int_{-1}^{1} f(x)\,dx - \sum_{j=0}^{k-1} c_j f^{(j)}(0) + cf(\lambda) + df(-\lambda) + R(f).$$

La formule est de degré $k+4$ d'exactitude.

Pour trouver les coefficients c, d prenons successivement $f=x^k$, $f=x^{k+1}$. Compte tenu du fait que k est impair, nous en déduisons

$$c = d = \frac{1}{(k+2)\lambda^{k+1}}. \tag{12}$$

En prenant $f = x^i$ où i est impair nous trouvons $(k > 1)$

$$c_1 = c_3 = \ldots = c_{k-2} = 0. \tag{13}$$

Enfin si nous prenons $f = x^{2j}(x^{k-2j+1} - \lambda^{k-2j+1})$, $j = 0, 1, \ldots, \frac{k-1}{2}$, nous obtenons

$$c_{2j} = \frac{2}{(2j+1)!} - \frac{2}{(2j)!\,(k+2)\lambda^{k-2j+1}}, \quad j = 0, 1, \ldots, \frac{k-1}{2}. \tag{14}$$

La formule de quadrature cherchée s'écrit donc

$$\int_{-1}^{1} f(x)\,dx = \sum_{j=0}^{\frac{k-1}{2}} \left(\frac{2}{(2j+1)!} - \frac{2}{(2j)!\,(k+2)\lambda^{k-2j+1}} \right) f^{(2j)}(0) \tag{15}$$

$$+ \frac{1}{(k+2)\lambda^{k+1}} (f(\lambda) + f(-\lambda)) + R(f).$$

Le reste $R(f)$ est donné par la formule (voir [3]),

$$R(f) = R(x^{k+5})[\xi_1, \xi_2, \ldots, \xi_{k+6}; f]$$

ou $\xi_1, \xi_2, \ldots, \xi_{k+6}$ sont $k+6$ points distincts de l'intérieur de l'intervalle $[-1, 1]$ et qui dépendent, en général, de la fonction (continue) $f(x)$. $[\xi_1, \xi_2, \ldots, \xi_{k+6}; f]$ est la différence divisée de la fonction f sur les noeuds ξ_i, $i = 1, 2, \ldots, k+6$.

On peut obtenir le coefficient $R(x^{k+5})$ en prenant dans la formule (15) $f = x^{k+3}(x^2 - \lambda^2)$. On obtient ainsi

$$R(x^{k+5}) = R(x^{k+3}(x^2 - \lambda^2)) = 2\left(\frac{1}{k+6} - \frac{\lambda^2}{k+4}\right) = \frac{8}{(k+4)^2(k+6)}.$$

BIBLIOGRAPHY

1. A. Ossicini, *Construzione di formule di quadratura di tipo Gaussiano*, Ann. Mat. Pura Appl. (4) 72 (1966), 213–237. MR 34 #2180.

2. ———, *Le funzioni di influenzá nel problema di Gauss sulle formule di quadratura*, Matematiche (Catania) 23 (1968), 7–30. MR 40 #3719.

3. T. Popoviciu, *Sur une généralisation de la formule d'intégration numérique de Gauss*, Acad. R. P. Romîne. Fil. Iaşi. Stud. Cerc. Şti. 6 (1955), 29–57. (Romanian) MR 19, 64.

4. ———, *La simplicité du reste dans certaines formules de quadrature*, Mathematica (Cluj) 6 (29) (1964), 157–184. MR 32 #4848.

5. L. Čakalov, *Über eine allgemeine Quadraturformel*, C. R. Acad. Bulgare Sci. Math. Nat. 1 (1948), no. 2–3, 9–12. MR 10, 743.

6. P. Turán, *On the theory of the mechanical quadrature*, Acta Sci. Math. (Szeged) 12 (1950), 29–37. MR 12, 164.

Trudy Mat. Inst. Steklov.
134 (1975)

Proc. Steklov Inst. Math.
134 (1975)

ON STRUCTURAL CHARACTERIZATIONS OF CLASSES OF FUNCTIONS WITH A GIVEN DEGREE OF BEST APPROXIMATIONS

UDC 517: 51

M. K. POTAPOV

Abstract. In this paper a structural characterization is given for a class of nonperiodic functions belonging to $L_p[-1, 1]$ and satisfying the condition that for any n there exists an algebraic polynomial $P_n(x)$ of degree no higher than n such that

$$\left\| [f(x) - P_n(x)](1-x)^\alpha (1+x)^\beta \left(1 - x + \frac{1}{n'}\right)^{\rho_1} \left(1 + x + \frac{1}{n'}\right)^{\rho_2} \right\|_p \leqslant Cn^{-(r+\gamma)}.$$

From the results also follows, in particular, an answer to the question of a structural characterization of the class of nonperiodic functions satisfying the condition

$$E_n(f)_{p,\alpha,\beta} \leqslant Cn^{-(r+\gamma)}.$$

Here $E_n(f)_{p,\alpha,\beta}$ is the best approximation of $f(x)$ by algebraic polynomials of degree no higher than n, in the L_p-metric with weight $(1-x)^\alpha(1+x)^\beta$.

Bibliography: 11 titles.

I. As far back as the beginning of the century it was established that if $f(t)$ is a continuous 2π-periodic function, then the condition $f \in \operatorname{Lip} \alpha$, $0 < \alpha < 1$, is equivalent to the condition $\epsilon_n(f) \leq Cn^{-\alpha}$, where $\epsilon_n(f)$ is the best approximation of $f(t)$ by trigonometric polynomials of degree not larger than n.

At the same time it was established that if $f(x)$ is a nonperiodic function which is continuous on $[a, b]$ and $E_n(f)$ is the best approximation of $f(x)$ on $[a, b]$ by algebraic polynomials of degree not larger than n, then

a) if $f(x) \in \operatorname{Lip} \alpha$, $0 < \alpha < 1$, on the whole segment $[a, b]$ then $E_n(f) \leq Cn^{-\alpha}$;

b) if $E_n(f) \leq Cn^{-\alpha}$, $0 < \alpha < 1$, then $f(x) \in \operatorname{Lip} \alpha$ only on every segment $[c, d] \subset (a, b)$.

Thus for periodic functions the condition $\epsilon_n(f) \leq n^{-\alpha}$ allows us to describe in a constructive way the class of functions satisfying a Lipschitz condition on a whole period, and for nonperiodic functions the condition $E_n(f) \leq Cn^{-\alpha}$ does not give a constructive characterization of the class of functions satisfying a Lipschitz condition on the whole segment $[a, b]$. Therefore the following questions arise:

AMS (MOS) subject classifications (1970). Primary 41A10, 41A25, 41A50.

1) What is a constructive characterization of the class of nonperiodic functions satisfying a Lipschitz condition?

2) What is a structural characterization of the class of nonperiodic functions satisfying the condition $E_n(f) \leq Cn^{-\alpha}$?

For periodic functions the result on the constructive characterization of the class Lip α was subsequently extended to the space L_p and to other classes of functions. The question of the constructive characterization of nonperiodic functions $f(x) \in \text{Lip}\ \alpha$ remained open for a long time.

But in the year 1946 it was proved S. M. Nikol'skiĭ [1] that if $f(x) \in \text{Lip}\ \alpha$ on the segment $[-1, 1]$, then for every n there exists an algebraic polynomial $P_n(x)$ with degree no larger than n such that

$$|f(x) - P_n(x)| \leqslant \frac{\pi}{2} \frac{\sqrt{1-x^2}}{n} + |x| O\left(\frac{\ln n}{n^2}\right).$$

This result shows that it is possible to approximate a function $f(x) \in \text{Lip}\ 1$ better at the end points of the interval than at interior points.

On the basis of this fact, Nikol'skiĭ conjectured that a constructive characterization of a nonperiodic function given on a closed interval must take into account the position of the points on the interval. This conjecture was confirmed in the work of A. F. Timan [2] and V. K. Dzjadyk [3], in which a constructive characterization of the class Lip α was obtained; that is, an answer to the first question was obtained. Namely, they proved that in order for a nonperiodic continuous function $f(x)$ to be in Lip α, $0 < \alpha < 1$, on the whole segment $[-1, 1]$, it is necessary and sufficient that for every n there is an algebraic polynomial $P_n(x)$ of degree no larger than n such that

$$\max_{-1 \leqslant x \leqslant 1} \frac{|f(x) - P_n(x)|}{\left(\sqrt{1-x^2} + \frac{1}{n}\right)^\alpha} \leqslant \frac{C}{n^\alpha}.$$

Naturally the following question arose. Is the condition

$$\left(\int_{-1}^{1} \left| \frac{f(x) - P_n(x)}{\left(\sqrt{1-x^2} + \frac{1}{n}\right)^\alpha} \right|^p dx \right)^{1/p} \leqslant \frac{C}{n^\alpha}, \quad 1 \leqslant p < \infty;\ n = 1, 2, \ldots, \qquad (*)$$

a constructive characterization of the class of nonperiodic functions satisfying a Lipschitz condition in the L_p metric? The answer to this question was obtained by V. P. Motornyĭ [4]. Using results of G. K. Lebed' [5] and M. K. Potapov [6], [7], he proved that condition $(*)$ is not a constructive characterization of the class $\text{Lip}(\alpha, p)$.

Thus the following questions arise:

3) What is a constructive characterization of the class of nonperiodic functions satisfying a Lipschitz condition in the L_p metric?

4) What is a structural characterization of the class of nonperiodic functions for which condition $(*)$ is fulfilled?

The third question is still open.

Questions 2 and 4 are special cases of the more general question:

5) What is a structural characterization of the class of nonperiodic functions

which satisfy the condition: for every n there exists an algebraic polynomial $P_n(x)$ with degree no larger than n such that

$$\left\|[f(x)-P_n(x)](1-x)^{\alpha}(1+x)^{\beta}\left(1-x+\frac{1}{n^2}\right)^{\rho_1}\left(1+x+\frac{1}{n^2}\right)^{\rho_2}\right\|_p \leqslant Cn^{-(r+\gamma)}?$$

In the case where $\alpha = \beta \geq -\frac{1}{2}p$ and $\rho_1 = \rho_2$, the answer to the fifth question was obtained in [8]. An answer to the fifth question is given in the present article in the case $\alpha > \beta > -\frac{1}{2}p$.

II. We now cite the definitions which are necessary in what follows, and formulate the fundamental result of this paper. We will write, as usual, $f(x) \in L_p[a, b]$ if $f(x)$ is given on an interval $[a, b]$, for $1 \leq p < \infty$ the function is measurable and

$$\|f\|_p = \left(\int_a^b |f(x)|^p\,dx\right)^{1/p} < \infty,$$

and for $p = \infty$ the function $f(x)$ is continuous and

$$\|f\|_\infty = \|f\|_C = \max_{a\leqslant x\leqslant b} |f(x)| < \infty.$$

In the case where $[a, b] = [-1, 1]$ we will, for simplicity, write L_p instead of $L_p[-1, 1]$.

We will also write $f \in L_{p,\alpha,\beta}$ if the function

$$\varphi(x) = f(x)(1-x)^{\alpha}(1+x)^{\beta} \in L_p,$$

and also

$$\|f\|_{p,\alpha,\beta} = \|f(x)(1-x)^{\alpha}(1+x)^{\beta}\|_p.$$

We will write $E_n(f)_{p,\alpha,\beta}$ for the best approximation of the function $f(x)$ by means of algebraic polynomials $P_n(x)$ with degree at most n in the metric $L_{p,\alpha,\beta}$; that is,

$$E_n(f)_{p,\alpha,\beta} = \inf_{P_n} \|f(x) - P_n(x)\|_{p,\alpha,\beta}.$$

We assume further that $\alpha > \beta > -\frac{1}{2}$, and set

$$f(x, t, \alpha, \beta) = \frac{1}{\mu(\alpha, \beta)} \int_0^1 \int_{-1}^1 f\Big[x\cos t + rz\sin t\sqrt{1-x^2} - (1-r^2)(1-x)\sin^2\frac{t}{2}\Big](1-r^2)^{\alpha-\beta-1} r^{2\beta+1}(1-z^2)^{\beta-1/2}\,dz\,dr,$$

where

$$\mu(\alpha, \beta) = \int_0^1 \int_{-1}^1 (1-r^2)^{\alpha-\beta-1} r^{2\beta+1}(1-z^2)^{\beta-1/2}\,dz\,dr.$$

We remark that the function $f(x, t, \alpha, \beta)$ (a generalized shift function) arises naturally in the solution of the problem of structurally characterizing the class of functions for which $E_n(f)_{p,\alpha,\beta} \leq Cn^{-r-\gamma}$. That is, the form which it has is connected with the properties of the Jacobi polynomials which are used in the solution of this problem.

We will say that the function $f(x)$ has an rth derivative $f^{(r)}(x)$ inside the interval $(-1, 1)$ if on any fixed segment $[a, b] \subset (-1, 1)$ the function $f(x)$ has an absolutely continuous derivative of order $r-1$.

If $f(x)$ has an rth derivative $f^{(r)}(x) = \psi(x)$ inside the interval $(-1, 1)$, then the function $\psi(x, t, \alpha, \beta)$ will be designated by $f^{(r)}(x, t, \alpha, \beta)$.

The purpose of this paper is to prove the following theorem.

Theorem 1. *Suppose we have numbers* $\alpha, \beta, \rho_1, \rho_2, \delta_1, \delta_2, \gamma$ *and* r *such that* r *is a nonnegative integer,* $0 < \gamma \leq 1$,

$$r+\gamma > 2\max(\rho_1, \rho_2), \quad \max(\rho_1+\delta_1, \delta_2+\rho_2) < \gamma/2 < r/2 + 1 + 1/p + \min(\rho_1+\alpha, \rho_2+\beta),$$

for $1 \leqslant p < \infty$: $\alpha > \beta > -1/2p$, $\alpha-\delta_1 > \beta-\delta_2 > -r/2 - 1/2p$, $\delta_1 < \alpha + r/2 + 1/p$;

for $p = \infty$: $\alpha > \beta \geqslant 0$, $\alpha - \delta_1 > \beta - \delta_2 \geqslant -r/2$.

In order that a function $f(x) \in L_{p,\alpha,\beta}$ *have an* r*th derivative* $f^{(r)}(x)$ *inside the interval* $(-1, 1)$ *satisfying the condition*

$$\Big\| \Big[f^{(r)}(x) - f^{(r)}\Big(x, t, \alpha+\frac{r}{2}-\delta_1, \beta+\frac{r}{2}-\delta_2\Big)\Big](1-x)^{\alpha+r/2-\delta_1} \times (1+x)^{\beta+r/2-\delta_2}\Big(1-x+\sin^2\frac{t}{2}\Big)^{\rho_1+\delta_1}\Big(1+x+\sin^2\frac{t}{2}\Big)^{\rho_2+\delta_2}\Big\|_p \leqslant C_1\Big|\sin\frac{t}{2}\Big|^{\gamma},$$

it is necessary and sufficient that there exist a sequence of algebraic polynomials $P_n(x)$ *of degree not exceeding* n *such that*

$$\Big\|[f(x) - P_n(x)](1-x)^{\alpha}(1+x)^{\beta}\Big(1-x+\frac{1}{n^2}\Big)^{\rho_1}\Big(1+x+\frac{1}{n^2}\Big)^{\rho_2}\Big\|_p \leqslant C_2 n^{-(r+\gamma)},$$

where the constants C_1 *and* C_2 *do not depend on* $f(x)$, n *or* t.

This theorem is the logical join of Theorems 2 and 3 which are proved below. We mention one consequence of Theorem 1.

Let r be a nonnegative integer and $0 < \gamma \leq 1$. Let α and $\beta < \alpha$ be numbers with $\beta > ½p$ or $\beta \geq 0$ according as $1 \leq p < \infty$ or $p = \infty$. If $f(x) \in L_{p,\alpha,\beta}$, then the condition

$$E_n(f)_{p,\alpha,\beta} \leqslant \frac{C_1}{n^{r+\gamma}}$$

is equivalent to the condition that $f(x)$ has an rth derivative $f^{(r)}(x)$ inside the interval $(-1, 1)$ which satisfies the condition

$$\Big\|f^{(r)}(x) - f^{(r)}\Big(x, t, \alpha+\frac{r}{2}, \beta+\frac{r}{2}\Big)\Big\|_{p,\alpha+r/2,\beta+r/2} \leqslant C_2\Big|\sin\frac{t}{2}\Big|^{\gamma}.$$

III. Let m and q be two natural numbers. It is well known that the function

$$\nu(t) = \Big(\frac{\sin(mt/2)}{\sin(t/2)}\Big)^{2q+4}$$

is an even trigonometric polynomial of degree $(q+2)(m-1)$. It is easy to see that the function

$$K(t, m, \alpha, \beta) = \frac{\nu(t)}{\nu(m, \alpha, \beta)},$$

where $q > \alpha - 1$ and

$$\nu(m, \alpha, \beta) = \int_0^\pi \nu(t) \sin^{2\alpha+1}(t/2) \cos^{2\beta+1}(t/2)\, dt,$$

satisfies, for $2q + 2 > \gamma + 2\alpha > -2$, the conditions

$$\int_0^\pi K(t, m, \alpha, \beta) \sin^{2\alpha+1}\frac{t}{2} \cos^{2\beta+1}\frac{t}{2}\, dt = 1,$$

$$\int_0^\pi t^\gamma K(t, m, \alpha, \beta) \sin^{2\alpha+1}\frac{t}{2} \cos^{2\beta+1}\frac{t}{2}\, dt \leqslant Cm^{-\gamma},$$

where the constant C does not depend on m.

Lemma 1. *If the function $f(x)$ is summable on $[-1, 1]$ with weight $(1-x)^\alpha(1+x)^\beta$, then the function*

$$Q(x) = \int_0^\pi K(t, m, \alpha, \beta) f(x, t, \alpha, \beta) \sin^{2\alpha+1}\frac{t}{2} \cos^{2\beta+1}\frac{t}{2}\, dt$$

is an algebraic polynomial of degree $(q+2)(m-1)$.

Proof. Since $K(t, m, \alpha, \beta)$ is an even trigonometric polynomial of degree $(q+2)(m-1)$, we have

$$K(t, m, \alpha, \beta) = \sum_{n=0}^{(q+2)(m-1)} b_n P_{n,\alpha,\beta}(\cos t),$$

where the $P_{n,\alpha,\beta}(x)$ are algebraic polynomials of degree n which are pairwise orthogonal with weight $(1-x)^\alpha(1+x)^\beta$. Therefore

$$Q(x) = C_{m,\alpha,\beta} \sum_{n=0}^{(q+2)(m-1)} b_n \varphi_n(x),$$

where

$$\varphi_n(x) = 2^{\alpha+\beta+1} \int_0^\pi \int_0^1 \int_{-1}^1 f\left[x \cos t + rz \sin t \sqrt{1-x^2} - (1-r^2)(1-x)\sin^2\frac{t}{2}\right]$$
$$\times P_{n,\alpha,\beta}(x)(1-r^2)^{\alpha-\beta-1} r^{2\beta+1}(1-z^2)^{\beta-1/2} \sin^{2\alpha+1}\frac{t}{2} \cos^{2\beta+1}\frac{t}{2}\, dr\,dz\,dt.$$

Making the change of variables $h = \cos t$ in this integral, we have

$$\varphi_n(x) = \int_{-1}^1 \int_0^1 \int_{-1}^1 P_{n,\alpha,\beta}(x) f\left[xh + rz\sqrt{1-x^2}\sqrt{1-h^2} - \frac{1}{2}(1-r^2)(1-x)(1-h)\right]$$
$$\times (1-r^2)^{\alpha-\beta-1} r^{2\beta+1} (1-z^2)^{\beta-1/2} (1-h)^\alpha (1+h)^\beta\, dz\,dr\,dh.$$

Now making the change of variables

$$h = xH + RV\sqrt{1-x^2}\sqrt{1-H^2} - \frac{1}{2}(1-R^2)(1-x)(1-H),$$

$$z = \frac{(1-R^2+H+HR^2)\sqrt{1-x^2} - 2RVx\sqrt{1-H^2}}{\sqrt{[(1-R^2+H+HR^2)\sqrt{1-x^2} - 2RVx\sqrt{1-H^2}]^2 + 4R^2(1-V^2)(1-H^2)}},$$

$$r = \sqrt{\frac{R^2(1+x)(1-H) - 2RV\sqrt{1-x^2}\sqrt{1-H^2} + (1+H)(1-x)}{4 - [R^2(1-x)(1-H) + 2RV\sqrt{1-x^2}\sqrt{1-H^2} + (1+H)(1+x)]}},$$

we obtain

$$\varphi_n(x) = \int_{-1}^{1}\int_{0}^{1}\int_{-1}^{1} f(H) P_{n,\alpha,\beta}\Big[xH + RV\sqrt{1-x^2}\sqrt{1-H^2} - \frac{1}{2}(1-R^2)(1-x)(1-H)\Big](1-R^2)^{\alpha-\beta-1}R^{2\beta+1}(1-V^2)^{\beta-1/2} \times (1-H)^\alpha(1+H)^\beta\, dR\,dV\,dH.$$

Applying the addition formula for Jacobi polynomials to $P_{n,\alpha,\beta}(x)$ (see [9], p. 188), we have

$$\varphi_n(x) = \int_{-1}^{1} f(H) P_{n,\alpha,\beta}(H) P_{n,\alpha,\beta}(x) d_{n,\alpha,\beta} dH = d_{n,\alpha,\beta} a_{n,\alpha,\beta}(f) P_{n,\alpha,\beta}(x),$$

where $d_{n,\alpha,\beta}$ is some constant and $a_{n,\alpha,\beta}(f)$ is the Fourier coefficient of $f(x)$ for the system of Jacobi polynomials $P_{n,\alpha,\beta}(x)$. The last equality shows that $\phi_n(x)$ is an algebraic polynomial of degree n, and therefore $Q(x)$ is an algebraic polynomial of degree $(q+2)(m-1)$. The proof of Lemma 1 is complete.

Lemma 2 (see [10]). *Let $Q_n(x)$ be an algebraic polynomial of degree n, ρ_1 and ρ_2 any real numbers, and suppose that the numbers α, β, δ_1, δ_2, σ_1 and σ_2 satisfy $\alpha > -1/p$ and $\beta > -1/p$ for $1 \le p < \infty$, $\alpha \ge 0$ and $\beta \ge 0$ for $p = \infty$, $\delta_1 + \delta_2 \ge 0$, and $\sigma_1 + \sigma_2 \ge 0$. Then*

$$\Big\| Q_n(x)(1-x)^{\alpha+1/2}(1+x)^{\beta+1/2}\Big(1-x+\frac{1}{n^2}\Big)^{\rho_1}\Big(1+x+\frac{1}{n^2}\Big)^{\rho_2}\Big\|_p \leqslant C_1 n \Big\| Q_n(x)(1-x)^\alpha(1+x)^\beta\Big(1-x+\frac{1}{n^2}\Big)^{\rho_1}\Big(1+x+\frac{1}{n^2}\Big)^{\rho_2}\Big\|_p,$$

$$\Big\| Q_n(x)(1-x)^\alpha(1+x)^\beta\Big(1-x+\frac{1}{n^2}\Big)^{\rho_1}\Big(1+x+\frac{1}{n^2}\Big)^{\rho_2}\Big\|_p \leqslant C_2 n^{2\max(\delta_1+\delta_2,\,\sigma_1+\sigma_2)}\Big\| Q_n(x)(1-x)^{\alpha+\delta_1}(1+x)^{\beta+\sigma_1}\Big(1-x+\frac{1}{n^2}\Big)^{\rho_1+\delta_2}\Big(1+x+\frac{1}{n^2}\Big)^{\rho_2+\sigma_2}\Big\|_p,$$

where the constants C_1 and C_2 do not depend on n.

Lemma 3. *Let $f(x) \in L_{p,\alpha,\beta}$, and suppose that the numbers α, β, δ_1, δ_2, σ_1, σ_2, ρ_1, ρ_2 and γ satisfy $\alpha > -1/p$ and $\beta > -1/p$ for $1 \le p < \infty$, $\alpha \ge 0$ and $\beta \ge 0$ for $p = \infty$, $\delta_1 + \delta_2 \ge 0$, $\sigma_1 + \sigma_2 \ge 0$, $\gamma > 2\max(\delta_1+\delta_2, \sigma_1+\sigma_2)$, and $\gamma > 2\max(\rho_1, \rho_2) + 2\max(\delta_1+\delta_2, \sigma_1+\sigma_2)$. If there exists a sequence of algebraic polynomials $P_n(x)$ such that*

$$\Big\|[f(x) - P_n(x)](1-x)^{\alpha+\delta_1}(1+x)^{\beta+\sigma_1}\Big(1-x+\frac{1}{n^2}\Big)^{\rho_1+\delta_2}\Big(1+x+\frac{1}{n^2}\Big)^{\rho_2+\sigma_2}\Big\|_p \leqslant \frac{C_1}{n^\gamma},$$

then there exists a sequence of algebraic polynomials $R_n(x)$ *such that*

$$\left\|[f(x)-R_n(x)](1-x)^\alpha(1+x)^\beta\left(1-x+\frac{1}{n^2}\right)^{\rho_1}\left(1+x+\frac{1}{n^2}\right)^{\rho_2}\right\|_p \leqslant C_2 n^{-\gamma+2\max(\delta_1+\delta_2,\,\sigma_1+\sigma_2)},$$

where the constants C_1 *and* C_2 *do not depend on* $f(x)$ *or* n*. In addition,* $P_{2^k}(x) = R_{2^k}(x)$.

Proof. We consider the sequence of polynomials $Q_{2^k}(x)$, where $Q_{2^k}(x) = P_{2^k}(x) - P_{2^{k-1}}(x)$ for $k \geq 1$ and $Q_1(x) = P_1(x)$. From the conditions of the lemma it follows that

$$\left\|Q_{2^k}(x)(1-x)^{\alpha+\delta_1}(1+x)^{\beta+\sigma_1}\left(1-x+\frac{1}{2^{2k}}\right)^{\rho_1+\delta_2}\left(1+x+\frac{1}{2^{2k}}\right)^{\rho_2+\sigma_2}\right\|_p \leqslant C_3 2^{-k\gamma}.$$

Applying Lemma 2 and this estimate, we obtain that

$$\left\|Q_{2^k}(x)(1-x)^\alpha(1+x)^\beta\left(1-x+\frac{1}{2^{2k}}\right)^{\rho_1}\left(1+x+\frac{1}{2^{2k}}\right)^{\rho_2}\right\|_p \leqslant C_4 2^{-k\gamma+2k\max(\delta_1+\delta_2,\,\sigma_1+\sigma_2)}.$$

Now it follows that the series

$$\sum_{k=0}^{\infty} Q_{2^k}(x)$$

converges in $L_{p[a,b]}$ to $f(x)$ for any segment $[a, b] \subset (-1, 1)$.

We now consider

$$I = \left\|[f(x)-P_{2^N}(x)](1-x)^\alpha(1+x)^\beta\left(1-x+\frac{1}{2^{2N}}\right)^{\rho_1}\left(1+x+\frac{1}{2^{2N}}\right)^{\rho_2}\right\|_p.$$

It follows from above that

$$I \leqslant \sum_{k=N+1}^{\infty}\left\|Q_{2^k}(x)(1-x)^\alpha(1+x)^\beta\left(1-x+\frac{1}{2^{2N}}\right)^{\rho_1}\left(1+x+\frac{1}{2^{2N}}\right)^{\rho_2}\right\|_p.$$

We consider two cases: $\rho_1 \geq \rho_2$ and $\rho_1 < \rho_2$.

Let $\rho_1 \geq \rho_2$, whence it is clear that

$$I \leqslant C_5 \sum_{k=N+1}^{\infty}\left\|Q_{2^k}(x)(1-x)^\alpha(1+x)^\beta\left(1-x^2+\frac{1}{2^{2N}}\right)^{\rho_1}\left(1+x+\frac{1}{2^{2N}}\right)^{\rho_2-\rho_1}\right\|_p,$$

where the constant C_5 does not depend on N.

If $\rho_1 \leq 0$, then

$$I \leqslant C_5 \sum_{k=N+1}^{\infty}\left\|Q_{2^k}(x)(1-x)^\alpha(1+x)^\beta\left(1-x^2+\frac{1}{2^{2k}}\right)^{\rho_1}\left(1+x+\frac{1}{2^{2k}}\right)^{\rho_2-\rho_1}\right\|_p$$

$$\leqslant C_6 \sum_{k=N+1}^{\infty}\left\|Q_{2^k}(x)(1-x)^\alpha(1+x)^\beta\left(1-x+\frac{1}{2^{2k}}\right)^{\rho_1}\left(1+x+\frac{1}{2^{2k}}\right)^{\rho_2}\right\|_p$$

$$\leqslant C_7 \sum_{k=N+1}^{\infty} 2^{-k[\gamma-2\max(\delta_1+\delta_2,\,\sigma_1+\sigma_2)]} \leqslant C_8 2^{-N\gamma+2N\max(\delta_1+\delta_2;\,\sigma_1+\sigma_2)},$$

where the constants C_6, C_7 and C_8 do not depend on N.

If $\rho_1 > 0$ and $k \geq N$, then

$$\left(1-x^2+\frac{1}{2^{2N}}\right)^{\rho_1}\leqslant 2^{2(k-N)\rho_1}\left(1-x^2+\frac{1}{2^{2k}}\right)^{\rho_1},$$

and therefore

$$I\leqslant C_9 2^{-2N\rho_1}\sum_{k=N+1}^{\infty}2^{2k\rho_1}\left\|Q_{2^k}(x)(1-x)^\alpha(1+x)^\beta\left(1-x+\frac{1}{2^{2k}}\right)^{\rho_1}\left(1+x+\frac{1}{2^{2k}}\right)^{\rho_2}\right\|_p$$

$$\leqslant C_{10}2^{-2N\rho_1}\sum_{k=N+1}^{\infty}2^{-k[\gamma-2\rho_1-2\max(\delta_1+\delta_2,\,\sigma_1+\sigma_2)]}\leqslant C_{11}2^{-N\gamma+2N\max(\delta_1+\delta_2,\,\sigma_1+\sigma_2)},$$

where the constants C_9, C_{10} and C_{11} do not depend on N.

The case $\rho_1<\rho_2$ is handled similarly.

Thus, under the conditions of Lemma 3 we always have

$$I\leqslant C_{12}2^{-N\gamma+2N\max(\delta_1+\delta_2,\,\sigma_1+\sigma_2)},$$

where the constant C_{12} does not depend on N.

Now, setting $R_n(x)=P_{2^k}(x)$ for $2^{k-1}<n\le 2^k$, we obtain Lemma 3.

Lemma 4. *If* $0\le r\le 1$, $-1\le x\le 1$, $-1\le z\le 1$, ρ *is an arbitrary real number and*

$$x=Y\cos t+RV\sqrt{1-Y^2}\sin t-(1-R^2)(1-Y)\sin^2\frac{t}{2},$$

$$z=\frac{(1-R^2+Y+YR^2)\sin t-2RV\cos t\sqrt{1-Y^2}}{\sqrt{[(1-R^2+Y+YR^2)\sin t-2RV\cos t\sqrt{1-Y^2}]^2+4R^2(1-V^2)(1-Y^2)}},$$

$$r=\sqrt{\frac{R^2(1-Y)\cos^2\frac{t}{2}-RV\sqrt{1-Y^2}\sin t+(1+Y)\sin^2\frac{t}{2}}{4-\left[R^2(1-Y)\sin^2\frac{t}{2}+RV\sin t\sqrt{1-Y^2}+(1+Y)\cos^2\frac{t}{2}\right]}},$$

then $0\le R\le 1$, $-1\le Y\le 1$, $-1\le V\le 1$ *and*

$$(1-r^2)(1-x)=(1-R^2)(1-Y),\quad r^2(1-z^2)(1-x^2)=R^2(1-V^2)(1-Y^2),$$

$$x\sin t-rz\cos t\sqrt{1-x^2}+\frac{1}{2}(1-r^2)(1-x)\sin t=RV\sqrt{1-Y^2},$$

$$x\cos t+rz\sin t\sqrt{1-x^2}+\frac{1}{2}(1-r^2)(1-x)\cos t=$$
$$=Y+\frac{1}{2}(1-R^2)(1-Y),$$

$$\left(1-x+\frac{1}{n^2}\right)^\rho\leqslant C\left[1+\left(n\left|\sin\frac{t}{2}\right|\right)^{2|\rho|}\right]\left(1-Y+\frac{1}{n^2}\right)^\rho,$$

$$\left(1+x+\frac{1}{n^2}\right)^\rho\leqslant C\left[1+\left(n\left|\sin\frac{t}{2}\right|\right)^{2|\rho|}\right]\left(1+Y+\frac{1}{n^2}\right)^\rho,$$

$$\left(1-x+\frac{1}{n^2}\right)^\rho\leqslant C\left[1+\left(n\left|\sin\frac{t}{2}\right|\right)^{-2\rho}\right]\left(1-x+\sin^2\frac{t}{2}\right)^\rho,$$

$$\left(1+x+\frac{1}{n^2}\right)^\rho\leqslant C\left[1+\left(n\left|\sin\frac{t}{2}\right|\right)^{-2\rho}\right]\left(1+x+\sin^2\frac{t}{2}\right)^\rho,$$

where the constant C *does not depend on* x, t, Y *or* n.

The proof of this lemma has a computational character and is omitted because of the awkwardness of the calculations.

Lemma 5. *Suppose the function* $f(x) \in L_{p,\alpha,\beta}$, *let* ρ_1 *and* ρ_2 *be arbitrary real numbers, and pick numbers* α, β, δ_1 *and* δ_2 *such that* $\alpha > \beta > -\frac{1}{2}p$ *for* $1 \le p < \infty$, $\alpha > \beta \ge 0$ *for* $p = \infty$, $\delta_1 > 0$ *and* $\delta_2 \ge 0$. *If* $f(x)$ *has a derivative* $f'(x)$ *satisfying*

$$\left\| f'(x)(1-x)^{\alpha+1/2-\delta_1}(1+x)^{\beta+1/2-\delta_2}\left(1-x+\frac{1}{n^2}\right)^{\rho_1+\delta_1}\left(1+x+\frac{1}{n^2}\right)^{\rho_2+\delta_2}\right\|_p \leqslant C_1 n^{-\gamma},$$

then there exists a sequence of polynomials $P_n(x)$ *such that*

$$\left\| [f(x)-P_n(x)](1-x)^{\alpha}(1+x)^{\beta}\left(1-x+\frac{1}{n^2}\right)^{\rho_1}\left(1+x+\frac{1}{n^2}\right)^{\rho_2}\right\|_p \leqslant C_2 n^{-1-\gamma},$$

where the constants C_1 *and* C_2 *do not depend on* $f(x)$ *or* n.

Proof. We select a natural number q such that $q > \alpha + |\rho_1| + |\rho_2|$. For every n we select a natural number m such that

$$\frac{n}{q+2} < m \leqslant \frac{n}{q+2} + 1. \tag{*}$$

We consider the polynomial $Q(x)$ defined in Lemma 1. Thus $Q(x)$ has degree $(q+2)(m-1) \le n$. For every n satisfying condition $(*)$ we take the polynomial $Q(x)$ for $P_n(x)$. Then

$$f(x)-P_n(x) = \int_0^{\pi} K(t, m, \alpha, \beta)[f(x)-f(x, t, \alpha, \beta)] \sin^{2\alpha+1}\frac{t}{2}\cos^{2\beta+1}\frac{t}{2}\,dt$$

$$= \frac{1}{\mu(\alpha, \beta)}\int_0^{\pi} K(t, m, \alpha, \beta)\int_0^1\int_{-1}^1\int_0^t f'(Y)\,Y_1(1-r^2)^{\alpha-\beta-1}r^{2\beta+1}(1-z^2)^{\beta-1/2}$$
$$\times \sin^{2\alpha+1}\frac{t}{2}\cos^{2\beta+1}\frac{t}{2}\,du\,dz\,dr\,dt,$$

where

$$Y = x\cos u + rz\sin u\sqrt{1-x^2} - (1-r^2)(1-x)\sin^2\frac{u}{2},$$
$$Y_1 = x\sin u - rz\cos u\sqrt{1-x^2} + \tfrac{1}{2}(1-r^2)(1-x)\sin u.$$

Furthermore, applying Minkowski's inequality, we have

$$I = \left\| [f(x)-P_n(x)](1-x)^{\alpha}(1+x)^{\beta}\left(1-x+\frac{1}{n^2}\right)^{\rho_1}\left(1+x+\frac{1}{n^2}\right)^{\rho_2}\right\|_p$$
$$\leqslant \frac{1}{\mu(\alpha, \beta)}\int_0^{\pi} K(t, m, \alpha, \beta)\int_0^{t} \sin^{2\alpha+1}t\cos^{2\beta+1}\frac{t}{2}$$
$$\times \left\| \int_{-1}^1\int_0^1 f'(Y)\,Y_1(1-x)^{\alpha}(1+x)^{\beta}\left(1-x+\frac{1}{n^2}\right)^{\rho_1}\left(1+x+\frac{1}{n^2}\right)^{\rho_2}\right.$$
$$\left. \times (1-r^2)^{\alpha-\beta-1}r^{2\beta+1}(1-z^2)^{\beta-1/2}\,dr\,dz\right\|_p du\,dt.$$

We first consider the case $p = \infty$. Selecting ϵ such that $0 < \epsilon \le \alpha - \beta$ and $0 < \epsilon \le \delta_1$ and applying Lemma 4, we obtain that

$$I \leqslant C_3 \int_0^\pi K(t, m, \alpha, \beta) \int_0^t \left[1 + \left(n \sin \frac{u}{2}\right)^{2|\rho_1|+2\varepsilon+2|\rho_2|}\right] \sin^{2\alpha+1} \frac{t}{2} \cos^{2\beta+1} \frac{t}{2}$$

$$\times \int_{-1}^{1} \int_0^1 \left\| f'(Y)(1-Y)^{\alpha+1/2-\varepsilon}(1+Y)^{\beta+1/2}\left(1-Y+\frac{1}{n^2}\right)^{\rho_1+\varepsilon}\left(1+Y+\frac{1}{n^2}\right)^{\rho_2}\right.$$

$$\times (1-R^2)^{\alpha-\beta-\varepsilon} R^{2\beta}(1-V^2)^\beta \Big\|_C (1-r^2)^{\varepsilon-1} r (1-z^2)^{-1/2} \, dr dz dV dt$$

$$\leqslant C_4 \int_0^\pi K(t, m, \alpha, \beta) t [1 + (nt)^{2(|\rho_1|+|\rho_2|+\varepsilon)}] \sin^{2\alpha+1} \frac{t}{2} \cos^{2\beta+1} \frac{t}{2}$$

$$\times \left\| f'(Y)(1-Y)^{\alpha+1/2-\delta_1}(1+Y)^{\beta+1/2-\delta_2}\left(1+Y+\frac{1}{n^2}\right)^{\rho_1+\delta_1}\left(1+Y+\frac{1}{n^2}\right)^{\rho_2+\delta_2}\right\|_C$$

$$\leqslant C_5 n^{-1-\gamma},$$

where the constants C_3, C_4 and C_5 do not depend on $f(x)$ or n.

Now let p be an element of the interval $1 < p < \infty$. Selecting ϵ such that $0 < \epsilon < \alpha - \beta$ and $0 < \epsilon < \delta_1$ and applying Hölder's inequality, we can write

$$I \leqslant C_6 \int_0^\pi K(t, m, \alpha, \beta) \int_0^t \sin^{2\alpha+1} \frac{t}{2} \cos^{2\beta+1} \frac{t}{2}$$

$$\times \left(\int_{-1}^{1} \int_0^1 \int_{-1}^{1} \left| f'(Y) Y_1 (1-x)^{\alpha-\varepsilon}(1+x)^\beta \left(1-x+\frac{1}{n^2}\right)^{\rho_1+\varepsilon}\left(1+x+\frac{1}{n^2}\right)^{\rho_2}\right.\right.$$

$$\times \left.\left.(1-r^2)^{\alpha-\beta-\varepsilon-1/p} r^{2\beta+1/p}(1-z^2)^{\beta-1/2p}\right|^p dz dr dx \right)^{1/p} du dt.$$

Substituting into the triple integral according to the formulas

$$x = Y \cos u + RV\sqrt{1-\gamma^2} \sin u - (1-R^2)(1-Y)\sin^2 \frac{u}{2},$$

$$z = \frac{(1-R^2+Y+YR^2)\sin u - 2RV\cos u\sqrt{1-Y^2}}{\sqrt{[(1-R^2+Y+YR^2)\sin u - 2RV\cos u\sqrt{1-Y^2}]^2 + 4R^2(1-V^2)(1-Y^2)}},$$

$$r = \sqrt{\frac{R^2(1-Y)\cos^2(u/2) - RV\sqrt{1-Y^2}\sin u + (1+Y)\sin^2(u/2)}{4 - [R^2(1-Y)\sin^2(u/2) + RV\sin u\sqrt{1-Y^2} + (1+Y)\cos^2(u/2)}} \qquad (**)$$

and applying Lemma 4, we obtain that

$$I \leqslant C_7 \int_0^\pi K(t, m, \alpha, \beta) \int_0^t \sin^{2\alpha+1} \frac{t}{2} \cos^{2\beta+1} \frac{t}{D}\left[1 + \left(n \sin \frac{u}{2}\right)^{2(|\rho_1|+|\rho_2|+\varepsilon)}\right]$$

$$\times \left\{ \int_{-1}^{1} \int_0^1 \int_{-1}^{1} \left| f'(Y)(1-Y)^{\alpha+1/2-\varepsilon}(1+Y)^{\beta+1/2}\left(1-Y+\frac{1}{n^2}\right)^{\rho_1+\varepsilon}\left(1+Y+\frac{1}{n^2}\right)^{\rho_2}\right|^p\right.$$

$$\times \left.(1-R^2)^{(\alpha-\beta-\varepsilon)p-1} R^{2\beta p+1}(1-V^2)^{\beta p-1/2} dY dR dV \right\}^{1/p} du dt$$

$$\leqslant \frac{C_8}{n} \left\| f'(Y)(1-Y)^{\alpha+1/2-\delta_1}(1+Y)^{\beta+1/2-\delta_2}\left(1-x+\frac{1}{n^2}\right)^{\rho_1+\delta_1}\left(1+x+\frac{1}{n^2}\right)^{\rho_2+\delta_2}\right\|_p$$

$$\leqslant C_9 n^{-1-\gamma},$$

where the constants C_6, C_7, C_8 and C_9 do not depend on $f(x)$ or n.

Finally, let $p = 1$. Then

$$I \leqslant \frac{1}{\mu(\alpha, \beta)} \int_0^{\pi} K(t, m, \alpha, \beta) \int_0^{t} \sin^{2\alpha+1}\frac{t}{2} \cos^{2\beta+1}\frac{t}{2}$$

$$\times \int_{-1}^{1}\int_{0}^{1}\int_{-1}^{1} |f'(Y)\, Y_1|\, (1-x)^{\alpha}(1+x)^{\beta}\left(1-x+\frac{1}{n^2}\right)^{\rho_1}\left(1+x+\frac{1}{n^2}\right)^{\rho_2}$$

$$\times (1-r^2)^{\alpha-\beta-1} r^{2\beta+1} (1-z^2)^{\beta-1/2} dx\,dz\,dr\,du\,dt.$$

Substituting into the triple integral according to formulas (**) and applying Lemma 4, we obtain, after some obvious calculations, that $I \leq C_{10} n^{-1-\gamma}$, where C_{10} does not depend on $f(x)$ or n. Lemma 5 is completely proved.

IV. Theorem 2. *Suppose that the numbers* α, β, γ, ρ_1, ρ_2, δ_1, δ_2 *and* r *are such that* r *is a natural number,* $\gamma > 0$, $r + \gamma > 2\max(\rho_1, \rho_2)$, $\gamma > 2\max(\rho_1 + \delta_1, \rho_2 + \delta_2)$,

for $1 \leq p < \infty$: $\alpha > \beta > -\tfrac{1}{2}p$, $\delta_1 < \alpha + r/2 + 1/p$, $\delta_2 < \beta + r/2 + 1/p$,

for $p = \infty$: $\alpha > \beta \geq 0$, $\delta_1 \leq \alpha + r/2$, $\delta_2 \leq \beta + r/2$,

and suppose $f(x) \in L_{p,\alpha,\beta}$. *Then the following two conditions are equivalent.*

1) *The function* $f(x)$ *has an rth derivative inside the interval* $(-1, 1)$ *for which there exists a sequence of algebraic polynomials* $R_n(x)$ *such that*

$$\Big\| [f^{(r)}(x) - R_n(x)]\,(1-x)^{\alpha+r/2-\delta_1}(1+x)^{\beta+r/2-\delta_2}\left(1-x+\frac{1}{n^2}\right)^{\rho_1+\delta_1}$$

$$\times\left(1+x+\frac{1}{n^2}\right)^{\rho_2+\delta_2}\Big\|_p \leqslant C_1 n^{-\gamma}.$$

2) *For the function* $f(x)$ *there is a sequence of polynomials* $P_n(x)$ *such that*

$$\Big\| [f(x) - P_n(x)]\,(1-x)^{\alpha}(1+x)^{\beta}\left(1-x+\frac{1}{n^2}\right)^{\rho_1}\left(1+x+\frac{1}{n^2}\right)^{\rho_2}\Big\|_p \leqslant C_2 n^{-r-\gamma}.$$

Here the constants C_1 *and* C_2 *do not depend on* $f(x)$ *or* n.

Proof. We will prove that condition 2) follows from condition 1).

a) We first suppose that $\delta_1 > 0$ and $\delta_2 \geq 0$. We set $\phi_1(x) = f^{(r-1)}(x) - \int_0^x R_{n-1}(t)\,dt$, $\alpha_1 = \alpha + (r-1)/2 - \delta_1'$, $\delta_1' = \delta_{1/2}$, $\rho_1' = \rho_1 + \delta_1'$, $\beta_1 = \beta - (r - 1/2)$; hence condition 1) takes the form

$$\Big\| \varphi_1'(x)\,(1-x)^{\alpha_1+1/2-\delta_1'}(1+x)^{\beta_1+1/2-\delta_2}\left(1-x+\frac{1}{(n-1)^u}\right)^{\rho_1'+\delta_1'}$$

$$\times\left(1+x+\frac{1}{(n-1)^2}\right)^{\rho_2+\delta_2}\Big\|_p \leqslant C_1 (n-1)^{-\gamma}.$$

Applying Lemma 5, we obtain that for every n there exists an algebraic polynomial $T_n^*(x)$ such that

$$\left\|[\varphi_1(x) - T_n^*(x)](1-x)^{\alpha_1}(1+x)^{\beta_1}\left(1-x+\frac{1}{n^2}\right)^{\rho_1'}\left(1+x+\frac{1}{n^2}\right)^{\rho_2}\right\|_p \leqslant C_3 n^{-1-\gamma},$$

where the constant C_3 does not depend on $f(x)$ or n. Since $T_n(x) = T_n^*(x) - \int_0^x R_{n-1}(t)dt$ is an algebraic polynomial of degree no larger than n, this inequality may be written in the form

$$\Big\|[f^{(r-1)}(x) - T_n(x)](1-x)^{\alpha+\frac{r-1}{2}-\delta_1'}(1+x)^{\beta+\frac{r-1}{2}} \times\left(1-x+\frac{1}{n^2}\right)^{\rho_1+\delta_1'}\left(1+x+\frac{1}{n^2}\right)^{\rho_2}\Big\|_p \leqslant C_3 n^{-1-\gamma}.$$

Repeating this $r-1$ times establishes the validity of condition 2) for every $n > r$. For $n \leq r$ condition 2) is obvious.

b) We now consider the case $\delta_1 \leq 0$ and $\delta_2 < 0$. Take a positive number δ_1^0 and set $\alpha^0 = \alpha_1 - \delta_2 + \delta_1^0$, $\rho_1^0 = \rho_1 + \delta_1 - \delta_1^0$, $\beta^0 = \beta - \delta_2$, $\rho_2^0 = \rho_2 + \delta_2$ and $\delta_2^0 = 0$. Then condition 1) may be written as

$$\Big\|[f^{(r)}(x) - R_n(x)](1-x)^{\alpha^0+r/2-\delta_1^0}(1+x)^{\beta^0+r/2-\delta_2^0}\left(1-x+\frac{1}{n^2}\right)^{\rho_1^0+\delta_1^0} \times\left(1+x+\frac{1}{n^2}\right)^{\rho_2^0+\delta_2^0}\Big\|_p \leqslant C_1 n^{-\gamma}.$$

We are now in the situation of case a); hence it follows from this inequality that for every n there exists a polynomial $Q_n(x)$ such that

$$\Big\|[f(x) - Q_n(x)](1-x)^{\alpha-\delta_1+\delta_1^0}(1+x)^{\beta-\delta_2}\left(1-x+\frac{1}{n^2}\right)^{\rho_1+\delta_1-\delta_1^0} \times\left(1+x+\frac{1}{n^2}\right)^{\rho_2+\delta_2}\Big\|_p \leqslant C_4 n^{-r-\gamma},$$

where the constant C_4 does not depend on $f(x)$ or n. Now an application of Lemma 3 establishes condition 2).

The cases $\delta_1 \leq 0, \delta_2 > 0$ and $\delta_1 > 0, \delta_2 < 0$ are handled similarly.

Thus we have proved that condition 2) follows from condition 1). We now prove the converse; that is, that condition 2) implies condition 1).

We consider the polynomials $Q_{2^k}(x)$, where $Q_1(x) = P_1(x)$ and $Q_{2^k}(x) = P_{2^k}(x) - P_{2^{k-1}}(x)$ for $k \geq 1$. It is clear that

$$\left\|Q_{2^k}(x)(1-x)^{\alpha}(1+x)^{\beta}\left(1-x+\frac{1}{2^{2k}}\right)^{\rho_1}\left(1+x+\frac{1}{2^{2k}}\right)^{\rho_2}\right\|_p \leqslant C_5 2^{-k(r+\gamma)},$$

where the constant C_5 does not depend on N. Applying Lemma 2, we obtain that

$$\left\|Q_{2^k}^{(r)}(x)(1-x)^{\alpha+r/2-\delta_1}(1+x)^{\beta+r/2-\delta_2}\left(1-x+\frac{1}{2^{2k}}\right)^{\rho_1+\delta_1}\left(1+x+\frac{1}{2^{2k}}\right)^{\rho_2+\delta_2}\right\|_p \leqslant C_6 2^{-k\gamma},$$

where the constant C_6 also does not depend on k. From this inequality it follows that for every segment $[a, b] \subset (-1, 1)$ the series

$$\sum_{k=0}^{\infty} Q_{2^k}(x)$$

converges in $L_p[a, b]$ to the function $f(x)$, and the series

$$\sum_{k=0}^{\infty} Q_{2^k}^{(r)}(x) \tag{*}$$

also converges in $L_p[a, b]$. But then by a lemma of S. M. Nikol'skiĭ ([11], Lemma 4.4.5) this series converges to $f^{(r)}(x)$. Thus we have proved that the function $f(x)$ has an rth derivative almost everywhere on $(-1, 1)$ and that the series $(*)$ converges to it in $L_p[a, b]$ for every segment $[a, b] \subset (-1, 1)$. We now consider

$$I = \left\| \left[f^{(r)}(x) - P_{2^N}^{(r)}(x)\right](1-x)^{\alpha+r/2-\delta_1}(1+x)^{\beta+r/2-\delta_2} \times \left(1-x+\frac{1}{2^{2N}}\right)^{\rho_1+\delta_1}\left(1+x+\frac{1}{2^{2N}}\right)^{\rho_2+\delta_2}\right\|_p .$$

It is clear that

$$I \leqslant \sum_{k=N+1}^{\infty} \left\| Q_{2^k}^{(r)}(x)(1-x)^{\alpha+r/2-\delta_1}(1+x)^{\beta+r/2-\delta_2} \times \left(1-x+\frac{1}{2^{2N}}\right)^{\rho_1+\delta_1}\left(1+x+\frac{1}{2^{2N}}\right)^{\rho_2+\delta_2}\right\|_p .$$

If we assume that $\rho_1 + \delta_1 \geq \rho_2 + \delta_2$, then

$$I \leqslant C_7 \sum_{k=N+1}^{\omega} \left\| Q_{2^k}^{(r)}(x)(1-x)^{\alpha+r/2-\delta_1}(1+x)^{\beta+r/2-\delta_2}\left(1-x^2+\frac{1}{2^{2N}}\right)^{\rho_1+\delta_1} \times \left(1+x+\frac{1}{2^{2N}}\right)^{\rho_2+\delta_2-\rho_1-\delta_1}\right\|_p ,$$

where the constant C_7 does not depend on N.

If $\rho_1 + \delta_1 \leq 0$, then

$$I \leqslant C_7 \sum_{k=N+1}^{\infty} \left\| Q_{2^k}^{(r)}(x)(1-x)^{\alpha+r/2-\delta_1}(1+x)^{\beta+r/2-\delta_2}\left(1-x^2+\frac{1}{2^{2k}}\right)^{\rho_1+\delta_1}\right.$$

$$\left.\times\left(1+x+\frac{1}{2^{2k}}\right)^{\rho_2+\delta_2-\rho_1-\delta_1}\right\|_p \leqslant C_8 \sum_{k=N+1}^{\infty} \left\| Q_{2^k}^{(r)}(x)(1-x)^{\alpha+r/2-\delta_1}(1+x)^{\beta+r/2-\delta_2}\right.$$

$$\left.\times\left(1-x+\frac{1}{2^{2k}}\right)^{\rho_1+\delta_1}\left(1+x+\frac{1}{2^{2k}}\right)^{\rho_2+\delta_2}\right\|_p \leqslant C_9 \sum_{k=N+1}^{\infty} 2^{-k\gamma} \leqslant C_{10} 2^{-N\gamma},$$

where the constants C_8, C_9 and C_{10} do not depend on N.

If $\rho_1 + \delta_1 > 0$, then

$$I \leqslant C_7 \sum_{k=N+1}^{\infty} 2^{2(\rho_1+\delta_1)(k-N)} \left\| Q_{2^k}^{(r)}(x)(1-x)^{\alpha+r/2-\delta_1}(1+x)^{\beta+r/2-\delta_2}\left(1-x^2+\frac{1}{2^{2k}}\right)^{\rho_1+\delta_1}\right.$$

$$\left.\times\left(1+x+\frac{1}{2^{2k}}\right)^{\rho_2+\delta_2-\rho_1-\delta_1}\right\|_p \leqslant C_{11} 2^{-2N(\rho_1+\delta_1)} \sum_{k=N+1}^{\infty} 2^{2k(\rho_1+\delta_1)-k\gamma} \leqslant C_{12} 2^{-N\gamma},$$

where the constants C_{11} and C_{12} do not depend on N.

The case $\rho_1 + \delta_1 < \rho_2 + \delta_2$ follows similarly.

Thus we have proved that $I \leq C_{13} 2^{-N\gamma}$, which completes the proof of Theorem 2.

V. **Theorem 3.** *Suppose that the numbers* α, β, ρ_1, ρ_2 *and* γ *satisfy* $0 < \gamma < 2$, $\max(\rho_1, \rho_2) < \gamma/2 < 1 + 1/p + \min(\rho_1 + \alpha, \rho_2 + \beta)$,

for $1 \le p < \infty$: $\alpha > \beta > -\frac{1}{2p}$, *and for* $p = \infty$: $\alpha > \beta \ge 0$,

and assume that $f(x) \in L_{p,\alpha,\beta}$. *Then the following two conditions are equivalent:*

1) *The function* $f(x)$ *satisfies the condition*

$$\left\| [f(x) - f(x, t, \alpha, \beta)](1-x)^{\alpha}(1+x)^{\beta}\left(1 - x + \sin^2\frac{t}{2}\right)^{\rho_1}\left(1 + x + \sin^2\frac{t}{2}\right)^{\rho_2}\right\|_p \leqslant C_1\left|\sin\frac{t}{2}\right|^{\gamma}.$$

2) *For* $f(x)$ *there is a sequence of algebraic polynomials* $P_n(x)$ *such that*

$$\left\|[f(x) - P_n(x)](1-x)^{\alpha}(1+x)^{\beta}\left(1 - x + \frac{1}{n^2}\right)^{\rho_1}\left(1 + x + \frac{1}{n^2}\right)^{\rho_2}\right\|_p \leqslant C_2 n^{-\gamma}.$$

The constants C_1 *and* C_2 *do not depend on* $f(x)$, t *or* n.

Proof. We will prove that condition 2) follows from condition 1). We select a natural number q so that $q > \alpha - 1 + \gamma/2$ and $q > \alpha + \gamma/2 - \rho_1 - \rho_2$. For every natural number n we select a natural number m so that

$$\frac{n}{q+2} < m \leqslant \frac{n}{q+2} + 1,$$

and we consider the polynomial $Q(x)$ defined in Lemma 1. $Q(x)$ has degree $(q+2)(m-1) \le n$. For every n satisfying condition $(*)$ we choose $P_n(x)$ to be $Q(x)$. Then

$$f(x) - P_n(x) = \int_0^{\pi} K(t, m, \alpha, \beta)[f(x) - f(x, t, \alpha, \beta)]\sin^{2\alpha+1}\frac{t}{2}\cos^{2\beta+1}\frac{t}{2}\,dt.$$

We consider

$$I = \left\|[f(x) - P_n(x)](1-x)^{\alpha}(1+x)^{\beta}\left(1 - x + \frac{1}{n^2}\right)^{\rho_1}\left(1 + x + \frac{1}{n^2}\right)^{\rho_2}\right\|_p.$$

Applying Minkowski's inequality, we have

$$I \leqslant \int_0^{\pi} K(t, m, \alpha, \beta)\sin^{2\alpha+1}\frac{t}{2}\cos^{2\beta+1}\frac{t}{2}$$
$$\times\left\|[f(x) - f(x, t, \alpha, \beta)](1-x)^{\alpha}(1+x)^{\beta}\left(1 - x + \frac{1}{n^2}\right)^{\rho_1}\left(1 + x + \frac{1}{n^2}\right)^{\rho_2}\right\|_p dt.$$

Next an application of Lemma 4 gives

$$I \leqslant C_3\int_0^{\pi} K(t, m, \alpha, \beta)\sin^{2\alpha+1}\frac{t}{2}\cos^{2\beta+1}\frac{t}{2}\left[1 + \left(n\sin\frac{t}{2}\right)^{-2(\rho_1+\rho_2)}\right]$$
$$\times\left\|[f(x) - f(x, t, \alpha, \beta)](1-x)^{\alpha}(1+x)^{\beta}\left(1 - x + \sin^2\frac{t}{2}\right)^{\rho_1}\left(1 + x + \sin^2\frac{t}{2}\right)^{\rho_2}\right\|_p dt,$$

where the constant C_3 does not depend on $f(x)$ or n. Using condition 1) of the theorem, we have

$$I \leqslant C_4 \int_0^\pi K(t,\ m,\ \alpha,\ \beta) \sin^{2\alpha+1}\frac{t}{2}\cos^{2\beta+1}\frac{t}{2}\left[1+\left(n\sin\frac{t}{2}\right)^{-2(\rho_1+\rho_2)}\right]\sin^\gamma\frac{t}{2}\,dt \leqslant C_5 n^{-\gamma},$$

where the constants C_4 and C_5 do not depend on $f(x)$ or n.

Thus we have proved that condition 2) follows from condition 1).

We now prove the converse; that is, that 1) follows from 2).

We consider the polynomials $Q_{2^k}(x)$, where $Q_1(x) = P_1(x)$ and $Q_{2^k}(x) = P_{2^k}(x) - P_{2^{k-1}}(x)$ for $k \geq 1$. It is clear that

$$\left\| Q_{2^k}(x)(1-x)^\alpha(1+x)^\beta\left(1-x+\frac{1}{2^{2k}}\right)^{\rho_1}\left(1+x+\frac{1}{2^{2k}}\right)^{\rho_2}\right\|_p \leqslant C_6 2^{-k\gamma},$$

where the constant C_6 does not depend on k.

We must estimate

$$I = \left\|[f(x)-f(x,t,\alpha,\beta)](1-x)^\alpha(1+x)^\beta\left(1-x+\sin^2\frac{t}{2}\right)^{\rho_1}\left(1+x+\sin^2\frac{t}{2}\right)^{\rho_2}\right\|_p.$$

Selecting a number N such that $2^{-1-N} < |\sin(t/2)| \leq 2^{-N}$, we have

$$\begin{aligned} I &\leqslant C_7\left\|[f(x)-f(x,t,\alpha,\beta)](1-x)^\alpha(1+x)^\beta\left(1-x+\frac{1}{2^{2N}}\right)^{\rho_1}\left(1+x+\frac{1}{2^{2N}}\right)^{\rho_2}\right\|_p \\ &\leqslant C_7\left\{\left\|[f(x)-P_{2^N}(x)](1-x)^\alpha(1+x)^\beta\left(1-x+\frac{1}{2^{2N}}\right)^{\rho_1}\left(1+x+\frac{1}{2^{2N}}\right)^{\rho_2}\right\|_p\right. \\ &+\left\|[f(x,t,\alpha,\beta)-P_{2^N}(x,t,\alpha,\beta)](1-x)^\alpha(1+x)^\beta\left(1-x+\frac{1}{2^{2N}}\right)^{\rho_1}\left(1+x+\frac{1}{2^{2N}}\right)^{\rho_2}\right\|_p \\ &+\left.\sum_{k=0}^N\left\|[Q_{2^k}(x)-Q_{2^k}(x,t,\alpha,\beta)](1-x)^\alpha(1+x)^\beta\left(1-x+\frac{1}{2^{2N}}\right)^{\rho_1}\left(1+x+\frac{1}{2^{2N}}\right)^{\rho_2}\right\|_p\right\} \\ &\leqslant C_7(I_1+I_2+I_3), \end{aligned}$$

where the constant C_7 does not depend on $f(x)$ or N.

It is clear that $I_1 \leq C_8 2^{-N\gamma} \leq C_9|\sin t/2|^\gamma$, where the constants C_8 and C_9 do not depend on N.

We now estimate I_2.

First let $p = 1$. Then

$$I_2 \leqslant \int_{-1}^1\int_0^1\int_{-1}^1 |f(Y)-P_{2^N}(Y)|(1-r^2)^{\alpha-\beta-1}r^{2\beta+1}(1-z^2)^{\beta-1/2}(1-x)^\alpha(1+x)^\beta$$
$$\times\left(1-x+\frac{1}{2^{2N}}\right)^{\rho_1}\left(1+x+\frac{1}{2^{2N}}\right)^{\rho_2} dz\,dr\,dx,$$

where $Y = x\cos t + rz\sin t\sqrt{1-x^2} - (1-r^2)(1-x)\sin^2(t/2)$. Substituting according to the formulas

$$x = Y\cos t + RV\sin t\sqrt{1-Y^2} - (1-R^2)(1-Y)\sin^2\frac{t}{2},$$

$$z = \frac{(1-R^2+Y+YR^2)\sin t - 2RV\cos t\sqrt{1-Y^2}}{\sqrt{[(1-R^2+Y+YR^2)\sin t - 2RV\cos t\sqrt{1-Y^2}]^2 + 4R^2(1-V^2)(1-Y^2)}}, \quad (**)$$

$$r = \sqrt{\frac{R^2(1-Y)\cos^2\frac{t}{2} - RV\sqrt{1-Y^2}\sin t + (1+Y)\sin^2(t/2)}{4-[R^2(1-Y)\sin^2(t/2) + RV\sin t\sqrt{1-Y^2} + (1+Y)\cos^2(t/2)]}}$$

and applying Lemma 4, we have

$$I_2 \leqslant C_{10} \int_{-1}^{1}\int_{0}^{1}\int_{-1}^{1} |f(Y)-P_{2^N}(Y)|(1-Y)^{\alpha}(1+Y)^{\beta}\left(1-Y+\frac{1}{2^{2N}}\right)^{\rho_1}$$

$$\times\left(1+Y+\frac{1}{2^{2N}}\right)^{\rho_2}(1-R^2)^{\alpha-\beta-1}R^{2\beta+1}(1-V^2)^{\beta-1/2}dY\,dR\,dV \leqslant C_{11}2^{-N\gamma} \leqslant C_{12}\left|\sin\frac{t}{2}\right|^{\gamma},$$

where the constants C_{10}, C_{11} and C_{12} do not depend on $f(x)$, N or t.

Now let $1 < p < \infty$. Then for any ϵ such that $0 < \epsilon < \alpha - \beta$ it is possible to write

$$I_2 \leqslant \Bigg\| \int_0^1\int_{-1}^1 \{[f(Y)-P_{2^N}(Y)](1-r^2)^{\alpha-\beta-\varepsilon-1/p}r^{2\beta+1/p}(1-z^2)^{\beta-1/2p}\}$$

$$\times\{r^{1/p_1}\sqrt{1-z^2}^{-1/p_1}(1-r^2)^{\varepsilon-1/p_1}\}\,dz\,dr\,(1-x)^{\alpha-\varepsilon}(1+x)^{\beta}$$

$$\times\left(1-x+\frac{1}{2^{2N}}\right)^{\rho_1+\varepsilon}\left(1+x+\frac{1}{2^{2N}}\right)^{\rho_2}\Bigg\|_p,$$

where $1/p + 1/p_1 = 1$. Applying Hölder's inequality to the double integral, we obtain

$$I_2 \leqslant C_{13}\left\{\int_{-1}^{1}\int_{-1}^{1}\int_{0}^{1} |f(Y)-P_{2^N}(Y)|^p(1-r^2)^{(\alpha-\beta-\varepsilon)p-1}r^{2\beta p-1}(1-z^2)^{\beta p-1/2}\right.$$

$$\left.\times(1-x)^{(\alpha-\varepsilon)p}(1+x)^{\beta p}\left(1-x+\frac{1}{2^{2N}}\right)^{(\rho_1+\varepsilon)p}\left(1+x+\frac{1}{2^{2N}}\right)^{\rho_2 p}dr\,dz\,dx\right\}^{1/p},$$

where the constant C_{13} does not depend on $f(x)$ or N. Substituting according to (**) and applying Lemma 4, we have

$$I_2 \leqslant C_{14}\left\{\int_{-1}^{1}\int_{0}^{1}\int_{-1}^{1} |f(Y)-P_{2^N}(Y)|^p(1-R^2)^{(\alpha-\beta-\varepsilon)p-1}R^{2\beta p+1}(1-V^2)^{\beta p-1/2}\right.$$

$$\left.\times(1-Y)^{(\alpha-\varepsilon)p}(1+Y)^{\beta p}\left(1-Y+\frac{1}{2^{2N}}\right)^{(\rho_1+\varepsilon)p}\left(1+Y+\frac{1}{2^{2N}}\right)^{\rho_2 p}dR\,dV\,dY\right\}^{1/p}$$

$$\leqslant C_{15}\left\|[f(Y)-P_{2^N}(Y)](1-Y)^{\alpha-\varepsilon}(1+Y)^{\beta}\left(1-Y+\frac{1}{2^{2N}}\right)^{\rho_1+\varepsilon}\left(1+Y+\frac{1}{2^{2N}}\right)^{\rho_2}\right\|_p,$$

where the constants C_{14} and C_{15} do not depend on $f(x)$ or N. Applying Lemma 3, we obtain

$$I_2 \leqslant C_{16}2^{-N\gamma} \leqslant C_{17}\left|\sin\frac{t}{2}\right|^{\gamma},$$

where the constants C_{16} and C_{17} do not depend on $f(x)$, N or t.

Finally, let $p = \infty$. Then for any ϵ such that $0 < \epsilon \leq \alpha - \beta$ Lemma 4 gives

$$I_2 \leqslant \Bigg\|\int_0^1\int_{-1}^1 |f(Y)-P_{2^N}(Y)|[(1-x^2)(1-r^2)]^{\alpha-\beta-\varepsilon}[r^2(1-z^2)(1-x^2)]^{\beta}(1-z^2)^{-1/2}$$

$$\times r(1-r^2)^{-1+\varepsilon}\left(1-x+\frac{1}{2^{2N}}\right)^{\rho_1+\varepsilon}\left(1+x+\frac{1}{2^{2N}}\right)^{\rho_2}dr\,dz\Bigg\|_C$$

$$\leqslant C_{18}\Big\|[f(Y)-P_{2^N}(Y)][(1-R^2)(1-Y)]^{\alpha-\beta-\varepsilon}[R^2(1-V^2)(1-Y^2)]^{\beta}\times$$

$$\times\left(1-Y+\frac{1}{2^{2N}}\right)^{\rho_1+\varepsilon}\left(1+Y+\frac{1}{2^{2N}}\right)^{\rho_2}\Bigg\|_C \int_0^1\int_{-1}^1 (1-z^2)^{-1/2}(1-r^2)^{-1+\varepsilon}\, r\,dr\,dz$$

$$\leqslant C_{19}\left\|[f(Y)-P_{2N}(Y)](1-Y)^{\alpha-\varepsilon}(1+Y)^{\beta}\left(1-Y+\frac{1}{2^{2N}}\right)^{\rho_1+\varepsilon}\left(1+Y+\frac{1}{2^{2N}}\right)^{\rho_2}\right\|_C,$$

where the constants C_{18} and C_{19} do not depend on $f(x)$ or N. Now, applying Lemma 3, we have

$$I_2\leqslant C_{20}2^{-N\gamma}\leqslant C_{21}\left|\sin\frac{t}{2}\right|^{\gamma},$$

where the constants C_{20} and C_{21} do not depend on $f(x)$, N or t.

Thus I_2 is estimated for all p in the interval $1\le p\le\infty$.

It remains to estimate I_3. It is easy to see that

$$Q_{2k}(x,t,\alpha,\beta)-Q_{2k}(x)$$

$$=\frac{1}{2\mu(\alpha,\beta)}\int_0^1\int_{-1}^1\int_0^t\int_{-u}^u [Q''_{2k}(Y)Y_1^2+Q'_{2k}(Y)Y_2](1-r^2)^{\alpha-\beta-1}r^{2\beta+1}(1-z^2)^{\beta-1/2}\,d\lambda\,du\,dr\,dz,$$

where

$$Y=x\cos\lambda+rz\sin\lambda\sqrt{1-x^2}-(1-r^2)(1-x)\sin^2\frac{\lambda}{2},$$

$$Y_1=-x\sin\lambda+rz\cos\lambda\sqrt{1-x^2}-\frac{1}{2}(1-r^2)(1-x)\sin\lambda,$$

$$Y_2=-x\cos\lambda-rz\sin\lambda\sqrt{1-x^2}-\frac{1}{2}(1-r^2)(1-x)\cos\lambda.$$

Applying Minkowski's inequality, we obtain

$$I_3\leqslant C_{22}\sum_{k=0}^{N}\int_0^t\int_{-u}^u I_4(\lambda)\,d\lambda\,du,$$

where the constant C_{22} does not depend on N and

$$I_4(\lambda)=\Bigg\|\int_0^1\int_{-1}^1[Q''_{2k}(Y)Y_1^2+Q'_{2k}(Y)Y_2](1-r^2)^{\alpha-\beta-1}(1-z^2)^{\beta-1/2}r^{2\beta+1}$$

$$\times(1-x)^{\alpha}(1+x)^{\beta}\left(1-x+\frac{1}{2^{2N}}\right)^{\rho_1}\left(1+x+\frac{1}{2^{2N}}\right)^{\rho_2}dr\,dz\Bigg\|_p.$$

Let $1\le p<\infty$. Choosing ϵ such that $0<\epsilon<\alpha-\beta$ and applying Hölder's inequality, we have

$$I_4(\lambda)\leqslant C_{23}\Bigg(\int_{-1}^1\int_0^1\int_{-1}^1|Q''_{2k}(Y)Y_1^2+Q_{2k}(Y)Y_2|^p(1-x)^{(\alpha-\varepsilon)p}(1+x)^{\beta p}$$

$$\times\left(1-x+\frac{1}{2^{2N}}\right)^{(\rho_1+\varepsilon)p}\left(1+x+\frac{1}{2^{2N}}\right)^{\rho_2 p}(1-r^2)^{(\alpha-\beta-\varepsilon)p-1}r^{2\beta p+1}(1-z^2)^{\beta p-1/2}d\lambda\,dr\,dz\Bigg)^{1/p},$$

where the constant C_{23} does not depend on k or N.

Substituting into the triple integral

$$x = Y\cos\lambda + RV\sin\lambda\sqrt{1-Y^2} - (1-R^2)(1-Y)\sin^2\frac{\lambda}{2},$$

$$z = \frac{(1-R^2+Y+YR^2)\sin\lambda - 2RV\cos\lambda\sqrt{1-Y^2}}{\sqrt{[(1-R^2+Y+YR^2)\sin\lambda - 2RV\cos\lambda\sqrt{1-Y^2}]^2 + 4R^2(1-V^2)(1-Y^2)}},$$

$$r = \sqrt{\frac{R^2(1-Y)\cos^2(\lambda/2) - RV\sqrt{1-Y^2}\sin\lambda + (1+Y)\sin^2(\lambda/2)}{4 - \left[R^2(1-Y)\sin^2\frac{\lambda}{2} + RV\sqrt{1-Y^2}\sin\lambda\right]^2 + (1+Y)\cos^2(\lambda/2)}},$$

and applying Lemma 4 we obtain

$$I_4(\lambda) \leqslant C_{24}\left(\int_{-1}^{1}\int_{0}^{1}\int_{-1}^{1}\left|Q''_{2^k}(Y)(1-Y^2)R^2V^2 - \left[Y + \frac{1}{2}(1-R^2)(1-Y)\right]Q'_{2^k}(Y)\right|^p\right.$$
$$\times(1-Y)^{(\alpha-\varepsilon)p}(1+Y)^{\beta p}\left(1+Y+\frac{1}{2^{2N}}\right)^{\rho_2 p}\left(1-Y+\frac{1}{2^{2N}}\right)^{(\rho_1+\varepsilon)p}$$
$$\left.\times(1-R^2)^{(\alpha-\beta-\varepsilon)p-1}(1-V^2)^{\beta p-1/2}R^{2\beta p+1}\,dV\,dR\,dY\right)^{1/p}$$
$$\leqslant C_{25}\left\{\left\|Q''_{2^k}(Y)(1-Y^2)(1-Y)^{\alpha-\varepsilon}(1+Y)^{\beta}\left(1-Y+\frac{1}{2^{2N}}\right)^{\rho_1+\varepsilon}\left(1+Y+\frac{1}{2^{2N}}\right)^{\rho_2}\right\|_p\right.$$
$$\left.+\left\|Q'_{2^k}(Y)(1-Y)^{\alpha-\varepsilon}(1+Y)^{\beta}\left(1-Y+\frac{1}{2^{2N}}\right)^{\rho_1+\varepsilon}\left(1+Y+\frac{1}{2^{2N}}\right)^{\rho_2}\right\|_p\right\}$$
$$= C_{25}(I_5 + I_6),$$

where the constants C_{24} and C_{25} do not depend on k or N.

We now estimate I_5 and I_6. We begin with the estimation of I_6. We select a number μ such that $\beta \geq \mu > -1/p$. Then, applying Lemma 2, we can write

$$I_6 \leqslant C_{26}\left\|Q'_{2^k}(Y)(1-Y^2)^{\mu}\left(1-Y+\frac{1}{2^{2N}}\right)^{\rho_1+\alpha-\mu}\left(1+Y+\frac{1}{2^{2N}}\right)^{\rho_2+\beta-\mu}\right\|_p.$$

Let $\rho_1 + \alpha \geq \rho_2 + \beta$. Then

$$I_6 \leqslant C_{27}\left\|Q'_{2^k}(Y)(1-Y^2)^{\mu}\left(1-Y^2+\frac{1}{2^{2N}}\right)^{\rho_2+\beta-\mu}\left(1-Y+\frac{1}{2^{2N}}\right)^{\rho_1+\alpha-\rho_2-\beta}\right\|_p.$$

If $\rho_2 + \beta - \mu \geq 0$, then for $k \leq N$

$$I_6 \leqslant C_{27}\left\|Q'_{2^k}(Y)(1-Y^2)^{\mu}\left(1-Y^2+\frac{1}{2^{2k}}\right)^{\rho_2+\beta-\mu}\left(1-Y+\frac{1}{2^{2k}}\right)^{\rho_1+\alpha-\rho_2-\beta}\right\|_p.$$

Applying Lemma 2, again, we obtain

$$I_6 \leqslant C_{28}2^{2k}\left\|Q_{2^k}(Y)(1-Y)^{\alpha}(1+Y)^{\beta}\left(1-Y+\frac{1}{2^{2k}}\right)^{\rho_1}\left(1+Y+\frac{1}{2^{2k}}\right)^{\rho_2}\right\|_p \leqslant C_{29}2^{k(2-\gamma)},$$

where the constants C_{28} and C_{29} do not depend on k.

If $\rho_2 + \beta - \mu < 0$, then

$$I_6 \leqslant C_{30}2^{2(k-N)(\rho_2+\beta-\mu)}\left\|Q'_{2^k}(Y)(1-Y^2)^{\mu}\left(1-Y^2+\frac{1}{2^{2k}}\right)^{\rho_2+\beta-\mu}\left(1-Y+\frac{1}{2^{2k}}\right)^{\rho_1+\alpha-\rho_2-\beta}\right\|_p$$
$$\leqslant C_{31}2^{2(k-N)(\rho_2+\beta-\mu)}2^{k(2-\gamma)},$$

where the constants C_{30} and C_{31} do not depend on k.

The estimation of I_5 is similar.

It is now easy to estimate I_3. For $\rho_2+\beta-\mu\geq 0$ and $\gamma<2$

$$I_3\leqslant C_{32}t^2\sum_{k=0}^{N}2^{k(2-\gamma)}\leqslant C_{33}t^2 2^{N(2-\gamma)},$$

and for $0>\rho_2+\beta-\mu>\gamma/2-1$

$$I_3\leqslant C_{34}t^2 2^{-2N(\rho_2+\beta-\mu)}\sum_{k=0}^{N}2^{k(2-\gamma+2\rho_2+2\beta-2\mu)}\leqslant C_{35}t^2 2^{N(2-\gamma)},$$

where the constants C_{32}, C_{33}, C_{34} and C_{35} do not depend on t or N.

The case $\rho_2+\beta>\rho_1+\alpha$ is handled in exactly the same way.

Thus if $\gamma<2$ and $\min(\rho_1+\alpha,\rho_2+\beta)>\gamma/2-1-\mu$, then $I_3\leq C_{36}t^2 2^{N(2-\gamma)}$, where the constant C_{36} does not depend on t or N.

Since it is possible to assume without loss of generality that $|t|\leq\pi$, it is easy to see that $t^2\leq C_{37}\sin^2 t/2$ and therefore $I_3\leq C_{38}|\sin(t/2)|^{\gamma}$, where the constants C_{37} and C_{38} do not depend on t. Since the last inequality is true for all μ such that $1/p<\mu<1-\gamma/2+\min(\rho_1+\alpha_1,\rho_2+\beta)$, the inequality $I_3\leq C_{39}|\sin(t/2)|^{\gamma}$ is true for all γ satisfying the condition

$$\frac{\gamma}{2}<1+\frac{1}{p}+\min(\rho_1+\alpha,\ \rho_2+\beta).$$

Thus we have proved that under the conditions of the theorem for $1\leq p<\infty$ the inequality $I_3\leq C_{39}|\sin t/2|^{\gamma}$ holds.

Now let $p=\infty$. We will estimate

$$I_7=|Q''_{2^k}(Y)Y_1^2+Q'_{2^k}(Y)Y_2|(1-x)^{\alpha-\varepsilon}(1+x)^{\beta}\left(1-x+\frac{1}{2^{2N}}\right)^{\rho_1+\varepsilon}\left(1+x+\frac{1}{2^{2N}}\right)^{\rho_2}$$
$$\times(1-r^2)^{\alpha-\beta-\varepsilon}r^{2\beta}(1-z^2)^{\beta}.$$

Applying Lemma 4, we obtain that

$$I_7\leqslant C_{40}\left\{\left\|Q''_{2^k}(Y)(1-Y^2)(1-Y)^{\alpha-\varepsilon}(1+Y)^{\beta}\left(1-Y+\frac{1}{2^{2N}}\right)^{\rho_1+\varepsilon}\left(1+Y+\frac{1}{2^{2N}}\right)^{\rho_2}\right\|_C\right.$$
$$\left.+\left\|Q'_{2^k}(Y)(1-Y)^{\alpha-\varepsilon}(1+Y)^{\beta}\left(1-Y+\frac{1}{2^{2N}}\right)^{\rho_1+\varepsilon}\left(1+Y+\frac{1}{2^{2N}}\right)^{\rho_2}\right\|_C\right\},$$

where the constant C_{40} does not depend on k.

In the same way as above, it is possible to prove that under the conditions of the theorem

$$I_7\leqslant C_{41}2^{k(2-\gamma)}[1+2^{2(k-N)\min(\rho_1+\alpha,\ \rho_2+\beta)}],$$

where the constant C_{41} does not depend on k or N. Reasoning as above, we obtain that for $p=\infty$ the inequality $I_3\leq C_{42}|\sin(t/2)|^{\gamma}$ holds. Considering the bounds for I_1, I_2 and I_3, we obtain $I\leq C_{43}|\sin(t/2)|^{\gamma}$, which completes the proof of Theorem 3.

BIBLIOGRAPHY

1. S. M. Nikol'skiĭ, *On the best approximation of functions satisfying Lipschitz conditions by polynomials*, Izv. Akad. Nauk SSSR Ser. Mat. **10** (1946), 295–322. (Russian) MR **8**, 153.

2. A. F. Timan, *The approximation of functions satisfying a Lipschitz condition by ordinary polynomials*, Dokl. Akad. Nauk SSSR 77 (1951), 969–972. (Russian) MR **12**, 823.

3. V. K. Dzjadyk, *Constructive characterization of functions satisfying the condition* Lip α $(0<\alpha<1)$ *on a finite segment of the real axis*, Izv. Akad. Nauk SSSR Ser. Mat. **20** (1956), 623–642. (Russian) MR **18**, 392.

4. V. P. Motornyĭ, *Approximation of functions by algebraic polynomials in the* L_p *metric*, Izv. Akad. Nauk SSSR Ser. Mat. 35 (1971), 874–899 = Math. USSR Izv. 5 (1971), 889–914. MR 44 #5665.

5. G. K. Lebed', *Certain problems of the approximation of functions of one variable by algebraic polynomials*, Dokl. Akad. Nauk SSSR **118** (1958), 239–242. (Russian) MR 21 #248.

6. M. K. Potapov, *On Jackson type theorems in the* L_p *metric*, Dokl. Akad. Nauk SSSR **111** (1956), 1185–1188. (Russian) MR **18**, 803.

7. ———, *On the approximation of nonperiodic functions by algebraic polynomials*, Vestnik Moskov. Univ. Ser. I Mat. Meh. **1960**, no. 4, 14–25. (Russian) MR 23 #A3406.

8. ———, *Structural and constructive characterizations of some classes of functions*, Trudy Mat. Inst. Steklov. **131** (1973), 211–231 = Proc. Steklov Inst. Math. **131** (1973), 219–241.

9. T. H. Koornwinder, *The addition formula for Jacobi polynomials*. I, Nederl. Akad. Wetensch. Proc. Ser. A 72 = Indag. Math. 34 (1972), 188–191. MR 46 #7590.

10. B. A. Hililova, *On some estimates for polynomials*, Scientific Conference of Aspirants, Abstracts of Reports, Akad. Nauk Azerbaidžan. SSR, Baku, 1973, pp. 20–21. (Russian)

11. S. M. Nikol'skiĭ, *Approximation of functions of several variables and imbedding theorems*, "Nauka", Moscow, 1969; English transl., Springer-Verlag, Berlin and New York, 1974. MR 46 #9714.

Translated by Le BARON O. FERGUSON

Trudy Mat. Inst. Steklov.
134 (1975)

Proc. Steklov Inst. Math.
134 (1975)

APPROXIMATIONS OF A SEMICIRCLE BY ALGEBRAIC POLYNOMIALS

UDC 517.51

BL. SENDOV

Abstract. In this paper we study the order of magnitude of the best approximation of a function $y_1 = \sqrt{1 - x^2}$ ($x \in [-1, 1]$) by algebraic polynomials in the Hausdorff metric. It is shown that the best approximations ϵ_n of this function relative to the Hausdorff distance satisfy the estimates $c_1 n^{-2} < \epsilon_n < c_2 n^{-2}$ ($n = 1, 2, \dots$), where c_1 and c_2 are positive absolute constants.

The proof of this assertion makes use of a method due to Bernstein for estimating the best approximations of the function $y = |x|$ in the ordinary uniform metric, and is based on the study of the properties of polynomial oscillators.

Bibliography: 3 titles.

One of the concrete problems on uniform approximation by algebraic polynomials is the problem of approximating the function $|x|$, which was solved by Bernstein. The fundamental significance of this result in the theory of approximation is well known.

The problem of approximating $|x|$ on the interval $[-1, 1]$ by algebraic polynomials is equivalent to the problem of approximating the semicircle $y = \sqrt{1 - x^2}$ on the same interval.

Let E_n denote the minimum positive number for which there exists an algebraic polynomial of degree no higher than n, whose graph on the interval $[-1, 1]$ lies in the strip between the semicircles

$$y_1 = \sqrt{1 - x^2} - E_n \text{ and } y_2 = \sqrt{1 - x^2} + E_n.$$

Then one of Bernstein's results [1] can be formulated as follows: There exist positive constants a and b such that $a/n < E_n < b/n$.

Let ϵ_n denote the minimum positive number for which there exists an algebraic polynomials $P(x)$ of degree no higher than n, whose graph on the interval $[-1, 1]$ lies in the strip between the semicircles

$$\eta_1 = \sqrt{(1 - \varepsilon_n)^2 - x^2}, \quad |x| \leqslant 1 - \varepsilon_n,$$

and

$$\eta_2 = \sqrt{(1 + \varepsilon_n)^2 - x^2}, \quad |x| \leqslant 1 + \varepsilon_n,$$

AMS (MOS) subject classifications (1970). Primary 41A10.

where $P(-1) = P(1) = 0$. The purpose of this note is to show that there exist positive constants c_1 and c_2 such that $c_1 n^{-2} < \epsilon_n < c_2 n^{-2}$. This result is the analogue of the above-mentioned result of Bernstein if, instead of approximating uniformly, we approximate relative to the Hausdorff distance [2].

To prove our assertion we use Bernstein's method [1]. We now give the definition and some properties of polynomial oscillators, a concept introduced by Bernstein.

Definition. We say that the polynomial

$$P(x) = A_0 x^{\alpha_0} + A_1 x^{\alpha_1} + \dots + A_n x^{\alpha_n}$$

is a *polynomial oscillator* on the interval $[0, 1]$, corresponding to the sequence of nonnegative exponents $\alpha_0 < \alpha_1 < \dots < \alpha_n$, if it achieves its maximum absolute value at $n+1$ points of the interval. The number n is the order of the polynomial oscillator.

The uniform norm of P on the interval $[0, 1]$ will be denoted by $\|P\|$:

$$\|P\| = \max_{0 \leqslant x \leqslant 1} |P(x)|.$$

We will need the polynomial oscillator

$$T_{2k}(x) = \cos(2k \arccos x) = (-1)^k \{1 - 2k^2 x^2 + \dots + (-1)^k 2^{k-1} x^{2k}\}, \qquad (1)$$

namely the Čebyšev polynomial, for which $\|T_{2k}\| = 1$.

It is not hard to show [1] that the coefficients of a polynomial oscillator alternate in sign and that consecutive extrema of a polynomial oscillator have opposite signs.

The following theorems of Bernstein [1] hold:

Theorem 1. *If* $P(x) = \sum_0^n A_i x^{\alpha_i}$ *is a polynomial oscillator and* $Q(x) = \sum_0^n B_i x^{\alpha_i}$ *is another polynomial with the same degree of* x *and with one common coefficient* $A_{i_0} = B_{i_0}$ *(where* $\alpha_{i_0} > 0$*), then* $\|P\| < \|Q\|$.

Theorem 2. *If*

$$P(x) = A_0 x^{\alpha_0} + \dots + A_{i-1} x^{\alpha_{i-1}} + x^m + A_{i+1} x^{\alpha_{i+1}} + \dots + A_n x^{\alpha_n}$$

and

$$Q(x) = B_0 x^{\beta_0} + \dots + B_{i-1} x^{\beta_{i-1}} + x^m + B_{i+1} x^{\beta_{i+1}} + \dots + B_n x^{\beta_n}$$

are two polynomial oscillators, and if

$$0 \leqslant \alpha_0 < \beta_0 < \dots < \alpha_{i-1} < \beta_{i-1} < m < \beta_{i+1} < \alpha_{i+1} < \beta_n < \alpha_n,$$

then $\|Q\| < \|P\|$.

Following the method of [1], we prove the following assertion:

Lemma 1. *The polynomial oscillator*

$$S_k(x) = x^2 + a_1 x^3 + \ldots + a_k x^{2k+1} \tag{2}$$

satisfies the inequalities

$$\frac{k^{-2}}{18} < \| S_k \| < (k+1)^{-2}. \tag{3}$$

Proof. If we let Q_{k+1} denote the polynomial oscillator

$$Q_{k+1}(x) = x^2 + b_1 x^4 + \ldots + b_k x^{2k+2},$$

then it follows from Theorem 2 that

$$\| S_k \| < \| Q_{k+1} \|.$$

On the other hand, from Theorem 1 and (1) we have

$$\| Q_{k+1} \| < \left\| \frac{1}{2}(k+1)^{-2} T_{2k+2} \right\| + \frac{1}{2}(k+1)^{-2} = (k+1)^{-2}.$$

This proves the right-hand side of (3).

It follows from Theorem 1 that for every $\mu \geq 0$ we have

$$\max_{0 \leqslant x \leqslant 1} \left| S_k \left(\frac{x}{1+\mu} \right) \right| \leqslant \| S_k \|$$

or

$$|(1+\mu) x^2 + a_1 x^3 + a_2' x^5 + \ldots + a_k' x^{2k+1}| \leqslant (1+\mu)^3 \| S_k \|.$$

Taking account of (2), we get

$$|\mu (x^2 + b_1 x^5 + \ldots + b_{k-1} x^{2k+1})| \leqslant (1 + (1+\mu)^3) \| S_k \|. \tag{4}$$

from the last inequality. Let $R_x(x)$ denote the polynomial

$$R_k(x) = x^2 + b_1 x^5 + \ldots + b_{k-1} x^{2k+1}.$$

Then it follows from Theorem 2 and (1) that

$$\| R_k \| > \left\| \frac{1}{2} k^{-2} T_{2k}(x) + (-1)^k \frac{1}{2} k^{-2} \right\| > \frac{1}{2} k^{-2} \| T_{2k} \| = \frac{1}{2} k^{-2}.$$

Therefore we deduce from (4) that

$$\| S_k \| \geqslant \frac{\mu}{2k^2 (1 + (1+\mu)^3)}, \qquad \mu \geqslant 0.$$

For $\mu = 1$ the last inequality reduces to the left-hand side of (3). The lemma is proved.

Let x_i, $i = 0, 1, \ldots, k$, be the points where $|S_k(x)|$ achieves its maximum absolute value on the interval $[0, 1]$. If they are arranged in increasing order, then

$$0 < x_0 < x_1 < \ldots < x_{k-1} < x_k = 1.$$

On the other hand, since

$$S_k(0) = 0, \qquad S_k'(0) = 0, \qquad S_k''(0) = 2 > 0,$$

we have $S_k(x_0) > 0$, and therefore

$$S_k(x_i) = (-1)^i \| S_k \|, \quad i = 0, 1, 2, \ldots, k. \tag{5}$$

Theorem 3. *Let ϵ_n denote the smallest positive number for which there exists an algebraic polynomial $P_n(x)$ of degree not exceeding n, with $P_n(-1) = P(1) = 0$, whose graph on the interval $[-1, 1]$ lies in the strip between the semicircles*

$$\eta_1 = \sqrt{(1-\varepsilon_n)^2 - x^2}, \quad |x| \leqslant 1 - \varepsilon_n,$$

and

$$\eta_2 = \sqrt{(1+\varepsilon_n)^2 - x^2}, \quad |x| \leqslant 1 + \varepsilon_n.$$

Then

$$\frac{n^{-2}}{9} < \varepsilon_n < 5n^{-2}. \tag{6}$$

Proof. We introduce the polynomial oscillator (2), replace x by $\sqrt{1-t^2}$ and write it in the following form:

$$\frac{S_k(\sqrt{1-t^2})}{\sqrt{1-t^2}} = \sqrt{1-t^2} - V_k(t^2), \tag{7}$$

where $V_k(t^2) = -a_1(1-t^2) - \cdots - a_k(1-t^2)^k$. As a result, we have

$$\sqrt{1-t^2} - \frac{\| S_k \|}{\sqrt{1-t^2}} \leqslant V_k(t^2) \leqslant \sqrt{1-t^2} + \frac{\| S_k \|}{\sqrt{1-t^2}}. \tag{8}$$

It can be verified at once that the inequality

$$\sqrt{1-t^2} - \frac{\alpha}{\sqrt{1-t^2}} \geqslant \sqrt{(1-\alpha)^2 - t^2}$$

holds for $|t| \leq 1 - \alpha$, and therefore

$$V_k(t^2) \geqslant \sqrt{(1 - \| S_k \|)^2 - t^2}, \quad |t| \leqslant 1 - \| S_k \|. \tag{9}$$

It follows from (5) and (7) that

$$V_k(t^2) \leqslant \sqrt{1-t^2} \text{ for } t_0 = \sqrt{1-x_0^2} \leqslant |t| \leqslant 1 \tag{10}$$

and

$$V_k(t_0^2) = \sqrt{1-t_0^2} - \frac{\| S_k \|}{\sqrt{1-t_0^2}}.$$

Since $S_k(t)$ is a polynomial oscillator, $V_k(t^2)$ decreases monotonically on the interval $[t_0, 1]$, and hence is nonnegative. Then, obviously,

$$t_0 < 1 - \| S_k \|.$$

On the other hand, it is not hard to verify that the inequality

$$\sqrt{1-t^2} + \frac{\alpha}{\sqrt{1-t^2}} \leqslant \sqrt{\left(1 + \frac{5}{4}\alpha\right)^2 - t^2}$$

holds for $|t| \leq 1 - \alpha$. It follows from the last inequality and (10) that

$$V_k(t^2) \leqslant \sqrt{\left(1+\frac{5}{4}\|S_k\|\right)^2 - t^2} \tag{11}$$

for $|t| \leq 1$. But (9), (11) and (3) imply

$$\varepsilon_n \leqslant \frac{5}{4}\|S_k\| < \frac{5}{4(k+1)^2} \leqslant 5n^{-2},$$

which proves the right-hand side of (6).

We now prove the left-hand side of (6). The following inequalities are easily verified:

$$\sqrt{1-t^2} - \frac{\alpha}{\sqrt{1-t^2}} \leqslant \sqrt{\left(1-\frac{1}{2}\alpha\right)^2 - t^2}, \qquad t^2 \leqslant 1-\alpha, \tag{12}$$

$$\sqrt{1-t^2} + \frac{\alpha}{\sqrt{1-t^2}} \geqslant \sqrt{(1+\alpha)^2 - t^2}, \qquad |t| \leqslant 1. \tag{13}$$

We note that $V_k(t^2)$ is an even polynomial of degree $2k$ and that the equalities in (8) are achieved alternately at the $k+1$ points $0 = t_k < t_{k-1} < \cdots < t_0 < 1$. Then, using (12) and (13), we can conclude that there does not exist a polynomial of degree $\leq 2k$ whose graph on the interval $[-1, 1]$ lies between the semicircles

$$y_1 = \sqrt{\left(1-\frac{1}{2}\|S_k\|\right)^2 - t^2}$$

and

$$y_2 = \sqrt{(1+\|S_k\|)^2 - t^2}$$

and which vanishes for $t = -1$ and $t = 1$. In fact, there would otherwise be such a polynomial of degree not exceeding $2k+1$ with the same value as $V_k(t^2)$ at $2k+2$ points, and this polynomial would therefore coincide with $V_k(t^2)$, which is impossible. Hence, in accordance with (3), we get

$$\varepsilon_n \geqslant \frac{1}{2}\|S_k\| > \frac{1}{36}k^{-2} \geqslant \frac{1}{9}n^{-2}$$

and the theorem is proved.

Corollary. *For every natural number n there exists an algebraic polynomial P of degree not exceeding n for which*

$$\int_{-1}^{1} \left|\sqrt{1-x^2} - P(x)\right| dx \leqslant 5\pi n^{-2}, \qquad P(-1) = P(1) = 0.$$

Remark. In [3] we proved the following assertion:

Theorem 4. *Let f be an analytic function in the disk $|z| < 1$ which is bounded in the closed disk $|z| \leq 1$ and takes real values on the interval $[-1, 1]$. Then the best approximation $E_{n,r}(f)$ of f by algebraic polynomials of degree not exceeding n relative to the Hausdorff distance on $[-1, 1]$ satisfies the inequality*

$$E_{n,r}(t) \leqslant \left(\frac{8\ln n}{n}\right)^2 + \frac{5M}{n^2}, \tag{14}$$

where $M = \sup_{|z|\leq 1} |f(z)|$.

It follows from Theorem 3 that the function $\phi(x) = \sqrt{1 - x^2}$ satisfies the inequality

$$E_{n,r}(\varphi) \leqslant \frac{5}{n^2},$$

i.e., for this function we can improve the order of the estimate, getting rid of $(\ln n)^2$. This can be done for all functions of the form $\phi_\alpha(x) = (1 - x^2)^\alpha$, $0 < \alpha < 1$. We note that the estimate (14) cannot be improved in the class of all functions of the type considered in Theorem 4.

The problem of narrowing the bounds in (6) and of finding the asymptotic value of ϵ_n remains open.

BIBLIOGRAPHY

1. Serge Bernstein, *Sur la meilleure approximation de $|x|$ par des polynomes des degrés donnés*, Acta Math. 37 (1914), 1–57.

2. Bl. Sendov, *Certain questions in the theory of approximations of functions and sets in the Hausdorff metric*, Uspehi Mat. Nauk 24 (1969), no. 5 (149), 141–178 = Russian Math. Surveys 24 (1969), no. 5, 143–183. MR 43 #2390.

3. ———, *Approximation of analytic functions in Hausdorff metric*, Proc. Internat. Conf. Functional Analysis Appl. (Madras, 1973), Lecture Notes in Math., vol. 399, Springer-Verlag, Berlin and New York, 1974, pp. 490–500. Zbl. 293 #41006.

Translated by R. A. SILVERMAN

Trudy Mat. Inst. Steklov.
134 (1975)

Proc. Steklov Inst. Math.
134 (1975)

ON MEAN VALUES OF THE MODULUS OF A TRIGONOMETRIC SUM

UDC 511

S. B. STEČKIN

Abstract. We set

$$S_n(X) = \sum_{1 \leqslant k \leqslant X} e\left(\sum_{\nu=1}^{n} \alpha_\nu k^\nu\right) \qquad (e(u) = e^{2\pi i u}),$$

$$J = J_n(X, l) = \int_{\mathbf{T}^n} |S_n(X)|^{2e} \, d\alpha_1 \ldots d\alpha_n.$$

The following common upper bound for the J can be found using the method of I. M. Vinogradov:

Theorem. *Let $n, r \in \mathbf{N}$, $n \geq 2$, $l \in \mathbf{R}$, $l \geq rn$, and $X \in \mathbf{R}$, $X \geq 1$. Then*

$$J_n(X, l) \leqslant D(r, n) X^{2l - \frac{n(n+1)}{2} + \frac{n^2}{2}\left(1 - \frac{1}{n}\right)^r}$$

where

$$D(r, n) = \exp(C \min(r, n) n^2 \ln n)$$

and C is an absolute constant.

This theorem is applied to estimate trigonometric sums of the form

$$T(Q) = \sum_{P < k \leqslant P+Q} e(F(k)),$$

where $F(u)$ is a sufficiently smooth function.

Bibliography: 21 titles.

Introduction

The following notation is used in this paper:

$\mathbf{N}$ is the set of all natural numbers,

$\mathbf{P}$ is the set of all prime numbers,

$\mathbf{R}$ is the set of all real numbers,

$\mathbf{Z}$ is the set of all integers,

$\mathbf{T} = \mathbf{R}/\mathbf{Z}$, $\mathbf{T}^n$ and $\mathbf{Z}^n$ have the usual meaning,

$C_1, \ldots, C_{61}$ are absolute positive constants.

In addition, as usual, $\exp(y) = e^y$ and $e(y) = e^{2\pi i y}$, $y \in \mathbf{R}$.

We set

AMS (MOS) subject classifications (1970). Primary 10G05; Secondary 10H15, 10B15, 10B45.

$$f(u)=f_n(u)=\sum_{\nu=1}^{n}\alpha_\nu u^\nu \quad (n\in \mathbf{N},\ n\geqslant 2,\ \alpha_\nu\in\mathbf{R},\ u\in\mathbf{R}),$$
$$S(X)=S_n(X)=\sum_{1\leqslant k\leqslant X} e(f_n(k)) \quad (k\in\mathbf{N},\ X\in\mathbf{R},\ X\geqslant 1). \tag{0.1}$$

The behavior of the trigonometric sums (0.1) and similar sums is closely related to the behavior of the mean values of the modulus of $S(X)$, i.e. to the behavior of the integral

$$J=J_n(X,\ l)=\int_{\mathbf{T}^n}|S_n(X)|^{2l}\,d\alpha_1\ldots d\alpha_n \qquad (l\in\mathbf{R},\ l\geqslant 0) \tag{0.2}$$

as a function of the parameters X, l and n. For applications to trigonometric sums we need rather precise estimates of J from above. The strongest estimates for J and the corresponding estimates for trigonometric sums have been obtained using the Vinogradov method. Two theorems about J are proved in Vinogradov's monograph [1]: one of them establishes a "common upper bound" for the J, and the other gives a "simplified upper bound" for the J (see Theorems 1 and 4 in Chapter 4).

Recently A. A. Karacuba [2] (see also [3]) obtained the following simplified upper bound for J, which is uniform in all the parameters:

Theorem A. *There exist absolute positive constants C_1 and C_2 such that for all $n\in\mathbf{N}$, $n\geq 2$, $l\in\mathbf{R}$, $l\geq C_1n^2\ln n$, and $X\in\mathbf{R}$, $X\geq 1$, the following inequality holds:*

$$J_n(X,\ l)\leqslant n^{C_2n^3}X^{2l-\frac{n(n+1)}{2}}. \tag{0.3}$$

In the present paper the domain of variation of the parameters in which the "common upper bound" for J holds is extended to its natural limits. Namely, we prove the following theorem.

Theorem 1. *Let*

$$n,\ r\in\mathbf{N},\ n\geqslant 2,\ l\in\mathbf{R},\ l\geqslant rn,\ X\in\mathbf{R},\ X\geqslant 1. \tag{0.4}$$

Then

$$J_n(X,\ l)\leqslant D(r,\ n)\,X^{2l-\frac{n(n+1)}{2}+\frac{n^2}{2}\left(1-\frac{1}{n}\right)^r}, \tag{0.5}$$

where

$$D(r,\ n)=\exp\left(C_3\min(r,\ n)\,n^2\ln n\right) \tag{0.6}$$

and C_3 is an absolute constant.

The proof applies the p-adic version of the Vinogradov method in [2] and [4], and uses ideas and methods from [1].

As usual, the proof is divided into two steps. We first derive a recursive estimate for J. Then, by repeated integration of this estimate, we derive an estimate from above for J. The first step is the hardest. It consists of an arithmetic and an analytic part. The basic arithmetic part is no different from that in [2]; the result we need is presented in the form of Lemma 1. In the analytic part of the proof, certain estimates must be computed more carefully (Lemma 2). The overall recursive estimate we obtain is more precise than those of previous authors, and we immediately reduce it to a convenient form (Lemma 4). The second step of the proof also requires some new considerations (Lemma 6). In addition, we use Theorem A in an essential way: it shows that a stronger assertion than Theorem 1 holds for $r \geq C_1 n \ln n$.

As an application, we derive estimates for trigonometric sums of the form

$$T(Q) = \sum_{P<k\leqslant P+Q} e(F(k)), \tag{0.7}$$

where $F(u)$ is an n times differentiable function. We follow the same plan of proof as Vinogradov in [5] and [6]. The sum $T(Q)$ is represented as a sum whose terms ard trigonometric sums (in our case this is the "Weyl representation" [7], [8]); these trigonometric sums are estimated in terms of Weyl sums of order $n-1$ whose coefficients are "spread out over" T^{n-1}; then these estimates are integrated over suitable regions in T^{n-1} and added together; after taking into account the multiplicity of the covering of T^{n-1}, we obtain an estimate of $\mathrm{T}(Q)$ in terms of J_{n-1} (Theorem 3). These results improve somewhat the known estimates (see [9]–[13]); in particular, they generalize and refine the estimates Rodosskiĭ obtained [14] for small n. We note that Rodosskiĭ's plan of proof was significantly more complicated, although this was actually not necessary.

We compare our estimates with those of van der Corput. It turns out that, for certain relations between the parameters, Theorem 3 gives better results, starting with $n = 9$.

We note that, in a manner analogous to the treatment of the mean value theorem in the present paper, it is possible to extend the domain of variation of the parameters and improve the constants in several of Karacuba's theorems estimating the number of solutions of certain systems of congruences and mixed systems of equations and congruences (see [15] and [16]).

§1. Fundamental properties of J

We list the fundamental properties of the interval J which will be used in what follows (see [9], [4]).

1) *If* $0 \leq l \leq l_1$, *then*

$$J_n(X, l_1) \leqslant X^{2(l_1-l)} J_n(X, l) \qquad (X \geqslant 1);$$

in particular, we have the "trivial estimate"

$$J_n(X, l) \leqslant X^{2l}.$$

Use the estimate $|S_n(X)| \leq X$.

2) *If* $0 \leq l_0 \leq l \leq l_1$ *and* $l = (1-t)l_0 + tl_1$, $0 \leq t \leq 1$, *then*

$$J_n(X, l) \leqslant \{J_n(X, l_0)\}^{1-t} \{J_n(X, l_1)\}^t \quad (X \geqslant 1).$$

Apply Hölder's inequality.

The most important properties of J are when $l \in \mathbf{N}$. This is related to the fact that in this case J has a simple arithmetic meaning:

3) *If* $l \in \mathbf{N}$, *then* J *is equal to the number of solutions of the system of Diophantine equations*

$$\sum_{j=1}^{l} x_j^\nu = \sum_{j=1}^{l} y_j^\nu \quad (\nu = 1, \ldots, n),$$
$$x_j, y_j \in \mathbf{N}, \ 1 \leqslant x_j \leqslant X, \ 1 \leqslant y_j \leqslant X \ (j = 1, \ldots, l). \tag{1.1}$$

Write J in the form

$$J = \int_{\mathbf{T}^n} S^l(X) \overline{S^l(X)} \, d\alpha_1 \ldots d\alpha_n$$

and compute the integral. For a survey of work relating to the system (1.1), see [17], Chapter 24, and [12], Chapter 4.

We shall call the following important property of (1.1) and similar systems the *shift property*.

4) *Let* $a \in \mathbf{Z}$. *Then the number of solutions of the system*

$$\sum_{j=1}^{l} (x_j - a)^\nu = \sum_{j=1}^{l} (y_j - a)^\nu,$$
$$x_j, y_j \in \mathbf{N}, \ 1 \leqslant x_j \leqslant X, \ 1 \leqslant y_j \leqslant X \ (j = 1, \ldots, l)$$

does not depend on a, *and equals* $J_n(X, l)$.

See [2] or [4].

5) *Let* $l, q \in \mathbf{N}$ *and* $\lambda_\nu \in \mathbf{Z}$, $\nu = 1, \ldots, n$, *and let* $J_n(q, l, \lambda_1, \ldots, \lambda_n)$ *be the number of solutions of the following system in* $x_j, y_j \in \mathbf{N}$:

$$\sum_{j=1}^{l} x_j^\nu - \sum_{j=1}^{l} y_j^\nu = \lambda_\nu \quad (\nu = 1, \ldots, n),$$
$$1 \leqslant x_j \leqslant q, \ 1 \leqslant y_j \leqslant q \ (j = 1, \ldots, n). \tag{1.2}$$

Then $J_n(q, l, \lambda_1, \ldots, \lambda_n) \leq J_n(q, l)$.

Write the number of solutions of (1.2) as an integral and estimate its magnitude.

6) *If* $l \in \mathbf{N}$ *and* $1 \leq X \leq X_1$, *then* $J_n(X, l) \leq J_n(X_1, l)$.

This follows from 3).

7) $J_n(X, n) \leq n!\, X^n$.

See [9] or [4].

Let $p, q \in \mathbf{N}$. We set

$$\int_{\mathbf{T}^n} |U(x)|^{2l} \, d\alpha_1 \ldots d\alpha_n = J_n(q, l). \tag{1.3}$$

8) *If* $l \in \mathbf{N}$ *and* $x \in \mathbf{Z}$, *then*

$$U(x) = U_n(x) = \sum_{u=0}^{q-1} e(f_n(x + pu)), \quad x \in \mathbf{Z}.$$

Write the left-hand side as the number of solutions of a system of Diophantine equations, and use the shift property (see [2] or [4]). This is the fundamental property on which the estimates of J are based.

§2. Recursive estimate for J

Let $n \in \mathbf{N}$, $n \geq 2$, $l \in \mathbf{Z}$, $l \geq 0$, and $p, q \in \mathbf{N}$. As in §1, we set

$$U(x) = U_n(x) = \sum_{u=0}^{q-1} e(f_n(x+pu)) \quad (x \in \mathbf{Z}). \tag{2.1}$$

Since

$$S_n(pq) = \sum_{x=1}^{p} U_n(x), \tag{2.2}$$

it follows that

$$J_n(pq, l+n) = \int_{\mathbf{T}^n} \left| \sum_{x=1}^{p} U_n(x) \right|^{2(l+n)} d\alpha_1 \ldots d\alpha_n. \tag{2.3}$$

We represent this integral in the following form:

$$\begin{aligned} J_n(pq, l+n) &= \int_{\mathbf{T}^n} \left| \sum_{x_1, \ldots, x_{l+n}} \prod_{j=1}^{l+n} U(x_j) \right|^2 d\alpha_1 \ldots d\alpha_n \\ &= \int_{\mathbf{T}^n} \left| \Sigma^{(1)} \prod_{j=1}^{l+n} U(x_j) + \Sigma^{(2)} \prod_{j=1}^{l+n} U(x_j) \right|^2 d\alpha_1 \ldots d\alpha_n, \end{aligned} \tag{2.4}$$

where $x_j \in \mathbf{N}$, $1 \leq x_j \leq p$ $(j = 1, \ldots, l+n)$ and the summation in $\sigma^{(1)}$ is taken over all systems $(x_1, \ldots, x_{l+n})_n$ containing at least n distinct numbers, and the summation in $\Sigma^{(2)}$ is taken over the remaining systems.

The basic part of J is estimated by means of the following lemma.

Lemma 1. *Let $l, n, q \in \mathbf{N}$ and $p \in \mathbf{P}$,*

$$n \geqslant 2, \ p > n, \ 1 \leqslant q \leqslant p^{n-1}. \tag{2.5}$$

Set

$$J^{(1)} - \int_{\mathbf{T}^n} \left| \Sigma^{(1)} \prod_{j=1}^{l+n} U(x_j) \right|^2 d\alpha_1 \ldots d\alpha_n, \tag{2.6}$$

where $\Sigma^{(1)}$ is defined as above. Then

$$J^{(1)} \leqslant n! \binom{l+n}{n}^2 p^{2l+\frac{n}{2}(3n-1)} J_n(q, l). \tag{2.7}$$

This lemma is essentially contained in Lemma 3 of [2] or Theorem 1 of [4], p. 94. For the sake of completeness, we give its proof.

The proof is divided into several steps.

1. The integral $J^{(1)}$ is the number of solutions of the following system of Diophantine equations in $x_j, y_j, u_j, v_j \in \mathbf{Z}$:

$$\sum_{j=1}^{l+n}(x_j+pu_j)^\nu=\sum_{j=1}^{l+n}(y_j+pv_j)^\nu \quad (\nu=1,\ldots,n),$$
$$1\leqslant x_j\leqslant p,\ 1\leqslant y_j\leqslant p, \tag{2.8}$$
$$0\leqslant u_j<q,\ 0\leqslant v_j<q \ (j=1,\ldots,l+n),$$

where each of the systems $(x_1,\ldots,x_{l+n})_n$ and $(y_1,\ldots,y_{l+n})_n$ contains no fewer than n distinct numbers.

Let N_n be the number of solutions of the system of Diophantine equations (2.8), in which each of the systems $(x_1,\ldots,x_n)_n$ and $(y_1,\ldots,y_n)_n$ consists of distinct numbers. We arbitrarily choose indices $j_1,\ldots,j_n$ from among the $l+n$ indices. This can be done in $\binom{l+n}{n}$ ways. Since the systems $(x_1,\ldots,x_n)_n$, $x_{n+1},\ldots,x_{n+1}$ and the systems in which the distinct numbers are in the indices $j_1,\ldots,j_n$ are in one-to-one correspondence, and similarly for the systems y, it follows that

$$J^{(1)}\leqslant\binom{l+n}{n}^2N_n. \tag{2.9}$$

2. We have

$$N_n=\int_{T^n}\left|\sum_{(x_1,\ldots,x_n)_n,\,x_{n+1},\ldots,x_{n+l}}\prod_{j=1}^{l+n}U(x_j)\right|^2d\alpha_1\ldots d\alpha_n,$$

where the systems $(x_1,\ldots,x_n)_n$ consist of n distinct numbers, and the systems $(x_{n+1},\ldots,x_{n+l})$ are arbitrary. Hence

$$N_n=\int_{T^n}\left|\sum_{(x_1,\ldots,x_n)_n}\prod_{j=1}^{n}U(x_j)\left\{\sum_{x=1}^{p}U(x)\right\}^l\right|^2d\alpha_1\ldots d\alpha_n$$

and by Hölder's inequality

$$N_n\leqslant\int_{T^n}\left|\sum_{(x_1,\ldots,x_n)_n}\prod_{j=1}^{n}U(x_j)\right|^2p^{2l-1}\sum_{x=1}^{p}|U(x)|^{2l}d\alpha_1\ldots d\alpha_n=p^{2l-1}\sum_{x=1}^{p}J(x), \tag{2.10}$$

where

$$J(x)=\int_{T^n}\left|\sum_{(x_1,\ldots,x_n)_n}\prod_{j=1}^{n}U(x_j)\,U^l(x)\right|^2d\alpha_1\ldots d\alpha_n\ (x\in\mathbf{Z}). \tag{2.11}$$

3. The integral $J(x)$ is equal to the number of solutions of the following system of Diophantine equations in $x_j, y_j, u_j, v_j, t_k, z_k\in\mathbf{Z}$:

$$\sum_{j=1}^{n}(x_j+pu_j)^\nu+\sum_{k=1}^{l}(x+pt_k)^\nu=\sum_{j=1}^{n}(y_j+pv_j)^\nu+\sum_{k=1}^{l}(x+pz_k)^\nu\ (\nu=1,\ldots,n),$$
$$1\leqslant x_j\leqslant p,\ 1\leqslant y_j\leqslant p,\ x_j\neq x_{j'},\ y_j\neq y_{j'}\ (j\neq j'),$$
$$0\leqslant u_j<q,\ 0\leqslant v_j<q\quad(j=1,\ldots,n), \tag{2.12}$$
$$0\leqslant t_k<q,\ 0\leqslant z_k<q\quad(k=1,\ldots,\eta).$$

If we make the substitution $a_j=x_j+pu_j$, $b_j=y_j+pv_j$, we see that this system is equivalent to the following system in $a_j, b_j, z_k, t_k\in\mathbf{Z}$:

$$\begin{gathered}\sum_{j=1}^{n} a_j^\nu - \sum_{j=1}^{n} b_j^\nu = \sum_{k=1}^{l} (x+pz_k)^\nu - \sum_{k=1}^{l} (x+pt_k)^\nu \quad (\nu=1,\ldots,n),\\ 1 \leqslant a_j \leqslant pq,\ 1 \leqslant b_j \leqslant pq \ (j=1,\ldots,n),\\ a_j \not\equiv a_{j'},\ b_j \not\equiv b_{j'} \pmod p \ (j \neq j'),\\ 0 \leqslant t_k < q,\ 0 \leqslant z_k < q \ (k=1,\ldots,l).\end{gathered} \tag{2.13}$$

Further, by the shift property, the system (2.13) is equivalent to the system in $a_j, b_j, z_k, t_k, \lambda_\nu \in \mathbf{Z}$:

$$\begin{gathered}\sum_{j=1}^{n} a_j^\nu - \sum_{j=1}^{n} b_j^\nu = p^\nu \left\{\sum_{k=1}^{l} z_k^\nu - \sum_{k=1}^{l} t_k^\nu\right\} \quad (\nu=1,\ldots,n),\\ 1-x \leqslant a_j \leqslant pq-x,\ 1-x \leqslant b_j \leqslant pq-x \ (j=1,\ldots,n),\\ a_j \not\equiv a_{j'},\ b_j \not\equiv b_{j'} \pmod p \ (j \neq j'),\\ 0 \leqslant t_k < q,\ 0 \leqslant z_k < q \ (k=1,\ldots,l),\end{gathered} \tag{2.14}$$

so that the number of solutions of (2.14) is equal to $J(x)$.

4. We transform the system (2.14) into the following system in $a_j, b_j, z_k, t_k, \lambda_\nu \in \mathbf{Z}$:

$$\sum_{j=1}^{n} a_j^\nu - \sum_{j=1}^{n} b_j^\nu = p^\nu \lambda_\nu, \tag{2.15}$$

$$\sum_{k=1}^{l} z_k^\nu - \sum_{k=1}^{l} t_k^\nu = \lambda_\nu \quad (\nu=1,\ldots,n) \tag{2.16}$$

with the same restrictions on the unknowns. We let $N_n(p\lambda_1, \ldots, p^n\lambda_n)$ denote the number of solutions of the system (2.15) for fixed $\lambda_1, \ldots, \lambda_n$. In §1 the number of solutions of (2.16) for fixed $\lambda_1, \ldots, \lambda_n$ was denoted by $J_n(q, l; \lambda_1, \ldots, \lambda_n)$, and there it was noted (property 5)) that

$$J_n(q, l, \lambda_1, \ldots, \lambda_n) \leqslant J_n(q, l) \qquad (\forall \lambda \in \mathbf{Z}^n). \tag{2.17}$$

Hence

$$\begin{aligned} J(x) &= \sum_{\lambda} N_n(p\lambda_1, \ldots, p^n\lambda_n) J_n(q, l, \lambda_1, \ldots, \lambda_n)\\ &\leqslant J_n(q, l) \sum_{\lambda} N_n(p\lambda_1, \ldots, p^n\lambda_n). \end{aligned} \tag{2.18}$$

5. Finally, we consider the sum

$$T' = \sum_{\lambda} N_n(p\lambda_1, \ldots, p^n\lambda_n). \tag{2.19}$$

This is the number of solutions of the following system of equations in $a_j, b_j, \lambda_\nu \in \mathbf{Z}$:

$$\begin{gathered}\sum_{j=1}^{n} a_j^\nu - \sum_{j=1}^{n} b_j^\nu = p^\nu \lambda_\nu \quad (\nu=1,\ldots,n),\\ 1-x \leqslant a_j \leqslant pq-x,\ 1-x \leqslant b_j \leqslant pq-x,\\ a_j \not\equiv a_{j'},\ b_j \not\equiv b_{j'} \pmod p \ (j \neq j').\end{gathered} \tag{2.20}$$

It is clear that $T' \leq T''$, where T'' is the number of solutions of the following *system of congruences* in a_j, $b_j \in \mathbf{Z}$:

$$\sum_{j=1}^{n} a_j^{\nu} - \sum_{j=1}^{n} b_j^{\nu} \equiv 0 \pmod{p^{\nu}} \quad (\nu = 1, \ldots, n),$$
$$1 - x \leqslant a_j \leqslant pq - x, \quad 1 - x \leqslant b_j \leqslant pq - x \quad (j = 1, \ldots, n), \tag{2.21}$$
$$a_j \not\equiv a_{j'} \pmod{p} \quad (j \neq j').$$

But, as proved by Linnik [18] (see also [10], Russian p. 484, Corollary, the case $m = 1$, or [2], Lemma 2), if $p \in \mathbf{P}$, $p > n$ and $q \leq p^{n-1}$, then

$$T'' \leqslant n!\, p^{2n^2 - \frac{n(n+1)}{2}}. \tag{2.22}$$

Comparing the estimates and formulas (2.9), (2.10), (2.18), (2.19) and (2.22), we obtain

$$J^{(1)} \leqslant \binom{l+n}{n}^2 p^{2l-1} \sum_{x=1}^{p} J_n(q,\, l)\, n!\, p^{2n^2 - \frac{n(n+1)}{2}},$$

which is equivalent to the assertion of the lemma.

Lemma 2. *Let l, n, $q \in \mathbf{N}$ and $p \in \mathbf{P}$ satisfy*

$$n \geqslant 2, \quad p > n, \quad 1 \leqslant q \leqslant p^{n-1}. \tag{2.23}$$

Then

$$J_n(pq,\, l+n) \leqslant (l+n)^{2n} p^{2l + \frac{n}{2}(3n-1)} J_n(q,\, l) + n^{2(l+n)} p^{2(n-1)} J_n(q,\, l+n). \tag{2.24}$$

Proof. We use the representation of J in the form (2.4). Since $|a + b|^2 \leq 2(|a|^2 + |b|^2)$, it follows that

$$J_n(pq,\, l+n) \leqslant 2(J^{(1)} + J^{(2)}), \tag{2.25}$$

where

$$J^{(m)} = \int_{\mathbf{T}^n} \left| \sum{}^{(m)} \prod_{j=1}^{l+n} U(x_j) \right|^2 d\alpha_1 \ldots d\alpha_n \qquad (m = 1,\, 2). \tag{2.26}$$

The estimate for $J^{(1)}$ is given by Lemma 1.

We estimate $J^{(2)}$. To do this, we note that if $x_j \in \mathbf{Z}$, then, by Hölder's inequality and property 8),

$$\int_{\mathbf{T}^n} \left| \prod_{j=1}^{l+n} U(x_j) \right|^2 d\alpha_1 \ldots d\alpha_n$$
$$\leqslant \prod_{j=1}^{l+n} \left\{ \int_{\mathbf{T}^n} |U(x_j)|^{2(l+n)}\, d\alpha_1 \ldots d\alpha_n \right\}^{\frac{1}{l+n}} = J_n(q,\, l+n). \tag{2.27}$$

But by the Cauchy inequality

$$\left|\sum^{(2)}\prod_{j=1}^{l+n}U(x_j)\right|^2\leqslant T\sum^{(2)}\left|\prod_{j=1}^{l+n}U(x_j)\right|^2,$$

where T is the numbers of terms in the sum $\Sigma^{(2)}$. From this and (2.27),

$$J^{(2)}\leqslant T\sum^{(2)}\int_{\mathbf{T}^n}\left|\prod_{j=1}^{l+n}U(x_j)\right|^2 d\alpha_1\ldots d\alpha_n\leqslant T^2J_n(q,\ l+n). \qquad (2.28)$$

We estimate T. We can choose $n-1$ distinct numbers from among the numbers $1,\ldots,p$ in $\binom{p}{n-1}$ ways; these numbers can be placed (with repetitions) in the $l+n$ places in no more than $(n-1)^{l+n}$ ways. Since each of the systems $(x_1,\ldots,x_{l+n})$ over which the summation $\Sigma^{(2)}$ is taken is obtained in that way, it follows that

$$T\leqslant\binom{p}{n-1}(n-1)^{l+n}\leqslant\frac{(n-1)^{l+n}}{(n-1)!}p^{n-1}\leqslant\frac{1}{2}n^{l+n}p^{n-1}$$

and consequently

$$J^{(2)}\leqslant\frac{1}{4}n^{2(l+n)}p^{2(n-1)}J_n(q,\ l+n). \qquad (2.29)$$

Comparing Lemma 1, (2.25) and (2.29), we find that

$$J_n(pq,\ l+n)\leqslant 2\left\{n!\binom{l+n}{n}^2p^{2l+\frac{n}{2}(3n-1)}J_n(q,\ l)\right.$$
$$\left.+\frac{1}{4}n^{2(l+n)}p^{2(n-1)}J_n(q,\ l+n)\right\}\leqslant(l+n)^{2n}p^{2l+\frac{n}{2}(3n-1)}J_n(q,\ l)$$
$$+n^{2(l+n)}p^{2(n-1)}J_n(q,\ l+n),$$

and the lemma is proved.

We transform the estimate (2.24) to a more convenient form. We shall need the following lemma on prime numbers.

Lemma 3. *For any $x\in\mathbf{R}$, $x\geq 2$, there exists a prime number p satisfying the conditions*

$$x<p\leqslant x\left(1+\frac{C_4}{\ln x}\right), \qquad (2.30)$$

where

$$C_4=\frac{4}{7}\ln 7=1,\ 112\ldots; \qquad (2.31)$$

the constant C_4 in this inequality is best possible, and is attained for $x=7$.

Proof. Rosser and Schoenfeld [19] have shown that the function

$$\pi(x)=\sum_{p\in\mathbf{P},\ p\leqslant x}1$$

satisfies the estimates

$$\frac{x}{\ln x-\frac{1}{2}}<\pi(x)<\frac{x}{\ln x}+\frac{3x}{2\ln^2 x}\quad(x\geqslant 67). \qquad (2.32)$$

We estimate $\pi(y)$ from below for

$$y = x\left(1 + \frac{10}{9\ln x}\right) \quad (x \geqslant e^{10}).$$

According to (2.32),

$$\pi(y) > \frac{y}{\ln y - \frac{1}{2}} = \frac{x\left(1+\frac{10}{9\ln x}\right)}{\ln x + \ln\left(1+\frac{10}{9\ln x}\right) - \frac{1}{2}}$$

$$> \frac{x\left(1+\frac{10}{9\ln x}\right)}{\ln x + \frac{10}{9\ln x} - \frac{1}{2}} = \frac{x\left(1+\frac{10}{9\ln x}\right)}{\ln x} \frac{1}{1-\left(\frac{1}{2}-\frac{10}{9\ln x}\right)\frac{1}{\ln x}}$$

$$> \frac{x}{\ln x}\left(1+\frac{10}{9\ln x}\right)\left(1+\left(\frac{1}{2}-\frac{10}{9\ln x}\right)\frac{1}{\ln x}\right).$$

But for $x \geq e^{10}$

$$\frac{10}{9\ln x} \leqslant \frac{1}{9}, \quad \frac{1}{2} - \frac{10}{9\ln x} \geqslant \frac{7}{18}$$

and consequently

$$\pi(y) \geqslant \frac{x}{\ln x}\left(1+\frac{10}{9\ln x}\right)\left(1+\frac{7}{18}\frac{1}{\ln x}\right)$$
$$> \frac{x}{\ln x}\left(1+\frac{3}{2\ln x}\right) > \pi(x) \quad (x \geqslant e^{10}). \tag{2.33}$$

Since $C_4 > 10/9$, it hence follows that there is a prime number p on the interval $(x, x(1 + C_4/\ln x))$ for $x \geq e^{10}$.

It remains to verify that for $2 \leq x \leq e^{10}$ the interval $(x, x(1 + C_4/\ln x)]$ also always has a prime number. We set $d_n = p_{n+1} - p_n$ and show that

$$d_n \ln p_n \leqslant p_n \qquad (11 \leqslant p_n \leqslant e^{10}). \tag{2.34}$$

If $p_n \leq e^{10} \approx 22026$, then we find from a table of primes that $d_n \leq 52$ (the pair 19609, 19661). Hence $d_n \ln p_n \leq 52 \cdot 10 \leq p_n$ for $520 \leq p_n \leq e^{10}$. Repeating this process, after several steps we show that the property (2.34) holds.

Now let $11 \leq x \leq e^{10}$; we find n from the condition $p_n \leq x \leq p_{n+1}$. Then, by (2.34),

$$x\left(1+\frac{1}{\ln x}\right) \geqslant p_n + \frac{p_n}{\ln p_n} \geqslant p_n + d_n = p_{n+1},$$

i.e.

$$p_{n+1} \in \left(x,\ x\left(1+\frac{1}{\ln x}\right)\right] \quad (11 \leqslant x \leqslant e^{10}).$$

Comparing this property with (2.33), we find that for all $x \geq 11$ there exists $p \in \mathbf{P}$, $p \in (x, x(1 + C_4/\ln x)]$.

Finally, an elementary investigation of the case $2 \leq x \leq 11$ shows that the most unfavorable value is $x = 7$, and we obtain the constant C_4 as stated in the lemma.

It would have been possible to prove this lemma with C_4 replaced by a larger constant, without all the computations (see [10]).

Lemma 4. *Let l, $n \in \mathbf{N}$ and $X \in \mathbf{R}$ satisfy*

$$n \geqslant 2, \quad l \geqslant n, \quad X \geqslant n^n. \tag{2.35}$$

Then

$$J_n(X, l+n) \leqslant A_l X^{2\frac{l}{n}+\frac{3n-1}{2}} J_n\left(X^{1-\frac{1}{n}}, l\right) + B_l X^{2\left(1-\frac{1}{n}\right)} J_n\left(X^{1-\frac{1}{n}}, l+n\right), \tag{2.36}$$

where

$$A_l = \exp\left(C_5 \frac{l+n^2}{\ln n}\right), \quad B_l = \exp\left((2l + C_6 n)\ln n\right). \tag{2.37}$$

Proof. We use Lemma 2, where we choose the parameters p and q in a suitable way.

We set

$$x = \frac{X^{\frac{1}{n}}}{1 - X^{\frac{1}{n}-1}},$$

so that

$$x > X^{\frac{1}{n}} \geqslant n \geqslant 2, \tag{2.38}$$

and we apply Lemma 3. We find that there exists a number $p \in \mathbf{P}$ which satisfies the conditions

$$x < p \leqslant x\left(1 + \frac{C_4}{\ln x}\right). \tag{2.39}$$

We fix such a p, and we set

$$q = \left[\frac{X}{p}\right] + 1. \tag{2.40}$$

For this choice of p and q we have

$$X \leqslant pq, \quad p > X^{\frac{1}{n}} \geqslant n, \tag{2.41}$$

$$q \leqslant \frac{X}{p} + 1 \leqslant \frac{X}{x} + 1 = X^{1-\frac{1}{n}}\left(1 - X^{\frac{1}{n}-1}\right) + 1 = X^{1-\frac{1}{n}} < p^{n-1}, \tag{2.42}$$

and, since

$$X^{\frac{1}{n}-1} \leqslant n^{1-n} \leqslant n^{-1} \quad \text{and} \quad \frac{\ln n}{n-1} \leqslant \ln 2 \quad (n \geqslant 2),$$

we have

$$p \leqslant \frac{X^{\frac{1}{n}}}{1 - X^{\frac{1}{n}-1}}\left(1 + \frac{C_4}{\ln x}\right) \leqslant X^{\frac{1}{n}} \frac{1 + \frac{C_4}{\ln n}}{1 - n^{-1}}$$

$$= X^{\frac{1}{n}}\left(1 + \frac{1}{n-1}\right)\left(1 + \frac{C_4}{\ln n}\right) \leqslant X^{\frac{1}{n}} \exp\left(\frac{1}{n-1} + \frac{C_4}{\ln n}\right)$$

$$= X^{\frac{1}{n}} \exp\left(\frac{1}{\ln n}\left(C_4 + \frac{\ln n}{n-1}\right)\right) \leqslant X^{\frac{1}{n}} \exp\left(\frac{C_4 + \ln 2}{\ln n}\right) \tag{2.43}$$

$$= X^{\frac{1}{n}} \exp\left(\frac{C_7}{\ln n}\right),$$

where $C_7 = C_4 + \ln 2 = 4/7 \ln 7 + \ln 2 < 2$.

Inequalities (2.41) and (2.42) show that all the conditions of Lemma 2 are fulfilled. Applying this lemma and using property 6), we obtain

$$J_n(X, l+n) \leqslant J_n(pq, l+n) \leqslant (l+n)^{2n} p^{2l+\frac{n}{2}(3n-1)} J_n(q, l)$$
$$+ n^{2(l+n)} p^{2(n-1)} J_n(q, l+n) \leqslant (l+n)^{2n} p^{2l+\frac{n}{2}(3n-1)} J_n\left(X^{1-\frac{1}{n}}, l\right) \tag{2.44}$$
$$+ n^{2(l+n)} p^{2(n-1)} J_n\left(X^{1-\frac{1}{n}}, l+n\right).$$

We estimate the coefficients of J_n in the right-hand side of this inequality. Using (2.43), we find

$$(l+n)^{2n} p^{2l+\frac{n}{2}(3n-1)} \leqslant (l+n)^{2n} X^{2\frac{l}{n}+\frac{3n-1}{2}} \exp\left(\frac{C_7}{\ln n}\left(\frac{n}{2}(3n-1) + 2l\right)\right)$$

$$\leqslant \exp\left\{2n \ln(l+n) + \frac{C_7}{\ln n}\left(2l + \frac{3}{2}n^2\right)\right\} X^{2\frac{l}{n}+\frac{3n-1}{2}}.$$

But, by Young's inequality $ab \leq C_8(a^{3/2} + b^3)$, $a, b > 0$, we have

$$n \ln n \ln(l+n) \leqslant C_9 n \ln n \,(l+n)^{\frac{1}{4}} \leqslant C_{10}\left\{(n \ln n)^{\frac{3}{2}} + (l+n)^{\frac{3}{4}}\right\} \leqslant C_{11}(l+n^2).$$

Hence

$$2n \ln(l+n) + \frac{C_7}{\ln n}\left(2l + \frac{3}{2}n^2\right) \leqslant C_5 \frac{l+n^2}{\ln n}$$

and consequently

$$(l+n)^{2n} p^{2l+\frac{n}{2}(3n-1)} \leqslant \exp\left(C_5 \frac{l+n^2}{\ln n}\right) X^{2\frac{l}{n}+\frac{3n-1}{2}}. \tag{2.45}$$

Similarly,

$$n^{2(l+n)} p^{2(n-1)} \leqslant n^{2(l+n)} X^{2\left(1-\frac{1}{n}\right)} \exp\left(2(n-1)\frac{C_7}{\ln n}\right)$$

$$\leqslant \exp(2l \ln n + C_6 n \ln n)\, X^{2\left(1-\frac{1}{n}\right)}. \tag{2.46}$$

Substituting these estimates in (2.44), we obtain the lemma.

We note the following two corollaries of this lemma.

Corollary 1 (the case of small l). *Let $l, n \in \mathbf{N}$ and $X \in \mathbf{R}$ satisfy*

$$n \geqslant 2, \quad n \leqslant l \leqslant n^2+1, \quad X \geqslant n^n. \tag{2.47}$$

Then

$$Jn(X, l+n) \leqslant \exp\left(C_{12}\frac{n^2}{\ln n}\right) X^{\frac{2l}{n}+\frac{3n-1}{2}} J_n\left(X^{1-\frac{1}{n}}, l\right)$$
$$+\exp(C_{13} l \ln n)\, X^{2\left(1-\frac{1}{n}\right)} J_n\left(X^{1-\frac{1}{n}}, l+n\right). \tag{2.48}$$

Corollary 2 (the case of large l). *Let $l, n \in \mathbf{N}$ and $X \in \mathbf{R}$ satisfy*

$$n \geqslant 2, \quad l \geqslant n^2, \quad X \geqslant n^{2n}. \tag{2.49}$$

Then

$$J_n(X, l+n) \leqslant \exp\left(C_{14}\frac{l}{\ln n}\right) X^{\frac{2l}{n}+\frac{3n-1}{2}} J_n\left(X^{1-\frac{1}{n}}, l\right). \tag{2.50}$$

In fact, by (2.36), for $l \geq n^2$ we have

$$J_n(X, l+n) \leqslant \exp\left(C_{15}\frac{l}{\ln n}\right) X^{\frac{2l}{n}+\frac{3n-1}{2}} J_n\left(X^{1-\frac{1}{n}}, l\right)$$
$$+\exp((2l+C_6 n)\ln n)\, X^{2\left(1-\frac{1}{n}\right)} J_n\left(X^{1-\frac{1}{n}}, l+n\right) = K_1 + K_2.$$

We estimate K_2. By property 1),

$$J_n\left(X^{1-\frac{1}{n}}, l+n\right) \leqslant X^{2n-2} J_n\left(X^{1-\frac{1}{n}}, l\right),$$

so that

$$K_2 \leqslant \exp((2l+C_6 n)\ln n)\, X^{2n-\frac{2}{n}} J_n\left(X^{1-\frac{1}{n}}, l\right)$$
$$\leqslant \exp((2l+C_6 n)\ln n)\, X^{\frac{n}{2}-\frac{2l}{n}+\frac{1}{2}} X^{\frac{2l}{n}+\frac{3n-1}{2}} J_n\left(X^{1-\frac{1}{n}}, l\right).$$

But for $l > n^2$ and $X > n^{2n}$ we have

$$\exp((2l+C_6 n)\ln n)\, X^{\frac{n}{2}-\frac{2l}{n}+\frac{1}{2}} \leqslant n^{2l+C_6 n+n^2-4l+n}$$

$$\leqslant n^{-2l+n^2+(C_6+1)n} \leqslant n^{-n^2+(C_6+1)n} \leqslant C_{16}.$$

Therefore

$$K_2 \leqslant C_{16} X^{\frac{2l}{n}+\frac{3n-1}{2}} J_n\left(X^{1-\frac{1}{n}}, l\right),$$

$$J_n(X, l+n) \leqslant \exp\left(C_{14}\frac{l}{\ln n}\right) X^{\frac{2l}{n}+\frac{3n-1}{2}} J_n\left(X^{1-\frac{1}{n}}, l\right),$$

and Corollary 2 is proved.

§3. Common upper bound for J

We shall need one technical lemma.

Lemma 5. *Let* $n \in \mathbf{N}$, $n \geq 2$, $d \in \mathbf{R}$, $0 \leq d < n$, *and*

$$\sigma_r = \sigma_r(d, n) = 2(rn + d) - \frac{n(n+1)}{2} + \eta\delta^r \quad (r = 1, 2, \ldots), \tag{3.1}$$

where

$$g_m = g_m(r, n) = 2n(1 - \delta^m) + \sigma_{r+1}\delta^m \quad (m = 0, 1, \ldots, n-1), \tag{3.2}$$

Set

$$\delta = 1 - \frac{1}{n}, \quad \eta \geqslant 0, \quad \frac{n^2}{2} - 2d\frac{n}{n-1} \leqslant \eta \leqslant \frac{n^2}{2}. \tag{3.3}$$

$$g_m \leqslant \sigma_{r+1} - C_{17} rm \quad (m = 0, 1, \ldots, n), \tag{3.4}$$

Then

$$g_m - \sigma_{r+1} = (2n - \sigma_{r+1})(1 - \delta^m) \quad (m = 0, 1, \ldots, n-1). \tag{3.5}$$

where $C_{17} = \frac{1}{2} - 1/e$.

Proof. We have

$$g_n = g_n(r) = 2n(1 - \delta^n) + \left(\sigma_{r+1} + \frac{\eta}{n}\delta^r\right)\delta^n.$$

But, by (3.2),

$$\sigma_1 = 2(n + d) - \frac{n(n+1)}{2} + \eta\frac{n-1}{n} \geqslant n$$

and

$$\sigma_{r+1} - \sigma_r = 2n - \frac{\eta}{n}\delta^r \geqslant 2n - \frac{n}{2} = \frac{3}{2}n,$$

so that

$$\sigma_{r+1} \geqslant \frac{3}{2}rn + \sigma_1 \geqslant \frac{3}{2}rn + n.$$

Hence

$$2n - \sigma_{r+1} \leqslant 2n - \frac{3}{2}rn - n \leqslant n - \frac{3}{2}rn \leqslant -\frac{1}{2}rn \tag{3.6}$$

and consequently

$$g_m - \sigma_{r+1} \leqslant -\frac{1}{2}rn(1 - \delta^m) \quad (m = 0, 1, \ldots, n-1). \tag{3.7}$$

But, since the function $\phi(x) = (1 - 1/n)^x$, $x \in \mathbf{R}$, is convex,

$$\varphi(x) \leqslant \left(1 - \frac{x}{n}\right)\varphi(0) + \frac{x}{n}\varphi(n) \quad (0 \leqslant x \leqslant n),$$

so that, because $(1 - 1/n)^n < 1/e$,

$$\left(1 - \frac{1}{n}\right)^m \leqslant 1 - \frac{m}{n} + \frac{m}{n}\left(1 - \frac{1}{n}\right)^n \leqslant 1 - \frac{m}{n} + e^{-1}\frac{m}{n} \quad (0 \leqslant m \leqslant n).$$

Thus

$$1-\delta^m \geqslant (1-e^{-1})\frac{m}{n} \qquad (m=0,\, 1, \ldots, n-1), \tag{3.8}$$

and hence

$$g_m - \sigma_{r+1} \leqslant -\frac{1}{2}(1-e^{-1})\, rm \qquad (m=0,\, 1, \ldots, n-1). \tag{3.9}$$

We now consider the case $m = n$. We have

$$g_n - \sigma_{r+1} = (2n - \sigma_{r+1})(1-\delta^n) + \frac{\eta}{n}\delta^{r+n},$$

so that, by (3.6), (3.2) and (3.8),

$$\begin{aligned} g_n - \sigma_{r+1} &\leqslant -\frac{1}{2} rn\,(1-\delta^n) + \frac{n}{2} e^{-1} \\ &\leqslant -\frac{1}{2} rn\,(1-e^{-1}) + \frac{1}{2e} rn = -\left(\frac{1}{2} - \frac{1}{e}\right) rn. \end{aligned} \tag{3.10}$$

Comparing (3.9) and (3.10), we find that

$$g_m - \sigma_{r+1} \leqslant -\left(\frac{1}{2} - \frac{1}{e}\right) rm \qquad (m=0,\, 1, \ldots, n),$$

and the lemma is proved.

We can now estimate $J_n(X, rn)$ for small r.

Lemma 6. *Let $n,\ r \in \mathbf{N}$ and $X \in \mathbf{R}$ satisfy*

$$n \geqslant 2, \quad 1 \leqslant r \leqslant n, \quad X \geqslant 1. \tag{3.11}$$

Then

$$J_n(X, rn) \leqslant D_2(r, n)\, X^{\sigma_r}, \tag{3.12}$$

where

$$\sigma_r = 2rn - \frac{n(n+1)}{2} + \frac{n^2}{2}\left(1-\frac{1}{n}\right)^r, \tag{3.13}$$

$$D_2(r, n) = \exp(C_{18} rn^2 \ln n). \tag{3.14}$$

The proof is by induction on r. Since $\sigma_1 = n$, it follows by property 7) that

$$J_n(X, n) \leqslant n!\, X^n \leqslant \exp(C_{19} n^2 \ln n)\, X^{\sigma_1} \qquad (X \geqslant 1), \tag{3.15}$$

i.e. the lemma holds for $r = 1$. Suppose that (3.12) is already established for some r, $1 \leq r \leq n-1$, in the form

$$J_n(X, rn) \leqslant D_r(a)\, X^{\sigma_r} \qquad (X \geqslant 1), \tag{3.16}$$

where $D_r(u) = \exp(arn^2 \ln n)$, and let us estimate $J_n(X, (r+1)n)$.

By Corollary 1 to Lemma 4,

$$\begin{aligned} J_n(X, (r+1)n) \leqslant A_r X^{2r + \frac{3n-1}{2}} J_n\left(X^{1-\frac{1}{n}}, rn\right) \\ + B_r X^{2\left(1-\frac{1}{n}\right)} J_n\left(X^{1-\frac{1}{n}}, (r+1)n\right), \end{aligned} \tag{3.17}$$

where

$$A_r = \exp(C_{20}n^2), \qquad B_r = \exp(C_{21}rn\ln n). \tag{3.18}$$

Let $b \geq e$ and $X \geq n^{bn}$, so that

$$X^{\left(1-\frac{1}{n}\right)^{n-1}} \geqslant X^{\frac{1}{e}} \geqslant n^n.$$

Then Corollary 1 to Lemma 4 can be applied $n-1$ times. Setting $\alpha = 2r + (3n-1)/2$, $\delta = 1 - 1/n$ and $\beta = 2\delta$, we obtain

$$J_n(X, (r+1)n) \leqslant \sum_{m=0}^{n} F_m, \tag{3.19}$$

where

$$\begin{aligned} F_m &= A_r B_r^m X^{\beta+\dots+\beta\delta^{m-1}+\alpha\delta^m} J_n(X^{\delta^{m+1}}, rn) \qquad (m = 0, \dots, n-1), \\ F_n &= B_r^n X^{\beta+\dots+\beta\delta^{n-1}} J_n(X^{\delta^n}, (r+1)n). \end{aligned} \tag{3.20}$$

We estimate each of the F_m, $m = 0, 1, \dots, n$. Let $0 \leq m \leq n-1$. Using (3.16), we find

$$F_m \leqslant A_r B_r^m D_r(a) X^{\beta\frac{1-\delta^m}{1-\delta}+\alpha\delta^m+\sigma_r\delta^{m+1}}.$$

But, by Lemma 5 with $d = 0$ and $\eta = n^2/2$, we have

$$\begin{aligned} &\beta\frac{1-\delta^m}{1-\delta} + \alpha\delta^m + \sigma_r\delta^{m+1} = 2(n-1)(1-\delta^m) \\ &+\left(2r + \frac{3n-1}{2} + \sigma_r\delta\right)\delta^m = 2(n-1)(1-\delta^m) \\ &+\sigma_{r+1}\delta^m \leqslant g_m \leqslant \sigma_{r+1} - C_{17}rm \qquad (m = 0, 1, \dots, n-1), \end{aligned}$$

so that

$$F_m \leqslant A_r D_r(a) X^{\sigma_{r+1}} (B_r X^{-C_{17}r})^m \qquad (0 \leqslant m \leqslant n-1). \tag{3.21}$$

Similarly, using property 1), we find

$$F_n \leqslant B_r^n X^{\beta\frac{1-\delta^n}{1-\delta}+2n\delta^n} J_n(X^{\delta^n}, rn) \leqslant B_r^n D_r(a) X^{2(n-1)(1-\delta^n)+2n\delta^n+\sigma_r\delta^n}.$$

But, by Lemma 5,

$$\begin{aligned} 2(n-1)(1-\delta^n) + 2n\delta^n + \sigma_r\delta^n &= 2(n-1)(1-\delta^n) \\ &+ \sigma_{r+1}\delta^n + \frac{\eta}{n}\delta^{r+n} \leqslant g_n \leqslant \sigma_{r+1} - C_{17}rn, \end{aligned}$$

so that

$$F_n \leqslant D_r(a) X^{\sigma_{r+1}} (B_r X^{-C_{17}r})^n.$$

Hence, since $B_r = \exp(C_{21}rn\ln n)$ and $X \geq \exp(bn\ln n)$, it follows that

$$\begin{aligned}J_n(X,(r+1)n) &\leqslant A_r D_r(a) X^{\sigma_{r+1}} \sum_{m=0}^{n} (B_r X^{-C_{17}r})^m \\ &= A_r D_r(a) X^{\sigma_{r+1}} \sum_{m=0}^{n} \exp(m(C_{21}rn\ln n - C_{17}brn\ln n)) \\ &= A_r D_r(a) X^{\sigma_{r+1}} \sum_{m=0}^{n} \exp(m(C_{21}-C_{17}b)rn\ln n).\end{aligned} \tag{3.22}$$

Here we set

$$b = C_{22} = \max\left(e, \frac{C_{21}+1}{C_{17}}\right).$$

For this choice of b,

$$\sum_{m=0}^{n} \exp(m(C_{21}-C_{17}b)rn\ln n) \leqslant \sum_{m=0}^{n} \exp(-mrn\ln n) \leqslant C_{23}$$

and consequently

$$J_n(X,(r+1)n) \leqslant C_{23} A_r D_r(a) X^{\sigma_{r+1}} \qquad (X \geqslant n^{C_{22}n}). \tag{3.23}$$

But

$$C_{23}A_r = C_{23}\exp(C_{20}n^2) \leqslant \exp(C_{24}n^2\ln n),$$
$$C_{23}A_r D_r(a) \leqslant \exp((C_{24}+ar)n^2\ln n).$$

Hence, if $a \geq C_{24}$, then

$$J_n(X,(r+1)n) \leqslant \exp(a(r+1)n^2\ln n) X^{\sigma_{r+1}} \qquad (X \geqslant n^{C_{22}n}).$$

Finally, if $1 \leq X \leq \exp(C_{22}n \ln n)$, then the trivial estimate gives

$$\begin{aligned}J_n(X,(r+1)n) &\leqslant X^{2(r+1)n} \\ &\leqslant \exp(2(r+1)nC_{22}n\ln n) \leqslant \exp(2C_{22}(r+1)n^2\ln n) X^{\sigma_{r+1}}.\end{aligned}$$

This implies that, if $a = C_{25} = \max(C_{19}, 2C_{22}, C_{24})$, then, whenever (3.16) holds, it follows that the same inequality holds for $r+1$, i.e. we may set $a = C_{25}$, and the lemma is proved.

Theorem 1. *Let*

$$n,\ r \in \mathbf{N},\ n \geqslant 2,\ l \in \mathbf{R},\ l \geqslant rn,\ X \in \mathbf{R},\ X \geqslant 1. \tag{3.24}$$

Then

$$J_n(X,l) \leqslant D(r,n) X^{2l - \frac{n(n+1)}{2} + \frac{n^2}{2}\left(1-\frac{1}{n}\right)^r}, \tag{3.25}$$

where

$$D(r,n) = \exp(C_3 \min(r,n) n^2 \ln n). \tag{3.26}$$

Proof. We begin by investigating the case $l = rn$ for large r. We show that, if $r \geq n$, then

$$J_n(X, rn) \leqslant \exp\left(C_{26}\frac{n}{\ln n}(r^2+n^2\ln^2 n)\right) X^{2rn-\frac{n(n+1)}{2}+\frac{n^2}{2}\left(1-\frac{1}{n}\right)^r}. \qquad (3.27)$$

Let $s \in \mathbf{Z}$, $s \geq -1$. We set

$$X_{-1}=1, \quad X_s = n^{2n\left(\frac{n}{n-1}\right)^s} \qquad (s=0, 1, \ldots).$$

We first consider the case $X_{s-1} \leq X \leq X_s$, $0 \leq s \leq r-n$, so that

$$X^{\left(1-\frac{1}{n}\right)^s} \leqslant n^{2n} \leqslant X^{\left(1-\frac{1}{n}\right)^{s-1}} \qquad (s \geqslant 1).$$

Writing Corollary 2 to Lemma 4 in the form

$$\frac{J_n(X, rn)}{X^{2rn-\frac{n(n+1)}{2}}} \leqslant \exp\left(C_{14}\frac{rn}{\ln n}\right)\frac{J_n\left(X^{1-\frac{1}{n}}, (r-1)n\right)}{X^{\left(1-\frac{1}{n}\right)\left(2(r-1)n-\frac{n(n+1)}{2}\right)}}$$

and applying it s times, we find

$$\frac{J_n(X, rn)}{X^{2rn-\frac{n(n+1)}{2}}} \leqslant \exp\left(C_{14}\frac{srn}{\ln n}\right)\frac{J_n\left(X^{\left(1-\frac{1}{n}\right)^s}, (r-s)n\right)}{X^{\left(1-\frac{1}{n}\right)^s\left(2(r-s)n-\frac{n(n+1)}{2}\right)}},$$

so that, by the trivial estimate,

$$\frac{J_n(X, rn)}{X^{2rn-\frac{n(n+1)}{2}}} \leqslant \exp\left(C_{14}\frac{r^2n}{\ln n}\right) X^{\left(1-\frac{1}{n}\right)^s\frac{n(n+1)}{2}} \leqslant \exp\left(C_{27}\frac{n}{\ln n}(r^2+n^2\ln^2 n)\right),$$

i.e.

$$J_n(X, rn) \leqslant \exp\left(C_{27}\frac{n}{\ln n}(r^2+n^2\ln^2 n)\right) X^{2rn-\frac{n(n+1)}{2}}. \qquad (3.28)$$

Hence

$$J_n(X, rn) \leqslant \exp\left(C_{27}\frac{n}{\ln n}(r^2+n^2\ln^2 n)\right) X^{2rn-\frac{n(n+1)}{2}+\frac{n^2}{2}\left(1-\frac{1}{n}\right)^r}$$
$$(1 \leqslant X \leqslant X_{r-n}). \qquad (3.29)$$

Now let $X \geq X_{r-n}$. Applying Corollary 2 to Lemma 4 $r-n$ times, we similarly find

$$\frac{J_n(X, rn)}{X^{2rn-\frac{n(n+1)}{2}}} \leqslant \exp\left(C_{14}\frac{r^2n}{\ln n}\right)\frac{J_n\left(X^{\left(1-\frac{1}{n}\right)^{r-n}}, n^2\right)}{X^{\left(1-\frac{1}{n}\right)^{r-n}\left(2n^2-\frac{n(n+1)}{2}\right)}},$$

so that, by Lemma 6,

$$J_n(X, rn) \leqslant \exp\left(C_{28}\frac{n}{\ln n}(r^2+n^2\ln^2 n)\right) X^{2rn-\frac{n(n+1)}{2}+\frac{n^2}{2}\left(1-\frac{1}{n}\right)^r} \quad (X \geqslant X_{r-n}).$$
$$(3.30)$$

The estimates (3.29) and (3.30) show that (3.27) holds for all $X \geq 1$.

It further follows by Theorem A that Theorem 1 is valid for $r \geq C_1 n \ln n$, and it remains to consider the case $r \leq C_1 n \ln n$.

Inequality (3.30) shows that

$$J_n(X, rn) \leqslant \exp(C_{29}n^3\ln n)\, X^{2rn-\frac{n(n+1)}{2}+\frac{n^2}{2}\left(1-\frac{1}{n}\right)^r}$$

for all $X \geq 1$ and all r, $n \leq r \leq C_1 n \ln n$, and Lemma 6 shows that

$$J_n(X, rn) \leqslant \exp(C_{18}rn^2\ln n)\, X^{2rn-\frac{n(n+1)}{2}+\frac{n^2}{2}\left(1-\frac{1}{n}\right)^r}$$

for all $X \geq 1$ and all r, $1 \leq r \leq n$.

Comparing these results, we see that

$$J_n(X, rn) \leqslant \exp(C_3 \min(r, n) n^2 \ln n) X^{2rn - \frac{n(n+1)}{2} + \frac{n^2}{2}\left(1-\frac{1}{n}\right)^r}$$

for all $r \geq 1$ and all $X \geq 1$.

Finally, if $l \geq rn$, it follows by property 1) that for all $X \geq 1$

$$J_n(X, l) \leqslant X^{2(l-rn)} J_n(X, rn) \leqslant \exp(C_6 \min(r, n) n^2 \ln n) X^{2l - \frac{n(n+1)}{2} + \frac{n^2}{2}\left(1-\frac{1}{n}\right)^r},$$

and Theorem 1 is completely proved.

The following corollary is sufficient for the usual applications.

Corollary. *Let $n, r \in \mathbf{N}$, $n \geq 2$, and $l, X \in \mathbf{R}$, $l \geq rn$, $X \geq 1$. Then*

$$J_n(X, l) \leqslant n^{C_3 n^3} X^{2l - \frac{n(n+1)}{2} + \frac{n^2}{2}\left(1-\frac{1}{n}\right)^r}. \tag{3.31}$$

We now show that, for fixed n, r and l, $l > n$, $l \geq rn$,

$$\frac{J_n(X, l)}{X^{2l - \frac{n(n+1)}{2} + \frac{n^2}{2}\left(1-\frac{1}{n}\right)^r}} \to 0 \quad (X \to \infty). \tag{3.32}$$

We shall need the following lemma, which was proved by Hua Lo-keng [11] (Lemma 4.5):

Lemma 7. *Let $n \in \mathbf{N}$, $n \geq 2$, $X \in \mathbf{R}$, $X \geq e$, and $L = \ln X$. Then*

$$J_n(X, n+1) \leqslant A(n) X^{n+1} L^{2^n - 1}. \tag{3.33}$$

Setting $X = e$ in (3.33), we find that

$$A(n) \geqslant e^{-C_{30} n}. \tag{3.34}$$

Theorem 2. *Let $n, r \in \mathbf{N}$, $n \geq 2$, $X \in \mathbf{R}$, $X \geq e$, $L = \ln X$, and $l \in \mathbf{R}$, $l = rn + 1 + tn$, $0 \leq t \leq 1$. Then*

$$J_n(X, l) \leqslant D_3(r, n) X^{2l - \frac{n(n+1)}{2} + \eta\left(1-\frac{1}{n}\right)^{r-1}\left(1-\frac{t}{n}\right)} L^{2^n - 1}, \tag{3.35}$$

where

$$\eta = \frac{1}{2}(n+1)(n-2), \tag{3.36}$$

$$D_3(r, n) = A(n) \exp(C_{31} \min(r, n) n^2 \ln n) \tag{3.37}$$

and $A(n)$ has the same meaning as in Lemma 7.

If we show that $A(n) \leq \exp(C_{32} n^2 \ln n)$, then the factor $A(n)$ in (3.37) can be omitted.

The proof repeats the proof of Theorem 1, and is divided into four steps:

1) the case $l = rn + 1$ with r small ($1 \leq r \leq n$),

2) the case $l = rn + 1$ with r large ($r \geq n$),

3) cutting off too large r,

4) passing to intermediate l.

The case of small r is encompassed in the following lemma.

Lemma 8. *Let* $n, r \in \mathbf{N}$, $n \geq 2$, $1 \leq r \leq n$, $X \in \mathbf{R}$, $X \geq e$ *and* $L = \ln X$. *Then*

$$J_n(X, rn+1) \leqslant D_4(r, n)\, X^{2(rn+1)-\frac{n(n+1)}{2}+\eta\left(1-\frac{1}{n}\right)^{r-1}} L^{2^n-1}, \tag{3.38}$$

where

$$D_4(r, n) = A(n)\exp(C_{33} r n^2 \ln n). \tag{3.39}$$

The proof is analogous to the proof of Lemma 6.

The remaining steps in the proof of Theorem 2 do not differ significantly from the corresponding steps in the proof of Theorem 1.

The case $l = rn + 1$ *and large* r $(r \geq n)$. If we take into account that $A(n) \geq e^{-C_{30}n}$ and $L \geq 1$, by (3.34), we find that in this case for $X \geq e$

$$J_n(X, rn+1) \leqslant D_5(r, n)\, X^{2(rn+1)-\frac{n(n+1)}{2}+\eta\left(1-\frac{1}{n}\right)^{r-1}} L^{2^n-1}, \tag{3.40}$$

where

$$D_5(r, n) = A(n)\exp\left(C_{34}\frac{n}{\ln n}(r^2+n^2\ln^2 n)\right).$$

Further, according to Theorem A, if $r \geq C_1 n \ln n$, then

$$J_n(X, rn+1) \leqslant \exp(C_2 n^3 \ln n)\, X^{2(rn+1)-\frac{n(n+1)}{2}} \leqslant$$
$$\leqslant \exp(C_2 n^3 \ln n)\, X^{2(rn+1)-\frac{n(n+1)}{2}+\eta\left(1-\frac{1}{n}\right)^{r-1}} L^{2^n-1} \quad (X \geqslant e).$$

But

$$\exp(C_2 n^3 \ln n) \leqslant A(n)\exp(C_{35} n^3 \ln n) = D_6(r, n),$$

so that for $r \geq C_1 n \ln n$

$$J_n(X, rn+1) \leqslant D_6(r, n)\, X^{2(rn+1)-\frac{n(n+1)}{2}+\eta\left(1-\frac{1}{n}\right)^{r-1}} L^{2^n-1} \quad (X \geqslant e),$$

and we finish the proof in the same way as for Theorem 1.

This theorem and Lemma 7 imply the property (3.32). We consider two cases:

1) $n < l < n + 1$. By Lemma 7 and property 2),

$$J_n(X, l) \leqslant \{A(n)\}^{\frac{l}{n+1}} X^l L^{\frac{l}{n+1}(2^n-1)}$$

and

$$l < 2l - \frac{n(n+1)}{2} + \frac{n^2}{2}\left(1-\frac{1}{n}\right) = 2l - n.$$

2) $l \geq n + 1$. Since the exponents under consideration are piecewise linear, it suffices to prove that for $l = rn$ $(r \geq 2)$ the exponent of X in Theorem 2 is less than the exponent of X in Theorem 1, i.e.

$$\eta\left(1-\frac{1}{n}\right)^{r-2}\left(1-\frac{t}{n}\right) < \frac{n^2}{2}\left(1-\frac{1}{n}\right)^r,$$

where η is defined by (3.36) and $t = (n-1)/n$, as is easy to verify.

§4. Estimates for trigonometric sums

Let

$$T(Q) = \sum_{k=1}^{Q} e(F(k)), \tag{4.1}$$

where $Q \in \mathbf{N}$ and $F(u)$ $(1 \leq u \leq Q)$ is a real-valued n times differentiable function satisfying the conditions

$$0 < \lambda \leqslant F^{(n)}(u) \leqslant 2\lambda \qquad (1 \leqslant u \leqslant Q). \tag{4.2}$$

Van der Corput obtained the following estimate for the sum $T(Q)$ ([20], p. 841; see also [21], p. 114, Theorem 20):

$$|T(Q)| \leqslant C_{36} Q \left\{ \lambda^{\frac{1}{2^{n}-2}} + Q^{-\frac{1}{2^{n-1}}} + (\lambda Q^{n})^{-\frac{1}{2^{n-1}}} \right\}. \tag{4.3}$$

In what follows we obtain an estimate for $T(Q)$ in terms of J_{n-1} and compare the results with (4.3).

Lemma 9. *Suppose that $l, n, q, Q, R \in \mathbf{N}$, $n \geq 2$, $1 \leq R \leq Q$, $\lambda > 0$, the function $F(u)$, $1 \leq u \leq Q$, satisfies the condition*

$$\left| \frac{F^{(n)}(u)}{n!} \right| \leqslant \lambda \qquad (1 \leqslant u \leqslant Q), \tag{4.4}$$

and $\Omega(m)$ is the region consisting of points $\alpha = (\alpha_1, \ldots, \alpha_{n-1}) \in \mathbf{R}^{n-1}$ for which

$$\left| \frac{F^{(\nu)}(m)}{\nu!} - \alpha_\nu \right| \leqslant \frac{1}{2} \lambda q^{n-\nu} \qquad (\nu = 1, \ldots, n-1). \tag{4.5}$$

Then, setting

$$T_m(R) = \sum_{k=1}^{R} e(F(k+m) - F(m)) \qquad (1 \leqslant R \leqslant q;\ 1 \leqslant m \leqslant Q - R), \tag{4.6}$$

we have for $\alpha \in \Omega(m)$

$$|T_m(R)|^{2l} \leqslant n^{C_{37} l} \left\{ (q^n \lambda)^{2l} R^{-1} \sum_{k=1}^{R-1} |S_{n-1}(k, \alpha)|^{2l} + |S_{n-1}(R, \alpha)|^{2l} \right\} \quad (R = 1, \ldots, q). \tag{4.7}$$

Proof. Let $\alpha \in \Omega(m)$. We set

$$u_k = u_k(m) = F(k+m) - F(m) - \sum_{\nu=1}^{n-1} \alpha_\nu k^\nu$$
$$(m - 1, \ldots, Q \quad k;\ k = 0, 1, \ldots, q). \tag{4.8}$$

By the Abel transform,

$$T_m(R) = \sum_{k=1}^{R} e(F(k+m) - F(m)) = \sum_{k=1}^{R} e\left(\sum_{\nu=1}^{n-1} \alpha_\nu k^\nu + u_k \right)$$
$$= \sum_{k=1}^{R-1} \{e(u_k) - e(u_{k+1})\} S_{n-1}(k, \alpha) + e(u_R) S_{n-1}(R, \alpha), \tag{4.9}$$

so that

$$|T_m(R)| \leqslant \sum_{k=1}^{R-1} |e(u_k) - e(u_{k+1})|\,|S_{n-1}(k, \alpha)| + |S_{n-1}(R, \alpha)|. \tag{4.10}$$

We estimate $|e(u_k) - e(u_{k+1})|$. By Taylor's formula

$$F(k+m) - F(m) = \sum_{\nu=1}^{n-1} \frac{F^{(\nu)}(m)}{\nu!} k^\nu + \frac{1}{(n-1)!} \int_m^{k+m} (k+m-n)^{n-1} F^{(n)}(u)\,du,$$

so that

$$u_{k+1} - u_k = F(k+m+1) - F(k+m) - \sum_{\nu=1}^{n-1} \alpha_\nu \{(k+1)^\nu - k^\nu\}$$
$$= \sum_{\nu=1}^{n-1} \left\{\frac{F^{(\nu)}(m)}{\nu!} - \alpha_\nu\right\} \{(k+1)^\nu - k^\nu\}$$
$$+ \frac{1}{(n-1)!} \int_m^{k+m} \{(k+m+1-u)^{n-1} - (k+m-u)^{n-1}\} F^{(n)}(u)\,du$$
$$+ \frac{1}{(n-1)!} \int_{k+m}^{k+m+1} (k+m+1-u)^{n-1} F^{(n)}(u)\,du.$$

Hence, by (4.5) and (4.4),

$$|u_{k+1} - u_k| \leqslant \frac{1}{2}\lambda \sum_{\nu=1}^{n-1} q^{n-\nu} \nu (k+1)^{\nu-1} + nk(n-1)(k+1)^{n-2}\lambda$$
$$+ n\lambda \leqslant C_{38} n^2 q^{n-1} \lambda \qquad (k = 1, \ldots, q-1)$$

and consequently

$$|e(u_k) - e(u_{k+1})| = |1 - e(u_{k+1} - u_k)| \leqslant 2\pi |u_{k+1} - u_k|$$
$$\leqslant C_{39} n^2 q^{n-1} \lambda \qquad (k = 1, \ldots, q-1). \tag{4.11}$$

Substituting this estimate in (4.10), we obtain

$$|T_m(R)| \leqslant C_{39} n^2 q^{n-1} \lambda \sum_{k=1}^{R-1} |S_{n-1}(k, \alpha)| + |S_{n-1}(R, \alpha)|,$$

so that, by Hölder's inequality,

$$|T_m(R)|^{2l} \leqslant 2^{2l} \left\{ (C_{39} n^2 q^{n-1} \lambda)^{2l} \left(\sum_{k=1}^{R-1} |S_{n-1}(k, \alpha)| \right)^{2l} + |S_{n-1}(R, \alpha)|^{2l} \right\}$$
$$\leqslant 2^{2l} \left\{ (C_{39} n^2 q^{n-1} \lambda)^{2l} R^{2l-1} \sum_{k=1}^{R-1} |S_{n-1}(k, \alpha)|^{2l} + |S_{n-1}(R, \alpha)|^{2l} \right\}$$
$$\leqslant (C_{40} n^2)^{2l} \left\{ (q^{n-1}\lambda)^{2l} q^{2l} R^{-1} \sum_{k=1}^{R-1} |S_{n-1}(k, \alpha)|^{2l} + |S_{n-1}(R, \alpha)|^{2l} \right\}$$
$$\leqslant n^{C_{37} l} \left\{ (q^n \lambda)^{2l} R^{-1} \sum_{k=1}^{R-1} |S_{n-1}(k, \alpha)|^{2l} + |S_{n-1}(R, \alpha)|^{2l} \right\} \quad (R = 1, \ldots, q),$$

and the lemma is proved.

Lemma 10. *Suppose that the conditions of the previous lemma are fulfilled, and that, in addition,*

$$0 < h \leqslant 1, \quad \lambda h \leqslant \frac{F^{(n)}(n)}{n!} \leqslant \lambda \; (0 \leqslant u \leqslant Q),$$
$$Q = Pq + R, \quad 0 \leqslant R < q, \quad P \in \mathbf{N}, \; \lambda q \leqslant 1. \tag{4.12}$$

Consider $\Omega(sq)$, $s = 0, \ldots, P-1$, *as regions in* T^{n-1}, *and let* G *be the maximal multiplicity of covering of points* $\alpha \in \mathbf{T}^{n-1}$ *by the regions* $\Omega(sq)$. *Then*

$$G \leqslant C_{41} n h^{-1} (\lambda Q + 1). \tag{4.13}$$

This lemma is well known (see [4], p. 100). For the sake of completeness, we give a simple proof of it.

Proof. If the regions $\Omega(sq)$ and $\Omega(s'q)$ in $\mathbf{T}^{n-1}$ overlap, it follows by condition (4.5) for $\nu = n-1$ that the numbers $F^{(n-1)}(sq)/(n-1)!$ and $F^{(n-1)}(s'q)/(n-1)$ belong to intervals which are congruent mod 1 and have length $\lambda q \leq 1$. Since

$$\frac{1}{(n-1)!}\{F^{(n-1)}(s+1)q) - F^{(n-1)}(sq)\} = n \int\limits_{sq}^{(s+1)q} \frac{F^{(n)}(u)}{n!}\,du,$$

it follows by (4.12) that

$$nh\lambda q \leqslant \frac{1}{(n-1)!}\{F^{(n-1)}((s+1)q) - F^{(n-1)}(sq)\} \leqslant n\lambda q.$$

Therefore, the overall variation in $F^{(n-1)}(sq)/(n-1)!$ over P steps does not exceed $nPq\lambda \leq n\lambda Q$, and there are no more than $(nh)^{-1}+1$ points on each interval $\Delta \subset \mathbf{R}$ of length λq. Hence

$$G \leqslant \left(\frac{1}{nh} + 1\right)(n\lambda Q + 2) \leqslant C_{42} h^{-1}(n\lambda Q + 2) \leqslant C_{41} n h^{-1}(\lambda Q + 1),$$

and the lemma is proved.

Theorem 3. *Suppose that* $l, n, Q \in \mathbf{N}$, $l \geq n-1 \geq 2$, $\lambda, h \in \mathbf{R}$, $Q^{-n} \leq \lambda \leq 1$, $0 \leq h < 1$, *and the function* $F(u)$, $1 \leq u \leq Q$, *satisfies the conditions* (4.12). *Then*

$$|T(Q)| \leqslant \exp(C_{43} n)\, Q \left\{ h^{-1}(\lambda + Q^{-1})\, q^{1-2l+\frac{n(n-1)}{2}} J_{n-1}(q, l) \right\}^{\frac{1}{2l}}, \tag{4.14}$$

where

$$q = \left[\lambda^{-\frac{1}{n}}\right]. \tag{4.15}$$

Proof. We use the notation of Lemma 9. By our choice of q, we have

$$q \in \mathbf{N},\ 1 \leqslant q \leqslant \lambda^{-\frac{1}{n}} \leqslant Q,\ 2^{-n} \leqslant \lambda q^n \leqslant 1, \tag{4.16}$$

so that we obtain the following estimate for the volume of $\Omega(m)$:

$$V = V_{n-1}(\Omega(m)) = (\lambda q^n)^{n-1} q^{-\frac{n(n-1)}{2}} \geqslant 2^{-n^2} q^{-\frac{n(n-1)}{2}}. \tag{4.17}$$

Let

$$Q = Pq + R, \tag{4.18}$$

where $P \in \mathbf{N}$ and $0 \leq R < q$.

We represent $T(Q)$ in the form

$$T(Q) = \sum_{s=0}^{P-1} \sum_{k=1}^{q} e(F(sq+k)) + \sum_{k=1}^{R} e(F(Pq+k)).$$

Then

$$|T(Q)| \leqslant \sum_{s=0}^{P-1} |T_{sq}(q)| + |T_{Pq}(R)|$$

and, by Hölder's inequality,

$$|T(Q)| \leqslant (P+1)^{1-\frac{1}{2l}} \left\{ \sum_{s=0}^{P-1} |T_{sq}(q)|^{2l} + |T_{Pq}(R)|^{2l} \right\}^{\frac{1}{2l}}$$
$$\leqslant 2P^{1-\frac{1}{2l}} \left\{ \sum_{s=0}^{P-1} |T_{sq}(q)|^{2l} + |T_{Pq}(R)|^{2l} \right\}^{\frac{1}{2l}}. \tag{4.19}$$

Further, if $\alpha \in \Omega(sq)$, then, by Lemma 9 and (4.16),

$$|T_{sq}(q)|^{2l} \leqslant n^{C_{37}l} \left\{ (\lambda q^n)^{2l} q^{-1} \sum_{k=1}^{q-1} |S_{n-1}(k, \alpha)|^{2l} + |S_{n-1}(q, \alpha)|^{2l} \right\}$$
$$\leqslant n^{C_{37}l} \left\{ q^{-1} \sum_{k=1}^{q-1} |S_{n-1}(k, \alpha)|^{2l} + |S_{n-1}(q, \alpha)|^{2l} \right\}.$$

Integrating over $\Omega(sq)$, we obtain

$$|T_{sq}(q)|^{2l} \leqslant n^{C_{37}l} V^{-1} \left\{ q^{-1} \sum_{k=1}^{q-1} \int_{\Omega(sq)} |S_{n-1}(k, \alpha)|^{2l} d\alpha_1 \dots d\alpha_{n-1} \right.$$
$$\left. + \int_{\Omega(sq)} |S_{n-1}(q, \alpha)|^{2l} d\alpha_1 \dots d\alpha_{n-1} \right\}.$$

If we let G denote the maximal multiplicity of covering of points $\alpha \in \mathrm{T}^{n-1}$ by the regions $\Omega(sq)$, $s = 0, 1, \dots, P-1$, and use property 6), we hence deduce that

$$\sum_{s=0}^{P-1} |T_{sq}(q)|^{2l} \leqslant n^{C_{37}l} G V^{-1} \left\{ q^{-1} \sum_{k=1}^{q-1} \int_{\mathrm{T}^{n-1}} |S_{n-1}(k, \alpha)|^{2l} d\alpha + \int_{\mathrm{T}^{n-1}} |S_{n-1}(q, \alpha)|^{2l} d\alpha \right\}$$
$$\leqslant n^{C_{44}l} G V^{-1} \int_{\mathrm{T}^{n-1}} |S_{n-1}(q, \alpha)|^{2l} d\alpha = n^{C_{44}l} G V^{-1} J_{n-1}(q, l).$$

In exactly the same way,

$$|T_{Pq}(R)|^{2l} \leqslant n^{C_{44}l} V^{-1} J_{n-1}(q, l) \qquad (0 \leqslant R < q).$$

Hence, taking into account Lemma 10 and (4.17), we have

$$\sum_{s=0}^{P-1} |T_{sq}(q)|^{2l} + |T_{Pq}(R)|^{2l} \leqslant n^{C_{45}l} G V^{-1} J_{n-1}(q, l)$$
$$\leqslant n^{C_{46}l} 2^{n^2} (\lambda Q + 1) h^{-1} q^{\frac{n(n-1)}{2}} J_{n-1}(q, l). \tag{4.20}$$

Comparing (4.19), (4.18) and (4.20), we finally obtain

$$|T(Q)| \leqslant 2P^{1-\frac{1}{2l}} \left\{ \sum_{s=0}^{P-1} |T_{sq}(q)|^{2l} + |T_{Pq}(R)|^{2l} \right\}^{\frac{1}{2l}}$$
$$\leqslant 2(Q/q)^{1-\frac{1}{2l}} \left\{ n^{C_{46}l} 2^{n^2} (\lambda Q + 1) h^{-1} q^{\frac{n(n-1)}{2}} J_{n-1}(q, l) \right\}^{\frac{1}{2l}} \leq$$

$$\leqslant n^{C_{47}} 2^{\frac{n^2}{2l}} Q\left\{h^{-1}(\lambda+Q^{-1}) q^{1-2l+\frac{n(n-1)}{2}} J_{n-1}(q, l)\right\}^{\frac{1}{2l}}$$

$$\leqslant \exp(C_{43}n)\, Q\left\{h^{-1}(\lambda+Q^{-1}) q^{1-2l+\frac{n(n-1)}{2}} J_{n-1}(q, l)\right\}^{\frac{1}{2l}},$$

and the theorem is proved.

We now obtain several more convenient estimates for the sum $T(Q)$ using this theorem and Theorems 1 and 2.

Proposition 1. *Suppose that* $n, q \in \mathbf{N}$, $n \geq 2$, $\lambda, h \in \mathbf{R}$, $\lambda > 0$, $0 < h \leq 1$, *and the function* $F(u)$, $0 \leq u \leq Q$, *satisfies the conditions*

$$h\lambda \leqslant \frac{F^{(n)}(u)}{n!} \leqslant \lambda \qquad (1 \leqslant u \leqslant Q). \tag{4.12}$$

Then there exist constants C_{48}, C_{49} *and* C_{50} *such that*

$$|T(Q)| \leqslant e^{C_{48}n} Q h^{-\tau_n}\left\{\lambda^{\rho_n} + \left(Q\lambda^{\frac{1}{n}}\right)^{-\rho_n}\right\}, \tag{4.21}$$

where

$$\rho_n = \frac{C_{49}}{n^2 \ln n}, \quad \tau_n = \frac{C_{50}}{n^2 \ln n}. \tag{4.22}$$

Proof. Without loss of generality, we may assume that $Q^{-n} \leq \lambda \leq 1$, since otherwise (4.21) is trivial. We make use of Theorem A. This shows that there exists

$$l_n \in \mathbf{N}, \quad C_{51}n^2 \ln n \leqslant l_n \leqslant C_{52}n^2 \ln n$$

such that

$$J_{n-1}(q, l_n) \leqslant n^{C_1 n^3} q^{2l_n - \frac{n(n-1)}{2}}.\text{[1]}$$

For this choice of $l = l_n$ the preceding theorem gives

$$|T(Q)| \leqslant e^{C_{43}n} Q\left\{h^{-1}(\lambda+Q^{-1}) n^{C_1 n^3} q\right\}^{\frac{1}{2l_n}} \leqslant e^{C_{48}n} Q\left\{h^{-1}(\lambda+Q^{-1})\lambda^{-\frac{1}{n}}\right\}^{\frac{1}{2l_n}}$$

$$\leqslant e^{C_{48}n} Q h^{-\frac{1}{2l_n}}\left\{\lambda^{\frac{n-1}{n \cdot 2l_n}} + \left(Q\lambda^{\frac{1}{n}}\right)^{-\frac{1}{2l_n}}\right\} \leqslant e^{C_{48}n} Q h^{-\tau_n}\left\{\lambda^{\rho_n} + \left(Q\lambda^{\frac{1}{n}}\right)^{-\rho_n}\right\},$$

and Proposition 1 is proved.

For large λ the first term in (4.21) is the dominating one, and for small λ (in particular, for $Q^{-n} \leq \lambda \leq Q^{-1}$) the second term dominates. In the statement of Proposition 1, condition (4.12) may be replaced by

$$h\lambda \leqslant F^{(n)}(u) \leqslant \lambda, \tag{4.23}$$

since this only effects the value of the constant C_{48}.

Since Theorem A does not give the asymptotic behavior of l_n,

---

(1) Strictly speaking, Theorem A only implies this property for $n \geq 3$. But the case $n = 2$ is trivial, since then $J_1(q - 1) = q$.

we cannot derive an asymptotic estimate for ρ_n from this theorem. If instead of Theorem A we make use of the simplified upper bound for J given by Vinogradov ([1], Chapter 4, Theorem 4), then we obtain

$$|T(Q)| \leqslant A(n) h^{-\rho_n} Q \left\{\lambda^{\frac{n-1}{n}\rho_n} + \left(Q\lambda^{\frac{1}{n}}\right)^{-\rho_n}\right\} \quad (n \geqslant 12), \tag{4.24}$$

where

$$\rho_n = \{2[n^2(2\ln n + \ln\ln n + 2{,}6)]\}^{-1} \approx \frac{1}{4n^2 \ln n}, \tag{4.25}$$

but we do not have an estimate for the constant $A(n)$.

We note that it is only necessary to use the simplified upper bound for J to estimate $T(Q)$ in the case when λ is very small, say, of order $Q^{-n}\ln Q$. But if $\lambda \geq Q^{1/n-1}$, then we can use Theorem 1 or 2, in fact, for values of the parameter l for which these theorems can be established without using Theorem A.

Proposition 2. *Suppose that all the conditions of Proposition 1 are fulfilled and that, in addition,*

$$Q^{\alpha-n} \leqslant \lambda \leqslant Q^{-1}, \tag{4.26}$$

where $1/n \leq \alpha \leq n-1$. *Then*

$$|T(Q)| \leqslant e^{C_{53}n} h^{-\frac{C_{54}}{n^2 \ln n}} Q^{1-\frac{C_{55}\alpha}{n^3 \ln n}}. \tag{4.27}$$

We use Theorem 1, where we set $r = [4n \ln n + 1]$. Then

$$r = [4n \ln n + 1],$$

and

$$\frac{(n-1)^2}{2}\left(1 - \frac{1}{n-1}\right)^r \leqslant \frac{(n-1)^2}{2} e^{-4\ln n} \leqslant \frac{1}{2n^2} \leqslant \frac{\alpha}{2n}.$$

But, by our choice of r,

$$J_{n-1}(X, rn) \leqslant e^{C_3 n^3 \ln n} X^{2rn - \frac{n(n-1)}{2} + \frac{\alpha}{2n}} \quad (X \geqslant 1).$$

Hence, by Theorem 3,

$$C_{56} n^2 \ln n \leqslant rn < C_{57} n^2 \ln n,$$

and this implies (4.27).

If $\lambda \leq Q^{-1}$ and is not too small, then similar estimates can also be obtained for $T(N)$, $N \leq Q$ (compare [4], p. 105, Theorem 5).

Proposition 3. *Suppose that all the conditions of Proposition 1 are fulfilled and that, in addition,*

$$N \in \mathbf{N},\ \gamma \in \mathbf{R},\ N \leqslant Q,\ \gamma > 1,\ n > 4\gamma,\ Q^{-\gamma} \leqslant \lambda \leqslant Q^{-1}. \tag{4.28}$$

Then

$$|T(N)| \leqslant e^{C_{58}n} Q\left(hQ^{\frac{1}{2}}\right)^{-\rho_n}, \tag{4.29}$$

where $\rho_n = C_{59}/n^2 \ln n$.

First let

$$N = Q, \quad n \geqslant 2\gamma. \tag{4.30}$$

Since the case $hQ^{1/2} \leq 1$ is trivial, without loss of generality we may assume that $hQ^{1/2} \geq 1$. By (4.28) and (4.30) we have $\gamma \leq n/2$ and $\alpha = n - \gamma \geq n/2$. Hence, as in the proof of Proposition 2,

$$|T(Q)| \leqslant e^{C_{58}n} Q\left(h^{-1}Q^{-1}\lambda^{-\frac{1}{n}}\right)^{\frac{1}{2rn}} \leqslant e^{C_{58}n} Q\left(hQ^{\frac{1}{2}}\right)^{-\frac{1}{2rn}} \leqslant e^{C_{58}n} Q\left(hQ^{\frac{1}{2}}\right)^{-\rho_n}. \tag{4.31}$$

The constant C_{59} in the definition of ρ_n can be taken sufficiently small; in particular, so that $\rho_n \leq 1$.

We now proceed to investigate the case when $N \leq Q$ and $n \geq 4\gamma$. If $N \leq Q^{1/2}$, then, since $\rho_n \leq 1$, we have

$$e^{C_{58}n} Q\left(hQ^{\frac{1}{2}}\right)^{-\rho_n} \geqslant Q^{\frac{1}{2}} \geqslant N,$$

and hence (4.29) is trivial. But if $N \geq Q^{1/2}$, then

$$N^{-2\gamma} \leqslant \lambda \leqslant Q^{-1} \leqslant N^{-1}, \quad n \geqslant 2(2\gamma),$$

and (4.31) is applicable with Q replaced by N. Hence

$$|T(N)| \leqslant e^{C_{58}n} N\left(hN^{\frac{1}{2}}\right)^{-\rho_n} \leqslant e^{C_{58}n} Q\left(hQ^{\frac{1}{2}}\right)^{-\rho_n},$$

and Proposition 3 is proved.

In the case of large λ, $Q^{-1} \leq \lambda \leq 1$, we can obtain an asymptotic estimate for ρ_n.

Proposition 4. *Suppose that all the conditions of Proposition* 1 *are fulfilled and that, in addition,*

$$n \geqslant 3, \quad Q^{-1} \leqslant \lambda \leqslant 1. \tag{4.32}$$

Then

$$|T(Q)| \leqslant e^{C_{60}n} h^{-\tau_n} Q \lambda^{\rho_n}, \tag{4.33}$$

where

$$\tau_n = \{2(n-1)[(n-1)(\ln n + \ln\ln n - \ln 2) + 1]\}^{1},$$

$$\rho_n = \frac{1}{2n^2 \ln n}\left(1 - \frac{\ln\ln n + 1 - \ln 2}{\ln n}\right). \tag{4.34}$$

Proof. Theorem 3 shows that for $\lambda \geq Q^{-1}$

$$|T(Q)| \leqslant e^{C_{43}n} Q\left\{h^{-1}\lambda q^{1-2l+\frac{n(n-1)}{2}} J_{n-1}(q, l)\right\}^{\frac{1}{2l}},$$

where $q=[\lambda^{-1/n}]$. To estimate J_{n-1} we use Theorem 1, according to which, if $l\geq r(n-1)$, then

$$J_{n-1}(q,\ l)\leqslant n^{C_3n^3}q^{2l-\frac{n(n-1)}{2}+\eta}, \tag{4.35}$$

where

$$\eta=\frac{(n-1)^2}{2}\left(1-\frac{1}{n-1}\right)^r.$$

Hence

$$|T(Q)|\leqslant e^{C_{43}n}Q\{h^{-1}\lambda n^{C_3n^3}q^{1+\eta}\}^{\frac{1}{2l}}\leqslant\exp\left(C_{43}n+C_3\frac{n^3\ln n}{2l}\right)h^{-\frac{1}{2l}}Q\lambda^{\frac{n-1-\eta}{2ln}}. \tag{4.36}$$

Here we set

$$r=[(n-1)(\ln n+\ln\ln n-\ln 2)+1],\ \ l=r(n-1). \tag{4.37}$$

Then $r\in\mathbf{N}$ and

$$(n-1)(\ln n+\ln\ln n-\ln 2)\leqslant r\leqslant(n-1)(\ln n+\ln\ln n-\ln 2)+1,$$

$$\left(1-\frac{1}{n-1}\right)^r\leqslant\left(1-\frac{1}{n-1}\right)^{(n-1)(\ln n+\ln\ln n-\ln 2)}\leqslant\frac{2}{n\ln n},$$

$$\eta=\frac{(n-1)^2}{2}\left(1-\frac{1}{n-1}\right)^r\leqslant\frac{(n-1)^2}{n\ln n},$$

$$\begin{aligned}\frac{n-1-\eta}{2ln}&\geqslant\frac{n-1-\dfrac{(n-1)^2}{n\ln n}}{2n(n-1)n\left(\ln n+\ln\ln n-\ln 2+\dfrac{1}{n}\right)}\\&=\frac{1}{2n^2\ln n}\,\frac{1-\dfrac{n-1}{n\ln n}}{1+\dfrac{\ln\ln n-\ln 2+1/n}{\ln n}}\\&>\frac{1}{2n^2\ln n}\left(1-\frac{n-1}{n\ln n}\right)\left(1-\frac{\ln\ln n-\ln 2+1/n}{\ln n}\right)\\&>\frac{1}{2n^2\ln n}\left(1-\frac{n-1}{n\ln n}-\frac{\ln\ln n-\ln 2+1/n}{\ln n}\right)\\&=\frac{1}{2n^2\ln n}\left(1-\frac{\ln\ln n+1-\ln 2}{\ln n}\right)=\rho_n,\end{aligned}$$

so that, because $l\geq C_{61}n^2\ln n$, we have

$$|T(Q)|\leqslant e^{C_{60}n}h^{-\frac{1}{2l}}\lambda^{\rho_n},$$

and the proposition is proved.

For $n\geq 10$ this proposition gives better estimates for $T(Q)$ than those which follow from (4.3).

We note that

$$\rho_n=\frac{1}{2n^2\ln n}\left(1+O\left(\frac{\ln\ln n}{\ln n}\right)\right) \tag{4.38}$$

and that the constant ½ in the dominant term of this formula is better than the known constants.

For small n this result can be further refined.

Proposition 5. *Suppose that all the conditions of Proposition 1 are fulfilled and that, in addition* $n \geq 3$, $r \in \mathbf{N}$ *and* $l = r(n-1)+1$. *Then for any* $\epsilon > 0$

$$|T(Q)| \leqslant A(n, \varepsilon) h^{-\frac{1}{2l}} Q \lambda^{\rho_n - \varepsilon} \quad (Q^{-1} \leqslant \lambda \leqslant 1), \tag{4.39}$$

$$|T(Q)| \leqslant A(n, \varepsilon) h^{-\frac{1}{2l}} Q^{1-\frac{1}{2l}} \lambda^{-\tau_n - \varepsilon} \quad (\lambda \leqslant Q^{-1}), \tag{4.40}$$

where

$$\rho_n = \frac{n-1-\frac{n(n-3)}{2}\left(1-\frac{1}{n-1}\right)^{r-1}}{2nl}, \quad \tau_n = \frac{1+\frac{n(n-3)}{2}\left(1-\frac{1}{n-1}\right)^{r-1}}{2nl}. \tag{4.41}$$

We use Theorems 2 and 3.

In conclusion, we make some remarks concerning Rodosskiĭ's results [14].

Clearly, Theorem 3 of the present paper is a generalization and refinement of the fundamental theorem in [14]. Theorem 3 is valid for a wider class of functions $F(u)$; the factor $(\ln q)^{n/2l}$ and the term $Q\lambda^{1/2n}$ are missing in the right-hand side of (4.14); and the theorem gives an estimate of the growth of the constant.

Proposition 5 refines Theorem A in [14]. Computations show that, if $n = 9$, then the max in the formula for ρ_n is reached at $r = 20$, and

$$\rho_9^* = \max_r \rho_9 > \frac{1}{495}, \tag{4.42}$$

i.e., for $n = 9$ Proposition 5 gives stronger estimates than (4.3) in the region $Q^{-1} \leq \lambda \leq 1$. We similarly obtain a strengthening of Theorem A (the cases $n = 10$ and $n = 11$).

Finally, the last Theorem B in [14] has a conditional character. It is obtained under the assumption of the following conjecture:

Conjecture H_n. *Let* $n \in \mathbf{N}$ *and* $X \geq 1$. *Then for any* $\epsilon > 0$

$$J_n\left(X, \frac{1}{2}n(n+1)\right) \leqslant A(n, \varepsilon) X^{\frac{n(n+1)}{2}+\varepsilon}.$$

Rodosskiĭ notes that, if Conjecture H_{n-1} holds, then the following estimate is fulfilled for the trigonometric sum $T(Q)$, where $F(u)$ satisfies conditions (4.12) and certain additional restrictions:

$$|T(Q)| \leqslant A(n, \varepsilon) h^{-\frac{1}{n(n-1)}} Q \left\{\lambda^{\frac{1}{n^2}-\varepsilon} + Q^{-\frac{1}{n(n-1)}} \lambda^{-\frac{1}{n^2(n-1)}-\varepsilon}\right\} \quad (\varepsilon > 0).$$

If we suppose that Conjecture H_{n-1} holds and set $l = \frac{1}{2}n(n+1)$ in Theorem 3, we see that this estimate then holds for any function $F(u)$ satisfying conditions (4.12), i.e., none of the additional restrictions on $F(u)$ are needed.

Thus, the present paper contains, actually in a stronger form, all of the results in [14].

BIBLIOGRAPHY

1. I. M. Vinogradov, *The method of trigonometrical sums in the theory of numbers*, rev. ed., "Nauka", Moscow, 1971. (Russian; cf. [5]).

2. A. A. Karacuba, *The mean value of the modulus of a trigonometric sum*, Izv. Akad. Nauk SSSR Ser. Mat. 37 (1973), 1203–1227 = Math. USSR Izv. 7 (1973), 1199–1224. MR **49** #2586.

3. ———, *On a system of indeterminate equations*, Mat. Zametki 4 (1968), 125–128 = Math. Notes 4 (1968), 565–567. MR **38** #1062.

4. K. Chandrasekharan, *Arithmetical functions*, Springer-Verlag, Berlin and New York, 1970. MR 43 #3223.

5. I. M. Vinogradov, *The method of trigonometrical sums in the theory of numbers*, Trudy Mat. Inst. Steklov. 23 (1947); English transl., Interscience, New York, 1954. MR **10**, 599; **15**, 941.

6. ———, *Selected works*, Izdat. Akad. Nauk SSSR, Moscow, 1952. (Russian) MR **14**, 610.

7. H. Weyl, *Zur Abschätzung von* $\zeta(1+ti)$, Math. Z. **10** (1921), 88–101.

8. ———, Reprint of [7], *Gesammelte Abhandlungen*. Band II, Springer-Verlag, New York, 1968, pp. 181–194. MR **37** #6157.

9. E. C. Titchmarsh, *Theory of the Riemann zeta function*, Cambridge, 1930.

10. K. Prachar, *Primzahlverteilung*, Springer-Verlag, Berlin, 1957; Russian transl., "Mir", Moscow, 1967. MR **19**, 393.

11. Hua Loo-keng, *Additive theory of prime numbers*, Trudy Mat. Inst. Steklov. 22 (1947); English transl., Amer. Math. Soc., Providence, R. I., 1965. MR **10**, 597; 33 #2614.

12. ———, *Die Abschätzung von Exponentialsummen und ihre Anwendung in der Zahlentheorie*, Enzyklopädie der math. Wissenschaften, Bände I, 2, Heft 13, Teil I, Teubner, Leipzig, 1959. MR **24** #A94.

13. ———, *Additive Primzahltheorie*, Teubner, Leipzig, 1959. [German transl. of [11]] MR **23** #A1620.

14. K. A. Rodosskiĭ, *On some aspects of the method of I. M. Vinogradov*, Izv. Vysš. Učebn. Zaved. Matematika **1965**, no. 1 (44), 123–132. (Russian) MR 31 #135.

15. A. A. Karacuba, *On systems of congruences*, Izv. Akad. Nauk SSSR Ser. Mat. 29 (1965), 935–944; English transl., Amer. Math. Soc. Transl. (2) **82** (1969), 73–84. MR 32 #7530.

16. ———, *Theorems on the mean and complete trigonometric sums*, Izv. Akad. Nauk SSSR Ser. Mat. 30 (1966), 183–206. (Russian) MR **33** #1296.

17. L. E. Dickson, *History of the theory of numbers*. Vol. II, Carnegie Inst. of Washington, Washington, D. C., 1920; reprinted by Chelsea, New York, 1966. MR 39 #6807b.

18. Ju. V. Linnik, *On Weyl's sums*, C. R. (Dokl.) Acad. Sci. URSS 34 (1942), 184–186. MR 4, 211.

19. J. B. Rosser and L. Schoenfeld, *Approximate formulas for some functions of prime numbers*, Illinois J. Math. 6 (1962), 64–94. MR 25 #1139.

20. J. G. van der Corput, *Sur la méthode de Weyl dans la théorie des nombres*. III, Nederl. Akad. Wetensch. Proc. **40** (1937), 836–845.

21. J. F. Koksma, *Diophantische Approximationen*, Springer-Verlag, Berlin, 1936.

Translated by N. KOBLITZ

Trudy Mat. Inst. Steklov.
134 (1975)

Proc. Steklov Inst. Math.
134 (1975)

ON THE CONVERGENCE IN THE L-METRIC OF FOURIER SERIES WITH QUASI-MONOTONE COEFFICIENTS

UDC 517.52

S. A. TELJAKOVSKIĬ AND G. A. FOMIN

Abstract. It is shown that if a cosine or sine series is a Fourier series and its coefficients a_n are quasi-monotone, then it converges in the L-metric if and only if $a_n \log n \to 0$ as $n \to \infty$.

Bibliography: 9 titles.

The problem of the convergence in the L-metric of the series

$$\frac{a_0}{2} + \sum_{k=1}^{\infty} a_k \cos kx \tag{1}$$

under the condition that its coefficients tend to zero and the sequence $\{a_k\}$ is convex (i.e. $\Delta^2 a_k = \Delta a_k - \Delta a_{k+1} = a_k - 2a_{k+1} + a_{k+2} \geq 0$ for all k) or quasi-convex (i.e. $\sum_{k=1}^{\infty} k|\Delta^2 a_{k-1}| < \infty$) was solved by A. N. Kolmogorov in [1]. He proved that in this case the series (1) is a Fourier series (for convex $\{a_k\}$ this result was obtained earlier by Young [2]) and that the condition

$$a_n \log n \to 0 \quad \text{as} \quad n \to \infty \tag{2}$$

is necessary and sufficient for convergence of the series in the L-metric.

After this, similar questions were considered by a number of authors [3]–[6].

Young [2] proved that a series

$$\sum_{k=1}^{\infty} a_k \sin kx \tag{3}$$

with monotone decreasing coefficients a_k is a Fourier series if and only if $\sum_1^{\infty} k^{-1} a_k < \infty$, and Hille and Tamarkin [3] established that under these conditions the series (3) converges in the L-metric. Results established in [1]–[3] were cited in Zygmund's monograph [7], §5.1.

G. A. Fomin [5], in the form of corollaries obtained from certain more general bounds, considered the problem of convergence in the L-metric of the series (1) under the assumption that it is a Fourier series and that its coefficients are monotone decreasing, and showed that in this case condition (2) is sufficient and the condition

---

AMS (MOS) subject classifications (1970). Primary 42A32, 42A20; Secondary 40C05.

$$\frac{a_{2n}^2}{a_n}\log n \to 0 \quad \text{for} \quad n\to\infty \tag{4}$$

is necessary. One can show that (2) follows from $a_k \downarrow 0$ and condition (4), and in this way condition (2) is shown to be necessary and sufficient in this problem. We shall not dwell on this, since this paper gives a simple proof of the fact that relation (2) is a necessary and sufficient condition for convergence in the L-metric of series (1) and (3) if they are Fourier series and their coefficients are quasi-monotone. A sequence $\{a_k\}$ is *quasi-monotone* if for some $\alpha \geq 0$ the numbers a_k/k^α are monotone decreasing. The monotone sequences are obtained when $\alpha = 0$.

Theorem 1. *If* (1) *is a Fourier series of a function f and its coefficients are quasi-monotone, then it converges in the L-metric if and only if condition* (2) *holds.*

Proof. *Sufficiency.* Let us consider the means of the series (1):

$$u_n(f,\ x) = \lambda_{n,0}\frac{a_0}{2} + \sum_{k=1}^{2n}\lambda_{n,k}a_k\cos kx, \tag{5}$$

where

$$\lambda_{n,k} = \begin{cases} 1 & \text{for} \quad 0\leqslant k\leqslant n, \\ \dfrac{2n-k}{n}\left(\dfrac{n}{k}\right)^\alpha & \text{for} \quad n+1\leqslant k\leqslant 2n, \end{cases}$$

and α is the number involved in the definition of quasi-monotonicity. Let

$$s_n(f,\ x) = \frac{a_0}{2} + \sum_{k=1}^{n} a_k\cos kx$$

be a partial sum of the series (1). We have

$$\int_0^\pi |f(x) - s_n(f,\ x)|\,dx \leqslant \int_0^\pi |f(x) - u_n(f,\ x)|\,dx + \int_0^\pi |u_n(f,\ x) - s_n(f,\ x)|\,dx. \tag{6}$$

Since $0 \leq \Delta\lambda_{n,k} \leq (1+\alpha)/n$, $k = 0, 1, \dots, 2n-1$, we obtain (cf. [8], for example) that the summability method defined by the coefficients $\lambda_{n,k}$ is regular in the C-metric, and thus according to a theorem of Lozinskiĭ [9] it is also regular in the L-metric. Therefore

$$\int_0^\pi |f(x) - u_n(f,\ x)|\,dx \to 0 \quad \text{as} \quad n\to\infty. \tag{7}$$

Now let us bound the second integral on the right of (6). We introduce the notation

$$b_k = \lambda_{n,k}a_k = \frac{2n-k}{n^{1-\alpha}}\frac{a_k}{k^\alpha}, \qquad k = n+1, \dots, 2n.$$

By the qussi-monotonicity of the sequence $\{a_k\}$ we have

$$b_{n+1} \geqslant b_{n+2} \geqslant \dots \geqslant b_{2n}.$$

Since

$$u_n(f, x) - s_n(f, x) = \sum_{k=n+1}^{2n-1} b_k \cos kx = -b_{n+1} D_n(x) + \sum_{k=n+1}^{2n-1} \Delta b_k D_k(x),$$

where $D_k(x)$ is the Dirichlet kernel, we find, denoting the Lebesgue constants by L_k, that

$$\begin{aligned} \frac{2}{\pi}\int_0^\pi |u_n(f, x) - s_n(f, x)|\, dx &\leqslant b_{n+1}L_n + \sum_{k=n+1}^{2n-1} \Delta b_k L_k \\ &= 2b_{n+1}L_n + \sum_{k=n+1}^{2n-1} b_k (L_k - L_{k-1}) \leqslant b_{n+1}(L_{2n-1} + L_n). \end{aligned} \tag{8}$$

But $b_{n+1} \leq a_{n+1}$. Thus from (2) and (8) we conclude that

$$\int_0^\pi |u_n(f, x) - s_n(f, x)|\, dx \to 0 \quad \text{as} \quad n \to \infty. \tag{9}$$

The sufficiency now follows from (6), (7), and (9).

Necessity. The polynomials

$$Q_n(x) = \sum_{k=1}^{n} \frac{\cos(n+k)x - \cos(n-k)x}{k} = -2 \sin nx \sum_{k=1}^{n} \frac{\sin kx}{k}$$

are uniformly bounded. Thus, for some positive constant c,

$$\begin{aligned} \frac{2}{\pi}\int_0^\pi |f(x) - s_n(f, x)|\, dx &\geqslant \frac{2c}{\pi}\int_0^\pi [f(x) - s_n(f, x)]\, Q_n(x)\, dx \\ &= c\sum_{k=1}^{n} \frac{a_{n+k}}{k} \geqslant cn^\alpha \sum_{k=1}^{n} \frac{1}{k}\, \frac{a_{n+k}}{(n+k)^\alpha} \geqslant cn^\alpha \frac{a_{2n}}{(2n)^\alpha} \log n. \end{aligned}$$

The necessity of condition (2) follows from this, and the theorem is proved.

A similar theorem holds for a series of sines.

Theorem 2. *If* (3) *is a Fourier series and its coefficients are quasi-monotone, then it converges in the L-metric if and only if condition* (2) *holds.*

We shall not give a proof of this theorem, since it does not differ from the proof of Theorem 1.

As was previously noted, a series (3) with quasi-monotone coefficients converges in the L-metric if it is a Fourier series. In connection with Theorem 2, let us show by an example that *there exist sine series with quasi-monotone coefficients (for $\alpha > 0$) which are Fourier series but do not converge in the L-metric.* It is clear that here we can restrict ourselves to small values of α.

Let $0 < \alpha \leq 1$ and let $m_j = 2^{j^3}$ and $c_j = 1/j^2 m_j^\alpha$ for $j = 1, 2, \ldots$. Set

$$d_k(j) = \begin{cases} c_j k^\alpha & \text{for } 0 \leqslant k \leqslant m_j, \\ c_j(2m_j^\alpha - k^\alpha) & \text{for } m_j < k < 2^{\frac{1}{\alpha}} m_j, \\ 0 & \text{for } 2^{\frac{1}{\alpha}} m_j \leqslant k \end{cases}$$

and

$$d_k = \max_j d_k(j).$$

Then, as $k \to \infty$, $d_k \to 0$ and $d_k/k^\alpha \downarrow 0$; i.e. the d_k are quasi-monotone and $|\Delta d_k| \leq \max_j |\Delta d_k(j)|$. Note that, by the condition $\alpha \leq 1$, the numbers $|\Delta d_k(j)|$ are monotone decreasing with increasing k and fixed j. Thus for $A_k = \max_{n \geq k} |\Delta d_n|$ we have

$$A_k \leqslant \sum_{j=1}^{\infty} |\Delta d_k(j)|,$$

from which we find that

$$\sum_{k=1}^{\infty} A_k \leqslant \sum_{j=1}^{\infty} \sum_{k=1}^{\infty} |\Delta d_k(j)| \leqslant \sum_{j=1}^{\infty} 2m_j^\alpha c_j < \infty. \tag{10}$$

Moreover,

$$\sum_{k=1}^{\infty} \frac{d_k}{k} \leqslant \sum_{k=1}^{\infty} \sum_{j=1}^{\infty} \frac{d_k(j)}{k} \leqslant \sum_{j=1}^{\infty} \sum_{k \leqslant 2^{\frac{1}{\alpha}} m_j} c_j k^{\alpha-1} < \infty. \tag{11}$$

By (10) and (11) the series

$$\sum_{k=1}^{\infty} d_k \sin kx \tag{12}$$

is a Fourier series (cf. Theorem 2 of [6]). But since

$$d_{m_j} = d_{m_j}(j) = c_j m_j^\alpha = \frac{1}{j^2},$$

we have that

$$d_{m_j} \log m_j = \frac{1}{j^2} \log 2^{j^3} \to \infty \quad \text{as} \quad j \to \infty,$$

and thus (12) does not converge in the L-metric. This assertion follows from our Theorem 2 as well as from Theorem 4 of [6].

BIBLIOGRAPHY

1. A. N. Kolmogoroff, *Sur l'ordre de grandeur des coefficients de la série de Fourier-Lebesgue*, Bull. Internat. Acad. Polon. Sci. Lett. Cl. Sci. Math. Nat. Sér. A Sci. Math. 1923, 83–86.

2. W. H. Young, *On the Fourier series of bounded functions*, Proc. London Math. Soc. (2) 2 (1913), 41–70.

3. E. Hille and J. D. Tamarkin, *On the summability of Fourier series*. II, Ann. of Math. (2) 34 (1933), 329–348.

4. S. A. Teljakovskiĭ, *On the question of the convergence of Fourier series in the L-metric*, Mat. Zametki 1 (1967), 91–98 = Math. Notes 1 (1967), 59–63. MR 34 #8067.

5. G. A. Fomin, *On convergence of Fourier series in the L-metric*, Applications of Functional Anal. in Approximation Theory., Materials for a Conference, Kalinin, 1970, pp. 170–173. (Russian)

6. S. A. Teljakovskiĭ, *A certain sufficient condition of Sidon for the integrability of trigonometric series*, Mat. Zametki 14 (1973), 317–328 = Math. Notes 14 (1973), 742–748 (1974). MR 48 #6798.

7. A. Zygmund, *Trigonometrical series*, PWN, Warsaw, 1935.

8. G. A. Fomin, *On linear methods of summability of Fourier series*, Mat. Sb. 65 (107) (1964), 144–152. (Russian) MR 29 #6243.

9. S. M. Lozinskiĭ, *On convegence and summability of Fourier series and interpolation processes*, Mat. Sb. 14 (56) (1944), 175–268. MR 6, 264.

Translated by W. L. BYNUM

Trudy Mat. Inst. Steklov.
134 (1975)

Proc. Steklov Inst. Math.
134 (1975)

ON THE ISOMETRIC MAPPING OF SOME ULTRAMETRIC SPACES INTO L_p-SPACES

UDC 517.5:513.88

A. F. TIMAN

Abstract. Countable ultrametric spaces $R(a_1, a_2, \dots)$ are considered in which the metric for an arbitrary monotone decreasing sequence of numbers $d_n \geq 0$ is defined by the equation $\rho(a_m, a_n) = d_n$ for $m > n$.

It is shown that each such space can be mapped isometrically into the space L_p, where $p \geq 1$.

It is shown that these spaces have certain dimensional properties, and an application is made to the solution of the inverse problem for the ϵ-entropy of compact sets in L_p.

Bibliography: 9 titles.

§1. Ultrametric spaces and their characteristic property

A metric space R with a metric ρ is said to be *ultrametric* (see [1], Chapter III, §8, Problem 4) if the triangle inequality $\rho(x, y) \leq \rho(x, z) + \rho(y, z)$ for any three points $x, y, z \in R$ is satisfied also in the stronger form

$$\rho(x, y) \leqslant \max\{\rho(x, z), \rho(y, z)\}. \tag{u}$$

As a trivial example of an ultrametric space we can take any discrete metric space R; that is, a space R in which $\rho(x, y) = 1$ for arbitrary distinct points $x, y \in R$ and $\rho(x, x) = 0$. One of the first nontrivial examples of an ultrametric space was the Baire null-space (see [2], §20.4). Among other known examples of ultrametric spaces an important place is held by the space $\mathbf{Z}_p$ of integers and the space $\mathbf{Q}_p$ of all real p-adic numbers, where p is a prime (see, for example, Chapter I in [3]).

A metric space R is ultrametric if and only if all triangles in the space are isosceles and the base of each is the smallest side. This geometrical property of ultrametric spaces, which characterizes such spaces, can be formulated in the following more general manner.

If R is an ultrametric space, then for any natural number $n \geq 2$ and system

AMS (MOS) subject classifications (1970). Primary 54E99, 54C25; Secondary 46E30, 47H15.

of points $x_1, \ldots, x_n \in R$, among the $n(n-1)/2$ numbers $d_{ij} = \rho(x_i, x_j)$, $i, j = 1, \ldots, n$, $i \neq j$, in the corresponding matrix of distances $\{\rho(x_i, x_j)\}$ the number $r = r(x_1, \ldots, x_n)$ of mutually distinct numbers does not exceed $n-1$ and the largest of these is of multiplicity $s = s(x_1, \ldots, x_n) \geq n-1$.

For $n = 3$ this assertion is known and is equivalent to inequality (u).

Suppose that for some natural number $n \geq 3$ the above assertion has already been proved. We consider an arbitrary system of $n+1$ points $x_1, \ldots, x_{n+1}$ and all of the distances $\delta_\nu = \rho(x_{n+1}, x_\nu)$, $\nu = 1, \ldots, n$.

If $\delta_1 = \cdots = \delta_n = d$, then by inequality (u) we know that d is not less than any of the $r \leq n-1$ positive numbers which represent the possible distances $\rho(x_i, x_j)$ in the system of points $x_1, \ldots, x_n$, and therefore it is clear that since the assertion is true by assumption for the system of points $x_1, \ldots, x_n$, it is also true for the system of points $x_1, \ldots, x_{n+1}$.

Now suppose that $\min_i \delta_i < \max_i \delta_i$. Without loss of generality we can assume that the smallest of the numbers δ_i $(i = 1, \ldots, n)$ is δ_n. For an arbitrary $i = 1, \ldots, n-1$ we consider the three points x_i, x_n and x_{n+1}; since $\delta_n < \delta_i$, by inequality (u) we must have $\rho(x_{n+1}, x_i) = \rho(x_n, x_i)$. Thus if $\delta_n < \delta_i$ for some $i = 1, \ldots, n-1$, then δ_i coincides with one of the pairwise distances for the system of points $x_1, \ldots, x_n$. Thus the inequality $r(x_1, \ldots, x_n) \leq n-1$ implies that $r(x_1, \ldots, x_{n+1}) \leq n$.

Further suppose that δ_j is the largest of the numbers δ_i $(i = 1, \ldots, n)$. Since in this case $\delta_n < \delta_j$, by inequality (u) we have $\rho(x_j, x_n) = \delta_j$. If for some pair of points x_ν, x_k, $1 \leq \nu < k$, we were to have $\rho(x_\nu, x_k) > \rho(x_j, x_n)$, then for the points x_ν, x_k and x_{n+1} we would have, by inequality (u), the relation $\delta_j = \rho(x_j, x_n) < \max(\delta_\nu, \delta_k)$, which would contradict the choice of δ_j. Therefore δ_j, the largest of the numbers δ_i $(i = 1, \ldots, n)$, coincides with the largest of the distances between the points $x_1, \ldots, x_n$. Thus, if $s(x_1, \ldots, x_n) \geq n-1$, we have

$$s(x_1, \ldots, x_n, x_{n+1}) \geqslant s(x_1, \ldots, x_n) + 1 \geqslant n.$$

We note that the resulting estimates

$$1 \leqslant r(x_1, \ldots, x_n) \leqslant n-1 \leqslant s(x_1, \ldots, x_n) \leqslant \frac{n(n-1)}{2}$$

are best possible for each natural number $n \geq 2$. For each $n \geq 2$ we can find an ultrametric space R consisting of n points $x_1, \ldots, x_n$ in which

$$r(x_1, \ldots, x_n) = n-1 = s(x_1, \ldots, x_n).$$

§2. Ultrametric sets in metric spaces and the statement of the problem

The effect of the exotic geometric structure of ultrametric spaces generated by the ultrametric inequality (u) (in connection with this, see [1], Chapter III, §8, Problem 4, where certain singular features of such spaces are pointed out) is dispersed when we consider in ordinary metric spaces those sets which satisfy (u).

Let R be a metric space. A set $A \subset R$ is said to be *ultrametric* if A as a metric space with the metric induced by R is an ultrametric space.

As an example of an ultrametric set in a metric space R we can take any finite or infinite sequence of points $a_1, a_2, \ldots \in R$ for which $\rho(a_m, a_n) = d_n$ when $m > n$, where $d = \{d_i\}$ $(d_1 \geq d_2 \geq \cdots \geq 0)$ is an arbitrary nonincreasing sequence of positive numbers. Such ultrametric sequences $A \subset R$ will be denoted by $A(d; R)$.

As a concrete example of an ultrametric set in the space C we can take an arbitrary set of functions with compact disjoint supports, and as an example of a sequence $A(d; C)$ we can take any sequence of such functions which converges uniformly to zero with an appropriate enumeration.

We note that not even every finite system of n points in a metric space R which constitutes an ultrametric set in R can be enumerated in such a way that the pairwise distances between the points $a_1, a_2, \ldots$ are defined by equations $\rho(a_i, a_j) = d_i$ for $j > i$, where $d_1 \geq d_2 \geq \cdots$. Counterexamples can easily be found for $n = 4$ in three-dimensional euclidean space.

Consideration of the simplest ultrametric sets in well-known spaces shows that the geometric singularities caused by inequality (u) are often brought about not by some special feature of the metric, which can turn out to be induced by some ordinary metric, but rather by a special distribution of the points. In this context there arises the natural question of whether it is possible to find an isometric imbedding of an ultrametric space into a space with a previously chosen metric. Certain other questions are connected with this one, and one of these questions is considered in §6.

The present work is devoted to the above question. The fundamental result is concerned with the possibility of realizing countable ultrametric spaces $R(d; R)$ in the form of sequences $A(d; L_p)$ with respect to an L_p-metric with $1 \leq p < \infty$. The possibility of such a realization in L_∞ is an immediate corollary of the universality property of this space and can easily be established as an independent result. In the transition from the case $p = \infty$ to the case where p is finite, the usual property of universality even with respect to a very simple metric space is lost. This fact is demonstrated by the following example of a metric space consisting of four points (a_1, a_2, a_3, a_4) with the metric

$$\rho(a_1, a_2) = \rho(a_1, a_3) = \rho(a_2, a_3) = 1,$$

$$\rho(a_1, a_4) = \rho(a_2, a_4) = \frac{1}{\sqrt{3}},$$

$$\rho(a_3, a_4) = d \quad \left(1 - \frac{1}{\sqrt{3}} < d < \frac{1}{\sqrt{3}}\right),$$

which clearly cannot be isometrically mapped into l_2. This circumstance shows that it is appropriate to investigate the problem which we have posed.

§3. Existence of a solution of certain systems of nonlinear equations

The question of existence of isometric imbeddings of ultrametric spaces into a given metric space is always connected with the question of existence of a solution of a corresponding system of nonlinear equations.

Let $\phi(t)$ be an arbitrary real, even function defined on $-\infty < t < \infty$ which satisfies the following conditions on the positive axis $t \geq 0$:

1) $\phi(t)$ is continuous,

2) $\phi(0) = 0$,

3) $\phi(t_1) < \phi(t_2)$ for $0 \leq t_1 < t_2$,

4) $\phi(t') + \phi(t'') \leq \phi(t' + t'')$ for $t', t'' \geq 0$.

We have the following result.

Theorem 1. *For each monotone decreasing sequence of nonnegative numbers* $d_1 \geq d_2 \geq \cdots \geq 0$ *and each function* $\phi(t)$ *which satisfies conditions* 1)–4) *there exists an infinite triangular matrix of numbers* $\{\xi_{i,j}\}$ $(i = 1, 2, \ldots; j = 1, \ldots i)$ *which satisfy the system of equations*

$$\begin{gathered}\sum_{j=1}^{k} \varphi(\xi_{k,j}) = \varphi(d_1),\\ \sum_{j=1}^{l-1} \varphi(\xi_{l-1,j} - \xi_{k,j}) + \sum_{j=l}^{k} \varphi(\xi_{k,j}) = \varphi(d_l) \qquad (l = 2, 3, \ldots, k)\end{gathered} \qquad (\alpha_k)$$

for all natural numbers k.

The proof of this assertion reduces to the proof of the following fundamental lemma.

Lemma 1. *For each monotone decreasing sequence of nonnegative numbers* $d_1 \geq d_2 \geq \cdots \geq 0$ *and each function* $\phi(t)$ *which satisfies conditions* 1)–4), *for an arbitrary natural number n there exists one and only one real nonnegative matrix* $\{\xi_{i,j}\}$, $1 \leq i \leq n$, $1 \leq j \leq i$, *which satisfies the system of equations* (α_k) *for all* $k \leq n$. *For each* $k \geq 2$ $(k \leq n)$ *this matrix satisfies the condition*

$$\xi_{j+1,j} = \xi_{j+2,j} = \xi_{j+3,j} = \cdots = \xi_{k,j} \leqslant \xi_{j,j} \quad (j = 1, \ldots, k-1). \qquad (\beta_{kj})$$

Proof of Lemma 1. In the case $n = 1$ the assertion in Lemma 1 means that the equation $\phi(\xi) = \phi(d_1)$ has a unique nonnegative solution $\xi_{1,1} = d_1$, and this follows from condition 3). Letting $\xi_{1,1} = d_1$, we now consider the case where $n = 2$. Here the system (α_n) is of the form

$$\begin{gathered}\varphi(\xi_{2,1}) + \varphi(\xi_{2,2}) = \varphi(d_1),\\ \varphi(\xi_{1,1} - \xi_{2,1}) + \varphi(\xi_{2,2}) = \varphi(d_2).\end{gathered} \qquad (\alpha_2)$$

Consequently the value of $\xi_{2,1}$ must be determined by the equation

$$\varphi(\xi_{2,1}) - \varphi(\xi_{1,1} - \xi_{2,1}) = \varphi(d_1) - \varphi(d_2). \qquad (1)$$

But since $0 \le d_2 \le d_1$ and the continuous function $\phi(x) - \phi(d_1 - x)$ is an injective mapping of $[d_1/2, d_1]$ onto $[0, \phi(d_1)]$, the equation has one and only one solution $\xi_{2,1}$ such that $d_1/2 \le \xi_{2,1} \le d_1$. Therefore from the first equation in (α_2) we can determine one and only one nonnegative value of $\xi_{2,2}$ which together with $\xi_{2,1}$ satisfies the system (α_2). The inequality $\xi_{2,1} \le \xi_{1,1}$ expresses condition (β_{21}).

Now we consider the case where $n = 3$. The system of equations (α_3), where $\xi_{1,1} = d_1$, and $\xi_{2,1}$ and $\xi_{2,2}$ are values already determined from (α_2), is of the form

$$\begin{aligned} \varphi(\xi_{3,1}) + \varphi(\xi_{3,2}) + \varphi(\xi_{3,3}) &= \varphi(d_1), \\ \varphi(\xi_{1,1} - \xi_{3,1}) + \varphi(\xi_{3,2}) + \varphi(\xi_{3,3}) &= \varphi(d_2), \\ \varphi(\xi_{2,1} - \xi_{3,1}) + \varphi(\xi_{2,2} - \xi_{3,2}) + \varphi(\xi_{3,3}) &= \varphi(d_3). \end{aligned} \qquad (\alpha_3)$$

Thus it is clear that the value of $\xi_{3,1}$ must satisfy the equation

$$\varphi(\xi_{3,1}) - \varphi(\xi_{1,1} - \xi_{3,1}) = \varphi(d_1) - \varphi(d_2),$$

which coincides with (1). Therefore its unique solution coincides with the previously determined value $\xi_{2,1}$, and consequently it satisfies condition (β_{31}); that is, we have

$$\xi_{2,1} = \xi_{3,1} \leqslant \xi_{1,1}, \qquad (\beta_{31})$$

and condition (β_{kj}) has been verified for $k = 2, 3;\ j = 1$.

By virtue of (β_{21}) it follows from (α_3) that we have the equation

$$\begin{aligned} \varphi(\xi_{3,2}) - \varphi(\xi_{2,2} - \xi_{3,2}) &= \varphi(d_2) - \varphi(d_3) - \varphi(\xi_{1,1} - \xi_{3,1}) \\ &= \varphi(d_2) - \varphi(d_3) - \varphi(\xi_{1,1} - \xi_{2,1}) \end{aligned}$$

for $\xi_{3,2}$, which by virtue of (α_2) can be rewritten in the form

$$\varphi(\xi_{3,2}) - \varphi(\xi_{2,2} - \xi_{3,2}) = \varphi(\xi_{2,2}) - \varphi(d_3). \qquad (2)$$

We now show that

$$|\varphi(\xi_{2,2}) - \varphi(d_3)| \leqslant \varphi(\xi_{2,2}).$$

In fact, from (α_2) it follows that

$$2\varphi(\xi_{2,2}) = \varphi(d_1) + \varphi(d_2) - \varphi(\xi_{2,1}) - \varphi(d_1 - \xi_{2,1}).$$

By condition 4) we know that

$$\varphi(\xi_{2,1}) + \varphi(d_1 - \xi_{2,1}) \leqslant \varphi(d_1),$$

and therefore $2\phi(\xi_{2,2}) \geq \phi(d_2) \geq \phi(d_3)$ or $\phi(\xi_{2,2}) - \phi(d_3) \geq -\phi(\xi_{2,2})$. If we consider the continuous function $\phi(x) - \phi(\xi_{2,2} - x)$, which is an injective mapping from $[0, \xi_{2,2}]$ onto $[-\phi(\xi_{2,2}), \phi(\xi_{2,2})]$, we see that equation (2) has one and only one solution $\xi_{3,2}$ such that $0 \leq \xi_{3,2} \leq \xi_{2,2}$, where this inequality expresses condition (β_{32}). By virtue of (β_{21}) and (α_2) we then have

$$\varphi(\xi_{3,1}) + \varphi(\xi_{3,2}) \leqslant \varphi(\xi_{2,1}) + \varphi(\xi_{2,2}) = \varphi(d_1).$$

Consequently there must exist one and only one nonnegative value of $\xi_{3,3}$ which together with $\xi_{3,1}$ and $\xi_{3,2}$ satisfies the system (α_3).

Now suppose that for all natural numbers $1, \dots, n-1$ the assertion in Lemma 1 is valid. We will prove that there exists a system of real nonnegative numbers $\xi_{n,1}, \dots, \xi_{n,n}$, which together with the nonnegative solution $\{\xi_{i,j}\}$ $(i = 1, \dots, n-1;\ j = 1, \dots, i)$ of the system of equations (α_{n-1}) satisfies the system of equations (α_n) and the conditions (β_{nj}).

By virtue of our assumptions and the conditions (β_{kj}), for an arbitrary $k = 1, \dots, n-1$ the system of equations (α_k) can be written in the form

$$\begin{aligned} &\sum_{j=1}^{k} \varphi(\xi_{k,j}) = \varphi(d_1), \\ &\varphi(\xi_{l-1,l-1} - \xi_{k,l-1}) + \sum_{j=l}^{k} \varphi(\xi_{k,j}) = \varphi(d_l) \qquad (l = 2, 3, \dots, k). \end{aligned} \qquad (\gamma_k)$$

We consider the system (α_n), in which the values $\{\xi_{i,j}\}$ $(i = 1, \dots, n-1;\ j = 1, \dots, i)$ have already been determined. As before, we obtain the equation

$$\varphi(\xi_{n,1}) - \varphi(\xi_{1,1} - \xi_{n,1}) = \varphi(d_1) - \varphi(d_2)$$

for $\xi_{n,1}$. But this equation coincides with the equation

$$\varphi(\xi_{k,1}) - \varphi(\xi_{1,1} - \xi_{k,1}) = \varphi(d_1) - \varphi(d_2),$$

from which we already determined the values $\xi_{k,1}$ in the system (α_k) for an arbitrary $k = 2, 3, \dots, n-1$. Consequently

$$\xi_{n,1} = \xi_{n-1,1} = \dots = \xi_{2,1} \leqslant \xi_{1,1}, \qquad (\beta_{n1})$$

and condition (β_{kj}) is satisfied for $k = n$, $j = 1$. By virtue of (β_{n1}) we obtain from the system (α_n) the following equation for determining $\xi_{n,2}$;

$$\varphi(\xi_{n,2}) - \varphi(\xi_{2,2} - \xi_{n,2}) = \varphi(d_2) - \varphi(d_3) - \varphi(\xi_{1,1} - \xi_{n,1}),$$

or

$$\varphi(\xi_{n,2}) - \varphi(\xi_{2,2} - \xi_{n,2}) = \varphi(d_2) - \varphi(d_3) - \varphi(\xi_{1,1} - \xi_{2,1}).$$

Since by assumption the values of $\xi_{1,1}$ and $\xi_{2,1}$ satisfy the system (α_2), the preceding equation is of the form

$$\varphi(\xi_{n,2})-\varphi(\xi_{2,2}-\xi_{n,2})=\varphi(\xi_{2,2})-\varphi(d_3).$$

This equation for determining the nonnegative value of $\xi_{n,2}$ coincides with equation (2), from which we determined $\xi_{3,2}$. Consequently

$$\xi_{n,2}=\xi_{n-1,2}=\dots=\xi_{3,2}\leqslant\xi_{2,2}, \qquad (\beta_{n2})$$

that is, condition (β_{kj}) is satisfied for $k=n$ and $j=2$.

Now suppose that for a given n we have already determined nonnegative values of $\xi_{n,1},\dots,\xi_{n,l-1}$ such that

$$\xi_{n,j}=\xi_{n-1,j}=\dots=\xi_{j+1,j}\leqslant\xi_{j,j}, \qquad (\beta_{nj})$$

where $j=1,\dots,l-1$. In the system (α_n) we consider the two equations

$$\sum_{j=1}^{l-1}\varphi(\xi_{l-1,j}-\xi_{n,j})+\sum_{j=l}^{n}\varphi(\xi_{n,j})=\varphi(d_l),$$

$$\sum_{j=1}^{l}\varphi(\xi_{l,j}-\xi_{n,j})+\sum_{j=l+1}^{n}\varphi(\xi_{n,j})=\varphi(d_{l+1}).$$

By virtue of (β_{nj}) for $j=1,\dots,l-1$ these equations can be written in the form

$$\varphi(\xi_{l-1,l-1}-\xi_{n,l-1})+\sum_{j=l}^{n}\varphi(\xi_{n,j})=\varphi(d_l),$$

$$\varphi(\xi_{l,l}-\xi_{n,l})+\sum_{j=l+1}^{n}\varphi(\xi_{n,j})=\varphi(d_{l+1}).$$

Therefore we have the equation

$$\varphi(\xi_{n,l})-\varphi(\xi_{l,l}-\xi_{n,l})=\varphi(d_l)-\varphi(d_{l+1})-\varphi(\xi_{l-1,l-1}-\xi_{n,l})$$

for determining $\xi_{n,l}$. But from (γ_l) it follows that

$$\varphi(d_l)-\varphi(\xi_{l-1,l-1}-\xi_{l,l-1})=\varphi(\xi_{l,l}).$$

Therefore the preceding equation can be written in the form

$$\varphi(\xi_{n,l})-\varphi(\xi_{l,l}-\xi_{n,l})=\varphi(\xi_{l,l})-\varphi(d_{l+1}). \qquad (3)$$

We will show that

$$|\varphi(\xi_{l,l})-\varphi(d_{l+1})|\leqslant\varphi(\xi_{l,l}).$$

To this end we note that by virtue of our assumption and (β_{kj}) and (γ_l) we have

$$2\varphi(\xi_{l,l})=\varphi(d_1)+\varphi(d_l)-\sum_{j=1}^{l-1}\varphi(\xi_{l,j})-\varphi(\xi_{l-1,l-1}-\xi_{l,l-1})$$

$$=\varphi(d_1)+\varphi(d_l)-\sum_{j=1}^{l-2}\varphi(\xi_{l,j})-\varphi(\xi_{l,l-1})-\varphi(\xi_{l-1,l-1}-\xi_{l,l-1}).$$

By condition 4) we have

$$\varphi(\xi_{l,l-1})+\varphi(\xi_{l-1,l-1}-\xi_{l,l-1})\leqslant\varphi(\xi_{l-1,l-1}).$$

Consequently

$$2\varphi(\xi_{l,l})\geqslant\varphi(d_1)+\varphi(d_l)-\sum_{j=1}^{l-2}\varphi(\xi_{l,j})-\varphi(\xi_{l-1,l-1}).$$

Since $l\le n-1$, by our assumption we have

$$\xi_{l,1}=\xi_{l-1,1},\ \ldots,\ \xi_{l,l-2}=\xi_{l-1,l-2}.$$

Therefore

$$2\varphi(\xi_{l,l})\geqslant\varphi(d_l)+\varphi(d_1)-\sum_{j=1}^{l-1}\varphi(\xi_{l,j})=\varphi(d_l)\geqslant\varphi(d_{l+1});$$

that is,

$$\varphi(\xi_{l,l})-\varphi(d_{l+1})\geqslant-\varphi(\xi_{l,l}).$$

Since $|\phi(\xi_{l,l})-\phi(d_{l+1})|\le\phi(\xi_{l,l})$, we know that equation (3), by virtue of the same arguments as for equation (2), has one and only one nonnegative solution $\xi_{n,l}$ such that $0\le\xi_{n,l}\le\xi_{l,l}$. We note that equation (3) for determining $\xi_{n,l}$ coincides with the equation derived in the same way from (γ_k) for determining $\xi_{k,l}$ for an arbitrary $k=l+1,\ldots,n-1$. Consequently

$$\xi_{n,l}=\xi_{n-1,l}=\ \ldots\ =\xi_{l+1,l}\leqslant\xi_{l,l}. \qquad (\beta_{nl})$$

Thus in a step-by-step fashion we have uniquely determined for a given n the values $\xi_{n,1},\ldots,\xi_{n,n-1}$ which satisfy the conditions (β_{nj}) for $j=1,\ldots,n-1$. By virtue of these conditions we have

$$\sum_{j=1}^{n-1}\varphi(\xi_{n,j})=\sum_{j=1}^{n-2}\varphi(\xi_{n-1,j})+\varphi(\xi_{n,n-1})\leqslant\sum_{j-1}^{n-1}\varphi(\xi_{n-1,j})=\varphi(d_1).$$

Thus it is clear that there must exist one and only one nonnegative value of $\xi_{n,n}$ which together with the previously determined values of $\xi_{n,1},\ldots,\xi_{n,n-1}$ satisfies the system (α_n). Here the equation $\xi_{n,n}=0$ is equivalent to the equation $\xi_{n,n-1}=\xi_{n-1,n-1}$. Since the assertion in the lemma has already been proved for $n=3$, the lemma is now proved for an arbitrary natural number n.

From the lemma it follows immediately that Theorem 1 is true.

§4. A theorem on the isometric mapping of spaces $R(d;\ R)$ into L_p

If we consider a function $\phi(t)=|t|^p$, $1\le p<\infty$, which satisfies conditions 1)–4), then by Theorem 1 we know that for each sequence of numbers $d_1\ge d_2\ge\cdots\ge 0$ and each $p\ge 1$ there exists a nonnegative infinite numerical matrix $\{\xi_{i,j}\}$, $i=1,2,\ldots;\ j=1,\ldots,i$, satisfying the system of equations

$$\left(\sum_{j=1}^{k} \xi_{k,j}^{p}\right)^{1/p} = d_1,$$

$$\left(\sum_{j=1}^{l-1} |\xi_{l-1,j} - \xi_{k,j}|^p + \sum_{j=l}^{k} \xi_{k,j}^{p}\right)^{1/p} = d_l \qquad (l = 2, 3, \ldots, k)$$

for all natural numbers k. This means that if $\{\xi_{i,j}\}$ is such a matrix, then the sequence of points $a_{i+1}(\xi_{i,1}, \ldots, \xi_{i,i}, 0, \ldots, 0)$ in the real l_p-space $(p \geq 1)$ together with the zero point $a_1(0, 0, \ldots)$ is an ultrametric sequence $A(d; l_p)$.

For an arbitrary given sequence of positive numbers μ_j, $j = 1, 2, \ldots$, we consider the space $l_p(u)$, $p \geq 1$, of sequences $\xi = \{\xi_j\}_1^\infty$ with the norm

$$\|\xi\| = \left(\sum_{j=1}^{\infty} |\xi_j|^p \mu_j\right)^{1/p} < \infty.$$

In this more general case the sequence of points

$$a_{i+1}(\xi_{i,1}\mu_1^{-1/p}, \ \ldots, \ \xi_{i,i}\mu_i^{-1/p}, \ 0, \ 0, \ 0, \ \ldots) \qquad (i = 1, 2, \ldots)$$

for the same matrix $\{\xi_{i,j}\}$ together with the zero point $a_1(0, 0, \ldots)$ is an ultrametric sequence $A(d; l_p(\mu))$.

Let (Y, S, μ) be a measure space, and let $L_p(\mu)$ $(p \geq 1)$ be the normed linear space of all measurable functions $F(y)$, $y \in Y$, such that

$$\|F\|_p = \left\{\int_Y |F(y)|^p \, d\mu\right\}^{1/p} < \infty.$$

We will assume that Y contains an infinite collection of disjoint measurable subsets of positive measure. Then we have the following result.

Theorem 2. *Each countable ultrametric space $R(d; R)$ can be mapped isometrically into $L_p(\mu)$, $p \geq 1$.*

If $p = 1$ or $p = 2$, we can describe a simple algorithm for the effective construction of an ultrametric sequence $A(d; l_p)$ for a given sequence of numbers $d_1 \geq d_2 \geq \cdots \geq 0$.

For $p = 1$ the sequence $\{a_i\}$ $(i = 2, 3, \ldots)$ consists of the points

$$a_2(d_1, \ 0, \ 0, \ 0, \ \ldots), \ \ldots,$$

$$a_i\left(d_1 - \frac{d_2}{2}, \ \frac{d_2 - d_3}{2}, \ldots, \ \frac{d_{i-2} - d_{i-1}}{2}, \ \frac{d_{i-1}}{2}, \ 0, \ 0, \ 0, \ \ldots\right) \ (i = 3, 4, \ldots).$$

For $p = 2$ the sequence $\{a_i\}$ $(i = 2, 3, \ldots)$ is of the form

$$a_2(d_1, \quad 0, \ 0, \ 0, \ \ldots),$$

$$a_3\left(d_1 - \frac{d_2^2}{2d_1}, \ \sigma_2, \ 0, \ \ldots\right),$$

$$a_4\left(d_1-\frac{d_2^2}{2d_1},\ \sigma_2-\frac{d_3^2}{2\sigma_2},\ \sigma_3,\ 0,\ 0,\ 0,\ \ldots\right),$$
$$a_5\left(d_1-\frac{d_2^2}{2d_1},\ \sigma_2-\frac{d_3^2}{2\sigma_2},\ \sigma_3-\frac{d_4^2}{2\sigma_3},\ \sigma_4,\ 0,0,0,\ldots\right),\cdots,$$

where the values of $\sigma_2, \sigma_3, \ldots$ are determined successively by the condition

$$\|a_m-a_n\|_{l_2}=\max(d_m,\ d_n)\qquad (m\neq n).$$

§5. The connection with linear dimensionality

If for some value of n we have $d_i = 0$ for $i > n$, then the ultrametric space $R(d;\ R)$ consists of n points, and, as we see from our previous arguments, it can be mapped isometrically into $(n-1)$-dimensional euclidean space.

We define the dimension $\dim R(d;\ R)$ of the ultrametric space $R(d;\ R)$ as the smallest value of n for which the space can be mapped isometrically into n-dimensional euclidean space, and we obtain the following result.

Theorem 3. *For each ultrametric space $R(d;\ R)$ which consists of a finite number of points, we have*

$$\dim R(d;\ R)=\operatorname{card} R-1.$$

In the case where $\rho(x, y) = 1$ for each pair of distinct points $x, y \in R$ and $\rho(x, x) = 0$, the equation in Theorem 3 was proved in [4], where an analogous definition was made for the dimension of a finite graph.

We note that, in contrast to an n-dimensional euclidean space, an n-dimensional space with the metric of l_1 or l_∞ contains ultrametric sets consisting of more than $n+1$ points.

The proof of Theorem 3 depends on the following lemma.

Lemma 2. *If $\phi(t)$ is a function such that in condition 4) it is the case that for all $t', t'' > 0$ the strict inequality holds, then each solution $\{\xi_{i,j}\}$ $(i = 1, \ldots, k;\ j = 1, \ldots, i)$ of the collection of systems of equations (α_k), $k = 1, 2, \ldots,$ satisfies the conditions*

$$\operatorname{sgn}\xi_{\nu,j}=\operatorname{sgn}\xi_{j,j}\qquad (1<\nu\leqslant k),$$
$$|\xi_{j+1,j}|=|\xi_{j+2,j}|=|\xi_{j+3,j}|=\ldots=|\xi_{\nu,j}|\leqslant|\xi_{j,j}|\qquad (j<k). \qquad (\beta_{kj}^*)$$

Proof of Lemma 2. The equation $\phi(\xi_{1,1}) = \phi(d_1)$ has two solutions $\xi_{1,1} = d_1$ and $\xi_{1,1} = -d_1$. The properties (β_{21}^*) and (β_{31}^*) in each of these two cases follow immediately from the considerations in §3 if we assume that $\phi(t)$ is an even function.

The property (β_{32}^*) in the case where $d_3 = 0$ becomes obvious if we revert to the system (α_3). For $d_3 > 0$ the difference $\phi(\xi_{2,2}) - \phi(d_3)$ always satisfies the following inequality by virtue of Lemma 2, as we see from the arguments in §3:

$$|\varphi(\xi_{2,2}) - \varphi(d_3)| < \varphi(\xi_{2,2}).$$

Therefore, if $\phi(t)$ satisfies the hypothesis of Lemma 2, the value of $\xi_{3,2}$, which is uniquely determined by equation (2), must satisfy condition (β^*_{32}).

Now suppose that for all $k \leq n-1$ the assertion in Lemma 2 holds. We will show that an arbitrary triangular matrix $\{\xi_{i,j}\}$ $(i = 1, \ldots, n;\ j = 1, \ldots, i)$ which satisfies the collection of systems of equations (α_k) for $k = 1, \ldots, n$ has property (β^*_{nj}). From our assumption and condition (β^*_{kj}) for each $k = 1, \ldots, n-1$ it follows that the system (α_k) can be written in the form (γ_k). We consider the system (α_n), in which the values $\{\xi_{i,j}\}$ $(i = 1, \ldots, n-1;\ j = 1, \ldots, i)$ constitute one of the solutions of the collection of systems (α_k) for $k = 1, \ldots, n-1$. From this system, as in §3, we obtain an equation for $\xi_{n,1}$ which coincides with the equation

$$\varphi(\xi_{k,1}) - \varphi(\xi_{1,1} - \xi_{k,1}) = \varphi(d_1) - \varphi(d_2),$$

from which we already uniquely determined the value of $\xi_{k,1}$ for each $k = 2, 3, \ldots, n-1$. Consequently we must have

$$\xi_{n,1} = \xi_{n-1,1} = \ldots = \xi_{2,1},$$

and by virtue of (β^*_{21}) the condition (β^*_{n1}) is satisfied. By virtue of this condition, as in §3, we obtain from (α_n) the equation

$$\varphi(\xi_{n,2}) - \varphi(\xi_{2,2} - \xi_{n,2}) = \varphi(\xi_{2,2}) - \varphi(d_3),$$

which must be satisfied by $\xi_{n,2}$. Since this equation coincides with the equation for determining $\xi_{3,2}$, we have

$$\xi_{n,2} = \xi_{n-1,2} = \ldots = \xi_{3,2},$$

and by virtue of (β^*_{32}) the condition (β^*_{n2}) is satisfied.

Suppose that for the above value of n and for some l $(3 \leq l \leq n-1)$ we have already verified (β^*_{nj}) for $j = 1, \ldots, l-1$. The arguments in §3 show that $\xi_{n,l}$ must satisfy (3). This equation must also be satisfied by $\xi_{k,l}$ for an arbitrary $k = l+1, \ldots, n-1$.

If $d_{l+1} = 0$, then by virtue of the assumptions in Lemma 2 the unique solution of (3) is $\xi_{n,l} = \xi_{l,l}$. Thus

$$\xi_{l+1,l} = \xi_{l+2,l} = \xi_{l+3,l} = \ldots = \xi_{n,l},$$

and the condition (β^*_{nl}) holds. For $d_{l+1} > 0$ the difference $\phi(\xi_{l,l}) - \phi(d_{l+1})$, by virtue of the considerations in §3, satisfies the inequality

$$|\varphi(\xi_{l,l}) - \varphi(d_{l+1})| < \varphi(\xi_{l,l}).$$

Therefore under the hypothesis of Lemma 2 the value of $\xi_{n,l}$ is always uniquely determined by equation (3), and thanks to property 3) it satisfies the inequality $|\xi_{n,l}| \le |\xi_{l,l}|$; that is, the condition (β^*_{nl}) holds also in the case where $d_{l+1} > 0$.

Thus we have shown that if the assertion in Lemma 2 holds for all $k \le n-1$, then the values of $\xi_{n,1}, \dots, \xi_{n,n-1}$ in an arbitrary matrix $\{\xi_{i,j}\}$ $(i = 1, \dots, n;$ $j = 1, \dots, i)$, which satisfies the collection of systems of equations (α_k) for $k = 1, \dots, n$, must satisfy condition (β^*_{nj}). Since for $k = 3$ the condition (β^*_{kj}) has already been verified, it also holds for an arbitrary natural number k.

Proof of Theorem 3. In order to prove Theorem 3 we have to show that if some system of points $a_1, \dots, a_{n+1}$ in an $(n-1)$-dimensional euclidean space l_2^{n-1} constitutes an ultrametric set $A(d;\, l_2^{n-1})$, then $d_n = 0$; that is, $a_{n+1} = a_n$. For $n = 2$ this is obvious. Suppose that the assertion has been proved for all euclidean spaces of dimension less than $n-1$ $(n \ge 2)$. Then for $2 \le k \le n$ the system of vectors $a_i - a_1$ $(i = 2, 3, \dots, k)$ is linearly independent. Therefore if we orthogonalize these vectors we can find in the space l_2^{n-1} an orthonormal basis $e_1, \dots, e_{n-1}$ such that

$$a_{i+1} = \sum_{j=1}^{i} \xi_{i,j} e_j$$

for all $i = 1, \dots, n-1$, and

$$a_{n+1} = \sum_{j=1}^{n-1} \xi_{n,j} e_j,$$

where a_1 serves as the zero of the space.

Since the system of points a_i $(i = 1, \dots, n+1)$ constitutes an ultrametric set $A(d;\, l_2^{n-1})$, the matrix $\{\xi_{i,j}\}$ $(i = 1, 2, 3, \dots, n;\ j = 1, \dots, i;\ \xi_{n,n} = 0)$ satisfies the system of equations (α_k), where $\phi(t) = t^2$, for an arbitrary natural number $k \le n$. By virtue of Lemma 2 condition (β^*_{kj}) holds for all $k \le n$. Therefore

$$d_1^2 = \sum_{i=1}^{n-1} \xi_{n,j}^2 = \sum_{j=1}^{n-2} \xi_{n-1,j}^2 + \xi_{n,n-1}^2 \leqslant \sum_{j=1}^{n-1} \xi_{n-1,j}^2 = d_1^2.$$

Consequently $\xi_{n,n-1}^2 = \xi_{n-1,n-1}^2$. Since condition $(\beta^*_{n,n-1})$ holds, we have $\xi_{n,n-1} = \xi_{n-1,n-1}$. Since according to (β^*_{kj}) we have $\xi_{n,j} = \xi_{n-1,j}$ for $j = 1, \dots, n-2$, we have $a_{n+1} = a_n$.

By carrying these arguments further we could show that the equation established in Theorem 3 is true for all ultrametric spaces R.

§6. Application to the inverse problem for the ϵ-entropy of compact sets

In this section we apply our fundamental result to the solution of the inverse problem for the ϵ-entropy of compact sets in L_p-spaces.

Suppose that $\overline{X}$ is a metric space, A is a compact subset of $\overline{X}$, and $H_\epsilon^{\overline{X}}(A)$, $H_\epsilon(A)$ and $C_\epsilon(A)$ respectively are the ϵ-entropy of A with respect to $\overline{X}$, the absolute ϵ-entropy of A, and the ϵ-capacity of A in the sense of Kolmogorov (see [5] and [6]). For an arbitrary nonempty compact set A and for all $\epsilon > 0$ the functions $H_\epsilon^A(A)$, $H_\epsilon(A)$ and $C_\epsilon(A)$ have the following properties: 1) they are not negative; 2) they do not increase; 3) they vanish beginning with some $\epsilon > 0$; 4) they are continuous from the right for all $\epsilon > 0$. It is natural that functions $f(\epsilon)$ which satisfy conditions 1)–4) and for which $2^{f(\epsilon)}$ is always an integer are called *entropy functions.* The inverse problem for ϵ-entropy (see [7], p. 690) consists in deciding whether in an infinite-dimensional Banach space $\overline{X}$ there exist compact sets A whose ϵ-entropy or ϵ-capacity for all $\epsilon > 0$ coincides with a given entropy function $f(\epsilon)$. For the space C this question was answered affirmatively for $H_\epsilon^C(A)$ in [8], and for $H_\epsilon^A(A)$, $H_\epsilon(A)$ and $C_\epsilon(A)$, in [9]. But if $\overline{X} = l_2$, then for $H_\epsilon^{\overline{X}}(A)$ the situation is different (see [8]). By means of Theorem 2 we can show that for $H_\epsilon^A(A)$, $H_\epsilon(A)$ and $C_\epsilon(A)$ the inverse problem has an affirmative solution for all L_p-spaces $(p \geq 1)$.

Theorem 4. *If* (Y, S, μ) *is a measure space such that* Y *contains infinitely many disjoint subsets of positive measure, then for any entropy function* $f(\epsilon)$ *there exists in the space* $L_p(\mu)$ $(p \geq 1)$ *a compact set* A *such that*

$$H_\epsilon^A(A) = C_\epsilon(A) = H_\epsilon(A) = f(\epsilon)$$

for all $\epsilon > 0$.

If $p = \infty$, then, since C is universal, this result is equivalent to a result of Kreĭnovič in [9] which proves the existence of a metric compact set A for which the above equation holds. Kreĭnovič noted that this property is possessed by the following compact set. To each point of discontinuity ϵ_i $(i = 1, 2, \dots)$ of the function $f(\epsilon)$ we associate a system of elements a_i^j, $1 \leq j \leq \Delta_i$, where

$$\Delta_i = 2^{f(\varepsilon_i)} - 2^{f(\varepsilon_i - 0)} \quad (\varepsilon_1 > \varepsilon_2 > \varepsilon_3 > \dots).$$

The desired compact set A is the set of all elements a_i^j which are complementary to some one element a_∞^0, using the metric

$$\rho(a_i^j, a_k^l) = \begin{cases} \varepsilon_i, & k > i, \\ \varepsilon_i, & k = i,\ j \neq l, \\ 0, & k = i,\ j = l. \end{cases}$$

Since this compact set is an ultrametric set of the type $R(d; R)$, the investigation of the inverse problem for ϵ-entropy in L_p, when $p \geq 1$ is finite, is connected with the formulation of the question presented in §2, and Theorem 2 enables us to assert the validity of Theorem 4.

BIBLIOGRAPHY

1. J. Dieudonné, *Foundations of modern analysis*, Academic Press, New York, 1960. MR 22 #11074.

2. F. Hausdorff, *Mengenlehre*, 3rd ed., de Gruyter, Berlin, 1937; English transl., Chelsea, New York, 1957. MR 19, 111.

3. A. Weil, *Basic number theory*, Springer-Verlag, Berlin and New York, 1967. MR 38 #3244.

4. P. Erdös, F. Harary and R. T. Tutte, *On the dimension of a graph*, Mathematika 12 (1965), 118–122. MR 32 #5537.

5. A. N. Kolmogorov, *On certain asymptotic characteristics of totally bounded metric spaces*, Dokl. Akad. Nauk SSSR 108 (1956), 385–388. (Russian) MR 18, 324.

6. A. N. Kolmogorov and V. M. Tihomirov, *ε-entropy and ε-capacity of sets in functional spaces*, Uspehi Mat. Nauk 14 (1959), no. 2 (86), 3–86; English transl., Amer. Math. Soc. Transl. (2) 17 (1961), 277–364. MR 22 #2890; 23 #A2031.

7. A. F. Timan, *Some investigations in the theory of approximation of functions of a real variable*, Proc. Fourth All-Union Math. Congr. (Leningrad, 1961), vol. II, "Nauka", Leningrad, 1964, pp. 683–693. (Russian) MR 37 #4463.

8. S. B. Babadžanov, *Diameters of compact sets in Banach spaces*, Dissertation, Moscow State Univ., Moscow, 1967. (Russian)

9. V. Ja. Kreĭnovič, *On the inverse problem of A. F. Timan for the ε-entropy of compacta*, Uspehi Mat. Nauk 30 (1975), no. 1 (181), 241–242. (Russian)

Translated by F. A. CEZUS

Trudy Mat. Inst. Steklov.
134 (1975)

Proc. Steklov Inst. Math.
134 (1975)

ON VARIOUS FORMS OF CONVERGENCE IN THE CLASSES $\phi(L)$

UDC 517:52

P. L. UL'JANOV

Abstract. An investigation is carried out of the function classes $\phi(L)$ and the topological spaces U_ϕ constructed with respect to them. An interdependence is established between the following forms of convergence: with respect to ϕ-distance, with respect to the topology of U_ϕ and with respect to the norm of the Orlicz space L^*_ϕ. An equivalent definition of the convergence of a sequence $\{f_i\}$ with respect to the topology of U_ϕ is found in terms of the behavior of corresponding integrals of $\phi(f_i - g)$. The class of convergent sequences in the topology of U_ϕ is completely described for the case when ϕ satisfies the condition $\phi(t+1) \neq O\{\phi(t)\}$ for $t \to \infty$.

Bibliography: 8 titles.

§1. Introduction

Let Φ be the totality of all finite, even, nonnegative and nondecreasing on $[0, \infty)$ functions $\phi(t)$ satisfying the conditions $\phi(0) = 0$ and

$$\lim_{t\to\infty} \varphi(t) = \varphi(\infty) = \infty.$$

If $\phi(t) \in \Phi$, let $\phi(L)$ denote the set of all measurable functions f on $[0, 1]$ for which $\int_0^1 \phi(f(x))dx < \infty$.

For any two functions $f_1(x)$ and $f_2(x)$ from $\phi(L)$ the not necessarily finite quantity

$$\rho_\varphi(f_1, f_2) \equiv \int_0^1 \varphi(f_1(x) - f_2(x))\,dx$$

is called the *generalized ϕ-distance* between them. We note that the function $\rho_\phi(f_1, f_2)$ does not necessarily satisfy the well known axioms of a metric space (see, for example, [1], Chapter 4); more precisely, it need not satisfy the triangle inequality.

A sequence $\{f_n\}_1^\infty$ with $f_n \in \phi(L)$, where $\phi \in \Phi$, is said to *converge in ϕ-distance* to an element $F \in \phi(L)$ if $f_n - F \in \phi(L)$ for $n \geq N$, where N is some natural number, and

---

AMS (MOS) subject classifications (1970). Primary 46E30, 54C35.

$$\lim_{n\to\infty} \rho_\varphi(f_n, F) = 0.$$

Further, let

$$A_\varphi(F, \varepsilon) \equiv \{f : f \in \varphi(L),\ \rho_\varphi(f, F) < \varepsilon\}$$

be an ϵ-neighborhood of an element $F \in \phi(L)$, where ϵ is a positive number.

The collection of neighborhoods $A_\phi(F, \epsilon)$ (for arbitrary numbers $\epsilon > 0$ and elements $F \in \phi(L)$) may be used as a subbase to generate a topology for $\phi(L)$ and thereby define a topological space U_ϕ. A more detailed discussion of this point can be found in [5] and [6] (see also [1], Chapter 1 and [2], Chapter II, §5). We have previously established (see [4]–[6]) some approximation and topological properties of $\phi(L)$ and U_ϕ. In particular, we have found necessary and sufficient conditions for the case when $\phi(L)$ is separable relative to ϕ-distance, the case when the topological space U_ϕ is separable, and the case when the set

$$\varphi^+(L) = \{f : f \in \varphi(L),\ f(x) \geqslant 0 \quad \text{for} \quad x \in [0, 1]\}$$

is separable relative to ϕ-distance. In [6] we sketched the historical background of the above line of investigation.

We note that one can replace the interval $[0, 1]$ by a measurable set E in the above definitions and thereby obtain analogous notions of ϕ-distance and of a topological space $U_\phi(E)$. In those parts of the analysis where the set E is not explicitly indicated, one should put $E = [0, 1]$.

The present article is a continuation of [4]–[6]. The main contents of this article were published without proof in [7].

§2. Auxiliary assertions

In this section we establish a series of lemmas needed for the proofs of the theorems.

Lemma 1. *Suppose $\phi \in \Phi$ and $\phi(+0) = 0$. If*

$$\varphi(t_0 + 0) > \varphi(t_0) \quad \textit{for some} \quad t_0 > 0, \tag{1}$$

then there exist functions $F \in \phi(L)$ and $f_n \in \phi(L)$ $(n = 1, 2, \ldots)$ such that

$$\lim_{n\to\infty} \int_0^1 \varphi(F - f_n)\, dt = 0, \tag{2}$$

whereas

$$\int_0^1 \varphi(f_n)\, dt \geqslant A + \int_0^1 \varphi(F)\, dt \qquad \textit{for all} \quad n = 1, 2, \ldots, \tag{3}$$

where $A = \phi(t_0 + 0) - \phi(t_0) > 0$.

Proof. We put $F(t) \equiv t_0$ and $f_n(t) = t_0 + 1/n$, $n = 1, 2, \ldots$. Clearly, the

f_n $(n = 1, 2, \ldots)$ and F belong to $\phi(L)$. Since $\phi(+0) = 0$, we have

$$\int_0^1 \varphi(F - f_n)\,dt = \int_0^1 \varphi\left(\frac{1}{n}\right)dt \to 0 \quad \text{for } n \to \infty,$$

i.e. (2) is satisfied. And since $\phi(t)$ is nondecreasing on $[0, \infty)$ and condition (1) holds, we have

$$\int_0^1 \varphi(f_n)\,dt = \int_0^1 \varphi\left(t_0 + \frac{1}{n}\right)dt \geqslant \varphi(t_0 + 0) = \varphi(t_0 + 0) - \varphi(t_0) + \int_0^1 \varphi(F)\,dt,$$

i.e. (3) is valid for all $n = 1, 2, \ldots$. Q.E.D.

Lemma 2. *Suppose $\phi \in \Phi$ and $\phi(+0) = 0$. If*

$$\varphi(t_0 - 0) < \varphi(t_0) \qquad \textit{for some} \quad t_0 > 0, \tag{4}$$

then there exist functions $F \in \phi(L)$ and $f_n \in \phi(L)$ $(n = 1, 2, \ldots)$ for which

$$\lim_{n\to\infty} \int_0^1 \varphi(F - f_n)\,dt = 0, \tag{5}$$

whereas

$$\int_0^1 \varphi(f_n)\,dt \leqslant -B + \int_0^1 \varphi(F)\,dt \qquad \textit{for all} \quad n \geqslant 1, \tag{6}$$

where $B = \phi(t_0) - \phi(t_0 - 0) > 0$.

Proof. Let n_0 be a natural number such that $n_0 > 1/t_0$. We put

$$F(t) \equiv t_0, \quad f_n(t) = t_0 - \frac{1}{n + n_0} \qquad (n = 1, 2, \ldots).$$

Since $\phi(+0) = 0$,

$$\int_0^1 \varphi(F - f_n)\,dt = \int_0^1 \varphi\left(\frac{1}{n + n_0}\right)dt \to 0 \qquad \text{for } n \to \infty,$$

i.e. (5) is satisfied. And since $\phi(t)$ is nondecreasing on $[0, \infty)$ and $1/(n + n_0) < t_0$ for $n = 1, 2, \ldots$ (see also (4)), we have

$$\int_0^1 \varphi(f_n)\,dt = \int_0^1 \varphi\left(t_0 - \frac{1}{n + n_0}\right)dt \leqslant \varphi(t_0 - 0) = [\varphi(t_0 - 0) - \varphi(t_0)] + \int_0^1 \varphi(F)\,dt,$$

i.e. (6) is satisfied. Q.E.D.

Lemma 3. *Suppose $\phi \in \Phi$ and $\phi(\eta) = 0$ for some $\eta > 0$. Then there exist functions F, Ψ, f_n, $\psi_n \in \phi(L)$ for $n \geq 1$ such that*

$$\lim_{n\to\infty} \int_0^1 \varphi(F - f_n)\,dt = \lim_{n\to\infty} \int_0^1 \varphi(\Psi - \psi_n)\,dt = 0, \tag{7}$$

whereas

$$\int_0^1 \varphi(f_n)\,dt \geqslant A + \int_0^1 \varphi(F)\,dt,$$
$$\int_0^1 \varphi(\psi_n)\,dt \leqslant -A + \int_0^1 \varphi(\Psi)\,dt \qquad (n = 1, 2, \ldots), \tag{8}$$

where A is a certain positive number.

The proof of this lemma (like Lemmas 1 and 2) is almost obvious. We put

$$\alpha = \sup_{\varphi(t)=0} \{t\}, \quad F(t) \equiv \frac{\alpha}{2}, \quad f_n(t) \equiv \frac{5}{4}\alpha, \quad \Psi(t) \equiv \frac{5}{4}\alpha,$$
$$\psi_n(t) \equiv \frac{\alpha}{2} \qquad (n = 1, 2, \ldots).$$

Clearly, $\alpha \geq \eta > 0$ and

$$\int_0^1 \varphi(F - f_n)\,dt = \int_0^1 \varphi(\Psi - \psi_n)\,dt = \int_0^1 \varphi\left(\frac{3}{4}\alpha\right)dt = 0 \qquad (n = 1, 2, \ldots),$$

i.e. (7) is satisfied. On the other hand,

$$\int_0^1 \varphi(f_n)\,dt = \int_0^1 \varphi\left(\frac{5}{4}\alpha\right)dt \geqslant \varphi\left(\frac{5}{4}\alpha\right) + \int_0^1 \varphi(F)\,dt,$$

and

$$\int_0^1 \varphi(\psi_n)\,dt = 0 \leqslant -\varphi\left(\frac{5}{4}\alpha\right) + \int_0^1 \varphi\left(\frac{5}{4}\alpha\right)dt = -\varphi\left(\frac{5}{4}\alpha\right) + \int_0^1 \varphi(\Psi)\,dt,$$

i.e. (8) is satisfied if one puts $A = \phi(5\alpha/4)$. The number A is positive by virtue of the choice of α. Q.E.D.

Lemma 4. *Suppose $\phi \in \Phi$ and*

$$\varphi(2t) \neq O\{\varphi(t)\} \quad \text{for} \quad t \to \infty. \tag{9}$$

Then there exist functions $F \in \phi(L)$ and $f_n \in \phi(L)$ for $n \geq 1$ such that

$$\lim_{n\to\infty} \int_0^1 \varphi(F - f_n)\,dt = 0, \tag{10}$$

whereas

$$\lim_{n\to\infty} \int_0^1 \varphi(f_n)\,dt = \infty. \tag{11}$$

Proof. Condition (9) implies the existence of a sequence $\{t_k\}_1^\infty$ such that

$$0 < t_k < t_{k+1}, \quad \varphi(t_k) > 2^{k+1}, \quad \varphi(2t_k) > 2^k\varphi(t_k) \quad \text{for } k = 1, 2, \ldots. \tag{12}$$

We put

$$c_k = 2^{-k}, \quad d_k = c_k + \frac{1}{k^2\varphi(t_k)} \quad (k = 1, 2, \ldots), \tag{13}$$

$$F(t) = \begin{cases} t_k & \text{for} \quad t \in [c_k, d_k] \quad (k \geqslant 1), \\ 0 & \text{for} \quad t \in [0, 1] - \sum_{k=1}^{\infty} [c_k, d_k] \end{cases} \tag{14}$$

and (for each $n = 1, 2, \ldots$)

$$f_n(t) = \begin{cases} 2t_n & \text{for} \quad t \in [c_n, d_n] \\ F(t) & \text{for} \quad t \in [0, 1] - [c_n, d_n]. \end{cases} \tag{15}$$

Clearly the intervals $[c_k, d_k]$ are mutually disjoint (see (12) and (13)), and hence definitions (14) and (15) are correct.

Since $\phi(0) = 0$ (see (13)–(15)), we have

$$\begin{aligned} \int_0^1 \varphi(f_n(t))\,dt &= \int_{c_n}^{d_n} \varphi(2t_n)\,dt + \sum_{\substack{k=1 \\ k\neq n}}^{\infty} \int_{c_k}^{d_k} \varphi(t_k)\,dt \\ &= \frac{\varphi(2t_n)}{n^2\varphi(t_n)} + \sum_{\substack{k=1 \\ k\neq n}}^{\infty} \varphi(t_k)\frac{1}{k^2\varphi(t_k)} = \frac{\varphi(2t_n)}{n^2\varphi(t_n)} + \sum_{\substack{k=1 \\ k\neq n}}^{\infty} \frac{1}{k^2}. \end{aligned} \tag{16}$$

It follows for every fixed n that $f_n(t) \in \phi(L)$. Further, since $\phi(t)$ is nondecreasing and $0 \leq F(t) \leq f_n(t)$ (see (14) and (15)) for all $t \in [0, 1]$, we have $F \in \phi(L)$. It therefore also follows from (16) (see (12)) that

$$\lim_{n\to\infty} \int_0^1 \varphi(f_n)\,dt = \infty,$$

i.e. (11) is valid.

On the other hand (see (13)–(15)),

$$\int_0^1 \varphi(F - f_n)\,dt = \int_{c_n}^{d_n} \varphi(-t_n)\,dt = \frac{1}{n^2} \to 0 \quad \text{for} \quad n \to \infty,$$

i.e. equality (10) is satisfied. Q.E.D.

We note that in Lemmas 1–4 the functions $f - f_n$ belong to $\phi(L)$ for all $n \geq 1$.

Lemma 5. *Suppose $\phi \subset \Phi$, $\phi(+0) = 0$ and the following conditions are satisfied:*

a) *$\phi(t)$ is continuous on $(0, \infty)$,*

b) *$\phi(t) > 0$ for all $t > 0$,* (17)

c) *$\phi(2t) = O\{\phi(t)\}$ for $t \to \infty$.*

If $f \in \phi(L)$ and $f_n \in \phi(L)$ for $n \geq 1$, then the equality

$$\lim_{n\to\infty} \int_0^1 \varphi(f - f_n)\,dt = 0 \tag{18}$$

implies the equality

$$\lim_{n\to\infty}\int_0^1 \varphi(f_n)\,dt = \int_0^1 \varphi(f)\,dt. \tag{19}$$

Proof. We will first establish that the family of integrals

$$\left\{\int_E \varphi(f_n(t))\,dt\right\}_{n=1}^{\infty} \tag{20}$$

is equicontinuous relative to measurable sets $E\subset[0,1]$.

Condition c) of (17) implies that $\phi(2t)\le C\phi(t)$ for $t\ge t_0$, where t_0 is some positive number. But then, if $\eta>0$, we have (see condition b) of (17))

$$\varphi(2t)\leqslant\left(C+\frac{\varphi(2t_0)}{\varphi(\eta)}\right)\varphi(t)\equiv A_\eta\varphi(t)\quad\text{for}\quad t\geqslant\eta$$

and therefore, if $\alpha\ge 0$ and $\beta\ge 0$,

$$\varphi(\alpha+\beta)\leqslant A_\eta(\varphi(\alpha)+\varphi(\beta))\qquad\text{for all }\{\alpha,\ \beta\}\geqslant\eta. \tag{21}$$

We take an arbitrary $\epsilon>0$ and find (since $\phi(+0)=0$) a number $\eta=\eta(\epsilon)>0$ such that

$$\varphi(2\eta)<\frac{\varepsilon}{3}. \tag{22}$$

We next choose (see (18)) a natural number $N=N(\eta)=N(\eta(\epsilon))$ such that

$$\int_0^1 \varphi(f-f_n)\,dt<\frac{\varepsilon}{3A_\eta}\qquad\text{for}\quad n\geqslant N. \tag{23}$$

Finally, we find a number $\delta=\delta(\epsilon)>0$ such that

$$\int_E \varphi(f)\,dt<\frac{\varepsilon}{3A_\eta},\quad \int_E \varphi(f_n)\,dt<\varepsilon\quad\text{for}\quad 1\leqslant n\leqslant N\text{ and }|E|<\delta. \tag{24}$$

For a set $E\subset[0,1]$ with measure $|E|<\delta$ let

$$E_1=E\cap B\cap D_n,\qquad E_2=E\cap\{([0,1]-B)\cup([0,1]-D_n)\},$$

where

$$B=\{t:t\in[0,1],\ |f(t)|<\eta\},\qquad D_n=\{t:t\in[0,1],\ |f_n(t)-f(t)|<\eta\}.$$

Then (see (21), (22), (24) and (23)) for a set E with $|E|<\delta$ and $n>N$ we have

$$\int_E \varphi(f_n(t))\,dt\leqslant\int_E \varphi(|f_n(t)-f(t)|+|f(t)|)\,dt\leqslant\int_{E_1}\varphi(|f-f_n|+|f|)\,dt$$

$$+\int_{E_2}\varphi(|f-f_n|+|f|)\,dt\leqslant\int_{E_1}\varphi(2\eta)\,dt+A_\eta\left\{\int_{E_2}\varphi(f(t))\,dt+\int_{E_2}\varphi(f_n(t)-f(t))\,dt\right\}$$

$$<\frac{\varepsilon}{3}+A_\eta\int_E\varphi(f(t))\,dt+A_\eta\int_0^1\varphi(f_n-f)\,dt<\frac{\varepsilon}{3}+\frac{\varepsilon}{3}+\frac{\varepsilon}{3}=\varepsilon,$$

i.e. $\int_E \phi(f_n(t))dt < \epsilon$ for $n > N$ and $|E| < \delta \equiv \delta(\epsilon)$. Thus (see also (24)) the family of integrals (20) is absolutely equicontinuous.

Let $\epsilon_n \equiv \int_0^1 \phi(f - f_n)dt$. By assumption, $\epsilon_n \to 0$ as $n \to \infty$. It follows from condition b) of (17) that for $\sigma > 0$

$$\varepsilon_n \geqslant \int_{\{t: |f(t)-f_n(t)| \geqslant \sigma\}} \varphi(f(t) - f_n(t))\,dt \geqslant \varphi(\sigma)\, m\{t: |f(t) - f_n(t)| \geqslant \sigma\},$$

where $\phi(\sigma) > 0$. Therefore

$$\overline{\lim_{n\to\infty}}\, m\{t: t \in [0, 1],\ |f(t) - f_n(t)| \geqslant \sigma\} \leqslant \overline{\lim_{n\to\infty}} \frac{\varepsilon_n}{\varphi(\sigma)} = 0,$$

i.e. the sequence $\{f_n(t)\}$ converges in measure on $[0, 1]$ to $f(t)$.

Consider now the sequence of functions $\{\phi(f_n(t))\}_{n=1}^\infty$, where $\phi(t)$ is a continuous function on the real line $(-\infty, \infty)$. This sequence has the property that every subsequence $\{\phi(f_{n_k}(t))\}_{k=1}^\infty$ of it contains a subsequence $\{\phi(f_{n_{k_i}}(t))\}_{i=1}^\infty$ that converges on $[0, 1]$ to $\phi(f(t))$. This follows from the fact that $\{f_{n_k}(t)\}_{k=1}^\infty$ converges in measure to $f(t)$, which implies the existence of a sequence $\{k_i\}$ such that $f_{n_{k_i}}(t) \to f(t)$ for almost all $t \in [0, 1]$ (it is then only necessary to make use of the continuity of $\phi(t)$). From what has been said it follows that every sequence $\{\phi(f_{n_k})\}_{k=1}^\infty$ contains a subsequence $\{\phi(f_{n_{k_i}})\}_{i=1}^\infty$ that converges in measure to $\phi(f)$. But under convergence in measure the space is metrizable, and it therefore follows from what has been proved that the sequence $\{\phi(f_n(t))\}_{n=1}^\infty$ converges in measure on $[0, 1]$ to $\phi(f(t))$. Since we have already proved that the family of integrals (20) is equicontinuous, it follows on the basis of Vitali's theorem (see [8], Chapter VI, §3, Theorem 4) that (19) is valid. Q.E.D.

Lemma 6. *Suppose $\phi \in \Phi$ satisfies the conditions*

a) $\phi(t + 0) = \phi(t)$ *for all* $t \geq 0$,

b) $\phi(t) > 0$ *for all* $t > 0$, (25)

c) $\phi(2t) = O\{\phi(t)\}$ *for* $t \to \infty$.

Then convergence in ϕ-distance is equivalent to convergence in the topology of U_ϕ.

Proof. It suffices for us to establish under conditions (25) that

$$f \in \varphi(L),\quad f_n \in \varphi(L) \quad (n \geqslant 1) \qquad \text{and} \qquad \lim_{n\to\infty} \int_0^1 \varphi(f - f_n)\,dt = 0 \tag{26}$$

imply the convergence of $\{f_n\}$ to f in the topology of U_ϕ.

A base B for the topology of U_ϕ is provided, in particular, by the finite intersections of neighborhoods $A_\phi(F, \epsilon)$, where $F \in \phi(L)$ and ϵ is an arbitrary positive number. It therefore suffices to prove that $\{f_n\}$ converges to f with respect to the base B if conditions (25) and (26) are satisfied.

Suppose $M = \bigcap_1^m A_\phi(F_i, \epsilon_i)$ is a member of B containing f, i.e. $f \in M$. We have previously proved (see [6], Theorem 6.2) that if conditions (25) are satisfied, then the space U_ϕ has property d, and hence the condition $f \in A_\phi(F_i, \epsilon_i)$ $(1 \le i \le m)$ implies the existence of numbers $\delta_i > 0$ $(1 \le i \le m)$ for which

$$A_\varphi(f, \delta_i) \subset A_\varphi(F_i, \varepsilon_i).$$

Let $\delta = \min_{1 \le i \le m} \delta_i$. Clearly, $\delta > 0$ and

$$A_\varphi(f, \delta) \subset M \in B. \tag{27}$$

But then, if N is a natural number such that $\int_0^1 \phi(f - f_n)dt < \delta$ for $n \ge N$ (such a number exists by virtue of conditions (26)), we have $f_n \in A_\phi(f, \delta)$ for all $n \ge N$, and a fortiori (see (27)) $f_n \in M$ for all $n \ge N$, which means that $\{f_n\}_1^\infty$ converges to f with respect to the base B. Q.E.D.

Lemma 7. *Suppose $\phi \in \Phi$ and*

a) $\phi(t + 0) = \phi(t)$ *for all* $t \in [0, \infty)$,

b) $\phi(2t) \ne O\{\phi(t)\}$ *for* $t \to \infty$. (28)

Then there exist functions f, g, $f_n \in \phi(L)$, $n \ge 1$, such that $f - g \in \phi(L)$ and

$$\rho_\varphi(f, f_n) \to 0 \quad \textit{for} \quad n \to \infty, \quad \lim_{n\to\infty} \int_0^1 \varphi(f_n)\,dt = \int_0^1 \varphi(f)\,dt, \tag{29}$$

whereas

$$\int_0^1 \varphi(f_n - g)\,dt = \infty \qquad \textit{for all} \quad n \geqslant 1. \tag{30}$$

Proof. Condition b) of (28) implies the existence of a strictly increasing sequence $\{t_n\}_1^\infty$ such that $t_1 > 1$ and

$$\varphi\left(\left(1 + \frac{1}{n}\right)t_n\right) > 2^n \varphi(t_n) > 2^{2n+1} \quad \text{for} \quad n \geqslant 1. \tag{31}$$

Let $c_n = 2^{-n}$, $d_n = c_n + 1/n^2\phi(t_n)$, and

$$F(t) = \begin{cases} t_n & \text{for} \quad t \in [c_n, d_n] \quad (n \geqslant 1), \\ 0 & \text{for} \quad t \in [0, 1] - \sum_{n=1}^{\infty} [c_n, d_n]. \end{cases}$$

Clearly, $F(t)$ is nonnegative on $[0, 1]$ and

$$\int_0^1 \varphi(F(t))\,dt = \sum_{n=1}^{\infty} \int_{c_n}^{d_n} \varphi(t_n)\,dt = \sum_{n=1}^{\infty} \frac{1}{n^2} < \infty,$$

i.e. $F \in \phi(L)$. On the other hand, for every number $\gamma > 1$ there exists an integer $n_0 > 1/(\gamma - 1)$, and hence (see (31))

$$\int_0^1 \varphi(\gamma F(t))\,dt = \sum_{n=1}^{\infty}\int_{c_n}^{d_n}\varphi(\gamma t_n)\,dt \geqslant \sum_{n=n_0}^{\infty}\varphi\left(\left(1+\frac{1}{n}\right)t_n\right)\frac{1}{n^2\varphi(t_n)} \geqslant \sum_{n=1}^{\infty}\frac{2^n}{n} = \infty,$$

i.e.

$$\int_0^1 \varphi(F)\,dt < \infty, \qquad \text{while} \quad \int_0^1 \varphi(\gamma F)\,dt = \infty \quad \text{for all } \gamma > 1. \tag{32}$$

We now put

$$f(t) = \frac{1}{2}F(t), \quad g(t) = -\frac{1}{2}F(t), \quad f_n(t) = \frac{1}{2}\left(1+\frac{1}{n}\right)F(t) \quad (n \geqslant 1). \tag{33}$$

Since $F(t) \in \phi(L)$ (see (32)), we get $f \in \phi(L)$, $g \in \phi(L)$ and $f_n \in \phi(L)$ for $n \geq 1$. Also, $f - g = F(t) \in \phi(L)$.

Further, by virtue of the finiteness of $f(t)$ we have

$$\lim_{n\to\infty}\frac{1}{n}f(t) = 0 \qquad \text{for all } t \in [0, 1]$$

and hence (since $\phi(+0) = 0$ (see (28)))

$$\lim_{n\to\infty}\varphi\left(\frac{1}{n}f(t)\right) = \varphi(+0) = \varphi(0) = 0 \qquad \text{for all } t \in [0, 1]. \tag{34}$$

In addition,

$$\varphi\left(\frac{1}{n}f(t)\right) \leqslant \varphi(F(t)) \in L(0, 1) \qquad \text{for all } n \geqslant 1. \tag{35}$$

Therefore, according to Lebesgue's theorem (see (33)–(35)),

$$\int_0^1 \varphi(f - f_n)\,dt = \int_0^1 \varphi\left(\frac{1}{n}f(t)\right)dt \to 0 \quad \text{for } n \to \infty,$$

i.e. $\rho_\phi(f, f_n) \to 0$ for $n \to \infty$. On the other hand, since $\phi(t)$ is continuous on the right on $[0, \infty)$ (see (28)) and $F(t)$ is nonnegative, we have

$$\lim_{n\to\infty}\varphi\left(\left(1+\frac{1}{n}\right)f(t)\right) = \varphi(f(t)) \qquad \text{for all } t \in [0, 1], \tag{36}$$

while

$$0 \leqslant \varphi\left(\left(1+\frac{1}{n}\right)f(t)\right) \leqslant \varphi(2f(t)) = \varphi(F(t)) \in \varphi(L) \qquad \text{for } n \geqslant 1. \tag{37}$$

Therefore, according to Lebesgue's theorem (see also (36) and (37)),

$$\lim_{n\to\infty}\int_0^1 \varphi(f_n)\,dt = \lim_{n\to\infty}\int_0^1 \varphi\left(\left(1+\frac{1}{n}\right)f(t)\right)dt = \int_0^1 \varphi(f(t))\,dt.$$

We have thus established that the f_n and f satisfy (29).

Finally (see (33) and (32)),

$$\int_0^1 \varphi(f_n - g)\,dt = \int_0^1 \varphi\left(\left(1+\frac{1}{2n}\right)F(t)\right)dt = \infty \qquad \text{for all} \quad n \geqslant 1,$$

i.e. (30) is satisfied. Q.E.D.

Lemma 8. *Suppose given a continuous on* $[0, \infty)$ *function* $\phi \in \Phi$ *that is positive on* $(0, \infty)$, *and a sequence* $\{f_n\}_1^\infty$ *with* $f_n \in \phi(L)$ *that converges in the topology of* U_ϕ *to an element* $f \in \phi(L)$. *If* $g \in \phi(L)$, *then*

a) *under the assumption* $f - g \in \phi(L)$ *there exists a natural number* $N = N(g, \{f_n\})$ *such that*

$$f_n - g \in \varphi(L) \quad \textit{for} \quad n \geqslant N \qquad \textit{and} \qquad \lim_{n\to\infty}\int_0^1 \varphi(f_n - g)\,dt = \int_0^1 \varphi(f-g)\,dt; \tag{38}$$

b) *under the assumption* $f - g \notin \phi(L)$

$$\lim_{n\to\infty}\int_0^1 \varphi(f_n - g)\,dt = \infty = \int_0^1 \varphi(f-g)\,dt. \tag{39}$$

Proof. Since $\{f_n\}$ converges to f in the topology of U_ϕ, we have, in particular,

$$\lim_{n\to\infty}\int_0^1 \varphi(f - f_n)\,dt = 0 \quad \text{and} \quad f - f_n \in \varphi(L) \quad \text{for} \quad n \geqslant N_1, \tag{40}$$

where N_1 is some natural number. By assumption, $\phi(t) > 0$ for $t > 0$, and hence (40) implies that $\{f_n\}$ converges to f in measure on $[0, 1]$.

Let $\{f_{n_k}(t)\}_{k=1}^\infty$ be an arbitrary subsequence of $\{f_n\}_1^\infty$ that converges to $f(t)$ almost everywhere on $[0, 1]$.

Consider now case a).

By assumption, $f - g \in \phi(L)$, and hence for every $\epsilon > 0$ there exists a number $\delta = \delta(\epsilon)$ such that

$$\int_E \varphi(f-g)\,dt < \frac{\varepsilon}{3} \quad \text{for any set} \quad E \subset [0, 1] \quad \text{with } |E| < \delta. \tag{41}$$

According to Luzin's and Egorov's theorems, there exists a perfect set $P \subset [0, 1]$ with measure $|P| > 1 - \delta$ such that the functions f, g and f_n $(n \geq 1)$ are continuous on P and the sequence $\{f_{n_k}(t)\}_{k=1}^\infty$ converges uniformly on P to $f(t)$. But ϕ is continuous on $(-\infty, \infty)$, and hence

$$|\varphi(f_{n_k}(t) - g(t)) - \varphi(f(t) - g(t))| < \frac{\varepsilon}{3} \quad \text{for} \quad t \in P \quad \text{and} \quad k \geqslant K_1, \tag{42}$$

where $K_1 = K_1(\epsilon, P)$ is some natural number.

Let

$$F(t) = \begin{cases} g(t) & \text{for} \quad t \in [0, 1] - P, \\ f(t) & \text{for} \quad P. \end{cases}$$

Clearly, $F \in \phi(L)$ since $f \in \phi(L)$ and $g \in \phi(L)$. Take the neighborhoods

$$A_1 = A_\varphi\left(f, \frac{\varepsilon}{3}\right), \quad A_2 = A_\varphi\left(F, \frac{\varepsilon}{3}\right) \text{ and } A = A_1 \cap A_2.$$

We have $f \in A_1$ by definition and $f \in A_2$ by virtue of the fact that (see (41))

$$\int_0^1 \varphi(f - F)\,dt = \int_{[0,1]-P} \varphi(t - g)\,dt < \frac{\varepsilon}{3}.$$

Therefore $f \in A$. Since $\{f_n\}$ converges to f in the topology of U_ϕ, it follows that $f_n \in A$ for all sufficiently large n, and consequently $f_{n_k} \in A_2$ for $k \geq K_2$, where K_2 is some natural number. The latter fact means that

$$\int_0^1 \varphi(F - f_{n_k})\,dt < \frac{\varepsilon}{3} \quad \text{for } k \geqslant K_2$$

and a fortiori

$$\int_{[0,1]-P} \varphi(g(t) - f_{n_k}(t))\,dt < \frac{\varepsilon}{3} \quad \text{for } k \geqslant K_2. \tag{43}$$

But then for $k \geq K = \max\{K_1, K_2\}$ we have (see (42), (43) and (41))

$$\left|\int_0^1 \varphi(f_{n_k} - g)\,dt - \int_0^1 \varphi(f - g)\,dt\right| \leqslant \int_P |\varphi(f_{n_k} - g) - \varphi(f - g)|\,dt$$

$$+ \int_{[0,1]-P} \varphi(f_{n_k} - g)\,dt + \int_{[0,1]-P} \varphi(f - g)\,dt < \frac{\varepsilon}{3} + \frac{\varepsilon}{3} + \frac{\varepsilon}{3} = \varepsilon$$

i.e.

$$\lim_{k\to\infty} \int_0^1 \varphi(f_{n_k} - g)\,dt = \int_0^1 \varphi(f - g)\,dt, \tag{44}$$

where $f_{n_k} - g \in \phi(L)$ for $k \geq K = K(\epsilon, \{f_{n_k}\})$.

Now take the neighborhoods

$$B_1 = A_\varphi(f, 1), \quad B_2 = A_\varphi\left(g, 1 + \int_0^1 \varphi(f - g)\,dt\right) \text{ and } B = B_1 \cap B_2.$$

Clearly, $f \in B$. From the convergence of $\{f_n\}$ to f in the topology of U_ϕ we have $f_n \in B \subset B_2$ for $n \geq N_2$, i.e.

$$\int_0^1 \varphi(f_n - g)\,dt < 1 + \int_0^1 \varphi(f - g)\,dt < \infty \quad \text{for } n \geqslant N_2.$$

Hence

$$f_n - g \in \varphi(L) \quad \text{for } n \geqslant N_2. \tag{45}$$

Let

$$d_n = \int_0^1 \varphi(f_n - g)\,dt \quad \text{and} \quad d = \int_0^1 \varphi(f - g)\,dt.$$

It is evident (see (45)) that the numbers d_n are finite for $n \geq N_2$, i.e. the first part of (38) is established. On the other hand, since $\{f_n\}$ converges in measure to f, for every sequence of natural numbers $\{p_i\}_1^\infty$ one can choose indices i_k such that the sequence $\{f_{p_{i_k}}(t)\}_{k=1}^\infty$ will converge to $f(t)$ almost everywhere on $[0, 1]$. But then (44) implies that

$$\lim_{k\to\infty} d_{p_{i_k}} = d.$$

And this implies that $d_n \to d$ for $n \to \infty$, i.e. assertion a) is established.

Consider case b).

By assumption, $f - g \notin \phi(L)$, and hence for every number $N > 0$ there exists a number $\delta_1 > 0$ such that

$$\int_E \varphi(f - g)\,dt > N \quad \text{for all sets} \quad E \subset [0, 1] \text{ with } |E| > 1 - \delta_1. \tag{46}$$

According to Luzin's and Egorov's theorems, there exists a perfect set $P \subset [0, 1]$ with $|P| > 1 - \delta_1$ such that the functions f, g and f_n $(n \geq 1)$ are continuous on P and the sequence $\{f_{n_k}(t)\}_{k=1}^\infty$ converges uniformly on P to $f(t)$ (here $\{f_{n_k}\}$ is the sequence chosen at the beginning of the proof of Lemma 8). But then, by virtue of the continuity of $\phi(t)$,

$$\varphi(f_{n_k}(t) - g(t)) \to \varphi(f(t) - g(t)) \qquad \text{uniformly on } P \text{ for } k \to \infty.$$

Therefore (see (46))

$$\lim_{k\to\infty} \int_P \varphi(f_{n_k}(t) - g(t))\,dt = \int_P \varphi(f(t) - g(t))\,dt > N$$

and a fortiori

$$\varliminf_{k\to\infty} \int_0^1 \varphi(f_{n_k}(t) - g(t))\,dt > N.$$

By virtue of the arbitrariness of N we get that

$$\lim_{k\to\infty} \int_0^1 \varphi(f_{n_k}(t) - g(t))\,dt = \infty. \tag{47}$$

And from this relation we get (39) in the same way as in the proof of assertion a) of Lemma 8, since the sequence $\{f_{n_k}(t)\}_{k=1}^\infty$ in (47) is an arbitrary subsequence of $\{f_n\}_1^\infty$ that converges almost everywhere on $[0, 1]$. Lemma 8 is completely proved.

Remark 1. In the proof of assertion b) of Lemma 8 we did not use the condition $g \in \phi(L)$ but only the fact that $g(t)$ is measurable and finite almost everywhere on $[0, 1]$. By the same token we have

Lemma 9. *Suppose $\phi(t) \in \Phi$, $\phi(t) \in C(0, \infty)$ and $\phi(t) > 0$ for $t > 0$. If a sequence $\{f_n\}$ converges to $f \in \phi(L)$ in the topology of U_ϕ and $g(t)$ is an arbitrary function that is measurable and finite almost everywhere on* $[0, 1]$, *then the equality*

$$\lim_{n\to\infty} \int_0^1 \varphi(f_n - g)\,dt = \int_0^1 \varphi(f - g)\,dt \tag{48}$$

holds in each of the following three cases:

a) $g \in \phi(L)$, $g - f \in \phi(L)$,

b) $g \in \phi(L)$, $g - f \notin \phi(L)$,

c) $g \notin \phi(L)$, $g - f \notin \phi(L)$.

We note that the integrals in (48) (as well as in (39)) can be infinite.

Lemma 10. *Suppose given a function ϕ satisfying the conditions of Lemma 8, a sequence $\{f_n\}_1^\infty$ and an element f such that $f \in \phi(L)$ and $f_n \in \phi(L)$ for $n \geq 1$. If*

$$\lim_{n\to\infty} \int_0^1 \varphi(f_n - g)\,dt = \int_0^1 \varphi(f - g)\,dt, \tag{49}$$

for every element $f \in \phi(L)$ with $f - g \in \phi(L)$, then $\{f_n\}$ converges to f in the topology of U_ϕ.

Proof. Since the set $\{A_\phi(g, \epsilon)\colon g \in \phi(L), \epsilon > 0\}$ of ϵ-neighborhoods is (by definition) a subbase of the topological space U_ϕ, it suffices to prove that $\{f_n\}$ converges to f with respect to the ϵ-neighborhoods (whose centers do not necessarily coincide with f). Suppose f is contained in an ϵ-neighborhood $A_\phi(g, \epsilon)$ of an element $g \in \phi(L)$. Then

$$\int_0^1 \varphi(f - g)\,dt < \varepsilon \tag{50}$$

and a fortiori $f - g \in \phi(L)$. From condition (49) we consequently have

$$\lim_{n\to\infty} \int_0^1 \varphi(f_n - g)\,dt = \int_0^1 \varphi(f - g)\,dt < \varepsilon$$

which implies $\int_0^1 \phi(f_n - g)dt < \epsilon$ for $n \geq N$, where N is some natural number. But the latter inequality means that $f_n \in A_\phi(g, \epsilon)$ for all $n > N$, i.e. $\{f_n\}_1^\infty$ converges to f with respect to the ϵ-neighborhoods, and hence $\{f_n\}$ converges to f in the topology of U_ϕ. Q.E.D.

Remark 2. The only property of ϕ that we used in the proof of Lemma 10 was the fact that $\phi \in \Phi$. Also, we need only have assumed that $f_n - f \in \phi(L)$ for all sufficiently large n (see (49)).

Let us recall a definition (see [6], Definition 6.1). A topological space U_ϕ with $\phi \in \Phi$ has *property d* if for every neighborhood of the form $A_\phi(f, \epsilon)$ and every element $g \in A_\phi(f, \epsilon)$ there exists a number $\delta > 0$ such that $A_\phi(g, \delta) \subset A_\phi(f, \epsilon)$.

We have (see [6], Theorem 6.2)

Lemma 11. *Suppose $\phi \in \Phi$ and $\phi(+0) = 0$. Then the topological space U_ϕ has property d if and only if $\phi(t+0) = \phi(t) > 0$ for $t > 0$ and $\phi(2t) = O\{\phi(t)\}$ for $t \to \infty$.*

For the next lemma we require the following definition (see [6], Definition 3.3): a function $\phi \in \Phi$ satisfies the *ω-condition* if $\phi(t+1) = O\{\phi(t)\}$ for $t \to \infty$.

Lemma 12. *Suppose a function $\phi \in \Phi$ does not satisfy the ω-condition. Then for every real number A there exists a nonnegative function f such that*

$$\int_0^1 \varphi(f(t) + A)\, dt \leqslant 1 + \varphi(A), \quad \textit{whereas} \quad \int_0^1 \varphi(f(t) + A + \delta)\, dt = \infty$$

for all $\delta > 0$.

Proof. Since $\phi(t+1) \neq O\{\phi(t)\}$ for $t \to \infty$, there exist numbers $1 < t_1 < t_2 < \cdots$ such that

$$t_n > |A|, \ \varphi(t_n) > 2^n \ \text{and} \ \varphi\left(t_n + \frac{1}{n}\right) > 2^n \varphi(t_n) \quad \text{for} \ n \geqslant 1.$$

Let $c_n = 2^{-n}$, $d_n = c_n + (2^n \phi(t_n))^{-1}$ for $n \geq 1$ and

$$f(t) = \begin{cases} t_n - A & \text{for} \quad t \in [c_n, d_n] \quad (n \geqslant 1), \\ 0 & \text{for} \quad t \in [0, 1] - \sum_{n=1}^{\infty} [c_n, d_n]. \end{cases}$$

Clearly, $f(t) \geq 0$ for $t \in [0, 1]$ and

$$\int_0^1 \varphi(f(t) + A)\, dt = \left\{ \int_{[0,1] - \sum_{n=1}^{\infty} [c_n, d_n]} \varphi(A)\, dt \right\} + \left\{ \sum_{n=1}^{\infty} \int_{c_n}^{d_n} \varphi(t_n)\, dt \right\}$$

$$\leqslant \varphi(A) + \sum_{n=1}^{\infty} 2^{-n} = 1 + \varphi(A).$$

On the other hand, for every $\delta > 0$ there exists an integer $m \geq 1$ with $\delta m > 1$, and hence

$$\int_0^1 \varphi(f(t) + A + \delta)\, dt \geqslant \sum_{n=m}^{\infty} \int_{c_n}^{d_n} \varphi(t_n + \delta)\, dt \geqslant \sum_{n=m}^{\infty} \frac{\varphi(t_n + 1/n)}{n^2 \varphi(t_n)} \geqslant \sum_{n=1}^{\infty} \frac{2^n}{n^2} = \infty.$$

Q.E.D.

We note that Lemma 12 is analogous to Lemma 3.5 of [6] and reduces to it if $A = 0$.

Lemma 13. *Suppose a function $\phi \in \Phi$ does not satisfy the ω-condition. Then if each of the functions $f_i(t) \in \phi(L)$ of a sequence $\{f_i\}_1^\infty$ is not equivalent to zero on the interval* [0, 1], *there exists a function $F \in \phi(L)$ such that*

$$\int_0^1 \varphi(f_i(t) - F(t))\, dt = \infty \quad \text{for all} \quad i \geqslant 1.$$

This assertion was established in [6] (see part 1) of the proof of Lemma 6.2). It clearly remains valid when the interval [0, 1] is replaced by any set E with $|E| > 0$.

Lemma 14. *Suppose given a sequence of sets $E_i \subset (-\infty, \infty)$ having measure $|E_i| = \alpha_i > 0$ for all $i \geq 1$. Then there exist sets A_i such that $A_i \subset E_i$ $(i \geq 1)$, $A_i \cap A_j = 0$ for $i \neq j$ and $|A_i| > 0$ for all $i \geq 1$.*

This lemma is obvious (it was actually proved in [6]; see part 1) of the proof of Lemma 6.2).

§3. On the interdependence of the convergences

Here we establish on the basis of the results of §2 several assertions concerning convergence in ϕ-distance and in the topology of U_ϕ.

Theorem 1. *Suppose $\phi \in \Phi$ and $\phi(+0) = 0$. Then in order for the conditions*

$$f_n \in \varphi(L) \quad \text{for} \quad n \geqslant 1, \; f \in \varphi(L) \; \text{and} \; \lim_{n \to \infty} \rho_\varphi(f, f_n) = 0$$

to always imply the equality

$$\lim_{n \to \infty} \int_0^1 \varphi(f_n)\, dt = \int_0^1 \varphi(f(t))\, dt, \tag{51}$$

it is necessary and sufficient that the following three conditions be satisfied:

a) $\phi(t) \in C(0, \infty)$,

b) $\phi(t) > 0$ *for* $t > 0$,

c) $\phi(2t) = O\{\phi(t)\}$ *for* $t \to \infty$.

This assertion is a direct consequence of Lemmas 1–3.

Theorem 2. *Suppose $\phi \in \Phi$ and $\phi(+0) = 0$. Then in order for convergence in ϕ-distance to be equivalent in $\phi(L)$ to convergence in the topology of U_ϕ it is necessary and sufficient that the following three conditions be satisfied:*

$$\begin{aligned} &\text{a)}\ \phi(t+0) = \phi(t) \text{ for all } t \in (0, \infty), \\ &\text{b)}\ \phi(t) > 0 \text{ for all } t > 0, \\ &\text{c)}\ \phi(2t) = O\{\phi(t)\} \text{ for } t \to \infty. \end{aligned} \tag{52}$$

Proof. The sufficiency of the conditions is a direct consequence of Lemma 6.

Let us verify the necessity of conditions (52). Suppose one of the conditions a), b) or c) of (52) is not satisfied. In each of these cases there exist a number $\alpha > 0$ and functions $f \in \phi(L)$ and $f_n \in \phi(L)$ $(n \geq 1)$ such that (see Lemmas 1, 3 and 4)

$$\lim_{n\to\infty}\int_0^1 \varphi(f-f_n)\,dt=0, \tag{53}$$

whereas

$$\int_0^1 \varphi(f_n)\,dt>\alpha+\int_0^1 \varphi(f)\,dt=\alpha+I \quad \text{for all} \quad n\geqslant 1. \tag{54}$$

It follows from (53) that $\{f_n\}$ converges to f in ϕ-distance. We will see, however, that $\{f_n\}$ does not converge to f in the topology of U_ϕ. In fact, take the neighborhoods $A_1 \equiv A_\phi(f, 1)$ and $A_2 \equiv A_\phi(0, \alpha + I)$. Clearly, $f \in A_1$ and, since $\rho_\phi(0, f) = I < \alpha + I$ (because $\alpha > 0$), $f \in A_2$, which implies $f \in A_1 \cap A_2$.

On the other hand, the neighborhood $A_1 \cap A_2$ does not contain any of the elements f_n with $n \geq 1$, since for $n \geq 1$ (see (54)) we have $\rho_\phi(0, f_n) > \alpha + I$, i.e. $f_n \notin A_2$. Thus we have found a neighborhood $A_1 \cap A_2$ that contains the element f but none of the elements f_n for $n \geq 1$, which means that the sequence $\{f_n\}$ does not converge to f in the topology of U_ϕ.

In this way we have proved that if one of the conditions (52) is violated, there exists a sequence $\{f_n\}$ that converges to f in ϕ-distance but not in the topology of U_ϕ. We note that $\{f_n\}$ cannot converge in the topology of U_ϕ to any element of $\phi(L)$, since if it converged in the topology of U_ϕ to some element $g \in \phi(L)$, then that element g would necessarily have to be equal to f.

From what has been said it follows that convergence in ϕ-distance is equivalent to convergence in the topology of U_ϕ only if conditions (52) are satisfied. Q.E.D.

In connection with Theorems 1 and 2 a question arises. If a function ϕ satisfies only conditions a) and b) of (52) but a sequence $\{f_n\}$ for which (51) holds converges in ϕ-distance to an element $f \in \phi(L)$, does it not then follow that $\{f_n\}$ converges to f in the topology of U_ϕ? The answer to this question is provided by

Theorem 3. *Suppose $\phi \in \Phi$ and $\phi(t + 0) = \phi(t)$ for $t \geq 0$. If $\phi(2t) \neq O\{\phi(t)\}$ for $t \to \infty$, then there exists a sequence $\{f_n\}$ for which (51) holds that converges to an element $f \in \phi(L)$ in ϕ-distance but not in the topology of U_ϕ.*

Proof. Take functions f, g and f_n $(n \geq 1)$ satisfying the conditions of Lemma 7, and let

$$\delta=1+\int_0^1 \varphi(f-g)\,dt. \tag{55}$$

Clearly, $0 < \delta < \infty$ since $f - g \in \phi(L)$. Take the neighborhoods $A_1 = A_\phi(f, 1)$, $A_2 = A_\phi(g, \delta)$ and $A = A_1 \cap A_2$. It is obvious that $f \in A$, since (see (55))

$$\int_0^1 \varphi(f-g)\,dt<\delta.$$

If $\{f_n\}$ converged in the topology of U_ϕ to some element $r(t) \in \phi(L)$, we would have $r = f$ (convergence in the topology of U_ϕ always implies convergence in ϕ-distance). But if $\{f_n\}$ converges to f in the topology of U_ϕ, then $f_n \in A$ for all $n \geq N$, where N is some natural number. Therefore $f_n \in A_2$ for all $n \geq N$, and hence (by definition of the neighborhood A_2) $\int_0^1 \phi(f_n - g)dt < \delta$ for all $n \geq N$, which contradicts (30).

In this way we have constructed a sequence $\{f_n\}$ satisfying (51) that converges in ϕ-distance (see (29)) but not in the topology of U_ϕ. Q.E.D.

Thus Theorem 3 provides a negative answer to the question posed above. Nevertheless, convergence in the topology of U_ϕ can be characterized in terms of the behavior of integrals of the form $\int_0^1 \phi(f_n(t) + g(t))dt$.

Namely, we have

Theorem 4. *Suppose $\phi \in \Phi$, $\phi \in C(0, \infty)$ and $\phi(t) > 0$ for $t > 0$. Then a sequence $\{f_n\}_1^\infty$ with $f_n \in \phi(L)$ for $n \geq 1$ converges in the topology of U_ϕ to an element $f \in \phi(L)$ if and only if*

$$\lim_{n\to\infty} \int_0^1 \varphi(f_n - g)\, dt = \int_0^1 \varphi(f - g)\, dt \tag{56}$$

for every $g \in \phi(L)$ with $g - f \in \phi(L)$.

This assertion immediately follows from Lemmas 8 and 10.

We note that Theorem 4 concerns the convergence in the topology of U_ϕ of individual sequences $\{f_n\}$.

Remark 3. Theorem 4 (and its proof) remain valid for convergence in the topology of $U_\phi(E)$, where $E \subset [0, 1]$ with $|E| > 0$. It is only necessary in this case to replace the interval $[0, 1]$ in (56) by the set E.

Corollary 1. *If ϕ satisfies the conditions of Theorem 4 and $\{f_n\}$ converges to f in the topology of U_ϕ, then*

$$\lim_{n\to\infty} \int_0^1 \varphi(f - f_n)\, dt = 0, \qquad \lim_{n\to\infty} \int_0^1 \varphi(f_n)\, dt = \int_0^1 \varphi(f)\, dt. \tag{57}$$

Corollary 1 is a special case of Theorem 4 (it is only necessary to first put $g = f$ and then put $g = 0$ in (56) in order to obtain (57)).

Remark 4. If ϕ satisfies the Δ_2-condition (i.e. $\phi(2t) = O\{\phi(t)\}$ for $t \to \infty$), then Corollary 1 has a converse, i.e. the fulfilment of (57) implies the convergence of $\{f_n\}$ in the topology of U_ϕ (see Theorem 2). If, on the other hand, ϕ does not satisfy the Δ_2-condition, then the fulfilment of (57) does not automatically imply the convergence of $\{f_n\}$ in the topology of U_ϕ (see Theorem 3).

Remark 5. Suppose $\phi(t)$ is a nondecreasing even convex function on $[0, \infty)$ such that $\phi(0) = 0$ and

$$\lim_{t\to\infty} \frac{\varphi(t)}{t} = \infty.$$

Then, as is well known (see [3]), it is possible to introduce a norm in the set $\phi(L)$ and thereby construct an Orlicz space L_ϕ^*. The elements of L_ϕ^* will be of the form $Cf(t)$, where $f \in \phi(L)$ and C is an arbitrary real number. With the use of Luxemburg's norm (see [3], §9.7), one can easily see that *if* $u_n \in L_\phi^*$ *and* $u \in L_\phi^*$, *then*

$$\alpha_n \equiv \| u_n - u \|_{(\varphi)} \to 0 \quad as \quad n \to \infty \tag{58}$$

if and only if there exist numbers $\lambda_n \uparrow \infty$ *for which*

$$\beta_n \equiv \int_0^1 \varphi(\lambda_n(u_n(t) - u(t)))\, dt \to 0 \quad as \quad n \to \infty. \tag{59}$$

In fact, if (58) holds, there exist $\lambda_n \uparrow \infty$ for which $1 > \alpha_n\lambda_n \to 0$ as $n \to \infty$, and hence (see [3], Theorem 9.5)

$$\int_0^1 \varphi(\lambda_n(u_n - u))\, dt \leqslant \lambda_n \| u_n - u \|_{(\varphi)} = \lambda_n \alpha_n \to 0 \quad \text{as} \quad n \to \infty,$$

i.e. (59) holds.

Conversely, suppose (59) holds for some sequence $\lambda_n \uparrow \infty$. Then, by definition of the Luxemburg norm (see [3], §9.7),

$$\| u_n - u \|_{(\varphi)} \leqslant \frac{1}{\lambda_n} \to 0 \quad \text{as} \quad n \to \infty,$$

i.e. (58) holds.

The above assertion can also be formulated as follows (see [3], (9.24)):

If Cf *and* $C_n f_n$ $(n \geq 1)$ *belong to* $\phi(L)$ *for certain constants* $C > 0$ *and* $C_n > 0$ $(n \geq 1)$, *then* $\{f_n\}_1^\infty$ *converges to* f *in the norm of the Orlicz space* L_ϕ^* *if and only if*

$$\lim_{n\to\infty} \int_0^1 \varphi(\lambda_n(f_n - f))\, dt = 0 \qquad \textit{for certain} \qquad \lambda_n \uparrow \infty. \tag{60}$$

We assume here, of course, that ϕ satisfies the conditions formulated at the beginning of Remark 5.

It is interesting to compare condition (56), ensuring convergence in the topology of U_ϕ, with condition (60), ensuring convergence in the norm of L_ϕ^*. Condition (56) is clearly applicable for wider function classes $\phi(L)$ (in the sense that in (56) the function $\phi(t)$ can grow arbitrarily slowly (like $\ln\ln(|t| + e)$, for example), whereas in (60) the function $\phi(t)$ grows at least as rapidly as $|t|$). It is evident that conditions (56) and (60) coincide for the case $\phi(t) = |t|^p$ with $1 < p < \infty$, whereas neither contains the other in general (see Remarks 6 and 7).

Remark 6. Theorem 4 indicates the "poorness" of the class of convergent sequences in U_ϕ when $\phi(t)$ does not grow sufficiently slowly or its changes are not sufficiently gradual. This will be seen from the assertions proved below.

a) *Suppose a continuous on* $[0, \infty)$ *function* $\phi \in \Phi$ *with* $\phi(t) > 0$ *for* $t > 0$ *does not satisfy the* ω*-condition while a sequence* $\{f_i\}_1^\infty$ *converges in the topology of* $U_\phi(E)$ *to an element* f, *where* $E \subset [0, 1]$ *is some measurable set with measure* $|E| > 0$. *If* $f(t) = 0$ *for* $t \in M \subset E$, *then* $f_i(t) = 0$ *for almost all* $t \in M$ *and* $i \geq N$, *where* N *is some natural number.*

Proof. Suppose the contrary, i.e. suppose there exists a sequence $\{f_i\}_1^\infty$ that converges in the topology of $U_\phi(E)$ to an element f with $f(t) = 0$ for $t \in M$ ($|M| > 0$) but $f_{i_k} \not\sim 0$ (i.e. f_{i_k} is not equivalent to zero on M) for some sequence $\{i_k\}_1^\infty$ with $i_k \to \infty$ for $k \to \infty$. According to Lemma 13 (see also the comments following it) applied to the set M and the sequence $\{f_{i_k}\}$, there exists a function $F(t)$ such that

$$\int_M \varphi(F(t))\, dt < \infty, \quad \text{whereas} \quad \int_M \varphi(f_{i_k}(t) - F(t))\, dt = \infty \quad \text{for all } k \geqslant 1.$$

Let

$$g(t) = \begin{cases} F(t), & t \in M, \\ 0, & t \in E - M. \end{cases}$$

Clearly,

$$\left.\begin{aligned} \int_E \varphi(g(t))\, dt &= \int_M \varphi(F)\, dt < \infty, \\ \int_E \varphi(f(t) - g(t))\, dt &= \int_{E-M} \varphi(f)\, dt + \int_M \varphi(-g)\, dt < \infty, \end{aligned}\right\} \tag{61}$$

whereas

$$\int_E \varphi(f_{i_k}(t) - g(t))\, dt \geqslant \int_M \varphi(f_{i_k} - F)\, dt = \infty \quad \text{for all} \quad k \geqslant 1.$$

From the latter relation it follows that

$$\lim_{k\to\infty} \int_E \varphi(f_{i_k} - g)\, dt = \infty. \tag{62}$$

But $\{f_i\}_1^\infty$ converges in the topology of $U_\phi(E)$ to f and hence $\{f_{i_k}\}_1^\infty$ also converges in the topology of $U_\phi(E)$ to f. Therefore (see (61), Theorem 4 and Remark 3)

$$\lim_{k\to\infty} \int_E \varphi(f_{i_k} - g)\, dt = \int_E \varphi(f - g)\, dt < \infty,$$

which contradicts (62). Assertion a) is proved.

Assertion a) suggests that if ϕ grows very rapidly, then the only convergent sequences in $U_\phi(E)$ will be stationary sequences $\{f_i\}_1^\infty$ (in which $f_i(t) = f(t)$ for almost all $t \in E$ and $i \geq N$). But this is not the case. In fact, we have the assertion

b) *Suppose a function* ϕ *satisfies the conditions of Theorem* 4. *If a sequence* $\{f_i\}_1^\infty$ *of nonnegative and measurable on a set* $E \subset [0, 1]$ (*with* $|E| > 0$) *functions* $f_i(t)$ *converges in measure on* E *to a function* $f(t)$ *having the properties*

$$\int_E \varphi(f(t))\,dt < \infty \text{ and } f_i(t) \leqslant f(t) \quad \text{for} \quad t \in E \text{ and } i \geqslant 1, \tag{63}$$

then

$$\lim_{i\to\infty} \int_E \varphi(f_i + g)\,dt = \int_E \varphi(f + g)\,dt \tag{64}$$

for every function $g(t)$ *with*

$$\int_E \varphi(g(t))\,dt < \infty \text{ and } \int_E \varphi(f + g)\,dt < \infty. \tag{65}$$

Proof. Let

$$E_1 = \{t : t \in E,\ g(t) > 0\}, \quad E_2 = \{t : t \in E,\ -f(t) \leqslant g(t) \leqslant 0\},$$
$$E_3 = \{t : t \in E,\ g(t) < -f(t)\}.$$

Clearly, E_1, E_2 and E_3 are mutually disjoint and $E = E_1 + E_2 + E_3$. Let

$$F(t) = \begin{cases} f(t) + g(t) & \text{for } t \in E_1, \\ f(t) & \text{for } t \in E_2, \\ |g(t)| & \text{for } t \in E_3. \end{cases}$$

Then $F(t) \in \phi(L)$ on E since (see (63) and (65)) $f \in \phi(L)$, $g \in \phi(L)$ and $f + g \in \phi(L)$ on E. It is easily seen that

$$|f_i(t) + g(t)| \leqslant F(t) \quad \text{for all} \quad t \in E \text{ and } i \geqslant 1. \tag{66}$$

In fact, for $t \in E_1$ we have (see (63))

$$|f_i(t) + g(t)| \leqslant |f_i(t)| + |g(t)| = f_i(t) + g(t) \leqslant f(t) + g(t) = F(t),$$

for $t \in E_2$ we have

$$|f_i(t) + g(t)| \leqslant f(t) = F(t)$$

and for $t \in E_3$ we have

$$|f_i(t) + g(t)| \leqslant |g(t)| = F(t).$$

From (66) and the fact that $F \in \phi(L)$ on E it follows that the integrals

$$\left\{\int_e \varphi(f_i(t) + g(t))\,dt\right\}_{i=1}^{\infty} \quad \text{with } e \in E$$

are absolutely equicontinuous.

On the other hand, by assumption, the sequence $\{f_i + g\}_1^\infty$ converges in measure on E to $f + g$, and hence $\{\phi(f_i + g)\}_1^\infty$ converges in measure on E to $\phi(f + g)$. But then, according to Vitali's theorem (see [8], Chapter VI, §3, Theorem 4), equality (64) must hold. Q.E.D.

Corollary 2. *Under the conditions of assertion* b) *the sequence* $\{f_i\}_1^\infty$ *converges in the topology of* $U_\phi(E)$ *to* f.

This immediately follows from (64) upon taking into account Theorem 4 and Remark 3.

c) *Suppose* ϕ *does not satisfy the* ω*-condition but the requirements of Theorem 4 are satisfied. If* $f \in \phi(L)$, $f_i \in \phi(L)$ *for* $i \geq 1$ *and* $f_i(t) > f(t) \geq 0$ *for* $t \in E_i \subset [0, 1]$ *with* $|E_i| > 0$ $(i \geq 1)$, *then the sequence* $\{f_i\}_1^\infty$ *cannot converge to* f *in the topology of* U_ϕ.

Suppose the contrary, i.e. suppose the conditions of assertion c) are satisfied for some sequence $\{f_i\}_1^\infty$ converging to f in the topology of U_ϕ. According to Lemma 14, there exist sets $A_i \subset E_i$ $(i \geq 1)$ such that $A_i \cap A_j = 0$ for all $i \neq j$ and $|A_i| > 0$ for $i \geq 1$. By virtue of Luzin's theorem, we can assume that $f_i(t)$ and $f(t)$ are continuous on A_i (we could otherwise go over to a subset). In each set A_i we choose a perfect subset B_i $(i \geq 1)$ such that

$$|B_i| > 0 \quad \text{and} \quad |B_i|\,\varphi(\alpha_i) < \frac{1}{2^i}, \quad \text{where} \quad \alpha_i = \sup_{t \in B_i} f(t). \tag{67}$$

Let $\beta_i = \inf_{t \in B_i} f_i(t)$. Clearly, $0 \leq \alpha_i < \beta_i$ for $i \geq 1$. In accordance with Lemma 12, we construct a function $\tau_i(t)$ such that

$$\tau_i(t) \geqslant 0 \quad \text{for } t \in [0, 1], \quad \int_0^1 \varphi(\tau_i(t) + \beta_i)\,dt = \infty,$$
$$\int_0^1 \varphi(\tau_i(t) + \alpha_i)\,dt \leqslant 1 + \varphi(\alpha_i). \tag{68}$$

By making an appropriate change of variable (see, for example, [6], the proof of Lemma 3.6), we can find a function $\psi_i(t)$ defined on B_i such that (see (68))

$$\psi_i(t) \geqslant 0 \quad \text{for } t \in B_i, \quad \int_{B_i} \varphi(\psi_i(t) + \beta_i)\,dt = \infty,$$
$$\int_{B_i} \varphi(\psi_i(t) + \alpha_i)\,dt \leqslant |B_i|\,(1 + \varphi(\alpha_i)). \tag{69}$$

Finally, let

$$g(t) = \begin{cases} \psi_i(t) & \text{for } t \in B_i \ (i \geqslant 1), \\ 0 & \text{for } t \in [0, 1] - \sum_{i=1}^{\infty} B_i. \end{cases}$$

Clearly, $g(t) \geq 0$ for $t \in [0, 1]$ and $g(t) \in \phi(L)$, since (see (67) and (69))

$$\int_0^1 \varphi(g(t))\,dt = \sum_{i=1}^{\infty} \int_{B_i} \varphi(\psi_i(t))\,dt \leqslant \sum_{i=1}^{\infty} |B_i|\,(1 + \varphi(\alpha_i)) \leqslant \sum_{i=1}^{\infty} \left(|B_i| + \frac{1}{2^i}\right) \leqslant 2.$$

Here we have also made use of the inequality $\alpha_i \geq 0$.

On the other hand (see (69) and (67)),

$$\int_0^1 \varphi(g(t)+f(t))\,dt = \int_{[0,1]-\sum_{i=1}^{\infty} B_i} \varphi(f(t))\,dt + \sum_{i=1}^{\infty} \int_{B_i} \varphi(\psi_i(t)+f(t))\,dt$$

$$\leqslant \int_0^1 \varphi(f)\,dt + \sum_{i=1}^{\infty} \int_{B_i} \varphi(\psi_i(t)+\alpha_i)\,dt \leqslant \int_0^1 \varphi(f)\,dt$$

$$+ \sum_{i=1}^{\infty} |B_i|(1+\varphi(\alpha_i)) \leqslant 2 + \int_0^1 \varphi(f)\,dt < \infty,$$

i.e. $g+f \in \phi(L)$. Thus $\{f_i\}_1^\infty$ converges in the topology of U_ϕ to $f \in \phi(L)$ while $g \in \phi(L)$ is such that $g+f \in \phi(L)$. But then, according to Theorem 4,

$$\lim_{i\to\infty} \int_0^1 \varphi(f_i+g)\,dt = \int_0^1 \varphi(f+g)\,dt$$

and hence for some natural number N

$$\int_0^1 \varphi(f_i+g)\,dt < \infty \quad \text{for} \quad i \geqslant N, \tag{70}$$

since $f+g \in \phi(L)$. On the other hand (see (69)),

$$\int_0^1 \varphi(f_i(t)+g(t))\,dt \geqslant \int_{B_i} \varphi(f_i(t)+\psi_i(t))\,dt \geqslant \int_{B_i} \varphi(\beta_i+\psi_i(t))\,dt = \infty$$

for all $i \geq 1$. This contradicts (70). Assertion c) is proved.

In almost the same way, we can establish the assertion

d) *Suppose a function $\phi \in \Phi$ (with $\phi(t) \in C(0,\infty)$ and $\phi(t)>0$ for $t>0$) does not satisfy the ω-condition. If $f \in \phi(L)$, $f_i \in \phi(L)$ for $i \geq 1$ and $f_i(t) < 0 \leq f(t)$ for $t \in E_i \subset [0,1]$ with $|E_i| > 0$ $(i \geq 1)$, then $\{f_i\}_1^\infty$ cannot converge to f in the topology of U_ϕ.*

In the same way as in the proof of assertion c), we find perfect sets $B_i \subset [0,1]$ such that f_i and f are continuous on B_i, the sets B_i $(i \geq 1)$ are mutually disjoint and $\beta_i = \inf_{t\in B_i}(-f_i(t)) > 0$ for $i \geq 1$. In accordance with Lemma 12 (when $A=0$), we construct a nonnegative function $\psi(t)$ such that

$$\int_0^1 \varphi(\psi(t))\,dt < \infty, \quad \text{whereas} \quad \int_0^1 \varphi(\psi(t)+\delta)\,dt = \infty \qquad \text{for all} \quad \delta > 0.$$

Making a change of variable, we find a function $\psi_i(t)$ defined on B_i such that

$$\psi_i(t) \geqslant 0 \quad \text{for} \quad t \in B_i, \quad \int_{B_i} \varphi(\psi_i(t) + \beta_i)\, dt = \infty,$$

$$\int_{B_i} \varphi(\psi_i(t))\, dt = |B_i| \int_0^1 \varphi(\psi(t))\, dt. \tag{71}$$

Let

$$g(t) = \begin{cases} -\psi_i(t) & \text{for} \quad t \in B_i, \quad (i \geqslant 1), \\ 0 & \text{for} \quad t \in [0, 1] - \sum_{i=1}^{\infty} B_i. \end{cases}$$

Clearly (see (71)), $g(t) \leq 0$ for $t \in [0, 1]$ and $g(t) \in \phi(L)$, since

$$\int_0^1 \varphi(g(t))\, dt = \sum_{i=1}^{\infty} \int_{B_i} \varphi(\psi_i(t))\, dt = \sum_{i=1}^{\infty} |B_i| \int_0^1 \varphi(\psi(t))\, dt \leqslant \int_0^1 \varphi(\psi)\, dt < \infty.$$

Further, $f(t) + g(t) \in \phi(L)$, since (see (71)) f and ψ_i have the same sign on B_i, and hence

$$\int_0^1 \varphi(f(t) + g(t))\, dt = \int_{[0,1] - \sum_{i=1}^{\infty} B_i} \varphi(f(t))\, dt + \sum_{i=1}^{\infty} \int_{B_i} \varphi(f(t) - \psi_i(t))\, dt$$

$$\leqslant \int_0^1 \varphi(f(t))\, dt + \sum_{i=1}^{\infty} \int_{B_i} [\varphi(f(t)) + \varphi(\psi_i(t))]\, dt$$

$$\leqslant 2 \int_0^1 \varphi(f)\, dt + \int_0^1 \varphi(\psi)\, dt < \infty.$$

Therefore, if $\{f_i\}$ were to converge to f in the topology of U_ϕ, it would follow from Theorem 4 that

$$\lim_{i \to \infty} \int_0^1 \varphi(f_i + g)\, dt = \int_0^1 \varphi(f + g)\, dt$$

and hence that

$$\int_0^1 \varphi(f_i + g)\, dt < \infty \quad \text{for} \quad i \geqslant N. \tag{72}$$

On the other hand, for each $i \geq 1$ (see (71))

$$\int_0^1 \varphi(f_i + g)\, dt \geqslant \int_{B_i} \varphi(f_i - \psi_i)\, dt$$

$$\geqslant \int_{B_i} \varphi(-\beta_i - \psi_i(t))\, dt = \int_{B_i} \varphi(\psi_i(t) + \beta_i)\, dt = \infty,$$

which contradicts (72). Assertion d) is proved.

For any function $f(t)$ defined on $[0, 1]$ let

$$E^{(0)}(f)=\{t:t\in[0,1],\quad f(t)=0\},\quad E^+(f)=\{t:t\in[0,1],\quad f(t)>0\},$$
$$E^-(f)=\{t:t\in[0,1],\quad f(t)<0\}.$$

We note two obvious facts. If a sequence $\{f_i\}_1^\infty$ converges to a function f in the topology of U_ϕ, then (i) any subsequence $\{f_{i_k}\}_{k=1}^\infty$ of it has the same property, and (ii) the sequence $\{-f_i\}_1^\infty$ will converge to $-f$ in the topology of U_ϕ.

From assertions c) and d) we obtain

Corollary 3. *Suppose a function ϕ satisfies the conditions of assertion* c). *Then, if a sequence $\{f_i\}_1^\infty$ converges in the topology of U_ϕ to an element f, there exists a natural number N such that*

$$0\leqslant f_i(t)\leqslant f(t)\qquad \textit{for almost all}\qquad t\in E^+(f)\quad \textit{and all}\quad i\geqslant N. \tag{73}$$

Proof. If (73) did not hold, there would exist a sequence $\{f_{n_k}\}_{k=1}^\infty$ for which either the inequalities

$$0\leqslant f(t)<f_{n_k}(t)\ \text{ for }\ t\in D_k\subset E^+(f)\qquad (|D_k|>0,\ k\geqslant 1) \tag{74}$$

or the inequalities

$$f_{n_k}(t)<0\leqslant f(t)\ \text{ for }\ t\in D_k\subset E^+(f)\qquad (|D_k|>0,\ k\geqslant 1) \tag{75}$$

would hold. But then, according to assertions c) and d) (see (74) and (75)), $\{f_{n_k}\}$ would not converge to f in the topology of U_ϕ, which would contradict the fact that $\{f_i\}_1^\infty$ converges on $[0,1]$ to f in the topology of U_ϕ. Corollary 3 is proved.

We can now establish

Theorem 5. *Suppose a continuous on $[0,\infty)$ function $\phi\in\Phi$ with $\phi(t)>0$ for $t>0$ does not satisfy the ω-condition. Then a sequence $\{f_i\}_1^\infty$ with $f_i\in\phi(L)$ $(i\geq 1)$ converges in the topology of U_ϕ to an element $f\in\phi(L)$ if and only if*

1°) *$\{f_i\}_1^\infty$ converges in measure on $[0,1]$ to f*

and there exists a natural number N such that

2°) *$f_i(t)=0$ for almost all $t\in E^{(0)}(f)$ and $i\geq N$,*

3°) *$0\leq f_i(t)\leq f(t)$ for almost all $t\in E^+(f)$ and $i\geq N$,*

4°) *$f(t)\leq f_i(t)\leq 0$ for almost all $t\in E^-(f)$ and $i\geq N$.*

Proof. Suppose conditions 1°)–4°) are satisfied, and let E denote one of the sets $E^{(0)}(f)$, $E^+(f)$ or $E^-(f)$. Then for every function $g(t)$ defined on E and satisfying the conditions

$$\int_E \varphi(g(t))\,dt<\infty,\qquad \int_E\varphi(f+g)\,dt<\infty$$

we have

$$\lim_{i\to\infty}\int_E\varphi(f_i+g)\,dt=\int_E\varphi(f+g)\,dt. \tag{76}$$

In fact, equality (76) (i) is obvious if $E = E^{(0)}(f)$, (ii) follows from assertion b) if $E = E^{+}(f)$, and (iii) also follows from assertion b) (in which $\{f_i\}$ and f are replaced by $\{-f_i\}$ and $-f$ respectively) if $E = E^{-}(f)$. But $E^{(0)}(f) + E^{+}(f) + E^{-}(f) = [0, 1]$, and hence (see (76))

$$\lim_{i\to\infty}\int_0^1 \varphi(f_i+g)\,dt=\int_0^1 \varphi(f+g)\,dt \quad \text{for all} \quad g\in\varphi(L) \text{ with } f+g\in\varphi(L)$$

which implies (see Theorem 4) that $\{f_i\}$ converges to f in the topology of U_ϕ.

Conversely, if $\{f_i\}_1^\infty$ converges to f in the topology of U_ϕ, then, in particular, $\{f_i\}_1^\infty$ converges to f in ϕ-distance. But $\phi(t) > 0$ for $t > 0$, and hence $\{f_i\}$ converges in measure on $[0, 1]$ to f, i.e. 1°) is satisfied.

Further, property 2°) follows directly from assertion a) while property 3°) follows from Corollary 3. As to property 4°), it also follows from Corollary 3 (if $\{f_i\}$ and f are replaced by $\{-f_i\}$ and $-f$ respectively). Theorem 5 is proved.

The "poorness" of the class of convergent sequences $\{f_i\}$ in U_ϕ for the functions ϕ considered in Theorem 5 suggests that for such functions ϕ it might sometimes be more expedient to consider in the set $\phi(L)$ convergence in ϕ-distance or else convergence in the norm of the corresponding Orlicz space (when it exists). At the same time it should be borne in mind that, although the class of convergent sequences indicated in Theorem 5 is very poor, one nevertheless cannot conclude in this case that the set of convergent sequences in the corresponding Orlicz space (when it exists) is wider than the class of convergent sequences in U_ϕ. This will be seen below.

Remark 7. In proving that the fulfilment of requirements 1°)–4°) in Theorem 5 implies the convergence of a sequence $\{f_i\}$ in the topology of U_ϕ, we did not make use of the fact that ϕ does not satisfy the ω-condition (see also assertion b) and Corollary 2).

From the assertions proved above we can now draw certain conclusions that adequately characterize the interrelationships between the various kinds of convergence. For simplicity, we will assume below that the function $\phi \in \Phi$ is continuous on $[0, \infty)$ and positive on $(0, \infty)$. In this connection, if we speak of the norm of an element of the Orlicz space L_ϕ^*, it is assumed in addition that $\phi(t)$ is convex and

$$\lim_{t\to\infty}\frac{\varphi(t)}{t}=\infty.$$

We will denote by D_ϕ, D_{U_ϕ} and $D_{L_\phi^*}$ the classes of sequences in $\phi(L)$ converging respectively in ϕ-distance, in the topology of U_ϕ and in the norm of the Orlicz space L_ϕ^*.

The following four assertions hold.

e) *The inclusion*

$$D_{U_\varphi}\subset D_\varphi \tag{77}$$

is valid for any of the functions ϕ. *It is a strict inclusion if* $\phi(2t) \neq O\{\phi(t)\}$ *for* $t \to \infty$, *while the equality sign holds in* (77) *if* $\phi(2t) = O\{\phi(t)\}$.

This assertion follows directly from the definition of convergence in the topology of U_ϕ and Theorem 2.

f) *The inclusion*

$$D_{L^*_\varphi} \subset D_\varphi \tag{78}$$

is valid for any of the functions ϕ. *It is a strict inclusion if* ϕ *does not satisfy the* Δ_2-*condition, while the equality sign holds in* (78) *if* ϕ *satisfies the* Δ_2-*condition.*

This assertion is well known. In fact, if ϕ satisfies the Δ_2-condition, then convergence in ϕ-distance is equivalent to convergence in the norm of the Orlicz space L^*_ϕ (see [3], Theorem 9.4). If, on the other hand, ϕ does not satisfy the Δ_2-condition, then the requirements $f \in \phi(L)$, $f_i \in \phi(L)$, $i \geq 1$, and $\|f - f_i\|_{L^*_\phi} \to 0$ as $i \to \infty$ imply (see [3], Lemma 9.2) that $\int_0^1 \phi(f - f_i)dt \to 0$ as $i \to \infty$, i.e. $\{f_i\}$ converges to f in ϕ-distance, and hence the inclusion (78) holds. This inclusion is always strict (see [3], §9.6).

g) *If* ϕ *is such that* $\phi(2t) = O\{\phi(t)\}$ *for* $t \to \infty$, *then* $D_{U_\phi} = D_{L^*_\phi} = D_\phi$.

This equality follows directly from e) and f).

h) *If* ϕ *is such that* $\phi(2t) \neq O\{\phi(t)\}$ *for* $t \to \infty$, *then*

$$D_{U_\varphi} - D_{L^*_\varphi} \neq 0 \text{ and } D_{L^*_\varphi} - D_{U_\varphi} \neq 0, \tag{79}$$

i.e. the classes D_{U_ϕ} *and* $D_{L^*_\phi}$ *are noncomparable.*

In fact, take an arbitrary nonnegative function $f(t) \in \phi(L)$ such that $\gamma f(t) \notin \phi(L)$ for all $\gamma > 1$. Such functions exist (see, for example, [6], Corollary 3.3). For $i = 1, 2, \ldots$ let

$$F(t) = \frac{f(t)}{2}, \quad \tau_i(t) = \left(\frac{1}{2} + \frac{1}{i}\right) f(t)$$

and

$$f_i(t) = \begin{cases} f(t) & \text{for } f(t) \leqslant i, \\ 0 & \text{for } f(t) > i. \end{cases}$$

Clearly, $0 \leq f_i(t) \leq f(t)$ for all $t \in [0, 1]$, and the sequence $\{f_i\}_1^\infty$ converges in measure to f. Therefore (see assertion b) and Corollary 2) $\{f_i\}_1^\infty$ converges in the topology of U_ϕ to f, i.e. $\{f_i\}_1^\infty \in D_{U_\phi}$. But $\{f_i\} \notin D_{L^*_\phi}$, since we would otherwise have

$$\|2(f - f_i)\|_{L^*_\varphi} \to 0 \quad \text{as} \quad i \to \infty, \tag{80}$$

which implies

$$\int_0^1 \varphi(2(f-f_i))\,dt<1 \quad \text{for} \quad i\geqslant i_0,$$

i.e.

$$\int_0^1 \varphi(2f(t))\,dt \leqslant \int_{E\{t:f(t)>i_0\}} \varphi(2(f(t)-f_{i_0}(t))\,dt + \int_{E\{t:f(t)\leqslant i_0\}} \varphi(2i_0)\,dt \leqslant 1+\varphi(2i_0)<\infty,$$

and hence $2f \in \phi(L)$, which contradicts the choice of f. Thus $\{f_i\} \in D_{U_\phi}$ and $\{f_i\} \notin D_{L^*_\phi}$, i.e. we have established the first of relations (79).

On the other hand, the sequence $\{\tau_i\}_1^\infty$ converges in the norm of the Orlicz space L^*_ϕ to F, since in this case, if $\lambda_i = \sqrt{i}$, then

$$\tau_i \in \varphi(L), \quad F\in\varphi(L) \text{ and } \int_0^1 \varphi(\lambda_i(F-\tau_i))\,dt = \int_0^1 \varphi\left(\frac{f(t)}{\sqrt{i}}\right)dt \to 0$$

as $i\to\infty$, which implies (see Remark 5) that $\|F-\tau_i\|_{L^*_\phi} \to 0$ as $t\to\infty$. But $\{\tau_i\}$ does not converge in the topology of U_ϕ. For it would otherwise follow from Theorem 4 that the function $g = f/2 \in \phi(L)$ with $F+g = f/2+f/2 = f \in \phi(L)$ would have to be such that

$$\lim_{i\to\infty}\int_0^1 \varphi(\tau_i+g)\,dt = \int_0^1 \varphi(F+g)\,dt = \int_0^1 \varphi(f)\,dt<\infty,$$

which is impossible since

$$\int_0^1 \varphi(\tau_i+g)\,dt = \int_0^1 \varphi\left(\left(1+\frac{1}{i}\right)f(t)\right)dt = \infty \quad \text{for all} \quad i\geqslant 1.$$

This means that $\{\tau_i\} \notin D_{U_\phi}$, although $\{\tau_i\} \in D_{L^*_\phi}$, and the second of relations (79) is thereby also valid. Assertion h) is completely proved.

Remark 8. It follows from Remark 5 and the proof of assertion h) that the difference between the convergence of a sequence $\{f_i\}$ to f in the topology of U_ϕ and its convergence in the norm of the Orlicz space L^*_ϕ consists in the fact that the functions f_i, when they converge in the norm of the Orlicz space, more closely approximate the function f in some sense (for sufficiently large i) than when they converge in the topology of U_ϕ.

BIBLIOGRAPHY

1. J. L. Kelley, *General topology*, Van Nostrand, Princeton, N. J., 1955. MR 16, 1136.

2. A. N. Kolmogorov and S. V. Fomin, *Elements of the theory of functions and of functional analysis*, 2nd ed., "Nauka", Moscow, 1968; English transl. of 1st ed., Vols. I, II, Graylock Press, Albany, N. Y., 1957, 1961. MR **19**, 44; **22** #9566a; **38** #2559.

3. M. A. Krasnosel'skiĭ and Ja. B. Rutickiĭ, *Convex functions and Orlicz spaces*, GITTL, Moscow, 1958; English transl., Noordhoff, Groningen, 1961. MR **21** #5144; **23** #A4016.

4. P. L. Ul'janov, *Weierstrass theorems for $\phi(L)$ class functions*, Dokl. Akad. Nauk SSSR **197** (1971), 1030–1033 = Soviet Math. Dokl. **12** (1971), 629–633. MR **43** #5232.

5. ———, *On some approximative and topological properties of the function space $\phi(L)$*, Dokl. Akad. Nauk SSSR **203** (1972), 534–536 = Soviet Math. Dokl. **13** (1972), 427–430. MR **45** #9124.

6. ———, *Representation of functions by series and the classes $\phi(L)$*, Uspehi Mat. Nauk **27** (1972), no. 2 (164), 3–52 = Russian Math. Surveys **27** (1972), no. 2, 1–42.

7. ———, *On some properties of the function space $\phi(L)$*, Dokl. Akad. Nauk SSSR **216** (1974), 278–281 = Soviet Math. Dokl. **15** (1974), 799–802.

8. I. P. Natanson, *Theory of functions of a real variable*, 2nd rev. ed., GITTL, Moscow, 1957; English transl., Ungar, New York, 1961. MR **26** #6309.

Translated by S. SMITH

Trudy Mat. Inst. Steklov.
134 (1975)

Proc. Steklov Inst. Math.
134 (1975)

ON WELL-POSED PROBLEMS FOR A CLASS OF PARTIALLY HYPOELLIPTIC EQUATIONS IN A HALFSPACE

UDC 517.947.5

S. V. USPENSKIĬ

Abstract. General boundary-value problems, subject to a Lopatinskiĭ condition for a class of partially hypoelliptic operators, are investigated. In particular, operators of the form

$$a_0 \frac{\partial}{\partial t_0} + \sum_{j=1}^{k} a_j \frac{\partial^{2k_j+1}}{\partial t_j^{2k_j+1}} + L[D]$$

occur in this class, where $L[D]$ is a positive definite differential operator. Conditions of solvability and uniqueness in the Sobolev classes W_2^l in a halfspace are obtained.

Bibliography: 15 titles.

In this paper we investigate general boundary-value problems, subject to a Lopatinskiĭ condition for a class of nonclassical equations. An ultraparabolic equation, for example, is a special case of such equations.

The Cauchy problem (cf. [3]–[7]) has been well studied for ultraparabolic (and more general) equations.

Well-posed problems for general equations (systems) with constant coefficients in a halfspace were formulated in the papers of G. E. Šilov and his pupils (see [8]). In this connection, it was assumed that in terms of Fourier transforms the boundary values were attained in special sets determined by the roots of the characteristic equation.

For the class of equations considered in this paper, we study the classical formulation of the problem, where the initial values are attained in the metric of $L_2(E_n)$. This is possible, as is shown, under certain restrictions connecting the dimension of the space and the order of the differential equations. If these restrictions are satisfied, then a solution of the general boundary-value problems exists in the Sobolev class W_2^r and is unique.

The Dirichlet problem was considered earlier for a wide class of partially hypoelliptic equations by S. M. Nikol'skiĭ. For regular (in particular, for convex) regions it was shown in [9]–[11] which boundary data determine a well-posed

---

AMS (MOS) subject classifications (1970). Primary 35H05; Secondary 35E99, 35K35, 46E35.

formulation of the boundary-value problem in the classes S_2^r generalizing the spaces W_2^r.

Let us also mention [12], where for general equations one considers boundary-value problems with nonlocal boundary conditions. A detailed bibliography of papers on these subjects is also supplied there.

Let us introduce some notation:

$$x^\beta = x_0^{\beta_0},\ x_1^{\beta_1}\dots x_n^{\beta_n},\quad E_{n+1}^+ = (x_0 \geqslant 0,\ x'(x_1,\dots,x_n) \in E_n),$$

$$r(x) = \left[\sum_{i=l+1}^n x_i^2\right]^{1/2},\quad 1 \leqslant l \leqslant n,\quad |x| = \left[\sum_{i=1}^n x_i^{2\alpha_i^{-1}}\right]^{1/2},$$

$$D^\beta = D^{\beta_0}\dots D^{\beta_n} = \frac{\partial^{\beta_0}}{\partial x_0^{\beta_0}}\cdots\frac{\partial^{\beta_n}}{\partial x_n^{\beta_n}},\quad l = (l_0,\ l_1,\dots,l_n),$$

$$\alpha = (\alpha_0,\ \alpha_1,\dots,\alpha_n),\quad l\alpha = \sum_{i=0}^n l_i\alpha_i,\quad |\alpha| = \sum_{i=1}^n \alpha_i,$$

$$\alpha_i > 0,\ i = 0,\ 1,\ \dots, n,\qquad 1/\alpha_i \text{ are integers},$$

$$G_0(x_i) = \exp - |x_i|^{2m},\qquad G_0(x) = \prod_{i=1}^n G_0(x_i),$$

$$f_h = h^{-|\alpha|}(2\pi)^{-n}\int_{E_n}\hat{G}_0\left(\frac{t-x}{h^\alpha}\right)f(t)\,dt,\ \text{where}$$

$$\hat{G}_0(x) = \int_{E_n} e^{-ixt}G_0(t)\,dt,\qquad \|f\|_{L_2(E_n)} = \left(\int_{E_n}|f|^2\,dx\right)^{1/2}.$$

Let us introduce a class of operators $L[D]$ with constant coefficients:

$$L[D] = L[D_{x_0},\dots,D_{x_n}] = D_{x_0}^{1/\alpha_0} + L_1[D_{x_0},\dots,D_{x_n}] + L_2[D_{x_1},\dots,D_{x_n}].$$

We shall assume that the characteristic polynomial $L[is] = L[is_0,\dots,is_n]$ of the operator $L[D]$ satisfies the following conditions:

1) There exists a real vector $\alpha = (\alpha_0,\dots,\alpha_n)$, $\alpha_i > 0$, $i = 0,\dots,n$, such that for any $s \in E_{n+1}$ and $\lambda > 0$ one has

$$L[\lambda^\alpha is] = \lambda L[is].$$

2) $L[is_0,\dots,is_n] \neq 0$ if $s \in E_{n+1}$ and

$$r(s) = \left[\sum_{j=l+1}^n s_j^2\right]^{1/2} \neq 0,$$

where the roots of the equation

$$L[i\tau,\ is_1,\dots,is_n] = 0$$

satisfy the estimate

$$|I_m\tau_k| > Cr(s)^\mu,$$

for any $s \in E_n$, $|s| = 1$, $r(s) \neq 0$, and the constant C does not depend on s, $\mu > 0$.

3) The polynomial $L_1[is_0, \ldots, is_n]$ has degree less than $1/\alpha_0$ with respect to s_0.

For $1/\alpha_0 > 1$ we require additional restrictions on $L[is]$.

4) If $1/\alpha_0$ is even, then $L[-is] = L[is]$; if $1/\alpha_0$ is odd, then

$$L_1[-is] = -L_1[is], \quad L_2[is] = \sum_{k=l+1}^{n} s_k Q_k[is_1, \ldots, is_n]$$

and one has the estimate

$$L_2[is] > C \sum_{k=l+1}^{n} s_k^{2\gamma_k},$$

where $\gamma_k > 0$ are integers.

Example. The operator

$$\left(a_0 D_{t_0}^{2k_0+1} + \sum_{j=1}^{l} a_j D_{t_j}^{2k_j+1}\right)^{2\nu+1} + (-1)^m L_{2m}[D_{l+1}, \ldots, D_n],$$

where L_{2m} is a homogeneous positive definite elliptic operator of order $2m$ which satisfies the conditions 1) 4).

Let us make yet one more assumption on the roots of the equation

$$L[i\tau, is_1, \ldots is_n] = 0$$

for $s \in E_n$, $|s| = 1$, and $r(s) \neq 0$. We select roots $\tau_1^+, \ldots, \tau_m^+$ having $I_m \tau_k^+ > 0$ and roots $\tau_{m+1}^-, \ldots, \tau_{1/\alpha_0}$ having $I_m \tau_m^- < 0$.

5) We shall assume that for all $s \in E_n$, $|s| = 1$, and $r(s) \neq 0$ the number of roots with $I_m \tau_k > 0$ is fixed. For definiteness, we shall assume that

$$m = \max\left(m, \frac{1}{a_0} - m\right).$$

Let us consider the differential operators

$$B_j[D] = \sum_{k=1}^{m} P_{k,j}[D_1, \ldots, D_n] D_{x_0}^{k-1}, \quad j = 1, \ldots, m,$$

where $P_{k,j}$ is a polynomial in $D_1, \ldots, D_n$. We shall assume that the operators $B_j[D]$ satisfy the following conditions:

6) Let $\alpha = (\alpha_0, \ldots, \alpha_n)$ be a vector satisfying the condition 1); then for any $\lambda > 0$ and $s \in E_{n+1}$ one has

$$B_j[is\lambda]^\alpha = \lambda^{\beta_j} B_j[is],$$

where $\beta_j \geq 0$. The number β_j will be called the *homogeneity coefficient* of the operator B_j.

7) Lopatinskiĭ condition: the polynomials $B_j[i\tau, is_1, \ldots, is_n]$, $j = 1, \ldots, m$, as polynomials in τ, are linearly independent for any fixed arguments. Let us define the Sobolev class $W_2^{2/\alpha}(E_n)$, where $\alpha = (\alpha_0, \ldots, \alpha_n)$, $\alpha_i > 0$, and $1/\alpha_i$ are integers. We shall say that a function $f \in W_2^{2/\alpha}(E_n)$ if f has generalized derivatives $D^\beta f$, $\beta\alpha \leq 2$, and

$$\overset{\bullet}{\|} f \|_{W_2^{2/\alpha}(E_n)} = \sup_{x_0>0} \sum_{0\leqslant\beta\alpha\leqslant 2} \| D^\beta U \|_{L_2(E_n)} < \infty, \qquad f \in W_2^{2/\alpha}(E_n) \cap \overset{\bullet}{M},$$

if $f \in W_2^{2/\alpha}(E_n)$ and $\sup_{x\in E_{n+1}^+} |U| < \infty$.

Let us pose the following boundary-value problem.

To find a function $U \in W_2^{2/\alpha}(E_n) \cap M$ satisfying the equation

$$L[D]U = 0 \tag{1}$$

in the halfspace E_{n+1}^+, and the boundary condition

$$\begin{gathered} B_j[D]U|_{x_0=0} = \varphi_j, \quad j=1,\ldots m, \\ \lim_{x_0\to\infty} \| D^\nu U \|_{L_2(E_n)} = 0, \quad 0 \leqslant \nu\alpha \leqslant 2, \qquad \text{for all} \quad \nu. \end{gathered} \tag{2}$$

Here the equalities (1) and (2) are understood in the sense of the metric $L_2(E_n)$. We shall prove below that a solution of the problem (1), (2) exists and is unique if the operator $L[D]$ and the boundary operators are subject to the conditions 1)–7). We shall carry out the proof by using the method of Agmon, Douglis, and Nirenberg [13]. Let us formulate a series of auxiliary lemmas.

By the condition 7) the polynomials

$$B_j[i\tau, is] = \sum_{k=1}^{m} b_{j,k}(s)\tau^{k-1},$$

as polynomials in τ for any fixed $s \in E_n$, are linearly independent. Then the determinant of the matrix B satisfies

$$|B| = \begin{vmatrix} b_{11} & \ldots & b_{1m} \\ \cdot & \cdot\cdot\cdot & \cdot \\ b_{m1} & \ldots & b_{mm} \end{vmatrix} \neq 0 \tag{3}$$

for any $s \in E_n$. Let us denote by B^{-1} the matrix inverse to B, and by b^{jk} the elements of B^{-1}.

For $s \in E_n$, $|s| = 1$, and $r(s) \neq 0$ let us define the polynomial

$$M^+(\tau, s) = \prod_{k=1}^{m} (\tau - \tau_k^+) = \sum_{k=0}^{m} a_k(s)\tau^{m-k}, \quad a_0 = 1,$$

where $\tau_1^+, \ldots, \tau_m^+$ are the roots of $L[i\tau, is_1, \ldots, is_n] = 0$ for $|s| = 1$, $r(s) \neq 0$, $I_m \tau_k^+ > 0$. Put

$$\begin{gathered} M_j^+(\tau, s) = \sum_{k=0}^{j} a_k(s)\tau^{j-k}, \\ N_k(\tau, s) = \sum_{j=1}^{m} b^{jk} M_{m-j}^+(\tau, s). \end{gathered} \tag{4}$$

Since it follows from 2) that $|I_m \tau_k^+| > Cr(s)^\mu$ for $|s| = 1$, one can construct for any $\epsilon > 0$ and $r(s) > \epsilon$ a bounded contour γ_ϵ which is situated in the upper complex halfplane and which contains all of the roots $\tau_1^+, \ldots, \tau_m^+$. In the sequel we shall assume that in the complex region $z = x + iy$ the contour γ_ϵ (for any $\epsilon >$ 0) is the boundary of the region

$$G_\varepsilon = \left(|z| \leqslant N \cap y \geqslant \frac{C\varepsilon}{3}\right),$$

where N is a number sufficiently large that the disk $|z| < N/2$ contains all the zeros τ_k^+ for $|s| = 1$, and the constant C satisfies the condition 2). Then for $r(s) \geq \epsilon$ and $\tau \in \gamma_\epsilon$ one has

$$|\tau - \tau_k^+| \geqslant c_1 r(s)^\mu. \tag{5}$$

Indeed, if τ lies on the line $y = C/3\epsilon$, then

$$|\tau - \tau_k^+| \geqslant |I_m(\tau - \tau_k^+)| \geqslant \frac{2}{3}|I_m \tau_k^+| \geqslant \frac{2}{3} Cr(s)^\mu,$$

where the last inequality in the chain follows from condition 2). If also $\tau \in |z| = N$, then

$$|\tau - \tau_k^+| \geqslant \frac{N}{2} \geqslant c_1 r(s)^\mu.$$

It is not difficult to see that for $r(s) \geq \epsilon$ the estimate (5) is preserved if one replaces the contour γ_ϵ by the contour γ_{ϵ_1}, $\epsilon_1 < \epsilon$.

Lemma 1. *Let* $|s| = 1$ *and* $r(s) \geq \epsilon$. *Put*

$$I_{k,j} = \frac{1}{2\pi i} \int_{\gamma_\varepsilon} \frac{B_j[i\tau, is_1, \ldots, is_n] N_k(\tau, s)}{M^+(\tau, s)} d\tau.$$

Then

$$I_{k,j} = \begin{vmatrix} 1, & k = j \\ 0, & k \neq j \end{vmatrix} = \delta_{k,j}.$$

The proof of Lemma 1 is carried out in a manner analogous to that used in [13], p. 634.

Using the definition of $N_k(\tau, s)$, we have

$$I_{k,j} = \sum_{p=1}^m \frac{1}{2\pi i} \int_{\gamma_\varepsilon} b_{jp} \tau^{p-1} \sum_{s=1}^m b^{sk} \frac{M_{m-s}^+}{M^+} d\tau = \frac{1}{2\pi i} \sum_{p=1}^m \sum_{s=1}^m b_{jp} b^{sk} \int_{\gamma_\varepsilon} \tau^{p-1} \frac{M_{m-s}^+}{M^+} d\tau.$$

Let us denote by $Q_{m-s+p-1}$ the polynomial $\tau^{p-1} M_{m-s}^+$ of degree $m - s + p - 1$ with respect to τ. We prove that

$$\int_{\gamma_\varepsilon} \frac{Q_{m-s+p-1}}{M^+} d\tau = \delta_{p,s}. \tag{6}$$

a) $s > p$; then $m - s + p - 1$. Deforming γ_ϵ to a circle S_R of sufficiently large radius, we obtain

$$\left| \int_{\gamma_\varepsilon} \frac{Q_{m-s+p-1}}{M^+} d\tau \right| = \left| \int_{S_R} \frac{Q_{m-s+p-1}}{M^+} d\tau \right| < \frac{C}{R} \to 0 \quad (R \to \infty).$$

b) $s = p$. Since the coefficient of the leading term of $Q_{m-s+p-1}$ is 1, it follows that

$$\frac{1}{2\pi i} \int_{\gamma_\varepsilon} \frac{Q_{m-s+p-1}}{M^+} d\tau = \frac{1}{2\pi i} \int_{\gamma_\varepsilon} \frac{\tau Q_{m-1}}{\tau M^+} d\tau$$

$$= \frac{1}{2\pi i} \int_{S_R} \frac{1}{\tau} d\tau - \frac{1}{2\pi i} \int_{S_R} \frac{Q_{m-1}^1}{M^+} d\tau = 1,$$

where the last equality in the chain is a consequence of a).

c) $s < p$; then

$$Q_{m-s+p-1} = M^+_{m-s}\tau^{p-1} = \sum_{k=0}^{m-s} a_k\tau^{m-s+k+p-1} = \sum_{k=0}^{m} a_k\tau^{m-s-k+p-1}$$

$$- \sum_{k=m-s+1}^{m} a_k\tau^{m-s-k+p-1} = \tau^{p-s-1}M^+ - R_\nu,$$

where R_ν is a polynomial of degree $\nu \leq p-2 \leq m-2$. We have

$$\frac{1}{2\pi i}\int_{\gamma_\varepsilon} \frac{Q_{m-s+p-1}}{M^+}\,d\tau = \frac{1}{2\pi i}\int_{\gamma_\varepsilon} \tau^{p-s-1}d\tau + \frac{1}{2\pi i}\int_{\gamma_\varepsilon} \frac{R_\nu}{M^+}\,d\tau = 0.$$

Taking (6) into account, we obtain

$$I_{k,j} = \sum_{p=1}^{m}\sum_{s=1}^{m} b_{jp}b^{sk}\delta_{p,s} = \sum_{p=1}^{m} b_{jp}b^{pk} = \delta_{j,k}.$$

Lemma 2. *Put*

$$G_0^\varepsilon(x) = \begin{vmatrix} G_0(x), & r(x) > \varepsilon, \\ 0, & r(x) \leqslant \varepsilon, \end{vmatrix}$$

$$G_\varepsilon(x) = |x|^{2m} G_0^\varepsilon(x).$$

If $f \in L_2(E_n)$, *then*

$$f = \lim_{h\to 0}\lim_{\varepsilon\to 0}\int_h^{h^{-1}} v^{-|\alpha|-1}\int_{E_n} \hat{G}_\varepsilon\left(\frac{t-x}{v^\alpha}\right) f(t)\,dt\,dv$$

in the norm of $L_2(E_n)$.

Proof. It follows from a paper of the author [14] that if $f \in L_2(E_n)$, then

$$f = \lim_{h\to 0}\int_h^{h^{-1}} v^{-|\alpha|-1}\int_{E_n}\int_{E_n} G_0(s)\,|s|^{2m} e^{\frac{i(x-t)s}{v^\alpha}}\,ds f(t)\,dt\,dv$$

in the norm of L_2. Therefore it suffices to prove that

$$\lim_{\varepsilon\to 0}\int_{E_n}\int_{E_l}\int_{r(s)\leqslant\varepsilon} e^{\frac{i(x-t)s}{v^\alpha}}\,|s|^{2m} G_0(s)\,ds f(t)\,dt = 0.$$

Denoting

$$\psi_\varepsilon(x) = \begin{vmatrix} G_0(x)\,|x|^{2m}, & r(x) \leqslant \varepsilon, \\ 0 & r(x) > \varepsilon \end{vmatrix}$$

and applying the Minkowski inequality, we obtain

$$\left|\int_{E_n} \hat{\psi}_\varepsilon\left(\frac{t-x}{v^\alpha}\right) f(t)\,dt\right| \leqslant v^{|\alpha|}\|f\|_{L_2(E_n)}\|\hat{\psi}_\varepsilon\|_{L_2(E_n)}$$

$$\leqslant c\|f\|_{L_2(E_n)}\left[\int_{E_l}\int_{r(s)\leqslant\varepsilon} (|s|^{2m} G_0(s))^2\,ds\right]^{1/2} \to 0 \quad (\varepsilon \to 0).$$

Let us put

$$u_{h,\varepsilon}=\sum_{k=1}^{m}\frac{1}{(2\pi)^n}\int_h^{h^{-1}} v^{-|\alpha|+\beta_k-1}\int_{E_n}\int_{E_n} e^{\frac{i(x-t)s}{v^\alpha}}\frac{G_\varepsilon(s)}{|s|^{\beta_k}}$$

$$\times\frac{1}{2\pi i}\int_{\gamma_\varepsilon} e^{\frac{ix_0\tau|s|^{\alpha_0}}{v^{\alpha_0}}}\frac{N_k\left(\tau,\frac{s}{|s|^\alpha}\right)}{M^+\left(\tau,\frac{s}{|s|^\alpha}\right)}d\tau\varphi_k(t)\,dsdtdv.$$

Here $\alpha=(\alpha_0,\ldots,\alpha_n)$ satisfies 1), β_k is the homogeneity coefficient satisfying 6), and M^+ and N_k are polynomials in τ which were defined in (4).

Let us denote

$$\frac{G_\varepsilon(s)}{|s|^{\beta_k}}\frac{1}{2\pi i}\int_{\gamma_\varepsilon} e^{ix_0\tau|s|^{\alpha_0}}\frac{N_k\left(\tau,\frac{s}{|s|^\alpha}\right)}{M^+\left(\tau,\frac{s}{|s|^\alpha}\right)}d\tau=\Phi_{\varepsilon,k}(s,x_0), \tag{7}$$

$$\frac{1}{(2\pi)^n}\int_{E_n} e^{-ixs}\Phi_{\varepsilon,k}(s,x_0)\,ds=\hat{\Phi}_{\varepsilon,k}(x,x_0). \tag{8}$$

Then one can represent $U_{h,\epsilon}$ in the form

$$U_{h,\varepsilon}=\sum_{k=1}^{m}\int_h^{h^{-1}} v^{-|\alpha|+\beta_k-1}\int_{E_n}\hat{\Phi}_{\varepsilon,k}\left(\frac{t-x}{v^\alpha},\frac{x_0}{v^{\alpha_0}}\right)\varphi_k dtdv. \tag{9}$$

Let us study the differential properties of $U_{h,\epsilon}$. We shall assume that

$$\varphi_k\in L_2(E_n)\cap L_1(E_n),\ \nu=(\nu_0,\nu_1,\ldots,\nu_n),\ \alpha=(\alpha_0,\alpha_1,\ldots,\alpha_n),\ 0\leqslant\nu\alpha\leqslant 2.$$

Applying the Hölder and Minkowski inequalities, we have

$$|x_0|\,\|D^\nu U_{h,\varepsilon}\|_{L_2(E_n)}\leqslant C_{\varepsilon,h}\sum_{k=1}^{m}\|\varphi_k\|_{L_1}\left\|s^\nu\frac{\Phi_{\varepsilon,k}(s,x_0)}{|s|^{\alpha_0}}\right\|_{L_2}$$

$$|U_{h,\varepsilon}|\leqslant C_{\varepsilon,h}\sum_{k=1}^{m}\|\varphi_k\|_{L_2}\|\Phi_{\varepsilon,k}(s,x_0\|_{L_2}.$$

Since

$$\left|\frac{\Phi_{\varepsilon,k}(s,x_0)s^\nu}{|s|^{\alpha_0}}\right|\leqslant C_\varepsilon\left|\frac{s^\nu G_\varepsilon(s)}{|s|^{\alpha_0}}\right|\in L_2(E_n),$$

we have

$$|U_{h,\varepsilon}|\leqslant c_{\varepsilon,h}\sum_{k=1}^{m}\|\varphi_k\|_{L_2(E_n)},$$

$$|x_0|\,\|U_{h,\varepsilon}\|_{L_2(E_n)}\leqslant C_{\varepsilon,h}\sum_{k=1}^{m}\|\varphi_k\|_{L_1(E_n)} \tag{10}$$

where the constants $C_{\epsilon,h}$ do not depend on x_0.

Lemma 3.

$$L[D]U_{h,\varepsilon}=0,\ h>0,\ \varepsilon>0. \tag{11}$$

Proof. Using the homogeneity of $L[D]$, we obtain·

$$L[D]\hat{\Phi}_{\varepsilon,h}\left(\frac{t-x}{v^{\alpha}}, \frac{x_0}{v^{\alpha_0}}\right)=v^{-1}\int_{E_n} e^{\frac{i(x-t)s}{v^{\alpha}}}\frac{G_\varepsilon(s)}{|s|^{\beta_k}}$$

$$\times\frac{1}{2\pi i}\int_{\gamma_\varepsilon} e^{\frac{ix_0\tau|s|^{\alpha_0}}{v^{\alpha_0}}}\frac{N_k\left(\tau, \frac{s}{|s|^{\alpha}}\right)}{M^+\left(\tau, \frac{s}{|s|^{\alpha}}\right)}L[i\tau|s|^{\alpha_0}, is_1, \dots is_n]\,d\tau ds$$

$$=v^{-1}\int_{E_n} e^{\frac{i(x-t)s}{v^{\alpha}}}G_\varepsilon(s)|s|^{1-\beta_k}\frac{1}{2\pi i}\int_{\gamma_\varepsilon} e^{\frac{ix_0|s|^{\alpha_0}\tau}{v^{\alpha_0}}}\frac{L\left[i\tau, \frac{is}{|s|^{\alpha}}\right]N_k\left(\tau, \frac{s}{|s|^{\alpha}}\right)}{M^+\left(\tau, \frac{s}{|s|^{\alpha}}\right)}d\tau ds=0,$$

since

$$L\left[i\tau, \frac{is}{|s|^{\alpha}}\right]=M^+\left(\tau, \frac{s}{|s|^{\alpha}}\right)M^-\left(\tau, \frac{s}{|s|^{\alpha}}\right).$$

Lemma A. *If* $\phi_k\in L_2(E_n)\cap L_1(E_n)$, *then*

$$\lim_{h\to 0}\lim_{\varepsilon\to 0}B_j[D]U_{h,\varepsilon}\big|_{x_0=0}=\varphi_j,\ j=1, \dots m, \tag{12}$$

in the norm of $L_2(E_n)$.

Proof. Using the homogeneity of $B_j[i\tau, is]$, we have

$$B_j[D]\hat{\Phi}_{\varepsilon,k}\left(\frac{t-x}{v^{\alpha}}, \frac{x_0}{v^{\alpha_0}}\right)\bigg|_{x_0=0}=v^{\beta_j}\int_{E_n} e^{\frac{(x-t)is}{v^{\alpha}}}\frac{G_\varepsilon(s)|s|^{\beta_j}}{|s|^{\beta_k}}\frac{1}{2\pi i}$$

$$\times\int_{\gamma_\varepsilon}\frac{B_j\left[i\tau, \frac{is}{|s|^{\alpha}}\right]N_k\left(\tau, \frac{s}{|s|^{\alpha}}\right)}{M^+\left(\tau, \frac{s}{|s|^{\alpha}}\right)}d\tau ds$$

$$=\begin{vmatrix} v^{-\beta_k}\hat{G}_\varepsilon\left(\frac{t-x}{v^{\alpha}}\right), & k=j, \\ 0, & k\neq j, \end{vmatrix}$$

where the integral is evaluated with the help of Lemma 1. Then

$$B_j[D]U_{h,\varepsilon}|_{x_0=0}=\frac{1}{(2\pi)^n}\int_h^{h^{-1}} v^{-|\alpha|-1}\int_{E_n}\hat{G}\left(\frac{t-x}{v^{\alpha}}\right)\varphi_j dt dv.$$

Applying Lemma 2, we obtain (12).

Lemma 5. *Suppose that the condition* 2) *is satisfied and* $\phi_k\in W_1^{r_k}(E_n)$, *where* $r_k=(r_1^k, \dots, r_n^k)$ *satisfies the conditions*

$$-\frac{|\alpha|}{2}+\beta_k-2+r_s^k\alpha_s>0,\ \alpha=(\alpha_0, \alpha_1, \dots, \alpha_n), k=1, \dots, m, s=1, \dots, n,$$
$$\frac{|\alpha|}{2}>\beta_k,\ 2m\mu<n-l. \tag{13}$$

Then there exist $\delta_1, \delta_2>0$ *such that for any* $\nu=(\nu_0, \nu_1, \dots, \nu_n)$, $0\leq\nu\alpha\leq 2$, *one has the estimate*

$$\| D^\nu U_{h_1, \varepsilon_1} - D^\nu U_{h_2, \varepsilon_2} \|_{L_2(E_n)} \leqslant C (\varepsilon^{\delta_1} + h^{\delta_2}) \sum_{k=1}^{m} \| \varphi_k \|_{W_1^{r_k}(E_n)}, \tag{14}$$

where the constant C does not depend on x_0, ϵ *or* h; *here* $\epsilon = \max(\epsilon_1, \epsilon_2)$ *and* $h = \max(h_1, h_2)$.

Proof. For definiteness, let us put $\epsilon_2 > \epsilon_1$ and $h_2 > h_1$. We have

$$\begin{aligned}
&\| D^\nu U_{h_1, \varepsilon_1} - D^\nu U_{h_2, \varepsilon_2} \|_{L_2(E_n)} \\
&\leqslant \left\| \sum_{k=1}^{m} \int_{h_1}^{1} v^{-|\alpha|+\beta_k-1} \int_{E_n} [D^\nu \hat{\Phi}_{\varepsilon_1, k} - D^\nu \hat{\Phi}_{\varepsilon_2, k}] \varphi_k \, dt dv \right\|_{L_2} \\
&+ \left\| \sum_{k=1}^{m} \int_{h_1}^{h_2} v^{-|\alpha|+\beta_k-1} \int_{E_n} D^\nu \hat{\Phi}_{\varepsilon_2, k} \varphi_k \, dt dv \right\|_{L_2} \\
&+ \left\| \sum_{k=1}^{m} \int_{h_2^{-1}}^{h_1^{-1}} v^{-|\alpha|+\beta_k-1} \int_{E_n} D^\nu \hat{\Phi}_{\varepsilon_2, k} \varphi_k \, dt dv \right\|_{L_2} \\
&+ \left\| \sum_{k=1}^{m} \int_{1}^{h_1^{-1}} v^{-|\alpha|+\beta_k-1} \int_{E_n} [D^\nu \hat{\Phi}_{\varepsilon_1, k} - D^\nu \Phi_{\varepsilon_2, k}] \varphi_k \, dt dv \right\|_{L_2} \\
&= \|I_1\| + \|I_2\| + \|I_3\| + \|I_4\|.
\end{aligned}$$

It suffices to estimate a term in each sum I_j, $j = 1, \ldots, 4$. Applying the generalized Minkowski inequality and denoting $|y|^{2l} = \Sigma a_s y^{\gamma_s}$, we obtain

$$\begin{aligned}
\|I_1\| &\leqslant \sum |a_s| \int_0^1 v^{-|\alpha|+\beta_k-1-\nu\alpha+r_s^k\alpha_s} \left\| \int_{E_n} e^{\frac{i(x-t)y}{v^\alpha}} [\Phi_{\varepsilon_1, k} - \Phi_{\varepsilon_2, k}] \right. \\
&\times \left. \frac{|y|^{\alpha_0 \nu_0} y^{\sigma_s}}{|y|^{2l}} dy \right\|_{L_2} dv \| \varphi_k \|_{W_1^{r_k}} \\
&\leqslant \sum |a_s| \int_0^1 v^{-\gamma+r_s^k\alpha_s} \left\| \int_{E_n} e^{-ity} |y|^p y^{\sigma_s} [G_{\varepsilon_1}(y) - G_{\varepsilon_2}(y)] \right. \\
&\times \left. \frac{1}{2\pi i} \int_{\gamma_{\varepsilon_1}} e^{\frac{i x_0 \tau |y|^{\alpha_0}}{v^{\alpha_0}}} \tau^{\nu_0} \frac{N_k\left(\tau, \frac{y}{|y|^\alpha}\right)}{M^+\left(\tau, \frac{y}{|y|^\alpha}\right)} d\tau dy \right\|_{L_2} dv \| \varphi_k \|_{W_1^{r_k}} \\
&\leqslant c \sum |a_s| \int_0^1 v^{-\gamma+r_s^k\alpha_s} dv \left\| \frac{|y|^p y^{\sigma_s} [G_{\varepsilon_1} - G_{\varepsilon_2}]}{r(y)^{m\mu}} \right\|_{L_2} \| \varphi_k \|_{W_1^{r_k}} \\
&\leqslant c \varepsilon_2^{\delta} \sum |a_s| \int_0^1 v^{-\gamma+r_s^k\alpha_s} dv \left\| \frac{|y|^{p_1} y^{\sigma_s} G_0(y)}{r(y)^{m\mu+\delta}} \right\|_{L_2} \| \varphi_k \|_{W_1^{r_k}},
\end{aligned}$$

wherein $\gamma = |\alpha|/2 + \beta_k - 1 - \nu\alpha$ and the third inequality follows from the estimate $|\tau - \tau_k^+| \geq c r(y)^\mu$. We have

$$-\gamma+r_s^k\alpha_s=\frac{-|\alpha|}{2}+\beta_k-1-\nu\alpha+r_s^k\alpha_s>-1.$$

From the conditions of the lemma follows $2m\mu < n - l$. Consequently there exist $\delta > 0$, $2(m\mu + \delta) < n - l$ and a function $r(y)^{-2(m\mu+\delta)}$ which is integrable in a neighborhood of the zero $y_j = 0$, $j = l + 1, \ldots, n$. Consequently

$$\|I_1\|_{L_2}\leqslant C\varepsilon_2^\delta\sum_{k=1}^m\|\varphi_k\|_{W_1^{rk}}. \tag{15}$$

Carrying out computations as in (15), we obtain

$$\begin{aligned}\|I_2\|_{L_2(E_n)}&\leqslant\sum|a_s|\int_{h_1}^{h_2}v^{-\frac{|\alpha|}{2}+\beta_k-1-\nu\alpha+r_s^k\alpha_s}dv\left\|\frac{|y|^p y^{\sigma_s}G_0(y)}{r(y)^{\mu m}}\right\|_{L_2}\|\varphi_k\|_{W_1^{rk}}\\&\leqslant C(h_1^\gamma+h_2^\gamma)\|\varphi_k\|_{W_1^{rk}},\end{aligned} \tag{16}$$

using $\gamma = -|\alpha|/2 + \beta_k - 2 + r_s^k\alpha_s > 0$;

$$\|I_3\|_{L_2(E_n)}\leqslant c\int_{h_2^{-1}}^{h_1^{-1}}v^{\frac{-|\alpha|}{2}+\beta_k-1-\nu\alpha}dv\left\|\frac{y^\sigma|y|^pG_0(y)}{r(y)^{m\mu}}\right\|_{L_2}\|\varphi_k\|_{W_1^{rk}}\leqslant Ch_2^{\frac{|\alpha|}{2}-\beta_k}\|\varphi_k\|_{W_1^{rk}}, \tag{17}$$

since $|\alpha|/2 > \beta_k$;

$$\begin{aligned}\|I_4\|_{L_2}&\leqslant c\int_1^\omega v^{\frac{-|\alpha|}{2}+\beta_k-1-\nu\alpha}dv\left\|\int_{E_n}e^{-ity}|y|^{\nu_0\alpha_0}y^\sigma[\Phi_{\varepsilon_1,k}-\Phi_{\varepsilon_2,k}]\,dy\right\|_{L_2}\|\varphi_k\|_{L_1}\\&\leqslant C\varepsilon^\delta\int_1^\infty v^{-\frac{|\alpha|}{2}+\beta_k-1-\nu\alpha}\left\|\frac{|y|^py^\sigma G_0(y)}{r(y)^{\mu m+\delta}}\right\|_{L_2}\|\varphi_k\|_{L_1}\leqslant C\varepsilon^\delta\|\varphi_k\|_{L_1}.\end{aligned} \tag{18}$$

We obtain the lemma from (18), (17), (16) and (15).

Corollary 1. *Under the conditions of the lemma,*

$$\|D^\nu U_{h,\varepsilon}\|_{L_2(E_n)}\leqslant C\sum_{k=1}^m\|\varphi_k\|_{W_1^{rk}(E_n)}, \tag{19}$$

where C does not depend on x_0, ϵ or h, and $0 \le \nu\alpha \le 2$.

The proof follows from (10) and (14).

Corollary 2. *Under the conditions of the lemma, the sequence $U_{h,\epsilon}$ converges in $W_2^{2/\alpha}(E_n)$ to a function U_0, and for U_0 the estimate* (19) *is satisfied.*

Corollary 3. *Under the conditions of the lemma*

$$\lim_{x_0\to\infty}\|D^\nu U_0\|_{L_2(E_n)}=0,$$

if $0 \le \nu\alpha \le 2$.

Proof. It follows from (10) that for all fixed $\epsilon > 0$ and $h > 0$

$$\|D^\nu U_{h,\varepsilon}\|_{L_2(E_n)}\to 0\quad(x_0\to\infty).$$

Since the sequence $D^{\nu}U_{h,\epsilon}$ converges to $D^{\nu}U_0$ uniformly with respect to x_0, we have

$$\|D^{\nu}U_0\|_{L_2(E_n)} \to 0 \quad (x_0 \to \infty).$$

Lemma 6. *Suppose that the condition* 2) *is satisfied, and*

$$\varphi_k \in W_2^{r_k}(E_n), \text{ where } r_k = (r_1^k, \dots r_n^k),$$

$$\frac{-|\alpha|}{2} + \beta_k - 2 + r_s^k \alpha_s > 0, \quad \frac{|\alpha|}{2} > \beta_k, \quad 2\mu m < n - l,$$

$$\alpha = (\alpha_0, \alpha_1, \dots \alpha_n), \; k = 1, \dots, m, \; s = 1, \dots, n.$$

Then there exist δ_1 *and* δ_2 *such that one has the estimate*

$$\sup_{x \in E_{n+1}^+} |U_{h_1,\varepsilon_1} - U_{h_2,\varepsilon_2}| \leqslant C(\varepsilon^{\delta_1} + h^{\delta_2}) \sum_{k=1}^{m} \|\varphi_k\|_{W_2^{r_k}}, \tag{20}$$

where the constant C *does not depend on* ϵ, h *or* ϕ_k; *here* $\epsilon = \max(\epsilon_1, \epsilon_2)$ *and* $h = \max(h_1, h_2)$.

Proof. Using the notation of Lemma 5, we have

$$|U_{h_1,\varepsilon_1} - U_{h_2,\varepsilon_2}| \leqslant |I_1| + |I_2| + |I_3| + |I_4|,$$

$$|I_1| \leqslant \sum |a_s| \left| \int_0^1 v^{-|\alpha|+\beta_k-1+r_s^k\alpha_s} \int_{E_n}\int_{E_n} e^{-\frac{(x-t)y}{v^{\alpha}}} y^{\sigma_s} |y|^{-2l} [\Phi_{\varepsilon_1,k} - \Phi_{\varepsilon_2,k}] D^{r_s^k}\varphi_k \, dy\, dt\, dv \right|$$

$$\leqslant \sum |a_s| \cdot \int_0^1 v^{\frac{-|\sigma|}{2}+\beta_k-1+r_s^k\alpha_s} dv$$

$$\times \left\| \int_{E_n} e^{-ity} y^{\sigma_s} |y|^{-2l} |G_{\varepsilon_1}(y) - G_{\varepsilon_2}(y)| r^{-m\mu}(y)\, dy \right\|_{L_2} \|\varphi_k\|_{W_2^{r_k}} \leqslant c\varepsilon^{\delta} \|\varphi_k\|_{W_2^{r_k}},$$

applying Hölder's inequality and then carrying out computations as in (15) for $\nu = 0$;

$$|I_2| \leqslant \sum |a_s| \int_{h_1}^{h_2} v^{-\frac{|\alpha|}{2}+\beta_k-1+r_s^k\alpha_s} dv \, \|y^{\sigma_s}\|y|^{-2l}\Phi_{\varepsilon_2,k}\|_{L_2} \|\varphi_k\|_{W_2^{r_k}}$$

$$\leqslant c \sum a_s h^{-\frac{|\alpha|}{2}+\beta_k+r_s^k\alpha_s} \|\varphi_k\|_{W_2^{r_k}},$$

$$|I_3| \leqslant C \int_{h_2^{-1}}^{h_1^{-1}} v^{-\frac{|\alpha|}{2}+\beta_k-1} dv \, \|\Phi_{\varepsilon_2,k}\|_{L_2} \|\varphi_k\|_{L_2(E_n)} \leqslant Ch^{-\frac{|\alpha|}{2}+\beta_k} \|\varphi_k\|_{L_2(E_n)},$$

$$|I_4| \leqslant c \int_1^{\infty} v^{-\frac{|\alpha|}{2}+\beta_k-1} \|\Phi_{\varepsilon_1,k} - \Phi_{\varepsilon_2,k}\|_{L_2} \|\varphi_k\|_{L_2} \leqslant c\varepsilon^{\delta} \|\varphi_k\|_{L_2(E_n)}.$$

Corollary 1. *Under the conditions of Lemma* 6,

$$|U_{h,\delta}| \leqslant C \sum_{k=1}^{m} \|\varphi_k\|_{W_2^{r^k}}, \tag{21}$$

for any h and ϵ, where the constant C does not depend on h, ϵ, x or ϕ_k.

The proof follows from (10) and (20).

Corollary 2. *The limit function $U_0 \in W_2^{2/\alpha}$ of the sequence $U_{h,\epsilon}$ with respect to the norm in $W_2^{2/\alpha}(E_n)$ satisfies the estimate*

$$|U_0| \leqslant C \sum_{k=1}^{m} |\varphi_k|_{W_2^{r^k}(E_n)}, \tag{22}$$

where C does not depend on x and ϕ_k.

Theorem 1 (existence). *Consider the problem* (1), (2). *Assume that the operators $L[D]$ and $B_j[D]$, $j = 1, \ldots, m$, satisfy the conditions* 1)–7). *Let $\phi \in W_1^{r^k}(E_n) \cap W_2^{r^k}(E_n)$, where $r^k = (r_1^k, \ldots, r_n^k)$,*

$$\frac{-|\alpha|}{2} + \beta_k - 2 + r_s^k \alpha_s > 0, \quad \frac{|\alpha|}{2} > \beta_k, \quad 2\mu m < n - l,$$

$$\alpha = (\alpha_0, \ \ldots \ \alpha_n), \ k = 1, \ \ldots, \ m, \ s = 1, \ \ldots, \ n, \ |\alpha| = \sum_{i=1}^{n} \alpha_i. \tag{23}$$

Then there exists a solution of the problem (1), (2) *$U \in W_2^{2/\alpha}(E_n) \cap M$, $\alpha = (\alpha_0, \alpha_1, \ldots, \alpha_n)$, and the estimates*

$$\|U\|_{W_2^{2/\alpha}(E_n)} \leqslant c \sum_{k=1}^{m} \|\varphi_k\|_{W_2^{r^k}(E_n)},$$

$$\sup_{x \in E_{n+1}^+} |U \leqslant c \sum_{k=1}^{m} \|\varphi_k\|_{W_2^{r^k}(E_n)} \tag{24}$$

are satisfied, where the constant C does not depend on ϕ_k.

Proof. Consider the sequence of functions $U_{h,\epsilon}$; $U_0 = \lim_{n\to 0} \lim_{\epsilon\to 0} U_{h,\epsilon}$ (with respect to the norm in $W_2^{2/\alpha}(E_n)$) satisfies, by virtue of Lemmas 3–6, the estimate (24) and

$$L[D]U_0 = 0$$
$$B_j[D]U_0|_{x_0=0} = \varphi_j, \ j = 1 \ldots m,$$
$$\|D^\nu U_0\|_{L_2(E_n)} \to 0, \ x_0 \to \infty,$$

for all $0 \leq \nu\alpha \leq 2$.

Theorem 2 (uniqueness). *The problem* (1)–(2) *has a unique solution in the class $W_2^{2/\alpha}(E_n) \cap M$.*

Proof. Let us assume the existence of two solutions U_1 and U_2. Then $U = U_1 - U_2$ satisfies the homogeneous problem (1)–(2). For the proof of the theorem it suffices to show that

$$\frac{\partial^k U}{\partial x_0^k}\Big|_{x_0=0}=0,\ k=0,\ \ldots,\ 1/\alpha_0-1.$$

Then $U\equiv 0$ by virtue of the uniqueness of the Cauchy problem in E^+_{n+1} (cf. [15]).

Since

$$B_j[D]U|_{x_0=0}=0,\ j=1\ldots m,$$

after a Fourier transform with respect to $x_1,\ldots,x_n$ we obtain

$$\sum_{j=0}^{m} b_{kj}D^j_{x_0}\tilde{U}|_{x_0=0}=0.$$

Since the matrix $B=(b_{kj})$ is nonsingular, for any $\phi\in E_n$

$$D^j_{x_0}\tilde{U}(y,\ x_0)|_{x_0=0}=0,\ j=0,\ \ldots,\ m-1,$$

or

$$D^j_{x_0}U(x,\ x_0)|_{x_0=0}=0.$$

If $1/\alpha_0=1$, the theorem is proved. Let us consider the case $1/\alpha_0>1$; $1/\alpha_0$ is odd. Put $E^+_n=(x_0=t,\ x(x_1,\ldots,x_n)\in E_n)$. Taking into account the condition at infinity for (1)–(2), we obtain

$$0=\lim_{t\to\infty}\int_0^t\int_{E_n}UL[D]U\,dx\,dx_0=\lim_{t\to\infty}\sum\int_{E^+_n}D^\gamma UD^\beta dx$$

$$+\lim_{t\to\infty}\int_0^t\int_{E_n}UL[-D]U\,dx\,dx_0=\int_{E^+_{n+1}}U[(-1)D^{1/\alpha_0}U+L_1[-D]U]$$

$$+UL_2[-D]U\,dx\,dx_0$$

$$=\int_{E^+_{n+1}}(-1)UL[D]+UL_2[D]+UL_2[-D]\,dx\,dx_0$$

$$=\int_{E^+_{n+1}}\bar{\tilde{U}}(L_2[\hat{D}]U+L_2[-\hat{D}]U)\,dx\,dx_0$$

$$=\int_{E^+_{n+1}}\bar{\tilde{U}}\tilde{U}(L_2[-is]+L_2[is])\,dx\,dx_0$$

$$\geqslant\sum_{j=l+1}^{n}2\int_{E^+_{n+1}}(is_j)^{\overline{\gamma_j}}\,\tilde{U}(is_j^{\gamma_j})\,\tilde{U}\,dx\,dx_0=2\sum_{j=l+1}^{n}\int_{E^+_{n+1}}(D^{\gamma_j}U)^2\,dx\,dx_0=0,$$

taking into account condition 3) for the operator $L_1[D]$ and condition 4) for $L_2[D]$. Since $U\in L_2(E_n)$, the function U does not depend on the variables $x_{l+1},\ldots,x_n$, and by the condition 4) on the operator $L_2[D]$

$$L_2[D]U=0.$$

Then we have

$$0=\lim_{t\to\infty}\int_0^t\int_{E_n} D_{x_0}^{m_1}U\,[D_{x_0}^{1/\alpha_0}U+L_1[D]\,U]\,dxdx_0. \tag{25}$$

Let $1/\alpha_0 \geq 3$, $m \geq 2$, $D_{x_0}^k U|_{x_0=0} = 0$, $k = 0, \ldots, m-1$. Considering that derivatives with respect to x_0 of order $\beta < 1/\alpha_0$ are contained in the operator $L_1[D]$, we shall have, by putting $m_1 = 2$ in (25) and using the fact that $1/\alpha_0 = m + m - 1$,

$$\begin{aligned} 0&=\lim_{t\to\infty}\int_0^t\int_{E_n} D_{x_0}^2 U\,[D_{x_0}^{1/\alpha_0}U+L_1[D]\,U]\,dxdx_0\\ &=\lim_{t\to\infty}\int_0^t\int_{E_n}(-1)^{m-2}D_{x_0}^m U D_{x_0}^{1/\alpha_0-m+2}U+L_1[-D]\,D_{x_0}^2 UU dxdx_0\\ &=(-1)^{m-1}\int_{E_n}\left(\frac{\partial^m U}{\partial x_0^m}\right)^2 dx\Big|_{x_0=0}+\int_{E_{n+1}^+}(-1)\,L[D]\,D_{x_0}^2\,UUdxdx_0\\ &=(-1)^{m-1}\int_{E_n}\left(\frac{\partial^m U}{\partial x_0^m}\right)^2 dx\Big|_{x_0=0}=0. \end{aligned} \tag{26}$$

From this we obtain

$$\frac{\partial^m U}{\partial x_0^m}\Big|_{x_0=0}=0.$$

Let us consider the case of even $1/\alpha_0$. In this case $L[D] = L[-D]$. For any m_1 we have

$$0=\lim_{t\to\infty}\int_0^t\int_{E_n} D_{x_0}^{m_1}UL[D]\,U\,dxdx_0. \tag{27}$$

Integrating by parts in (27), we have (for $m_1 = 1$)

$$0=(-1)^{m-2}\int_{E_n}\left(\frac{\partial^m U}{\partial x_0^m}\right)^2 dx\Big|_{x_0=0}+\int_{E_{n+1}^+} UD_{x_0}^2 L[D]\,dxdx_0$$

or

$$\frac{\partial^m U}{\partial x_0^m}\Big|_{x_0=0}=0.$$

Putting $m_1 = 3, 5, \ldots$, we obtain

$$\frac{\partial^k U}{\partial x_0^k}\Big|_{x_0=0}=0,\ k=0,\ldots,1/\alpha_0-1.$$

Theorem 2 is proved completely.

BIBLIOGRAPHY

1. N. S. Piskunov, *Problèmes limites pour les équations du type elliptico-parabolique*, Mat. Sb. 7 (49) (1940), 385–424. (Russian) MR 2, 203.

2. J.-L. Lions, *Sur certaines équations aux dérivées partielles à coefficients opérateurs non bornés*, J. Analyse Math. 6 (1958), 333–355. MR 23 #A2623.

3. T. G. Genčev, *Ultraparabolic equations*, Dokl. Akad. Nauk SSSR 151 (1963), 265–268 = Soviet Math. Dokl. 4 (1963), 979–982. MR 27 #1715.

4. A. M. Il′in, *On a class of ultraparabolic equations*, Dokl. Akad. Nauk SSSR **159** (1964), 1214–1217 = Soviet Math. Dokl. **5** (1964), 1673–1676. MR **30** #1315.

5. V. S. Vladimirov and Ju. N. Drožžinov, *Generalized Cauchy problem for an ultraparabolic equation*, Izv. Akad. Nauk SSSR Ser. Mat. 31 (1967), 1341–1360 = Math. USSR Izv. **1** (1967), 1285–1304. MR **37** #1799.

6. L. R. Volevič and S. G. Gindikin, *Pseudodifferential operators and the Cauchy problem for differential equations with variable coefficients*, Funkcional. Anal. i Priložen. 1 (1967), no. 4, 8–25 = Functional Anal. Appl. **1** (1967), 262–277. MR **36** #6769.

7. S. G. Gindikin, *Correctness of the Cauchy problem for differential equations with variable coefficients*, Funkcional. Anal. i Priložen. 4 (1970), no. 2, 83–84 = Functional Anal. Appl. 4 (1970), 160–162. MR 42 #2158.

8. G. E. Šilov, *Mathematical analysis: Second special course*, "Nauka", Moscow, 1965; English transl., *Generalized functions and partial differential equations*, Gordon and Breach, New York, 1968. MR **36** #2943; **37** #5694.

9. S. M. Nikol′skiĭ, *The first boundary problem for a general linear equation*, Dokl. Akad. Nauk SSSR **146** (1962), 767–769 = Soviet Math. Dokl. **3** (1962), 1388–1390. MR **25** #4237.

10. ———, *Uniqueness of the solution to a boundary value problem for a convex region*, Dokl. Akad. Nauk SSSR **148** (1963), 1022–1025 = Soviet Math. Dokl. **4** (1963), 218–221. MR **26** #2711.

11. ———, *Functions with dominant mixed derivative. Boundary value problems for equations of hypoelliptic and more general type*, Outlines Joint Sympos. Partial Differential Equations (Novosibirsk, 1963), Acad. Sci. USSR Siberian Branch, Moscow, 1963, pp. 185–190. MR **35** #3231.

12. A. A. Dezin and V. N. Maslennikova, *Nonclassical boundary problems*, Partial Differential Equations (Proc. Sympos. S. L. Sobolev on the Occasion of His Sixtieth Birthday), "Nauka", Moscow, 1970, pp. 81–95; English transl., Amer. Math. Soc. Transl. (2) **105** (1976), 117–138.

13. S. Agmon, A. Douglis and L. Nirenberg, *Estimates near the boundary for solutions of elliptic partial differential equations satisfying general boundary conditions*. I, Comm. Pure Appl. Math. **12** (1959), 623–727. MR **23** #A2610.

14. S. V. Uspenskiĭ, *On the representation of functions defined by a class of hypoelliptic operators*, Trudy Mat. Inst. Steklov. **117** (1972), 292–299 = Proc. Steklov Inst. Math. **117** (1972), 343–352.

15. I. M. Gel′fand and G. E. Šilov, *Fourier transforms of rapidly increasing functions and questions of the uniqueness of the solution of Cauchy's problem*, Uspehi Mat. Nauk 8 (1953), no. 6 (58), 3–54; English transl., Amer. Math. Soc. Transl. (2) 5 (1957), 221–274. MR **15**, 867; **18**, 736.

Translated by J. D. FABREY

Trudy Mat. Inst. Steklov. 134 (1975)
Proc. Steklov Inst. Math. 134 (1975)

THE $C(I)$ NORMS OF ORTHOGONAL PROJECTIONS ONTO SUBSPACES OF POLYGONALS*

UDC 517.51

Z. CIESIELSKI

Abstract. In the space $C(I)$ of continuous functions on the interval $I = \langle 0, 1\rangle$, consider the orthogonal projection P_n onto the subspace of polygonal functions with n equally spaced nodes. This paper studies the asymptotic behavior of the norm of P_n as $n \to \infty$. From the results it follows, in particular, that $2(1 - \delta_n) < \|P_n\| < 2$, where $\delta_n \to 0$ as $n \to \infty$, i.e. in any case $\lim_{n\to\infty}\|P_n\| = 2$.

This is a refinement of the estimate $\|P\| \le 3$, found earlier by the author and valid for the norm of the projection onto the subspace of polygonal functions with arbitrarily distributed (nonuniform, in general) nodes.

In the same direction, the author has established in previous papers that the supremum of the norms of the partial sum operators for Fourier series in the Franklin system is $2 + (2 - \sqrt{3})^2$, i.e. it is larger than the norm of P_n.

Bibliography: 2 titles.

1. Introduction. We denote by $C(I)$ the space of all continuous real functions on $I = \langle 0, 1\rangle$, and we define $\|f\| = \sup\{|f(t)| : t \in I\}$. For given partition $\pi : 0 = s_0 < \ldots < s_n = 1$ we denote by $S_\pi(I)$ the set of all $f \in C(I)$ such that f is linear in each $I_j = \langle s_{j-1}, s_j\rangle$, $j = 1, \ldots, n$. The notion of orthogonality is being introduced in $C(I)$ by means of the usual scalar product

$$(f, g) = \int_I f(t)\, g(t)\, dt.$$

To each $f \in C(I)$ there is unique $P_\pi f \in S_\pi(I)$ such that

$$\|f - P_\pi f\|_2 = \inf\{\|f - g\|_2 : g \in S_\pi(I)\},$$

where

$$\|f\|_2 = \left(\int_I |f(t)|^2\, dt\right)^{1/2}.$$

Thus, the mapping $P_\pi : C(I) \to S_\pi(I)$ is an orthogonal projection of $C(I)$ onto $S_\pi(I)$, and its $C(I)$ norm is given by the formula

$$\|P_\pi\| = \sup\{\|P_\pi f\| : \|f\| = 1,\ f \in C(I)\}.$$

It has been proved by the author in [1] that for each π

$$\|P_\pi\| \le 3. \tag{1.1}$$

AMS (MOS) subject classifications (1970). Primary 46E15; Secondary 42A56.

**Editor's note.* The body of this article is an unaltered reprinting of the original paper. Details of the bibliography have been checked and put in standard form, and MR citations have been added.

It is not known to the author whether the constant 3 in (1. 1) is best possible for the class of all partitions π. In this note it is shown that for the uniform partitions $\pi_n : s_j = j/n,\ j=0, \ldots, n,\ n > 0$, we have

$$2(1-\delta_n) < \|P_n\| < 2,\ n>1,\ \|P_1\| = \frac{3}{2}, \tag{1. 2}$$

where $P_n = P_{\pi_n}$ and $\delta_n = (2-3^{1/2})\sinh^{-1}[n/2]\alpha$ with α as the positive solution of the equation $\cosh\alpha = 2$, $[x]$ as the greatest integer not exceeding x.

In the case of the sequence of dyadic partitions corresponding to the Franklin orthonormal system (c. f. [1]) one can prove that $2+(2-3^{1/2})^2$ is the basis constant for the Franklin system i. e. it is the lowest upper bound for the orthogonal projections corresponding to the partial sums of the Franklin — Fourier series. The tedious prove will be not given here.

2. Preliminaries. Let us start with fixed partition π. The partition of unity, corresponding to π, of polygonals is defined as follows

$$N_{\pi, j}(t) = (s_{j+1} - s_{j-1})[s_{j-1},\ s_j,\ s_{j+1};\ (s-t)_+], \tag{2. 1}$$

where the square bracket denotes the divided difference of $(s-t)_+ = \max(s-t, 0)$ as a function of s taken at the points $s_{j-1},\ s_j,\ s_{j+1}$. It follows that $\operatorname{supp} N_{\pi, j} = \langle s_{j-1},\ s_{j+1}\rangle$, $N_{\pi, j} \geqslant 0$ and for all t

$$\sum_j N_{\pi, j}(t) = 1.$$

The orthogonal projection P_π defined as above is an integral operator corresponding to the kernel

$$K_\pi(t, s) = \sum_{j=0}^{n} N_{\pi, j}(t)\, N_{\pi, j}(s). \tag{2. 2}$$

It is well known that

$$\|P_\pi\| = \max_{t\in I} \int_I |K_\pi(t, s)|\, ds.$$

Notice that for each $t \in I$

$$\int_{I_j} |K_\pi(t, s)|\, ds \leqslant |I_j| \max(|K_\pi(t, s_{j-1})|,\ |K_\pi(t, s_j)|),$$

where $|I_j| = s_j - s_{j-1}$. Therefore it follows that

$$\|P_\pi\| = \max_{0\leqslant j\leqslant n} \int_I |K_\pi(s_j, s)|\, ds. \tag{2. 3}$$

Using the fact that

P_π is the orthogonal projection of $C(I)$ onto $S_\pi(I)$ we find that $a_{i, j} = K_\pi(s_i, s_j)$ satisfy the following equations $(i = 0, 1, \ldots, n)$

$$\begin{gathered} 2a_{0, i} + a_{1, i} = \frac{6}{\delta_1}\delta_{0, i}, \\ \delta_j(a_{j-1, i} + 2a_{j, i}) + \delta_{j+1}(2a_{j, i} + a_{j+1, i}) = 6\delta_{ij}, \\ a_{n-1, i} + 2a_{n, i} = \frac{6\delta_{n, i}}{\delta_n}, \end{gathered} \tag{2. 4}$$

where $j = 1, \ldots, n$, and $\delta_j = |I_j|$.

In the case of uniform partition i. e. for $\delta_j = 1/n$ the solution of equations (2.4) is given by the formula (c. f. [2])

$$a_{i,j} = (-1)^{i+j} \frac{6n}{\sinh\alpha \sinh\alpha n} \cosh\alpha(i \wedge j) \cosh\alpha[(n-i) \wedge (n-j)], \qquad (2.5)$$

where $i \wedge j = \min(i, j)$ and α is the positive solution of $\operatorname{ch}\alpha = 2$.

Now, according to (2.5) we have $a_{i,j} = (-1)^{i+j} |a_{i,j}|$ and therefore it follows that

$$\int_{I_j} |K_\pi(s_i, s)|\, ds = \frac{\delta_j}{2} \frac{|a_{j-1,i}|^2 + |a_{j,i}|^2}{|a_{j-1,i}| + |a_{j,i}|},$$

whence we infer

$$\|P_\pi\| = \max_{0 \leqslant i \leqslant n} \frac{1}{2} \sum_{j=1}^{n} \delta_j \frac{|a_{j-1,i}|^2 + |a_{j,i}|^2}{|a_{j-1,i}| + |a_{j,i}|}. \qquad (2.6)$$

It is useful in this place to introduce the sequences $\{A_n\}$ and $\{B_n\}$ of positive integers as the unique solution of the following equations

$$\begin{aligned} A_{n+1} &= 2A_n + 3B_n, & A_0 &= 1 \\ B_{n+1} &= A_n + 2B_n, & B_0 &= 0. \end{aligned} \qquad (2.7)$$

It can be easily seen that

$$A_n = \cosh \alpha n, \quad 3^{\frac{1}{2}} B_n = \sinh \alpha n, \qquad (2.8)$$

These sequences are used to simplify formula (2.6). The combination of (2.5), (2.6) and (2.8) gives

$$\|P_n\| = \max_{0 \leqslant i \leqslant n} \frac{1}{B_n} (A_{n-i} S_i + A_i S_{n-i}), \qquad (2.9)$$

where

$$S_n = \sum_{j=1}^{n} \frac{A_{j-1}^2 + A_j^2}{A_{j-1} + A_j}, \qquad S_0 = 0, \quad n > 0. \qquad (2.10)$$

3. The proof of (1.2). According to (2.8) we have

$$B_n = B_i A_{n-i} + B_{n-i} A_i, \qquad 0 \leqslant i \leqslant n. \qquad (3.1)$$

To prove the right hand side of (1.2) it is sufficient, in view of (2.9), to show that

$$S_j < 2B_j, \qquad j > 0. \qquad (3.2)$$

The proof of (3.2) goes as follows. According to equations (2.7) we have

$$\begin{aligned} A_j - A_{j-1} &= B_j + B_{j-1}, \\ A_j + A_{j-1} &= 3(B_j - B_{j-1}), \end{aligned}$$

whence we infer

$$\frac{A_j^2 + A_{j-1}^2}{A_j + A_{j-1}} = 2(B_j - B_{j-1}) \left[\frac{1}{12} \frac{(B_j + B_{j-1})^2}{(B_j - B_{j-1})^2} + \frac{3}{4} \right]. \qquad (3.3)$$

Now, by (2.8) $B_{j-1} < (2-3^{1/2}) B_j$ and since $(2-3^{1/2})(3^{1/2}+1) = (3^{1/2}-1)$ we have $B_j + B_{j-1} < 3^{1/2}(B_j - B_{j-1})$ and the square bracket in (3.3) is less than 1. Thus,

$$\frac{A_j^2 + A_{j-1}^2}{A_j + A_{j-1}} < 2(B_j - B_{j-1}), \qquad j > 0,$$

and this gives (3.2).

The left hand side of (1.2) is being proved as follows. Let $a = 2+3^{1/2}$, $b = 2-3^{1/2}$, and let

$$g(t) = \frac{(1+b)-(a+1)t}{(1-b)+(a-1)t}.$$

Then, $g(0) = 3^{1/2}$ and

$$g(b^{2j}) = \frac{B_j + B_{j-1}}{B_j + B_{j-1}}.$$

Applying the mean value theorem we obtain

$$g(b^{2j}) > 3^{\frac{1}{2}}(1-2ab^{2j}).$$

Thus,

$$S_j = 2B_j - 2\sum_{k=1}^{j}(B_k - B_{k-1})b^{2k-1} > 2B_j\left[1 - \frac{b}{3^{1/2}B_j}\right] = 2B_j\left[1 - \frac{b}{\sinh aj}\right]. \quad (3.4)$$

Formula (2.9) gives for $1 < i < n$, $n > 2$

$$\|P_n\| \geqslant \frac{1}{B_n}(A_{n-i}S_i + A_iS_{n-i}) > \frac{2}{B_n}(A_{n-i}B_i + A_iB_{n-i})\min\left(\frac{S_i}{2B_i}, \frac{S_{n-i}}{2B_{n-i}}\right),$$

and therefore by (3.4) and (3.1)

$$\|P_n\| > 2\left(1 - \frac{b}{\sinh a(i \wedge (n-i))}\right), \qquad 1 < i < n.$$

However $i \wedge (n-i) = [n/2]$ for $i = [n/2]$ and therefore the required inequality follows. For $n = 2$ $\|P_2\| = 17/9$ and inequalities (1.2) hold too.

BIBLIOGRAPHY

1. Z. Ciesielski, *Properties of the orthonormal Franklin system*, Studia Math. 23 (1963), 141–157. MR 28 #419.

2. ———, *Properties of the orthonormal Franklin system. II*, Studia Math. 27 (1966), 289–323. MR 34 #3202.

Trudy Mat. Inst. Steklov.
134 (1975)

Proc. Steklov Inst. Math.
134 (1975)

THE VARIATIONAL PRINCIPLE IN GAME THEORY AND OPTIMAL CONTROL PROBLEMS

UDC 518.9:519.3

V. M. ŠALOV

Abstract. In the paper, the connection is established between game theory and optimal control problems on the one hand and the classical calculus of variations in functional spaces on the other.

It is proved that the existence of a so-called quasi-saddle point of a certain functional on the admissible set is necessary and sufficient for the existence of an optimal solution of the game, and that it is a fortiori sufficient if such a point exists for the same functional on the entire space, without any constraints.

The variational formulation of the game is given for controlled systems which are described by partial differential equations, ordinary differential equations, integral equations, and relations in finite-dimensional Euclidean spaces.

In conclusion, an example is considered of a differential game with a second-order partial differential equation from the domain of optimal design. In the one-dimensional case, the optimal solution of this game is written out in an explicit form.

Bibliography: 20 titles.

At present, thanks to the Pontrjagin maximum principle [1], [2] the solution of nonclassical extremal problems is far advanced. The main success has been achieved here in the area of problems described by ordinary differential equations.

The ideas of the Pontrjagin maximum principle have been extended to a broader class of problems (see [11], [13], [16]–[18]), and, first of all, to the problems described by partial differential equations (see [12], [14], [19] and [20]).

Differential games [3]–[6] are of particular interest. They often arise in practical problems, and, in this connection, optimal control problems constitute a particular case with only a single player.

However, it should be noted that the problems of optimal control and game theory, although they constitute a development of the classical calculus of variations, formed in fact a new direction in mathematics, which has been developing, in essence, independently of variational principles.

On the other hand, thanks to the works of Sobolev [7], Nikol'skii [8], [9], Kudrjavcev [10], and many other authors, variational principles for studying problems of mathematical physics, which make wide use of the theory of functions apparatus and embedding theorems, have been greatly developed.

AMS (*MOS*) *subject classifications* (1970). Primary 49B30; Secondary 90D25.

In this paper, an attempt is made to consider the problems of game theory and optimal control from the viewpoint of the modern variational principles in mathematical physics.

The work consists of three parts. In the first part, the optimality principle is formulated in abstract form in linear spaces, in order to describe as wide a set of extremal problems as possible. Here we prove the equivalence of the problem of obtaining an optimal solution of a game and the problem of finding a so-called quasi-saddle point of a certain functional on the admissible set. It is shown that, if one succeeds in finding a quasi-saddle point of the functional on the entire space, without any constraints, then the coordinates of this point a fortiori constitute an optimal solution of the game.

In the second part, the variational optimality principle is formulated for the class of extremal problems which are described by partial differential equations, ordinary differential equations, integral equations, and for mathematical programming problems. The existence of a solution is assumed in all cases, and solvability conditions of extremal problems are not considered in this work.

In the third part, the application of the results obtained to some simplest problems from the domain of optimal design is considered. The purpose of this part of the work is to formulate the optimal designing problem, in particular examples, as a mathematical game of the designer with nature.

It is shown in an example that, for this type of problems with "phase" constraints, the solutions appear in a natural way in the domain of generalized functions (the δ-function appears in the right-hand side of the equation). For simplest cases, the optimal solution is constructed in an explicit form on the basis of the variational principle presented.

§1. The variational optimality principle in abstract form

Let there be given linear spaces E_1, E_2 and E_3. We consider a controlled system whose behavior can be changed by two players U and V. We agree that (x, u, v) is an element (a point) of the space E_1, that x characterizes the state of the system, u denotes a strategy for the player U, v denotes a strategy for the player V, ψ is an element of E_2, and ω is an element of E_3.

Moreover, we shall assume the following: $E_{xuv} \subset E_1$ is a subset of E_1 on which the state of the system is defined; $E_\psi \subset E_2$ is a linear set which is the range of elements ψ; $E_\omega \subset E_3$ is a linear set which is the range of elements ω; and R is the space of real numbers.

Our purpose is to find an optimal (in a certain sense) behavior of the system under consideration by studying the corresponding mathematical game.

This problem is formulated as follows: find

$$\max_v \min_u f(x, u, v) - \text{the goal of the game.} \tag{1}$$

under the condition that the following "equations", which describe the behavior of the system, are satisfied:

$$p(x, u, v, \psi)=0, \quad \forall \psi \in E_{\psi}, \tag{2}$$

and that the following inequalities, which impose constraints upon the behavior of the system, are also satisfied:

$$h(x, u, v, \omega) \geqslant 0, \quad \forall \omega \in E_{\omega}, \tag{3}$$

where

$$f: E_{xuv} \rightarrow R, \quad p: E_{xuv} \times E_{\psi} \rightarrow R, \quad h: E_{xuv} \times E_{\omega} \rightarrow R$$

are given mappings.

The mappings f, p and h are functionals; therefore we shall clarify several of their properties.

Firstly, we introduce the necessary notation:

$$\operatorname{Ker} p=\{(x, u, v): p(x, u, v, \psi)=0, \ \forall \psi \in E_{\psi}\} \tag{4}$$

and, similarly,

$$\operatorname{Ker} h_{+}=\{(x, u, v): h(x, u, v, \omega) \geqslant 0, \ \forall \omega \in E_{\omega}\}. \tag{5}$$

Further, we denote

$$K_{ph_{+}}=\operatorname{Ker} p \cap \operatorname{Ker} h_{+}, \tag{6}$$

and, moreover,

$$K_{v}=\{v: \exists (x, u) \mid (x, u, v) \in K_{ph_{+}}\} \tag{7}$$

and accordingly

$$E_{v}=\{v: \exists (x, u) \mid (x, u, v) \in E_{xuv}\}. \tag{8}$$

The properties of the functionals p and h. We assume by definition that the functionals p and h have the following properties:

1°. The sets $\operatorname{Ker} p$, $\operatorname{Ker} h_+$ and K_{ph_+} are not empty.

2°. $p(x, u, v, \alpha\psi' + \beta\psi'') = \alpha p(x, u, v, \psi') + \beta p(x, u, v, \psi'')$ for any ψ', $\psi'' \in E_{\psi}$ and any $\alpha, \beta \in R$.

3°. $h(x, u, v, 0) = 0$, $\forall (x, u, v) \in E_{xuv}$.

4°. $0 = \inf_{\omega \in E_{\omega}} h(x, u, v, \omega)$, $(x, u, v) \in K_{ph_+}$.

Thus, the relation (2) determines a generalized (in some sense) solution x_{uv}, which depends on the elements u and v, and (3) is a generalized inequality which, in fact, corresponds to a number of inequalities (see, for example, (61) and (62)).

We now introduces a number of definitions.

Definition 1. An element $(\bar{x}, \bar{u}, \bar{v}) \in E_{xuv}$ will be called an *optimal solution of the game* (1)–(3) if the following relations are satisfied:

1) The element $(\bar{x}, \bar{u}, \bar{v})$ satisfies (2) and (3), i.e.,

$$p(\bar{x}, \bar{u}, \bar{v}, \psi)=0, \ \forall\psi\in E_\psi, \ h(\bar{x}, \bar{u}, \bar{v}, \omega)\geqslant 0, \ \forall\omega\in E_\omega. \tag{9}$$

2) For any fixed element $v\in K_v$, there exist elements $\bar{x}_v$ and $\bar{u}_v$ which determine the point $(\bar{x}_v, \bar{u}_v, v)\in K_{ph_+}$ and are such that

$$f(\bar{x}_v, \bar{u}_v, v)\leqslant f(x', u', v), \ \forall (x', u', v)\in K_{ph_+}. \tag{10}$$

3) There exists an element $\bar{v}\in K_v$ such that

$$f(\bar{x}_v, \bar{u}_v, v)\leqslant f(\bar{x}, \bar{u}, \bar{v}), \ \forall (\bar{x}_v, \bar{u}_v, v)\in K_{ph_+}, \tag{11}$$

where $(\bar{x}, \bar{u}, \bar{v}) = (\bar{x}_{\bar{v}}, \bar{u}_{\bar{v}}, \bar{v})$.

For the mathematical game (1)–(3), we consider the following functional on the set $\mathfrak{E} = E_{xuv}\times E_\psi \times E_\omega$:

$$\Phi(x, u, v, \psi, \omega)=f(x, u, v)+p(x, u, v, \psi)+h(x, u, v, \omega), \tag{12}$$

which, in a sense, is similar to the corresponding functionals used in employing indeterminate Lagrange multipliers or generalized multipliers in the Kuhn-Tucker theorem for mathematical programming problems.

Definition 2. An element $(\hat{x}, \hat{u}, \hat{v}, \hat{\psi}, \hat{\omega})\in\mathfrak{E}$ will be called a *quasi-saddle point of the functional* (12) *on the set* $\mathfrak{E}$, if the following conditions are satisfied:

1) For any fixed element $v\in E_v$, there exist elements $\hat{x}_v, \hat{u}_v, \hat{\psi}_v$ and $\hat{\omega}_v$ which determine the point $(\hat{x}_v, \hat{u}_v, v, \hat{\psi}_v, \hat{\omega}_v)\in\mathfrak{E}$, and are such that the following relation holds:

$$\begin{aligned} f(\hat{x}_v, \hat{u}_v, v)+p(\hat{x}_v, \hat{u}_v, v, \hat{\psi}_v)+h(\hat{x}_v, \hat{u}_v, v, \hat{\omega}_v)& \\ \leqslant f(x, u, v)+p(x, u, v, \psi)+h(x, u, v, \omega),& \end{aligned} \tag{13}$$

which is true for a given fixed $v\in E_v$ for any $(x, u, v, \psi, \omega)\in\mathfrak{E}$.

2) There exists an element $\hat{v}\in E_v$ such that the following relation holds for the point $(\hat{x}, \hat{u}, \hat{v}, \hat{\psi}, \hat{\omega}) = (\hat{x}_{\hat{v}}, \hat{u}_{\hat{v}}, \hat{v}, \hat{\psi}_{\hat{v}}, \hat{\omega}_{\hat{v}})$:

$$\begin{aligned} f(\hat{x}_v, \hat{u}_v, v)+p(\hat{x}_v, \hat{u}_v, v, \hat{\psi}_v)+h(\hat{x}_v, \hat{u}_v, v, \hat{\omega}_v)& \\ \leqslant f(\hat{x}, \hat{u}, \hat{v})+p(\hat{x}, \hat{u}, \hat{v}, \hat{\psi})+h(\hat{x}, \hat{u}, \hat{v}, \hat{\omega}),& \end{aligned} \tag{14}$$

which is true for any $(\hat{x}_v, \hat{u}_v, v, \hat{\psi}_v, \hat{\omega}_v)\in\mathfrak{E}$.

Along with these, we shall need one more definition.

Definition 3. An element $(\bar{x}, \bar{u}, \bar{v}, \bar{\psi}, \bar{\omega})\in K_{ph_+}\times E_\psi\times E_\omega$ will be called a *quasi-saddle point of the functional* (12) *on the admissible set* $K_{ph_+}\times E_\psi\times E_\omega$ if the following conditions hold:

1) For any fixed $v\in K_v$, there exist elements $\bar{x}_v, \bar{u}_v, \bar{\psi}_v$ and $\bar{\omega}_v$ which determine the point $(\bar{x}_v, \bar{u}_v, v, \bar{\psi}_v, \bar{\omega}_v)\in K_{ph_+}\times E_\psi\times E_\omega$, and are such that the following inequality holds:

$$\begin{aligned} f(\bar{x}_v, \bar{u}_v, v)+p(\bar{x}_v, \bar{u}_v, v, \bar{\psi}_v)+h(\bar{x}_v, \bar{u}_v, v, \bar{\omega}_v)& \\ \leqslant f(x', u', v)+p(x', u', v, \psi)+h(x', u', v, \omega),& \end{aligned} \tag{15}$$

which is true for all $(x', u', v, \psi, \omega) \in K_{ph_+} \times E_\psi \times E_\omega$.

2) There exists an element $\bar{v} \in K_v$ such that the inequality

$$f(\bar{x}_v, \bar{u}_v, v) + p(\bar{x}_v, \bar{u}_v, v, \bar{\psi}_v) + h(\bar{x}_v, \bar{u}_v, v, \bar{\omega}_v) \\ \leqslant f(\bar{x}, \bar{u}, \bar{v}) + p(\bar{x}, \bar{u}, \bar{v}, \bar{\psi}) + h(\bar{x}, \bar{u}, \bar{v}, \bar{\omega}), \tag{16}$$

which is true for all $(\bar{x}_v, \bar{u}_v, v, \bar{\psi}_v, \bar{\omega}_v) \in K_{ph_+} \times E_\psi \times E_\omega$, holds for the point

$$(\bar{x}, \bar{u}, \bar{v}, \bar{\psi}, \bar{\omega}) = (\bar{x}_{\bar{v}}, \bar{u}_{\bar{v}}, \bar{v}, \bar{\psi}_{\bar{v}}, \bar{\omega}_{\bar{v}}) \in K_{ph_+} \times E_\psi \times E_\omega$$

We shall now prove that the following theorem holds.

Theorem 1. *For an optimal solution* $(\bar{x}, \bar{u}, \bar{v}) \in E_{xuv}$ *of the game* (1)–(3) *to exist, it is necessary and sufficient that there exist a quasi-saddle point* $(\bar{x}, \bar{u}, \bar{v}, \bar{\psi}, \bar{\omega})$ *of the functional* (12) *on the admissible set* $K_{ph_+} \times E_\psi \times E_\omega$.

Necessity. Let an element $(\bar{x}, \bar{u}, \bar{v}) \in E_{xuv}$ be an optimal solution of the game (1)–(3). We shall show that then there exist elements $\bar{\psi} \in E_\psi$ and $\bar{\omega} \in E_\omega$ such that the element $(\bar{x}, \bar{u}, \bar{v}, \bar{\psi}, \bar{\omega})$ is a quasi-saddle point of the functional (12) on the admissible set. Indeed, if $(\bar{x}, \bar{u}, \bar{v})$ is an optimal solution of the game (1)–(3), then by virtue of Definition 1 there exist, for any $v \in K_v$, elements $\bar{x}_v$ and $\bar{u}_v$ which determine the point $(\bar{x}_v, \bar{u}_v, v) \in K_{ph_+}$ and an element $\bar{v} \in K_v$ such that (10) and (11) hold. Further, for the elements $(\bar{x}_v, \bar{u}_v, v)$ and $(x', u', v) \in K_{ph_+}$ the following equality holds for any $v \in K_v$:

$$0 = p(\bar{x}_v, \bar{u}_v, v, \bar{\psi}_v) = p(x', u', v, \psi), \quad \forall \psi \in E_\psi. \tag{17}$$

Also, the inequality

$$0 = h(\bar{x}_v, \bar{u}_v, v, \bar{\omega}_v) \leqslant h(x', u', v, \omega), \quad \forall \omega \in E_\omega, \tag{18}$$

holds, where $\bar{\psi}_v$ is a fixed element of E_ψ, and $\bar{\omega}_v$ is a fixed element of E_ω at which the functional h vanishes (such an element exists by virtue of properties 3° and 4° of the functional h).

Now, adding up the left- and right-hand sides of (10), (17) and (18), we obtain (15).

Considering the relations (17) and (18) with $v = \bar{v}$, we have

$$0 = p(\bar{x}_v, \bar{u}_v, v, \bar{\psi}_v) = p(\bar{x}_{\bar{v}}, \bar{u}_{\bar{v}}, \bar{v}, \bar{\psi}_{\bar{v}}), \tag{19}$$

and

$$0 = h(\bar{x}_v, \bar{u}_v, v, \bar{\omega}_v) = h(\bar{x}_{\bar{v}}, \bar{v}_{\bar{v}}, \bar{v}, \bar{\omega}_{\bar{v}}). \tag{20}$$

Hence, taking into account the condition $(\bar{x}_{\bar{v}}, \bar{u}_{\bar{v}}, \bar{v}) = (\bar{x}, \bar{u}, \bar{v})$ and inequality (11), we obtain (16), in which $\bar{\psi} = \bar{\psi}_{\bar{v}}$ and $\bar{\omega} = \bar{\omega}_{\bar{v}}$.

Thus, we have established that the existence of an optimal solution of the game (1)–(3) necessarily implies the existence of a quasi-saddle point of the functional (12) on the admissible set.

Sufficiency. Now, let it be given that the functional (12) has a quasi-saddle point $(\bar{x}, \bar{u}, \bar{v}, \bar{\psi}, \bar{\omega})$ on the admissible set. We shall show that then $(\bar{x}, \bar{u}, \bar{v})$ is a fortiori an optimal solution of the game (1)–(3).

By virtue of Definition 3, there exist, for any $v \in K_v$, a point $(\bar{x}_v, \bar{u}_v, v, \bar{\psi}_v, \bar{\omega}_v) \in K_{ph_+} \times E_\psi \times E_\omega$ and an element $\bar{v} \in K_v$ such that (15) and (16) hold.

By definition, for elements $(\bar{x}_v, \bar{u}_v, v) \in K_{ph_+}$ we have

$$p(\bar{x}_v, \bar{u}_v, v, \psi) = 0, \quad \forall \psi \in E_\psi, \tag{21}$$

and

$$h(\bar{x}_v, \bar{u}_v, v, \omega) \geqslant 0, \quad \forall \omega \in E_\omega, \tag{22}$$

which hold for any $v \in K_v$.

Setting $x' = \bar{x}_v$, $u' = \bar{u}_v$ and $\psi = \bar{\psi}_v$ in (15), we obtain

$$h(\bar{x}_v, \bar{u}_v, v, \bar{\omega}_v) \leqslant h(\bar{x}_v, \bar{u}_v, v, \omega), \ \forall \omega \in E_\omega;$$

hence, by virtue of property 4° of the functional h, it follows that

$$h(\bar{x}_v, \bar{u}_v, v, \bar{\omega}_v) = 0, \quad \forall v \in K_v. \tag{23}$$

Inequality (15) with $\psi = 0$ and $\omega = 0$, with relations (21), (22) and property 3° of the functional h taken into account, implies that (10) holds.

By virtue of property 2) of a quasi-saddle point on the admissible set, there exists an element $\bar{v} \in K_v$ for which (16) is satisfied. Hence and from (21), (23) and the condition $(\bar{x}, \bar{u}, \bar{v}) = (\bar{x}_{\bar{v}}, \bar{u}_{\bar{v}}, \bar{v})$ follows inequality (11). Moreover, it follows from (21) and (22) with $v = \bar{v}$ that (9) holds.

Thus, it has been proved that the coordinates $(\bar{x}, \bar{u}, \bar{v})$ of the quasi-saddle point $(\bar{x}, \bar{u}, \bar{v}, \bar{\psi}, \bar{\omega})$ of the functional (12) on the admissible set are an optimal solution of the game (1)–(3). The sufficiency is proved.

This proves Theorem 1 in its entirety.

Theorem 1 has been proved under very general requirements on the functionals f, p and h, and it yields necessary and sufficient optimality conditions. The following theorem gives wider possibilities for the search for an optimal solution of the game, although the conditions of this theorem in the general case are not necessary.

Theorem 2. *For an element* $(\bar{x}, \bar{u}, \bar{v}) \in E_{xuv}$ *to be an optimal solution of the game* (1)–(3), *it is sufficient that* $\bar{x}$, $\bar{u}$ *and* $\bar{v}$ *be the coordinates of a quasi-saddle point of the functional* (12) *on the set* $E_{xuv} \times E_\psi \times E_\omega$.

Proof. Let there exist a quasi-saddle point $(\hat{x}, \hat{u}, \hat{v}, \hat{\psi}, \hat{\omega})$ of the functional (12) in the sense of Definition 2. We shall show that then the coordinates $(\bar{x}, \bar{u}, \bar{v}) = (\hat{x}, \hat{u}, \hat{v})$ form an optimal solution of the game (1)–(3).

By definition, the existence of a quasi-saddle point means that, for any $v \in E_v$, there exist a point $(\bar{x}_v, \bar{u}_v, v, \bar{\psi}_v, \bar{\omega}_v) \in E_{xuv} \times E_\psi \times E_\omega$ and an element $\bar{v} \in E_v$ such that (13) and (14) are satisfied.

We now consider (13) with $(\hat{x}_v, \hat{u}_v, v, \psi, \hat{\omega}_v)$. We have

$$p(\hat{x}_v, \hat{u}_v, v, \hat{\psi}_v) \leqslant p(\hat{x}_v, \hat{u}_v, v, \psi), \qquad \forall\psi \in E_\psi. \tag{24}$$

Similarly, substituting the element $(\hat{x}_v, \hat{u}_v, v, \hat{\psi}_v, \omega)$ into the right-hand side of (13), we obtain

$$h(\hat{x}_v, \hat{u}_v, v, \hat{\omega}_v) \leqslant h(\hat{x}_v, \hat{u}_v, v, \omega), \qquad \forall\omega \in E_\omega. \tag{25}$$

By virtue of the linearity of the functional p and of the set E_ψ, we obtain from (24) that

$$p(\hat{x}_v, \hat{u}_v, v, \psi) = 0, \qquad \forall\psi \in E_\psi. \tag{26}$$

By virtue of the properties of the functional h, the condition (25) implies the equality

$$h(\hat{x}_v, \hat{u}_v, v, \hat{\omega}_v) = 0, \tag{27}$$

and the inequality

$$h(\hat{x}_v, \hat{u}_v, v, \omega) \geqslant 0, \qquad \forall\omega \in E_\omega. \tag{28}$$

Setting $v = \hat{v}$ in (26) and (28), and denoting $(\hat{x}_{\hat{v}}, \hat{u}_{\hat{v}}, \hat{v})$ by $(\bar{x}, \bar{u}, \bar{v})$, we obtain that the relations (9) hold. Further, making use of (26) and (27), we obtain inequality (10) with $(\bar{x}_v, \bar{u}_v, v) = (\hat{x}_v, \hat{u}_v, v)$ from inequality (13) considered only on the set of elements $(x', u', v, 0, 0) \in K_{ph_+} \times E_\psi \times E_\omega$.

Similarly, inequality (14) with (26) and (27) taken into account implies (11).

Thus we have shown that the coordinates $(\hat{x}, \hat{u}, \hat{v})$ of the quasi-saddle point $(\hat{x}, \hat{u}, \hat{v}, \hat{\psi}, \hat{\omega})$ of the functional (12) on the set $E_{xuv} \times E_\psi \times E_\omega$ are an optimal solution of the game (1)–(3) in the sense of Definition 1. Theorem 2 is proved.

Corollary 1. *If there exists a quasi-saddle point of the functional* (12) *on the set* $\mathfrak{S}$, *if the linear spaces* E_1, E_2 *and* E_3 *are Banach spaces, and if the functionals* f, p *and* h *are Gâteaux or Fréchet differentiable on a neighborhood of the saddle point, then the optimal solution* $(\bar{x}, \bar{u}, \bar{v})$ *of the game* (1)–(3) *and the elements* $\bar{\psi}$ *and* $\bar{\omega}$ *satisfy the following variational equations*:

$$\begin{aligned}
&\langle \partial_x\Phi, \delta x\rangle = \langle f'_x + p'_x + h'_x, \delta x\rangle = 0, && \forall\delta x : (x + \delta x, \bar{u}, \bar{v}) \in E_{xuv},\\
&\langle \partial_u\Phi, \delta u\rangle = \langle f'_u + p'_u + h'_u, \delta u\rangle = 0, && \forall\delta u : (x, u + \delta u, v) \in E_{xuv},\\
&\langle \partial_v\Phi, \delta v\rangle = \langle f'_v + p'_v + h'_v, \delta v\rangle = 0, && \forall\delta v : (x, u, v + \delta v) \in E_{xuv},\\
&\langle \partial_\psi\Phi, \delta\psi\rangle = p(\bar{x}, \bar{u}, \bar{v}, \delta\psi) = 0, && \forall\delta\psi \in E_\psi,\\
&\langle \partial_\omega\Phi, \delta\omega\rangle = \langle h'_\omega, \delta\omega\rangle = 0, && \forall\delta\omega : \bar{\omega} + \delta\omega \in E_\omega,
\end{aligned} \tag{29}$$

where the usual notation of nonlinear functional analysis is used.

In order to find the element $(\bar{x}_v, \bar{u}_v, v, \bar{\psi}_v, \bar{\omega}_v)$, the equation $(\partial_v\Phi, \delta v) = 0$ is not considered, and the solution of the remaining equations, which depends on an element v, is found.

If we have an element $(\bar{x}, \bar{u}, \bar{v}) \in K_{ph_+}$ and are required to verify whether it is an optimal solution of the game (1)–(3), the following theorem is useful.

Theorem 3. *For an element* $(\bar{x}, \bar{u}, \bar{v}) \in K_{ph_+}$ *to be an optimal solution of the game* (1)–(3), *it is necessary and sufficient that, for every* $v \in K_v$, *there exist elements* $(\bar{x}_v, \bar{u}_v, v) \in K_{ph_+}$ *and an element* $\bar{v} \in K_v$ *such that the set*

$$\operatorname{Ker} \Delta f_+ = \{(x, u, v) : f(\bar{x}_v, \bar{u}_v, v) \leqslant f(x, u, v) \,\forall (x, u, v) \in E_{xuv} \text{ and } f(\bar{x}_v, \bar{u}_v, v) \leqslant f(\bar{x}, \bar{u}, \bar{v})\}, \tag{30}$$

where $(\bar{x}, \bar{u}, \bar{v}) = (\bar{x}_v, \bar{u}_v, v)$, *has the property that*

$$K_{ph_+} \subset \operatorname{Ker} \Delta f_+. \tag{31}$$

Necessity. Assume that there exists an optimal solution of the game (1)–(3). Then, by definition, there exist elements $(\bar{x}_v, \bar{u}_v, v) \in K_{ph_+}$ and an element $\bar{v} \in K_v$ such that (10) and (11) hold for all $(x', u', v) \in K_{ph_+}$, i.e., indeed $K_{ph_+} \subset \operatorname{Ker} \Delta f_+$.

Sufficiency. Suppose that, for every $v \in K_v$, there exist elements $(\bar{x}_v, \bar{u}_v, v) \in K_{ph_+}$ and an element $\bar{v} \in K_v$ such that (31) is satisfied.

By definition, the relations (9) hold for the element $(\bar{x}, \bar{u}, \bar{v}) = (\bar{x}_{\bar{v}}, \bar{u}_{\bar{v}}, \bar{v}) \in K_{ph_+}$. Developing the condition (31), we obtain that (10) and (11) are also satisfied, i.e. we obtain indeed that the element $(\bar{x}, \bar{u}, \bar{v})$ is in this case an optimal solution of the game (1)–(3) in the sense of Definition 1. Theorem 3 is proved.

Thus Theorems 1–3, together with Corollary 1, yield a certain method of finding and investigating an optimal solution in game theory and control problems. Subsequently this method will be called the *variational optimality principle.*

We shall now consider the application of the variational optimality principle to certain concrete classes of extremal problems.

§2. The variational formulation of game theory and optimal control problems for various classes of equations

The variational optimality principle considered above can be applied in practice if the game is given in the form (1)–(3), i.e. if the functionals p and h are constructed in an explicit form (the functional f is usually given).

Below, we shall write out in an explicit form the functionals f, p and h of form (1)–(3) for controlled systems which are described by differential equations, integral equations, or finite algebraic relations.

If the functional $\Phi = f + p + h$ constructed for these problems in the appropriate spaces has a quasi-saddle point in the sense of Definition 2, then, by virtue of the corollary to Theorem 2, an optimal solution of such a game is determined from the variational equations (29).

For the game problems presented below the solvability question is not considered, the existence of an optimal solution of the game is assumed, and all the operations carried out are assumed to be correct.

1°. *Systems described by second-order partial differential equations.* Let $\xi = (\xi^0, \xi^1, \dots, \xi^n)$ be a point in the $(n+1)$-dimensional Euclidean space E^{n+1}, let $\Omega \subset E^{n+1}$ be a bounded domain with the boundary $\partial\Omega$, and let the vector-valued functions $x = (x^1, \dots, x^m)$, $u = (u^1, \dots, u^p)$, and $v = (v^1, \dots, v^q)$

be defined on the domain Ω and characterize the state of a controlled system, the strategy for the player U, and the strategy for the player V, respectively.

Further, we set

$$D_\alpha = \frac{\partial}{\partial \xi^\alpha}, \quad D = (D_0, D_1, \ldots D_n)$$

and consider the following differential game: find

$$\max_v \min_u \left\{ \int_\Omega f(x, Dx, u, v, \xi)\, d\Omega + \int_{\partial\Omega} \varphi(x, Dx, u, v, \xi)\, d\Gamma \right. \tag{32}$$

under the condition that

$$\begin{gathered} \sum_{\alpha=0}^{n} D_\alpha A^{\alpha\beta}(x, Dx, u, v, \xi) = f^\beta(x, Dx, u, v, \xi), \quad \beta \in \{1, 2, \ldots m\} = M; \\ (x^\beta - x_0^\beta)_{|\partial\Omega_0} = 0, \ \beta \in M_0 \subset M, \ \left(\sum A^{\alpha\beta} n_\alpha - \varphi_0^\beta\right)_{|\partial\Omega_0} = 0, \ \beta \in N_0 \subset M; \\ (x^\beta - x_1^\beta)_{|\partial\Omega_1} = 0, \ \beta \in M_1 \subset M, \ \left(\sum A^{\alpha\beta} n_\alpha - \varphi_1^\beta\right)_{|\partial\Omega_1} = 0, \ \beta \in N_1 \subset M, \end{gathered} \tag{33}$$

and that, also,

$$\begin{gathered} h^i(x, Dx, u, v, \xi) \geqslant 0, \ \xi \in \Omega \ (i = 1, 2, \ldots r), \\ h^j(x, Dx, u, v, \xi)_{|\partial\Omega} \geqslant 0 \ (j = r+1, r+2, \ldots s), \ s \geqslant r, \end{gathered} \tag{34}$$

$$\int_\Omega h^k(x, Dx, u, v, \xi)\, d\Omega + \int_{\partial\Omega} \Lambda^k(x, Dx, u, v, \xi)\, dr \geqslant 0 \ (k = s+1, s+2 \ldots l), \ l \geq s,$$

where $\partial\Omega_0$ and $\partial\Omega_1$ are parts of the boundary $\partial\Omega$; M_0, N_0, M_1 and N_1 are the subsets of indices of the set M which correspond to the correctly posed boundary problem for the given controlled system; n_α are the components of the outward unit normal to the boundary; and x_0^β, ϕ_0^β, x_1^β and ϕ_1^β are given functions on the boundary which depend, generally speaking, on x, u and v.

In our case the functionals f, p and h in the game (1)–(3) have the form

$$f(x, u, v) = \int_\Omega f(x, Dx, u, v, \xi)\, d\Omega + \int_{\partial\Omega} \varphi(x, Dx, u, v, \xi)\, d\Gamma,$$

$$p(x, u, v, \psi) = \int_\Omega \sum_{\beta=1}^{m} \left\{ \sum_{\alpha=0}^{n} A^{\alpha\beta}(x, Dx, u, v, \xi)\, D_\alpha \psi_\beta \right.$$

$$\left. + f^\beta(x, Dx, u, v, \xi)\, \psi_\beta \right\} d\Omega + \int_{\partial\Omega_0} \sum_{\beta \in N_0} \varphi_0^\beta \psi_\beta\, d\Gamma + \int_{\partial\Omega_1} \sum_{\beta \in N_1} \varphi_1^\beta \psi_\beta\, d\Gamma,$$

$$\begin{aligned} h(x, u, v, \omega) = {} & \int_\Omega \sum_{i=1}^{r} \omega_i^2 h^i(x, Dx, u, v, \xi)\, d\Omega + \int_{\partial\Omega} \sum_{j=r+1}^{s} \omega_j^2 h^j(x, Dx, u, v, \xi)\, d\Omega \\ & + \sum_{k=s+1}^{l} \mathring{\omega}_k^2 \left\{ \int_\Omega h^k(x, Dx, u, v, \xi)\, d\Omega + \int_{\partial\Omega} \Lambda^k(x, Dx, u, v, \xi)\, d\Gamma \right\}, \end{aligned} \tag{35}$$

where the following notation is used:

$$\psi = (\psi_1, \psi_2, \ldots \psi_m), \quad \omega = (\omega_1, \ldots \omega_r|_\Omega, \omega_{r+1}, \ldots \omega_s|_{\partial\Omega}, \mathring{\omega}_{s+1}, \ldots \mathring{\omega}_l). \tag{36}$$

Here $\overset{\circ}{\omega}{}^2_{s+1}, \overset{\circ}{\omega}{}^2_{s+2}, \ldots, \overset{\circ}{\omega}{}^2_l$ are numbers.

For the functionals p and h thus constructed, the requirement that property 1° hold means the solvability of the system of differential equations for fixed elements u and v, the consistency of the system of inequalities (34), and the fact that the class of functions which satisfy both the differential equations and the system of inequalities is nonempty.

Property 2° is satisfied for the functional p. Properties 3° and 4° hold for the functional h.

The set of functions on which the functional $\Phi = f + p + h$ determined by (35) is considered must satisfy the condition

$$x \in \{x : (x^\beta - x_0^\beta)|_{\partial\Omega_0} = 0, \ \beta \in M_0, \ (x^\beta - x_1^\beta)|_{\partial\Omega_1} = 0, \ \beta \in M_1\},$$

$$\psi \in \{\psi : \psi_\beta|_{\partial\Omega_0} = 0, \ \beta \in M \setminus N_0, \ \psi_\beta|_{\partial\Omega_1} = 0, \ \beta \in M \setminus N_1\},$$

and must belong to the corresponding functional spaces in which an optimal solution of the game is being sought.

2°. *Systems described by first-order partial differential equations.* Let us be given the following differential game: find

$$\max_v \min_u \left\{ \int_\Omega f(x, u, v, \xi)\, d\Omega + \int_{\partial\Omega} \varphi(x, u, v, \xi)\, d\Gamma \right\} \tag{37}$$

subject to the conditions

$$\sum_{\alpha=0}^{n} D_\alpha A^{\alpha\beta}(x, u, v, \xi) = f^\beta(x, u, v, \xi), \quad \beta \in \{1, 2, \ldots m\} = M;$$

$$(x^\beta - x_0^\beta)|_{\partial\Omega_0} = 0, \quad \beta \in M_0 \subset M, \quad \left(\sum_{\alpha=0}^{n} A^{\alpha\beta} n_\alpha - \varphi_0^\beta\right)\Big|_{\partial\Omega_0} = 0, \quad \beta \in N_0 \subset M; \tag{38}$$

$$(x^\beta - x_1^\beta)|_{\partial\Omega_1} = 0, \quad \beta \in M_1 \subset M, \quad \left(\sum_{\alpha=0}^{n} A^{\alpha\beta} n_\alpha - \varphi_1^\beta\right)\Big|_{\partial\Omega_1} = 0, \quad \beta \in N_1 \subset M,$$

and

$$\begin{gathered} h^i(x, u, v, \xi) \geqslant 0, \quad \xi \in \Omega \quad (i = 1, 2, \ldots r), \\ h^j(x, u, v, \xi)|_{\partial\Omega} \geqslant 0 \quad (j = r+1, \ r+2, \ldots s), \quad s \geqslant r, \\ \int_\Omega h^k(x, u, v, \xi)\, d\Omega + \int_{\partial\Omega} \Lambda^k(x, u, v, \xi)\, d\Gamma \geqslant 0 \\ (k = s+1, \ s+2, \ldots l), \quad l \geqslant s, \end{gathered} \tag{39}$$

where the same notation is used as in the preceding problem.

In this case the functionals f, p and h (1)–(3) have the form

$$f(x, u, v) = \int_\Omega f(x, u, v, \xi)\, d\Omega + \int_{\partial\Omega} \varphi(x, u, v, \xi)\, d\Gamma,$$

$$\begin{aligned} p(x, u, v, \psi) = &\int_\Omega \sum_{\beta=1}^m \left\{ \sum_{\alpha=0}^n A^{\alpha\beta}(x, u, v, \xi)\, D_\alpha \psi_\beta + f^\beta(x, u, v, \xi)\, \psi_\beta \right\} d\Omega \\ &+ \int_{\partial\Omega_0} \sum_{\beta\in N_0} \varphi_0^\beta \psi_\beta d\Gamma + \int_{\partial\Omega_1} \sum_{\beta\in N_1} \varphi_1^\beta \psi_\beta d\Gamma, \end{aligned} \tag{40}$$

$$\begin{aligned} h(x, u, v, \omega) = &\int_\Omega \sum_{i=1}^r \omega_i^2 h^i(x, u, v, \xi)\, d\Omega + \int_{\partial\Omega} \sum_{j=r+1}^s \omega_j^2 h^j(x, u, v, \xi)\, d\Gamma \\ &+ \sum_{k=s+1}^l \mathring{\omega}_k^2 \left\{ \int_\Omega h^k(x, u, v, \xi)\, d\Omega + \int_{\partial\Omega} \Lambda^k(x, u, v, \xi)\, d\Gamma \right\}, \end{aligned}$$

where the notation (36) is used for ψ and ω, and f, p and h have the same properties as in the preceding problem.

3°. *Systems described by ordinary differential equations.* The differential games (32)–(34) and (37)–(39) presented above represent, in the particular case of $n = 0$, games for controlled systems which are described by ordinary differential equations of the second and first order, respectively.

Let us consider this case in more detail. If we set $\xi^0 = t$, $\Omega = \{t : 0 \leq t \leq T\}$ and $\partial\Omega = \{t : t = 0,\ t = T\}$, then the differential game (32)–(34) takes on the following form: find

$$\max_v \min_u \left\{ \int_0^T f\left(x, \frac{dx}{dt}, u, v, t\right) dt + \varphi\left(x, \frac{dx}{dt}, u, v, t\right) \Big|_0^T \right\} \tag{41}$$

subject to the condition that

$$\begin{aligned} &\frac{d}{dt} A^\beta\left(x, \frac{dx}{dt}, u, v, t\right) = f^\beta\left(x, \frac{dx}{dt}, u, v, t\right), \quad \beta \in \{1, 2, \dots m\} = M; \\ &(x^\beta - x_0^\beta)|_{t=0} = 0, \quad \beta \in M_0 \subset M, \quad (A^\beta - \varphi_0^\beta)|_{t=0} = 0, \quad \beta \in N_0 \subset M; \\ &(x^\beta - x_1^\beta)|_{t=T} = 0, \quad \beta \in M_1 \subset M, \quad (A^\beta - \varphi_1^\beta)|_{t=T} = 0 \quad \beta \in N_1 \subset M, \end{aligned} \tag{42}$$

and that

$$\begin{aligned} &h^i\left(x, \frac{dx}{dt}, u, v, t\right) \geqslant 0, \quad t \in [0, T] \quad (i = 1, 2, \dots r), \\ &h^j\left(x, \frac{dx}{dt}, u, v, t\right) \Big|_0^T \geqslant 0 \quad (j = r+1, r+2, \dots s), \quad s \geqslant r \\ &\int_0^T h^k\left(x, \frac{dx}{dt}, u, v, t\right) dt + \Lambda^k\left(x, \frac{dx}{dt}, u, v, t\right) \Big|_0^T \geqslant 0 \quad (k = s+1, s+2, \dots l), \quad l \geqslant s, \end{aligned} \tag{43}$$

where the notation has the same sense as before.

In this case f, p and h are written down as follows:

$$f(x, u, v) = \int_0^T f(x, \dot{x}, u, v, t)\,dt + \varphi(x, \dot{x}, u, v, t)\Big|_0^T,$$

$$\begin{aligned} p(x, u, v, \psi) = \int_0^T \sum_{\beta=1}^{m} \{A^\beta(x, \dot{x}, u, v, t)\dot{\psi}_\beta + f^\beta(x, \dot{x}, u, v, t)\psi_\beta\}\,dt \\ + \sum_{\beta\in N_0} (\varphi_0^\beta \psi_\beta)\big|_{t=0} - \sum_{\beta\in N_1} (\varphi_1^\beta \psi_\beta)\big|_{t=T}, \end{aligned} \tag{44}$$

$$\begin{aligned} h(x, u, v, \omega) = \int_0^T \sum_{i=1}^{r} \omega_i^2 h^i(x, \dot{x}, u, v, t)\,dt + \sum_{j=r+1}^{s} \omega_j^2 h^j(x, \dot{x}, u, v, t)\Big|_0^T \\ + \sum_{k=s+1}^{l} \mathring{\omega}_k^2 \left\{\int_0^T h^k(x, \dot{x}, u, v, t)\,dt + \Lambda^k(x, \dot{x}, u, v, t)\Big|_0^T\right\}. \end{aligned}$$

The differential game (37)–(39) with $n = 0$ can be considered in a similar way. In this case, this game will be formulated as follows: find

$$\max_v \min_u \left\{\int_0^T f(x, u, v, t)\,dt + \varphi(x, u, v, t)\Big|_0^T\right\} \tag{45}$$

under the conditions

$$\begin{aligned} &\frac{d}{dt}A^\beta(x, u, v, t) = f^\beta(x, u, v, t), \quad \beta \in \{1, 2, \dots m\} = M; \\ &(x^\beta - x_0^\beta)\big|_{t=0} = 0, \quad \beta \in M_0 \subset M, \quad (A^\beta - \varphi_0^\beta)\big|_{t=0} = 0, \quad \beta \in N_0 \subset M; \\ &(x^\beta - x_1^\beta)\big|_{t=T} = 0, \quad \beta \in M_1 \subset M, \quad (A^\beta - \varphi_1^\beta)\big|_{t=T} = 0, \quad \beta \in N_1 \subset M, \end{aligned} \tag{46}$$

and

$$\begin{aligned} &h^i(x, u, v, t) \geqslant 0, \quad t \in [0, T] \quad (i = 1, 2, \dots r), \\ &h^j(x, u, v, t)\Big|_0^T \geqslant 0 \quad (j = r+1, r+2, \dots s), \quad s \geqslant r, \\ &\int_0^T h^k(x, u, v, t)\,dt + \Lambda^k(x, u, v, t)\Big|_0^T \geqslant 0 \quad (k = s+1, s+2, \dots l), \quad l \geqslant s. \end{aligned} \tag{47}$$

Then f, p and h have the following form:

$$f(x, u, v)=\int_0^T f(x, u, v, \xi)\,dt+\varphi(x, u, v, t)\Big|_0^T,$$

$$\begin{aligned} p(x, u, v, \psi)=&\int_0^T \sum_{\beta=1}^{m}\{A^\beta(x, u, v, t)\dot{\psi}_\beta+f^\beta(x, u, v, t)\psi_\beta\}\,dt \\ &+\sum_{\beta\in N_0}(\varphi_0^\beta\psi_\beta)\big|_{t=0}-\sum_{\beta\in N_1}(\varphi_1^\beta\psi_\beta)\big|_{t=T}, \end{aligned} \tag{48}$$

$$\begin{aligned} h(x, u, v, \omega)=&\int_0^T\sum_{i=1}^{r}\omega_i^2 h^i(x, u, v, t)\,dt+\sum_{j=r+1}^{s}\omega_j^2 h^j(x, u, v, t)\Big|_0^T \\ &+\sum_{k=s+1}^{l}\mathring{\omega}_k^2\left\{\int_0^T h^k(x, u, v, t)\,dt+\Lambda^k(x, u, v, t)\Big|_0^T\right\}. \end{aligned}$$

The set of functions on which the functional $\Phi = f + p + h$, determined by (44) or (48), is considered must satisfy the conditions

$$x\in\{x:(x^\beta-x_0^\beta)\big|_{t=0}=0,\ \beta\in M_0,\ (x^\beta-x_1^\beta)\big|_{t=T}=0,\ \beta\in M_1\},$$
$$\psi\in\{\psi:\psi_\beta\big|_{t=0}=0,\ \beta\in M\setminus N_0,\ \psi_\beta\big|_{t=T}=0,\ \beta\in M\setminus N_1\}$$

and belong to the corresponding functional spaces in which an optimal solution of the game is sought.

4°. *Systems described by integral equations.* Let there be given the following integral game: find

$$\max_v\min_u\int_\Omega\left\{\int_\Omega K(x, u, v, \xi, \eta)\,d\eta+f(x, u, v, \xi)\right\}d\xi \tag{49}$$

subject to the conditions

$$x^\beta(\xi)=\int_\Omega K^\beta(x, u, v, \xi, \eta)\,d\eta+f^\beta(x, u, v, \xi)\quad(\beta=1, 2, \dots m), \tag{50}$$

and

$$\int_\Omega H^i(x, u, v, \xi, \eta)\,d\eta+h^i(x, u, v, \xi)\geqslant 0\quad \xi\in\Omega\quad(i=1, 2, \dots, r),$$

$$\int_\Omega\left\{\int_\Omega H^j(x, u, v, \xi, \eta)\,d\eta+h^j(x, u, v, \xi)\right\}d\eta\geqslant 0 \tag{51}$$

$$(j=r+1, r+2, \dots s),\quad s\geqslant r,$$

where x, u and v denote the same vector-valued functions as before, $\xi = (\xi^0, \xi^1, \dots, \xi^n)$ is a point of the space E^{n+1}, $\xi\in\Omega$, $\Omega\subset E^{n+1}$ is the integration domain, $\eta=(\eta^0, \eta^1, \dots, \eta^n)$ is the variable point of integration, and K, f, H and h are given functions of their arguments.

By assumption, all the operations in (49)–(51) and below make sense.

In this case the functionals f, p, and h of the game (1)–(3) have the following form:

$$
\begin{aligned}
f(x, u, v) &= \int_{\Omega} \left\{ \int_{\Omega} K(x, u, v, \xi, \eta)\, d\eta + f(x, u, v, \xi) \right\} d\xi, \\
p(x, u, v, \psi) &= \sum_{\beta=1}^{m} \int_{\Omega} \left\{ \int_{\Omega} K^{\beta}(x, u, v, \xi, \eta)\, d\eta + f^{\beta}(x, u, v, \xi) - x^{\beta}(\xi) \right\} \psi_{\beta} d\xi, \\
h(x, u, v, \omega) &= \sum_{i=1}^{r} \int_{\Omega} \left\{ \int_{\Omega} H^{i}(x, u, v, \xi, \eta)\, d\eta + h^{i}(x, u, v, \xi) \right\} \omega_i^2 d\xi \\
&\quad + \sum_{j=r+1}^{s} \overset{\circ}{\omega}{}_j^2 \int_{\Omega} \left\{ \int_{\Omega} H^{j}(x, u, v, \xi, \eta)\, d\eta + h^{j}(x, u, v, \xi) \right\} d\xi.
\end{aligned}
\tag{52}
$$

If the functional $\Phi = f + p + h$ determined by (52) has a quasi-saddle point in the sense of Definition 2, then, by virtue of Theorem 2, an optimal solution of the integral game is determined from the variational equations (29).

5°. *Systems described by relations in finite-dimensional Euclidean spaces.* Let there be given the following game: find

$$\max_{v} \min_{u} f^0(x, u, v) \tag{53}$$

subject to the conditions

$$f^{\beta}(x, u, v) = 0 \quad (\beta = 1, 2, \dots m) \tag{54}$$

and

$$h^{i}(x, u, v) \geqslant 0 \quad (i = 1, 2, \dots r), \tag{55}$$

where

$$x = (x^1, x^2, \dots x^m), \quad u = (u^1, u^2, \dots u^p), \quad v = (v^1, v^2, \dots v^q)$$

denote points in the Euclidean spaces E^m, E^p, and E^q.

In this case f, p and h are written down as follows:

$$
\begin{aligned}
f(x, u, v) &= f^0(x, u, v), \\
p(x, u, v, \psi) &= \sum_{\beta=1}^{m} \psi_{\beta} f^{\beta}(x, u, v), \\
h(x, u, v, \omega) &= \sum_{i=1}^{r} \omega_i^2 h^{i}(x, u, v),
\end{aligned}
\tag{56}
$$

where

$$\psi = (\psi_1, \psi_2, \dots \psi_m) \in E_m, \quad \omega = (\omega_1, \omega_2, \dots \omega_r) \in E_r$$

are points with covariant coordinates in E_m and E_r, respectively.

If there exists a quasi-saddle point of the functional

$$\Phi = f^0(x, u, v) + \sum_{\beta=1}^{m} \psi_\beta f^\beta(x, u, v) + \sum_{i=1}^{r} \omega_i^2 h^i(x, u, v) \tag{57}$$

on the set $E^m \times E^p \times E^q \times E_m \times E_r$, then an optimal solution of the game (53)–(55) is determined from the variational equations (29) for this functional. In the case of smooth functions f^0, f^β and h^i, these equations have the form

$$\begin{aligned}
&\frac{\partial f^0}{\partial x^\lambda}(x, u, v) + \sum_{\beta=1}^{m} \psi_\beta \frac{\partial f^\beta}{\partial x^\lambda}(x, u, v) + \sum_{i=1}^{r} \omega_i^2 \frac{\partial h^i}{\partial x^\lambda}(x, u, v) = 0 \quad (\lambda = 1, 2, \ldots m),\\
&\frac{\partial f^0}{\partial u^\omega}(x, u, v) + \sum_{\beta=1}^{m} \psi_\beta \frac{\partial f^\beta}{\partial u^\omega}(x, u, v) + \sum_{i=1}^{r} \omega_i^2 \frac{\partial h^i}{\partial u^\omega}(x, u, v) = 0 \quad (\omega = 1, 2, \ldots p),\\
&\frac{\partial f^0}{\partial v^\mu}(x, u, v) + \sum_{\beta=1}^{m} \psi_\beta \frac{\partial f^\beta}{\partial v^\mu}(x, u, v) + \sum_{i=1}^{r} \omega_i^2 \frac{\partial h^i}{\partial v^\mu}(x, u, v) = 0 \quad (\mu = 1, 2, \ldots q),\\
&f^\beta(x, u, v) = 0 \quad (\beta = 1, 2, \ldots m),\\
&\omega_\nu h^\nu(x, u, v) = 0 \quad (\nu = 1, 2, \ldots r).
\end{aligned} \tag{58}$$

The existence of a quasi-saddle point of the functional (57) is closely related to the solvability of the first equation in $\psi_1, \ldots, \psi_m$ and, in general, does not constitute a necessary condition for the existence of an optimal solution of the game (53)–(55).

Necessary and sufficient conditions for the existence of an optimal solution of the game are given by Theorem 1.

Theorem 1 can be also made use of in subsections 1°–4° for finding an optimal solution of the game.

We shall now consider a concrete example of a differential game for a partial differential equation from the domain of optimal designing.

§3. The optimal design problem as a mathematical game between the designer and nature. An example

The statement of correct mathematical problems in the theory of optimal design (finding elastic constructions with the least weight or cost, obtaining aerodynamic forms which have the least drag while providing useful payloads, controllability, and stability along the trajectory, and other problems) is a difficult task, and, at present, it is in the initial state of development.

It seems to us that the application of game theory in this domain is a very promising direction, since finding, e.g., an optimal elastic construction (the choice of dimensions, cross-section areas, material, etc.) is also connected with the necessity of considering the worst external conditions (all loads, heat streams, the effects of turbulent atmosphere), which can be suitably treated as the the influence of the second player, namely nature, on the game.

We present below a simple illustrative example of such a problem (from the mathematical viewpoint, the problem is quite complicated; it was formulated by

J.-L. Lions in his monograph [15], Chapter 2, Remark 7.6, as an unsolved problem in a simpler case). Under the assumption of the existence of a quasi-saddle point of a certain functional, the variational equations of form (2), which the optimal solution being sought satisfies, are written down in a concrete form. In the one-dimensional case, the optimal solution is written down in an explicit form.

We consider the following differential game: find

$$\max_{v} \min_{u} \int_{\Omega} u \, d\Omega \tag{59}$$

subject to the conditions

$$\frac{\partial}{\partial \xi^1}\left(u \frac{\partial x}{\partial \xi^1}\right)+\frac{\partial}{\partial \xi^2}\left(u \frac{\partial x}{\partial \xi^2}\right)+v=0, \quad x\,|_{\partial\Omega}=0 \tag{60}$$

and

$$u \geqslant 0, \quad 0 \leqslant v \leqslant 1, \quad \int_{\Omega} v\, d\Omega \leqslant p \quad (p>0), \quad x \leqslant a. \tag{61}$$

The differential game can be viewed as a problem of optimal design which has the following physical meaning: there is a membrane with variable "thickness" u, fastened along its contour $\partial\Omega$. The load v acts on the membrane. Two players, namely the designer, who chooses the nonnegative thickness u, and the nature, which applies a load v from a certain class, can influence the state of the membrane. The aim of the designer is to choose a nonnegative membrane "thickness" $\bar{u}$ such that the membrane deflection $\bar{x}$ does not exceed a given value a under the "worst" load $\bar{v}$ from the given class, and that the weight (volume) of the membrane remains the smallest possible. The rules of the game are given by Definition 1.

Here the functionals f, p and h have the following form:

$$f(x, u, v)=\int_{\Omega} u\, d\Omega,$$

$$p(x, u, v, \psi)=\int_{\Omega}\left(u \frac{\partial x}{\partial \xi'} \frac{\partial \psi}{\partial \xi'}+u \frac{\partial x}{\partial \xi^2} \frac{\partial \psi}{\partial \xi^2}-v \psi\right) d\Omega, \tag{62}$$

$$h(x, u, v, \omega)=\int_{\Omega}\left\{\omega_1^2 u+\omega_2^2 v+\omega_3^2(1-v)+\omega_4^2(a-x)\right\} d\Omega+\mathring{\omega}_5^2\left(p-\int_{\Omega} v\, d\Omega\right),$$

where $\mathring{\omega}_5^2$ is a number.

The set of functions on which the functional $\Phi=f+p+h$ determined by (62) is considered must satisfy the conditions

$$x \in \{x: x\,|_{\partial\Omega}=0\}=E_x, \quad \psi \in \{\psi: \psi\,|_{\partial\Omega}=0\}=E_\psi. \tag{63}$$

We write down the variational equations (29) for the functional Φ in a formal way:

$$
\begin{aligned}
&\langle \partial_x\Phi, x\rangle = \int_\Omega \left(\bar u \frac{\partial\bar\psi}{\partial\xi^1}\frac{\partial x}{\partial\xi^1} + \bar u \frac{\partial\bar\psi}{\partial\xi^2}\frac{\partial x}{\partial\xi^2} - \omega_4^2 x\right) d\Omega = 0, \quad \forall x \in E_x,\\
&\langle \partial_u\Phi, u\rangle = \int_\Omega \left(1 + \frac{\partial\bar x}{\partial\xi^1}\frac{\partial\bar\psi}{\partial\xi^1} + \frac{\partial\bar x}{\partial\xi^2}\frac{\partial\bar\psi}{\partial\xi^2} + \bar\omega_1^2\right) u\, d\Omega = 0, \quad \forall v \in E_u,\\
&\langle \partial_v\Phi, v\rangle = \int_\Omega (-\bar\psi + \bar\omega_2^2 - \bar\omega_3^2 - \bar\omega_5^2)\, v\, d\Omega = 0, \quad \forall v \in E_v,\\
&\langle \partial_\psi\Phi, \psi\rangle = \int_\Omega \left(\bar u \frac{\partial\bar x}{\partial\xi^1}\frac{\partial\psi}{\partial\xi^1} + \bar u \frac{\partial\bar x}{\partial\xi^2}\frac{\partial\psi}{\partial\xi^2} - \bar v\psi\right) d\Omega = 0, \quad \forall \psi \in E_\psi,\\
&\langle \partial_\omega\Phi, \delta\omega\rangle = 2\int_\Omega \{\bar\omega,\ \bar u\delta\omega_1 + \bar\omega_2\bar v\delta\omega_2 + \bar\omega_3(1-\bar v)\delta\omega_3 + \bar\omega_4(a-\bar x)\delta\omega_4\}\, d\Omega\\
&\qquad + \ldots + 2\bar\omega_5\left(p - \int_\Omega v\, d\Omega\right)\delta\bar\omega_5 = 0 \quad \forall\delta\omega : \bar\omega + \delta\omega \in E_\omega,
\end{aligned}
\tag{64}
$$

where $(\bar x, \bar u, \bar v, \bar\psi, \bar\omega)$ is an optimal solution of the game, and $(x, u, v, \psi, \delta\omega)$ are variations of the solutions in the corresponding classes of functions E_x, E_u, E_v, E_ψ and E_ω, which contain at least infinitely differentiable functions.

The variational equations (64) allow us to establish many properties of the solution without actually finding it.

In particular, the following theorem holds.

Theorem 4. *An optimal solution of the game* (59)–(61) *which satisfies the variational equations* (64) *has the following properties in the case of sufficient smoothness*:

$$
\int_\omega \bar u\, d\Omega = \int_\omega \bar v\bar\psi\, d\Omega - \int_{\partial\omega} \bar u \frac{\partial\bar x}{\partial n}\bar\psi\, d\Gamma \tag{65}
$$

and

$$
\int_\omega \bar u\, d\Omega = -\int_\omega \bar\omega_4^2\bar x\, d\Omega - \int_{\partial\omega} \bar u \frac{\partial\bar\psi}{\partial n}\bar x\, d\Gamma, \tag{66}
$$

where ω *is an arbitrary subdomain of* Ω.

Proof. Let there exist an optimal solution of the game (59)–(61) which satisfies the variational equations (64) and has the necessary smoothness. From the second equation of (64) we have

$$
1 + \frac{\partial\bar x}{\partial\xi^1}\frac{\partial\bar\psi}{\partial\xi^1} + \frac{\partial\bar x}{\partial\xi^2}\frac{\partial\bar\psi}{\partial\xi^2} + \omega_1^2 = 0, \tag{67}
$$

and from the last equation we obtain that

$$
\bar\omega_1\bar u = 0, \quad \bar\omega_1^2\bar u = 0. \tag{68}
$$

Multiplying (67) by $\bar u$ and integrating over the domain $\omega \subset \Omega$ with the condition (68) taken into account, we obtain

$$\int\limits_{\omega}\left(\bar{u}+\bar{u}\frac{\partial\bar{x}}{\partial\xi^1}\frac{\partial\bar{\psi}}{\partial\xi^1}+\bar{u}\frac{\partial\bar{x}}{\partial\xi^2}\frac{\partial\bar{\psi}}{\partial\xi^2}\right)d\Omega=0. \tag{69}$$

On the other hand, it follows from the equation $\langle\partial_{\psi}\Phi, \psi\rangle = 0$ of system (64) for smooth functions that

$$\int\limits_{\omega}\left(\bar{u}\frac{\partial\bar{x}}{\partial\xi^1}\frac{\partial\bar{\psi}}{\partial\xi^1}+\bar{u}\frac{\partial\bar{x}}{\partial\xi^2}\frac{\partial\bar{\psi}}{\partial\xi^2}-\bar{v}\bar{\psi}\right)d\Omega-\int\limits_{\partial\omega}\bar{u}\frac{\partial\bar{x}}{\partial n}\bar{\psi}d\Gamma=0 \tag{70}$$

and, similarly, from the equation $\langle\partial_x\Phi, x\rangle = 0$ of (64) that

$$\int\limits_{\omega}\left(\bar{u}\frac{\partial\bar{x}}{\partial\xi^1}\frac{\partial\bar{\psi}}{\partial\xi^1}+\bar{u}\frac{\partial\bar{x}}{\partial\xi^2}\frac{\partial\bar{\psi}}{\partial\xi^2}-\bar{\omega}_4^2\bar{x}\right)d\Omega-\int\limits_{\partial\omega}\bar{u}\frac{\partial\bar{\psi}}{\partial n}\bar{x}d\Gamma=0. \tag{71}$$

Combining (69)–(71), we obtain (65) and (66). Theorem 4 is proved.

We shall now try to obtain the solution of system (64) in an explicit form for some particular cases, since in the general case the system (64) is sufficiently complicated. Analyzing the variational equations (64), we see that the inequality $x \le a$ in (61) cannot be transformed into equality on a set of points of Ω which contains an entire circle of arbitrarily small radius $r > 0$, i.e. the inequality can hold with the equality sign only at a single point, on a line, and on a more complicated set of points of Ω.

Let Ω be such that the inequality $x \le a$ holds with the equality sign only at a single point $\mathring{\xi} \in \Omega$. Then it follows from the last equation of (64) that $\omega_4^2 = \mathring{\omega}_4^2\delta(\xi - \mathring{\xi})$, where $\mathring{\omega}_4^2$ is a number and $\delta(\xi - \mathring{\xi})$ is the Dirac δ-function.

Apparently, in this case an optimal solution of the game (59)–(61) is piecewise smooth and satisfies almost everywhere the following system of nonlinear partial differential equations, which are obtained from the variational equations (64):

$$\begin{gathered}\frac{\partial}{\partial\xi^1}\left(u\frac{\partial\psi}{\partial\xi^1}\right)+\frac{\partial}{\partial\xi^2}\left(u\frac{\partial\psi}{\partial\xi^2}\right)+\mathring{\omega}_4^2\delta(\xi-\mathring{\xi})=0, \quad \psi|_{\partial\Omega}=0,\\ 1+\frac{\partial x}{\partial\xi^1}\frac{\partial\psi}{\partial\xi^1}+\frac{\partial x}{\partial\xi^2}\frac{\partial\psi}{\partial\xi^2}+\omega_1^2=0,\\ \psi+\omega_2^2-\omega_3^2-\mathring{\omega}_5^2=0,\\ \frac{\partial}{\partial\xi^1}\left(u\frac{\partial x}{\partial\xi^1}\right)+\frac{\partial}{\partial\xi^2}\left(u\frac{\partial x}{\partial\xi^2}\right)+v=0, \quad x|_{\partial\Omega}=0,\\ \omega_1u=\omega_2v=\omega_3(1-v)=\mathring{\omega}_4(a-x)|_{\xi=\mathring{\xi}}=\mathring{\omega}_5\left(p-\int\limits_{\Omega}vd\Omega\right)=0.\end{gathered} \tag{72}$$

Additional relations must be satisfied on the surface of discontinuities of an optimal solution or discontinuities of the derivatives. These relations are obtained from the variational equations (64).

Let us consider a one-dimensional analogue of the game (59)–(61), which describes the problem of the optimal design of a membrane which has the form of an infinite strip, or the problem of finding an optimal tension-bearing rod for an arbitrary external load from a given class. The problem is formulated as follows: find

$$\max_{v} \min_{u} \int_0^1 u\,d\xi \tag{73}$$

subject to the conditions

$$\frac{d}{d\xi}\left(u\,\frac{dx}{d\xi}\right)+v=0, \qquad x(0)=x(1)=0 \tag{74}$$

and

$$u\geqslant 0, \qquad 0\leqslant v\leqslant 1, \quad \int_0^1 v\,d\xi\leqslant p \ \ (p>0), \ \ x\leqslant a, \tag{75}$$

where the notation has the same sense as in (59)–(61). In this case the equations (72) for finding an optimal solution have the form

$$\begin{gathered}
(ux'_\xi)'_\xi+\mathring{\omega}_4^2\delta(\xi-\mathring{\xi})=0, \quad \psi(0)=\psi(1)=0,\\
1+x'_\xi\cdot\psi'_\xi+\omega_1^2=0,\\
-\psi+\omega_2^2-\omega_3^2-\mathring{\omega}_5^2=0,\\
(ux'_\xi)'_\xi+v=0, \quad x(0)=x(1)=0,\\
\omega_1 u=\omega_2 v=\omega_3(1-v)=\mathring{\omega}_4(a-x)_{|\xi=\mathring{\xi}}=\mathring{\omega}_5\left(p-\int_0^1 v\,d\xi\right)=0.
\end{gathered} \tag{76}$$

Now we introduce the notation

$$M_0=\left\{\xi: 0\leqslant\xi<\frac{1-p}{2}\right\}, \quad M=\left\{\xi:\frac{1-p}{2}\leqslant\xi\leqslant\frac{1+p}{2}\right\}, \quad M_1=\left\{\xi:\frac{1+p}{2}<\xi\leqslant 1\right\}$$

and, moreover,

$$\alpha=\frac{2}{p}, \qquad u_0=\frac{1}{\alpha}\left|\frac{p}{4}-\frac{p^2}{12}\right|. \tag{77}$$

Then an optimal solution of the game (73)–(75) can be written down in the form

$$\begin{aligned}
x&=\left(\frac{\xi}{\alpha u_0},\ -\frac{2}{3u_0}\left|\xi-\frac{1}{2}\right|^{\frac{3}{2}}+a,\ \frac{1-\xi}{\alpha u_0}\right),\\
u&=\left(u_0,\ \alpha^{\frac{1}{2}}u_0\left|\xi-\frac{1}{2}\right|^{\frac{1}{2}},\ u_0\right),\\
v&=(0,\ 1,\ 0),
\end{aligned} \tag{78}$$

where the components of the vectors x, u and v denote, respectively, the values of the optimal solution on the sets M_0, M and M_1, and the numbers α and u_0 are determined by (77). For the sake of brevity, we do not present the expressions for ψ and ω.

Finally, let us consider the simplest finite-dimensional analogue of the game (59)–(61), which corresponds to finding an optimal spring.

The problem is formulated as follows: find

$$\max_{v} \min_{u} u \tag{79}$$

under the condition that

$$xu - v = 0 \tag{80}$$

and that

$$u \geqslant 0, \ 0 \leqslant v \leqslant 1, \ x \leqslant a. \tag{81}$$

In this case f, p and h have the form

$$f(x, u, v) = u, \ p(x, u, v, \psi) = (xu - v)\psi,$$
$$h(x, u, v, \omega) = \omega_1^2 u + \omega_2^2 v + \omega_3^2 (1 - v) + \omega_4^2 (a - x).$$

The system of equations for finding an optimal solution is obtained by equating to zero the corresponding derivatives of the functional $\Phi = f + p + h$:

$$\begin{aligned}
&\partial_x \Phi = u\psi - \omega_4^2 = 0,\\
&\partial_u \Phi = 1 + x\psi + \omega_1^2 = 0,\\
&\partial_v \Phi = -\psi + \omega_2^2 - \omega_3^2 = 0,\\
&\partial_\psi \Phi = xu - v = 0,\\
&\frac{1}{2}\partial_\omega \Phi = \omega_1 u = \omega_2 v = \omega_3 (1 - v) = \omega_4 (a - x) = 0.
\end{aligned}$$

This system of equations has the following two solutions:

$$\begin{aligned}
&(x, u, v, \psi, \dot{\omega})_1 = (x_0, 0, 0, \omega_2^2, -1 - x_0\omega_2^2, \omega_2^2, 0, 0),\\
&(x, u, v, \psi, \omega)_2 = \left(a, \frac{1}{a}, 1, -\frac{1}{a}, 0, 0, \frac{1}{a}, \frac{1}{a^2}\right),
\end{aligned} \tag{82}$$

where x_0 and ω_2^2 are arbitrary constants.

Making use of Theorem 3, we shall verify the optimality of these solutions. In our case, we have for the second solution

$$\operatorname{Ker} \Delta f_+ = \left\{(x, u, v) \colon \frac{v}{a} \leqslant u, \ \frac{v}{a} \leqslant \frac{1}{a}\right\},$$

and, on the other hand,

$$K_{ph_+} = \{(x, u, v) \colon xu - v = 0, \ u \geqslant 0, \ 0 \leqslant v \leqslant 1, \ x \leqslant a\}$$
$$= \left\{(x, u, v) \colon u \geqslant 0, \ 0 \leqslant v \leqslant 1, \ \frac{v}{u} \leqslant a\right\}$$

and therefore $K_{ph_+} \subset \operatorname{Ker} \Delta f_+$. Then, by virtue of Theorem 3, the solution $(a, 1/a, 1, -1/a, 0, 0, 1/a, -1/a^2)$ is optimal for the game (79)–(81).

According to Theorem 3, the first solution (82) is not optimal.

In conclusion, the author would like to take this opportunity to express his gratitude to Academician N. N. Krasovskiĭ, corresponding Member E. F. Miščenko of the Academy of Sciences of the USSR, Professor D. V. Anosov, and Professor Ju. V. Egorov, for useful discussions of the work.

BIBLIOGRAPHY

1. L. S. Pontrjagin et al., *The mathematical theory of optimal processes*, Fizmatgiz, Moscow, 1961; English transl., Wiley, New York, 1962; Macmillan, New York, 1964. MR 29 #3316a, b; 32 #3896.

2. V. G. Boltjanskiĭ, R. V. Gamkrelidze and L. S. Pontrjagin, *On the theory of optimal processes*, Dokl. Akad. Nauk SSSR **110** (1956), 7–10. (Russian) MR 18, 859.

3. L. S. Pontrjagin, *On the theory of differential games*, Uspehi Mat. Nauk **21** (1966), no. 4 (130), 219–274 = Russian Math. Surveys 21 (1966), no. 4, 193–246. MR 34 #4033.

4. R. Isaacs, *Differential games*, Wiley, New York, 1965. MR 35 #1362.

5. N. N. Krasovskiĭ, *Game problems on the encounter of motions*, "Nauka", Moscow, 1970; English transl., *Differential games: Problems on the meeting of motions*, Dekker, New York (to appear). MR 44 #1452.

6. P. B. Gusjatnikov and M. S. Nikol'skiĭ, *Optimality of pursuit time*, Dokl. Akad. Nauk SSSR **184** (1969), 518–521 = Soviet Math. Dokl. **10** (1969), 103–106. MR 38 #6853.

7. S. L. Sobolev, *Applications of functional analysis in mathematical physics*, Izdat. Leningrad. Gos. Univ., Leningrad, 1950; English transl., Amer. Math. Soc., Providence, R. I., 1963. MR 14, 565; 29 #2624.

8. S. M. Nikol'skiĭ, *On the solution of the polyharmonic equation by a variational method*, Dokl. Akad. Nauk SSSR **88** (1953), 409–411. (Russian) MR 15, 425.

9. ———, *Approximation of functions of several variables and embedding theorems*, "Nauka", Moscow, 1969; English transl., Springer-Verlag, Berlin and New York, 1974. MR 46 #9714.

10. L. D. Kudrjavcev, *Direct and inverse imbedding theorems*, Trudy Mat. Inst. Steklov. 55 (1959); English transl., Amer. Math. Soc., Providence, R. I., 1974. MR 33 #7838.

11. A. Filippov, *On certain questions in the theory of optimal control*, Vestnik Moskov. Univ. Ser. Mat. Meh. Astronom. Fiz. Him. **1959**, no. 2, 25–32; English transl., SIAM J. Control 1 (1962), 76–84. MR 22 #13373; 26 #7469.

12. Ju. V. Egorov, *Necessary conditions for the optimality of control in Banach spaces*, Mat. Sb. 64 (106) (1964), 79–101; English transl., Amer. Math. Soc. Transl. (2) 49 (1965), 63–85. MR 31 #6716.

13. A. Ja. Dubovickiĭ and A. A. Miljutin, *Extremal problems with constraints*, Ž. Vyčisl. Mat. i Mat. Fiz. 5 (1965), 395–453 = USSR Comput. Math. and Math. Phys. 5 (1965), no. 5, 1–80. MR 32 #9093.

14. A. I. Egorov, *Necessary optimality conditions for distributed-parameter systems*, Mat. Sb. 69 (111) (1966), 371–421; English transl., SIAM J. Control 5 (1967), 352–408. MR 33 #2443; 36 #760.

15. J.-L. Lions, *Contrôle optimal de systèmes gouvernés par des équations aux dérivées partielles*, Dunod; Gauthier-Villars, Paris, 1968; English transl., Springer-Verlag, New York and Berlin, 1971. MR 39 #5920; 42 #6395.

16. B. N. Pšeničnyĭ, *Necessary conditions for an extremum,* "Nauka", Moscow, 1969; English transl., Dekker, New York, 1971. MR 43 #2584; #2585.

17. V. F. Dem'janov, *Minimax: Directional differentiability*, Izdat. Leningrad. Univ., Leningrad, 1974. (Russian)

18. A. V. Balakrishnan, *Introduction to optimization theory in Hilbert space*, Lecture Notes in Operations Research and Math. Systems, vol. 42, Springer-Verlag, Berlin and New York, 1971. MR 43 #7990.

19. A. G. Butkovskiĭ, *Optimal processes in systems with distributed parameters*, Avtomat. i Telemeh. 22 (1961), 17–26 = Automat. Remote Control 22 (1961), 13–21.

20. K. A. Lur′e, *Optimal problems for distributed systems*, Optimal Systems, Statistical Methods, "Nauka", Moscow, 1967, pp. 120–129. (Russian) RŽMat. 1968, #4Б571.*

Translated by K. MAKOWSKI

**Editor's note.* The one-line review refers the reader to an earlier paper by the same author, namely, *The Mayer-Bolza problem for multiple integrals*, Prikl. Mat. Meh. 27 (1963), 842–853 = J. Appl. Math. Mech. 27 (1963), 1284–1299 (MR 28 #5863).

Trudy Mat. Inst. Steklov.
134 (1975)

Proc. Steklov Inst. Math.
134 (1975)

SOME PROPERTIES OF SOLUTIONS OF QUASILINEAR ELLIPTIC EQUATIONS

UDC 517.946

G. N. JAKOVLEV

Abstract. In the paper a study is made of the differential properties of certain functions of solutions of quasilinear equations of second order that are elliptic but not uniformly elliptic. For a quite general class of such equations, the properties of the generalized solutions are studied both interior to the domain of definition and right up to the boundary. It is shown, for example, that a generalized solution of the second boundary value problem satisfies (i) the differential equation almost everywhere in the domain of definition, and (ii) the corresponding boundary condition in the mean and almost everywhere on the boundary.

Bibliography: 13 titles.

In this paper we study the differential properties of certain functions of solutions of quasilinear equations of second order that are elliptic but not uniformly elliptic. Such equations include, in particular, Euler's equations for certain elementary functionals, e.g. for a functional given in the form of an integral of the sum of the pth powers of the absolute values of the first derivatives of the desired function. Questions of the existence in W_p^1 of generalized solutions of appropriate boundary value problems for such equations have been considered in many works (see, for example, [2] or [3]). But the study of the differential properties of generalized solutions of nonuniformly elliptic equations remains incomplete. Papers [4]–[13] are devoted directly to this question. In the present paper we study the properties of generalized solutions for a more general class of equations not only interior to the domain of definition but also right up to the boundary. The latter question is especially interesting for the second boundary value problem, since one is able to show that a generalized solution of it satisfies (i) the differential equation almost everywhere in the domain of definition, and (ii) the corresponding boundary condition almost everywhere on the boundary.

§1. Definitions and auxiliary lemmas

In the paper we study the generalized solutions $u(x) \in W_p^1(G)$, $p \geq 2$, of quasilinear elliptic equations of the form

AMS (MOS) subject classifications (1970). Primary 35J60, 35B99, 35D99; Secondary 35J25, 35J67.

$$\sum_{i=1}^{n} D_i a^i(x, Du) = a^0(x, Du), \tag{1.1}$$

where $D_i a^i(x, Du)$ is the total derivative with respect to x_i of the function $a^i(x, Du) = a^i(x, u(x), D_1 u(x), \ldots, D_n u(x))$.

As usual $W_p^1(G)$, $1 < p < \infty$, denotes the space of functions $u(x) \in L_p(G)$ having first order generalized derivatives $D_j u(x)$, $j = 1, \ldots, n$, belonging to $L_p(G)$. The norm of a function $u(x) \in W_p^1(G)$ will be denoted by $\|u\|_{W_p^1(G)}$, or, briefly, by $\|u\|_{1,p}$, and is defined by the equality

$$\|u\|_{1,p} = \left(\sum_{j=0}^{n} \| D_j u \|_p^p \right)^{\frac{1}{p}}, \tag{1.2}$$

where $D_0 u = u$ and $\|D_j u\|_p$ is the norm in $L_p(G)$ of $D_j(u)(x)$. Let $\overset{\circ}{W}{}_p^1(G)$ denote the closure in the norm (1.2) of the set of infinitely differentiable functions with compact support in G. A function $u(x) \in W_p^1(G)$ is then called a generalized solution of equation (1.1) if

$$\sum_{i=0}^{n} \int_G a^i(x, Du) D_i v\, dx = 0, \quad \forall v \in \overset{\circ}{W}{}_p^1(G). \tag{1.3}$$

It is assumed everywhere below that (i) the domain G is bounded, (ii) the functions $a^i(x, y)$, $i = 0, 1, \ldots, n$, are defined for all $x \in G$ and all $y = (y_0, y_1, \ldots, y_n) \in E^{n+1}$, and (iii) the functions $a^i(x, Du(x))$ are measurable in G and all of the integrals in (1.3) are finite for any function $u(x) \in W_p^1(G)$. More concrete conditions on the domain G and the functions $a^i(x, y)$ will be indicated in the appropriate sections.

We will denote by $\Delta_j(f(x); h)$ the first order difference with step h of a function $f(x)$ with respect to x_j, i.e. $\Delta_j(f(x); h) = f(x + he_j) - f(x)$, where e_j is the unit vector in the direction of the x_j axis, and by G_δ the subdomain of G whose points lie at a distance of more than $\delta > 0$ from the boundary ∂G of G. If $f(x) \in W_p^1(G)$, $1 < p < \infty$, it is easily seen that $\Delta_j(f; h)$ is defined in G_δ for $|h| < \delta$, and that

$$\| \Delta_j(f; h) \|_{p, G_\delta} \leqslant |h| \| f_{,j} \|_p, \tag{1.4}$$

where $\| \cdots \|_{p, G_\delta}$ is the norm in $L_p(G_\delta)$.

In the sequel $f_{,j}(x)$ denotes the derivative with respect to the jth variable of a function $f(x)$, i.e. $f_{,j}(x) = D_j f(x)$. In the case of a composite function $a^i(x, Du(x))$ the expression $a^i_{,j}(x, Du(x))$ denotes the value of the derivative with respect to x_j of $a^i(x, y)$ at the point $(x, Du(x))$. For convenience, we put $f_{,0}(x) = D_0 f(x) = f(x)$.

The following proposition is used everywhere in the present paper.

Lemma 1. *Suppose given a function* $f(x) \in L_q(G)$, $1 < q < \infty$. *If there exists a constant* C *such that*

$$\|\Delta_j(f;\ h)\|_{q,\ G_\delta} \leqslant C|h|,\quad \forall h,\ |h| < \delta,$$

then the derivative $f_{,j}(x)$ *exists and belongs to* $L_q(G_\delta)$, *with*

$$\|f_{,j}\|_{q,\ G_\delta} \leqslant C.$$

This lemma is well known; a proof of it can be found in the book by S. M. Nikol'skii [1], Chapter 4, §4.8.

In conclusion, we formulate an elementary assertion.

Lemma 2. *Let* $X(h)$ *and* $Y(h)$ *be nonnegative functions of* h *for* $|h| < \delta$. *If there exist constants* A, B, A_1, B_1 *such that*

$$Y(h) \leqslant Ah^2 + B|h|X(h), \tag{1.5}$$

$$X(h) \leqslant A_1|h| + B_1\sqrt{Y(h)}, \tag{1.6}$$

then $X(h) < C|h|$, *where* $C = 2A_1 + 2\sqrt{A}B_1 + B_1^2B$.

§2. Properties of generalized solutions interior to the domain of definition

In this section we will assume that for each $y \in E^{n+1}$ the functions $a^i = a^i(x, y)$, $i = 1, \ldots, n$, as functions of x, are locally summable in G and have generalized (in the Sobolev sense) first order derivatives $a^i_{,j}$, $j = 1, \ldots, n$, and second order derivatives $a^i_{,jj}$, the functions a^i, $a^i_{,j}$ and $a^i_{,jj}$ being continuous in E^{n+1} as functions of y for almost all $x \in G$. The function $a^0 = a^0(x, y)$ has the same properties but without second derivatives.

In addition, we will assume that the following conditions are satisfied.

I. *Growth conditions*: there exist numbers $\mu > 0$, $p \geq 2$ and a function $\alpha(x) \in L_q(G)$, $1/p + 1/q = 1$, such that if one puts $b = b(x, y) \equiv |\alpha(x)|^q + \Sigma_{j=0}^n |y_j|^p$, then the inequalities

$$|a^0|,\ |a^0_{,j}|,\ |a^i|,\ |a^i_{,j}|,\ |a^i_{,jj}| \leqslant \mu b^{\frac{1}{q}},\quad i,\ j = 1,\ \ldots,\ n, \tag{2.1}$$

are valid for all $y \in E^{n+1}$ and almost all $x \in G$.

For what follows, we introduce the notation

$$\Delta_y a^i = a^i(x,\ y + \Delta y) - a^i(x,\ y), \tag{2.2}$$

$$b^0 = b^0(x,\ y,\ \Delta y) \equiv \{b(x,\ y) + b(x,\ y + \Delta y)\}^{1-\frac{2}{p}}. \tag{2.3}$$

II. *Ellipticity* (*monotonicity*) *condition*: there exists a number $\mu_0 > 0$ such that

$$B = B(x,\ y;\ \Delta y) \equiv \sum_{i=0}^n \Delta_y a^i \cdot \Delta y_i + \mu_0 b^0 |\Delta y_0|^2 \geqslant 0 \tag{2.4}$$

for all $y \in E^{n+1}$, all $\Delta y \in E^{n+1}$ and almost all $x \in G$.

III. *Subordination conditions*: there exists a number $\mu_1 > 0$ such that

$$|\Delta_y a^i|, \ |\Delta_y a^i_{,j}| \leqslant \sqrt{\mu_1 b^0 B} + \mu_1 b^0 |\Delta y_0|, \quad i, \ j = 1, \ \ldots, \ n, \tag{2.5}$$

for all $y \in E^{n+1}$, all $\Delta y \in E^{n+1}$ and almost all $x \in G$.

We note that it can be assumed without loss of generality that $\alpha(x)$ is the same function in each of the indicated conditions. It can also be assumed that $\mu = \mu_0 = \mu_1$.

Theorem 1. *Suppose the functions* $a^i(x, y)$, $i = 0, 1, \ldots, n$, *satisfy the growth conditions* (2.1), *the ellipticity condition* (2.4) *and the subordination conditions* (2.5) *with* $p \geq 2$. *If a function* $u(x) \in W^1_p(G)$ *is a generalized solution of equation* (1.1), *then each function* $a^i(x, Du(x))$, $i = 1, \ldots, n$, *belongs to* $W^1_q(G_\delta)$ *for any* $\delta > 0$, *and each function* $a^i_{,j}(x, Du(x))$, $i, j = 1, \ldots, n$, *belongs to* $L_q(G)$ *and has a generalized derivative* $D_j a^i_{,j}(x, Du(x)) \in L_q(G_\delta)$, $\delta > 0$. *In addition, there exists a constant* C *depending only on* δ, μ, μ_1, μ_0 *and* n *such that*

$$\sum_{i=1}^n \|a^i(x, \ Du)\|_{W^1_q(G_\delta)} + \sum_{i, \ j=1}^n \|D_j a^i_{,j}(x, \ Du)\|_{L_q(G_\delta)} \leqslant C\{\|\alpha\|_q + \|u\|^{p-1}_{1, p}\}.$$

Proof. Let $\eta = \eta(x)$, $x \in E^{n+1}$, denote a nonnegative continuously differentiable function that is equal to unity in G_δ, $\delta > 0$, is equal to zero outside $G_{\delta/2}$ and satisfies the condition $\eta \leq 1$ in G. Such a function is easily constructed.

It is not difficult to see that the function

$$v = \Delta_j \{\eta^2(x) \Delta_j (u(x); \ h); \ -h\}$$

belongs to $\overset{\circ}{W}{}^1_p(G)$ for $|h| < \delta/4$ if, for example, the function $u(x)$ is extended by zero outside G. Substituting it into (1.3), we obtain, after some elementary transformations, the equality

$$\sum_{i=0}^n \int_G \Delta_j (a^i(x, \ Du); \ h) D_i \{\eta^2 \Delta_j (u(x); \ h)\} \, dx = 0, \tag{2.6}$$

where $\Delta_j(a^i(x, Du); h)$ is the total difference of $a^i(x, Du(x))$ with respect to x_j with step h, i.e.

$$\Delta_j (a^i(x, \ Du); \ h) = a^i(x + he_j, \ Du(x + he_j)) - a^i(x, \ Du(x)).$$

The index j and step h will not vary in the sequel. Therefore in place of $\Delta_j(a^i(x, Du); h)$ and $\Delta_j(u(x); h)$ we will write $\Delta a^i(x, Du)$ and $\Delta u(x)$ or, more briefly, Δa^i and Δu. In this notation equality (2.6) (after a differentation) takes the form

$$\sum_{i=0}^n \int_G \eta^2 \Delta a^i \Delta u_{,i} dx = -2 \sum_{i=1}^n \int_G \eta \Delta a^i \eta_{,i} \Delta u dx. \tag{2.7}$$

It follows from the growth conditions (2.1) that the functions $\eta \Delta a^i$, $i = 0, 1, \ldots, n$, belong to $L_q(G)$, and hence that all of the integrals in (2.7) are finite. Suppose $|\eta_{,i}(x)| \le \kappa$, $i = 1, \ldots, n$. It then follows from (2.7) that a generalized solution $u(x) \in W_p^1(G)$ of (1.1) satisfies the inequality

$$\sum_{i=0}^{n} \int_G \eta^2 \Delta a^i \Delta u_{,i} dx \leqslant 2\kappa |h| \|u_{,j}\|_p \sum_{i=1}^{n} \|\eta \Delta a^i\|_q \tag{2.8}$$

for $|h| < \delta/4$. Since

$$\Delta a^i (x, Du) = \Delta_y a^i (x, Du(x)) + \Delta_x a^i (x, Du(x+he_j)), \tag{2.9}$$

where

$$\Delta_y a^i (x, Du(x)) = a^i (x, Du(x+he_j)) - a^i (x, Du(x)),$$
$$\Delta_x a^i (x, Du(x)) = a^i (x+he_j, Du(x)) - a^i (x, Du(x)),$$

we obtain from (2.8) upon setting $B = B(x, Du; \Delta Du)$ and $b^0 = b^0(x, Du, \Delta Du)$ (see the monotonicity condition) the inequality

$$\int_G \eta^2 B(x, Du; \Delta Du)\, dx$$
$$\leqslant \mu \int_G \eta^2 b^0 |\Delta u|^2 dx + 2\kappa |h| \|u_{,j}\|_p \sum_{i=1}^{n} \|\eta \Delta a^i\|_q + \sum_{i=0}^{n} |A_i|, \tag{2.10}$$

where

$$A_i = \int_G \eta^2 \Delta_x a^i (x, Du(x+he_j)) \Delta u_{,i} dx.$$

From the definition of b^0 and from (1.4) it follows that

$$\int_G \eta^2 b^0 |\Delta u|^2 dx \leqslant 2h^2 \|u_{,j}\|_p^2 D_p^{1-\frac{2}{p}}, \tag{2.11}$$

where $D_p = D_p(\alpha; u) = \|\alpha\|_q^q + \|u\|_{1,p}^p$.

The function $a^0(x, y)$ has a generalized derivative with respect to x_j that satisfies the growth condition (2.1); hence

$$|A_0| \leqslant \mu h^2 \|u_{,j}\|_p D_p^{\frac{1}{q}}. \tag{2.12}$$

Further, since $\eta(x) = 0$ outside $G_{\delta/2}$ and $|h| < \delta/4$, we have

$$A_i = \int_G \Delta_j (\eta^2(x); -h) \Delta_x a^i (x - he_j, Du(x)) u_{,i}(x)\, dx$$
$$- \int_G \eta^2(x) \Delta\Delta_x a^i (x - he_j, Du(x)) u_{,i}(x)\, dx = A_i' + A_i'' \tag{2.13}$$

for $i = 1, \ldots, n$. The functions $\eta(x)$ and $a^i(x, y)$ have derivatives with respect to x_j, which implies by virtue of (1.4) and the growth condition (2.1) that

$$|A_i'| \leqslant 2\kappa\mu h^2 \|u_{,i}\|_p D_p^{\frac{1}{q}}. \tag{2.14}$$

For A_i'' we have

$$|A_i''| \leqslant \|u_{,i}\|_p \|\eta^2(x) \Delta\Delta_x a^i (x - he_j, Du(x))\|_q. \tag{2.15}$$

We estimate the norms of the compound differences:

$$\| \eta^2(x)\,\Delta\Delta_x a^i(x - he_j,\ Du(x)) \|_q \leqslant |h| \left\| \eta^2 \int_0^1 \Delta a^i_{,j}(x - hte_j,\ Du(x))\,dt \right\|_q$$

$$\leqslant |h|\,\| \eta^2 \Delta a^k_{,j} \|_q + \mu h^2 D_p^{\frac{1}{q}}. \qquad (2.16)$$

It now follows from (2.10)–(2.16) that a generalized solution of (1.1) satisfies the inequality

$$\int_G \eta^2 B(x,\ Du;\ \Delta Du)\,dx$$

$$\leqslant C_1 h^2 D_p(\alpha;\ u) + C_2 |h|\,\|u\|_{1,p} \sum_{i=1}^n (\|\eta\Delta a^i\|_q + \|\eta^2 \Delta a^i_{,j}\|_q), \qquad (2.17)$$

where $C_1 = \mu(3 + n + 2\kappa n)$ and $C_2 = \max\{1, 2\kappa\}$. This inequality has the form of (1.5).

The norms $\|\eta\Delta a^i\|_q$ and $\|\eta^2\Delta a^i_{,j}\|_q$ have the same estimates. Let us estimate the first of them.

From (2.9) we have

$$\|\eta\Delta a^i\|_q \leqslant \|\eta\Delta_y a^i\|_q + \|\eta\Delta_x a^i(x,\ Du(x + he_j))\|_q. \qquad (2.18)$$

We have already estimated the norm of the difference $\Delta_x a^i$ (see (2.14)) and shown that

$$\|\eta\Delta_x a^i\|_q \leqslant \mu\,|h|\,D_p^{\frac{1}{q}}. \qquad (2.19)$$

The norm of $\Delta_y a^i$ is estimated with the use of the subordination condition (2.5):

$$\|\eta\Delta_y a^i\|_q \leqslant \mu\,\|\eta b^0 \Delta u\|_q + \sqrt{\mu}\,\|\eta\sqrt{b^0 B}\|_q. \qquad (2.20)$$

From the definition of b^0 and (1.4) we obtain, as in (2.11), the estimate

$$\|\eta b^0 \Delta u\|_q \leqslant 2\,|h|\,D_p^{\frac{1}{q}} \qquad (2.21)$$

for the first summand in (2.20). Applying Hölder's inequality, we obtain

$$\|\eta\sqrt{b^0 B}\|_q \leqslant 2 D_p^{\frac{1}{2} - \frac{1}{p}} \left(\int_G \eta^2 B\,dx \right)^{\frac{1}{2}} \qquad (2.22)$$

for the second summand in (2.20). It follows from (2.18)–(2.22) that

$$\|\eta\Delta a^i\|_q \leqslant 3\mu\,|h|\,D_p^{\frac{1}{q}} + 2 D_p^{\frac{1}{2} - \frac{1}{p}} \left(\mu \int_G \eta^2 B\,dx \right)^{\frac{1}{2}}$$

for $|h| < \delta/4$. The same estimate holds for $\|\eta^2 \Delta a^i_{,j}\|_q$.

Thus any function $u(x) \in W^1_p(G)$, $p \geq 2$, satisfies the estimate

$$\sum_{i=1}^{n} (\| \eta \Delta a^i \|_q + \| \eta^2 \Delta a^i_{,j} \|_q) \leqslant 6n\mu |h| D_p^{\frac{1}{q}} + 4\sqrt{\mu} n D_p^{\frac{1}{2}-\frac{1}{p}} \left(\int_G \eta^2 B dx \right)^{\frac{1}{2}}$$

for $|h| < \delta/4$. This inequality has the form of (1.6).

Applying Lemma 2 of §1, we get that a generalized solution $u(x) \in W_p^1(G)$, $p \geq 2$, of equation (1.1) satisfies the inequality

$$\sum_{i=1}^{n} (\| \eta \Delta a^i \|_q + \| \eta^2 \Delta a^i_{,j} \|_q) \leqslant C |h| D_p^{\frac{1}{q}} (\alpha; u)$$

for $|h| < \delta/4$, where $C = 72n^2\mu(1+\kappa)$.

By virtue of Lemma 1 of §1, the functions $a^i(x, Du(x))$ and $a^i_{,j}(x, Du(x))$, $i = 1, \ldots, n$, have generalized (in the Sobolev sense) derivatives $D_j a^i(x, Du(x)) \in L_q(G_\delta)$ and $D_j a^i_{,j}(x, Du(x)) \in L_q(G_\delta)$, such that

$$\sum_{i=1}^{n} (\| D_j a^i \|_{L_q(G_\delta)} + \| D_j a^i_{,j} \|_{L_q(G_\delta)}) \leqslant C D_p^{\frac{1}{q}} (\alpha; u).$$

Theorem 1 is proved.

§3. Properties of generalized solutions of the first boundary value problem

In this section we study the generalized solutions $u(x) \in W_p^1(G)$ of equation (1.1) satisfying the condition

$$u(x) - f(x) \in \mathring{W}_p^1(G), \tag{3.1}$$

where $f(x)$ is a given function. As usual (see, for example, [2]), such a function $u(x)$ is called a generalized solution of problem (1.1), (3.1).

In regard to equation (1.1) it will be assumed that the functions $a^i(x, y)$ satisfy the conditions of §2.

It will be assumed that the domain G is bounded and has a Lipschitz boundary ∂G. We recall that the latter assumption means the following: for any point $M \in \partial G$ there exists an n-dimensional neighborhood $O(M)$ such that, in a coordinate system $\tilde{x} = (\tilde{x}_1, \ldots, \tilde{x}_n)$ obtained from the given coordinate system $x = (x_1, \ldots, x_n)$ by a rotation, the surface $S^M = O(M) \cap \partial G$ has the representation

$$\tilde{x}_n = \varphi(\tilde{x}'), \quad \tilde{x}' = (\tilde{x}_1, \ldots, \tilde{x}_{n-1}) \in g', \tag{3.2}$$

where g' is an $(n-1)$-dimensional ball on the hyperplane $\tilde{x}_n = 0$ with center at the projection of M. In addition, the following conditions must be satisfied: the function (3.2) satisfies a Lipschitz condition, and there exists a number $\delta > 0$ such that the set $\{\tilde{x}' \in g', 0 < \tilde{x}_n - \phi(\tilde{x}') < \delta\}$ lies in G while the set $\{\tilde{x}' \in g', -\delta < \tilde{x}_n - \phi(\tilde{x}') < 0\}$ and G are mutually disjoint.

We will say that the bounded domain G has an l-times continuously differentiable boundary ∂G, and write $\partial G \in C^l$, if ∂G is a Lipschitz boundary and the function (3.2) is l-times continuously differentiable for each point $M \in \partial G$.

In regard to the domain G, it will be assumed in this section that it is bounded and $\partial G \in C^4$.

Let $\mathbf{n}$ denote the inward unit normal to ∂G. In the coordinate system $\widetilde{x}$ this vector has at the point on S^M corresponding to the point $x' = \xi' \in g'$ the coordinates $\nu_i = -\phi_{,i}\Phi^{-1}$, $i = 1, \dots, n-1$, and $\nu_n = \Phi^{-1}$, where $\Phi^2 = \Phi^2(\xi') = 1 + |\nabla\phi(\xi')|^2$, and $\phi_{,i}$ is the value of the partial derivative of $\phi(\widetilde{x}')$ with respect to $\widetilde{x}_i$ at ξ'. Let $\Pi(S^M; \delta)$ denote the set of all points $x \in E^n$ that lie on the normals to S^M at a distance of less than $\delta > 0$ from S^M. It is easily seen that $\Pi(S^M; \delta)$ is the image of the finite cylinder $Ц(g'; \delta) = \{\xi' \in g', |\xi_n| < \delta\}$ under the mapping T given by the formulas

$$\begin{aligned} \tilde{x}_i &= \nu_i(\xi')\xi_n + \xi_i, \quad i = 1, \dots, n-1, \\ \tilde{x}_n &= \nu_n(\xi')\xi_n + \varphi(\xi'). \end{aligned} \tag{3.3}$$

By assumption, $x = A\widetilde{x}$, where the matrix $A = (a_{ij})$ is orthogonal and such that $\det A = \pm 1$. Consequently, if the coordinates of the vector $\mathbf{n}(\xi')$ are denoted in the coordinate system x by $n_i(\xi')$, $i = 1, \dots, n$, the mapping $T\colon Ц(g'; \delta) \to \Pi(S^M; \delta)$ is given in this coordinate system by the formulas

$$x_i = n_i(\xi')\xi_n + \sum_{j=1}^{n-1} a_{ij}\xi_j + a_{in}\varphi(\xi'), \quad i = 1, \dots, n. \tag{3.4}$$

From (3.3) one sees that the mapping T is continuously differentiable if the boundary ∂G is twice continuously differentiable. For sufficiently small $\delta > 0$ this mapping has an inverse that is also continuously differentiable. In fact, for the Jacobians $\widetilde{J}(\xi)$ and $J(\xi)$ of the mappings given by formulas (3.3) and (3.4) respectively we have $J(\xi) = \det A\widetilde{J}(\xi)$. Clearly, $\widetilde{J}(\xi) \to \Phi(\xi')$ as $\xi_n \to 0$. There therefore exists a $\delta > 0$ such that $|J(\xi)| \geq ½$ in $Ц(g'; \delta)$, and hence the mapping T is one-to-one and bicontinuously differentiable from $Ц(g'; \delta)$ into $\Pi(S^M; \delta)$.

Let $\Pi^+(S^M; \delta)$ and $\Pi^-(S^M; \delta)$ denote the images of the sets $Ц^+(g'; \delta) = \{\xi' \in g'; 0 < \xi_n < \delta\}$ and $Ц^-(g'; \delta) = \{\xi' \in g'; -\delta < \xi_n < 0\}$ under the mapping T. The fact that $\partial G \in C^4$ implies the existence of a $\delta > 0$ such that $\Pi^+(S^M; \delta)$ lies in G while $\Pi^-(S^M; \delta)$ and G are mutually disjoint. For each point $M \in \partial G$ we choose a $\delta = \delta^M > 0$ so small that all of the enumerated properties are fulfilled.

Let g'_τ denote the open $(n-1)$-dimensional ball having a common center with g' and a radius smaller by $\tau > 0$ than the radius of g'. Let S^M_τ denote the part of S^M for which $\xi' \in g'_\tau$.

The family of sets $\Pi(S^M_\tau; \delta^M)$, $M \in \partial G$, forms a covering of the bounded closed set ∂G. By virtue of the Borel covering theorem it is possible to select from this family a finite subfamily that also covers ∂G. Suppose the sets $\Pi(S^{M_j}_\tau; \delta^{M_j})$, $j = 1, \dots, m$, form such a finite subfamily. Let G_δ denote the set of points of G lying at a distance of more than $\delta > 0$ from ∂G, and let δ_0 denote the least of the δ^{M_j}, $j = 1, \dots, m$. Then the sets $G_{\delta_0/2}$, $\Pi^+(S^{M_j}_\tau; \delta_0)$, $j = 1, \dots, m$, form a covering of G.

In any domain of the form G_δ, $\delta > 0$, the differential properties of the generalized solutions $u(x) \in W_p^1(G)$ do not depend on the boundary conditions; these properties were studied in §2. Thus it remains to study such properties in $G - G_{\delta_0}$, and it suffices for this purpose to consider one of the sets $\Pi^+(S_\tau^{M_j}; \delta_0)$.

We note that the following function is defined for any $x \in G - G_{\delta_0}$ and any $y \in E^{n+1}$:

$$N(x, y) = \sum_{i=1}^{n} a^i(x, y) n_i(x),$$

where the $n_i(x)$ are the coordinates of the unit normal to ∂G that passes through x and whose part from x to ∂G belongs to $G - G_{\delta_0}$.

Theorem 2. *Suppose equation* (1.1) *satisfies all of the conditions of Theorem* 1, *the domain* G *is bounded and* $\partial G \in C^4$. *If* $u(x) \in W_p^1(G)$ *is a generalized solution of problem* (1.1), (3.1) *with* $f(x) \in W_p^2(G)$, *then the following assertions are valid*:

1) $a^i(x, Du(x)) \in W_q^1(G_\delta)$, $i = 1, \ldots, n$, $\delta > 0$.

2) *There exists a* $\delta > 0$ *such that* $N(x, Du(x)) \in W_q^1(G - G_\delta)$.

3) *The function* $u(x)$ *satisfies* (1.1) *for almost all* $x \in G$.

In addition, in any $\Pi^+(S_\tau^M; \delta)$, $0 < \delta < \delta_0$ *(for the notation, see above), the functions* $a^i(x, Du(x))$, $i = 1, \ldots, n$, *considered as functions of the local coordinates* ξ, *have the generalized derivatives*

$$\frac{\partial a^i}{\partial \xi_j} \in L_q(\Pi^+(g'_\tau; \delta)), \quad j = 1, \ldots, n-1.$$

Proof. Consider one of the sets $\Pi^+(S^{M_j}; \delta_0)$. For the functions $v(x) \in \overset{\circ}{W}{}_p^1(G)$ that vanish outside this set, condition (1.3) takes (in local coordinates) the form

$$\sum_{i=0}^{n} \int_g a^i(x, Du(x)) v_{,i}(x) J(\xi) d\xi = 0, \tag{3.5}$$

where $g = \Pi^+(g'; \delta_0)$.

For simplicity, the functions $u_{,j}(x(\xi))$ will be denoted by $u^j(\xi)$, $j = 0, 1, \ldots, n$. The vector-valued function with components $u^j(\xi)$, $j = 0, 1, \ldots, n$, will be denoted by $\mathbf{u}(\xi)$, while the functions $a^i(x(\xi), Du(x(\xi)))$, $i = 0, 1, \ldots, n$, will be denoted by $b^i = b^i(\xi, \mathbf{u}(\xi))$. Then (3.5) takes the form

$$\sum_{i=0}^{n} \int_g b^i v^i(\xi) J(\xi) d\xi = 0. \tag{3.6}$$

Let $\eta = \eta(\xi)$ denote a continuously differentiable function in E^n that is equal to unity on the set $\{\xi' \in g'_\tau, |\xi_n| < \delta - \tau\}$, vanishes outside the set $\{\xi' \in g'_{\tau/2}, |\xi_n| < \delta - \tau/2\}$ and satisfies the condition $0 \le \eta(\xi) \le 1$ on $\Pi(g'; \delta)$. Such a function is easily constructed.

In (3.5) or, equivalently, in (3.6) we set

$$v = v(x) = \frac{1}{J(\xi(x))} \Delta_j [\eta^2 (\xi(x)) \Delta_j (w;\ h);\ -h],$$

where $w = u - f$ while $\Delta_j(w, h)$ and $\Delta_j[\ldots; -h]$ are the first order differences with steps h and $-h$ respectively for $|h| < r/4$ of the considered functions with respect to ξ_j, $j = 1, \ldots, n-1$. As above, the values of the derivatives $w_{,i}(x)$ for $x = x(\xi)$ will be briefly denoted by $w^i(\xi)$, $i = 0, 1, \ldots, n$. Then

$$v^0(\xi) = \frac{1}{J(\xi)} \Delta_j [\eta^2 \Delta_j (w^0;\ h);\ -h],$$

$$v^i(\xi) = \frac{1}{J(\xi)} \Big\{ \Delta_j [\eta^2 \Delta_j (w^i;\ h);\ -h] + 2\Delta_j \Big[\eta \frac{\partial \eta}{\partial x_i} \Delta_j (w^0;\ h);\ -h \Big] \Big\} - J_i(\xi)\, v^0(\xi),$$

where

$$J_i(\xi) = \frac{1}{J(\xi)} \frac{\partial J(\xi)}{\partial x_i}, \quad i = 1, \ldots, n.$$

Consequently, a generalized solution $u(x)$ of problem (1.1), (3.1) satisfies the equality

$$\sum_{i=0}^{n} \int_g \eta^2 \Delta b^i \Delta w^i d\xi + 2 \sum_{i=1}^{n} \int_g \eta \frac{\partial \eta}{\partial x_i} \Delta b^i \Delta w^0 d\xi = \sum_{i=1}^{n} \int_g \eta^2 \Delta \{J_i(\xi) b^i\} \Delta w^0 d\xi, \tag{3.7}$$

where, for example, $\Delta w^i = \Delta_j(w^i(\xi);\ h)$, i.e. Δ denotes the first-order difference with step h with respect to ξ_j.

Since

$$|\Delta \{J_i(\xi) b^i\}| \leqslant |J_i(\xi + h e_j)|\, |\Delta b^i| + |\Delta J_i(\xi)|\, |b^i|,$$

it follows by virtue of the continuous differentiability of $J_i(\xi)$ that there exists a constant C_1 such that

$$|\Delta \{J_i b^i\}| \leqslant C_1 |\Delta b^i| + C_1 |h|\, |b^i|.$$

The function $\partial\eta/\partial x_i$ is, by construction, bounded in g. Suppose $|\partial\eta/\partial x_i| \le C_2$. Then (3.7) implies that a generalized solution satisfies the inequality

$$\sum_{i=0}^{n} \int_g \eta^2 \Delta b^i \Delta u^i d\xi \leqslant \sum_{i=0}^{n} \int_g \eta^2 \Delta b^i \Delta f^i d\xi + \sum_{i=1}^{n} \int_g \eta (|\Delta u^0| + |\Delta f^0|)(C_1 |h|\, |b^i| + C_3 |\Delta b^i|)\, d\xi, \tag{3.8}$$

where $C_3 = C_1 + 2C_2$. For $i = 1, \ldots, n$ we have

$$\left| \int_g \eta^2 \Delta b^i \Delta f^i d\xi \right| \leqslant |h| \, \| f^i_{,j} \|_p \, \| \eta \Delta b^i \|_q, \tag{3.9}$$

$$\begin{aligned} \int_g \eta \, (|\Delta u^0| + |\Delta f^0|) \, (C_1 |h| \, |b^i| + C_3 |\Delta b^i|) \, d\xi \\ \leqslant |h| \, (\| u^0_{,j} \|_p + \| f^0_{,j} \|_p) \, (C_1 |h| \, \| b^i \|_q + C_3 \| \Delta b^i \|_q), \end{aligned} \tag{3.10}$$

where $\| \ \|_p$ and $\| \ \|_q$ are the norms of $L_p(g)$ and $L_q(g)$. Further,

$$\int \eta^2 \Delta b^0 \Delta f^0 d\xi = \int_g b^0 \Delta \{ \eta^2 \Delta (f^0; \ h); \ -h \} \, d\xi$$

and hence

$$\left| \int_g \eta^2 \Delta b^0 \Delta f^0 d\xi \right| \leqslant |h| \, \| b^0 \|_q \left\| \frac{\partial}{\partial \xi_j} \{ \eta^2 \Delta f^0 \} \right\|_p .$$

The function $\eta(\xi)$ has bounded derivatives. Suppose $|\eta_{,j}(\xi)| \le \kappa$, $j = 1, \ldots, n$. Then

$$\left| \int_g \eta^2 \Delta b^0 \Delta f^0 d\xi \right| \leqslant h^2 \| b^0 \|_q \, (2\kappa \| f^0_{,j} \|_p + \| f^0_{,jj} \|_p). \tag{3.11}$$

Inequalities (3.9)–(3.11) provide estimates of all of the integrals in the right side of (3.8). The norms $\|f^0_{,j}\|_p$, $\|f^0_{,jj}\|_p$, $\|f^i_{,j}\|_p$, $\|u^0_{,j}\|_p$ can be estimated in terms of the corresponding norms of the functions $f(x) \in W^2_p(G)$ and $u(x) \in W^1_p(G)$ by virtue of the smoothness of the transformation T. There therefore exists a constant C_4, not depending on $u(x)$ or h, such that

$$\sum_{i=0}^{n} \int_g \eta^2 \Delta b^i \Delta u^i d\xi \leqslant C_4 \, (\| f \|_{2,p} + \| u \|_{1,p}) \left(h^2 \sum_{i=0}^{n} \| b^i \|_q + |h| \sum_{i=1}^{n} \| \eta \Delta b^i \|_q \right). \tag{3.12}$$

Inequality (3.12) is analogous to (2.8). The proof now follows the plan of the proof of Theorem 1.

We will assume, if g_τ denotes a subdomain of g, where $\eta \equiv 1$, that we have proved

$$D_j b^i (\xi, \ \mathbf{u}(\xi)) \in L_q (g_\tau), \tag{3.13}$$

$$D_j b^i_{,j} (\xi, \ \mathbf{u}(\xi)) \in L_q (g_\tau), \tag{3.14}$$

where $i = 1, \ldots, n$ but $j = 1, \ldots, n-1$.

Let us prove the second assertion of our theorem.

We consider one of the sets $\Pi^+(S^M; \delta_0)$ and go over to local coordinates ξ in it. Then

$$D_i a^i (x, \ Du(x)) = \sum_{j=1}^{n} D_j b^i (\xi, \ \mathbf{u}(\xi)) \frac{\partial \xi_j}{\partial x_i}, \tag{3.15}$$

where, as everywhere, D_i denotes the total derivative with respect to the ith variable, i.e. with respect to x_i or with respect to ξ_i.

From (3.4), as is easily shown, it follows that

$$\frac{\partial \xi_n}{\partial x_i} = n_i(\xi'), \quad i = 1, \ldots, n. \tag{3.16}$$

From (3.15), (3.16) and the fact that $u(x)$ satisfies (1.1) almost everywhere in G it now follows that

$$\sum_{i=1}^{n} \sum_{j=1}^{n-1} D_j b^i(\xi, \mathbf{u}) \frac{\partial \xi_j}{\partial x_i} + D_n N(x(\xi), \mathbf{u}) = b^0(\xi, \mathbf{u}).$$

We have already proved that $b^0(\xi, \mathbf{u})$ and $D_j b^i(\xi, \mathbf{u})$ ($i = 1, \ldots, n$ and $j = 1, \ldots, n-1$) belong to $L_q(g_\tau)$, and, since the $\partial\xi_j/\partial x_i$ are bounded,

$$D_n N(x(\xi), \mathbf{u}(\xi)) \in L_q(g_\tau). \tag{3.17}$$

It follows from (3.13) and (3.17) that $N(x(\xi), \mathbf{u}(\xi)) \in W_q^1(g_\tau)$, and this means that condition 2) of our theorem is satisfied in the δ-layer $G - G_\delta$ adjoining ∂G. The theorem is proved.

§4. Properties of generalized solutions of the second boundary value problem

In this section we study the differential properties of the generalized solutions $u(x)$ of the following boundary value problem: $u(x)$ satisfies equation (1.1) in G and the boundary condition

$$N(x, Du) = \varphi(x, u) \tag{4.1}$$

on ∂G, where $N(x, y)$ is defined in §3 and $\phi(x, u)$ is a given function of the variables $x \in \partial G$ and $u \in E^1$.

It will be assumed that the functions $a^i(x, y)$, $i = 0, 1, \ldots, n$, satisfy all of the conditions of §3, that the domain G is bounded and that $\partial G \in C^4$.

In regard to the function $\phi(x, u)$, it will be assumed that it is summable over ∂G as a function of x for each $u \in E^1$ and continuous on E^1 as a function of u for almost all $x \in \partial G$. It will also be assumed that there exist a number $\mu > 0$ and a function $\beta(x) \in L_q(\partial G)$ such that the following three inequalities are satisfied for almost all $x \in \partial G$, all $u \in E^1$ and all $\Delta u \in E^1$.

I. $$|\varphi(x, u)| \leqslant \mu(|\beta(x)|^q + |u|^p)^{\frac{1}{q}}, \tag{4.2}$$

where the numbers p and q are the same as in the growth conditions for the functions $a^i(x, y)$.

II. $$\Delta_u \varphi(x, u)\Delta u \geqslant 0, \tag{4.3}$$

where $\Delta_u \phi(x, u) = \phi(x, u + \Delta u) - \phi(x, u)$.

III. $$|\Delta_u \varphi(x, u)| \leqslant \mu V(x, u, \Delta u) |\Delta u|, \tag{4.4}$$

where $V = (|\beta(x)|^q + |u|^p + |u + \Delta u|^p)^{1-2/p}$. The last inequality implies the condition

$$|\Delta_u \varphi(x, u)| \leqslant \sqrt{\mu V(x, u, \Delta u)\, \Delta_u \varphi(x, u)\, \Delta u}. \tag{4.5}$$

For the formulation of additional properties of $\phi(x, u)$ it is necessary to introduce local coordinates (see §3). Consider one of the sets $\Pi^+(S^M; \delta)$; let $\xi = (\xi', \xi_n) = (\xi_1, \dots, \xi_{n-1}, \xi_n)$ be the local coordinate system in this set constructed in §3 and let $x = \chi(\xi')$, $\xi' \in g'$, be the representation of S^M. The function $\phi(\chi(\xi'), u)$, $\xi' \in g'$, $u \in E^1$, will be denoted by $\psi(\xi', u)$.

It will be assumed that for each u the function $\psi(\xi', u)$ has as a function of ξ' first and second order generalized derivatives $\psi_{,i}$, $\psi_{,ij}$, $i, j = 1, \dots, n-1$, in the Sobolev sense that are continuous on E^1 as functions of u for almost all $\xi' \in g'$. It will also be assumed that the following inequalities are satisfied for all u and Δu in E^1 and almost all $\xi' \in g'$:

$$|\psi_{,j}|,\ |\psi_{,jj}| \leqslant \mu (|\beta(\chi(\xi'))|^q + |u|^p)^{\frac{1}{q}}, \tag{4.6}$$

$$|\Delta_u \psi_{,j}| \leqslant \sqrt{\mu V(\chi(\xi'), u, \Delta u)\, \Delta_u \psi \Delta u}, \quad j = 1, \dots, n-1, \tag{4.7}$$

where $\beta(x)$, μ, p and q are the same as in (4.2) and (4.4).

As usual [2], a function $u(x) \in W_p^1(G)$ is called a generalized solution of the boundary value problem (1.1), (4.1) if for any function $v(x) \in W_p^1(G)$

$$\sum_{i=0}^{n} \int_G a^i(x, Du) v_{,i} dx + \int_{\partial G} \varphi(x, f) w dS = 0, \tag{4.8}$$

where f and w are the traces of $u(x)$ and $v(x)$ respectively on ∂G. Under our conditions, the integral over ∂G in (4.8) is finite.

Theorem 3. *Suppose the domain G and equation* (1.1) *satisfy all of the conditions of Theorem* 2, *and suppose the function $\phi(x, u)$ satisfies conditions* (4.2)–(4.4), (4.6) *and* (4.7). *If $u(x) \in W_p^1(G)$, $p \geq 2$, is a generalized solution of problem* (1.1), (4.1), *then the following assertions are valid*:

1) $u^i(x, Du(x)) \in W_q^1(G_\delta)$, $\forall \delta > 0$, $i = 1, \dots, n$.

2) *There exists a* $\delta > 0$ *such that* $N(x, Du(x)) \in W_q^1(G - G_\delta)$.

3) *The function $u(x)$ satisfies equation* (1.1) *almost everywhere in* G, *and the trace of $N(x, Du(x))$ on ∂G is equal to $\phi(x, f)$ if f is the trace of $u(x)$ on ∂G.*

Furthermore, in any $\Pi^+(S_r^M; \delta)$, $0 < \delta < \delta_0$ (for the notation, see §3), *the functions $a^i(x, Du(x))$, $i = 1, \dots, n$, have, as functions of the local coordinates ξ, generalized derivatives in L_q with respect to ξ_j, $j = 1, \dots, n-1$, while the function $\phi(x, f(x))$, $x \in \partial G$, as a function of the local coordinates ξ', is such that*

$$\frac{\partial\varphi}{\partial\xi_j}\in L_q(g'_\tau),\quad j=1,\ \ldots,\ n-1. \tag{4.9}$$

Proof. Consider one of the sets $\Pi(S^M;\ \delta_0)$. For the functions $v(x)\in W_p^1(G)$ that vanish outside the closure of this set, condition (4.8) takes (after going over to local coordinates in the notation introduced in §3 and under the conditions imposed on $\phi(x,\ u)$) the form

$$\sum_{i=0}^{n}\int_g b^i v^i J\,d\xi=-\int_{g'}\psi(\xi',\ f)\,w(\xi')\,J'(\xi')\,d\xi',$$

where $J'(\xi')d\xi'$ is the area element of the surface S^M. Setting

$$v=v(x)=\frac{1}{J(\xi(x))}\Delta_j\{\eta^2(\xi(x))\,\Delta_j[u(x);\ h];\ -h\},$$

where the function $\eta=\eta(\xi)$ and the operators $\Delta_j[\ldots;\ h]$ and $\Delta_j\{\ldots;\ -h\}$ were defined in §3, we obtain the equality

$$\sum_{i=0}^{n}\int_g \eta^2\Delta b^i\Delta u^i d\xi+\int_{g'}\sigma^2\Delta(\psi J^*)\,\Delta f d\xi'=\sum_{i=1}^{n}\int_g\left\{\frac{\partial\eta^2}{\partial x_i}\Delta b^i-\eta^2\Delta(J_{,i}b^i)\right\}\Delta u^0 d\xi, \tag{4.10}$$

where $\sigma=\eta(\xi',\ 0)$, $\psi=\psi(\xi',\ f(\xi'))$, and $J^*=J'(\xi')/J(\xi',\ 0)$. All of the integrals in (4.10), except the integral over g', have already been considered in §3 in the proof of Theorem 2. We transform the last integral. Let

$$\Delta_u\psi=\psi(\xi',\ f(\xi'+he_j))-\psi(\xi',\ f(\xi')),$$
$$\Delta_\xi\psi=\psi(\xi'+he_j,\ f(\xi'))-\psi(\xi',\ f(\xi')).$$

Then

$$\Delta(\psi J^*)=J^*\Delta_u\psi+\Delta_\xi\{\psi(\xi',\ f(\xi'+he_j))\,J^*(\xi')\}. \tag{4.11}$$

It now follows from (4.10) and the estimates carried out in §3 that

$$\sum_{i=0}^{n}\int_g \eta^2\Delta b^i\Delta u^i d\xi+\int_{g'}\sigma^2 J^*\Delta_u\psi\Delta f d\xi'\leqslant I+C_3\|u^0_{,j}\|_p\sum_{i=1}^{n}(h^2\|b^i\|_q+|h|\,\|\Delta b^i\|_q), \tag{4.12}$$

where

$$I=\left|\int_{g'}\sigma^2\Delta_\xi\Psi\Delta f d\xi'\right|,\qquad \Psi=\psi(\xi',\ f(\xi'+he_j))\,J^*(\xi').$$

Since $\sigma^2\Delta f=\Delta(\sigma^2 f)-f(\xi'+he_j)\Delta\sigma^2$, we have

$$I\leqslant\left|\int_{g'}\Delta_\xi\Psi\Delta(\sigma^2 f)\,d\xi'\right|+\left|\int_{g'}f(\xi'+he_j)\,\Delta_\xi\Psi\Delta\sigma^2 d\xi'\right|.$$

In the first integral we transfer the difference from $\sigma^2 f$ to $\Delta_\xi\Psi$, while in the second integral we note that $|\Delta\sigma^2|\le 2\kappa|h|$ in g' and $\Delta\sigma^2=0$ outside $g'_{\tau/4}$. Therefore

$$I \leqslant \|f\|_p \|\sigma^2\Delta\{\Delta_\xi\Psi;\ -h\}\|_q + 2\varkappa|h|\|f\|_p\|\Delta_\xi\Psi\|_q^*, \tag{4.13}$$

where $\|\ \|_p$, $\|\ \|_q$ and $\|\ \|_q^*$ denote the norms of the corresponding functions in $L_p(g')$, $L_q(g')$ and $L_q(g'_{T/4})$.

By assumption, the functions $\psi(\xi', u)$ and $J^*(\xi')$ have derivatives with respect to ξ_j. Therefore $\Psi_{,j}$ exists and

$$\Delta_\xi\Psi = h\int_0^1 \tilde{\Psi}_{,j}dt, \text{ where } \tilde{\Psi} = \psi(\xi' + the_j,\ f(\xi' + he_j))\, J^*(\xi' + the_j).$$

The functions $J^*(\xi')$ and $J^*_{,j}(\xi')$ are bounded, i.e. $|J^*| \le C_4$ and $|J^*_{,j}| \le C_4$. Consequently

$$\|\Delta_\xi\Psi\|_q^* \leqslant C_4|h|\int_0^1 \{\|\tilde{\psi}_{,j}\|_q^* + \|\tilde{\psi}\|_q^*\}\, dt. \tag{4.14}$$

The norms of $\tilde{\psi}_{,j}$ and $\tilde{\psi}$ here are estimated in identical fashion with the use of conditions (4.6) and (4.2). Namely,

$$\|\tilde{\psi}\|_q^* \leqslant \mu\gamma_p^{\frac{1}{q}}, \tag{4.15}$$

where $\gamma_p = \gamma_p(\beta;\ f) \equiv \|\beta\|_q^q + \|f\|_p^p$. The same estimate holds for $\tilde{\psi}_{,j}$. From (4.13)–(4.15) we now obtain the inequality

$$I \leqslant \|f\|_p\|\sigma^2\Delta\{\Delta_\xi\Psi;\ -h\}\|_q + 4\mu\varkappa C_4h^2\gamma_p. \tag{4.16}$$

We estimate the norm of the compound difference by noting that $\Delta\{\Delta_\xi\Psi;\ -h\} = \Delta_\xi\{\Delta(\Psi;\ -h)\}$, which implies

$$\|\sigma^2\Delta\{\Delta_\xi\Psi;\ -h\}\|_q \leqslant |h|\int_0^1 \|\sigma^2\Delta\{\tilde{\Psi}_{,j};\ -h\}\|_q\, dt, \tag{4.17}$$

where $\tilde{\Psi}_{,j} = \tilde{\psi}_{,j}\tilde{J}^* + \tilde{\psi}\tilde{J}^*_{,j}$. Clearly,

$$|\Delta\{\tilde{\psi}_{,j}\tilde{J}^*;\ -h\}| \leqslant C_4|\Delta\tilde{\tilde{\psi}}_{,j}| + C_4|h|(1-t)\|\tilde{\tilde{\psi}}_{,j}\| + C_4|h|t|\tilde{\psi}_{,j}|, \tag{4.18}$$

where $\tilde{\tilde{\psi}}_{,j} = \psi_{,j}(\xi' + (t-1)he_j,\ f(\xi'))$, and an analogous estimate holds for $\Delta\{\tilde{\tilde{\psi}}\tilde{J}^*_{,j};\ -h\}$, under the assumption that $|J^*_{,jj}(\xi')| \le C_4$. From (4.18) and an inequality of form (4.15) it follows that

$$\|\sigma^2\Delta\{\tilde{\psi}_{,j}\tilde{J}^*;\quad h\}\|_q \leqslant C_4\|\sigma^2\Delta\tilde{\tilde{\psi}}_{,j}\|_q + \mu C_4|h|\gamma_p^{\frac{1}{q}}. \tag{4.19}$$

Analogously,

$$\|\sigma^2\Delta\{\tilde{\psi}\tilde{J}^*_{,j};\ -h\}\|_q \leqslant C_4\|\sigma^2\Delta\tilde{\tilde{\psi}}\|_q + \mu C_4|h|\gamma_p^{\frac{1}{q}}. \tag{4.20}$$

Let us estimate the norms of the differences $\Delta\tilde{\tilde{\psi}}_{,j}$ and $\Delta\tilde{\tilde{\psi}}$ in (4.19) and (4.20). Since $|\Delta\tilde{\tilde{\psi}}| \le |\Delta\psi| + |\Delta(\tilde{\tilde{\psi}} - \psi)|$, it follows by (4.6) that

$$\|\sigma^2\Delta\tilde{\tilde{\varphi}}\|_q \leqslant \|\sigma^2\Delta\psi\|_q + \mu|h|\gamma_p^{\frac{1}{q}}. \tag{4.21}$$

Analogously,

$$\|\sigma^2\Delta\tilde{\tilde{\varphi}}_{,j}\|_q \leqslant \|\sigma^2\Delta\psi_{,j}\|_q + \mu|h|\gamma_p^{\frac{1}{q}}. \tag{4.22}$$

This completes the estimate of the integral I. In fact, it follows from (4.16), (4.17) and (4.19)–(4.22) that

$$I \leqslant C_5 h^2\gamma_p + C_4|h|\{\|\sigma^2\Delta\psi\|_q + \|\sigma^2\Delta\psi_{,j}\|_q\}\|f\|_p, \tag{4.23}$$

where $C_5 = 4(1+\kappa)C_3\mu$.

The conditions imposed on G imply the existence of a constant $\nu > 0$ such that $J^*(\xi') \geq \nu$. It therefore follows from (4.12) and (4.23) that for $|h| < \tau/4$

$$\begin{aligned}\sum_{i=0}^{n}\int_g \eta^2\Delta b^i\Delta u^i d\xi + \nu\int_{g'}\sigma^2\Delta_h\psi\Delta f d\xi' &\leqslant C_5 h^2\gamma_p \\ &+ C_3\|u^0_{,j}\|_p\sum_{i=1}^{n}(h^2\|b^i\|_q + |h|\|\eta\Delta b^i\|_q) \\ &+ C_4|h|\{\|\sigma^2\Delta\psi\|_q + \|\sigma^2\Delta\psi_{,j}\|_q\}\|f\|_p,\end{aligned} \tag{4.24}$$

where $\|\ \|_p$ and $\|\ \|_q$ are the norms in L_p and L_q of the corresponding functions defined in either g or g'.

Let us estimate the last summand in (4.24). From (4.11) we have

$$\|\sigma^2\Delta\psi\|_q \leqslant \|\sigma^2\Delta_u\psi\|_q + \|\sigma^2\Delta_\xi\psi(\xi', f(\xi'+he_j))\|_q$$

and hence (see (4.14))

$$\|\sigma^2\Delta\psi\|_q \leqslant \|\sigma^2\Delta_u\psi\|_q + \mu|h|\gamma_p^{\frac{1}{q}}. \tag{4.25}$$

The norm of the difference $\Delta_u\psi$ is estimated with the use of condition (4.5) and Hölder's inequality with the appropriate exponents. After simplifying, we get

$$\|\sigma^2\Delta_u\psi\|_q \leqslant \sqrt{2\mu I_0(f)}\,\gamma_p^{\frac{1}{2}-\frac{1}{p}}, \tag{4.26}$$

where $I_0(f) = \int_{g'}\sigma^2\Delta_u\psi\Delta f d\xi'$.

Finally, from (4.25) and (4.27) we see that the inequality

$$\|\sigma^2\Delta\psi\|_q \leqslant \mu|h|\gamma_p^{\frac{1}{q}} + \sqrt{2\mu I_0(f)}\,\gamma_p^{\frac{1}{2}-\frac{1}{p}}$$

is satisfied for any function $f(\xi') \in L_p(g')$, $p \geq 2$. The same estimate can be obtained for $\|\sigma^2\Delta\psi_{,j}\|_q$ with the use of conditions (4.6) and (4.7). Thus

$$C_3|h|\{\|\sigma^2\Delta\psi\|_q + \|\sigma^2\Delta\psi_{,j}\|_q\}\|f\|_p \leqslant \varepsilon I_0(f) + C_\varepsilon h^2\gamma_p,$$

where $C_\epsilon = 2\mu C_4(1 + C_4\epsilon^{-1})$ and ϵ is any positive number. Setting $\epsilon = \nu/2$ and substituting this estimate in (4.24), we get

$$\sum_{i=0}^{n}\int_{g}\eta^2\Delta b^i\Delta u^i d\xi+\frac{\nu}{2}I_0(f)$$
$$\leqslant C_6 h^2\gamma_p+C_3\|u^0_{;j}\|_p\sum_{i=1}^{n}(h^2\|b^i\|_q+|h|\|\eta\Delta b^i\|_q), \tag{4.27}$$

where $C_6 = 4\mu C_4(1+\kappa+C_4\nu^{-1})$.

All of the assertions of our theorem concerning $a^i(x, Du)$ follow, as in the proof of Theorem 2, from (4.27).

Let us prove the assertions of the theorem concerning ψ.

From (4.27) we have

$$I_0(f)\leqslant C_6 h^2(\gamma_p+D_p),\quad |h|<\frac{\tau}{4}.$$

It follows from this result and (4.17) that

$$\|\sigma^2\Delta\psi\|_q\leqslant C_7|h|(\gamma_p+D_p)^{\frac{1}{q}},$$

which implies by virtue of Lemma 1 that $D_j\psi(\xi', f(\xi'))\in L_q(\xi'_\tau)$, i.e. assertion (4.9) of the theorem is proved.

Let us show, finally, that the function $N(x, Du(x))$ has a trace on ∂G equal to $\psi(x, f)$.

A generalized solution $u(x)\in W^1_p(G)$, $p\geq 2$, of problem (1.1), (4.1) satisfies the condition

$$\lim_{\delta\to 0}\int_{\partial G_\delta} vN(x, Du)\,dS=\int_{\partial G} v\varphi(x, f)\,dS \tag{4.28}$$

since it satisfies (1.1) almost everywhere in G and each function $a^i(x, Du(x))$ has a trace on ∂G_δ, $\delta>0$.

On the other hand, we already know that $N(x, Du(x))$ belongs to $W^1_q(G-G_\delta)$ and hence has a trace on ∂G. Let us show that it is equal to $\phi(x, f)$.

Consider one of the sets $\Pi(S^M;\delta_0)$. For a function $v(x)\in W^1_p(G)$ that vanishes outside $\Pi(S^M_\tau;\delta_0)$, condition (4.28) has the form

$$\lim_{\delta\to 0}\int_{g'_\tau} v^0(\xi',\delta)\tilde N(\xi',\delta)J'(\xi',\delta)\,d\xi'=\int_{g'_\tau} v^0(\xi', 0)\psi(\xi', f)J'(\xi')\,d\xi', \tag{4.29}$$

where $\tilde N(\xi',\delta)$ is the trace of $\tilde N(\xi) = N(x(\xi), \mathbf{u}(\xi))$ for $\xi_n=\delta$, while $J'(\xi',\delta)d\xi'$ and $J'(\xi')d\xi'$ are the area elements of the surfaces $\partial G_\delta\cap\Pi^+(S^M_\tau;\delta_0)$ and S^M_τ.

We have already shown that the function $\tilde N(\xi)$ belongs to $W^1_q(g_\tau)$ and hence has the trace $\tilde N(\xi', 0)$. Also, it is easily shown that for any function $v(\xi)\in W^1_p(g_\tau)$ the function $v(\xi)\tilde N(\xi)J'(\xi)$ belongs to $W^1_1(g_\tau)$ and has a trace $v(\xi', 0)\tilde N(\xi', 0)J'(\xi')$ for $\xi_n=0$ such that

$$\lim_{\delta\to 0}\int_{g'_\tau} v(\xi',\delta)\tilde N(\xi',\delta)J'(\xi',\delta)\,d\xi'=\int_{g'_\tau} v(\xi', 0)\tilde N(\xi', 0)J'(\xi')\,d\xi'. \tag{4.30}$$

It follows from (4.29) and (4.30) that the equality

$$\int_{g'} w(\xi')\{\psi(\xi', f)-\tilde N(\xi', 0)\}J'(\xi')\,d\xi'=0$$

holds for any function $u(\xi')$ that is the trace for $\xi_n = 0$ of a function belonging to $W_p^1(g_T)$. The set of traces for $\xi_n = 0$ of the functions belonging to $W_p^1(g_T)$ is dense in $L_p(g_T')$, the difference in braces belongs to $L_q(g_T')$, and the function $J'(\xi')$ is greater than unity. Therefore $\widetilde{N}(\xi', 0) = \psi(\xi', f)$ almost everywhere in g'. The theorem is proved.

BIBLIOGRAPHY

1. S. M. Nikol'skiĭ, *Approximation of functions of several variables and embedding theorems*, "Nauka", Moscow, 1969; English transl., Springer-Verlag, Berlin and New York, 1974. MR 46 #9714.

2. O. A. Ladyženskaja and N. N. Ural'ceva, *Linear and quasi-linear equations of elliptic type*, "Nauka", Moscow, 1964; English transl., Academic Press, New York, 1968. MR 35 #1955; 39 #5941.

3. M. I. Višik, *Quasi-linear strongly elliptic systems of differential equations in divergence form*, Trudy Moskov. Mat. Obšč. 12 (1963), 125–184 = Trans. Moscow Math. Soc. 1963, 140–208. MR 27 #6017.

4. Ju. D. Kaščenko, *A nonlinear differential equation with a polynomial functional generating it*, Differencial'nye Uravnenija 3 (1967), 1157–1165 = Differential Equations 3 (1967), 607–611. MR 36 #512.

5. G. N. Jakovlev, *A certain variational problem*, Differencial'nye Uravnenija 5 (1969), 1303–1312 = Differential Equations 5 (1969), 960–966. MR 41 #889.

6. ———, *On weak solutions of second order quasi-linear elliptic systems*, Differencial'nye Uravnenija 6 (1970), 157–163 = Differential Equations 6 (1970), 125–129. MR 40 #7629.

7. A. D. Džabrailov, *An investigation of certain classes of second-order quasilinear elliptic equations*. I, II, III, Differencial'nye Uravnenija 5 (1969), 2245–2257; 6 (1970), 889–899, 1279–1288 = Differential Equations 5 (1969), 1683–1693; 6 (1970), 684–693, 977–982. MR 41 #7292; 42 #6418.

8. ———, *Differential properties of generalized solutions of certain classes of second-order quasilinear parabolic equations*, Differencial'nye Uravnenija 6 (1970), 1467–1480 = Differential Equations 6 (1970), 1118–1128. MR 43 #719.

9. G. N. Jakovlev, *Weak solutions of quasilinear parabolic equations*, Differencial'nye Uravnenija 6 (1970), 1643–1651 = Differential Equations 6 (1970), 1249–1255. MR 43 #2362.

10. A. D. Džabrailov, *Estimates near the boundary of generalized solutions of certain classes of second order quasilinear elliptic equations*, Differencial'nye Uravnenija 7 (1971), 1459–1472 = Differential Equations 7 (1971), 1108–1117. MR 44 #5616.

11. G. N. Jakovlev, *On the solutions of a class of second order quasilinear elliptic equations*, Dokl. Akad. Nauk SSSR 202 (1972), 1020–1023 = Soviet Math. Dokl. 13 (1972), 255–258. MR 45 #715.

12. ———, *The first boundary value problem for quasilinear elliptic equations of second order*, Trudy Mat. Inst. Steklov. 117 (1972), 321–340 = Proc. Steklov Inst. Math. 117 (1972), 381–403.

13. ———, *Properties of solutions of a class of second-order quasilinear elliptic equations in divergence form*, Trudy Mat. Inst. Steklov. 131 (1974), 232–242 = Proc. Steklov Inst. Math. 131 (1974), 242–252.

Translated by S. SMITH